The Human Microbiota and Chronic Disease

The Human Microbiota and Chronic Disease

Dysbiosis as a Cause of Human Pathology

EDITED BY

Luigi Nibali

Queen Mary University of London, London, United Kingdom

Brian Henderson

University College London, London, United Kingdom

WILEY Blackwell

Published by John Wiley & Sons, Inc., Hoboken, New Jersey
Published simultaneously in Canada

For general information on our other products and services or for technical support, please contact our Customer
Care Department within the United States at (800) 762-2974, outside the United States at (317) 572-3993 or
fax (317) 572-4002.

Wiley also publishes its books in a variety of electronic formats. Some content that appears in print may not be
available in electronic formats. For more information about Wiley products, visit our web site at www.wiley.com.

Library of Congress Cataloging-in-Publication Data

Names: Henderson, Brian (Professor), editor. | Nibali, Luigi, 1978– editor.
Title: The human microbiota and chronic disease : dysbiosis as a cause of human pathology /
    edited by Luigi Nibali and Brian Henderson.
Description: Hoboken, New Jersey : John Wiley & Sons, 2016. | Includes bibliographical references and index.
Identifiers: LCCN 2016016110 (print) | LCCN 2016025000 (ebook) | ISBN 9781118982877 (cloth) |
    ISBN 9781118982884 (pdf) | ISBN 9781118982891 (epub)
Subjects: LCSH: Human body–Microbiology. | Chronic diseases.
Classification: LCC QR46 .H83 2016 (print) | LCC QR46 (ebook) |
    DDC 616.9/041–dc23
LC record available at https://lccn.loc.gov/2016016110

Cover credit: Gettyimages/STEVE GSCHMEISSNER/SPL

10 9 8 7 6 5 4 3 2 1

Printed and bound in Malaysia by Vivar Printing Sdn Bhd

Contents

Section 3 Dysbioses and bacterial diseases: Metchnikoff's legacy, 215

List of contributors

Luis G. Bermúdez-Humarán
AgroParisTech; UMR1319 Micalis; F-78350 Jouy-en-Josas, France;
INRA, UMR1319 Micalis, Commensal and Probiotics-Host Interactions Laboratory,
Domaine de Vilvert, 78352 Jouy-en-Josas Cedex, France

Aadil Bharwani
The Brain-body Institute and Firestone Institute for Respiratory Health, Ontario,
Canada

Hervé M. Blottière
Micalis Institute, INRA, AgroParisTech, Universitè Paris-Saclay, Paris, France

Katharina Brandl
Skaggs School of Pharmacy, University of California, San Diego, United States

Holger Brüggemann
Department of Biomedicine, Aarhus University, Aarhus, Denmark

Eugenia Bruzzese
University of Naples, Naples, Italy

Vittoria Buccigrossi
University of Naples, Naples, Italy

Marie-José Butel
Université Paris Descartes, Sorbonne Paris, Paris, France

John D. Carter
University of South Florida Morsani College of Medicine, Tampa, FL, United States

Séverine Couffin
UPEC, Université Paris Est Créteil Val de Marne-Equipe Universitaire EC2M3,
Paris, France

Mike Curtis
Institute of Dentistry, Queen Mary University of London

Jacqueline Detert
Charité-Universitätsmedizin Berlin, Berlin, Germany

Nik Ding
St. Mark's Hospital, London, United Kingdom

Joël Doré
Micalis Institute, INRA, AgroParisTech, Universitè Paris-Saclay, Paris, France

Alan Ebringer
King's College London, London, United Kingdom

Mehrbod Estaki
The University of British Columbia, Kelowna, Canada

Frida Fåk
Lund University, Lund, Sweden

Paul Forsythe
McMaster University, Hamilton, Ontario, Canada

Ralph Francescone
Fox Chase Cancer Center, Cancer Prevention and Control, Philadelphia, United States

Lionel Fry
Imperial College, London, United Kingdom

Markus B. Geuking
Mucosal Immunology Lab, University of Bern, Switzerland

Deanna L. Gibson
The University of British Columbia, Kelowna, Canada

Alfredo Guarino
University of Naples,Naples, Italy

George Hajishengallis
School of Dental Medicine, University of Pennsylvania, Philadelphia, United States

Ailsa Hart
St. Mark's Hospital, London, United Kingdom

Phillip Hay
St. George's, University of London, United Kingdom

Almut Heinken
Luxembourg Centre for Systems Biomedicine, University of Luxembourg, Belval, Luxembourg

Brian Henderson
University College London, London, United Kingdom

Anne-Judith Waligora-Dupriet
Université Paris Descartes, Sorbonne paris, Paris, France

Richard J. Lamont
School of Dentistry, University of Louisville, Louisville, KY, United States

Benjamin J. Marsland
Service de Pneumologie, CHUV, Faculty of Biology and Medicine, University of Lausanne, Lausanne, Switzerland

Luigi Nibali
Centre for Oral Clinical Research, Queen Mary University of London, London, United Kingdom

Candice Quin
The University of British Columbia, Kelowna, Canada

Taha Rashid
King's College London, London, United Kingdom

Giusy Ranucci
University of Naples, Naples, Italy

Dmitry A. Ravcheev
Luxembourg Centre for Systems Biomedicine, University of Luxembourg, Belval, Luxembourg

Frank Ryan
The Academic Unit of Medical Education, University of Sheffield, United Kingdom

Olawale Salami
Service de Pneumologie, CHUV, Faculty of Biology and Medicine, University of Lausanne, Lausanne, Switzerland

S. Tariq Sadiq
St. George's, University of London, United Kingdom

Joost Schalkwijk
Department of Dermatology, Radboud University Nijmegen Medical Centre, Nijmegen, The Netherlands

Bernd Schnabl
University of California, San Diego, United States

Boris A. Shenderov
Laboratory of Biology of Bifidobacteria, Head of Research Group Probiotics and Functional Foods, Gabrichevsky Research Institute of Epidemiology and Microbiology, Moscow, Russia

Jessica Snowden
University of Nebraska Medical Center, Omaha, Nebraska United States

Iradj Sobhani
Centre Hospitalier Universitaire Henri Mondor-Assistance Publique Hôpitaux, de Paris, Paris, France

Ines Thiele
Luxembourg Centre for Systems Biomedicine, University of Luxembourg, Belval, Luxembourg

Andrea Lo Vecchio
University of Naples, Naples, Italy

Débora B. Vendramini-Costa
Institute of Chemistry, University of Campinas, Campinas-SP, Brazil

William G. Wade
Centre for Immunobiology, Blizard Institute, Barts and The London School of Medicine and Dentistry, Queen Mary University of London, London, United Kingdom

Clyde Wilson
King Edward VII Memorial Hospital, Bermuda

Michael Wilson
UCL Eastman Dental Institute, University College London, United Kingdom

Kazuhisa Yamazaki
Division of Oral Science for Health Promotion, Niigata University Graduate School of Medical and Dental Sciences, Niigata, Japan

Patrick L.J.M. Zeeuwen
Department of Dermatology, Radboud University Nijmegen Medical Centre, Nijmegen, The Netherlands

Preface

The human organism comprises 10^{13} eukaryotic cells divided into a large number of distinct organs and tissues, with unimaginable requirements for inter- and intra-cellular communication. Malfunction in such communication inevitably results in the state we define as human disease. The emergent properties of the eukaryotic cellular complexity in *Homo sapiens* were beginning to be suspected in the 1950s and 1960s, when it was becoming clear that the bacteria that actually existed within the healthy human could have a major influence on many of its cellular and tissue systems, including innate and adaptive immunity. The development of antibiotic resistance in the 1970s produced a renaissance in microbiology that revealed just how heavily colonised healthy vertebrates were with bacteria. The human appears to be the acme of this colonisation process and it is now a familiar expression that 'for every human cell in our bodies there are ten bacteria'. Not only are we colonised by around 10^{14} bacteria, but the human population carries round with it a diversity of bacterial phylotypes that swamps the diversity of all the species in the aggregate of the world's zoological collections. Thus we can no longer think of bacteria in terms of 'us' and 'them'. *Homo sapiens*, like most vertebrates, must be viewed as a supra-organism colonised, on its mucosal surfaces and on the skin (and who knows where else) with complex populations of bacteria; each individual has a unique mixture of these bacteria, presumably a result of genetic (and/or epigenetic) factors controlling commensal bacterial colonisation and the stability of such colonisation.

Not only are we colonised by a large and diverse collection of bacteria (this volume will ignore colonisation by single-celled eukaryotes and by Archaea), but these bacteria generally take the form of dynamic multi-species biofilms that, like the comparison of human tissues to the disaggregated cells of these tissues, have emergent properties. Thus the collection of microbes in our bodies, which we call the microbiota, is a dynamically complex collection of multi-species biofilms. The formation of these biofilms requires an inordinate amount of intercellular signalling and this signalling must reciprocate with the cellular surfaces on which these biofilms co-exist. These cellular surfaces are 'us'.

In the 21st century, the concept of human health and disease has to take into account our intimate relationship with our microbiota. The regional complexity of the human microbiota is only now being revealed with the application of bacterial phylogenetic analyses and next-generation sequencing (NGS) methodologies. This overcomes the problem that only around 50% of the bacteria colonising the human can be cultivated and studied. Each of us is colonised with hundreds of bacterial phylotypes, each phylotype itself being composed of a varied range of strains, each containing different populations of genes. This generates the concept of the pan-genome in which each bacterial pan-genome perhaps has as many protein-coding genes as its host. This means that the individual bacterial population colonising each human has 10–100 (or more) times the number of genes utilised by the host. Every human host is colonised by a different combination of microbes, making him/her more or less susceptible to disease. Host genetic variants are largely responsible for determining the composition of human microbial biofilms. This creates a level of complexity that is difficult to comprehend but must be fully explored if we are to understand the healthy human and the diseases s/he is susceptible to.

Modern medicine, as a successful practice, can largely be dated from the late-19th-century discovery of the role of the bacterium in human infectious disease. At this stage it was assumed that humans were largely sterile and that infection was an aberrant state. For several decades after this monumental discovery, the paradigm of human disease was founded on bacterial or other infections as the causation of all disease, and it was only in the 1940s onwards that other mechanisms began to be sought for human disease pathology. The identification of monogenic diseases generated a successful paradigm for a proportion of human ailments, and this has morphed into our current belief that all idiopathic, and even infectious, disease has a genetic component. This paradigm has further developed with the identification of the effects of chemical modifications of our DNA on DNA function and has introduced the role of epigenetics in human diseases. However, the determination, starting in the 1980s, of how enormously colonised we are by bacteria, and the potential that these bacteria have for interfering with all aspects of our cellular homeostasis, has brought the bacterium to centre stage as a causative factor in maintaining human health and disease and even playing a role in our ageing processes.

Readers of this book live in a time when a major paradigm shift is in the offing about the causation of all human disease. There is a growing realisation that, in addition to directly causing 'infectious' disease, the bacteria that colonise us may generate other forms of pathology and that these will be dependent on our genetic/epigenetic constitution and on the composition of the bacteria colonising us. Microbiota-associated pathology can be a direct result of changes in general bacterial composition, such as might be found in periodontitis and bacterial vaginosis, and/or as the result of colonisation and/or overgrowth of so-called keystone species, such as the oral organism *Porphyromonas gingivalis* or the gastrointestinal bacterium *Helicobacter hepaticus*. This introduces the concept of *dysbiosis*, defined as a disruption in the composition of the normal microbiota.

This volume discusses the role of the microbiota in maintaining human health and introduces the reader to the biology of bacterial dysbiosis and its potential role in both bacterial disease and idiopathic chronic disease states. The current book is divided into five sections, starting from the concept of the human bacterial microbiota (chapter 1) with particular attention paid to the microbiotae of the gut, oral cavity and skin. A key methodology for exploring the microbiota, metagenomics, is also described. The second section attempts to show the reader the cellular, molecular and genetic complexities of the bacterial microbiota, its myriad connections with the host and how these can maintain tissue homeostasis. Section 3 begins to consider the role of dysbioses in human disease states, dealing with two of humanity's commonest bacterial diseases, periodontitis and bacterial vaginosis. In section 4 the discussion moves to the major chronic diseases of *Homo sapiens* and the potential role of dysbiosis in their induction and chronicity. This is a rapidly growing area where major discoveries are expected. The composition of some if not all microbiotas can be controlled by the diet and this is will be discussed in the final section, section 5. This last section will also take the reader to the therapeutic potential of manipulating the microbiota, introducing the concepts of probiotics, prebiotics and the administration of healthy human faeces (faecal microbiota transplantation), then to gaze into the crystal ball and imagine the future of medical treatment viewed from a microbiota-centric position.

This book should be of interest to a very wide audience ranging from clinicians interested in infectious and idiopathic diseases to pathologists interested in patho-mechanisms of disease and on to immunologists, molecular biologists, micro-biologists, cell biologists, biochemists, systems biologists, and so forth, who are attempting to understand the cellular and molecular bases of human diseases.

Luigi Nibali
Brian Henderson

An introduction to the human tissue microbiome

CHAPTER 1

The human microbiota: an historical perspective

Michael Wilson

UCL Eastman Dental Institute, University College London, United Kingdom

1.1 Introduction: the discovery of the human microbiota: why do we care?

The discovery by Antony van Leeuwenhoek in 1683 that we have a microbiota was very surprising and undoubtedly of great interest to 17th-century scientists. However, as modern-day researchers know only too well, this alone is not sufficient to ensure continued investigation of a subject. Further research into the microbes that inhabit humans proceeded at a very slow pace until it was realized that these microbes were able to cause disease and, much later, that they contribute to human health (i.e., in modern-day research parlance the research would be recognized as having "impact"). Our knowledge of those microbes with which we coexist has increased enormously during the last few years. An indication of the effort that has been devoted to determining the nature and function of the microbial communities inhabiting the various body sites of humans can be gleaned from the number of publications in this field listed in PubMed: in 2013 more than 2500 papers were published, nearly four times as many as in 2000.

What accounts for this recent huge growth of interest in the human microbiota? There appear to be two main driving forces: (a) increasing awareness of its importance in human disease, development, nutrition, behavior and wellbeing; (b) the development of technologies that enable us not only to identify which microbes are present but also to determine what these microbes are up to. In this chapter these two driving forces are described from a historical perspective.

1.2 The importance of the indigenous microbiota in health and disease

It has long been known that members of the indigenous microbiota of humans are responsible for a variety of infections, but only relatively recently has it been recognized that these microbes play an important role in maintaining human health and wellbeing.

The Human Microbiota and Chronic Disease: Dysbiosis as a Cause of Human Pathology, First Edition.
Edited by Luigi Nibali and Brian Henderson.
© 2016 John Wiley & Sons, Inc. Published 2016 by John Wiley & Sons, Inc.

1.2.1 The indigenous microbiota and human disease

In the late 19th and early 20th centuries many members of what we now recognize as the indigenous microbiota of humans were found to be the causative agents of a number of human infections (Table 1). However, at that time there was little understanding of what constituted the indigenous microbiota and therefore it was not realized that these newly recognized, disease-causing microbes were in fact regularly present on some, if not all, healthy humans and that, for the most part, they lived in harmony with their host (Table 1).

Subsequently, as knowledge of the indigenous microbiota improved, the involvement of members of these communities in disease processes became of great interest and was the subject of more intense research. Other members of the indigenous microbiota now known to cause human disease are shown in Table 2. More recently, it has become apparent not only that individual members of the microbiota are able to cause disease, but that shifts in the overall composition of the microbiota at a site can result in disease (Table 3). Such "dysbioses" are discussed in greater detail in subsequent chapters of this book. Recognition of the disease-inducing potential of the indigenous microbiota became an important stimulus to research into the characterization of the microbial communities associated with humans.

1.2.2 The indigenous microbiota and human health

Towards the end of the 19th century it became evident to many researchers that the intestinal microbiota was important in intestinal physiology, and Pasteur in 1885 went even further by suggesting that animal life would not be possible in the absence of the indigenous microbiota[19]. In the second half of the 20th century it became evident that the indigenous microbiota not only contributed to mammalian health and wellbeing in a number of ways but that it also played an important

Table 1 Early discoveries of the involvement of members of the indigenous microbiota in human infections.

Year	Researcher	Organism	Disease	Reference
1881	Alexander Ogston	staphylococci	abscesses	1
1884	Friedrich Rosenbach	*Strep. pyogenes*	Wound infections	2
1884	Friedrich Rosenbach	*Staphylococcus aureus*	Wound infections	2
1884	Friedrich Rosenbach	*Staphylococcus albus (i.e. Staph. epidermidis)*	Wound infections	2
1884	Albert Fraenkel	*Diplococcus pneumoniae (i.e. Strep. pneumoniae)*	Lobar pneumonia	3
1890s	Theodor Escherich	*Bacterium coli commune (i.e. Escherichia coli)*	Colicystitis (i.e. urinary tract infection)	—
1892	George Nuttall and William Welch	*Bacillus aerogenes capsulatus (i.e. Clostridium perfringens)*	gangrene	4
1898	Veillon and Zuber	A variety of anaerobic species including *Bacteroides fragilis*, *Fusobacterium nucleatum*	gangrene	5
1906	Thomas Horder	*Strep. salivarius*	infective endocarditis	6
1891	Albert Fraenkel	*Bacillus coli communis (i.e. Escherichia coli)*	peritonitis	7

Table 2 Diseases caused by members of the indigenous microbiota (in addition to those listed in Table 1).

Organism	Disease
Enterococcus faecalis	Urinary tract infections, endocarditis, meningitis, wound infections
Moraxella catarrhalis	Bronchopneumonia, sinusitis, otitis media
Haemophilus influenzae	Meningitis, pneumonia, sinusitis, otitis media, epiglottitis
Proteus mirabilis	Urinary tract infections
Helicobacter pylori	Gastritis, ulcers, carcinoma
Streptococcus mutans	Dental caries, endocarditis
Porphyromonas gingivalis	periodontitis
Actinomyces israelii	actinomycosis
Staphylococcus saprophyticus	Urinary tract infections
Neisseria meningitidis	meningitis
Malassezia spp.	Atopic dermatitis, seborrhoeic dermatitis, folliculitis
Gardnerella vaginalis	Bacterial vaginosis
Corynebacterium minutissimum	erythrasma

Table 3 Diseases resulting from dysbiosis.

Disease	Microbiota involved	Reference
obesity	intestinal tract	8
rhinosinusitis	nasal cavity	9
chronic obstructive pulmonary disorder	lungs	10
asthma	lungs	11
autism	intestinal tract	12
inflammatory bowel diseases	intestinal tract	13
multiple sclerosis	intestinal tract	14
arthritis	intestinal tract	15
periodontitis	oral cavity	16
colorectal cancer	intestinal tract	17
type II diabetes	intestinal tract	18

role in mammalian development (Table 4). While many of these discoveries were made in animals such as mice and rats, in some cases these effects have also been demonstrated in humans.

That the indigenous microbiota exerted a protective effect by preventing colonization of exogenous pathogens was demonstrated in 1962 when it was found that mice were 100,000-fold more susceptible to infection with *Salmonella enteritidis* following the administration of a single dose of streptomycin[20]. This was attributed to the disruptive effect of the antibiotic on the composition of the intestinal microbiota, thereby destroying its barrier function. This protective effect was termed "colonization resistance".

Most information regarding the role of microbes in mammalian development has been obtained by comparative studies involving germ-free animals, animals with a normal microbiota and those colonized with particular microbial species. Such studies became possible following the successful breeding of germ-free animal colonies (mainly rats and mice) in the 1950s and were well underway by the 1970s. The absence of an indigenous microbiota can have dramatic effects on

Table 4 Beneficial effects of the human microbiota.

Role of microbiota	Body site involved
colonization resistance (i.e. exclusion of exogenous pathogens)	all
development of immune functions	all
tissue and organ differentiation and development	intestinal tract
development of nutritional capabilities	intestinal tract
provision of nutrients	intestinal tract
provision of vitamins	intestinal tract
detoxification of harmful dietary constituents	intestinal tract
prevention of bowel cancer	intestinal tract

Table 5 Attributes of germ-free animals compared to their counterparts with an indigenous microbiota[22].

decreased mass of heart, lung and liver
increased size of cecum (may be eight times larger)
decreased water absorption by large intestine
increased redox potential of large intestine
decreased concentration of deconjugated bile salts
increased pH of stomach
altered surface epithelial mucins
decreased mass of small intestine
absence of immune cells in lamina propria
decreased intestinal peristalsis
prolonged intestinal epithelial cell cycle time
shorter, more slender villi in intestinal epithelial cells
increased length of microvilli
decreased surface area of intestinal mucosal
decreased number of lymphocytes in lamina propria
decreased size of Peyer's patches, mesenteric lymph nodes, spleen and thymus
decreased macrophage chemotaxis and phagocytic activity
decrease in plasma cells in lamina propria and Peyer's patches
decreased production of IgG and IgA
decreased number of TCRαβ+ intra-epithelial lymphocytes

the anatomy and physiology of an animal and examples of these are listed in Table 5; these are often termed "germ-free animal characteristics" (GACs)[21]. Those aspects of the host's anatomy, immunology, physiology or biochemistry that are influenced by the indigenous microbiota have been termed "microbiota-associated characteristics" (MACs) and, as is evident from Table 5, these are many and varied. Many of the abnormalities observed in germ-free animals can be reversed by inoculation with the indigenous microbiota or constituents of the microbiota.

A number of studies in the early 2000s involving the gut symbiont *Bacteroides thetaiotaomicron* and germ-free animals revealed the multiple contributions of the gut microbiota to host development[22,23]. Colonization of germ-free mice with *B. thetaiotaomicron* results in changes in the expression of several host genes involved in the processing and absorption of carbohydrates, lipids and micronutrients and thereby contributes to the development of the host's nutritional capabilities. Germ-free adult mice have a greatly reduced capillary network in their intestinal villi compared with conventional mice and it has been shown that in conventional mice the development of the capillary network coincides with the establishment of a complex intestinal microbiota. Inoculation of germ-free mice with *B. thetaiotaomicron* induces the formation of a normal capillary network, thereby greatly increasing the host's ability to absorb nutrients.

One of the major roles of the intestinal microbiota is in stimulating the growth and differentiation of intestinal epithelial cells[22,24]. Germ-free rodents have fewer crypt cells than conventional animals and the rate of production of such cells is reduced. Hence, in conventional rodents the rate of enterocyte turnover is almost twice that found in germ-free animals. The microbially induced proliferation and differentiation of epithelial cells is mediated by the short chain fatty acids (SCFAs) produced by fermentation of carbohydrates and amino acids. Although all three major SCFAs (i.e. butyrate, acetate and propionate) are able to induce this trophic effect, butyrate is the most potent in this respect and has been shown to alter the expression of a number of genes in epithelial cells *in vitro*. Butyrate can also inhibit DNA synthesis in, and proliferation of, neoplastic cells, and it has been suggested that this may account for the protective effect that dietary fibre exhibits against bowel cancer.

It is well established that the indigenous microbiota plays a key role in the development of a competent immune system[22,25]. Because the gut-associated lymphoid tissue (GALT) contains the largest collection of immunocompetent cells in the human body, most studies have involved the gastrointestinal tract and its microbiota. As shown in Table 5, the immune system of germ-free animals has a number of structural and functional abnormalities including low densities of lymphoid cells in the gut mucosa, low concentrations of circulating antibodies, specialized follicle structures are small, etc. However, exposure of the gut mucosa to the indigenous microbiota has a dramatic effect on the GALT. Hence, the number of intraepithelial lymphocytes expands greatly, germinal centres with antibody-producing cells appear in follicles and in the lamina propria, the levels of circulating antibodies increase and increased quantities of IgA are secreted into the gut lumen. Many studies have shown that the indigenous microbiota stimulates the secretory IgA system and B lymphocyte function in general[22].

The colonic microbiota functions as an effective scavenger of dietary constituents that the stomach and small intestine are unable to digest (mainly complex carbohydrates), have failed to digest (carbohydrates, proteins, peptides), or have failed to absorb (amino acids and monosaccharides). The colonic microbiota degrades these materials to assimilable molecules that can serve as nutrients for the host as well as for resident microbes[22,26]. Although a variety of microbes can digest the complex carbohydrates reaching the colon, the most effective species are those belonging to the genera *Bacteroides* and *Bifidobacterium*. In addition to carbohydrates, the colon also receives proteins and peptides from the diet, exfoliated epithelial cells and pancreatic enzymes. These are rapidly degraded by microbial proteases and peptidases and these may be of significant nutritional value to the host. Furthermore, many colonic microbes can ferment these amino acids to generate a range of products, including SCFAs, that are of great nutritional value to the host, providing up to 9% of the host's energy requirements[22]. Colonocytes can utilize each of the three SCFAs as an energy source, with butyrate being the most important and acetate the least important in this respect. It has been estimated that the colonic epithelium derives up to 70% of its energy from these SCFAs[22]. The acids are also used as precursors for the synthesis of mucosal lipids. Apart from acting as a major energy source and its involvement in lipid synthesis, butyrate has a number of effects on the colonic epithelium. Hence, it can stimulate cell growth and proliferation, induce differentiation, alter gene expression, induce apoptosis, stimulate tight junctions, increase mucus production and reduce inflammation[27]. SCFAs, therefore, appear to play a key role in maintaining gut integrity. There is also some evidence to suggest that the butyrate produced by colonic microbes exerts a protective effect against large bowel cancer.

The colonic microbiota, therefore, constitutes a means by which the host can achieve maximum recovery of the nutrients present in its diet without it having to elaborate the vast range of enzymes that would be needed to degrade a wide range of dietary constituents. It also plays a major role in energy harvest, storage and expenditure, and consequently is an important factor in human obesity[8].

A number of vitamins are present in the colon and many are derived from the colonic microbiota — particularly *Bifidobacterium* spp., *Bacteroides* spp., *Clostridium* spp. and enterobacteria. Vitamins produced by colonic bacteria include biotin, vitamin K, nicotinic acid, folate, riboflavin, pyridoxine, vitamin B12 and thiamine[22,28]. A number of studies have shown that the tremendous metabolic capabilities of the colonic microbiota can achieve detoxification of potentially harmful dietary constituents[22,29]. Heterocyclic aromatic amines (HAAs) are pyrolysis products of amino acids found in cooked meat and fish products and may have a role in the etiology of colon cancer. *Lactobacillus* spp., *Clostridium* spp. and *Bifidobacterium* spp. are able to reduce the mutagenicity of HAAs by binding to them and/or by altering their structure. Studies in humans have shown that the consumption of *Lactobacillus casei* or *Lactobacillus acidophilus* results in a greatly reduced urinary and fecal mutagenicity following the ingestion of meat.

More recently (i.e. during the early years of the 21st century), it has become increasingly evident that the gut microbiota is able to communicate with the central nervous system (via neural, hormonal, immunological and metabolic pathways) and influence brain function and behavior[30] (see chapter 27). Hence, there is considerable evidence (from both animal and human studies) that the composition of the gut microbiota can play a role in regulating memory, cognition, anxiety, sleep, mood and pain[31]. It is becoming clear, therefore, that the influence of the microbiota on human wellbeing extends far beyond what was recognized in the latter years of the 20th century (summarised in Table 4).

1.3 The development of technologies for characterising the indigenous microbiota

The above section has outlined how our knowledge of the role of the indigenous microbiota in health and disease has progressed. The establishment of the indigenous microbiota as a reservoir of disease-causing microbes, the finding that changes in its overall composition (i.e. dysbiosis) can result in a range of chronic diseases, the realization that it has a profound role in the development and health of humans and, finally, the recent recognition of its ability to influence brain function and behavior have all provoked an enormous drive to characterize the microbial communities that live on us. The above-mentioned revelations would not have been possible without the development and application of a range of new techniques for identifying what organisms are present in these complex communities and for establishing exactly what they are doing there. In this section, the historical emergence of these techniques will be outlined.

1.3.1 Light microscopy

The discovery of the microbial world and, indeed, the realization that we are colonized by microbes arose from the use of a simple light microscope. Not only was Antony van Leeuwenhoek the first person to report that he had seen microbes (in stored rainwater in 1676) but he was the first to report (in 1683) the presence of microbes in humans. On 17th September 1683, he wrote a letter to the Royal Society in which he reported

> Tho my teeth are kept usually very clean, nevertheless when I view them in a magnifying glass, I find growing between them a little white matter as thick as wetted flower: in this substance tho I could not see any motion I judged there might probably be living Creatures. I therfore took some of this flower and mixt it either with pure rain water wherein there were no Animals: or else with some of my Spittle (having no air bubbles to cause a motion in it) and then to my great surprize perceived that the aforesaid matter contained very many small living Animals which moved themselves very extravagantly. Their motion was strong and nimble, and they darted themselves thro the water or spittle as a Jack or Pike does thro water.

In a letter to Robert Hooke in 1719, he also reported the presence of "animalcules" in his feces.

Light microscopy continues to be an important tool for detecting and enumerating the microbes that colonize humans, although a number of developments since 1683 have greatly increased its usefulness and versatility. These include:

- the production of more powerful microscopes (increasing the magnification from X300 to more than X1000)
- dark-field microscopy
- phase-contrast microscopy
- the use of simple and differential staining techniques (e.g. Gram stain)
- the use of fluorescent probes
- the use of stains that distinguish between live and dead microbes
- confocal microscopy

These various techniques have not necessarily been used to investigate all of the microbial communities inhabiting humans. Some have proved to be more useful for studying particular body sites than others. For example, dark field and phase-contrast microscopy have been used extensively to investigate the oral microbiota whereas their use in studying microbial communities at other body sites has been more limited. A brief overview of these techniques is provided below:

- Light microscopy is one of the simplest and most direct approaches used to study microbial communities. One of its advantages is that it can reveal details of the physical structure of a community and the spatial arrangement of the constituent organisms. It also serves as a "gold standard" with respect to the total number of microbes that are present within a sample; this is often used as a yardstick for assessing the ability of other, less direct techniques to detect all of the organisms present in a community. Hence, it has been shown that culture-based analysis of feces may detect as few as 20% of the organisms that can be observed microscopically. Differential counts of the various morphotypes in a sample give an indication of the diversity of the microbiota, and this has proved useful to ascertain whether the composition of the vaginal and subgingival microbiotas in an individual are indicative of health or disease.

- Dark field microscopy (invented in 1830) is useful for examining unstained, living microbial communities. The resolution (0.02 μm) is approximately 10X higher than that obtained with a traditional bright field microscope, which means that thin and fragile microbes (e.g. spirochetes) can be visualized. It has been used to provide information on the main bacterial morphotypes present in biofilms found in the gingival crevice of healthy adults[32]. The relative proportions of these morphotypes (cocci, straight rods, curved rods, filaments, fusiforms and spirochetes) were shown to be similar in biofilms present in healthy adults. Shifts in their relative proportions (e.g. a decrease in cocci accompanied by an increase in rods and spirochetes) were found to be indicative of disease (gingivitis or periodontitis).

- Phase-contrast microscopy, first described in 1934, depends on differences in refractive indices between microbes and their surroundings and, like dark field microscopy, enables microbes to be examined in their living state. It has been used to study the various bacterial morphotypes present in biofilms from the fissures on teeth in adults[33].

- Fluorescent probes have long proved useful in determining which organisms are present in a microbial community. Initially, antibodies were used to identify the organisms present, but nowadays oligonucleotide-based probes are increasingly being used in a technique known as fluorescent *in situ* hybridization (FISH). This involves the use of fluorescent-labeled oligonucleotide probes to target specific regions of bacterial DNA. Most of the probes currently used are those that recognize genes encoding 16S ribosomal RNA (16S rRNA). The gene encoding 16S rRNA in a bacterium consists of both constant and variable regions. Within the molecule there are regions that are highly specific for a particular bacterial species as well as regions that are found in all bacteria, in only one bacterial genus, or in closely-related groups of bacteria. Probes, therefore, can be designed to identify an individual species, a particular genus, certain related microbial groups, or even all bacteria. Oligonucleotide probes can, of course, also be used to detect microbes other than bacteria. An important advantage of this approach is that it can be automated and the resulting data can be processed using computerized image analysis software. One of the earliest studies to use fluorescent antibodies to investigate the indigenous microbiota was that of Ritz[34], who used this technique to detect *Nocardia* spp. in dental plaque. FISH has also been used to detect *Bifidobacterium* spp. in fecal samples[35]. Since then, it has been used to analyse microbial communities at other sites including the vagina[36], the oral cavity[37] and the skin[38].

- Staining procedures that can distinguish between live and dead cells can provide useful information about the physiological status of members of microbial communities. One such procedure involves treating the specimen with a mixture of two DNA-binding dyes — propidium iodide (which fluoresces red) only enters cells with a damaged cytoplasmic membrane, while SYTO9 (fluoresces green) enters all cells. Live cells (strictly speaking, those with an intact membrane) appear green, while dead cells (strictly speaking, those with a damaged membrane) appear red. This has been employed to study biofilms in the oral cavity[39] but is more frequently used in conjunction with a confocal laser scanning microscope (CLSM) as described below.

- CLSMs, which became available in the early 1990s, produce a series of very thin optical sections through the object under examination. These sections can be built up to produce a 3D image of the object. A CLSM can be used to investigate communities in their living, hydrated state and so provides valuable information concerning the true spatial organization of the constituent cells as well as the overall shape and dimensions of the community. It is a technique that has revolutionized our understanding of the structures of biofilm communities. Additional information can be obtained by using vital stains, fluorescent-labeled antibodies and labeled oligonucleotide probes. Furthermore, information about the nature of the environment within the biofilm (e.g. pH, Eh, etc.) can be obtained using appropriate probes. It is also possible to monitor gene expression within biofilms using reporter genes such as green fluorescent protein. Because it is so useful for studying biofilms, CLSM has been widely used by researchers investigating the various types of dental plaque[40-43]. It has also been used to study biofilms formed on particulate matter in the human colon[44] and to study the microbial communities associated with the intestinal mucosa[45].

1.3.2 Electron microscopy

The electron microscope is capable of much higher magnifications (up to several million times) and has a greater resolving power than a light microscope, allowing it to see much smaller objects in finer detail. Consequently it can provide information that is not obtainable by ordinary light microscopy, and the high magnifications that are possible can be used to reveal details of microbial adhesins and adhesive structures. The organisms that are present can be identified using antibodies conjugated to electron-dense markers (e.g. gold or ferritin). However, a major disadvantage of electron microscopy is that specimen processing and the accompanying dehydration alters the structure of the sample. There are two basic types of electron microscope: the transmission electron microscope (TEM) and the scanning electron microscope (SEM).

The TEM became commercially available in 1939 and has been used to study microbial communities inhabiting most body sites. In 1969 the location of microbes within epidermal samples taken from various body sites was investigated by TEM[46]. The TEM has also been used to demonstrate the presence of viruses in feces[47], the development of biofilms on the tooth surface[48], bacterial attachment to the oral mucosa[49], the distribution of bacteria on the vaginal mucosa[50], the formation of bacterial microcolonies on the tonsillar epithelium[51] and the attachment of bacteria to the urethral epithelium[52]. The SEM, while capable of only lower magnification than the TEM, has the great advantage of being able to produce three-dimensional images. Hence, the overall shape and structure of microbial communities can be visualized. It became commercially available in 1965 and has been used to demonstrate the formation of dental plaque[53], the attachment of bacteria to the pharyngeal epithelium[54], the presence of bacteria in the urethra[55], the association of bacteria with the colonic mucosa[56], bacteria adhering to the oral mucosa[57] and bacterial microcolonies on skin[58].

1.3.3 Culture-based approaches to microbial community analysis
1.3.3.1 Techniques

In 1881 Robert Koch demonstrated the use of solid culture media (with gelatin as the solidifying agent) on glass plates to isolate pure cultures of bacteria. This was a major breakthrough in practical microbiology and forms the basis of all subsequent culture-based methods of isolating, purifying and identifying microbes from the mixed communities that inhabit humans. Shortly afterwards, other members of Koch's laboratory improved on this approach by replacing gelatin with agar (in 1882) and by introducing petri dishes in place of glass plates (in 1887). In the early 1900s, selective media were developed for the isolation of specific groups of microbes from mixed communities. One of the first of these (1905) was MacConkey agar, which incorporated bile salts to inhibit the growth of all but lactose-fermenting bacteria from faecal samples[59]. Since then a huge variety of media have been developed including elective, diagnostic and chromogenic media in addition to selective media — in 1930 a total of 2,543 different culture media formulations were recognized[60]. As well as nutrients, microbes also require appropriate environmental conditions for growth. Until 1861, it was thought that all living creatures needed oxygen for growth but in 1861 Pasteur discovered microbes that could grow in the absence of air — these he called "anaerobies". He discovered the anaerobe *Clostridium butyricum* and showed that it produced butyric acid under anaerobic conditions. In 1878, he recognized that microbes could be divided into three groups depending on their relationship to air: "either exclusively aerobic, at once aerobic and anaerobic, or exclusively anaerobic". Pasteur removed oxygen from his culture media by boiling, but subsequently many techniques were developed to enable the cultivation of anaerobic microbes[61]. Between 1888 and 1918 over 300 different methods for producing anaerobic conditions were described in the literature[62]. Subsequent developments led (in the 1960s) to the anaerobic cabinets widely used today for the isolation and cultivation of anaerobic microbes. The ability to grow and identify anaerobic microbes represented a huge step forward in characterising the indigenous microbiota, as these organisms comprise significant and, indeed, dominant proportions of the communities inhabiting many body sites.

Most of our knowledge of the composition of indigenous microbial communities has come from using quantitative culture techniques. This involves some form of sample dispersion, plating out the sample (and usually dilutions of it) on various media, incubation, subculture of isolated colonies and then identification of each isolate. Given the complexity of the communities at most body sites, this is a very labor-intensive and painstaking task. There are also a number of problems with the technique. First of all, if a non-selective medium is to be used, then one must be chosen that can support the growth of all of the species likely to be present; this is virtually impossible given the disparate (and often very exacting) nutritional requirements of the members of such communities. Furthermore, it is difficult to provide the optimum environmental conditions (e.g. pH, oxygen content, CO_2 content, etc.) necessary to enable the growth of all the physiological types of microbes present. Problems arise as a result of some organisms growing faster than others, resulting in overgrowth of plates and failure to isolate slow-growing organisms. In samples taken from sites with a very dense microbiota (e.g. the colon, vagina, dental plaque), it is essential to use dilutions of the sample to obtain isolated colonies for subsequent identification. This means that organisms present in low proportions are "diluted out" and so are rarely isolated. Many studies have used selective media instead of, or in addition to, non-selective media. These can be useful, but analysis of a complex microbiota requires the use of a number of media selective for the various groups of organisms present. However, no medium can be relied upon to be truly selective and the inhibitory constituents may also have some adverse effect on the organisms for which the medium is supposedly selective. These problems all contribute to a greater workload, which inevitably results in an increase in the number of errors, a decrease in the number of samples that can be processed and hence a decrease in the statistical reliability of the data obtained[63].

Comparison of samples analysed by culture and by microscopy has revealed that even the best culture methods seriously underestimate the number of organisms present in the microbiotas of certain body sites — particularly those from the gastrointestinal tract (GIT) and oral cavity. The reasons for this are many and include: (i) the failure to satisfy the nutritional and/or environmental requirements for some of the organisms present, (ii) the presence in the community of organisms in a "viable but not cultivable" state, (iii) the failure to disrupt chains or clusters of organisms prior to plating out — this results in the production of only one "colony forming unit" from a cluster or chain consisting of many viable bacteria, (iv) the death of viable cells during transportation and processing of the sample[63]. Collectively, these difficulties have resulted in a serious underestimate of the number and variety of organisms in a sample taken from any environment, and it has been estimated that we are able to culture in the laboratory no more than 1-2% of the microbial species present on planet Earth (of which there are thought to be at least 10^{12}). Once individual isolates have been obtained, the next task is to identify each one. Traditionally this has involved the use of a battery of morphological, physiological and metabolic tests that is very labor-intensive and often not very discriminatory. The use of commercially available kits for this purpose has made the process less technically demanding. Other phenotypic tests that have been used for identification purposes include cell wall protein analysis, serology and fatty acid methyl ester analysis.

During the last few years, there has been an increasing trend to use molecular techniques for identifying the organisms isolated and one of these is based on the sequencing of genes encoding 16S rRNA[63]. The gene is amplified by PCR and the sequence of the resulting DNA determined and then compared with the sequences of the 16S rRNA genes of organisms that have been deposited in databases. If the sequence is >98% similar to that of one already in the database, then it is assumed that the gene is from the same species and hence the identity of an unknown organism can be established. The procedure is much simpler to perform than a battery of phenotypic identification tests and has the great advantage that it enables phylogenetic comparisons of the isolated organisms. However, some taxa are recalcitrant to PCR and some (e.g. many viridans streptococci) are so closely related that they cannot be differentiated using this approach. Alternative gene targets for speciation have been used including *recA, rpoB, tuf, gyrA, gyrB* and *cpn60* family proteins.

In fungi, the rRNA gene complex consists of four ribosomal genes, 18S (small subunit), 5.8S, 28S (large subunit) and 5S genes. Within this region, the internal transcribed spacer (ITS) and an approximately 600 bp D1/D2 region of the 28S subunit are the most phylogenetically variable regions and have been widely used for fungal taxonomy and identification.

A variety of other techniques are used for the identification and further characterization of isolated colonies. These include pulsed field gel electrophoresis (PFGE), ribotyping, multiplex PCR, arbitrary-primed PCR, matrix-assisted laser desorption/ionization time-of-flight mass spectrometry (MALDI-TOF-MS), Raman spectroscopy and Fourier transform-infrared (FT-IR) spectroscopy.

1.3.3.2 Outcomes

Early culture-based studies of the oral cavity revealed the presence of a number of species and what was known of the indigenous microbiota of the mouth in 1875 was summarised in an essay by Peirce[64]. In this review, reference is made to the presence of bacteria, yeasts and protozoa including *Oidium albicans* (*Candida albicans*), *Cryptococcus cerevisiae*, *Leptothrix buccalis* (*Leptotrichia buccalis*), *Leptomitus oculi*, vibrios and paramecia. Williams (1899)[65] isolated a variety of microbes from the mouth including diphtheroids, actinomyces, *Staphylococcus pyogenes albus*, *Staphylococcus pyogenes aureus* (*Staphylococcus aureus*), *Sarcina lutea* and *Bacillus buccalis maximus* (*Leptotrichia buccalis*).

In 1884, Theodor Escherich isolated a number of bacteria from the feces of infants including *Bacterium coli commune* (*Escherichia coli*), *Proteus vulgaris*, *Streptococcus coli gracilis*, *Bacillus subtilis*, *Bacterium lactis aerogenes* (*Klebsiella pneumoniae*) and *Micrococcus ovalis*. He also cultured four different yeasts including *Monilia candida* (*Candida albicans*) and a *Torula* species. In 1886, he published his work on the intestinal microbiota as a monograph entitled "The Intestinal Bacteria of the Infant and Their Relation to the Physiology of Digestion" (this was republished in English in 1988[66]). In 1896, Harris isolated *Proteus vulgaris* and *Bacterium coli commune* from the human duodenum[67] and Strauss isolated the Boas-Oppler bacillus (i.e. *Lactobacillus acidophilus*) from the stomach of healthy individuals[68].

In the late 1880s and early 1890s a number of investigators, including Maggiora Vergano (1889), Preindlsberger (1891) and Welch (1892), isolated and identified large numbers of different types of bacteria from the skin of healthy individuals[69-71]. The predominant organism was found to be a staphylococcus of low pathogenicity, which was named *Staphylococcus epidermis albus* (i.e. *Staphylococcus epidermidis*). *Staphylococcus pyogenes aureus* was isolated from the skin by Bockhart in 1887[72]. Bordoni-Uffreduzzi cultured five different species of micrococci and two bacilli from the skin including the foul-smelling *Bacterium graveolens*, which was found between the toes[73].

In a book published in 1892, Sternberg reviewed those microbes that had been cultivated from various body sites[74] and these are shown in Table 6. However, at this stage, investigators were concerned exclusively with determining the identities of those microbes present at a body site. No quantitative studies were undertaken and therefore there was no attempt to define the relative proportions of the various microbes present in a particular community. Culture-based approaches to analysing the indigenous microbiota of humans reached their zenith in the early 2000s, but since then they are increasingly being replaced by culture-independent methods. The "state-of-play" of our knowledge of the cultivable microbes present in these communities prior to the large-scale use of culture-independent techniques will now be summarised.

Table 6 List of microbes that had been cultivated from various body sites as reviewed by Sternberg in 1892[74].

Body site	Organisms cultivated from the site
Skin	*Diplococcus albicans tardus, Diplococcus citreus liquefaciens, Diplococcus flavus liquefaciens tardus, Staphylococcus viridis flavescens, Bacillus graveolens, Bacillus epidermidis, Ascobacillus citreus, Bacillus fluorescens liquefaciens minutissimus, Bacillus aureus, Bacillus ovatus minutissimus, Bacillus albicans pateriformis, Bacillus spiniferus, Micrococcus tetragenus versatilis, Bacillus Havaniensis liquefaciens*
Feces	*Streptococcus coli gracilis, Hicrococcus ahogenes, Micrococcus ovalis, Porzellanococcus, Bacillus subtilis, Bacillus aerogenes, Bacterium aerogenes, Bacterium lactis erythrogenes, Clostridium foetidum, Bacillus muscoides, Bacillus putrificus coli, Bacillus subtilis similis, Bacillus zopfii, Bacillus liquefaciens communis, Bacillus intestinus liquefaciens, Bacillus intestinus motilis, Bacillus fluorescens liquefaciens, Bacillus mesentericus vulgatus, Staphylococcus pyogenes aureus, Bacillus septicaemise haemorrhagicae, Bacillus enteritidis, Bacillus pseudomurisepticus, Bacillus coli communis, Bacillus lactis aerogenes, Bacillus cavicida, Bacillus coprogenes foetidus, Bacillus leporis lethalis, Bacillus acidiformans, Bacillus cuniculicida, Bacillus cadaveris, Bacillus cavicida Havaniensis, Proteus vulgaris, Helicobacterium aerogenes*
Conjunctiva	various micrococci and occasional bacilli
Nose	*Micrococcus nasalis, Diplococcus coryzae, Micrococcus albus liquefaciens, Micrococcus cumulates tenuis, Micrococcus tetragenus subflavus, Diplococcus fluorescens foetidus, Micrococcus totidus, Vibrio nasalis, Bacillus striatus flavus, Bacillus striatus albus, Staphylococcus pyogenes aureus, Staphylococcus pyogenes albus, Streptococcus pyogenes, Bacillus Of Friedlander, Bacillus foetidus ozaenae, Bacillus mallei, Bacillus smaragdinus foetidus, Diplococcus fluorescens foetidus, Micrococcus albus liquefaciens, Vibrio nasalis*
Mouth	*Micrococcus roseus, Sarcina pulmonum, Sarcina lutea, Micrococcus candicaiis, Bacillus virescens, Vibrio rugula, Vibrio lingualis, Pseudo-diphtheria bacillus, Bacillus mesentericus vulgatus, Bacillus subtilis, Bacillus subtilis similis, Bacillus radiciformis, Bacillus luteus, Bacillus fluorescens non-liquefaciens, Bacillus ruber, Bacillus viridiflavus, Proteus zenkeri, Vibrio viridans, Micrococcus nexifer, Iodococcus magnus, Ascococcus buccalis, Bacillus fuscans, Staphylococcus pyogenes albus, Staphylococcus pyogenes aureus, Staphylococcus salivarius septicus, Streptococcus pyogenes, Micrococcus salivarius septicus, Micrococcus tetragenus, Micrococcus gingivae pyogenes, Streptococcus septo-pysemicus, Streptococcus articulorum, Micrococcus pneumoniae crouposae, Bacillus of Friedlander, Bacillus bronchitidse putridae, Bacillus septicaemias haemorrhagicae, Bacillus gingivae pyogenes, Bacillus pulpse pyogenes, Bacillus dentalis viridans, Bacillus crassus sputigenus, Bacillus saprogenes, Bacillus pneumoniae agilis, Bacillus pneumoniae of Klein, Bacillus pneumosepticus, Spirillum sputigenum, Spirillum dentium, Vibrio rugula, Vibrio lingualis*
Stomach	*Sarcina ventriculi, Bacillus pyocyaneus, Bacillus lactis aerogenes, Bacillus subtilis, Bacillus mycoides, Bacillus amylobacter, Vibrio rugula, Micrococcus tetragenus mobilis ventriculi*
Vagina	*Diplococcus albicans amplus, Diplococcus albicans tardissimus*
Urethra	*Streptococcus giganteus urethrae, Bacillus nodosus parvus*

1.3.3.2.1 The skin

Gram-positive species (belonging to one or more of the genera *Propionibacterium, Staphylococcus* and *Corynebacterium*) are usually the numerically dominant organisms at any skin site[75]. As can be seen in Figure 1, in general, propionibacteria are the predominant organisms of sebum-rich regions (e.g. scalp, forehead).

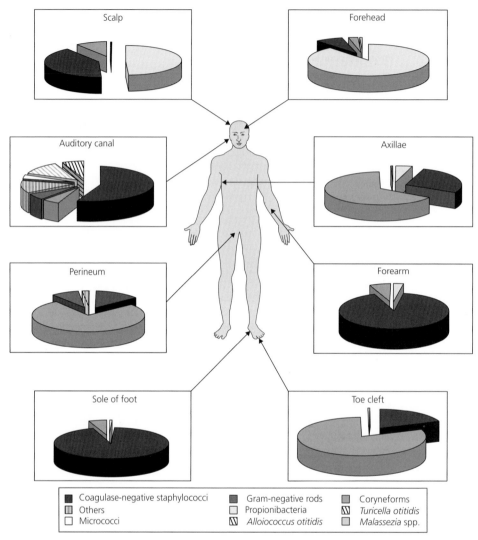

Figure 1 Relative proportions of the various organisms comprising the cultivable microbiota of a number of skin sites (reproduced with permission from Wilson M. *Bacteriology of humans: an ecological perspective.* Oxford: Wiley-Blackwell, 2008). (*see color plate section for color details*).

Staphylococci dominate in dry regions (e.g. arms, legs), while corynebacteria comprise the highest proportions of microbes in communities inhabiting moist regions (e.g. axillae, perineum). Apart from *Acinetobacter* spp., few Gram-negative species are present on the skin surface. As well as bacteria, fungi (Malassezia spp.) are found at many sites. Transients are often present on the skin surface, and these are derived from the environment and from other body sites that have openings onto the skin surface, e.g. the rectum, vagina, etc.

1.3.3.2.2 The conjunctiva
The conjunctival surfaces of a large proportion of individuals appear to be free of cultivable microbes and, when a microbial community is found, it tends to have a low population density and a simple composition — usually no more

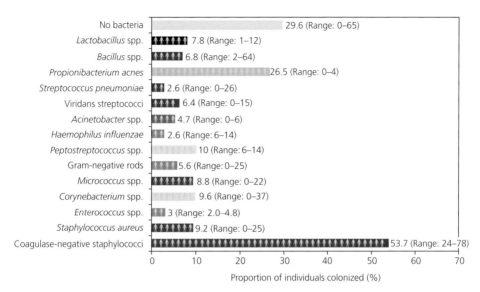

Figure 2 Frequency of detection of various microbes on the conjunctivae of healthy adults. The data shown are mean values (and ranges) derived from the results of 17 culture-based studies involving 4623 individuals from a number of countries (reproduced with permission from Wilson M. *Bacteriology of humans: an ecological perspective.* Oxford: Wiley-Blackwell, 2008).

than two species[76]. Coagulase-negative staphylococci (CNS) and, to a lesser extent, *Propionibacterium acnes* are the most frequently isolated organisms (Figure 2). Other organisms occasionally found include *Staph. aureus* and species belonging to the genera *Corynebacterium, Streptococcus, Lactobacillus, Peptostreptococcus, Bacillus* and *Micrococcus*. Gram-negative species are infrequently isolated. The eyelid margins have a similar cultivable microbiota to that found on the conjunctiva, but the population density tends to be higher.

1.3.3.2.3 The respiratory tract
Only the upper regions of the respiratory tract (the nose and pharynx) have resident microbial communities; the lower regions appear to be largely devoid of cultivable microbes[77]. Although the dominant cultivable organisms in the microbial communities at each site within the tract are known, the exact composition of each community is complex and poorly defined. This is not only because of the complexity of these communities, but is also attributable to the fact that the various regions of the respiratory tract are carriage sites of several very important human pathogens (*Streptococcus pyogenes, Neisseria meningitidis, Streptococcus pneumoniae, Haemophilus influenzae, Moraxella catarrhalis* and *Staph. aureus*) and, consequently, most bacteriological studies have focused on the detection of these organisms. Other more numerous members of the microbial communities have received little attention. In the nasopharynx (Figure 3) and oropharynx (Figure 4), the most frequently detected organisms include species belonging to the genera *Streptococcus* (mainly viridans streptococci), *Haemophilus, Neisseria, Staphylococcus* (mainly CNS), *Corynebacterium, Prevotella, Propionibacterium, Bacteroides, Porphyromonas* and *Veillonella*. Mollicutes are frequently present, but little is known regarding their identity or their exact prevalence.

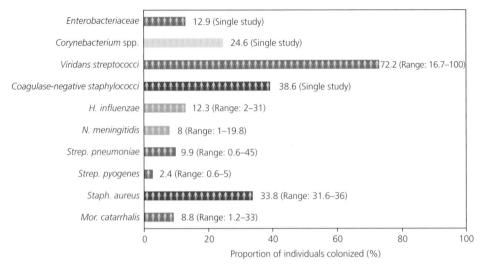

Figure 3 Organisms most frequently detected in the nasopharynx of adults (reproduced with permission from Wilson M. *Bacteriology of humans: an ecological perspective.* Oxford: Wiley-Blackwell, 2008).

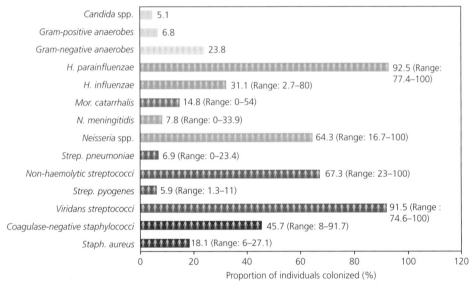

Figure 4 Organisms most frequently detected in the oropharynx (reproduced with permission from Wilson M. *Bacteriology of humans: an ecological perspective.* Oxford: Wiley-Blackwell, 2008).

The microbiotas of the nasal vestibule (Figure 5) and cavity (Figure 6) differ from those of the pharyngeal regions and the most frequently detected organisms are *Corynebacterium* spp., CNS and *Propionibacterium* spp.

1.3.3.2.4 The urinary tract of females

Only the urethra of the female urinary tract has a resident microbiota and, as this is relatively short, microbes can be detected along its entire length[78]. Remarkably few studies have been directed at ascertaining the composition of the urethral

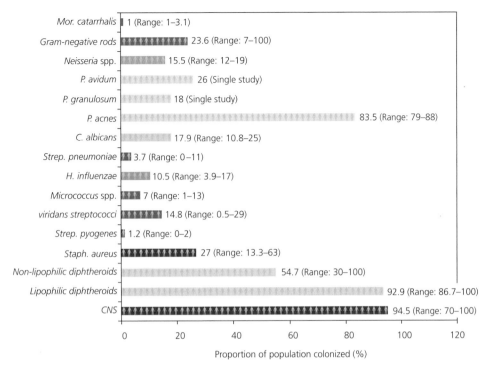

Figure 5 Frequency of detection (mean value and range) of various microbes in the nasal vestibule (reproduced with permission from Wilson M. *Bacteriology of humans: an ecological perspective*. Oxford: Wiley-Blackwell, 2008).

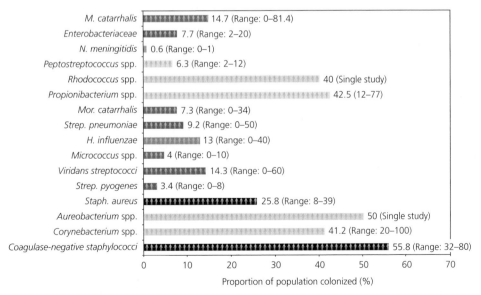

Figure 6 Frequency of detection (mean value and range) of various microbes in the nasal cavity (reproduced with permission from Wilson M. *Bacteriology of humans: an ecological perspective*. Oxford: Wiley-Blackwell, 2008).

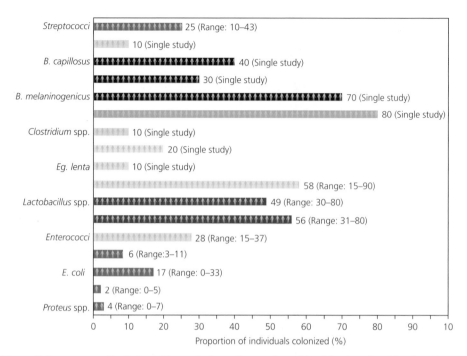

Figure 7 Frequency of isolation of bacteria from the urethra of healthy females. The data shown are the means (and ranges) based on the results of six studies involving 219 pre-menopausal females (reproduced with permission from Wilson M. *Bacteriology of humans: an ecological perspective*. Oxford: Wiley-Blackwell, 2008).

microbiota. The organisms most frequently isolated include *Corynebacterium* spp., Gram-positive anaerobic cocci (GPAC), *Bacteroides* spp., CNS and lactobacilli (Figure 7). However, the sexual maturity of the individual has a profound effect on microbial community composition. *Corynebacterium* spp., CNS, streptococci, and lactobacilli dominate the microbiota of pre-menarcheal girls and pre-menopausal women, whereas Gram-negative anaerobic bacilli and lactobacilli dominate that of post-menopausal individuals.

1.3.3.2.5 The reproductive system of females

Regions of the female reproductive system that are colonized by microbes are the vulva, vagina, and cervix[79]. In addition to being affected by the usual inter-individual variations (e.g. age, socioeconomic factors, etc.), the composition of the microbial communities at these sites is also profoundly influenced by the sexual maturity of the individual and, in females of reproductive age, the menstrual cycle. Although a wide variety of species have been detected within each of these communities in the population as a whole, in an individual female, each microbial community is generally dominated by a limited number of species. The species most frequently isolated from the vagina (Figure 8) and cervix (Figure 9) in females of reproductive age include lactobacilli, CNS, GPAC, Gram-negative anaerobic bacilli (GNAB), coryneforms and Mollicutes. Lactobacilli are generally the numerically dominant organisms in both the vaginal and cervical microbiotas — but this is not the case in pre-menarcheal girls and postmenopausal women who

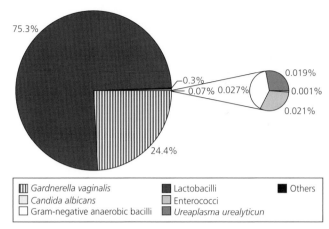

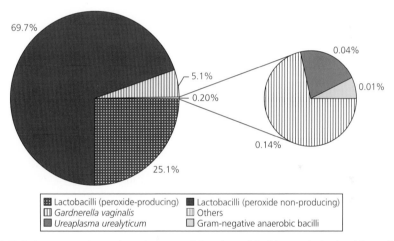

Figure 8 Relative proportions of the predominant organisms constituting the vaginal microbiota of 21 postmenarcheal/pre-menopausal, healthy, non-pregnant females (reproduced with permission from Wilson M. *Bacteriology of humans: an ecological perspective.* Oxford: Wiley-Blackwell, 2008).

Figure 9 Relative proportions of species comprising the cultivable cervical microbiota of 21 healthy, pre-menopausal females (reproduced with permission from Wilson M. *Bacteriology of humans: an ecological perspective.* Oxford: Wiley-Blackwell, 2008).

are not on hormone replacement therapy. The microbiota of the labia minora is similar to that of the vagina, whereas that of the labia majora consists of both vaginal and cutaneous species (Figure 10).

1.3.3.2.6 The reproductive and urinary systems of males

Of the various regions of the urinary and reproductive tracts in males, only the distal portion of the urethra and the glans penis appear to be colonized by microbes[80]. Unlike in females, where microbes are found along the whole length of the urethra, microbes can be detected only in the distal 6 cm of the male urethra. CNS, viridans streptococci, *Corynebacterium* spp., GPAC and GNAB are the most frequently encountered organisms (Figure 11), and the composition of the micro-biota varies along the urethra. Engaging in sexual activity has a dramatic effect on

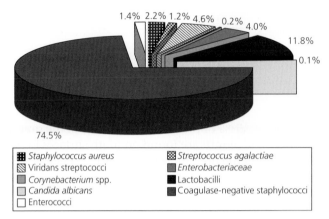

Figure 10 Relative proportions of the various microbes that comprise the cultivable microbiota of the labia majora of post-menarcheal/pre-menopausal females. Data represent the mean values obtained in a study involving 102 individuals (reproduced with permission from Wilson M. *Bacteriology of humans: an ecological perspective*. Oxford: Wiley-Blackwell, 2008). (*see color plate section for color details*).

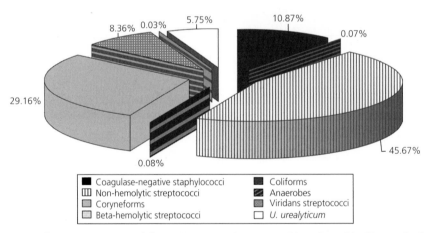

Figure 11 Relative proportions of the various organisms comprising the cultivable urethral microbiota of adult males. Data are derived from an analysis of 60 adult males (reproduced with permission Wilson M. *Bacteriology of humans: an ecological perspective*. Oxford: Wiley-Blackwell, 2008).

the urethral microbiota — it becomes more complex and contains organisms derived from the vagina. The microbiota of the glans penis differs substantially between circumcised and uncircumcised individuals, the population density and species diversity being greater in the latter. *Malassezia* spp., anaerobes, and facultative Gram-negative bacilli are frequently encountered in uncircumcised individuals, whereas CNS, *Propionibacterium* spp., and *Corynebacterium* spp. dominate in circumcised individuals.

1.3.3.2.7 The oral cavity

Because of its complex anatomy, the oral cavity has a large variety of habitats available for microbial colonization[81]. Uniquely, it also has non-shedding surfaces, the teeth, that make possible the formation of substantial and complex biofilms. The oral cavity harbors a variety of microbial communities, most of which have a high species diversity. As many as 700 phylotypes have been detected in the oral cavity, and approximately 50% of these have not yet been cultivated. Although mucosal surfaces comprise 80% of the total surface area of the oral cavity, most of the microbes present in the mouth are found on tooth surfaces in biofilms known as dental plaques. The microbial composition of these plaques is complex and is dependent on their anatomical location. In supragingival plaques, viridans streptococci and *Actinomyces* spp. are usually the dominant organisms, but anaerobes such as *Veillonella* spp. and *Fusobacterium* spp. are also invariably present (Figure 12). The microbial composition of plaque alters with time and is also affected by the host's diet. The microbiota of the plaque found in the gingival crevice is more diverse than that of supragingival plaques and, although streptococci are usually the dominant organisms, the proportion of anaerobes is greater than in supragingival plaques (Figure 13). Anaerobic organisms frequently detected include *Veillonella* spp., Gram-positive anaerobic cocci, *Prevotella* spp., *Fusobacterium* spp., *Selenomonas* spp., *Eubacterium* spp. and spirochetes.

The tongue is densely colonized by microbes, and the composition of the resident communities varies with the anatomical location (Figure 14). Streptococci, again, are generally the dominant organisms, and a variety of anaerobes are frequently present, including species belonging to the genera *Prevotella, Veillonella, Eubacterium* and *Fusobacterium*. Other mucosal surfaces are relatively sparsely populated compared with the tongue. The community composition varies with the anatomical location, but facultative anaerobes and capnophiles are usually the dominant organisms, e.g. streptococci, *Gemella* spp., *Neisseria* spp., *Haemophilus* spp. and *Capnocytophaga* spp. However, anaerobes such as *Fusobacterium* spp., *Veillonella* spp. and *Prevotella* spp. are also often present.

1.3.3.2.8 The gastro-intestinal tract

The gastro-intestinal tract (GIT) has a number of distinct regions, each harboring a characteristic microbial community or communities[82]. In the upper GIT (oral cavity, pharynx and esophagus), the resident microbiota is associated with surfaces, and because material (food, secretions, etc.) passes rapidly through these regions, microbial communities cannot become established in their lumen. As the passage of material becomes slower in the lower regions of the GIT, there is an opportunity for communities to develop within the lumen as well as on the mucosal surface — in the distal ileum, cecum, colon and rectum such communities are substantial. Very few studies have investigated the oesophageal microbiota, but these few have shown that it is dominated by staphylococci, lactobacilli and *Corynebacterium* spp. (Figure 15).

Because of its low pH, the stomach is a hostile environment for a wide range of organisms. Organisms detected in the lumen are mainly acid-tolerant species of streptococci and lactobacilli together with staphylococci, *Neisseria* spp., and various anaerobes. These organisms are also present in the mucosa-associated community, which in addition often contains the important pathogen *Helicobacter pylori* (Figure 16).

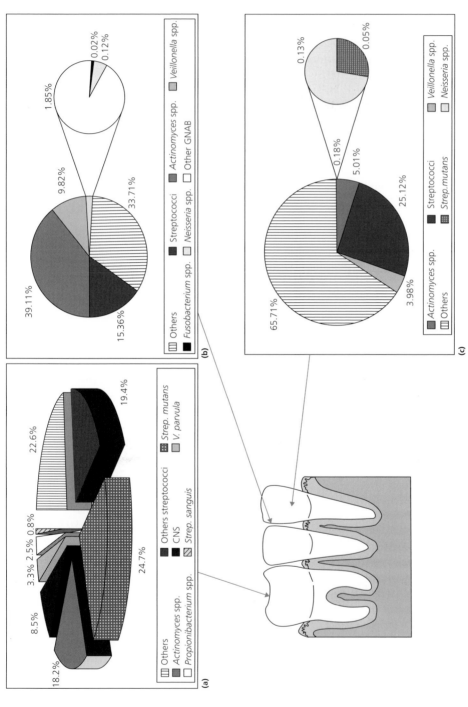

Figure 12 The predominant cultivable microbiota of the three main types of supragingival plaque: (a) fissure, (b) approximal, and (c) smooth surface. Data are derived from three studies involving a total of 40 healthy adults (reproduced with permission from Wilson M. *Bacteriology of humans: an ecological perspective.* Oxford: Wiley-Blackwell, 2008).

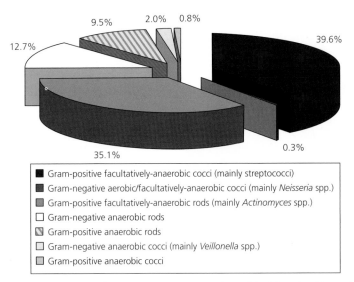

- ■ Gram-positive facultatively-anaerobic cocci (mainly streptococci)
- ■ Gram-negative aerobic/facultatively-anaerobic cocci (mainly *Neisseria* spp.)
- ■ Gram-positive facultatively-anaerobic rods (mainly *Actinomyces* spp.)
- □ Gram-negative anaerobic rods
- ◨ Gram-positive anaerobic rods
- □ Gram-negative anaerobic cocci (mainly *Veillonella* spp.)
- ▨ Gram-positive anaerobic cocci

Figure 13 Relative proportions of organisms comprising the cultivable microbiota of the gingival crevice. Data are derived from a study involving seven healthy adults (reproduced with permission from Wilson M. *Bacteriology of humans: an ecological perspective*. Oxford: Wiley-Blackwell, 2008). (*see color plate section for color details*).

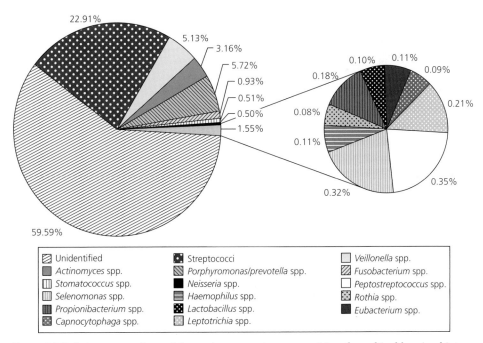

▨ Unidentified	▨ Streptococci	□ *Veillonella* spp.
■ *Actinomyces* spp.	▨ *Porphyromonas/prevotella* spp.	▨ *Fusobacterium* spp.
▥ *Stomatococcus* spp.	■ *Neisseria* spp.	□ *Peptostreptococcus* spp.
▥ *Selenomonas* spp.	▬ *Haemophilus* spp.	▨ *Rothia* spp.
▥ *Propionibacterium* spp.	▨ *Lactobacillus* spp.	■ *Eubacterium* spp.
▨ *Capnocytophaga* spp.	▨ *Leptotrichia* spp.	

Figure 14 Relative proportions of the various organisms comprising the cultivable microbiota of the tongue. Data are derived from a study involving 17 healthy adults (reproduced with permission from Wilson M. *Bacteriology of humans: an ecological perspective*. Oxford: Wiley-Blackwell, 2008).

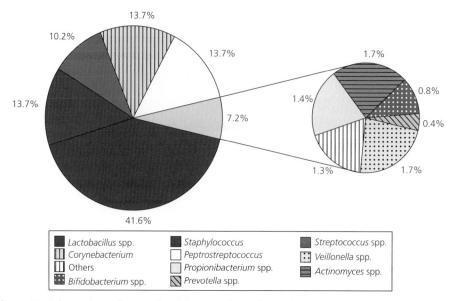

Figure 15 Culture-dependent study of the oesophageal microbiota. Relative proportions of the organisms present. Data are mean values derived from the results of two studies involving 17 healthy adults (reproduced with permission from Wilson M. *Bacteriology of humans: an ecological perspective*. Oxford: Wiley-Blackwell, 2008).

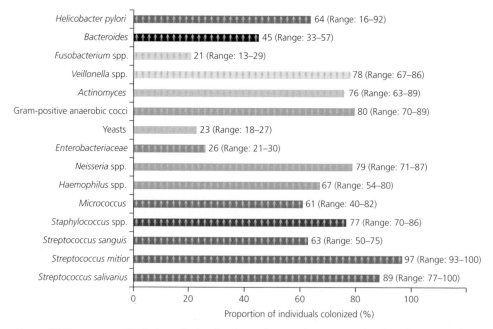

Figure 16 Frequency of isolation of microbes from the gastric mucosa. The data shown are the means (and ranges) based on the results of two studies involving 58 adults (reproduced with permission from Wilson M. *Bacteriology of humans: an ecological perspective*. Oxford: Wiley-Blackwell, 2008).

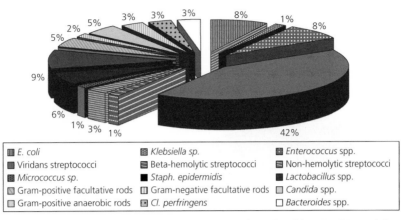

Figure 17 Relative proportions of organisms comprising the cultivable microbiota of the duodenal mucosa of 26 healthy adults (reproduced with permission from Wilson M. *Bacteriology of humans: an ecological perspective.* Oxford: Wiley-Blackwell, 2008). (*see color plate section for color details*).

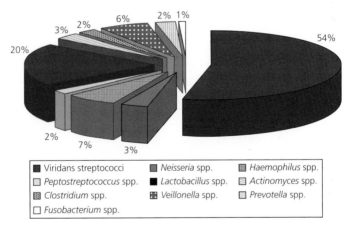

Figure 18 Relative proportions of organisms comprising the cultivable microbiota of the jejunal mucosa of 20 healthy adults (reproduced with permission from Wilson M. *Bacteriology of humans: an ecological perspective.* Oxford: Wiley-Blackwell, 2008).

The environments within the duodenum and jejunum are also largely inimical to many microbes because of the low pH, the presence of bile (and other antimicrobial compounds), and the rapid transit of material. Consequently, the mucosa and the lumen of both of these regions have sparse microbiotas consisting mainly of acid-tolerant streptococci and lactobacilli (Figures, 17 and 18).

In the ileum, especially the terminal region, conditions are less hostile to microbes, and the microbiotas within the lumen and on the mucosa are more substantial. Streptococci, enterococci and coliforms are the dominant organisms in the lumen, but the microbiota of the mucosa is very different and consists of high proportions of anaerobes including *Bacteroides* spp., *Clostridium* spp., GPAC and *Bifidobacterium* spp. The cecum has a lower pH and a higher content of easily fermentable compounds than the more distal regions of the GIT, and consequently

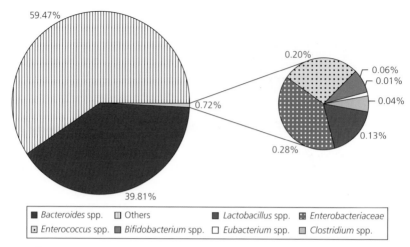

Figure 19 Relative proportions of organisms comprising the cultivable microbiota of the caecal mucosa in 19 healthy adults (reproduced with permission from Wilson M. *Bacteriology of humans: an ecological perspective.* Oxford: Wiley-Blackwell, 2008).

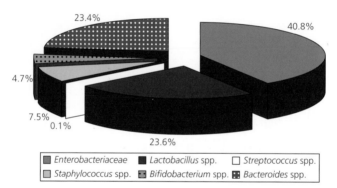

Figure 20 Relative proportions of organisms comprising the cultivable microbiota of the contents of the cecum in 21 healthy adults (reproduced with permission from Wilson M. *Bacteriology of humans: an ecological perspective.* Oxford: Wiley-Blackwell, 2008).

it harbors microbial communities that are very different from those in the rest of the large intestine. The lumen is dominated by facultative organisms (mainly Enterobacteriaceae and lactobacilli), although substantial proportions of anaerobes (*Bacteroides* spp. and *Clostridium* spp.) are also present (Figure 19). The mucosal microbiota appears to be dominated by *Bacteroides* spp. (Figure 20).

The colon is colonized by a very large and diverse microbial population. Up to 80% of the organisms present have not yet been grown in the laboratory, and many of these are novel phylotypes. Most of our knowledge of the colonic microbiota has come from analysis of feces in which obligate anaerobes are 1000-fold greater in number than facultative organisms — the predominant genera being *Bacteroides, Eubacterium* (and related genera), *Bifidobacterium* and *Clostridium* (Figure 21). The microbiota of the colonic mucosa is dominated by *Bacteroides* spp. (Figure 22).

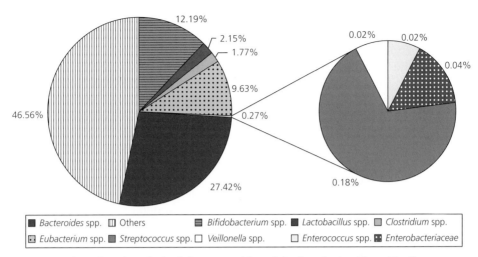

Figure 21 Culture-based analysis of the composition of the faecal microbiota. The figures represent mean values for the relative proportions of the various genera —— these have been derived from the results of ten studies involving 212 healthy adults from several countries (reproduced with permission from Wilson M. *Bacteriology of humans: an ecological perspective.* Oxford: Wiley-Blackwell, 2008).

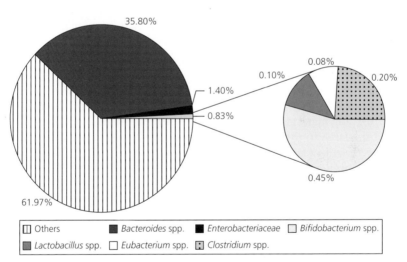

Figure 22 Cultivable microbiota of the colonic mucosa. Figures denote mean values for the relative proportions of the various organisms isolated, and are derived from the results of three studies involving 61 healthy adults (reproduced with permission from Wilson M. *Bacteriology of humans: an ecological perspective.* Oxford: Wiley-Blackwell, 2008).

1.4 Culture-independent approaches to microbial community analysis

Many of the problems inherent in culture-based approaches to analysing microbial communities can be circumvented by the use of molecular techniques, the main advantage being that organisms that have not yet been cultivated in the laboratory can be identified in these communities. However, it must be pointed out that such approaches are not without their own problems[63]. The first stage in the analysis of a microbial community by a molecular technique is to isolate either DNA or RNA from the sample. Extraction of nucleic acids from microbes requires that the cells are lysed and the ease of lysis varies significantly among different organisms. Numerous protocols for the lysis of microbes present in samples have been devised and include the use of enzymes, chemicals and mechanical methods. Care has to be taken that, once lysed, the nucleic acids do not undergo shearing and are not degraded by nucleases. The extracted nucleic acids can then be used in a variety of ways to reveal the identity of the microbes originally present in the sample and/or to produce a "profile" or "fingerprint" of the microbial community.

Universal primers can be used to amplify all of the 16S rRNA genes present in the DNA extracted from the sample and the amplified sequences are then cloned and sequenced. The sequences of the clones are then determined and compared to sequences in databases. If a sequence is >98% similar to one already in the database, then it is regarded as identical and so the corresponding organism can be assumed to have been present in the sample[83,84]. In this way the sequences of the 16S rRNA genes of all organisms present in the community can be determined and, if these sequences match those of known organisms in databases, then the identities of all the organisms present will be revealed. However, not all of the sequences of the 16S rRNA genes obtained correspond to sequences in databases. A significant limitation of 16S rRNA sequencing is the introduction of biases by PCR primer design, which may select for or against particular groups of organisms. Furthermore, bacterial contamination of reagents may occur, and therefore extensive controls are required. The 16S rRNA operon is also present in between one and fifteen copies in bacterial genomes and this can influence the apparent relative abundance of an organism.

Another useful approach is to separate the amplicons produced by either temperature-gradient gel electrophoresis or denaturing-gradient gel electrophoresis[85]. Staining of the DNA in the resulting gel reveals a banding pattern or "fingerprint" that is characteristic of that particular community. The individual bands can be cut out and each amplicon eluted, re-amplified, sequenced and identified using databases as described above. Alternatively, the fingerprints produced from samples from the same individual obtained on different occasions can be compared and analysed for differences. Hence, bands appearing or disappearing with time can be sequenced to determine the gain or loss of an organism from the community. The method is also useful for comparing the microbiotas present at the same body site in different individuals and this is facilitated by computer and statistical analysis of the banding patterns obtained.

Microarrays offer yet another possible approach. A number of different phylogenetic microarrays, consisting of multiple probes designed to discriminate specific organisms or subgroups of organisms, have been developed[86]. However, the probes and the organisms targeted must be preselected, which limits their usefulness in detecting novel organisms in the community under investigation. A microarray known as the Human Intestinal Tract Chip (HITChip) has been developed to detect target 1140 different phylotypes in the human intestinal microbiota[87].

Quantitative PCR can be used to quantify species, or groups of organisms, in a community[88]. However, this is technically demanding and time-consuming.

Metagenomics, the analysis of all the genomes present in an ecosystem, was first described in 1998 and has become an extremely powerful tool for analysing microbial communities both in terms of what species are present as well as the activities of which they are capable[89]. New sequencing technologies (e.g. pyrosequencing, Illumina) have made possible shotgun sequencing of the metagenomic DNA of a microbial community inhabiting a body site in a rapid and cost-effective manner. For example, in a study of the faecal microbiota of 124 subjects, metagenomic analysis revealed the presence of approximately 1150 bacterial species of which about 160 species were associated with each individual[90]. 57 species were common to more than 90% of individuals.

The above approaches let microbial community structure be determined in terms of species richness, community evenness and diversity. Such measures can reveal a great deal about the dynamics and selection pressures experienced by the system. Increased richness, evenness and diversity can be associated with stable, longer-established or less-active ecosystems[91]. Furthermore, association of these measures with environmental and clinical parameters can give valuable insights into states of health and disease[92,93].

1.5 Determination of microbial community functions

Metagenomics not only enables us to determine what organisms are present in a community but, as it provides a catalogue of the genes present, also furnishes us with information concerning the range of activities that the community has the potential to undertake. Such studies have revealed that the intestinal microbiota is enriched in gene categories (i.e. clusters of orthologous groups of proteins; http://www.ncbi.nlm.nih.gov/COG; Kyoto Encyclopedia of Genes and Genomes; http://www.genome.jp/kegg) involved in carbohydrate metabolism, energy metabolism and storage, generation of SCFAs, amino acid metabolism, biosynthesis of secondary metabolism and metabolism of cofactors and vitamins[94].

Other technologies that have been developed to provide information about community function rather than composition include metatranscriptomics, metaproteomics and metabolomics.

Metatranscriptomics involves extracting and sequencing the mRNA molecules present in the microbial community, thereby identifying which of the genes present are actually being expressed[95]. For example, a metatranscriptomic analysis of the faecal microbiota has demonstrated that the main functionalities expressed were carbohydrate metabolism, energy production and synthesis of cellular components[96].

Metaproteomics involves analysing all of the proteins present in a microbial community and is an approach that is developing very rapidly due to improvements in protein separation techniques combined with highly accurate, high-throughput mass spectrometry[97]. In a study of the feces from three healthy females, a core metaproteome (i.e. those proteins shared between different individuals) was identified[98]. This consisted of 1216 proteins among which glutamate dehydrogenases, pyruvate formate lyases, phosphoenolpyruvate carboxykinases, GroEL chaperonins and a NifU protein with chaperone function were highly abundant. These core functions have also been identified in other intestinal metaproteome studies. Current studies of the intestinal metaproteome emphasize the importance of proteins involved in carbohydrate metabolism, in maintaining protein integrity and in coping with the low redox potential in the gut. Specifically, flagellins, glutamate dehydrogenase and glyceraldehydes-3-phosphate dehydrogenase appear to be essential proteins for maintaining bacterial life in the intestinal tract.

Metabolomics involves analysing (usually by gas chromatography-mass spectrometry) the metabolites produced by a microbial community. In a recent metabolomics study of feces obtained from healthy and cirrhotic patients, a total of 9,215 metabolites were detected[99]. Six major groups of metabolites (bile acids, bile pigments, lysophosphatidylcholines, aromatic amino acids, fatty acids and acylcarnitines) were found to be significantly altered in the cirrhotic patients compared with healthy controls.

Data from studies using the above culture-independent techniques, along with data from culture-dependent studies such as those described previously in this chapter, has contributed enormously to our understanding of the composition and function of the indigenous microbiota of humans. Progress in the field has been rapid thanks to the efforts of research consortia such as the MetaHIT (Metagenomics of the Human Intestinal Tract; www.metahit.eu) project supported by the European Union and the Human Microbiome Project funded by the National Institutes of Health in the USA (http://www.hmpdacc.org). Contemporary views of the nature of the microbial communities found at various body sites will be described in subsequent chapters (see chapters 3, 4, 5, 16 and 24).

1.6 Closing remarks

Imagine the surprise and excitement, as well as an undoubted feeling of trepidation, three and a half centuries ago at the discovery of little "animalcules" wriggling around inside our mouths and intestines. Since then, our microbial companions have never ceased to surprise and amaze us. First they became something to be feared as we discovered that they can cause a range of diseases, some of which (e.g. meningitis, pneumonia) are life-threatening. Then they became something to be cherished as we found that they are essential for our development, are involved in protecting us from exogenous pathogens and supply us with nutrients, energy and vitamins. Now, in the second decade of the 21st century, we are learning that they are also involved in determining our physique, behavior and mood. An historical appreciation of the development of our relationship with our microbiota can only teach us that we are in store for many more surprises.

TAKE-HOME MESSAGE

- A major driving force behind the resurgence of interest in the human microbiota is an increasing awareness of its importance in human development, nutrition, behavior, wellbeing and disease.

- Knowledge of the human microbiota has increased dramatically in the 21st century due to the development of technologies that can identify which microbes are present in a community, as well as determining their possible functions, without needing to grow them in the laboratory.

- We are only just beginning to understand the composition and functions of the microbial communities that inhabit humans.

References

1 Ogston A. Report upon micro-organisms in surgical diseases. *Br Med J* 1881; **1**: 369.

2 Rosenbach FJ. *Mikro-organismen bei den wund-infections-krankheiten des menschen.* Weisbaden: JF Bergman. 1884. p. 18.

3 Fraenkel A. 1884 Ueber die genuine Pneumonie, Verhandlungen des Congress für innere Medicin. *Dritter Congress* **3**, 17–31, April 21.

4 Welch WH, Nuttall GHF. A gas-producing bacillus (*Bacillus aerogenes capsulatus,* Nov. Spec.) capable of rapid development in the body after death. *Bull Johns Hopkins Hosp Baltimore* 1892; **3**: 81–91.

5 Veillon A, Zuber A. Recherches sur quelques microbes strictement anaérobies et leur rôle en pathologie. *Arch Med Exper d'Anat Path* 1898; **10**: 517–545.

6 "Report on a Study of Micro-organisms associated with Rheumatic Fever and Malignant Endocarditis," Report of the Medical Officer, Local Govt. Board, London, 1906–7, p. 279.

7 Fraenkel A. Ueber peritoneal Infektion. Wiener klin. Wochenschr., 1891 Nos. 13–15.

8 Turnbaugh PJ, Hamady M, Yatsunenko T *et al.* A core gut microbiome in obese and lean twins. *Nature* 2008; **457**: 480–4.

9 Suzaki H, Watanabe S, Pawankar R. Rhinosinusitis and asthma — microbiome and new perspectives. *Curr Opin Allergy Clin Immunol* 2013; **13**: 45–9.

10 Dickson RP, Erb-Downward JR, Huffnagle GB. The role of the bacterial microbiome in lung disease. *Expert Rev Respir Med* 2013; **7**: 245–57.

11 Ege MJ, Mayer M, Normand A-C *et al.* Exposure to environmental microorganisms and childhood asthma. *N Engl J Med* 2011; **364**: 701–9.

12 Hsiao EY, McBride SW, Hsien S *et al.* Microbiota modulate behavioral and physiological abnormalities associated with neurodevelopmental disorders. *Cell* 2013; **155**: 1451–63.

13 Bringiotti R, Lerardi E, Lovero R *et al.* Intestinal microbiota: The explosive mixture at the origin of inflammatory bowel disease? *World J Gastrointest Pathophysiol* 2014; **5**: 550–9.

14 Berer K, Mues M, Koutrolos M *et al.* Commensal microbiota and myelin autoantigen cooperate to trigger autoimmune demyelination. *Nature* 2011; **479**: 538–41.

15 Wu HJ, Ivanov II, Darce J *et al.* Gut-residing segmented filamentous bacteria drive autoimmune arthritis via T helper 17 cells. *Immunity* 2010; **32**: 815–827.

16 Zaura E, Nicu EA, Krom BP, Keijser BJ. Acquiring and maintaining a normal oral microbiome: current perspective. *Front Cell Infect Microbiol.* 2014; **4**: 85.

17 Louis P, Hold GL, Flint HJ. The gut microbiota, bacterial metabolites and colorectal cancer. *Nature Revs Microbiol* 2014: **12**; 661–672.

18 Wu X, Ma C, Han L *et al.* Molecular characterisation of the faecal microbiota in patients with type II diabetes. *Curr Microbiol* 2010; **61**: 69–78.

19 Pasteur L. Observations relatives à la note précédente de M. Duclaux. *C R Acad Sci (Paris).* 1885; **100**: 68.

20 Bohnhoff M, Miller CP. Enhanced susceptibility to salmonella infection in streptomycin-treated mice. *J Infect Dis* 1962; **111**: 117–127.

21 Gordon HA, Pesti L. The gnotobiotic animal as a tool in the study of host microbial relationships. *Bacteriol Rev* 1971; **35**: 390.

22 Wilson M. Role of the indigenous microbiota in maintaining human health. In: *Microbial Inhabitants of Humans: their ecology and role in health and disease* (Wilson M ed.). Cambridge: Cambridge University Press, 2005; 375–394.

23 Zocco MA, Ainora ME, Gasbarrini G, Gasbarrini A. *Bacteroides thetaiotaomicron* in the gut: molecular aspects of their interaction. *Dig Liver Dis* 2007; **39**: 707–12.

24 Sommer F, Bäckhed F. The gut microbiota — masters of host development and physiology. *Nat Rev Microbiol* 2013; **11**: 227–238.

25 Bengmark S. Gut microbiota, immune development and function. *Pharmacol Res* 2013; **69**: 87–113.

26 Flint HJ, Scott KP, Louis P, Duncan SH. *Nat Rev Gastroenterol Hepatol* 2012; **9**: 577–589.

27 Joyce SA, Gahan CGM. The gut microbiota and the metabolic health of the host. *Curr Opin Gastroenterol* 2014; **30**: 120–127.

28 LeBlanc JG, Milani C, de Giori GS *et al.* Bacteria as vitamin suppliers to their host: a gut microbiota perspective. *Curr Opin Biotechnol;.* 2013; **24**: 160–8.

29 Knasmüller S, Steinkellner H, Hirschl AM *et al.* Impact of bacteria in dairy products and of the intestinal microflora on the genotoxic and carcinogenic effects of heterocyclic aromatic amines. *Mutat Res* 2001; **480**: 129–38.

30 Cryan JF, Dinan TG. Mind-altering microorganisms: the impact of the gut microbiota on brain and behaviour. *Nat Rev Neurosci* 2012; **13**: 701–12; Moloney RD, Desbonnet L, Clarke G *et al.* The microbiome: stress, health and disease. *Mamm Genome* 2014; **25**: 49–74.

31 Galland L. The gut microbiome and the brain. *J Med Food* 2014; **17**: 1–12.

32 Listgarten MA, Helldén L. Relative distribution of bacteria at clinically healthy and periodontally diseased sites in humans. *J Clin Periodontol* 1978; **5**: 115–132.

33 Theilade E, Fejerskov O, Prachyabrued W, Kilian M. Microbiologic study on developing plaque in human fissures. *Scand J Dent Res* 1974; **82**: 420–7.

34 Ritz HL. Localization of nocardia in dental plaque by immunofluorescence. *Proc Soc Exp Biol Med* 1963; **113**: 925–9.

35 Langendijk PS, Schut F, Jansen GJ *et al.* Quantitative fluorescence *in situ* hybridization of *Bifidobacterium* spp. with genus-specific 16S rRNA-targeted probes and its application in fecal samples. *Appl Environ Microbiol* 1995; **61**: 3069–75.

36 Fredricks DN, Fiedler TL, Marrazzo JM. Molecular identification of bacteria associated with bacterial vaginosis. *N Engl J Med* 2005; **353**: 1899–911.

37 Moter A, Hoenig C, Choi BK *et al.* Molecular epidemiology of oral treponemes associated with periodontal disease. *J Clin Microbiol* 1998; **36**: 1399–403.

38 Jahns AC, Oprica C, Vassilaki I *et al.* Simultaneous visualization of *Propionibacterium acnes* and *Propionibacterium granulosum* with immunofluorescence and fluorescence *in situ* hybridization. *Anaerobe* 2013; **23**: 48–5.

39 Netuschil L, Reich E, Brecx M. Direct measurement of the bactericidal effect of chlorhexidine on human dental plaque. *J Clin Periodont* 1989; **16**: 484–88.

40 Netuschil L, Reich E, Unteregger G *et al.* A pilot study of confocal laser scanning microscopy for the assessment of undisturbed dental plaque vitality and topography. *Arch Oral Biol* 1998; **43**: 277–85.

41 Wood SR, Kirkham J, Marsh PD *et al.* Architecture of intact natural human plaque biofilms studied by confocal laser scanning microscopy. *J Dent Res* 2000; **79**: 21–7.

42 Zaura-Arite E, van Marle J, ten Cate JM. Conofocal microscopy study of undisturbed and chlorhexidine-treated dental biofilm. *J Dent Res* 2001; **80**: 1436–40.

43 Palmer RJ Jr, Gordon SM, Cisar JO, Kolenbrander PE. Coaggregation-mediated interactions of streptococci and actinomyces detected in initial human dental plaque. *J Bacteriol* 2003; **185**: 3400–9.

44 Macfarlane S, Macfarlane GT. Composition and metabolic activities of bacterial biofilms colonizing food residues in the human gut. *Appl Environ Microbiol.* 2006; **72**: 6204–11.

45 Ahmed S, Macfarlane GT, Fite A *et al.* Mucosa-associated bacterial diversity in relation to human terminal ileum and colonic biopsy samples. *Appl Environ Microbiol* 2007; **73**: 7435–42.

46 Montes LF, Wilborn WH. Location of bacterial skin flora. *Br J Dermatol* 1969; **81**: **Suppl 1**: 23–26.

47 Flewett TH, Bryden AS, Davies H. Diagnostic electron microscopy of faeces. I. The viral flora of the faeces as seen by electron microscopy. *J Clin Pathol* 1974; **27**: 603–8.

48 Frank RM, Brendel A. Ultrastructure of the approximal dental plaque and the underlying normal and carious enamel. *Arch Oral Biol* 1966; **11**: 883–912.

49 Vitkov L, Krautgartner WD, Hannig M, Fuchs K. Fimbria-mediated bacterial adhesion to human oral epithelium. *FEMS Microbiol Lett* 2001; **202**: 25–30.

50 Sadhu K, Domingue PAG, Chow AW *et al.* A morphological study of the in situ tissue-associated autochthonous microflora of the human vagina. *Microbial Ecol Health Dis* 1989; **2**: 99–106.

51 Fredriksen F, Räisänen S, Myklebust R, Stenfors LE. Bacterial adherence to the surface and isolated cell epithelium of the palatine tonsils. *Acta Otolaryngol* 1996; **116**: 620–6.

52 Zaviacic M, Jakubovský J, Polák S *et al.* Rhythmic changes of human female uroepithelial squamous cells during menstrual cycle. Transmission and scanning electron microscopic study. *Int Urol Nephrol* 1984; **16**: 301–9.

53 Connor JN, Schoenfeld CM, Taylor RL. Study of in vivo plaque formation. *J Dent Res* 1976; **55**: 481–8.

54 Ebenfelt A, Geterud A, Granström G, Lundberg C. Imprints from the oropharyngeal mucosa: a novel method for studies of cell-kinetics and spatial relations between leukocytes, epithelial cells and bacteria in the secretion on the surface of the mucosa. *Acta Otolaryngol* 1995; **115**: 106–11.

55 Colleen S, Myhrberg H, Mårdh PA. Bacterial colonization of human urethral mucosa. I. Scanning electron microscopy. *Scand J Urol Nephrol* 1980; **14**: 9–15.

56 Macfarlane S, Macfarlane GT. Bacterial growth on mucosal surfaces and biofilms in the large bowel. In: *Medical Implications of Biofilms* (Wilson M, Devine D, eds). Cambridge: Cambridge University Press, 2003; 262–286.

57 Vitkov L, Krautgartner WD, Hannig M, Fuchs K. Fimbria-mediated bacterial adhesion to human oral epithelium. *FEMS Microbiol Lett* 2001; **202**: 25–30.

58 Katsuyama M, Kobayashi Y, Ichikawa H *et al.* A novel method to control the balance of skin micro-flora, Part 2. A study to assess the effect of a cream containing farnesol and xylitol on atopic dry skin. *J Dermatol Sci* 2005; **38**: 207–13.

59 MacConkey A. Lactose-fermenting bacteria in faeces. *J Hyg* 1905; **8**: 333–379.

60 Levine M, Schoenlein HW (1930) A compilation of culture media. Baltimore, Williams and Williams.

61 Roux E. Sur la culture des microbes anaérobies. *Ann de l'Inst Past* 1887; **1**; 49.

62 Bulloch W. *The History of Bacteriology*. London and New York: Oxford University Press, 1938.

63 Wilson M. The human-microbe symbiosis. In: *Bacteriology of humans: an ecological perspective*. Oxford: Wiley-Blackwell, 2008; 1–55.

64 Peirce CN. An essay on the lower forms of life found within the oral cavity. *Dental Cosmos* 1875; **17**: 449–464.

65 Williams JL. A contribution to the bacteriology of the human mouth. *Dental Cosmos* 1899; **41**: 317–349.

66 Escherich, T. The intestinal bacteria of the neonate and breast-fed infant: 1884. *Rev Infect Dis* 1988; **10**: 1220–25.

67 Harris VD. The mycological processes of the intestines. *J Path Bacteriol* 1896; **3**: 310–321.

68 Strauss. Uber die Abhangigkeit der Milchsaure Gahrung vom HC1 Gehalt des Magensafts. *Zeitschr f klin Med* 1895; **28**: 567–578.

69 Maggiora Vergano, A. Contributo allo studio dei microfiti della pelle umana normale e special-mente del piede. *Giorn. della Soc. italiana d'igiene* 1889; **XI**: 335–366.

70 Preindlsberger J. Zur Kenntnis der Bacterien des Unternagelraumes und zur Desinfection der Hande, Wien, A. Holder, 1891.

71 Welch W. Some considerations concerning antiseptic surgery. *Maryland M J* 1891; **26**: 45.

72 Bockhart. Ueber eine neue Art der Zubereitung von Fleisch als fester Nahrboden filr Mikroorganismen. Tagebl. d. 60. Versammlung deutscher Naturforscher und Aerzte in Wiesbaden, 1887, p. 347.

73 Bordoni-Ufpreduzzi G. Ueber die biologischen Eigenschaften der normalen Hautmikrophyten. *Fortschr der Med* 1886; No. 5.

74 Sternberg GM. *Manual of Bacteriology*. New York: W. Wood & Company, 1892.

75 Wilson M. The indigenous microbiota of the skin. In: *Bacteriology of Humans: An Ecological Perspective* (Wilson M, ed). Oxford: Wiley-Blackwell, 2008; 56–94.

76 Wilson M. The indigenous microbiota of the eye. In: *Bacteriology of Humans: An Ecological Perspective* (Wilson M, ed). Oxford: Wiley-Blackwell, 2008; 95–112.

77 Wilson M. The indigenous microbiota of the respiratory tract. In: *Bacteriology of Humans: An Ecological Perspective* (Wilson M, ed), Oxford: Wiley-Blackwell, 2008; 113–158.

78 Wilson M. The indigenous microbiota of the urinary system of females. In: *Bacteriology of Humans: An Ecological Perspective* (Wilson M, ed). Oxford: Wiley-Blackwell, 2008; 159–169.

79 Wilson M. The indigenous microbiota of the reproductive system of females. In: *Bacteriology of Humans: An Ecological Perspective* (Wilson M, ed). Oxford: Wiley-Blackwell, 2008; 170–206.

80 Wilson M. The indigenous microbiota of the urinary and reproductive systems of males. In: *Bacteriology of Humans: An Ecological Perspective* (Wilson M, ed). Oxford: Wiley-Blackwell, 2008; 207–221.

81 Wilson M. The indigenous microbiota of the oral cavity. In: *Bacteriology of Humans: An Ecological Perspective* (Wilson M, ed). Oxford: Wiley-Blackwell, 2008; 222–265.

82 Wilson M. The indigenous microbiota of the gastrointestinal tract. In: *Bacteriology of Humans: An Ecological Perspective* (Wilson M, ed). Oxford: Wiley-Blackwell, 2008; 266–326.

83 Olsen GJ, Lane DJ, Giovannoni SJ *et al.* Microbial ecology and evolution: a ribosomal RNA approach. *Annu Rev Microbiol* 1986; **40**: 337–365.

84 Lane DJ. (1991) 16S/23S rRNA sequencing. In Stackebrandt E. and Goodfellow M. (eds), *Nucleic Acid Techniques in Bacterial Systematics*. John Wiley and Sons, Chichester.

85 Muyzer G, de Waal EC, Uitterlinden AG. Profiling of complex microbial populations by denaturing gradient gel electrophoresis analysis of polymerase chain reaction-amplified genes coding for 16SrRNA. *Appl Environ Microbiol* 1993; **59**: 695–700.

86 Palmer C, Bik EM, Eisen MB *et al.* Rapid quantitative profiling of complex microbial populations. *Nucleic Acids Res* 2006; **10**; 34: e5.

87 Rajilic-Stojanovic M, Heilig HG, Molenaar D *et al.* Development and application of the human intestinal tract chip, a phylogenetic microarray: analysis of universally conserved phylotypes in the abundant microbiota of young and elderly adults. *Environ Microbiol* 2009; **11**: 1736–51.

88 Furet JP, Firmesse O, le Gourmelon M *et al.* Comparative assessment of human and farm animal faecal microbiota using real-time quantitative PCR. *FEMS Microbiol Ecol* 2009; **68**: 351–62.

89 Handelsman J, Rondon MR, Brady SF *et al.* Molecular biological access to the chemistry of unknown soil microbes: a new frontier for natural products. *Chem Biol* 1998; **5**: R245–9.

90 Qin J, Li R, Raes J, *et al.* A human gut microbial gene catalogue established by metagenomic sequencing. *Nature* 2010; **464**: 59–65.

91 Legendre P, Legendre L. *Numerical Ecology*, 3rd edn. Amsterdam: Elsevier. 2012.

92 Cox MJ, Allgaier M, Taylor B *et al.* Airway microbiota and pathogen abundance in age-stratified cystic fibrosis patients. *PLoS One* 2010; **5**: e11044.

93 Ley RE, Backhed F, Turnbaugh P *et al.* Obesity alters gut microbial ecology. *Proc Natl Acad Sci USA* 2005; **102**: 11070–75.

94 Maccaferri S, Biagi E, Brigidi P. Metagenomics: key to human gut microbiota. *Dig Dis* 2011; **29**: 525–30.

95 Moran MA. Metatranscriptomics: eavesdropping on complex microbial communities. *Microbe* 2009; **4**: 329–35.

96 Gosalbes MJ, Durban A, Pignatelli M *et al.* Metatranscriptomic approach to analyze the functional human gut microbiota. *PLoS One* 2011; **6**: e17447.

97 Kolmeder CA, de Vos WM. Metaproteomics of our microbiome — developing insight in function and activity in man and model systems. *J Proteomics* 2014; **97**: 3–16.

98 Kolmeder CA, de Been M, Nikkilä J *et al.* Comparative metaproteomics and diversity analysis of human intestinal microbiota testifies for its temporal stability and expression of core functions. *PLoS One* 2012; **7**: e29913.

99 Huang HJ, Zhang AY, Cao HC *et al.* Metabolomic analyses of faeces reveals malabsorption in cirrhotic patients. *Dig Liver Dis* 2013; **45**: 677–82.

CHAPTER 2

An introduction to microbial dysbiosis

Mike Curtis

Institute of Dentistry, Queen Mary University of London

2.1 Definition of dysbiosis

Our mucosal surfaces are colonized by complex communities of micro-organisms, or microbiota, which are uniquely adapted to the different environmental niches in the human body. These microbiota are composed of distinct and specialized consortia of micro-organisms which are characteristic of the respective niche, for example, the mouth, the gastrointestinal or the genito-urinary tract. Collectively, the microbiota on our mucosal surfaces and other anatomical locations in the body comprise the human microbiome, which has become an area of intensive investigation in recent years because of the recognition that the balance between these organisms and the human host plays a fundamentally important role in our biology, the maintenance of our health and the development of disease. The intestinal microbiota, for example, has a profound impact on human physiology, metabolism, local tissue organization and the development of the immune system (see chapters 3, 19, 22, 23 and 26).

An unfavorable alteration of the microbiota composition is called dysbiosis[1] and it has become increasingly evident that this phenomenon, characterized by alterations to the normally balanced microbial populations at different sites of the human body, may have profound effects on human health[2]. Conceptually, the potential importance of a balanced relationship between the microbiome and its respective host to the maintenance of the health of that host and, conversely, the potential deleterious effects of perturbation of this system to the development of disease has been recognized for decades[2]. The previous chapter gave an historical perspective into techniques for analysis of the microbial communities colonizing humans. From this it has clearly emerged that only in relatively recent years has the application of the newer high-throughput molecular methods to the in-depth characterization of complex microbial communities made possible an understanding, or at least a more detailed and expanded analysis, of these systems. The aim of this chapter is to provide some introductory thoughts on the concept of microbial dysbiosis by addressing three related issues. What should be regarded as

The Human Microbiota and Chronic Disease: Dysbiosis as a Cause of Human Pathology, First Edition.
Edited by Luigi Nibali and Brian Henderson.
© 2016 John Wiley & Sons, Inc. Published 2016 by John Wiley & Sons, Inc.

normal in the context of a benign commensal microbiota? What are the characteristics of dysbiosis? Lastly, what is the functional importance of dysbiosis or, more precisely, what is the current evidence that microbial dysbiosis is the cause rather than the consequence of disease?

2.2 The 'normal' microbiota

What is normal? The famous remark of President John F. Kennedy, 'everything changes but change itself,'[3] is a helpful reminder that biological systems do not necessarily remain stable and indeed that change is normal. We are well accustomed to the concept that the host-associated microbiota change throughout normal human development. For example, the intestinal microbiota at birth is initially colonized by *Enterobacteria*. However, in the following days, strict anaerobes dominate the microbial community. During the first month, bifidobacterial species predominate but then, following the introduction of solid foods, clostridial species expand. By 2–3 years, the microbiota consists mainly of Bacteroidaceae, Lachnospiraceae and Ruminococcaceae, which then remain relatively stable to adulthood[4]. However, it is now apparent that there is significant flexibility in the microbiota following this developmental period. This is becoming increasingly evident through the application of deep sequencing technologies to analyse the complex communities of bacteria that populate the different environmental niches of the body. Furthermore, it is clear that variability is evident not only at the level of the site but also at the level of the individual, between different individuals and across populations over time.

In one of the earliest comprehensive studies to exploit the radically cheaper sequencing and analysis technologies to the comprehensive analysis of human microbiota, Costello *et al.*[5] examined the variation in the microbiota sampled from different sites in the body from 13 healthy individuals over four different time points. In this in-depth study, over 800 samples were analysed from the gut (stool), oral cavity, external auditory canal, the nostrils, hair on the head and up to sixteen different skin sites (see Figure 1).

Differences in the overall bacterial community structure were assessed using a phylogeny-based metric, UniFrac[6]. In this analysis system, a relatively small UniFrac difference implies that two bacterial communities are similar, comprising lineages sharing a common evolutionary history. As would be predicted from the established microbiological literature, the largest variation in bacterial community structure was observed when comparing different sites in the body, thereby confirming the critical influence of the nature of the ecological niche on the composition of the resident microbiota. Thus, hierarchical structuring revealed that microbiota were grouped first on the basis of sample site, second on the basis of the individual and finally on the basis of time: the composition varied significantly less within the same sample site than between different sites and within-sample site variation was significantly less between the same sample sites over time than between different individuals sampled on the same day. However, this study also demonstrated that the degree of variation in composition of the microbiota was closely linked to the nature of the sample site. Skin, hair, nostril and external auditory canal microbiota had the highest levels of within-person variation over time, comparable, for example, to the level of variability seen in gut samples from

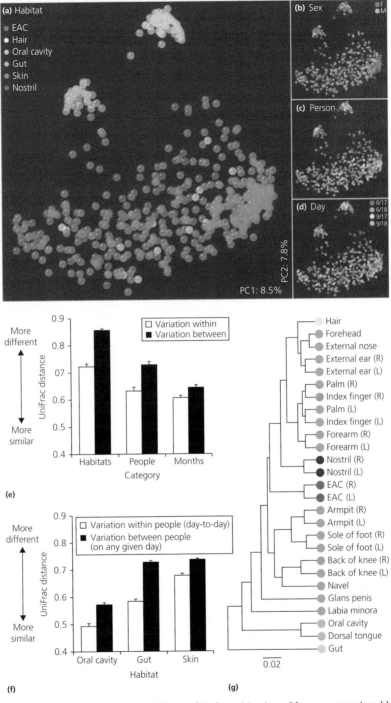

Figure 1 16S rRNA gene surveys reveal hierarchical partitioning of human-associated bacterial diversity. (a to d) Communities clustered using principal component analysis of the unweighted UniFrac distance matrix. Each point corresponds to a sample colored by (a) body habitat, (b) host sex, (c) host individual, or (d) collection date. The same plot is shown in each panel. F, female; M, male. (e and f) Mean (± SEM) unweighted UniFrac distance between communities. In (e) habitats are weighted equally and in (f) skin comparisons are within sites. (g) UPGMA (Unweighted Pair Group Method with Arithmetic Mean) clustering of composite communities from the indicated locales. Leaves are colored according to body habitat as in (a). R, right; L, left. Reproduced with permission from Costello EK, Lauber CL, Hamady M, Fierer N, Gordon JI, Knight R. Bacterial community variation in human body habitats across space and time. *Science* 2009; **326**: 1694–1697. (*see color plate section for color details*).

different individuals. The oral microbiota appeared to be the most stable over time of all sample sites, including the gut. Overall, this study emphasises that the microbiota of humans not only differs significantly by location in the body but appears to be very highly influenced by the individual host and can vary over time. In a follow-up study by the same group[7], the gut, mouth and two sites on the skin were examined daily in two individuals for up to 15 months. Once again, despite relatively stable differences between body sites and between individuals, they recorded pronounced variation within an individual's microbiota across months, weeks and even days.

More recent investigations by other workers have yielded similar conclusions. For example, David et al.[8] examined the microbial community structure of gut (stool) and saliva samples from two individuals over the course of one year and linked their findings to multiple measures of the lifestyles and wellness of the individuals over the sampling period (see Figure 2).

This outcome reinforced the findings of earlier investigations but also demonstrated which features of the host lifestyle may be instrumental in eliciting change in their associated microbiota. For example, whilst the overall pattern of the intestinal microbiota appeared relatively stable over periods of several months, a period spent living in the developing world by one individual coincided with a very marked shift in the community structure of the gut microbiota of that individual. Importantly, this did not appear to be through acquisition of novel bacteria but through changes in the abundance of pre-existing microbial groups, characterized most obviously by a twofold increase in the *Bacteroides* to *Firmicutes* ratio. Following return to their residence in the developed world this change was reversed to the original status. Other notable factors associated with measurable fluctuations in the microbiota included dietary change. Dietary fibre ingestion positively correlated with next-day abundance of clusters of bacteria representing over 15% of the total community. These included *Bifidobacteria, Roseburia* and *Eubacterium rectale* species, all of which have been shown in previous investigations to be sensitive to fibre intake. Similarly, the ingestion of citrus fruits was positively correlated with the level of *Faecalibacterium prausnitzii,* an organism suggested to have a therapeutic role in colitis, and other members of the *Clostridiales.* Significantly, pectin is a major carbohydrate component in citrus fruits, which has been shown to act as a food source for *F. prausnitzii* via *in vitro* investigations.

Perhaps the most dramatic demonstration of massive differences in the microbiota of humans has again come from analysis of the gut, in this instance from distinct populations based in the developed world versus those in the developing world. For example, in the study by De Filippo et al.[9], the fecal microbiota of European children was compared with that of children from a rural African village in Burkina Faso. In the latter population, the diet mainly consists mainly of cereals, legumes and vegetables, with a correspondingly high content of carbohydrate, fibre and non-animal protein, and is thought still to resemble the diet of Neolithic subsistence famers of approximately 10,000 years ago. Conversely, the diet of the European children comprised a typical Western diet high in animal protein, sugar, starch and fat and low in fibre. The results of the microbiota analyses were striking. The fecal microbiota of African children demonstrated a highly significant enrichment of Bacteroidetes and reductions in Firmicutes and Proteobacteria phyla. Within the Bacteroidetes in the African population, there

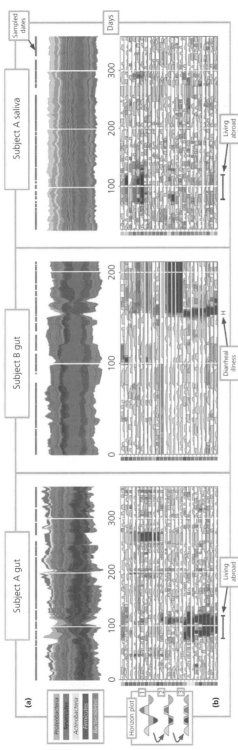

Figure 2 Gut and salivary microbiota dynamics in two subjects over one year. (a) Stream plots showing OTU (Operational Taxonomic Unit) fractional abundances over time. Each stream represents an OTU and streams are grouped by phylum: Firmicutes (purple), Bacteroidetes (blue), Proteobacteria (green), Actinobacteria (yellow), and Tenericutes (red). Stream widths reflect relative OTU abundances at a given time point. Sampled time points are indicated with gray dots over each stream plot. (b) Horizon graphs of most common OTUs abundance over time. Warmer regions indicate date ranges where a taxon exceeds its median abundance and cooler regions denote ranges where a taxon falls below its median abundance. Colored squares on the vertical axis correspond to stream colors in (A). Lower black bars span Subject A's travel abroad (days 71 to 122) and Subject B's Salmonella infection (days 151 to 159). For further details see David LA, Materna AC, Friedman J *et al.* Host lifestyle affects human microbiota on daily timescales. *Genome Biol* 2014; **15**: R89. Reproduced with permission. (*see color plate section for color details*).

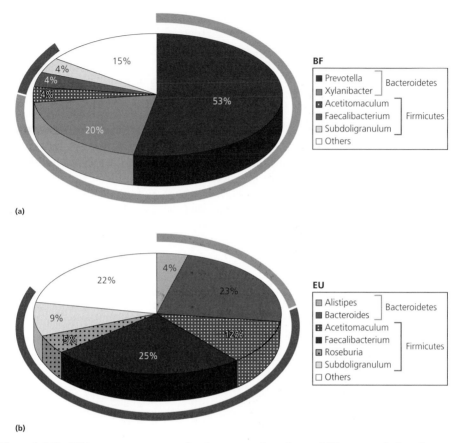

Figure 3 16S rRNA gene surveys reveal a clear separation of two children populations in Burkina Faso (a) and Europe (b). Pie charts of median values of bacterial genera present in fecal samples of Burkina Faso and European children (>3%). Rings represent corresponding phylum (Bacteroidetes in green and Firmicutes in red) for each of the most frequently represented genera. Adapted from De Filippo C, Cavalieri D, Di Paola M *et al.* Impact of diet in shaping gut microbiota revealed by a comparative study in children from Europe and rural Africa. *Proc Natl Acad Sci USA* 2010; **107**: 14691–14696. Reproduced with permission. (*see color plate section for color details*).

was a remarkable abundance of the Prevotella and Xylanibacter genera, which were completely absent in the European children. The likely critical importance of diet to these changes is amply demonstrated when one considers that both the Prevotella and Xylanibacter contain the genetic machinery for the hydrolysis and metabolism of cellulose and xylan, both of which constituents were over-represented in the African diets (see Figure 3).

Similar comparisons have now been undertaken on other populations drawn from rural communities in the developing world versus western societies. A comparison of the gut microbiota of children and adults in rural communities in Malawi and Venezuela versus urban populations in the United States also revealed a predominance of Prevotella within the Bacteroidetes phylum of the rural communities as opposed to Bacteroides in the North American study group[10]. Furthermore, this altered Prevotella versus Bacteroides signature in the gut microbiota has also been reported in a study comparing healthy children living in an urban slum in Bangladesh with that of children of the same age range in an upper-middle-class suburban community in the United States[11].

These investigations into diet-induced changes in the human gut microbiome are complemented by extensive studies in animals both in the wild[12,13,14] and under laboratory conditions. For example, Turnbaugh *et al.*[15] developed a humanized gnotobiotic mouse model to examine this phenomenon. Adult human fecal microbiota were transplanted into previously germ-free mice that were shown to be stably colonized with much of the microbial diversity of the human donor microbiome. Altering the diet fed to these animals from a low-fat, plant-polysaccharide-rich diet to a high-fat, high-sugar diet characteristic of Westernized human populations changed the gut microbial population structure within a single day. As in the comparative studies of different human populations described above, the microbiota changes were accompanied by alterations to the major metabolic pathways represented in the microbiome. Hence we can conclude that the gut microbiome is highly responsive to dietary changes, and that the composition can alter dramatically over short timescales and generate community structures that are phylogenetically and metabolically most appropriate for the altered nutrient supply in the gastrointestinal tract (discussed in chapter 29 by Gibson).

A final example of the flexibility of human associated microbiota comes from the relatively new field of paleo-microbiology, the investigation of ancient microbiomes from discrete locations and time points, which helps provide a view into the coevolution of microbes and their hosts through time. Dental calculus is one such time capsule of historical data that has proven to be not only a rich source of human bacterial communities of the past but also a dietary record of the individual from whom the sample originated[16]. The calcified biofilm that forms of the surface of teeth is present in all human populations, both present-day and in prehistory, as well as in extinct species of primates including Neanderthal and Sivapithecus apes. The biomolecular and structural preservation properties of dental calculus are remarkable: archeological calculus retains much of the organic and inorganic composition and remnant organic structures of mature dental calculus sampled from present-day individuals. As a result, analysis of the microbial population structures of calculus provides a unique insight into the changes that have occurred to the oral microbiome of human populations through history (see Figure 4).

Adler *et al.*[17] performed phylum-level comparisons of bacteria in dental calculus from Mesolithic, Neolithic, Bronze Age, Medieval and modern sources from globally dispersed sites. In order to rule out the potential contribution of environmental contamination of their samples they also examined the microbial content of freshwater, sediment and soil. The results demonstrated that the bacterial composition of ancient calculus was similar to that of modern oral samples and sequences from the Human Oral Microbiome database but different to potential environmental sources of contamination and different also to the samples from within the sampled teeth. Although dental calculus samples had a characteristic oral microbiome signature, the data did suggest changes in the microbiota over time. A significant change was observed between Mesolithic and Neolithic samples, coincident with the change from hunter-gather to farming cultures indicative of the potential influence of dietary alterations in the population at that time boundary. Perhaps surprisingly, there was no major change from the Neolithic

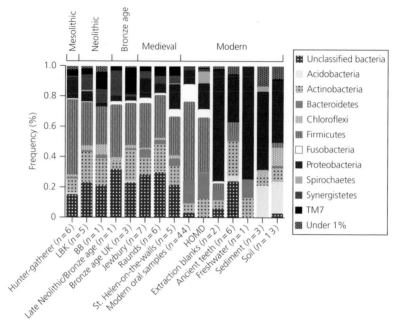

Figure 4 Phylum-level microbial composition of ancient dental calculus deposits. The distribution is similar to that of modern oral samples and distinct from those of non-template controls, ancient human teeth and environmental samples. The phylum frequencies for the V3 region of 16S rRNA are presented for the ancient calculus samples (BB, Bell Beaker), modern oral samples, which included pyro-sequenced (calculus, plaque and saliva) and cloned (plaque) data, non-template controls (or extraction blanks), ancient human teeth and environmental samples (freshwater, sediments and soils). Phylum frequencies from the Human Oral Microbiome Database (HOMD) were generated from partial and full-length sequences of the 16S rRNA gene. Reproduced with permission from Adler CJ, Dobney K, Weyrich LS *et al*. Sequencing ancient calcified dental plaque shows changes in oral microbiota with dietary shifts of the Neolithic and Industrial revolutions. *Nat Genet* 2013; **45**: 450–455.

period through the Bronze Age and into the Medieval Periods. Indeed, the next major shift was recorded at the end of the Medieval period corresponding to the dawn of the Industrial Revolution and the development of modern day diets. Overall, they observed a reduction in the diversity of the oral microbiota in modern samples coupled to an increase in the levels of potentially cariogenic species and concluded that this less diverse community structure may be a contributory factor to the prevalence of chronic microbially mediated oral diseases of the present day.

Whilst one conclusion from this study may be to suggest that changes in diet over time are responsible for these shifts in the oral microbial communities in these historical samples, it is important not to automatically extrapolate findings from the study of the gut to other regions of the alimentary canal. Indeed, there is good evidence to suggest that oral microbial communities are far less responsive to dietary change than those in the intestine. For example, in a study examining the oral microbiology of fed and fasted primates, Beighton *et al*.[18] found no influence of dietary intake on the total number of bacteria in dental plaque from developmental grooves on molar teeth and few changes in the nature of the bacteria present nor in the growth rate of some of these organisms. Exceptions were significant increases in *Streptococcus mitior* and *Neisseria mucosa* in the fasted animals. These authors concluded that the bacterial communities in the mouth gain a significant component of their nutritional needs not from dietary intake but from saliva. This conclusion is reinforced by the demonstration that the hydrolytic enzyme repertoire and metabolic capacity of oral bacteria present on newly cleaned tooth surfaces appear ideally tailored to the breakdown and metabolism of complex salivary glycoproteins. Hence, particularly when considering the change in the oral microbial communities in dental calculus samples from medieval to modern times, it is necessary to consider alternative causative factors in addition to diet. One such factor is likely to be modern practices of oral hygiene, which will have a significant influence on the microbial community structure present on tooth surfaces: daily debridement of the surfaces of the teeth will lead to continuous disruption of the successional development of biofilms and may as a consequence prevent or limit the establishment of a climax community organization represented by oral biofilms allowed to develop undisturbed for significant periods of time.

The overall conclusions of the studies listed above are that the community organization and structure of human associated microbiota are highly flexible. Of course, they are profoundly influenced by the nature of the site in the human body and the stage of development. Equally, they are responsive to the lifestyle of the individual and to environmental effects that can elicit changes in composition and abundance of different species over both extremely short and prolonged timescales. Large shifts or differences in the population structure of human-associated microbiota do not necessarily equate to a dysbiosis; rather, they are representative of the flexibility of these populations that is entirely compatible with health. Accumulating evidence suggests that there is an intimate relationship between our microbiota and our immune and non-immune systems and that these relationships have defining roles on our development and on the maintenance of homeostasis and health. Given the apparent flexibility of the microbiota associated with humans, there must be equal levels of flexibility in the recognition and response systems which interface with these microbiota, the full details of which have yet to be elucidated.

2.3 Main features of dysbiosis

If flexibility represents the norm, how is dysbiosis defined? The critical differentiating factor is the response of the host. Dysbiosis is now recognized as a definitive change in the microbiota at a given site in the body, crucially accompanied by a breakdown of host-microbial mutualism. Dysbiosis of human-associated microbiota is now thought to be the defining event of multiple inflammatory and systemically driven pathologies, summarised in part in other chapters in this volume. Whilst no single organism or collection of organisms has been identified as a consistent marker in any human-associated microbiota, several defining features of dysbiosis have emerged through investigation of these pathology-associated microbial populations in the last decade.

First, in several examples, dysbiosis is reflected in a reduction in the overall microbial diversity of the corresponding symbiotic community. A reduction in taxonomic diversity and species membership of the microbiota has been observed in multiple studies of the human gut microbiota in disease as well as in animal infection models[19]. Dysbiosis generally leads to a depletion of obligate anaerobic bacteria such as Bacteroides and Ruminococcus spp., and conversely an increase in facultative anaerobes, including the family Enterobacteriaceae (e.g. *E. coli, Klebsiella* spp., *Proteus* spp.). Manichanh *et al.*[20] reported a decrease in microbial diversity in the fecal microbiota in Crohn's disease. Lepage *et al.*[21] reported a similar reduction in ulcerative colitis and Chang *et al.*[22] described decreased diversity of the fecal microbiome in recurrent *Clostridium difficile*-associated diarrhoea. A functional consequence of a less diverse gut microbiota appears to be a reduced metabolic capacity, exemplified by a decline in short-chain fatty acid production. These physiological by-products of carbohydrate fermentation by the microbiota are important energy sources for the host and also enhance the mucosal barrier and inhibit intestinal inflammation and oxidative stress. Hence a reduction in the metabolic capacity of the microbiota may contribute to the impairment of host defenses and thereby promote the stability of a dysbiotic community[23] (see Figure 5).

The second emerging feature of dysbiotic communities is a preferential loss of organisms considered beneficial to human health and a corresponding increase in pathobionts, members of the normal commensal microbiota with the potential to cause pathology. In the case of dysbiosis of the gut, these changes are exemplified by well documented reductions in the levels of obligate anaerobic members of the Firmicutes phylum and increased representation of members of the Proteobacteria, in particular the family *Enterobactericeae* (recently reviewed in Walker and Lawley[24] and Hold *et al.*[25]). Most human pathogens belong to the Proteobacterium phylum including members of the Enterobacteriaceae, which contains a number of frank and opportunistic pathogens including *Salmonella* spp, *Shigella* spp, *Klebsiella* spp, *Proteus* spp and *Eschericia coli*. Similarly, in periodontal disease, a chronic inflammatory condition of the tooth-supporting tissues, marked shifts in the microbial population structure of subgingival dental plaque are observed. These community-wide changes have been described over the last 50 years, initially on the basis of microbial culture, followed by targeted molecular techniques typically involving DNA:DNA hybridization studies and more recently by 16S rRNA profiling (discussed in chapters 13 and 14).

In a series of landmark investigations based on cultural microbiology, Moore *et al.*[26] examined the bacteriology of severe generalized periodontitis in 21 individuals. They described the isolation and characterization by biochemical techniques of 2,723 individual isolates representing 190 bacterial species, subspecies, or serotypes. Of these, 11 species exceeded 1% of the subgingival flora and were most closely associated with disease, and 11 others were also sufficiently frequently isolated to be deemed suspected agents of tissue destruction. This study highlights the marked difference between the overall microbiota in periodontally diseased subgingival sites compared to the adjacent supragingival microbiota. Whilst the supragingival microbiota is dominated by the *Actinomyces* spp, *Streptococcus* spp and *Veillonella* spp, which comprised some 40% of the total cultivatable bacteria, the same genera represent only approximately 10% of the

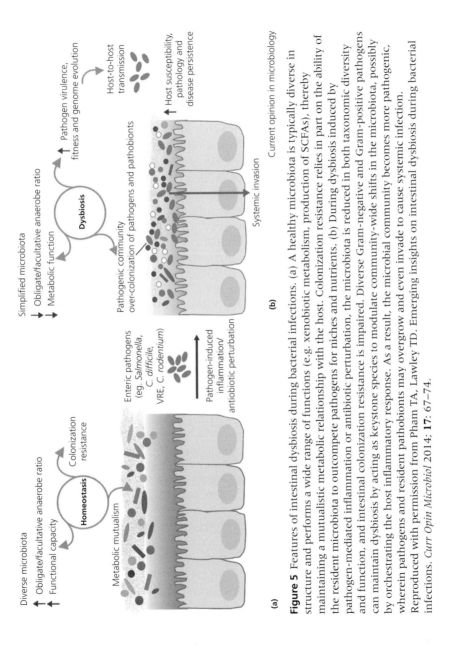

Figure 5 Features of intestinal dysbiosis during bacterial infections. (a) A healthy microbiota is typically diverse in structure and performs a wide range of functions (e.g. xenobiotic metabolism, production of SCFAs), thereby maintaining a mutualistic metabolic relationship with the host. Colonization resistance relies in part on the ability of the resident microbiota to outcompete pathogens for niches and nutrients. (b) During dysbiosis induced by pathogen-mediated inflammation or antibiotic perturbation, the microbiota is reduced in both taxonomic diversity and function, and intestinal colonization resistance is impaired. Diverse Gram-negative and Gram-positive pathogens can maintain dysbiosis by acting as keystone species to modulate community-wide shifts in the microbiota, possibly by orchestrating the host inflammatory response. As a result, the microbial community becomes more pathogenic, wherein pathogens and resident pathobionts may overgrow and even invade to cause systemic infection. Reproduced with permission from Pham TA, Lawley TD. Emerging insights on intestinal dysbiosis during bacterial infections. *Curr Opin Microbiol* 2014; **17**: 67–74.

organisms at subgingival diseased sites. Conversely, members of the *Bacteroides* and *Fusobacteria* are present in approximately 20% of the subgingival microbiota but in only approximately 5% supragingivally.

In a subsequent and hugely influential study, Socransky *et al.*[27] analysed approximately 13,000 plaque samples from 185 subjects using whole genomic DNA probes to 40 bacterial species. Associations were sought among species using cluster analysis and community ordination techniques. One of the key and fundamentally important findings of this study, which has shaped our under-standing of periodontal infections ever since, was the definition of bacterial com-plexes, as opposed to individual bacterial species, that were associated with either periodontal health or periodontal disease. This finding led to the concept that there may be a co-dependency or synergy between different bacterial species act-ing in concert as a specific complex. The complex most strongly associated with periodontal disease, the "red complex", was composed of three bacterial species that subsequently became the focus of intense investigation: *Porphyromonas gin-givalis, Treponema denticola* and *Tannerella forsythia*. Other complexes, for example the yellow complex, which comprised predominantly different streptococcal species, and the green complex, which contained a preponderance of capnocy-tophaga species, represented early colonizers of dental plaque that were more closely associated with health. The orange complex contained those organisms generally considered to colonize dental plaque later: fusobacteria species, mem-bers of the prevotella and the campylobacter. The presence of these organisms is thought to facilitate colonization of mature dental plaque by the red complex organisms either through the presentation of appropriate binding sites or by the creation of a suitable environment for the growth of these more fastidious species.

More recent 16S rRNA-based microbial population analyses have expanded this view of community-wide changes in periodontal disease characterized by loss of beneficial species and increase in pathobionts. In a very recent example, Kirst *et al.*[28] used 16S rRNA sequencing to survey the subgingival microbiota in 25 individuals with periodontal disease and 25 healthy controls. Consistent with earlier investigations, they described a significantly altered microbial structure with decreased diversity in disease. In addition, through cluster analysis techniques, they demonstrated the presence of two predominant clusters. One cluster, char-acterized by high levels of *Fusobacterium* and *Porphyromonas* bacterial species, was strongly associated with the diseased population, whereas a second cluster, this time dominated by *Rothia* and *Streptococci,* was representative of periodontal health. Analogous to different gut enterotypes, based on the abundance of key bacteria in the gut microbiota, the authors suggested that these discrete clusters in periodontal health and disease may represent different periodonto-types that differentiate between periodontal health and disease.

The overriding consensus of investigations of dysbiotic microbiota in human disease is that rather than seeking to describe a single bacterium or even group of organism responsible for the disease in a hugely complex microbial system, the microbiota as a whole should be viewed as the pathological determinant. As elegantly expressed by Hold *et al.*[29], instead of seeking to identify a needle in a haystack, the entire haystack is faulty. Whilst this may be the case, it is not to diminish the potentially pivotal role played by individual organisms in driving dysbiosis in a host-associated microbiota. Recent observations using the mouse model of periodontal disease have demonstrated that one of the presumed important pathobionts of the disease, *Porphyromonas gingivalis* — a member of the previously described "red complex" of periodontal bacteria — may contribute to disease by altering the normal oral microbiota to a dysbiotic state[30]. These studies in mice demonstrated that the oral commensal microbiota is responsible for the tissue and bone destruction associated with periodontitis when just low numbers of *P. gingivalis* are present. Colonization of mice with *P. gingivalis* led to a very significant change in the commensal microbiota: the total load of the microbiota rose by approximately 2 $\log_{10}$ units, a qualitative shift in the population structure of this commensal microbiome and elevated hard-tissue destruction. The requirement of the normal oral microbiota for the induction of disease was demonstrated by colonization of germ-free mice with *P. gingivalis*, which had no demonstrable effect on periodontal health. This organism was accordingly described as a keystone pathogen able to orchestrate the normal benign periodontal microbiota into a dysbiotic community structure. The concept of a keystone species derives from ecological studies and is defined as a species that is present in low abundance yet provides a major supporting role for an entire ecological community[31,32] (see Figure 6).

The data help explain an apparent paradox in describing *P. gingivalis* as a key agent in periodontal disease in that several lines of investigation have demonstrated that this bacterium is not a potent inducer of inflammation[33,34]. The lipopolysaccharide of *P. gingivalis* has an unusually low inflammatory potency, in stark contrast to the highly inflammatory properties of LPS produced by most other bacteria[35]. Furthermore, *P. gingivalis* is unusual in that it does not induce secretion of IL-8 by gingival epithelial cells, unlike a variety of other oral bacteria. Instead, it inhibits the secretion of this potent chemokine for neutrophil recruitment through a phenomenon termed local chemokine paralysis[36]. Such an anti-inflammatory property, in this case down-regulation of phagocytic cell trafficking into the periodontium, is contrary to the functions one would anticipate to be a prerequisite of an organism that drives the disease process through up-regulation of the inflammatory response. These apparent paradoxes can be explained if periodontitis is viewed as a disease driven by an entire dysbiotic microbiota in which *P. gingivalis* exerts its role though the orchestration of community disturbance (see Figure 7). This concept will be discussed further in chapter 14.

2.4 Conclusions

To summarize, dysbiosis may be defined as a deleterious perturbation to the health-associated microbial population structures of human microbiota and an accompanying breakdown in the normal homeostatic balance between the host and the resident microbes. Key features include a decrease in microbial population diversity, reductions in beneficial microbes and increased levels of pathobionts. Multiple factors are likely to be able to affect this change, including alterations in the efficiency of the immune and inflammatory status of the individual, infection or medical interventions such as antibiotics. Whilst the recent upsurge in interest in the analysis of dysbiotic microbial populations using deep sequencing

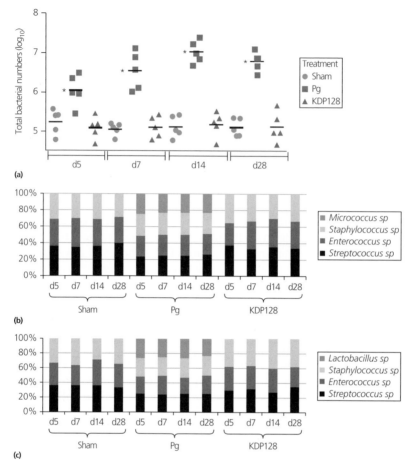

Figure 6 Oral inoculation of mice with *P. gingivalis* but not with an isogenic gingipain-deficient mutant (KDP128) causes elevation of the oral bacterial load and an altered microbiota composition. Mice were orally inoculated with *P. gingivalis* ATCC3327 (Pg), its isogenic gingipain-deficient mutant (KDP128), or vehicle only (Sham). At the indicated timepoints (d = day), maxillary periodontal tissue was harvested to determine total bacterial numbers using real-time PCR of the 16S rRNA gene (a). Changes to the qualitative composition of the cultivatable microbiota detected by aerobic (b) or anaerobic culture (c) were determined following sub-culture of the predominant organisms, based on colonial morphology, followed by identification by 16S rRNA sequence analysis. Five mice were used per group at each timepoint. *$p < 0.01$ vs. both Sham and KDP128. Mice inoculated with Pg but not KDP128 developed accelerated periodontal bone loss compared to sham animals. Reproduced with permission from Maekawa T, Krauss JL, Abe T *et al. Porphyromonas gingivalis* manipulates complement and TLR signaling to uncouple bacterial clearance from inflammation and promote dysbiosis. *Cell Host Microbe* 2014; **15**: 768–778.

technologies and metagenomic investigations continues to shed further light on the nature of these communities, their temporal changes and differences at the individual and population level, we are only at the beginning of the journey to understand the functional significance of microbial dysbiosis.

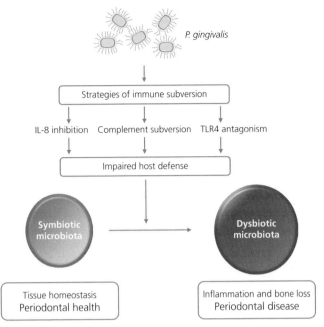

Figure 7 The red complex bacterium *P. gingivalis* causes periodontal inflammation and bone loss by remodeling the oral commensal microbiota. *P. gingivalis* modulates innate host defense functions that can have global effects on the oral commensal community. Immune subversion of IL-8 secretion, complement activity, or TLR4 activation can result in an impaired host defense. The inability of the host to control the oral commensal microbial community in turn results in an altered oral microbial composition and an increased microbial load. This alteration from a symbiotic to a dysbiotic microbiota is responsible for pathologic inflammation and bone loss. Reproduced with permission from Darveau RP, Hajishengallis G, Curtis MA. *Porphyromonas gingivalis* as a potential community activist for disease. *J Dent Res* 2012; **91**: 816–820. (*see color plate section for color details*).

Using the definition of dysbiosis presented above, changes in the microbial population structure in dysbiosis are intimately correlated with disease. Correlations, however, do not equate to causality. It is feasible to argue that alterations to the inflammatory status of a site, triggered perhaps by an autoimmune reaction, will lead to the selection of those members of the microbial community best adapted to survive the increased inflammatory pressure and/or take maximal advantage of the changed nutritional environment presented by such an alteration. Bacteria less well adapted to the injurious properties of the innate response or ill equipped to compete metabolically in this new environment will diminish in their overall proportions in the population. In this scenario, the dysbiosis is consequential, not causative. The repertoire of immune evasive strategies practiced by *P. gingivalis* in periodontal disease and the reduced susceptibility of Salmonella spp to effective elimination by heightened host defense systems in the gut compared to symbionts provide appropriate examples in support of this scenario. Further evidence that the environmental changes induced by the diseased status of the host drive and sustain the alterations to the microbiota rather than the reverse comes from the efficacy, in certain instances, of immuno-modulatory drugs that target the inflammatory systems of the host as opposed to direct intervention at the level of the microbiota[37].

Nonetheless, there is now accumulating evidence to support the counter argument that dysbiosis in itself is sufficient to both initiate and drive disease. Perhaps the most compelling data have come from studies in animals over the last decade that have demonstrated that transfer of a dysbiotic microbiome from a diseased animal into a healthy recipient can be sufficient to recapitulate both the dysbiosis and the disease phenotype. For example, transplantation experiments where the disease-associated gut microbiota is transferred into healthy germ-free animals have demonstrated that the disease phenotype can be reproduced in the recipients in a variety of conditions including adiposity, metabolic syndrome and colitis[38].

Equally, the restoration of a normal healthy microbiome following successful of fecal transplantation for the treatment of *Clostridium difficile* infection suggests that restoration of a normal balanced microbiota in the gut is sufficient to restore intestinal health. *C. difficile* infection, which occurs most frequently after prolonged antibiotic therapy, is a relapsing condition characterized by diarrhoea and disturbed gut microbiota and is notoriously difficult to treat. In the randomized trial reported by van Nood et al.,[39] patients with *C. difficile* infection were treated with an initial vancomycin regimen, followed by bowel lavage and subsequent infusion of a solution of healthy donor feces through a nasoduodenal tube. They reported an 81% success rate in resolution of C. difficile-associated diarrhoea in this patient group compared to 31% receiving vancomycin alone. After donor feces-infusion, the patients showed an increased fecal microbial diversity, similar to that in the healthy donors, a corresponding increase in Bacteroidetes and Clostridium spp and a decrease in the Proteobacterium phylum. Fecal transplantation treatment of the other major inflammatory bowel diseases, ulcerative colitis and Crohn's disease, has so far yielded less successful outcomes. However, the principle of treating inflammatory diseases associated with a perturbed microbiome by modalities aimed entirely at conversion of the dysbiotic microbiota to the previously harmonious and balanced system is now firmly established. Alternative strategies aimed at conversion of a microbial community back to a healthy state, or rebiosis, is accordingly now an area of intensive investigation. These aspects will be discussed in the last section of this book.

In conclusion, the remarkable advances in microbial population analysis over the last decade have uncovered a hitherto unseen level of complexity and, to a variable extent flexibility, in the bacterial communities associated with humans. Additionally we are now gaining a detailed inventory of how these population structures differ in diseased states of the mouth, the gut and other sites in and on the human body. Defining the mechanisms through which these community structures interface with host tissues to either maintain health or elicit disease will be a significant challenge given the complexity on both sides of the host:bacterium boundary. However, it should be anticipated that even incremental increases in our understanding of microbial dysbiosis have the potential to reap significant benefits in the diagnosis and treatment of human disease.

TAKE-HOME MESSAGE

- The human microbiome displays significant variation not only between different sites in the human body, but also at the same site in different individuals and in the same individual over time. Shifts or differences in the population structure of human associated microbiota do not necessarily equate to a dysbiosis; rather, they are representative of the flexibility of these populations and are entirely compatible with health.

- Dysbiosis can be defined as a substantive change in the microbiota at a given site in the body that, crucially, is accompanied by a breakdown of host-microbial mutualism at that site. The drivers of dysbiosis can include environmental alterations, including the status of the host immune and inflammatory response, and in some instances the activities of individual low-abundance bacteria — keystone pathogens.

- Understanding the mechanism of conversion of bacterial communities to a dysbiotic state and, conversely, the reversal of this process or rebiosis, is likely to have a significant impact on the treatment of many bacterially-driven inflammatory conditions with complex microbial etiologies.

Acknowledgment

The author thanks Juliet Ellwood for her significant contributions to the preparation and finalization of this chapter.

References

1 Hill DA, Artis D. Intestinal bacteria and the regulation of immune cell homeostasis. *Annu Rev Immunol* 2010; **28**: 623–667.
2 Frank DN, Zhu W, Sartor RB, Li E. Investigating the biological and clinical significance of human dysbioses. *Trends Microbiol* 2011; **19**: 427–434.
3 Kennedy JF. *From a speech to celebrate the 90th anniversary of Vanderbilt University on May 18.* 1963.
4 Arrieta MC, Stiemsma LT, Amenyogbe N, Brown EM, Finlay B. The intestinal microbiome in early life: health and disease. *Front Immunol* 2014; **5**: 427.
5 Costello EK, Lauber CL, Hamady M, Fierer N, Gordon JI, Knight R. Bacterial community variation in human body habitats across space and time. *Science* 2009; **326**: 1694–1697.
6 Lozupone C, Knight R. UniFrac: a new phylogenetic method for comparing microbial communities. *Appl Environ Microbiol* 2005; **71**: 8228–8235.
7 Caporaso JG, Lauber CL, Costello EK, *et al.* Moving pictures of the human microbiome. *Genome Biol* 2011; **12**: R50.
8 David LA, Materna AC, Friedman J, *et al.* Host lifestyle affects human microbiota on daily time-scales. *Genome Biol* 2014; **15**: R89.
9 De Filippo C, Cavalieri D, Di Paola M, *et al.* Impact of diet in shaping gut microbiota revealed by a comparative study in children from Europe and rural Africa. *Proc Natl Acad Sci U S A* 2010; **107**: 14691–14696.
10 Yatsunenko T, Rey FE, Manary MJ, *et al.* Human gut microbiome viewed across age and geography. *Nature* 2012; **486**: 222–227.
11 Lin A, Bik EM, Costello EK, *et al.* Distinct distal gut microbiome diversity and composition in healthy children from Bangladesh and the United States. *PLoS One* 2013; **8**: e53838.
12 Muegge BD, Kuczynski J, Knights D, *et al.* Diet drives convergence in gut microbiome functions across mammalian phylogeny and within humans. *Science* 2011; **332**: 970–974.
13 Amato KR, Yeoman CJ, Kent A, *et al.* Habitat degradation impacts black howler monkey (*Alouatta pigra*) gastrointestinal microbiomes. *ISME J* 2013; **7**: 1344–1353.

14 Amato KR, Leigh SR, Kent A, *et al*. The gut microbiota appears to compensate for seasonal diet variation in the wild black howler monkey (*Alouatta pigra*). *Microb Ecol* 2015; **69**: 434–443.

15 Turnbaugh PJ, Ridaura VK, Faith JJ, Rey FE, Knight R, Gordon JI. The effect of diet on the human gut microbiome: a metagenomic analysis in humanized gnotobiotic mice. *Sci Transl Med* 2009; **1**: 6ra1.

16 Warinner C, Speller C, Collins MJ, Lewis CM Jr. Ancient human microbiomes. *J Hum Evol* 2015; **79**: 125–136.

17 Adler CJ, Dobney K, Weyrich LS, *et al*. Sequencing ancient calcified dental plaque shows changes in oral microbiota with dietary shifts of the Neolithic and Industrial revolutions. *Nat Genet* 2013; **45**: 450–455.

18 Beighton D, Smith K, Hayday H. The growth of bacteria and the production of exoglycosidic enzymes in the dental plaque of macaque monkeys. *Arch Oral Biol* 1986; **31**: 829–835.

19 Pham TA, Lawley TD. Emerging insights on intestinal dysbiosis during bacterial infections. *Curr Opin Microbiol* 2014; **17**: 67–74.

20 Manichanh C, Rigottier-Gois L, Bonnaud E, *et al*. Reduced diversity of faecal microbiota in Crohn's disease revealed by a metagenomic approach. *Gut* 2006; **55**: 205–211.

21 Lepage P, Häsler R, Spehlmann ME, *et al*. Twin study indicates loss of interaction between microbiota and mucosa of patients with ulcerative colitis. *Gastroenterology* 2011; **141**: 227–236.

22 Chang JY, Antonopoulos DA, Kalra A, *et al*. Decreased diversity of the fecal Microbiome in recurrent *Clostridium difficile*-associated diarrhea. *J Infect Dis* 2008; **197**: 435–438.

23 Pham TA & Lawley TD *op cit*.

24 Walker AW, Lawley TD. Therapeutic modulation of intestinal dysbiosis. *Pharmacol Res* 2013; **69**: 75–86.

25 Hold GL, Smith M, Grange C, Watt ER, El-Omar EM, Mukhopadhya I. Role of the gut microbiota in inflammatory bowel disease pathogenesis: what have we learnt in the past 10 years? *World J Gastroenterol* 2014; **20**: 1192–1210.

26 Moore WEC, Holdeman LV, Smibert RM, Hash DE, Burmeister JA, Ranney RR. Bacteriology of severe periodontitis in young adult humans. *Infect Immun* 1982; **38**: 1137–1148.

27 Socransky SS, Haffajee AD, Cugini MA, Smith C, Kent RL Jr. Microbial complexes in subgingival plaque. *J Clin Periodontol* 1998; **25**: 134–144.

28 Kirst ME, Li EC, Alfant B, *et al*. Dysbiosis and alterations in predicted functions of the subgingival microbiome in chronic periodontitis. *Appl Environ Microbiol* 2015; **81**: 783–793.

29 Hold GL *et al., op cit*.

30 Hajishengallis G, Liang S, Payne MA, *et al*. Low-abundance biofilm species orchestrates inflammatory periodontal disease through the commensal microbiota and complement. *Cell Host Microbe* 2011; **10**: 497–506.

31 Darveau RP, Hajishengallis G, Curtis MA. *Porphyromonas gingivalis* as a potential community activist for disease. *J Dent Res* 2012; **91**: 816–820.

32 Hajishengallis G, Darveau RP, Curtis MA. The keystone-pathogen hypothesis. *Nat Rev Microbiol* 2012; **10**: 717–725.

33 Maekawa T, Krauss JL, Abe T, *et al*. *Porphyromonas gingivalis* manipulates complement and TLR signaling to uncouple bacterial clearance from inflammation and promote dysbiosis. *Cell Host Microbe* 2014; **15**: 768–778.

34 Curtis MA, Zenobia C, Darveau RP. The relationship of the oral microbiotia to periodontal health and disease. *Cell Host Microbe* 2011; **10**: 302–306.

35 Darveau RP. Periodontitis: a polymicrobial disruption of host homeostasis. *Nat Rev Microbiol* 2010; **8**: 481–490.

36 Darveau RP, Belton CM, Reife RA, Lamont RJ. Local chemokine paralysis, a novel pathogenic mechanism for *Porphyromonas gingivalis*. *Infect Immun* 1998; **66**: 1660–1665.

37 Delima AJ, Karatzas S, Amar S, Graves DT. Inflammation and tissue loss caused by periodontal pathogens is reduced by interleukin-1 antagonists. *J Infect Dis* 2002; **186**: 511–516.

38 Spor A, Koren O, Ley R. Unravelling the effects of the environment and host genotype on the gut microbiome. *Nat Rev Microbiol* 2011; **9**: 279–290.

39 van Nood E, Vrieze A, Nieuwdorp M, *et al*. Duodenal infusion of donor feces for recurrent *Clostridium difficile*. *N Engl J Med* 2013; **368**: 407–415.

CHAPTER 3

The gut microbiota: an integrated interactive system

Hervé M. Blottière and Joël Doré

Micalis Institute, INRA, AgroParisTech, Université Paris-Saclay, Paris, France

3.1 Introduction

Over the last years, our view of human physiology has been extensively challenged. Indeed, after a huge effort to sequence and analyse the human genome, the scientific community thought that it would explain most diseases besides infectious diseases. Susceptibility genes have been associated with chronic diseases, but their respective weight was insufficient. Evidence coming from germ-free animals started to provide information on a potential role of the microflora, as it was called at that time. In parallel, progress in sequencing technologies allowed easy molecular-based analysis of our inhabitants, and the realization that humans were not only the results of our 23,000 genes, but that our other genome, the microbiome, was a crucial component of human beings. Definitely, the human organism is composed of 100 trillion microorganisms, surpassing by 10 times the number of cells in the human body[1]. Due to the huge number of microbes colonizing the intestine, researchers obviously started to analyse the gut microbiota, but other organs are also of importance, including the skin, the vagina, the lung or the oral cavity.

Thus, ignored for long, the intestinal ecosystem is complex and results from a long co-evolution between the microbiota and its host or between the microbiome and the human genome. Although this notion is somehow controversial, the microbiota is considered as an organ with its metabolic functions and as such, it could be transplanted. One of the major difficulties is that it is very heterogeneous. Its composition varies among individuals, and it is in permanent interaction with food, host cells and ingested bacteria from food, and also with potential pathogens. The autochthonous microbiota exerts a barrier effect to prevent colonization by pathogens and also by transiting bacteria. In the present chapter, we describe what is known about this integrated interactive ecosystem. This system is also subject to dysbiosis associated with diverse pathologies[2]. This was introduced in chapter 2 by Curtis and will be discussed at length in section 3 of this book (from chapter 13 onwards).

The Human Microbiota and Chronic Disease: Dysbiosis as a Cause of Human Pathology, First Edition.
Edited by Luigi Nibali and Brian Henderson.
© 2016 John Wiley & Sons, Inc. Published 2016 by John Wiley & Sons, Inc.

3.2 Who is there, how is it composed?

The baby makes its first encounter with microbes in the perinatal period and its gut is very rapidly colonized within the first minutes of life. Then the ecosystem follows dynamic fluctuations in microbial composition during the first three years of life, with major increase in diversity, to reach a relatively stable equilibrium. Although this is still a matter of debate, Aagaard and collaborators proposed that infant might be first exposed *in utero* to a low-abundance microbiota present in the placenta and that the length of gestation may be a significant factor influencing the seeding[3]. A deep characterization of the placental microbiota was undertaken, revealing that it is composed of members of the Firmicutes, Tenericutes, Proteobacteria, Bacteroidetes, and Fusobacteria phyla, and its closest community is the oral microbiome (further discussed by Marsland and Salami in chapter 24). It is noteworthy that germ-free mice can be obtained by cesarean section derivation and live in isolators with germ-free foster mothers, thus calling into question the placenta hypothesis.

Colonization is affected by the mode of delivery[4,5]. Indeed, vaginal delivery leads to a baby bacterial community resembling its mother's vaginal microbiota and dominated by *Lactobacillus, Prevotella,* or *Sneathia* spp. In contrast, C-section-born babies acquire a microbiota resembling the skin surface bacterial community, including *Staphylococcus, Corynebacterium,* and *Propionibacterium* spp. Interestingly, it seems that the C-section-born baby bacterial community is no closer to its mother's skin than to other women's skin community[4,5], corresponding to the fact that the surgery-room environment is impacted by several individuals other than the mother. This observation is important to take into account if we consider the evolution of our modern living habits and that C-section is often considered economically convenient and probably less troubling. The impact of C-section on reducing richness or diversity is documented[6], and C-section is associated with higher risk of allergies, asthma and atopic diseases[7,8]. Antibiotic administration in early life is also a critical factor impacting colonization[9]. Obviously, maternal nutrition is also a factor impacting the baby microbiota, and the importance of the mode of feeding, i.e. breast-feeding versus formula feeding, is related to the presence of particular oligosaccharides in the mother's milk and is significant, although the environment will play a major role[10,11]. Finally, the switch to the first complementary food and weaning is required for a full microbiota maturation[11], while a recent report argues that the cessation of breast-feeding is more important than introduction of solid food[12].

During the first three years of life, the infant's intestinal microbiota is highly instable[13], its richness and diversity increasing over time to reach a plateau by three years of age. It remains mostly stable until adulthood in healthy conditions[14] (Figure 1). In adults, the gut microbiota is composed of hundreds of different species belonging to three major phyla, namely Firmicutes, Bacteroidetes and Actinobacteria, that are well adapted to this ecological niche[15,16]. The proportion of each phylum is individual specific, some being mainly dominated by Firmicutes, others presenting more Bacteroidetes. Other phyla are also present but to a lesser extent, including Proteobacteria, Verrucomicrobia and also Fusobacteria. Archaea are also present, especially *Methanobrevibacter smithii* in individuals excreting methane. Extensive metagenomic analysis revealed the presence of a common

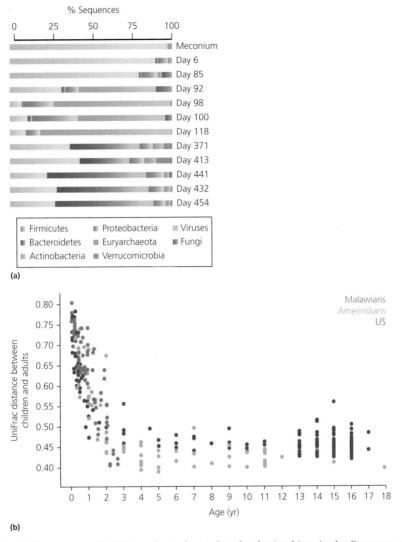

Figure 1 (a) Metataxonomic DNA analysis of an infant fecal microbiota in the first year and a half of life showing a chaotic evolution impacted by weaning and antibiotic consumption, from Koenig et al[13]. (b) Fecal microbial differences between children and adults from Amazonia (Amerindians: green dots), Malawi (red dots) and USA (blue dots), showing reduction of distance to adult microbiota during the first three years of life and then stability until adulthood, from Yatsunenko et al[14]. (*see color plate section for color details*).

core of bacterial species[15,17] that vary in amount between individuals. It is noteworthy that about 70% of the dominant species are considered unculti-vable[18], although calling these "not yet cultured" would seem more appropriate. In healthy conditions it is considered a stable community that is influenced by the dietary habits and the physiology of its host. Whole-metagenomic sequencing of 124 persons in the frame of the European Project MetaHIT revealed that each individual microbiome is composed of about 600,000 genes from a catalog ini-tially established to amount to 3.3 million non-redundant genes, indicating that the number of bacterial genes in the human gut surpassed the human genome by 150-fold[16]. This catalog increased with the number of sequenced individuals to reach 10 million genes based on 1267 sequenced microbiomes[19]. This effort towards deep characterization of the gut microbiome is essential to allow quantitative metagenomic, metatranscriptomic and metaproteomic analysis and to permit apprehending differences across healthy populations in the context of various nutritional behaviors or environments and understanding perturbation in diseases.

Interestingly, further analysis of individual microbiome revealed that it is possible to cluster people in at least three enterotypes corresponding to species and functional composition that appeared to be independent of gender and nationality[20]. These enterotypes correspond to balanced ecological arrangements dominated by *Bacteroides*, *Prevotella* and *Rumminococcus*, respectively. The latter enterotype is also associated to the presence of *M. smithii*. This description raised debate in the scientific community, but later other studies reinforced the concept[21,22]. Intriguingly, this concept seemed not limited to the gut but also to be true of other microbial communities across the human body. Similar clusters have also been described in mice and pigs[23,24]. Long-term dietary habits may be a key factor contributing to the structuration of these enterotypes[25]. Indeed, the *Bacteroides* enterotype was associated with consumption of protein and animal fat, whereas the *Prevotella* was related to carbohydrates. A follow up of volunteers submitted to high-fat/low-fibre or low-fat/high-fibre diet showed fast microbiome composition modifications (within 24h) but enterotypes were stable during the 10 days of dietary intervention. Enterotypes may also be correlated with physiological parameters of the host such as transit time[26]. Thus, these enterotype community structures may also be correlated to functional differences, methane production being for instance typical of the *Rumminococcus* enterotype. As in any other balanced complex systems, intestinal ecology may be subject to hysteresis[27].

3.3 A system in interaction with food

The first source of carbon and therefore energy for our microbiota comes from plant polysaccharides, especially dietary fibres, but also from mucus and animal-derived sugars. These complexes of various monosaccharides, linked to each other in different ways and modified with chemical radicals including acetyl and sulphate groups, necessitate a huge repertoire of enzymes for their degradation[28]. The human genome is poorly equipped with glycoside hydrolases involved in carbohydrate digestion, only allowing the breakdown of simple sugars (e.g. lactose, sucrose) and starch. But the gut microbiome encodes a wide variety of carbohydrate-active enzymes (CAZymes). Dietary fibres undergo partial or complete microbial fermentation in the large intestine that results in the production of short chain fatty acids available for the host and excreted or exhaled gas.

Gut microbial composition is highly impacted by food and food habits. Studying mouse gut microbiota, L. Zhao and his group reported that 57% of the variations in the composition of the intestinal microbiota were related to dietary variation, whereas genetic differences only accounted for 12%[29]. Food habits and lifestyle have changed greatly over the last centuries, starting with the "Neolithic revolution" where humans evolved from a hunting and gathering lifestyle to agriculture, animal husbandry and settlement. This evolution took centuries. However, over the last 100 years, changes have accelerated dramatically in Western countries with the discovery and use of antibiotics that are also used in livestock animals, the reduction of raw food and the development of processed food, and also with changes in hygiene. Thus, as also discussed in chapter 6 by Bermudez and 29 by Gibson, it has been proposed that these major lifestyle changes profoundly impacted the intestinal ecosystem and its microbial composition and diversity[30]. Such profound changes have been proven by studying various populations, including individuals from a rural African village in Burkina Faso[31], Guahibo Amerindians in the Amazonas and members of rural communities in Malawi[14] and Hadza of Tanzania[32], and comparing their microbiota to population from Western countries. In each study, a profound difference in gut microbiota was observed. Burkina Faso children, who consume more dietary fibres, showed exceptional abundance of *Prevotella* and *Xylanibacter,* which were almost absent in Italian children[31]. On the other hand, these EU children showed increased abundance in *Enterobacteriaceae* that were under-represented in the African children. Similarly, in the Amerindians and Malawi cohort described by Yatsunenko *et al.*, the average observed number of OTUs was high (about 1600) whereas in the US cohort it was around 1000[14]. The findings from the study on Hadza were that their microbiota displayed higher levels of richness and biodiversity than the EU citizens[32].

The notion of richness of the microbiome has been underlined by Le Chatelier and collaborators in a study on an obese population[33]. Applying a quantitative metagenomic approach on a Danish cohort of lean and obese individuals, the MetaHIT consortium observed a bimodal distribution of gene richness belonging to known and previously unknown microbial species (Figure 2). People with a low bacterial richness (about a quarter of the cohort) displayed higher adiposity, insulin resistance, dyslipidemia and a more pronounced low-grade inflammatory status than individuals with higher gene richness. Interestingly, the obese individuals with low gene richness had gained more weight over the preceding nine years than obese individuals with higher gene richness[33]. A similar finding was observed in a French cohort[34]. Interestingly, the French overweight and obese subjects who for six weeks ate a hyper low-calorie diet enriched in dietary fibres and proteins responded differently to the intervention diet depending on their microbiome gene richness. The group with low gene richness displayed increased bacterial richness after the intervention, but their body weight and other parameters did not improve so much (Figure 2). On the other hand, the group with higher gene richness did not show major modification of their richness, but the diet was more efficient[34]. Mechanisms of association between the gut microbiota and obesity are covered extensively in chapter 23.

The current view is that the human gut microbiota is relatively subject-specific, over and above the common core, and that it is stable and resistant to modification over time[35]. This concept has been illustrated by the resilience observed after short-term antibiotic treatment[36]; although in some individuals resilience was slower to achieve. This stability has been challenged by different dietary interventions, confirming, at least in part, the relative stability of the human gut microbiota[25,37]. Even at the level of strains, it seems that residents of the intestine

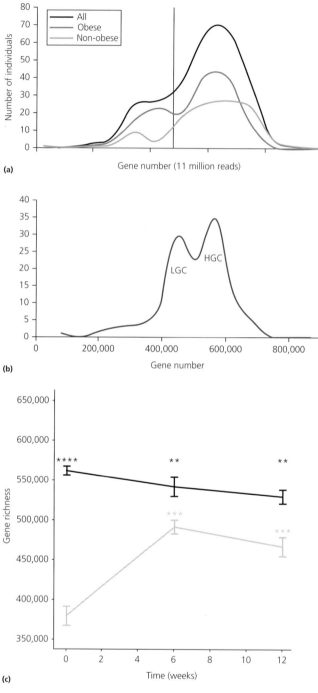

(a)

(b)

(c)

Figure 2 Bimodal distribution of microbial genes showing individual with low and high gene counts in a Danish cohort of lean and obese (a)[33] and in a French cohort of overweight and obese subjects (b)[34]. Gene richness evolution (c) of French subjects who underwent a six-week hyper-low-calorie diet depending on gene count at the beginning of the study.

are present for several years and not easily replaced[37,38]. Interestingly, Turnbaugh *et al. followed* the gut microbiota of individuals exposed to short-term diets composed either of animal- or plant-derived food products[39]. They showed that the animal-based diet had a positive impact on bile-tolerant bacteria from the *Alistipes, Bilophila* and *Bacteroides* genera and decreased Firmicutes usually associated with fibre degradation such as *Roseburia, Eubacterium rectale* and *Ruminococcus bromii*. They also confirmed that food-associated microbes occupied the gut transiently. However, using culture techniques, it was shown that rats exposed to a long term-diet of enriched fructo-oligosaccharides showed an impact of microbial composition only during the first few weeks of dietary intervention[40]. Resilience occurred after 8 to 27 weeks of the diet, and although the level of butyrate production was still elevated, gut microbiota composition returned to its initial structure. This notion of stability and resilience is a very challenging one that deserves to be studied more deeply. Understanding the mechanisms driving the stability of the ecosystem is essential in order to develop strategies of ecological modulation, especially in people with poor microbial richness or patients with dysbiosis. Moreover, aging is associated with major changes in the ecology of the gut[41], raising fundamental questions on how to modulate the microbiota to prevent disease.

3.4 A system highly impacted by the host

Although evidence shows that the composition of gut microbiota is less dependent on the host genotype than on dietary habits[29,42], host physiology is essential in conditioning gut microbiota. The innate immune system through the secretion of defensins and antimicrobial peptides has been revealed as essential in shaping the intestinal microbiota[43]. Among those, the role of RegIIIγ, a c-type lectin, has been shown to be key in promoting host-bacterial mutualism and regulating the special relationship between bacteria and host[44]. Comparably, mucins secreted by intestinal goblet cells act as a physical barrier, separating bacteria from the human mucosa, and are essential for antimicrobial protein efficacy[45]. Moreover, the complex sugar structure of mucins is a source of carbohydrates for bacteria and the capability to metabolize these oligosaccharides is expected to be a key factor in defining which bacterial species colonize the mucosal surface[46]. The identification of *Akkermansia muciniphila*, a member of the Verrucomicrobia, for its capacities to degrade mucus, later shown to play an essential role in obesity, underline the importance of mucins for the intestinal microbiota[47,48]. Furthermore, although the impact of water and electrolytes fluxes on the gut microbiota has not been investigated in detail, it is obvious that consistency of the luminal content will influence the ecosystem.

Similarly, transit time is evidently a key factor influencing gut microbiota composition. Twenty years ago, Galmiche and collaborators demonstrated in healthy volunteers that slowing down or accelerating transit time and gut motility with drugs impacted gut microbiota structure and functions[49]. A fast transit time resulted in enhanced presence of sulphate-reducing bacteria, increased SCFAs in feces and reduction in methanogenic Archaea and methane production in methane-excreting subjects. On the other hand, slowing-down transit time induced a rise in counts of methanogenic Archaea and methane production and a decrease in fecal SCFAs content with a fall in sulphate-reducing bacteria[49]. Recently, a meta-taxonomic analysis associated stool consistency, using the Bristol stool chart as a proxy for individual transit time, with species richness, enterotypes and community composition[26]. Among the findings, stool consistency was linked to the abundance of *Akkermansia* and *Methanobrevibacter*, reinforcing the findings of the 1996 study.

3.5 A system in interaction with human cells

Bacterial enzymes and metabolites are also important for the host. These comprise SCFAs, conjugated linoleic acids, choline derivatives, bile acids, aromatic amino acids, ammoniac (NH3), hydrogen sulphide, exopolysaccharides, and nutrients including vitamins B and K[50,51]. Among the various metabolites, SCFAs, in particular acetate, propionate and butyrate, have been extensively studied. Interestingly, advances in knowledge and tools led to a rediscovery of these essential metabolites. SCFAs are the end products of dietary fibre fermentation. They are the main energy source for colonocytes[52]. These SCFAs contribute to maintaining intestinal homeostasis and their action is linked to their activity as histone deacetylase inhibitors (HDACi) or less specifically as lysine deacetylase inhibitors (KDAC)[53]. Depending on the diet, the ratio between the three main acids varies, but is often reported at a 60:20:20 molar ratio for acetate, propionate and butyrate, respectively, for a total of 100-150 millimolar depending on the colonic segment considered and the diet[54]. Moreover, SCFAs also act as signaling molecules via the interaction with G-protein coupled receptors including free fatty acid receptors 2 (FFAR2/GPR43) and 3 (FFAR3/GPR41), even though GPR109A has also been proposed as a receptor[55]. The role of these receptors in the activation of entero-endocrine cells[56] and on immune cells including Treg lymphocytes has been recently highlighted[57].

Choline and carnithine have been shown to be metabolized by the intestinal gut microbiota to produce trimethylamine (TMA), acetate and ethanol[50]. Then, TMA is absorbed by intestinal epithelial cells and transformed into trimethylamine-*N*-oxide (TMAO) by flavin mono-oxygenase enzymes in the liver, and demethylated into dimethylamine and monomethylamine, or excreted in the urine[58]. This metabolism is important for the host and has been associated with cardiovascular diseases[59]. Similarly, dietary tryptophan and phenylalanine from cauliflower, cabbage, and broccoli are converted in kinurenine, an aryl hydrocarbon receptor (AhR) ligand. The activation of this pathway regulates immune and non-immune cells and, among other things, stimulates IL-22 secretion by innate lymphoid cells[60], which results in the production of antimicrobial peptides and mucus secretion[61]. Finally, our knowledge of potential metabolites is being increased considerably by the development of high-throughput analysis, including metabolomics[62,63] and functional metagenomics[64–66]. Indeed, several bacteria including *Faecalibacterium prausnitzii* have been shown strongly to influence host physiology by mechanisms that are yet to be deciphered[67].

3.6 Conclusion: an intriguing integrated interactive system deserving further study

When studying human pathophysiology, it is critical to consider not only each entity in the ecosystem, but also that a symbiosis is occurring in the gut and in other ecological niches. A recent paper by Pamer and collaborators demonstrates the importance of our other genome[68]. Indeed, in allogeneic hematopoietic stem cell transplantation the diversity of the gut microbiota at engraftment was revealed as an independent predictor of vital prognosis. More work is needed to clarify how man-microbe symbiosis impacts human physiology, health and well-being, taking into account the microbiome. Holistic studies must be conducted associating whole metagenome characterization, i.e. quantitative metagenomics, and not only metataxonomy or targeted metagenomics (16S DNA), metatranscriptomic, metaproteomic, metabolomics, and complete exhaustive patient phenotyping. Long-term follow-up with well selected large cohorts of individuals is also essential in gaining better understanding of the complexity of the intestinal ecosystem. Finally, it is crucial to consider the microbiota when defining strategic modulations of the diet for health benefits.

TAKE-HOME MESSAGE

- Gene richness of the intestinal ecosystem is a health stratifier.
- Bacterial metabolites are essential mediators of human physiology.
- Understanding and modeling the complexity of the intestinal ecosystem is an important future challenge.

References

1 Savage DC. Microbial ecology of the gastrointestinal tract. *Annu Rev Microbiol* 1977; **31**: 107–133.

2 de Vos WM, de Vos EA. Role of the intestinal microbiome in health and disease: from correlation to causation. *Nutr Rev* 2012; **70**: S45–56.

3 Aagaard K, Ma J, Antony KM, Ganu R, Petrosino J, Versalovic J. The placenta harbors a unique microbiome. *Sci Transl Med* 2014; **6**: 237ra65.

4 Dominguez-Bello MG, Costello EK, Contreras M *et al.* Delivery mode shapes the acquisition and structure of the initial microbiota across multiple body habitats in newborns. *Proc Natl Acad Sci USA* 2010; **107**: 11971–11975.

5 Dominguez-Bello MG, Blaser MJ, Ley RE, Knight R. Development of the human gastrointestinal microbiota and insights from high-throughput sequencing. *Gastroenterology* 2011; **140**: 1713–1719.

6 Jakobsson HE, Abrahamsson TR, Jenmalm MC *et al.* Decreased gut microbiota diversity, delayed Bacteroidetes colonisation and reduced Th1 responses in infants delivered by caesarean section. *Gut* 2014; **63**: 559–566.

7 Negele K, Heinrich J, Borte M *et al.* Mode of delivery and development of atopic disease during the first 2 years of life. *Pediatr Allergy Immunol* 2004; **15**: 48–54.

8 Bager P, Wohlfahrt J, Westergaard T. Caesarean delivery and risk of atopy and allergic disease: meta-analyses. *Clin Exp Allergy* 2008; **38**: 634–642.

9 Vangay P, Ward T, Gerber JS, Knights D. Antibiotics, pediatric dysbiosis, and disease. *Cell Host Microbe* 2015; **17**: 553–564.

10 Fallani M, Young D, Scott J *et al.* Intestinal microbiota of 6-week-old infants across Europe: geographic influence beyond delivery mode, breast-feeding, and antibiotics. *J Pediatr Gastroenterol Nutr* 2010; **51**: 77–84.

11 Fallani M, Amarri S, Uusijarvi A *et al.* Determinants of the human infant intestinal microbiota after the introduction of first complementary foods in infant samples from five European centres. *Microbiology* 2011;**157**: 1385–1392.

12 Bäckhed F, Roswall J, Peng Y *et al.* Dynamics and stabilization of the human gut microbiome during the first year of life. *Cell Host Microbe* 2015; **17**: 690–703.

13 Koenig JE, Spor A, Scalfone N *et al.* Succession of microbial consortia in the developing infant gut microbiome. *Proc Natl Acad Sci USA* 2011; **108**: 4578–4585.

14 Yatsunenko T, Rey FE, Manary MJ *et al.* Human gut microbiome viewed across age and geography. *Nature* 2012; **486**: 222–227.

15 Tap J, Mondot S, Levenez F *et al.* Towards the human intestinal microbiota phylogenetic core. *Environ Microbiol* 2009; **11**: 2574–2584.

16 Human Microbiome Project Consortium. Structure, function and diversity of the healthy human microbiome. *Nature* 2012; **486**: 207–214.

17 Qin J, Li R, Raes J *et al.* A human gut microbial gene catalogue established by metagenomic sequencing. *Nature* 2010; **464**: 59–65.

18 Suau A, Bonnet R, Sutren M *et al.* Direct analysis of genes encoding 16S rRNA from complex communities reveals many novel molecular species within the human gut. *Appl Environ Microbiol* 1999; **65**: 4799–4807.

19 Li J, Jia H, Cai X *et al.* An integrated catalog of reference genes in the human gut microbiome. *Nat Biotechnol* 2014; **32**: 834–841.

20 Arumugam M, Raes J, Pelletier E *et al.* Enterotypes of the human gut microbiome. *Nature* 2011; **473**: 174–180.

21 Ding T and Schloss PD. Dynamics and associations of microbial community types across the human body. *Nature* 2014; **509**: 357–360.

22 Zupancic ML, Cantarel BL, Liu Z *et al.* Analysis of the gut microbiota in the old order Amish and its relation to the metabolic syndrome. *PLoS One* 2012; **7**: e43052.

23 Wang J, Linnenbrink M, Künzel S *et al.* Dietary history contributes to enterotype-like clustering and functional metagenomic content in the intestinal microbiome of wild mice. *Proc Natl Acad Sci USA* 2014; **111**: E2703–E2710.

24 Mach N, Berri M, Estellé J *et al.* Early-life establishment of the swine gut microbiome and impact on host phenotypes. *Environ Microbiol Rep* 2015; **7**: 554–569.

25 Wu GD, Chen J, Hoffmann C *et al.* Linking longterm dietary patterns with gut microbial enterotypes. *Science* 2011; **334**: 105–108.

26 Vandeputte D, Falony G, Vieira-Silva S, Tito RY, Joossens M, Raes J. Stool consistency is strongly associated with gut microbiota richness and composition, enterotypes and bacterial growth rates. *Gut* 2015 Jun 11. pii: gutjnl-2015-309618 [Epub ahead of print].

27 Scheffer M, Carpenter SR, Lenton TM *et al.* Anticipating critical transitions. *Science* 2012; **338**: 344–348.

28 El Kaoutari A, Armougom F, Gordon JI, Raoult D, Henrissat B. The abundance and variety of carbohydrate-active enzymes in the human gut microbiota. *Nat Rev Microbiol* 2013; **11**: 497–504.

29 Zhang CH, Zhang MH, Wang SY *et al.* Interactions between gut microbiota, host genetics and diet relevant to development of metabolic syndromes in mice. *ISME J* 2010; **4**: 232–241.

30 Von Hertzen L, Beutler B, Bienenstock J *et al.* Helsinki alert of biodiversity and health. *Ann Med* 2015; **47**: 218–225.

31 De Filippo C, Cavalieri D, Di Paola M *et al.* Impact of diet in shaping gut microbiota revealed by a comparative study in children from Europe and rural Africa. *Proc Natl Acad Sci USA* 2010; **107**: 14691–14696.

32 Schnorr SL, Candela M, Rampelli S *et al.* Gut microbiome of the Hadza hunter-gatherers. *Nat Commun* 2014; **5**: 3654.

33 Le Chatelier E, Nielsen T, Qin J *et al.* Richness of human gut microbiome correlates with metabolic markers. *Nature* 2013; **500**: 541–546.

34 Cotillard A, Kennedy SP, Kong LC *et al.* Dietary intervention impact on gut microbial gene richness. *Nature* 2013; **500**: 585–588.

35 Lozupone CA, Stombaugh JI, Gordon JI, Jansson JK, Knight R. Diversity, stability and resilience of the human gut microbiota. *Nature* 2012, **489**: 220–230.

36 De la Cochetière MF, Durand T, Lepage P, Bourreille A, Galmiche JP, Doré J. Resilience of the dominant human fecal microbiota upon short-course antibiotic challenge. *J Clin Microbiol* 2005; **43**: 5588–5592.

37 Faith JH, Guruge JL, Mark Charbonneau M *et al.* The long-term stability of the human gut microbiota. *Science* 2013; **341**: 1237439.

38 Schloissnig S, Arumugam M, Sunagawa S *et al.* Genomic variation landscape of the human gut microbiome. *Nature* 2013; **493**: 45–50.

39 David LA, Maurice CF, Carmody RN *et al.* Diet rapidly and reproducibly alters the human gut microbiome. *Nature* 2014; **505**: 559–563.

40 Le Blay G, Michel C, Blottière HM, Cherbut C. Prolonged intake of fructo-oligosaccharides induces a short-term elevation of lactic acid-producing bacteria and a persistent increase in cecal butyrate in rats. *J Nutr* 1999; **129**: 2231–2235.

41 Claesson MJ, Jeffery IB, Conde S *et al.* Gut microbiota composition correlates with diet and health in the elderly. *Nature* 2012; **488**: 178–184.

42 Carmody RN, Gerber GK, Luevano JM Jr *et al.* Diet dominates host genotype in shaping the murine gut microbiota. *Cell Host Microbe* 2015; **17**: 72–84.

43 Mukherjee S, Hooper LV. Antimicrobial defense of the intestine. *Immunity* 2015; **42**: 28–39.

44 Vaishnava S, Yamamoto M, Severson KM *et al.* The antibacterial lectin RegIIIgamma promotes the spatial segregation of microbiota and host in the intestine. *Science* 2011; **334**: 255–258.

45 Birchenough GMH, Johansson MEV, Gustafsson JK, Bergström JH, Hansson GC. New developments in goblet cell mucus secretion and function. *Mucosal Immunol* 2015; **8**: 712–719.

46 Tailford LE, Crost EH, Kavanaugh D, Juge N. Mucin glycan foraging in the human gut microbiome. *Front Genet* 2015; **6**: 81.

47 Belzer C, de Vos WM. Microbes inside — from diversity to function: the case of *Akkermansia*. *ISME J* 2012; **6**: 1449–1458.

48 Dao MC, Everard A, Aron-Wisnewsky J *et al. Akkermansia muciniphila* and improved metabolic health during a dietary intervention in obesity: relationship with gut microbiome richness and ecology. *Gut* doi: 10.1136/gutjnl-2014-308778 [Epub ahead of print].

49 El Oufir L, Flourié B, Bruley des Varannes S *et al.* Relations between transit time, fermentation products, and hydrogen consuming flora in healthy humans. *Gut* 1996; **38**: 870–877.

50 Russell WR, Hoyles L, Flint HJ, Dumas ME. Colonic bacterial metabolites and human health. *Curr Opin Microbiol* 2013; **16**: 246–254.

51 Patterson E, Cryan JF, Fitzgerald GF, Ross RP, Dinan TG, Stanton C. Gut microbiota, the pharmabiotics they produce and host health. *Proc Nutr Soc* 2014; **73**: 477–489.

52 Astbury SM, Corfe BM. Uptake and metabolism of the short-chain fatty acid butyrate, a critical review of the literature. *Curr Drug Metab* 2012; **13**: 815–821.

53 Blottière HM, Buecher B, Galmiche JP, Cherbut C. Molecular analysis of the effect of short-chain fatty acids on intestinal cell proliferation. *Proc Nutr Soc* 2003; **62**: 101–106.

54 Cummings JH. Short chain fatty acids in the human colon. *Gut* 1981; **22**: 763–779.

55 Tilg H, Moschen AR. Food, immunity, and the microbiome. *Gastroenterology* 2015; **148**: 1107–1119.

56 Kaji I, Karaki S, Kuwahara A. Short-chain fatty acid receptor and its contribution to glucagon-like peptide-1 release. *Digestion* 2014; **89**: 31–36.

57 Smith PM, Howitt MR, Panikov N *et al.* The microbial metabolites, short-chain fatty acids, regulate colonic Treg cell homeostasis. *Science* 2013; **341**: 569–573.

58 Lang DH, Yeung CK, Peter RM *et al.* Isoform specificity of trimethylamine N-oxygenation by human flavin-containing monooxygenase (FMO) and P450 enzymes: selective catalysis by FMO3. *Biochem Pharmacol* 1998; **56**: 1005–1012.

59 Wang Z, Klipfell E, Bennett BJ *et al.* Gut flora metabolism of phosphatidylcholine promotes cardiovascular disease. *Nature* 2011; **472**: 57–63.

60 Kiss EA, Vonarbourg C. Aryl hydrocarbon receptor: a molecular link between postnatal lymphoid follicle formation and diet. *Gut Microbes* 2012; **3**: 577–582.

61 Zelante T, Iannitti RG, Cunha C *et al.* Tryptophan catabolites from microbiota engage aryl hydrocarbon receptor and balance mucosal reactivity via interleukin-22. *Immunity* 2013; **39**: 372–385.

62 Rezzonico E, Mestdagh R, Delley M *et al*. Bacterial adaptation to the gut environment favors successful colonization: microbial and metabonomic characterization of a simplified microbiota mouse model. *Gut Microbes* 2011; **2**: 307–318.

63 Nicholson JK, Holmes E, Kinross J *et al*. Host-gut microbiota metabolic interactions. *Science* 2012; **336**: 1262–1267.

64 Lakhdari O, Cultrone A, Tap J *et al*. Functional metagenomics: a high throughput screening method to decipher microbiota-driven NF-κB modulation in the human gut. *PLoS One* 2010; **5**: e13092.

65 Gloux K, Berteau O, El Oumami H, Béguet F, Leclerc M, Doré J. A metagenomic β-glucuronidase uncovers a core adaptive function of the human intestinal microbiome. *Proc Natl Acad Sci USA* 2011; **108 Suppl 1**: 4539–4546.

66 Lepage P, Leclerc MC, Joossens M *et al*. A metagenomic insight into our gut's microbiome. *Gut* 2013; **62**: 146–158.

67 Sokol H, Pigneur B, Watterlot L *et al*. *Faecalibacterium prausnitzii* is an anti-inflammatory commensal bacterium identified by gut microbiota analysis of Crohn disease patients. *Proc Natl Acad Sci USA* 2008; **105**: 16731–18736.

68 Taur Y, Jenq RR, Perales MA *et al*. The effects of intestinal tract bacterial diversity on mortality following allogeneic hematopoietic stem cell transplantation. *Blood* 2014; **124**: 1174–1182.

CHAPTER 4
The oral microbiota

William G. Wade

Centre for Immunobiology, Blizard Institute, Barts and The London School of Medicine and Dentistry, Queen Mary University of London, London, United Kingdom

4.1 Introduction

The human body is heavily colonized by microorganisms, known collectively as the human microbiome. In general, the microbiome lives in harmony with its human host, although there is increasing evidence that some chronic diseases can result from an inappropriate interaction between the microbiome and the immune system (several examples are provided later in this book). The mouth is an unusual ecosystem in that the oral microbiome must be controlled, normally by brushing with toothpaste, in order to prevent the common diseases dental caries (tooth decay) and the periodontal (gum) diseases. Both diseases are associated with bacteria but the primary risk factors are a sugar-rich diet for caries and host susceptibility and smoking for periodontitis. The oral microbiota is important for health because it prevents colonization by pathogens. It is also now recognized that oral bacteria play an important role in reducing nitrate to nitrite which is then converted to nitric oxide, which is critical for cardiovascular health. Oral bacteria also play a role in non-oral diseases either by directly causing infections or via the chronic inflammation that is a feature of periodontitis.

It is often stated that 90% of the cells in the human body are bacterial. A more recent re-evaluation of this estimate has revised the proportion down to around 1:3[1]. Whatever the true figure, there is no doubt that bacteria play an important role in the human body in preventing colonization by exogenous pathogens, in the digestion of food, providing enzymes that humans lack and providing an appropriate context to the immune system. Conversely, the commensal microbiota has also been linked to a number of chronic diseases and conditions such as obesity. The human microbiome is extremely complex and only with the introduction of next-generation sequencing methods has it become possible to study the composition of the microbiome, primarily by community profiling targeting the small sub-unit (16S) ribosomal RNA gene. The genetic potential of the microbiome can now be studied by metagenomic analyses using random shotgun sequencing and genome reconstruction[2]. In addition, microbiome gene expression is being investigated by metatranscriptomic techniques[3–5].

The Human Microbiota and Chronic Disease; Dysbiosis as a Cause of Human Pathology, First Edition.
Edited by Luigi Nibali and Brian Henderson.

Dental caries and the periodontal diseases are the commonest bacterial diseases of man and are complex diseases arising from an interaction between the oral microbiota and its human host, influenced by environmental factors. The aim of this chapter is to review what is currently known regarding the composition of the oral microbiome and its role in health and disease.

4.2 Composition of the oral microbiome

The oral microbiota includes representatives of all types of microorganisms: archaea, bacteria, fungi, protozoa and viruses and, after the colon, is the most heavily colonized body site.

4.2.1 Archaea

Archaea make up only a minor component of the oral microbiota and the representatives that are found are methanogens that are primarily found in subgingival plaque and are significantly associated with periodontitis[6]. Oral methanogens utilize short-chain fatty acids and CO_2 and hydrogen produced by other proteolytic anaerobes to produce methane and thus act as terminal degraders in the ecosystem; interestingly, their numbers are negatively correlated with another hydrogenotrophic group, the sulphate-reducing bacteria[7]. The most frequently detected species is *Methanobrevibacter oralis*, although the presence of *Methanobrevibacter smithii*, more commonly found in the colon, of representatives of the genus *Methanosarcina*, and of an as yet un-identified methanogenic lineage related to *Thermoplasmatales* has also been reported in oral samples[8,9].

4.2.2 Fungi

Candida is the principal fungal genus found in the mouth, with *C. albicans*, *C. dubliniensis*, *C. guillermondii*, *C. krusei*, *C. glabrata*, *C. parapsilosis* and *C. tropicalis* the predominant species responsible for a range of infections[10]. A culture-independent analysis targeting the fungal internal transcribed spacer (ITS) region revealed the presence of 85 fungal genera in the mouths of healthy individuals[11]. The most frequently detected taxa were *Candida*, *Cladosporium*, *Aureobasidium*, *Saccharomycetales*, *Aspergillus*, *Fusarium* and *Cryptococcus*. A similar study using an modified alogorithm to assign ITS sequences to genera refined and extended this list and also found that *Malasezzia*, a fungus commonly found on skin, was the most prevalent[12]. Whether all of these fungal genera are oral residents is debatable. Many are spore-forming organisms found in the air and food and may only be transient in the mouth, although mycobiome profiles assessed by culture have been shown to be stable within individuals over a 30-week period[13].

4.2.3 Protozoa

Two protozoa are commonly found in the mouth. Around half of individuals are colonized with *Entamoeba gingivalis*, while *Trichomonas tenax* is resident in around 20% of the population[14]. Their numbers are raised in subjects with poor oral hygiene and those with periodontitis and it was once thought that the protozoa might play a causative role. The current consensus is that they are harmless saprophytes that feed on food debris and bacteria; these are obviously more abundant in individuals with poor oral hygiene, who consequently have gingival disease[15].

4.2.4 Viruses

The viruses found in the mouth are those that infect either human or bacterial cells. *Herpes simplex* causes a variety of oral infections, and is thus frequently detected in the mouth[16]. Primary infections most often occur in children under 6 years of age and are typically subclinical. Older children may have a more serious infection, gingivostomatitis, which can affect the lips, tongue and buccal mucosa. Following the primary infection, herpes simplex enters a dormant state within the trigeminal nerve. Various factors including sunlight, cold weather, and stress can cause reactivation of the virus, which leads to encrustations of the lips known as herpes labialis, commonly known as coldsores[17]. Papilloma viruses are responsible for a range of oral infections[18] and have been linked with oropharyngeal cancer[19]. In individuals infected with rabies, the virus infects a variety of non-neural tissues, including the tongue and salivary glands[20], whilst mumps is the commonest viral infection of the salivary glands[21].

Other human viruses can be found in the mouth although the primary site of infection is elsewhere in the body. For example, the viruses responsible for upper respiratory tract infections, such as the common cold, are frequently detected in the mouth. Because there is a direct route from the bloodstream to the mouth via gingival crevicular fluid which flows around the teeth as they extrude from the body, blood-borne viruses, such as HIV and the hepatitis viruses, which cause systemic infections, can be found in the mouths of infected subjects[22-24].

A metagenomic analysis of double-stranded DNA viruses in healthy individuals performed as part of the Human Microbiome Project[25], found an average of 5.5 viral genera in each subject. Representatives of 13 viral genera were detected in the mouth with roseoloviruses, which includes the herpes viruses, the most frequently isolated, being present in 98% of subjects. Individuals have been found to harbor a personalized oral virome[26], which may be shared to some extent with members of the same household[27]. Different intra-oral habitats have characteristic viral profiles and the viral community found in subgingival plaque in subjects with periodontitis was significantly associated with disease status[28]. The majority of viruses found in dental plaque are bacteriophages[29] and it has been predicted that most oral bacteriophage follow a predominantly lysogenic lifestyle, thereby increasing the genetic potential of the host bacterial community.

4.2.5 Bacteria

Bacteria are the predominant type of microorganism in the mouth. Saliva contains around 100 million bacteria per ml and every surface of the mouth is covered by a bacterial biofilm. Around 700 species have been detected, 95% of which belong to the phyla *Firmicutes, Bacteroidetes, Proteobacteria, Actinobacteria, Fusobacteria* and *Spirochaetes*[30]. Other phyla consistently detected are *Synergistes, Chloroflexi* and the unnamed phylum-level Divisions GN02, SR1 and TM7[31]. The Human Oral Microbiome Database (www.homd.org) lists the bacterial taxa found in the mouth and provides descriptions of their phenotypes, where available, with links to genome sequence data as well as a 16S rRNA gene sequence identification tool[32]. The predominant phyla and genera found in the mouth are listed in Table 1.

Table 1 Predominant bacterial phyla and genera found in the human mouth.

Phylum/Division	Genera	Phylum/Division	Genera
Actinobacteria	Actinobaculum		Mogibacterium
	Actinomyces		Moryella
	Alloscardovia		Oribacterium
	Atopobium		Parvimonas
	Bifidobacterium		Peptococcus
	Corynebacterium		Peptoniphilus
	Cryptobacterium		Peptostreptococcus
	Eggerthella		Pseudoramibacter
	Olsenella		Selenomonas
	Parascardovia		Shuttleworthia
	Propionibacterium		Solobacterium
	Rothia		Staphylococcus
	Scardovia		Streptococcus
	Slackia		Veillonella
Bacteroidetes	Alloprevotella	Fusobacteria	Fusobacterium
	Bacteroides		Leptotrichia
	Bergeyella	GN02	"GN02"
	Capnocytophaga	Proteobacteria	Aggregatibacter
	Porphyromonas		Campylobacter
	Prevotella		Cardiobacterium
	Tannerella		Desulfobulbus
Chloroflexi	"Chloroflexi"		Desulfovibrio
Firmicutes	Abiotrophia		Eikenella
	Anaerococcus		Haemophilus
	Anaeroglobus		Kingella
	Bulleidia		Lautropia
	Catonella		Neisseria
	Dialister		Ottowia
	Enterococcus		Simonsiella
	Eubacterium	Spirochaetes	Treponema
	Filifactor	SR1	"SR1"
	Finegoldia	Synergistetes	Fretibacterium
	Gemella		Jonquetella
	Granulicatella		Pyramidobacter
	Johnsonella	Tenericutes	Mycoplasma
	Lachnoanaerobaculum	TM7 (Candidatus	"TM7"
	Lactobacillus	Saccharibacteria)	
	Megasphaera		

Only about two thirds of oral bacteria can be cultivated *in vitro*. There are numerous reasons for this, but one of the most important appears to be that bacteria naturally live in multi-species biofilms where community members communicate and supply each other with growth factors. A previously uncultivated lineage of the phylum *Synergistetes* was successfully cultured by co-culturing organisms from plaque samples and enriching the community for the target taxa by in-situ hybridization[33]. Members of Division TM7 (Candidatus *Saccharibacteria*[34]) are widely distributed in nature, including the human GI tract, and many attempts have been made to culture them *in vitro*. They can be detected readily in laboratory co-cultures but have proved difficult to grow in pure culture[35]. It has been shown recently that a TM7 phylotype is an obligate episymbiont of other bacteria, and is also able to enter the cells of other bacterial species[36]. The TM7 strain isolated has a small genome (c 0.7 MB)[36] and the genes for the synthesis of essential amino acids are missing, which explains its dependence on the host bacterium. It will be important to investigate if all members of this large Division/phylum are episymbionts or if any are free living.

4.3 The oral microbiota in health

The various habitats found in on different sites of the body are each colonized by a characteristic microbiota[37]. The oral cavity is host to a collection of microorganisms which, in general, are not found elsewhere in the body. Within the mouth, there are a variety of habitats with surfaces varying from the hard tissues of the teeth to the soft surface of the buccal mucosa and differing levels of oxygen and types of nutrients available. *Streptococcus* and *Gemella* are the most frequently detected genera[38]. The teeth harbor a bacterial community distinct from that of other surfaces with numbers of genera of obligate anaerobes such as *Fusobacterium* and *Prevotella* higher at subgingival sites than at supragingival sites, where facultative anaerobes including *Actinomyces* and *Capnocytophaga* were more prevalent[39].

The primary factor determining the precise composition of the oral microbiota is the individual, i.e. individuals have their own characteristic oral microbial community that is also stable over time[39]. It is often assumed that the human diet must be important in determining the composition of the oral bacterial community. In fact, when food is taken into the mouth and chewed, the flow of saliva is stimulated markedly and the food passes rapidly into the stomach via the oesophagus, giving the resident bacteria only a limited opportunity to use its constituents as nutrients. Some oral bacteria can rapidly take up sugars from the diet, particularly if it is rich in fermentable carbohydrate and/or it is frequently ingested. The acid produced by bacteria from fermentable carbohydrate can lead to dental caries, as discussed below. The primary source of nutrition for oral bacteria, then, is saliva and gingival crevicular fluid, the serum-like exudate that emanates from the gingival crevice around the teeth. Oral bacteria work together as a consortium to degrade glycoproteins from saliva and gingival crevicular fluid for the nutritional benefit of the whole bacterial community[40].

No differences in the composition of the salivary microbiota were seen in individuals who were omnivores or consumed ovo-lacto-vegetarian or vegan diets[41].

It is not surprising therefore that few differences in the composition of the oral bacterial community have been found at different geographical sites[42]. Where differences have been reported, they may be the result of differences in the oral disease status of the subjects being compared or in methodology, or the result of difficulties in sample collection and transport from remote locations. More definitive studies that address these issues are required, including the influence of host genotype.

4.3.1 Evolution of the oral microbiota
A number of interesting recent studies have determined the bacterial composition of dental calculus from archeological specimens. Calculus is formed by the mineralization of dental plaque and thus provides a protective layer for plaque bacteria from which intact DNA can be isolated and sequenced[43]. 16S rRNA gene

community profiling was performed on samples collected from 34 early European skeletons from the Mesolithic to the Medieval periods[44]. Oral bacterial communities similar to those seen in contemporary subjects were found, although a difference in composition between samples from the earliest and later periods was observed that was attributed to the introduction of farming in the Neolithic. The relative lack of caries-associated bacteria, members of the genus *Streptococcus*, in the ancient samples was assumed to be because the subjects studied pre-dated the increase in refined carbohydrates during and after the Industrial Revolution. A combination of 16S rRNA-based shotgun metagenomic analyses were performed on samples from four skeletons from the Medieval period[45] and it proved possible to assemble the genes coding for a range of bacterial and host proteins, and make inferences regarding their interactions. Periodontal disease-associated bacteria and products were found in the samples, suggesting that the subjects studied suffered from gingival disease; overall, the community found was similar to that seen in modern samples.

The data from these studies should be interpreted with caution. Calculus is formed from mature plaque, which is by definition a thick layer of plaque that would be expected to harbor an anaerobic bacterial community, with reduced levels of facultative anaerobes such as streptococci. Secondly, oral hygiene practices have changed enormously over time. Brushing with toothpaste is a relatively new phenomenon, being introduced towards the end of the 19th century in developed countries. The result of regular brushing is to render plaque in a permanently immature state. It is know that the primary colonizers of tooth enamel are streptococci and other facultative anaerobes[46], so it is not surprising that these organisms predominate in plaque from healthy sites in contemporary subjects.

4.3.2 Role of oral bacteria in health

All surfaces within the oral cavity are coated in a bacterial biofilm that inhibits colonization of the oral tissues by exogenous pathogens[47]. Evidence for this phenomenon is provided by the observation that opportunistic infections result from disturbance of the normal microbiota by antibiotic treatment, which can result in candidal infections such as antibiotic-associated glossitis.

It was stated above that diet has only a limited influence on oral microbial composition because food is rapidly washed from the mouth by saliva. Interestingly, there is an entero-salivary circuit for nitrate whereby, following nitrate ingestion and absorption in the upper intestine, nitrate is selectively taken up by the salivary glands and returned to the mouth[48,49], where it is reduced to nitrite by a range of oral bacteria[50]. Metagenomic analyses have identified 14 bacterial taxa with nitrate reductase activity, seven of which had not previously been detected by cultural investigation[51]. A proportion of the swallowed nitrite is directly acidified to nitric oxide in the acidic conditions of the stomach, whilst some nitrite enters the circulation, where it is converted to nitric oxide. Nitric oxide is a potent vasodilator and anti-atherogenic molecule[52]. Supplementation of the diet with nitrate lowers the blood pressure[53] and it has been demonstrated that short-term use of an antimicrobial mouthrinse raises blood pressure[54,55], presumably either by inhibiting the growth of oral nitrate-reducing bacteria or by direct inhibition of bacterial nitrate reductases. Humans do not possess innate nitrate reductase activity and thus the bacteria provide a mechanism for nitrite generation in the body that would not otherwise be possible and that appears to be important for cardiovascular health.

4.4 Role of oral microbiome in disease

It is well established that the microbiota at oral disease sites differs from that seen in health. The central, and long-debated, issue is whether disease-associated organisms play an active role in the disease process or whether their numbers rise in disease because the habitat has changed as a result of disease to one that they prefer. The role of the oral microbiota in periodontitis is a particular focus of this book; it will be dealt with in chapters 14 and 15 and is not covered here.

4.4.1 Dental caries

The microbiota found at carious lesions has raised proportions of acid-tolerating (aciduric) bacteria. Most attention has focused on *Streptococcus mutans* and lactobacilli[56] but a range of other aciduric organisms including bifidobacteria, *Propionibacterium acidifaciens* and *Scardovia wiggsiae* are present in significant numbers[57-60], and certain *Prevotella* species can be present in numbers inverse to those of lactobacilli[61,62]. pH gradients exist within carious lesions and low-pH regions have been found to have low diversity, with lactobacilli predominant and neutral pH regions harboring a richer community with representatives from a number of genera[63].

Much of the original evidence supporting a primary role for *S. mutans* has come from animal models of dental caries[64]. A deficiency of such models is that they have typically included a high-sucrose diet that selects for *S. mutans*, because *S. mutans* is naturally resistant to the antimicrobial properties of sucrose[65]. The experimental animals become colonized with a microbiota with a high proportion of *S. mutans* as a result, higher than would be seen in humans with active caries, and it is therefore not surprising, in particular, that anti-*S. mutans* strategies are effective in these models, which are thus poor predictors of efficacy in humans.

Caries is then an example of a disease arising from an interaction among environmental factors that cause a shift in the bacterial community to one that is dysbiotic and contributes to the disease. The processes involved are best explained by the ecological plaque hypothesis and its extension[66,67]. The primary risk factor for dental caries is a diet rich in fermentable carbohydrate[68]. Bacteria in plaque produce acid from the carbohydrate and lower the pH locally. The expression of genes responsible for acid production has been demonstrated by metatranscriptomics in individuals before and after eating a meal[69], and functional networks have been revealed which coordinate the functions involved in acid production and subsequent acid stress[5]. If carboydrate ingestion is infrequent then the pH will rise again, as a result of diffusion of the acid, the buffering activity of saliva and production of alkali by other bacteria, principally the formation of ammonia from arginine[70]. If sugar intake is excessive and/or frequent, however, the buffering capacity of saliva will be overcome and the pH will become lower continuously, changing the environment to one that favors the proliferation of aciduric bacteria that will produce additional acid and demineralize the enamel.

If caries is left untreated, further complications can arise. Endodontic infections occur when bacteria enter the pulp chamber and the host mounts an immune response in the periapical region. Endodontic infections are mixed infections of primarily anaerobic bacteria, but only a subset of the oral microbiota is found, suggesting that the habitat selects for those species that are able to flourish there[71].

Eventually, bacteria can pass through the periapical foramen into the surrounding alveolar bone, resulting in the formation of a dentoalveolar abscess.

4.4.2 Gingivitis

Plaque forms on the teeth in a structured way. Relatively few species can colonize a clean tooth surface and they do so by interactions with salivary glycoproteins adsorbed onto the tooth surface forming the layer known as pellicle[72]. The pioneer species are typically obligate aerobes such as *Neisseria* and *Rothia* and facultative anaerobes including *Streptococcus* and *Actinomyces*[73]. When the pioneer species have attached to the surface then other bacteria can join the biofilm by co-aggregating with them by means of adhesin-receptor interactions that can either be protein-protein or protein-carbohydrate (lectin-like)[74]. These organisms rapidly consume the available oxygen and the biofilm becomes anaerobic within a few days[75]. In addition, the proportion of Gram-negative oral species increases, particularly members of the genera *Veillonella*, *Fusobacterium* and *Prevotella*. Gram-negative bacteria naturally contain endotoxin (lipopolysaccharide) as part of their cell wall which passes into the tissues, causing inflammation of the gingivae, or gingivitis. The primary correlate with gingivitis severity is the amount of plaque present but there is also an intrinsic relationship with plaque maturity, anaerobiosis and the proportion of Gram-negatives present. A pyrosequencing study of the microbial changes in experimental gingivitis, where oral hygiene was withdrawn for 14 days, showed that *Fusobacterium nucleatum subsp. polymorphum*, *Lachnospiraceae* [G-2] HOT100, *Lautropia* HOTA94, and *Prevotella oulorum* were associated with gingivitis[76]. Interestingly, initial evidence seems to suggest that host susceptibility may influence the severity of experimental gingivitis[77].

4.4.3 Oral bacteria and non-oral disease

Oral bacteria have been implicated directly and indirectly in a wide range of non-oral diseases. They can cause infections when they gain access to non-oral sites, such as abscesses and actinomycosis in the soft tissues surrounding the mouth secondary to caries[78], or by aspiration into the lungs, where they can cause ventilator-assisted pneumonia[79] and are part of the bacterial community found in cystic fibrosis patients[80]. Human bites and clenched-fist injuries can also result in infections caused by oral bacteria[81].

Oral bacteria can gain access to the bloodstream relatively easily via the tissues surrounding the teeth during everyday activities such as chewing and toothbrushing as well as dental treatment[82]. Dentoalveolar abscesses secondary to odontogenic infections also provide a route into the blood. In susceptible individuals, they can cause endocarditis[83] and abscesses in other organs, particularly the brain and liver[84,85]. They have also been shown to colonize the placenta, where they have been linked to adverse pregancy outcomes[86]. *Fusobacterium nucleatum*, in particular, appears to be able to invade tissues in non-oral sites and has been linked both to stillbirth and colorectal cancer[87,88].

The indirect effects of oral bacteria primarily relate to diseases linked to the chronic inflammation that is a feature of periodontitis. Subjects with periodontitis are at increased risk of cardiovascular disease[89], and oral bacteria themselves have been detected in, and isolated from, atheromatous plaques in blood vessels[90] (more details are given in chapter 13 by Nibali). Rheumatoid arthritis has been shown to be associated with periodontitis[91] (discussed by Detert in chapter 18), and periodontal treatment found to reduce disease severity[92]. There is also some evidence that periodontitis, but not tooth loss, may be associated with mental impairment and dementia[93].

4.5 Future outlook

Community profiling using next-generation sequencing methods has revealed the diversity of the oral microbiota in both health and disease and confirmed that the composition of the oral microbiota is primarily determined by the local habitat. Concomitant genomic studies have revealed the extent of genetic variability among strains of the same species. Future studies need to take account of this diversity and need to model intra-microbiota microbial interactions in addition to host-microbiome relationships. In order to benefit from advances in metatranscriptomic techniques, a greatly expanded database of genome sequences is required, with appropriate numbers of sequences for each species depending on the natural genetic variation among strains of the species.

Microbiomic profiling is now straightforward and cheap to perform and has potential for use as a diagnostic/prognostic tool. It is known that the oral microbiome is different in subjects with caries and periodontal disease compared to those with health and it could be used in diagnosis, although whether there would be any additional benefit over clinical examination is unclear. Samples collected from supragingival plaque and the dorsum of the tongue have been shown to be as sensitive as subgingival samples for the detection of periodontal disease-associated microbial biomarkers[94]. If microbiome composition is altered before symptoms are manifest, then regular microbiome screening could be useful. There is some evidence that the oral microbiome is altered in non-oral diseases, e.g. pancreatic cancer[95] and inflammatory bowel disease[96], which would make salivary microbiome profile screening particularly valuable. Saliva is a readily available body fluid and long-term monitoring of individuals' microbiome profiles would be feasible and potentially clinically useful.

TAKE-HOME MESSAGE

- The human oral cavity is colonized by bacteria, archaea, viruses, protozoa and fungi; this normal oral microbiota is important for health, helping to prevent infection by exogenous pathogens, and is also important in nitrate metabolism, which is essential for cardiovascular health.

- Many oral bacteria cannot be grown in pure culture in the laboratory. This is because they have evolved as part of a multi-species biofilm and are dependent on interactions with other community members. In-vitro biofilm models have now been developed that enable complex oral biofilms to be grown, including previously uncultivated bacterial species.

- The oral microbiota is associated with the commonest bacterial diseases of man — dental caries and periodontal disease. These are complex diseases arising from an interaction among the human host, the microbiota and the environment which can lead to dysbiosis, a disease-associated microbiota.

References

1 American Academy of Microbiology. FAQ: Human Microbiome. 2014.
2 Belda-Ferre P, Alcaraz LD, Cabrera-Rubio R *et al*. The oral metagenome in health and disease. *ISME J.* 2012; **6**: 46–56.
3 Duran-Pinedo AE, Chen T, Teles R *et al*. Community-wide transcriptome of the oral microbiome in subjects with and without periodontitis. *Isme J* 2014; **8**: 1659–72.
4 Jorth P, Turner KH, Gumus P *et al*. Metatranscriptomics of the human oral microbiome during health and disease. *MBio* 2014; **5**: e01012–14.
5 Peterson SN, Meissner T, Su AI *et al*. Functional expression of dental plaque microbiota. *Front Cell Infect Microbiol* 2014; **4**: 108.
6 Nguyen-Hieu T, Khelaifia S, Aboudharam G *et al*. Methanogenic archaea in subgingival sites: a review. *APMIS* 2013; **121**: 467–77.
7 Vianna ME, Holtgraewe S, Seyfarth I *et al*. Quantitative analysis of three hydrogenotrophic microbial groups, methanogenic archaea, sulfate-reducing bacteria, and acetogenic bacteria, within plaque biofilms associated with human periodontal disease. *J. Bacteriol.* 2008; **190**: 3779–85.
8 Horz HP, Seyfarth I, Conrads G. McrA and 16S rRNA gene analysis suggests a novel lineage of Archaea phylogenetically affiliated with Thermoplasmatales in human subgingival plaque. *Anaerobe* 2012; **18**: 373–7.
9 Robichaux M, Howell M, Boopathy R. Growth and activities of sulfate-reducing and methanogenic bacteria in human oral cavity. *Curr. Microbiol.* 2003; **47**: 12–6.
10 Samaranayake LP, Keung Leung W, Jin L. Oral mucosal fungal infections. *Periodontol. 2000* 2009; **49**: 39–59.
11 Ghannoum MA, Jurevic RJ, Mukherjee PK *et al*. Characterization of the oral fungal microbiome (mycobiome) in healthy individuals. *PLoS Pathog.* 2010; **6**: e1000713.
12 Dupuy AK, David MS, Li L *et al*. Redefining the human oral mycobiome with improved practices in amplicon-based taxonomy: discovery of Malassezia as a prominent commensal. *PLoS One* 2014; **9**: e90899.
13 Monteiro-da-Silva F, Araujo R, Sampaio-Maia B. Interindividual variability and intraindividual stability of oral fungal microbiota over time. *Med. Mycol.* 2014; **52**: 498–505.
14 Wantland WW, Wantland EM, Remo JW *et al*. Studies on human mouth protozoa. *J. Dent. Res.* 1958; **37**: 949–50.
15 Wantland WW, Lauer D. Correlation of some oral hygiene variables with age, sex, and incidence of oral protozoa. *J. Dent. Res.* 1970; **49**: 293–7.
16 Scott DA, Coulter WA, Lamey PJ. Oral shedding of herpes simplex virus type 1: a review. *J. Oral Pathol. Med.* 1997; **26**: 441–7.
17 Arduino PG, Porter SR. Herpes Simplex Virus Type 1 infection: overview on relevant clinicopathological features. *J. Oral Pathol. Med.* 2008; **37**: 107–21.
18 Kumaraswamy KL, Vidhya M. Human papilloma virus and oral infections: an update. *J Cancer Res Ther* 2011; **7**: 120–7.
19 Khode SR, Dwivedi RC, Rhys-Evans P *et al*. Exploring the link between human papilloma virus and oral and oropharyngeal cancers. *J Cancer Res Ther* 2014; **10**: 492–8.
20 Li Z, Feng Z, Ye H. Rabies viral antigen in human tongues and salivary glands. *J. Trop. Med. Hyg.* 1995; **98**: 330–2.
21 Cascarini L, McGurk M. Epidemiology of salivary gland infections. *Oral Maxillofac. Surg. Clin. North Am.* 2009; **21**: 353–7.
22 Heiberg IL, Hoegh M, Ladelund S *et al*. Hepatitis B virus DNA in saliva from children with chronic hepatitis B infection: implications for saliva as a potential mode of horizontal transmission. *Pediatr. Infect. Dis. J.* 2010; **29**: 465–7.
23 Hermida M, Ferreiro MC, Barral S *et al*. Detection of HCV RNA in saliva of patients with hepatitis C virus infection by using a highly sensitive test. *J. Virol. Methods* 2002; **101**: 29–35.
24 Moore BE, Flaitz CM, Coppenhaver DH *et al*. HIV recovery from saliva before and after dental treatment: inhibitors may have critical role in viral inactivation. *J. Am. Dent. Assoc.* 1993; **124**: 67–74.
25 Wylie KM, Mihindukulasuriya KA, Zhou Y *et al*. Metagenomic analysis of double-stranded DNA viruses in healthy adults. *BMC Biol.* 2014; **12**: 71.
26 Abeles SR, Robles-Sikisaka R, Ly M *et al*. Human oral viruses are personal, persistent and gender-consistent. *Isme j* 2014; **8**: 1753–67.

27 Robles-Sikisaka R, Ly M, Boehm T *et al*. Association between living environment and human oral viral ecology. *Isme j* 2013; **7**: 1710–24.

28 Ly M, Abeles SR, Boehm TK *et al*. Altered oral viral ecology in association with periodontal disease. *MBio* 2014; **5**: e01133–14.

29 Naidu M, Robles-Sikisaka R, Abeles SR *et al*. Characterization of bacteriophage communities and CRISPR profiles from dental plaque. *BMC Microbiol.* 2014; **14**: 175.

30 Dewhirst FE, Chen T, Izard J *et al*. The human oral microbiome. *J. Bacteriol.* 2010; **192**: 5002–17.

31 Camanocha A, Dewhirst FE. Host-associated bacterial taxa from Chlorobi, Chloroflexi, GN02, Synergistetes, SR1, TM7, and WPS-2 Phyla/candidate divisions. *J Oral Microbiol* 2014; **6**.

32 Chen T, Yu WH, Izard J *et al*. The Human Oral Microbiome Database: a web accessible resource for investigating oral microbe taxonomic and genomic information. *Database (Oxford)* 2010; **2010**: baq013.

33 Vartoukian SR, Palmer RM, Wade WG. Cultivation of a Synergistetes strain representing a previously uncultivated lineage. *Environ. Microbiol.* 2010; **12**: 916–28.

34 Albertsen M, Hugenholtz P, Skarshewski A *et al*. Genome sequences of rare, uncultured bacteria obtained by differential coverage binning of multiple metagenomes. *Nat. Biotechnol.* 2013; **31**: 533–8.

35 Hugenholtz P, Tyson GW, Webb RI *et al*. Investigation of candidate division TM7, a recently recognized major lineage of the domain Bacteria with no known pure-culture representatives. *Appl. Environ. Microbiol.* 2001; **67**: 411–9.

36 He X, McLean JS, Edlund A *et al*. Cultivation of a human-associated TM7 phylotype reveals a reduced genome and epibiotic parasitic lifestyle. *Proc. Natl. Acad. Sci. U. S. A.* 2015; **112**: 244–9.

37 Ding T, Schloss PD. Dynamics and associations of microbial community types across the human body. *Nature* 2014; **509**: 357–60.

38 Aas JA, Paster BJ, Stokes LN *et al*. Defining the normal bacterial flora of the oral cavity. *J. Clin. Microbiol.* 2005; **43**: 5721–32.

39 Segata N, Haake SK, Mannon P *et al*. Composition of the adult digestive tract bacterial microbiome based on seven mouth surfaces, tonsils, throat and stool samples. *Genome Biol.* 2012; **13**: R42.

40 Bradshaw DJ, Homer KA, Marsh PD *et al*. Metabolic cooperation in oral microbial communities during growth on mucin. *Microbiology* 1994; **140 (Pt 12)**: 3407–12.

41 De Filippis F, Vannini L, La Storia A *et al*. The same microbiota and a potentially discriminant metabolome in the saliva of omnivore, ovo-lacto-vegetarian and Vegan individuals. *PLoS One* 2014; **9**: e112373.

42 Nasidze I, Li J, Quinque D *et al*. Global diversity in the human salivary microbiome. *Genome Res.* 2009; **19**: 636–43.

43 Warinner C, Speller C, Collins MJ. A new era in palaeomicrobiology: prospects for ancient dental calculus as a long-term record of the human oral microbiome. *Philos. Trans. R. Soc. Lond. B Biol. Sci.* 2015; **370**.

44 Adler CJ, Dobney K, Weyrich LS *et al*. Sequencing ancient calcified dental plaque shows changes in oral microbiota with dietary shifts of the Neolithic and Industrial revolutions. *Nat. Genet.* 2013; **45**: 450–5, 5e1.

45 Warinner C, Rodrigues JF, Vyas R *et al*. Pathogens and host immunity in the ancient human oral cavity. *Nat. Genet.* 2014; **46**: 336–44.

46 Nyvad B, Kilian M. Microbiology of the early colonization of human enamel and root surfaces in vivo. *Scand. J. Dent. Res.* 1987; **95**: 369–80.

47 Vollaard EJ, Clasener HA. Colonization resistance. *Antimicrob. Agents Chemother.* 1994; **38**: 409–14.

48 Qin L, Liu X, Sun Q *et al*. Sialin (SLC17A5) functions as a nitrate transporter in the plasma membrane. *Proc. Natl. Acad. Sci. U. S. A.* 2012; **109**: 13434–9.

49 Tannenbaum SR, Weisman M, Fett D. The effect of nitrate intake on nitrite formation in human saliva. *Food Cosmet. Toxicol.* 1976; **14**: 549–52.

50 Doel JJ, Benjamin N, Hector MP *et al*. Evaluation of bacterial nitrate reduction in the human oral cavity. *Eur. J. Oral Sci.* 2005; **113**: 14–9.

51 Hyde ER, Andrade F, Vaksman Z *et al*. Metagenomic analysis of nitrate-reducing bacteria in the oral cavity: implications for nitric oxide homeostasis. *PLoS One* 2014; **9**: e88645.

52 Cosby K, Partovi KS, Crawford JH *et al*. Nitrite reduction to nitric oxide by deoxyhemoglobin vasodilates the human circulation. *Nat. Med.* 2003; **9**: 1498–505.

53 Kapil V, Milsom AB, Okorie M *et al*. Inorganic nitrate supplementation lowers blood pressure in humans: role for nitrite-derived NO. *Hypertension* 2010; **56**: 274–81.

54 Kapil V, Haydar SM, Pearl V *et al*. Physiological role for nitrate-reducing oral bacteria in blood pressure control. *Free Radic. Biol. Med.* 2013; **55**: 93–100.

55 Bondonno CP, Liu AH, Croft KD *et al*. Antibacterial mouthwash blunts oral nitrate reduction and increases blood pressure in treated hypertensive men and women. *Am. J. Hypertens.* 2014.

56 Gibbons RJ, van Houte J. Dental caries. *Annu. Rev. Med.* 1975; **26**: 121–36.

57 Beighton D, Al-Haboubi M, Mantzourani M *et al*. Oral Bifidobacteria: caries-associated bacteria in older adults. *J. Dent. Res.* 2010; **89**: 970–4.

58 Downes J, Mantzourani M, Beighton D *et al*. Scardovia wiggsiae sp. nov., isolated from the human oral cavity and clinical material, and emended descriptions of the genus Scardovia and Scardovia inopinata. *Int. J. Syst. Evol. Microbiol.* 2011; **61**: 25–9.

59 Downes J, Wade WG. Propionibacterium acidifaciens sp. nov., isolated from the human mouth. *Int. J. Syst. Evol. Microbiol.* 2009; **59**: 2778–81.

60 Kaur R, Gilbert SC, Sheehy EC *et al*. Salivary levels of Bifidobacteria in caries-free and caries-active children. *Int. J. Paediatr. Dent.* 2013; **23**: 32–8.

61 Nadkarni MA, Caldon CE, Chhour KL *et al*. Carious dentine provides a habitat for a complex array of novel Prevotella-like bacteria. *J. Clin. Microbiol.* 2004; **42**: 5238–44.

62 Schulze-Schweifing K, Banerjee A, Wade WG. Comparison of bacterial culture and 16S rRNA community profiling by clonal analysis and pyrosequencing for the characterization of the dentine caries-associated microbiome. *Front Cell Infect Microbiol* 2014; **4**: 164.

63 Kianoush N, Adler CJ, Nguyen KA *et al*. Bacterial profile of dentine caries and the impact of pH on bacterial population diversity. *PLoS One* 2014; **9**: e92940.

64 Larson RH, Theilade E, Fitzgerald RJ. The interaction of diet and microflora in experimental caries in the rat. *Arch. Oral Biol.* 1967; **12**: 663–8.

65 Ikeda T, Sandham HJ. A high-sucrose medium for the identification of Streptococcus mutans. *Arch. Oral Biol.* 1972; **17**: 781–3.

66 Marsh PD. Are dental diseases examples of ecological catastrophes? *Microbiology* 2003; **149**: 279–94.

67 Takahashi N, Nyvad B. The role of bacteria in the caries process: ecological perspectives. *J. Dent. Res.* 2011; **90**: 294–303.

68 Sheiham A. Dietary effects on dental diseases. *Public Health Nutr.* 2001; **4**: 569–91.

69 Benitez-Paez A, Belda-Ferre P, Simon-Soro A *et al*. Microbiota diversity and gene expression dynamics in human oral biofilms. *BMC Genomics* 2014; **15**: 311.

70 Burne RA, Marquis RE. Alkali production by oral bacteria and protection against dental caries. *FEMS Microbiol. Lett.* 2000; **193**: 1–6.

71 Munson MA, Pitt-Ford T, Chong B *et al*. Molecular and cultural analysis of the microflora associated with endodontic infections. *J. Dent. Res.* 2002; **81**: 761–6.

72 Douglas CW. Bacterial-protein interactions in the oral cavity. *Adv. Dent. Res.* 1994; **8**: 254–62.

73 Diaz PI, Chalmers NI, Rickard AH *et al*. Molecular characterization of subject-specific oral microflora during initial colonization of enamel. *Appl. Environ. Microbiol.* 2006; **72**: 2837–48.

74 Kolenbrander PE. Oral microbial communities: biofilms, interactions, and genetic systems. *Annu. Rev. Microbiol.* 2000; **54**: 413–37.

75 Kenney EB, Ash MM, Jr. Oxidation reduction potential of developing plaque, periodontal pockets and gingival sulci. *J. Periodontol.* 1969; **40**: 630–3.

76 Kistler JO, Booth V, Bradshaw DJ *et al*. Bacterial community development in experimental gingivitis. *PLoS One* 2013; **8**: e71227.

77 Trombelli L, Scapoli C, Tatakis DN *et al*. Modulation of clinical expression of plaque-induced gingivitis: response in aggressive periodontitis subjects. *J. Clin. Periodontol.* 2006; **33**: 79–85.

78 Lypka M, Hammoudeh J. Dentoalveolar infections. *Oral Maxillofac Surg Clin North Am* 2011; **23**: 415–24.

79 Heo SM, Haase EM, Lesse AJ *et al*. Genetic relationships between respiratory pathogens isolated from dental plaque and bronchoalveolar lavage fluid from patients in the intensive care unit undergoing mechanical ventilation. *Clin. Infect. Dis.* 2008; **47**: 1562–70.

80 Filkins LM, Hampton TH, Gifford AH *et al*. Prevalence of streptococci and increased polymicrobial diversity associated with cystic fibrosis patient stability. *J. Bacteriol.* 2012; **194**: 4709–17.

81 Mennen U, Howells CJ. Human fight-bite injuries of the hand. A study of 100 cases within 18 months. *J Hand Surg Br* 1991; **16**: 431–5.

82 Silver JG, Martin AW, McBride BC. Experimental transient bacteraemias in human subjects with varying degrees of plaque accumulation and gingival inflammation. *J. Clin. Periodontol.* 1977; **4**: 92–9.

83 Pierce D, Calkins BC, Thornton K. Infectious endocarditis: diagnosis and treatment. *Am. Fam. Physician* 2012; **85**: 981–6.

84 Antunes AA, de Santana Santos T, de Carvalho RW *et al.* Brain abscess of odontogenic origin. *J. Craniofac. Surg.* 2011; **22**: 2363–5.

85 Schiff E, Pick N, Oliven A *et al.* Multiple liver abscesses after dental treatment. *J. Clin. Gastroenterol.* 2003; **36**: 369–71.

86 Aagaard K, Ma J, Antony KM *et al.* The placenta harbors a unique microbiome. *Sci Transl Med* 2014; **6**: 237ra65.

87 Han YW, Fardini Y, Chen C *et al.* Term stillbirth caused by oral Fusobacterium nucleatum. *Obstet. Gynecol.* 2010; **115**: 442–5.

88 Han YW. Fusobacterium nucleatum: a commensal-turned pathogen. *Curr. Opin. Microbiol.* 2015; **23c**: 141–7.

89 Beck JD, Offenbacher S. Systemic effects of periodontitis: epidemiology of periodontal disease and cardiovascular disease. *J. Periodontol.* 2005; **76**: 2089–100.

90 Serra e Silva Filho W, Casarin RC, Nicolela EL, Jr. *et al.* Microbial diversity similarities in periodontal pockets and atheromatous plaques of cardiovascular disease patients. *PLoS ONE* 2014; **9**: e109761.

91 Rutger Persson G. Rheumatoid arthritis and periodontitis – inflammatory and infectious connections. Review of the literature. *J Oral Microbiol* 2012; **4**.

92 Kaur S, Bright R, Proudman SM *et al.* Does periodontal treatment influence clinical and biochemical measures for rheumatoid arthritis? A systematic review and meta-analysis. *Semin. Arthritis Rheum.* 2014; **44**: 113–22.

93 Gil-Montoya JA, Sanchez-Lara I, Carnero-Pardo C *et al.* Is Periodontitis a Risk Factor for Cognitive Impairment and Dementia? A Case-Control Study. *J. Periodontol.* 2014: 1–14.

94 Galimanas V, Hall MW, Singh N *et al.* Bacterial community composition of chronic periodontitis and novel oral sampling sites for detecting disease indicators. *Microbiome* 2014; **2**: 32.

95 Farrell JJ, Zhang L, Zhou H *et al.* Variations of oral microbiota are associated with pancreatic diseases including pancreatic cancer. *Gut* 2012; **61**: 582–8.

96 Said HS, Suda W, Nakagome S *et al.* Dysbiosis of salivary microbiota in inflammatory bowel disease and its association with oral immunological biomarkers. *DNA Res.* 2014; **21**: 15–25.

CHAPTER 5

The skin microbiota

Patrick L.J.M. Zeeuwen and Joost Schalkwijk

Department of Dermatology, Radboud University Nijmegen Medical Centre, Nijmegen, The Netherlands

5.1 Normal skin

Skin is the interface of our body with the environment, and its uppermost layer, the stratum corneum, is subject to continuous abrasion by chemical and physical injury. To protect the body against invasion of microorganisms and toxic agents, as well as against loss of essential body fluids, the skin has evolved an elaborate differentiation process that results in a tough, water-impermeable outer covering that is constantly renewing[1]. The skin is sterile *in utero*, but almost immediately after birth, environmental microbes rapidly start to colonize the stratum corneum, eventually developing into a complex microbial ecosystem that is in homeostasis with its host[2,3]. Chapter 2 defined the total microbial community that lives in association with the human body collectively as the human microbiome, while the human skin microbiota includes all microbes found in a particular cutaneous body site habitat[4]. The diversity of the human microbiome is determined by various factors such as transmission of transient microbes, genetic predisposition, lifestyle, and host demographic and environmental characteristics[5,6]. Furthermore, certain physical and chemical factors influence the growth and survival of skin microbes, like moisture, acidic pH, salinity and sebum content[7,8]. Permanent or resident microbes on the skin are considered to be members of the 'normal' commensal skin microbiota, whereas transient micro-organisms are derived from the environment and live just temporarily on our skin. The resident microbial communities on human skin control colonization by potentially pathogenic micro-organisms[9,10], can modulate the cutaneous immune system[11–13], and are necessary for optimal skin immune fitness[14]. Altogether this indicates that preservation of our skin microbiota is beneficial for our health[15].

Several studies, using DNA-based next-generation sequencing (NGS) technologies for the detection and identification of microbial genes, have catalogued the microbiota of the surface of human skin under nonchallenged, static conditions in healthy volunteers. In agreement with Wilson's discussion in chapter 1, these culture-independent studies, using 16S rRNA gene sequencing, revealed a

The Human Microbiota and Chronic Disease: Dysbiosis as a Cause of Human Pathology, First Edition.
Edited by Luigi Nibali and Brian Henderson.

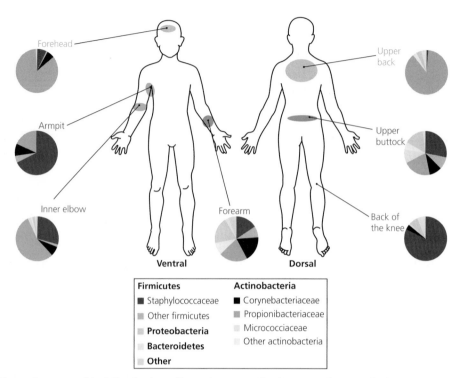

Figure 1 Topographical distribution of bacteria on specific skin sites. The microbial composition is shown at phylum and family level. Moist sites are labelled in green, sebaceous sites are labelled in blue, and dry surfaces in red. Data from Zeeuwen *et al.*[38] except for volar forearm, back of the knee, and the back (from Human Microbiome Project). (*see color plate section for color details*).

significantly greater diversity of micro-organisms than presumed earlier from culture-based methods[16]. It was shown that bacterial diversity largely depends on the topographical location on the body, that the dominant types of bacteria remain relatively stable over time, and that our skin microbiota has a high degree of interpersonal variation[17,18]. Furthermore, these studies in healthy volunteers showed that in general the most dominant resident skin bacteria could be categorized into four different phyla: Actinobacteria, Firmicutes, Proteobacteria and Bacteroidetes. In addition, it was demonstrated that specific bacteria were associated with dry, moist and/or sebaceous microenvironments (Figure 1). For example, *Propionibacterium* spp. are organisms that abundantly colonize sebaceous areas as the forehead, the back, and the piece of skin behind the ear, whereas *Staphylococcus* spp. and *Corynebacterium* spp. are found in large quantities in moist areas like the armpit, the inner elbow, and the skin behind the knee[17,18]. The dry areas of the skin like the buttock, the forearm, and parts of the hand are colonized by a microbiome of the highest diversity but lowest absolute numbers. With reference to numbers, on average 100 distinct species, making up a total of one million micro-organisms, reside on each square centimetre of our skin. However, many more microbes are found in the moist environment of the armpit ($\sim 10 \times 10^6/cm^2$), whereas lower numbers are found in the dry region of the upper buttock skin ($\sim 1 \times 10^5/cm^2$) (Figure 1).

5.2 Skin diseases

Under normal healthy conditions, our skin peacefully co-exists with commensal bacteria while fending off potential pathogenic invaders. Disturbance of these homeostatic relations between the host and its microbiota can result from changes in the composition of our skin microbiota (dysbiosis), an altered immune response to the microbiota itself, or both, and may be a driving factor in inflammatory skin diseases such as psoriasis and atopic dermatitis (AD)[19–21] (see also chapter 21 by Fry). Both are common chronic inflammatory skin diseases, characterized by various different clinical and histological features. Although AD and psoriasis are generally regarded as immune-mediated conditions, genetic studies of the last decade have indicated the importance of abnormalities in epithelium-expressed genes as a primary cause. Host susceptibility factors for AD and psoriasis, such as mutations in genes encoding epidermal proteins like filaggrin[22] and LCE3B/C[23], can result in variation of the skin barrier and could provoke a consequent cutaneous sensitization by penetration of microbes and bacterial products, which causes inflammation. Furthermore, polymorphisms in innate or adaptive immune elements[24–26] can also lead to excess inflammation before any significant microbial shift, which can lead to a state where both the microbiota and host immune response are altered and contribute to a vicious cycle leading to disease. It is speculated that in some patients the colonizing microbiota of the skin may elicit and maintain the skin disease. Therefore the identification of such 'causative' microbiota potentially could lead to early diagnostics, disease-modifying or, perhaps, curative therapies. To understand the role of microbes in inflammatory and allergic skin diseases, a few groups have recently analysed the skin microbiome of diseased and injured skin using NGS, as outlined in the following paragraphs.

5.2.1 Atopic dermatitis

Atopic dermatitis, or eczema, is a very common chronic inflammatory skin disease. One of the hallmarks of this disease is that patients have increased bacterial colonization, specifically *Staphylococcus aureus* (*S. aureus*), which is found on the lesional skin in more than 90% of AD patients[27]. The presence of *S. aureus* is thought to play an important role in the pathogenesis of AD[26], as treatments to reduce *S. aureus* colonization proved to decrease severity of disease in these patients[28]. In a NIH-funded Human Microbiome Project (HMP) study, the microbiota composition of lesional skin of moderate to severe pediatric AD patients was analysed at baseline, during disease flare, and after treatment[29]. Skin samples for microbiome analyses were taken from the back of the knee and the inner elbow (two commonly affected predilection sites) and as a control from the volar forearm. Overall, the AD skin of the inner elbow and the back of the knee showed a much lower microbiota diversity than the same skin area of healthy individuals and the control site. Furthermore, temporal shifts in skin microbiota composition were associated with disease flares and treatment. Examination and bacterial classifications of the samples revealed that the relative abundance of *S. aureus* and also the skin commensal *Staphylococcus epidermidis* (*S. epidermidis*) was increased with clinical disease activity, whereas increases of specific bacterial genera, such as *Corynebacterium*, *Streptococcus*, and *Propionibacterium*, were observed following therapy. These data suggest that treatment promotes microbial diversity, which leads to improvement in the clinical severity of disease. On the other hand, one could argue that the changing milieu of lesional AD skin might be a perfect microenvironment and substrate for specific (staphylococcal) species. Furthermore, abnormal skin biology is linked to genetically programmed differences in epidermal host defence[30,31], which might influence disease-associated changes in the microbiota composition as observed in inflammatory skin disorders as AD (and psoriasis). Another probability that could be extracted from these data is that the increased abundance of *S. epidermidis* in lesional AD skin reflects a microbial response to overgrowth of *S. aureus*, since *S. epidermidis* is able to selectively inhibit *S. aureus*[9]. The authors hypothesized that, as untreated disease flares have reduced diversity and high *Staphylococcus* proportions, increases in the proportion of *Staphylococcus* and reductions in microbial diversity precede worsening of AD disease severity[29]. To prove this hypothesis, however, longitudinal microbiome studies must be performed to examine the influence of microbiota on health and disease. Involvement of *S. aureus* in AD was further supported by a study that identified *S. aureus* derived deltatoxin as a potent inducer of mast cell degranulation and production of IgE and IL-4, both hallmarks of AD, and suggests a mechanistic link between *S. aureus* colonization and allergic skin disease[32]. However, this study used mice that were shaved and tape-stripped before *S. aureus* was applied to the skin, which means that deltatoxin might only penetrate the skin in the setting of genetic defects associated with disease (*e.g. filaggrin* mutations in AD).

5.2.2 Psoriasis

Psoriasis is a chronic inflammatory skin disease characterized by abnormal keratinocyte differentiation, hyperproliferation of keratinocytes, and recruitment of inflammatory cells in the epidermis[33] (discussed in detail in chapter 21). It is generally accepted that a combination of environmental and genetic factors contribute to its etiology[34]. As the pathophysiology of psoriasis in part suggests an inappropriately activated cutaneous immune response directed against hitherto unidentified antigens (possibly derived from skin microbiota), micro-organisms have been implicated in the immunopathogenesis of psoriasis[35]. The first, relatively small NGS-study (*n*=6 patients) of microbial communities in psoriatic skin revealed that the bacterial composition of the lesional skin was much more diverse than that observed in skin from healthy subjects or non-lesional skin from psoriatic patients[36]. Firmicutes were the most abundant and diverse phylum populating the lesional psoriatic skin, whereas the genus *Propionibacterium* was significantly underrepresented compared to healthy controls and non-lesional skin samples from psoriasis patients. These results contradict another subsequent small study (*n*=5 patients) that showed that microbiota diversity in psoriatic lesions was reduced relative to normal skin, and also the genera that were present at discriminating abundance levels were different[37]. These contradictions may be due to different sampling techniques: the first study determined the microbiota composition using superficial swab samples, while the subsequent study analysed the microbiota from the complete epidermis and dermis. As micro-organisms are also present in deeper (sub)epidermal compartments of the skin[38,39], one might suppose that the microbiota of the superficial layer is just a part of the total host indigenous skin microbiota. A similarity between both studies is the ratio *Streptococcus* versus

Propionibacterium, which was significantly increased in lesional psoriatic skin compared to healthy skin.

A third study of the skin microbiota of psoriasis (performed by the same authors as the first study, but now with a larger group of 51 patients) reported this time a trend to reduced taxonomic diversity in lesional skin[40]. Furthermore, they found that the psoriatic microbiota is associated with a cutaneotype enriched for Firmicutes and Actinobacteria, compared to healthy skin that showed an increased relative abundance of Proteobacteria. However, the overall composition in the controls differs from the main HMP report[41] and our own microbiome data[38], which yielded only low levels of cutaneous Proteobacteria. Another remark concerning all three studies on the psoriasis microbiome is the fact that no comments were given on the low sequence-depth of the samples. Our own pilot studies using samples of psoriasis patients revealed that these are seriously contaminated with eukaryotic DNA, which is due to human nuclei in the parakeratotic corneocytes of psoriatic lesional skin. However, none of the three above-mentioned studies indicates this observable fact. As mentioned earlier, longitudinal studies that explore the dynamics of the microbiota profile of psoriatic plaques before, during and after treatment will facilitate the identification of markers related to disease onset, progression and outcome.

5.2.3 Acne

Acne vulgaris is a common chronic inflammatory skin disorder of the face, neck, chest, and back, in which bacterial colonization of hair follicles by *Propionibacterium acnes* (*P. acnes*) is thought to play an important pathogenic role[42]. It is, however, most remarkable that *P. acnes* is also a common and abundant commensal bacterial species found on normal healthy human skin[17,18,38]. In a relatively large culture-independent microbiome study of the pilosebaceous units of the nose it was demonstrated that *P. acnes* was indeed the most abundant microbe; however, the strain populations of *P. acnes* were distinct between acne patients and healthy controls[43]. Certain dominant *P. acnes* strains, which were similar to each other, were associated with acne patients, whereas more diverse and other strains were found on healthy individuals. These data suggest that in addition to host-specific features such as hormone levels, sebum production, and physical changes in pilosebaceous units, specific pathogenic *P. acnes* strains may contribute to the development of acne vulgaris. In a follow-up study, a comparative genome analysis of 82 *P. acnes* strains was performed[44]. Specific genetic elements were found that differ among lineages, which may explain why *P. acnes* functions as a commensal microbe in healthy skin or as a pathogen in disease (acne). If these diversities at the strain level really exist between the healthy state and *P. acnes*-associated diseases, this will provide perspectives for new strain-specific and personalized therapies.

5.2.4 Rosacea

Rosacea is a common chronic inflammatory skin condition characterized by facial flushing, persistent redness, papules and pustules. The exact pathogenesis of rosacea is, however, not known[45]. Abnormalities in the pilosebaceous units, dysregulation of the neurovascular and immune system, digestive system (diet), environmental factors (heat, sun exposure, cosmetics), micro-organisms, and a genetic component are mentioned as a possible cause[46]. The micro-organisms associated with rosacea include *Helicobacter pylori*, *S. epidermidis*, *Chlamydophila pneumoniae*, and the microbiota of *Demodex folliculorum* mites[47], which are found in larger numbers on the skin of rosacea patients. However, most studies are controversial and compelling evidence for a role of micro-organisms in the pathogenesis of this condition is not provided. Until now, the only 16S rRNA gene sequencing study that has been performed relates to the analysis of microbiota from *Demodex* mites that were isolated from the skin of rosacea patients[48]. This preliminary study shows that the *Demodex* microbiota of different subtypes of rosacea (erythematol-angiectatic rosacea and rosacea papulopustular) are distinguishable from each other. This diversity might be caused by host-specific differences (innate immune response, nutrient composition of skin niches), but this needs to be investigated in future studies.

5.2.5 Seborrheic dermatitis and dandruff

The scalp is a unique piece of human skin because of the high follicular density and high sebum production. The environment of the scalp is relatively warm and dark, which makes it ideal for superficial fungal infections that are often associated with diseases of the scalp[49]. The etiology of dandruff and seborrheic dermatitis is linked to *Malassezia restricta* and *Malassezia globosa*, which have been identified as the predominant fungi on both normal skin and scalp with dandruff or seborrheic dermatitis[50]. The extent of sebum production and individual susceptibility are also believed to play a role in these conditions. However, an exact mechanism for how *Malassezia* spp., presumably nourished by sebum, do cause this skin disorder of the scalp is not yet clear. A first small pyrosequencing study to the fungus microbiota (mycobiome) in healthy scalp and those afflicted with dandruff showed that the phyla Ascomycota (*Acremonium* spp.) and Basidiomycota (*Filobasidium* spp.) were the predominant fungi, respectively[51]. However, a larger mycobiome study in healthy volunteers showed again that *Malassezia globosa* and *Malassezia restricta* (phylum Basidiomycota) are the most common fungi on the back of the head[52].

5.2.6 Primary immunodeficiencies

It is generally accepted now that genetically determined variation of stratum corneum properties and variants of the host immune system are linked to disturbed homeostatic relations between the host and its microbiota (dysbiosis). Skin alterations and inappropriate immune responses could cause changes in microbial community composition[29,30,53,54], and conversely, it has been shown that skin microbiota can modulate cutaneous immunity[12,14,55]. A perfect illustration of a genetically based dysbiosis, in which human genetic variants have an effect on microbial recognition and host response, is the situation in patients with chronic mucocutaneous candidiasis (CMC) and hyper-IgE syndrome (HIES), which are primary immunodeficiencies mainly caused by *STAT1* and *STAT3* mutations, respectively[56–59]. Both diseases are characterized by increased risk for skin and mucosal infections with different bacterial and fungal species due to impaired Th1 and Th17 responses[60]. It was reported that the bacterial communities of

CMC and HIES patients contained increased levels of Gram-negative bacteria, in particular *Acinetobacter* spp. (phylum Proteobacteria), and decreased levels of Gram-positive bacteria like the 'normal' commensal *Corynebacterium* spp. (phylum Actinobacteria) compared to healthy controls[61]. Furthermore, it was shown that *Acinetobacter* (*baumannii*), and not *Corynebacterium* (*jeikeium*), was able to suppress the cytokine response of primary leukocytes to *Candida albicans* (*C. albicans*) and *S. aureus*, which are two pathogens that frequently cause cutaneous infections in CMC and HIES patients. This means that the genetics-based dysbiosis in these patients impairs the host-innate immune response to *C. albicans* and *S. aureus*[61]. Another culture-independent pyrosequencing study among immunocompromised individuals (predominantly HIES) also revealed that the microbial communities in these patients are distorted when compared to a healthy state[62]. For example, HIES patients showed increased abundance of *Serratia marcencens*, a Gram-negative bacterium that belongs to the phylum Proteobacteria, in agreement with *Acinetobacter* spp. that were found at high levels in the previous study. Furthermore, the skin commensal *Propionibacterium* (phylum Actinobacteria) was significantly depleted in HIES individuals. However, no further experiments were performed to study the functional consequence of the shift from Gram-positive to Gram-negative microorganisms for the immune system. In addition to the first paper, the authors have performed an analysis to the fungal communities on the skin of HIES patients. A trend toward increased fungal richness in HIES patients was observed compared to healthy controls, which includes a reduction in the relative abundance of the commensal skin fungi *Malassezia* and an increase of *Aspergilles* and *Candida* organisms, both of which belong to the phylum Ascomycota[62]. Overall, one could conclude that these studies have demonstrated that human genetic variants could have an effect on cutaneous microbial communities and thereby on the antimicrobial epidermal host defence response, which both contribute to disease.

5.3 Experimental studies

Up to now, most studies of the human skin microbiota in health and disease are inventory and descriptive. Possibly with the exception of monogenic skin diseases, causality remains problematic. E.g. for polygenic diseases like AD and psoriasis, the question remains whether the microbial changes found in lesional skin are the cause or the consequence of the disease. Longitudinal studies are required to link the dynamics of the microbiome to the dynamics of the disease in time. A few experimental studies in human and animals have been performed to address such issues.

5.4 Dynamics of the skin microbiome

To study the potential contribution of stratum corneum structure and/or quality to microorganisms present on human skin, we have investigated the effect of skin barrier quality on cutaneous microbiota in a more dynamic situation[38].

This involves analysis of recolonization of normal healthy skin that has been mechanically damaged by removal of the stratum corneum. To study the micro-biome of human epidermis following skin barrier disruption we used the well-described 'tape stripping' method[63,64]. This method involves the repeated application of adhesive tape to the skin surface, thereby removing stratum cor-neum layers. In a pilot study we took samples from healthy upper buttock skin and from skin that was tape-stripped 1, 10, 15, and 30 times on this part of the body. This experiment revealed that after 15 times tape stripping, the lower epider-mal compartment is practically devoid of detectable bacterial DNA. Surprisingly, it was shown by others that the microbiota may even extend to the dermal com-partment[39]. This study, however, reported a striking difference with skin surface microbiomes in all other studies, with unusually high abundance of Proteobacteria and a relative underrepresentation of Actinobacteria and Firmicutes. The results of this study await confirmation by other investigators. As we showed in our pilot study that the lower epidermal compartment contained no detectable DNA, we tape-stripped upper buttock skin 5 and up to 10 times to analyze the microbiota profile of the deeper layers of the stratum corneum ($n=12$, 6 male and 6 female). We compared these data with the microbial communities present at the super-ficial layer of the skin on the same location. Significant differences between bac-terial amounts and composition were observed[38]. Furthermore, analyses using bioinformatics revealed inter-individual variability and a strong effect of gender on microbiota composition, a phenomenon that was also reported by others[8,65]. Then we investigated the temporal changes in the microbiota composition of the skin surface, following mechanical removal of the stratum corneum. We tape-stripped four areas on the upper buttock skin and these areas were sampled in time up to two weeks. This experiment revealed that the host seems to be an important denominator of microbiota composition. Initially, we observed that the microbiota profile of injured skin is considerably disturbed up to day 14 when compared to the microbiome of the superficial skin layer before tape stripping. However, after tape stripping, the microbiota composition on day 1 reflects the superficial layer, which then diverges within 14 days in the direction of the state of 10 times tape-stripped skin, the microbial composition of the deeper stratum corneum layer. Additional experiments showed a substantial inter-individual var-iation in epidermal antimicrobial protein (AMP) expression following disturbance of the skin barrier by tape-stripping, which probably accounts for the host-specific microbiota on skin of each individual. Based on these data, we have suggested that there is a short-lived recolonization of injured skin with microbes from the surrounding superficial skin layer, and that these temporary microbiota are replaced by microbiota that reside in the deeper layers of stratum corneum. During the subsequent recolonization process the microbiota of the host and invading bacteria from the environment trigger the skin to express AMPs and inflamma-tory molecules (e.g., cytokines). These host-specific innate immune responses may help the skin to repair the wound, resulting in a re-established barrier function in which epidermal keratinocytes are in homeostasis with the local microbiome. In conclusion, since bacteria are also present in deeper parts of the skin, it is obvious that the microbial constituents of the superficial layer are just a part of the host indigenous skin microbiome.

5.5 Axillary skin microbiome transplantation

Development and validation of novel therapeutics that specifically inhibit skin pathogens or restore bacterial communities to the normal 'healthy' state could be future strategies to treat or prevent disease. Selective modulation of skin microbiota by pre- and/or probiotics could be an option[66], but microbial transplantation is also a promising tool to manipulate the microbiome. Ground-breaking studies showed that transplantation of feces derived from healthy donors could restore normal distal gut microbial community structure in patients with recurrent infections with *Clostridium difficile*[67,68]. Other researchers have transplanted axillary skin microbiota from non-odorous persons to the armpits of individuals who suffer from bromhidrosis (odorous armpits)[69], which showed a dominance of *Corynebacterium* spp. in the axillary region. Earlier studies have reported that these bacteria are responsible for axillary malodour, while other species like *Staphylococcus* spp. reveal only low levels of odour[70]. Axillary bacterial transplantations revealed significant improvements in the hedonic values. Further analysis showed decreased abundance of the 'odorous' *Corynebacterium* spp. and enrichment with 'non-odorous' *Staphylococcus* spp. in the axillary region of these individuals[69]. These preliminary data show that transplantation of skin microbiota could be a valuable tool to improve bad body odour, and might possibly also be applicable to skin diseases with microbial involvement in which replacement of 'bad' bacteria with 'good' bacteria is desired.

5.6 Mouse skin microbiome studies

So far, just a few studies have been performed that link barrier disturbance to dysbiosis. In one of these studies the microbiome of wild-type mice versus *caspase-14*-deficient mice was compared[71]. Caspase-14, a cysteine-aspartic acid protease, is required for the generation of Natural Moisturizing Factors (NMFs) by processing of filaggrin protein[72] and, as previously mentioned, patients with filaggrin mutations are prone to develop AD[22]. The authors argued that one of the reasons for dysbiosis in AD could be a reduction in filaggrin breakdown products. They indeed demonstrated that *caspase-14* deficiency causes an imbalance of skin-resident bacterial communities. These *caspase-14*-deficient mice also showed an enhanced AMP response compared to wild-type mice. Another knockout mouse model for a protease (matriptase) involved in filaggrin processing also showed increased AMP expression[73]. Furthermore, these *matriptase*-deficient mice, which have an ichthyotic-like phenotype, showed a selective shift in skin microbiota. Furthermore, proteins like hornerin and filaggrin-2, which are both part of the epidermal barrier, also possess antimicrobial modes of action (Dr. Jens Schröder, personal communication)[74]. This suggests that the epidermal barrier as such could modulate microbiota communities. Another example of a study of the dynamics of host-microbe interactions was performed using diabetic mice[75]. It was shown by 16S rRNA gene sequencing that diabetic mouse skin contains quantitatively and qualitatively different microbial populations, which are accompanied by different host gene expression profiles. The authors conclude that integration and correlation of microbiome diversity and gene expression might lead to a better understanding of the dynamics of wound healing in diabetic ulcers.

5.7 Concluding remarks

Future microbial studies in skin diseases should focus on the identification of markers related to disease onset, progression, and outcome, which will need longitudinal cohorts of well characterized disease phenotypes. Recurrence of lesions in, for example, AD and psoriasis patients could be preceded by early shifts in microbiota composition. Such data will give us valuable clues on how to intervene to force the bacterial communities towards a 'healthy' composition. Development of antibiotic therapies that do not disturb beneficial species and host-microbiota homeostasis (which happens by the use of broad-spectrum antibiotics), but selectively target specific pathogens and restore natural commensal population structures, will be essential to keep away from clinical problems secondary to dysbiosis. This might be achieved by selective killing of harmful pathogens, by stimulating beneficial strains (using pre- and/or probiotics), perhaps by intervening in metabolic pathways that are essential for interactions between microorganisms and the host immune system[76], and, as previously mentioned, by transplantation of skin microbiota.

TAKE-HOME MESSAGE

- The microbiota of normal human skin has a high diversity and high interpersonal variation.
- Genetically determined variation of stratum corneum properties and variants of the host immune system are linked to disturbed homeostatic relations (dysbiosis) between the host and its microbiota.
- Skin alterations and inappropriate immune responses could cause changes in microbial community composition.
- Skin microbiota can modulate cutaneous immunity.
- Longitudinal studies are required to link the dynamics of the microbiome to the dynamics of a skin disease in time.
- As bacteria are also present in deeper layers of the stratum corneum, the microbial constituents of the superficial layer are just a part of the host indigenous skin microbiome.
- Development of therapies that do not disturb beneficial cutaneous species and host-microbiota homeostasis, but selectively target specific 'bad bugs' (pathogens) or stimulate 'good' bacteria, might be future strategies to restore natural commensal population structures.

References

1 Eady RA, Leigh IM, Pope FM. Anatomy and organization of human skin. In: *Textbook of Dermatology* (Champion RH, Burton JL, Burns DA *et al.*, eds), Sixth edition: Blackwell Science, Inc., Malden. 1998; 37–111.

2 Dominguez-Bello MG, Costello EK, Contreras M *et al.* Delivery mode shapes the acquisition and structure of the initial microbiota across multiple body habitats in newborns. *Proc Natl Acad Sci U S A* 2010; **107**: 11971–5.

3 Capone KA, Dowd SE, Stamatas GN *et al*. Diversity of the human skin microbiome early in life. *J Invest Dermatol* 2011; **131**: 2026–32.

4 Cho I, Blaser MJ. The human microbiome: at the interface of health and disease. *Nat Rev Genet* 2012; **13**: 260–70.

5 Rosenthal M, Goldberg D, Aiello A *et al*. Skin microbiota: microbial community structure and its potential association with health and disease. *Infect Genet Evol* 2011; **11**: 839–48.

6 Ursell LK, Clemente JC, Rideout JR *et al*. The interpersonal and intrapersonal diversity of human-associated microbiota in key body sites. *J Allergy Clin Immunol* 2012; **129**: 1204–8.

7 Sanford JA, Gallo RL. Functions of the skin microbiota in health and disease. *Semin Immunol* 2013; **25**: 370–7.

8 Fierer N, Hamady M, Lauber CL *et al*. The influence of sex, handedness, and washing on the diversity of hand surface bacteria. *Proc Natl Acad Sci U S A* 2008; **105**: 17994–9.

9 Iwase T, Uehara Y, Shinji H *et al*. Staphylococcus epidermidis Esp inhibits Staphylococcus aureus biofilm formation and nasal colonization. *Nature* 2010; **465**: 346–9.

10 Shu M, Wang Y, Yu J *et al*. Fermentation of Propionibacterium acnes, a commensal bacterium in the human skin microbiome, as skin probiotics against methicillin-resistant Staphylococcus aureus. *PLoS One* 2013; **8**: e55380.

11 Lai Y, Di NA, Nakatsuji T *et al*. Commensal bacteria regulate Toll-like receptor 3-dependent inflammation after skin injury. *Nat Med* 2009; **15**: 1377–82.

12 Wanke I, Steffen H, Christ C *et al*. Skin commensals amplify the innate immune response to pathogens by activation of distinct signaling pathways. *J Invest Dermatol* 2011; **131**: 382–90.

13 Li D, Lei H, Li Z *et al*. A Novel Lipopeptide from skin commensal activates TLR2/CD36-p38 MAPK signaling to increase antibacterial defense against bacterial infection. *PLoS One* 2013; **8**: e58288.

14 Naik S, Bouladoux N, Wilhelm C *et al*. Compartmentalized control of skin immunity by resident commensals. *Science* 2012; **337**: 1115–9.

15 Blaser M. Antibiotic overuse: Stop the killing of beneficial bacteria. *Nature* 2011; **476**: 393–4.

16 Grice EA, Segre JA. The skin microbiome. *Nat Rev Microbiol* 2011; **9**: 244–53.

17 Costello EK, Lauber CL, Hamady M *et al*. Bacterial community variation in human body habitats across space and time. *Science* 2009; **326**: 1694–7.

18 Grice EA, Kong HH, Conlan S *et al*. Topographical and temporal diversity of the human skin microbiome. *Science* 2009; **324**: 1190–2.

19 Scharschmidt TC, Fischbach MA. What lives on our skin: ecology, genomics and therapeutic opportunities of the skin microbiome. *Drug Discov Today. Dis Mech* 2013; **10**: e83–e9.

20 Kuo IH, Yoshida T, De Benedetto A *et al*. The cutaneous innate immune response in patients with atopic dermatitis. *J Allergy Clin Immunol* 2013; **131**: 266–78.

21 Zeeuwen PL, Kleerebezem M, Timmerman HM *et al*. Microbiome and skin diseases. *Curr Opin Allergy Clin Immunol* 2013; **13**: 514–20.

22 Palmer CN, Irvine AD, Terron-Kwiatkowski A *et al*. Common loss-of-function variants of the epidermal barrier protein filaggrin are a major predisposing factor for atopic dermatitis. *Nat Genet* 2006; **38**: 441–6.

23 de Cid R, Riveira-Munoz E, Zeeuwen PL *et al*. Deletion of the late cornified envelope LCE3B and LCE3C genes as a susceptibility factor for psoriasis. *Nat Genet* 2009; **41**: 211–5.

24 Hollox EJ, Huffmeier U, Zeeuwen PL *et al*. Psoriasis is associated with increased beta-defensin genomic copy number. *Nat Genet* 2008; **40**: 23–5.

25 Strange A, Capon F, Spencer CC *et al*. A genome-wide association study identifies new psoriasis susceptibility loci and an interaction between HLA-C and ERAP1. *Nat Genet* 2010; **42**: 985–90.

26 Boguniewicz M, Leung DY. Atopic dermatitis: a disease of altered skin barrier and immune dysregulation. *Immunol Rev* 2011; **242**: 233–46.

27 Matsui K, Nishikawa A, Suto H *et al*. Comparative study of Staphylococcus aureus isolated from lesional and non-lesional skin of atopic dermatitis patients. *Microbiol Immunol* 2000; **44**: 945–7.

28 Huang JT, Abrams M, Tlougan B *et al*. Treatment of Staphylococcus aureus colonization in atopic dermatitis decreases disease severity. *Pediatrics* 2009; **123**: e808–14.

29 Kong HH, Oh J, Deming C et al. Temporal shifts in the skin microbiome associated with disease flares and treatment in children with atopic dermatitis. *Genome Res* 2012; **22**: 850–9.

30 de Jongh GJ, Zeeuwen PL, Kucharekova M et al. High expression levels of keratinocyte antimicrobial proteins in psoriasis compared with atopic dermatitis. *J Invest Dermatol* 2005; **125**: 1163–73.

31 Zeeuwen PL, de Jongh GJ, Rodijk-Olthuis D et al. Genetically programmed differences in epidermal host defense between psoriasis and atopic dermatitis patients. *PLoS ONE* 2008; **3**: e2301.

32 Nakamura Y, Oscherwitz J, Cease KB et al. Staphylococcus delta-toxin induces allergic skin disease by activating mast cells. *Nature* 2013; **503**: 397–401.

33 Griffiths CE, Barker JN. Pathogenesis and clinical features of psoriasis. *Lancet* 2007; **370**: 263–71.

34 Bergboer JG, Zeeuwen PL, Schalkwijk J. Genetics of psoriasis: evidence for epistatic interaction between skin barrier abnormalities and immune deviation. *J Invest Dermatol* 2012; **132**: 2320–1.

35 Fry L, Baker BS. Triggering psoriasis: the role of infections and medications. *Clin Dermatol* 2007; **25**: 606–15.

36 Gao Z, Tseng CH, Strober BE et al. Substantial alterations of the cutaneous bacterial biota in psoriatic lesions. *PLoS ONE* 2008; **3**: e2719.

37 Fahlen A, Engstrand L, Baker BS et al. Comparison of bacterial microbiota in skin biopsies from normal and psoriatic skin. *Arch Dermatol Res* 2012; **304**: 15–22.

38 Zeeuwen PL, Boekhorst J, van den Bogaard EH et al. Microbiome dynamics of human epidermis following skin barrier disruption. *Genome biol* 2012; **13**: R101.

39 Nakatsuji T, Chiang HI, Jiang SB et al. The microbiome extends to subepidermal compartments of normal skin. *Nat commun* 2013; **4**: 1431.

40 Alekseyenko AV, Perez-Perez GI, De Souza A et al. Community differentiation of the cutaneous microbiota in psoriasis. *Microbiome* 2013; **1**: 31.

41 Consortium THMP. Structure, function and diversity of the healthy human microbiome. *Nature* 2012; **486**: 207–14.

42 Williams HC, Dellavalle RP, Garner S. Acne vulgaris. *Lancet* 2012; **379**: 361–72.

43 Fitz-Gibbon S, Tomida S, Chiu BH et al. Propionibacterium acnes strain populations in the human skin microbiome associated with acne. *J Invest Dermatol* 2013; **133**: 2152–60.

44 Tomida S, Nguyen L, Chiu BH et al. Pan-genome and comparative genome analyses of propionibacterium acnes reveal its genomic diversity in the healthy and diseased human skin microbiome. *mBio* 2013; **4**: e00003–13.

45 Lazaridou E, Giannopoulou C, Fotiadou C et al. The potential role of microorganisms in the development of rosacea. *J Dtsch Dermatol Ges* 2011; **9**: 21–5.

46 Steinhoff M, Schauber J, Leyden JJ. New insights into rosacea pathophysiology: a review of recent findings. *J Am Acad Dermatol* 2013; **69**: S15–26.

47 Holmes AD. Potential role of microorganisms in the pathogenesis of rosacea. *J Am Acad Dermatol* 2013; **69**: 1025–32.

48 Murillo N, Aubert J, Raoult D. Microbiota of Demodex mites from rosacea patients and controls. *Microb Pathog* 2014; **71–72**: 37–40.

49 Grimalt R. A practical guide to scalp disorders. *J Investig Dermatol Symp Proc* 2007; **12**: 10–4.

50 Gemmer CM, DeAngelis YM, Theelen B et al. Fast, noninvasive method for molecular detection and differentiation of Malassezia yeast species on human skin and application of the method to dandruff microbiology. *J Clin Microbiol* 2002; **40**: 3350–7.

51 Park HK, Ha MH, Park SG et al. Characterization of the fungal microbiota (mycobiome) in healthy and dandruff-afflicted human scalps. *PLoS One* 2012; **7**: e32847.

52 Findley K, Oh J, Yang J et al. Topographic diversity of fungal and bacterial communities in human skin. *Nature* 2013; **498**: 367–70.

53 Gallo RL, Nakatsuji T. Microbial symbiosis with the innate immune defense system of the skin. *J Invest Dermatol* 2011; **131**: 1974–80.

54 Nibali L, Henderson B, Sadiq ST et al. Genetic dysbiosis: the role of microbial insults in chronic inflammatory diseases. *J Oral Microbiol* 2014; **6**.

55 Lai Y, Cogen AL, Radek KA *et al.* Activation of TLR2 by a small molecule produced by Staphylococcus epidermidis increases antimicrobial defense against bacterial skin infections. *J Invest Dermatol* 2010; **130**: 2211–21.

56 van de Veerdonk FL, Plantinga TS, Hoischen A *et al.* STAT1 mutations in autosomal dominant chronic mucocutaneous candidiasis. *N Engl J Med* 2011; **365**: 54–61.

57 Liu L, Okada S, Kong XF *et al.* Gain-of-function human STAT1 mutations impair IL-17 immunity and underlie chronic mucocutaneous candidiasis. *J Exp Med* 2011; **208**: 1635–48.

58 Holland SM, DeLeo FR, Elloumi HZ *et al.* STAT3 mutations in the hyper-IgE syndrome. *N Engl J Med* 2007; **357**: 1608–19.

59 Minegishi Y, Saito M, Tsuchiya S *et al.* Dominant-negative mutations in the DNA-binding domain of STAT3 cause hyper-IgE syndrome. *Nature* 2007; **448**: 1058–62.

60 Puel A, Cypowyj S, Marodi L *et al.* Inborn errors of human IL-17 immunity underlie chronic mucocutaneous candidiasis. *Curr Opin Allergy Clin Immunol* 2012; **12**: 616–22.

61 Smeekens SP, Huttenhower C, Riza A *et al.* Skin microbiome imbalance in patients with stat1/stat3 defects impairs innate host defense responses. *J Innate Immun* 2014; **6**: 253–62.

62 Oh J, Freeman AF, Program NCS *et al.* The altered landscape of the human skin microbiome in patients with primary immunodeficiencies. *Genome Res* 2013; **23**: 2103–14.

63 Pinkus H. Examination of the epidermis by the strip method. II. Biometric data on regeneration of the human epidermis. *J Invest Dermatol* 1952; **19**: 431–47.

64 de Koning HD, Kamsteeg M, Rodijk-Olthuis D *et al.* Epidermal expression of host response genes upon skin barrier disruption in normal skin and uninvolved skin of psoriasis and atopic dermatitis patients. *J Invest Dermatol* 2011; **131**: 263–6.

65 Callewaert C, Kerckhof FM, Granitsiotis MS *et al.* Characterization of Staphylococcus and Corynebacterium clusters in the human axillary region. *PLoS One* 2013; **8**: e70538.

66 Krutmann J. Pre- and probiotics for human skin. *J Dermatol Sci* 2009; **54**: 1–5.

67 Khoruts A, Dicksved J, Jansson JK *et al.* Changes in the composition of the human fecal microbiome after bacteriotherapy for recurrent Clostridium difficile-associated diarrhea. *J Clin Gastroenterol* 2010; **44**: 354–60.

68 van Nood E, Vrieze A, Nieuwdorp M *et al.* Duodenal infusion of donor feces for recurrent Clostridium difficile. *N Engl J Med* 2013; **368**: 407–15.

69 Callewaert C, Kerckhof F, Van Keer T *et al.* Characterisation of the human malodorous axillary microbiome and a novel treatment to obtain a better body odour. *J Invest Dermatol* 2014; **134**: S78–S.

70 Leyden JJ, McGinley KJ, Holzle E *et al.* The microbiology of the human axilla and its relationship to axillary odor. *J Invest Dermatol* 1981; **77**: 413–6.

71 Kubica M, Hildebrand F, Brinkman BM *et al.* The skin microbiome of caspase-14-deficient mice shows mild dysbiosis. *Exp Dermatol* 2014; **23**: 561–7.

72 Denecker G, Hoste E, Gilbert B *et al.* Caspase-14 protects against epidermal UVB photodamage and water loss. *Nat Cell Biol* 2007; **9**: 666–74.

73 Scharschmidt TC, List K, Grice EA *et al.* Matriptase-deficient mice exhibit ichthyotic skin with a selective shift in skin microbiota. *J Invest Dermatol* 2009; **129**: 2435–42.

74 Hansmann B, Gerstel U, Schroder JM. C-terminal filaggrin-2 targets the bacterial replication: a new antimicrobial mode of action? *J Invest Dermatol* 2014; **134**: S76–S.

75 Grice EA, Snitkin ES, Yockey LJ *et al.* Longitudinal shift in diabetic wound microbiota correlates with prolonged skin defense response. *Proc Natl Acad Sci U S A* 2010; **107**: 14799–804.

76 Brestoff JR, Artis D. Commensal bacteria at the interface of host metabolism and the immune system. *Nature Immunol* 2013; **14**: 676–84.

CHAPTER 6

Metagenomic analysis of the human microbiome

Luis G. Bermúdez-Humarán[1,2]

[1] *AgroParisTech, UMR1319 Micalis, F-78350 Jouy-en-Josas, France*
[2] *INRA, UMR1319 Micalis, Commensal and Probiotics-Host Interactions Laboratory, Domaine de Vilvert, 78352 Jouy-en-Josas Cedex, France*

6.1 Introduction

The human microbiome (community of microbes and collection of genomes found in the human body) directly influences many aspects of human physiology, including metabolism, drug interactions and numerous diseases. Its study plays an important role because many diseases are reported to be associated with microbiome imbalance or dysbiosis. To explore variation in the human microbiome and to understand how gut bacteria (microbiota) may have co-evolved with humans, new technologies (such as metagenomics) are beginning to reveal important insights of host-microbiota interactions. Hence, the microbiome has been shown to interact with the host in several ways in diseases such inflammatory bowel diseases (IBD), irritable bowel syndrome (IBS), colorectal cancer (CRC), obesity and diabetes, among others. The aim of this chapter is to summarize the link between the human microbiome and health, highlighting the contribution of metagenomics in the recent advances in this field.

The term *microbiome* refers to the genetic material of the catalog of the microbial taxa associated with humans and is sometimes referred to as our second genome[1,2]. Frequently confused with the term microbiota (which is the microbial population present in numerous parts of the human body), the human microbiome was first defined by Joshua Lederberg in 2001[3] as an ecological community of commensal, symbiotic, and pathogenic microorganisms that literally share our body space. These indigenous microbial communities explain critical features of human biology and also play an important role in human health and disease. Factors that enable bacteria to colonize the gut are not yet fully explained, but there is a general acceptance of the mutual benefits for both the host and bacteria that is the key for this successful partnership.

It is estimated that the human gut harbors approximately 10^{14} microorganisms (literally 10 times more than the cells in the human body)[4]. This microorganism diversity is estimated to contain 3.3 million microbial genes, representing around 150-fold more than the human genome and suggesting a co-evolutionary

The Human Microbiota and Chronic Disease: Dysbiosis as a Cause of Human Pathology, First Edition.
Edited by Luigi Nibali and Brian Henderson.

relationship. However, because of the abundance of microhabitats in the human body and the numerous interactions among different species with the host and the external environment, the microbiome can also be conceptualized as a dynamic ecological community[5]. As in all ecosystems, the microbiota reaches a dynamic equilibrium in the human body, which can be altered by environmental factors and external stimuli such as the use of antibiotics[6] (see also chapter 2). These alterations frequently result in microbial imbalances on or inside the body, a phenomenon called dysbiosis introduced by Curtis in chapter 2. Thus, in some ecosystems, such as the gut, a high biodiversity is associated with a healthy status, while low biodiversity is more likely to be linked to pathological conditions[7].

Initially, to achieve the study of microbiota biodiversity, traditional methods were based on culture-dependent techniques. However, although the use of these techniques provided a large and interesting set of data, they also resulted in an erroneous view of the composition of the human microbiota in certain cases. Certainly, many microorganisms need special growth conditions, such as the extremely oxygen-sensitive (EOS) bacteria, that makes their culturing and even their detection difficult[8], whereas others have never been grown in culture and may require special, as yet unknown, growth conditions preventing their identification by culture-dependent methods[9]. Recently, several culture-independent techniques have been developed that allow for a qualitative and quantitative means of identification and are mostly based on PCR and DNA hybridization techniques. These simple methods have completely changed the notion of the human microbiome, opening the door to new and more complete fields such as metagenomics[10]. Metagenomics is a relatively new field of study (first introduced in 1998)[11] of microbial genomes within diverse environmental samples that is of increasing importance in microbiology. The introduction of this ecological perception of microbiology is the key to achieving real knowledge about the influence of the microbiota in human health and disease.

6.2 The human microbiome

Humans did not evolve as a unified species; instead, they evolved with a complex associated microbiome, resulting in a sort of "super-organism" or holobiont[12]. Thus, the human microbiome determines and defines the immune system and it is an integral part of fundamental processes in the human body, such as vitamin production, digestion, energy homeostasis, angiogenesis and maintenance of the intestinal barrier integrity. 16S ribosomal RNA gene (rRNA)[13] and whole-genome sequencing (WGS)[14] has provided a general view of the commensal microbial communities revealing that the human microbiome is comprised (in total) of around 1000 different bacterial phylotypes[15,16]. Although these studies revealed that two bacterial divisions, Bacteroidetes and Firmicutes, constitute over 90% of the known phylogenetic categories and dominate the gut microbiota[17], taxonomic composition of the microbiomes between healthy subjects can differ significantly[18]. However, in spite of this interpersonal variability, some bacterial groups share functionalities[14,19]. Thus, a 'core' set of specific bacterial taxa (although such a classification has yet to be unequivocally demonstrated) has been proposed[20]. This fact shows the importance of a particular pool of microbial genes, which can be found in a set of microbial species that yield similar functions. The human genome diversity pales in comparison to the diversity of the microbiome[15]. In one recent study, previously sequenced data of 22 fecal metagenomes of individuals from different countries were analyzed and three robust clusters or "enterotypes" (which are not nation- or continent-specific) were identified[21]. These enterotypes vary with respect to the associated microbial species and their functional potential. However, the abundance of molecular functions is not necessarily provided by the presence of abundant species, highlighting the importance of a functional analysis to understand microbial communities[21].

6.3 Changes in microbiota composition during host life cycles

Differences in microbiota composition exist across body sites and among individuals. However, changes are also evident across the human lifespan. The establishment of the human microbiotas begins at birth and reaches its maximum complexity in adolescence to finally remain relatively stable in healthy adults. In the later stages of life, the microbiome becomes comparatively less diverse, with reduced stability[22].

Today there is strong evidence to support the inheritance character of the microbiota. Although inheritance of the microbiota from the father has been poorly studied, increasing evidence supports inheritance from the mother[23,24]. The microbiota acquired by an infant depends on the mode of delivery (discussed in chapter 3), being similar to the mother's vaginal microbiota in case of natural birth and to the mother's skin microbiota in case of C-section delivery[25,26]. Because lactobacilli dominate the mother's vagina, the initial flourishing of these bacteria in the baby's GIT cannot be considered fortuitous. Another important factor for the establishment of the microbiome is the mode of feeding, as the microbiota of breast- and formula-fed babies is significantly different, both in composition and diversity. Bifidobacteria, known for their beneficial properties, are present in the microbiota of breast-fed babies, while with formula-fed babies *Escherichia coli, Clostridium difficile, Bacteroides fragilis,* and lactobacilli predominate[27]. Pregnant women of normal body weight and health-associated microbiota profiles (gut and breast-milk microbiota) have greater opportunities to pass on compounds, antigens modified by the mother's gut, and other agents that promote the development of a healthy immune system in the breast-fed infant[28]. Age-related changes in microbiota results in reduced ecosystem stability, making it more dynamic, with the dominating phyla at this stage shifting from Firmicutes to Bacteroides, accompanied by an increase in the number of Proteobacteria and reduction in Bifidobacteria[22]. The core microbiota of elderly individuals significantly differed from young adults, with a characteristic shift toward a Clostridium-dominated community[29].

6.4 The human microbiome and the environment

Chapter 2 by Curtis introduced the dynamic interactions between the human microbiome and its environments. The most important factor influencing the composition of the microbiome is the use of antibiotics. Indeed, the use of these drugs to treat infections also affects the host innate microbiota and impact the stability of the microbiome. The responses to antibiotic treatment are individualized and are influenced by prior exposure of an individual to the same antibiotic. For example, only one week after treatment with ciprofloxacin, the gut microbiota can regain its composition, returning to its pre-treatment state[30]. In contrast, clindamycin treatment, an antibiotic with strong anti-anaerobic activity, reduces Bacteroides diversity, after a week of consecutive use, and its impact can last up to two years[31].

Genetic diversity found in human intestinal microbiota also affects the microbiome, allowing digestion of different compounds using various metabolic pathways that are not explicitly encoded by the mammalian genome[32]. Each individual human microbiome offers new genes to digest new dietary products[33]. A number of studies support the observation that the human gut microbiota gains energy from polysaccharides and peptides that are indigestible by human enzymes[34]. For example, the human genome lacks genes encoding enzymes required for degrading plant polysaccharides with high carbohydrate containing xylan, pectin, arabinose, which humans usually consume. However, this capability is provided by the microbiome, allowing humans to utilize sucrose, glucose, galactose, fructose and mannose, and participates also in the synthesis of essential amino acids and vitamins[35,36]. In fact, the transformation of sugars into butyryl-CoA butyrate, a short- chain fatty acid (SCFAs) that is the main energy source of colonocytes and establishes a barrier that maintains a healthy intestinal condition, is carried out by the microbiome[37] (this is discussed in detail in chapter 23 by Fak). Metagenomic, metatranscriptiomic and metaproteomic analyses suggests that Firmicutes are useful in the metabolism of carbohydrates, while Bacteroidetes are reported to be involved in amino acid transport and metabolism[38].

Eating habits also play a key role in the establishment of the microbiome. For instance, Firmicutes and Proteobacteria dominates the microbiome of European individuals as European food is usually rich in animal protein, sugar, starch and fat and low in fibre, in contrast to the vegetarian diet rich in carbohydrates, fibre and non-animal protein consumed in Africa, resulting in a microbiome dominated by Actinobacteria and Bacteroidetes[39]. Despite the fact that certain bacterial phyla always dominate the human gut, variations in the relative percentages of Firmicutes and Bacteroidetes exist within the human population, reflecting the effects of diet and consumption of macronutrients[40]. The microbiome of individuals also depends on genetic background, as the microbiome of twins has been reported to be more similar, compared with those of parents or a third individual[14] (this is discussed more extensively in chapter 28). Moreover, recent studies have also shown that prebiotics (non-digestible food ingredients that stimulate the growth and/or activity of specific microorganisms) can influence the composition of the gut microbiota species, both in short-term dietary interventions and in response to the usual long-term dietary intakes (see also chapter 30)[41].

6.5 Disease and health implications of microbiome

Microbial communities reside on abundant sites of our body including the skin, the respiratory tract, the conjunctiva, the oral cavity, the vagina and the GIT. A description of microbial communities in these body sites based on culture techniques has been provided by Wilson in chapter 1. The composition and diversity of these bacterial communities is characteristic of their location. Current studies focus on the description of the microbiome modification that take place in specific disease states, or microbial temporal changes observed over the course of a disease. Chapters 3 and 4 of this book review the associations between the microbiome and specific human diseases. This chapter provides a brief overview of these associations, with a particular focus on gut microbiota.

6.5.1 The skin microbiota

The skin is the human body's largest organ, colonized by a diverse collection of microorganisms (including bacteria, fungi and viruses), most of which are harmless or even beneficial to their host[42]. Genomic approaches to characterize skin microbiome have revealed a greater diversity of organisms than that shown by culture-dependent methods (see chapter 1). Thus, metagenomic analyses revealed that most skin bacteria fall into four different phyla: Actinobacteria, Firmicutes, Bacteroidetes and Proteobacteria[42]. Chapter 5 introduced evidence for a shift in Firmicutes and Actinobacteria in psoriatic skin lesions[43]. Chapter 12 by Bruggemann will discuss evidence on the commensal skin bacterium *Propionibacterium acnes*, implicated in the common dermatological condition acne[44,45]. Atopic dermatitis (AD) is another chronic inflammatory condition of the skin that is also associated with microbial colonization and infection. Its prevalence has increased in incidence approximately three-fold over the last 30 years in industrialized countries, suggesting a potential role for microbiome alterations[46]. Chronic skin ulcers, which are often secondary to venous stasis or diabetes, lead to substantial morbidity. Skin microbiome shifts have been observed in these conditions, such as an increased abundance of Pseudomonadaceae in patients with chronic ulcers that were treated with antibiotics, and an increased abundance of Streptococcaceae in diabetic ulcers[47]. Such shifts may interact with aberrantly expressed host cutaneous defence response genes[48], thereby increasing disease risk.

6.5.2 The airway microbiome

The human respiratory system is the ecological niche for numerous commensal microorganisms. However, in contrast to the GIT, the respiratory tract of healthy individuals harbors a homogeneous microbiota that decreases in biomass from upper to lower tract[49]. Although colonization by potential pathogens of the upper respiratory tract (especially the nasopharyngeal microbiome) induces identifiable disease in only a small percentage of individuals, colonization represents an important source of secretions containing bacteria that spread between individuals[50]. For example, in the case of children the impact of age, season, type of day care, number of siblings, acute respiratory illness, diet, and sleeping position have been described, whereas in adults, other factors have been also implicated, such as contact with children, chronic obstructive pulmonary disease, obesity, immunosuppression, allergic conditions, acute sinusitis, etc.[50]. For instance, a metagenomic study on the detailed composition and variability in nasopharyngeal microbiota in samples from young children revealed that it differs among seasons[51]. During fall and winter, which tend to be associated with increased incidence of respiratory and invasive infections, a predominance of Proteobacteria and Fusobacteria was observed. However, in spring, Bacteroidetes and Firmicutes were more abundant. Another component of nasopharyngeal microbiome are viruses, and ~30% of all suspected viral cases fail diagnostic tests for etiologic agents[52]. Therefore, metagenomics could allow detection of known viruses in this specific environment as well as the detection of new[53]. Similarly, some authors propose to define 'the human virome project' as the systematic exploration of viruses that infect humans for the investigation of a new pathogen, and provide a model for comprehensive diagnosis of unexplained acute illnesses or outbreaks in clinical and public settings[54,55]. For example, in the case of influenza virus H_1N_1 in 2009, this strategy has proven to have the potential to replace conventional diagnostic tests[55]. Little is known about the lower respiratory tract microbiome recently described, even if it is likely to provide important pathogenic insights (cystic fibrosis, respiratory disease of the newborn, chronic obstructive pulmonary disease, and asthma)[56]. Moreover, infectious agents are known to be or suspected of having a key role in a number of chronic lung conditions. The amount of published data shows that phylogenetically diverse microbial communities in the lungs of healthy humans may be detected using high-throughput sequencing[57–59]. A better characterization of lung microbiome could help in understanding their role in preserving health or causing disease mainly in specific patient groups, for example in smokers[60].

6.5.3 Vaginal microbiome

The vaginal microbiome has an important influence on human development, physiology, and immunity. This ecosystem of mutualistic bacteria constitutes the first line of defense for the host by excluding non-indigenous microbes that may cause sickness[61]. Although the vaginal ecosystem is dynamic as a result of its physiological function (menstrual cycle) and personal habits (contraception and hygiene practices), it remains stable over the long term as a result of physiological and microbiological factors. A mature microbiota is already established in early adolescence after the hormonal changes typical of this period[62] and it includes some microorganisms also present in the GIT, even if the relative frequencies are different. The first culture-dependent studies of the human vagina reported lactobacilli as the dominant microorganisms of this organ, being more than 70% of all microorganisms isolated from vaginal exudates of healthy and fertile women[61,63,64]. On the other hand, the metagenomics analyses undertaken in the context of the 'Vaginal Human Microbiome Project' that aims to investigate the complex vaginal microbiome and its relation to human health and disease (as well as its variability with different physiological conditions) also describe the vagina as an ecosystem rich in lactobacilli[65]. Although the paradigm of the association between lactobacilli abundance and vaginal health seems to be true for the majority of women, it does not necessarily apply to all (see also chapter 16). For example, vaginal ecosystems dominated by other lactic-acid-producing bacteria, such as *Bifidobacterium* spp., *Megasphaera* spp., and *Leptotrichia* spp.

have also been described[66–68]. A higher vaginal pH has been reported in some racial and ethnic groups[69–71] as well as the absence of lactobacilli and the predominance of *Gardnerella vaginalis*, *Prevotella* spp., *Pseudomonas* spp., and/or *Streptococcus* spp.[66]. In fact, several studies have demonstrated the presence of different microbiome profiles (named vagitypes), many of which are dominated by a single bacterial taxon[65]. However, caution should be taken in this interpretation, since it is not possible completely to rule out a transition state between illness and health in these atypical microbiomes[72,73]. The vaginal microbiota, primarily lactobacilli, asserts its beneficial effects against pathogens through two main mechanisms: i) exclusion, driven by competition for epithelial cell receptors, and ii) growth inhibition due to the generation of antimicrobial compounds[74]. The first mechanism results from the ability of lactobacilli to compete for receptors against urogenital pathogens[75,76]. The second mechanism refers to the ability of lactobacilli to produce several antimicrobial compounds that are mainly organic acids produced from the fermentation of sugars, typically leading to the low pH of the vagina that inhibits the growth of most pathogens[77]. Furthermore, vaginal lactobacilli are also able to produce bacteriocins, biosurfactants[78,79], and hydrogen peroxide (H_2O_2)[80]. Abnormal vaginal microbiome can occur because of sexually transmitted pathogens or overgrowth of resident organisms. The most common pathologies are bacterial vaginosis, the proliferation of *Candida albicans* (candidiasis) and *Trichomonas vaginalis* (trichomoniasis)[61]. Bacterial vaginosis is the most frequent vaginal imbalance and was shown by molecular methods to be associated with a high microbiome diversity[81] and the presence of unfamiliar bacteria such as *Mobiluncus* spp., *Atopobium* spp., *Megasphaera* spp., and *Ureaplasma urealyticum*[82].

6.5.4 Gut microbiota and disease

Although the composition of the human gut microbiota is less diverse than other microbial ecosystems (such as those found in soil or water), probably reflecting the harsh physicochemical conditions of this niche, most microorganisms that colonize humans are bacteria present in the GIT at a density of approximately $10^{13}-10^{14}$ cells/g of stool (particularly in the colon, 70% of the total microbiota)[83]. The intestinal microbiota also contains a minority population of eukaryotic microorganisms (fungi, yeast), viruses, and archae.[9] The gut microbiota is considered to be beneficial for the host due to its key role in the maturation and stimulation of the immune system, promoting mucosal structure and function, and providing resistance against pathogen attack and colonization[84–86]. Intestinal epithelial cells (IECs) constitute the first line of contact of the gut microbiota and act as a barrier to prevent the translocation of substances systemically[87]. IECs are also the interface between the external environment and the most extensive host lymphoid compartment, the gut-associated lymphoid tissue (GALT), a tissue rich in cells of the innate and adaptive immune system. To maintain intestinal homeostasis, GALT must process large quantities of information at the interface between the luminal side and the host, distinguishing between commensal bacteria and pathogenic microorganisms[88]. GALT interacts with intestinal bacteria, which are sampled by dendritic cells (DCs) and IECs through pattern recognition receptors such as toll-like receptors (TLRs) and nucleotide-binding oligomerization domain (NOD)[89,90]. These innate immune receptors recognize conserved microbial structures found on microorganisms referred as microbial associated molecular patterns (MAMPs) both on pathogens and commensal bacteria. Also, Th17 cell differentiation in the mouse lamina propria requires the presence of segmented filamentous bacteria (SFB)[91], and polysaccharide A produced by *Bacteroides fragilis* mediates the conversion of CD4+ T cells into regulatory T cells[92]. An essential symbiotic relationship thus exists between our intestinal epithelium and commensal bacteria and, although some intestinal bacteria are potential pathogens, the relationship between the intestinal microbiota and the human host is mostly symbiotic in healthy individuals[93] (see also chapter 3). Disruption of this interaction and/or alterations of the microbiome could potentially affect human health and promote disease states or dysbiosis[94,95]. The GIT is covered by a protective mucus layer containing predominantly mucin glycoproteins that are synthesized and secreted by goblet cells. Intestinal microbes may directly affect goblet cell functions through the local release of bioactive factors, such as those released during the fermentation with beneficial microbes[96]. There are many diseases associated with gut microbiota and their imbalance, such as inflammatory bowel disease (IBD), inflammatory bowel syndrome (IBS), colorectal cancer (CRC), obesity, diabetes, antibiotic-associated diarrhea, etc. Advances in the analysis of the composition of the gut microbiota and the implications for health and disease have recently emerged. In particular, early infancy exposure to antibiotics has been associated with a significantly increased risk for Crohn's disease[97], suggesting that gut microbiome perturbations are important for disease risk. Microbial diversity is significantly diminished in Crohn's disease[7], suggesting a decreased gut microbiome resilience that could affect immune interactions. For example, decrease in Firmicutes, such as *Faecalibacterium prausnitzii,* is well documented in Crohn's disease patients[98,99]. On the other hand, metegenomic studies revealed a clear division of microbial composition in healthy and ulcerative colitis patients[100]. Another study reported an analysis of fecal microbiota of ulcerative colitis patients and confirmed the reduction of bacterial diversity in these individuals[101]. Specific bacteria of the Enterobacteriaceae family may act together with a disordered microbiome to increase the risk of ulcerative colitis[102]. Between twins that are discordant for ulcerative colitis, those affected had significantly reduced bacterial diversity but increased proportions of Actinobacteria and Proteobacteria[103].

The dominating four bacterial phyla, Firmicutes, Bacteroidetes, Actinobacteria, and Proteobacteria are the best-studied groups in the microbiome; however, archaea, viruses, fungi and eukarya, including helminths, are also constituents of this community; their study is necessary for better understanding of the relationship between the microbiome and host, as a whole in addition to their complex intertwined metabolic networks. Chapter 19 provides further details on the current status of the evidence for the association between dysbiosis and inflammatory bowel disease.

The gut microbiota has also emerged as an important factor that may contribute to its pathophysiology of irritable bowel syndrome (IBS)[104], a common functional gut disorder with an estimated prevalence of up to 20% of the population and with a negative social and economic impact on patients[105,106]. Symptoms include abdominal pain, altered stool consistency and frequency, bloating and distension. Some of the proposed beneficial effects of the gut microbiota include maintenance of intestinal homoeostasis, maintenance of peristalsis, intestinal mucosal integrity, and protection against pathogens[107–109]. For instance, animal studies have shown that alterations in the gut microbiota can lead to changes in GI functions such as an altered intestinal motility and visceral hypersensitivity (two disorders often observed in IBS patients)[107–109]. In addition, epidemiological observations linking bacterial gastroenteritis and small intestinal bacterial overgrowth to IBS, microbiology studies describing quantitative and qualitative alterations in the gut microbiota in patients with IBS, and clinical studies demonstrating the efficacy of antimicrobial and probiotic treatments in patients with IBS clearly demonstrate the role of an altered intestinal microbiota in the pathogenesis of IBS[110]. The onset of IBS symptoms after a gastroenteritis episode or the occurrence of an enteric infection represents one of the strongest points of evidence for the importance of gut microbiota in IBS[111]. It has also been demonstrated that subgroups of IBS patients may have an altered microbiota composition compared to healthy individuals (based on the analysis of fecal microbiota)[112]. Both an increase and a decrease in the diversity of the microbiota have been reported in IBS patients[113,114]. From this perspective, the abnormal variation likely reflects a loss of homeostasis, in which the bacterial community is unable to maintain its normal composition. Finally, recent studies have demonstrated the important role of the microbiome-brain-gut communication in the etiology of IBS, and altered central nervous system (CNS) control of visceral pain and inflammatory responses are now considered as integral pathophysiological features of this syndrome[115].

Recent studies show that gut microbiota may increase or decrease cancer susceptibility and progression by various mechanisms, such as by modulating inflammation, affecting genomic stability of the host cells, and producing metabolites that function as histone deacetylase inhibitors to regulate host gene expression[116]. The gut microbiota has been suspected for a long time to be involved in the development of colorectal cancer (CRC)[117], possibly by synthesizing SCFAs and other metabolites. SCFAs, in particular butyrate, may induce apoptosis, cell cycle arrest and differentiation[118]. Microorganisms may also be genotoxic to colonic epithelial cells or even promote CRC by eliciting host responses, for example, by stimulating exacerbated immune responses, potentially through Th17 cells[119]. Recent studies have associated members of anaerobic genus *Fusobacterium* with CRC[120,121]. *Fusobacterium nucleatum* is a mucosally adherent, pro-inflammatory microbe that was first identified in the mouth[122]. In CRC samples, *F. nucleatum* sequences were significantly enriched compared with samples obtained from control tissue, while both Bacteroidetes and Firmicutes were depleted relative to other bacteria in *Fusobacterium*-rich malignancies[121]. The enrichment of *Fusobacterium* species was confirmed when evaluating the microbiome of CRC compared to adjacent normal tissues in an expanded collection of biopsies[120]. However, the causal direction of the association has not yet been confirmed (see chapter 26).

6.5.5 Metabolic disorders (obesity/diabetes)

Obesity and its associated disorders, such as diabetes mellitus type 2, are a major public health problem, mainly caused by a combination of overeating and sedentary lifestyle, with contributions of host genetics, environment and inflammation of adipose tissue. Furthermore, in recent years an increasing number of studies have also shown that an altered composition and diversity of the intestinal microbiota might play an important role in the development of metabolic disorders such as obesity and diabetes[123]. From this perspective, recent studies on the distal gut microbiome have proposed that changes in gut microbiota are associated with obesity, metabolic syndrome and the Western diet. In fact, changes have been reported in the composition of the intestinal microbiota in humans and obese mice with a relative decrease in the proportion of Bacteroidetes in obese versus lean individuals[14,40,124], suggesting a potential link between microbial diversity and obesity. For example, genetically obese (*ob/ob*) mice have decreased Bacteroidetes/Firmicutes ratios compared with their lean counterparts[40]. In addition, transplantation of gut microbiota from *ob/ob* to GF mice conferred an obese phenotype, demonstrating the transmissibility of metabolic phenotypes[32]. Further support for a causal link between microbiota and obesity is provided by GF mice studies in which these animals have reduced total body fat relative to their conventionally reared counterparts, and are resistant to diet-induced obesity[125,126]. Interestingly, colonization of GF mice with commensal bacteria harvested from conventionally-reared mice produces a 60% increase in body weight within two weeks despite decreases in food consumption[127]. In humans, the relative proportions of members of the Bacteroidetes phylum increase with weight loss[124]. In studies of monozygotic and dizygotic twins, obesity was associated with smaller populations of Bacteroidetes, diminished bacterial diversity and enrichment of genes related to lipid and carbohydrate metabolism. Despite substantial taxonomic variation, functional metagenomic differences were minor[14]. Modern lifestyles that change the selection pressures on microbiomes could alter exposures to bacteria during the early lives of hosts and thus may contribute to the development of obesity. Antibiotic use in human infancy (before the age of six months) was significantly associated with obesity development[128]. By contrast, probiotic treatment has also been demonstrated to have beneficial anti-obesity effects in both mouse and human studies[129-131]. Perinatal administration of a *Lactobacillus rhamnosus* probiotic decreased excessive weight gain during childhood[131]. These early studies provide support for the concept that perturbations in microbiota could lead to childhood-onset obesity, which might be modifiable. Likewise, a single dose of *Lactobacillus plantarum* in GF mice maintained on high-fat/high-sugar diet resulted in the up-regulation of genes involved in carbohydrate transport and metabolism[130]. Similarly, the presence of the probiotic strain of *Bifidobacterium longum* elicits an expansion in the diversity of polysaccharides targeted for degradation by a prominent component of the adult human gut microbiota, *Bacteroides thetaiotaomicron*, when simultaneously colonized in GF mice[132]. Alterations in the gut microbiota also occur when interventions are used to treat obesity. Roux-en-Y surgery significantly increases levels of Proteobacteria and alters specific metabolic markers, such as the production of urinary amines and cresols[133]. Therefore, although composition of the gut microbiota varies between individuals, a 'core gut microbiome' has been identified that confers greater risk of obesity[14]. In addition, preclinical studies have shown that a seven-week exposure in mice to low doses of the antibiotics induced taxonomic changes in the microbiome characterized by reductions in the Bacteroidetes/Firmicutes ratio, which resembles the microbiota profile in obese humans, and was paralleled by increased body fat[134]. A link between the metabolic disease diabetes and bacterial populations in the gut has been also established. Specifically, the proportions of Firmicutes and class Clostridia were significantly reduced in the diabetic group compared to the control group[135]. More recently a correlation between gut microbiome profiles and diabetes mellitus type 2 was demonstrated in a Chinese population to the extent that composition of the gut microbiota could predict the disorder in a second cohort[136]. Low bacterial richness has been associated with obesity, low-grade inflammation and dysmetabolism richness[137,138]. Altogether, these data suggest that high microbial richness of the gut microbiome may be an indicator for metabolic homeostasis and a protecting factor against metabolic deviation and disease.

6.6 Conclusions

Today, we recognize the need to study the human microbiome as a whole ecosystem to better understand the relationship between microbiota and host health or illness. Maintaining mucosal homeostasis is a fine balance between the host and the gut microbiome. The host uses several mechanisms to contain the intestinal microbiome and prevent the development of unsuitable inflammation, for example. In parallel, the host intestinal immune system needs microbial stimulation for proper development. The end result is a steady state where microbial stimulation promotes normal immune function in the gut, which in turn allows the intestinal microbioma to prosper in the absence of unnecessary inflammation. Hence, due to the close relationship between the microbiome and health and the existence of biomarkers typical of different pathologies (IBD, CRC, obesity, etc.) the suggestion to modulate our microbiota seems logical from a therapeutic point of view. Furthermore, the inter-individual differences and physiological parameters suggest personal medicine as a future treatment. Thus, new powerful approaches are needed to analyse these ecosystems, with metagenomic methodologies being key for a more detailed analysis of the human microbiome. These kinds of approaches would identify biomarkers of well-being that correspond to a general and more balanced microbiota. In conclusion, human microbiome (which is a direct consequence of the mutualism between the host and its microbiota) is fundamental for the maintenance of the homeostasis of a healthy individual.

TAKE-HOME MESSAGE

- The human microbiome must be considered as a whole ecosystem for better understanding of its dynamic role in health and diseases.
- Resolution of gut dysbiosis by maintaining a mucosal homeostasis could be the key to improved treatments for several chronic diseases (such as IBD).
- Large-scale metagenomic analyses are the key to explore the relationship between the human microbiota and diseases.

References

1 Ursell LK, Metcalf JL, Parfrey LW *et al.* Defining the human microbiome. *Nutr Rev* 2012; **70 Suppl 1**: S38–44.

2 Bruls T, Weissenbach J. The human metagenome: our other genome? *Human Molecular Genetics* 2011; **20**: R142–R8.

3 Lederberg J, McCray AT. 'Ome sweet 'omics — A genealogical treasury of words. *Scientist* 2001; **15**: 8–.

4 Turnbaugh PJ, Ley RE, Hamady M *et al.* The human microbiome project. *Nature* 2007; **449**: 804–10.

5 Foxman B, Goldberg D, Murdock C *et al.* Conceptualizing human microbiota: from multicelled organ to ecological community. *Interdiscip Perspect Infect Dis* 2008; **2008**: 613979.

6 Dethlefsen L, Huse S, Sogin ML *et al.* The pervasive effects of an antibiotic on the human gut microbiota, as revealed by deep 16S rRNA sequencing. *PLoS Biol* 2008; **6**: e280.

7 Manichanh C, Rigottier-Gois L, Bonnaud E *et al.* Reduced diversity of faecal microbiota in Crohn's disease revealed by a metagenomic approach. *Gut* 2006; **55**: 205–11.

8 Miquel S, Martin R, Rossi O *et al.* Faecalibacterium prausnitzii and human intestinal health. *Curr Opin Microbiol* 2013; **16**: 255–61.

9 Weinstock GM. Genomic approaches to studying the human microbiota. *Nature* 2012; **489**: 250–6.

10 Dimon MT, Wood HM, Rabbitts PH *et al.* IMSA: integrated metagenomic sequence analysis for identification of exogenous reads in a host genomic background. *PLoS One* 2013; **8**: e64546.

11 Handelsman J, Rondon MR, Brady SF *et al.* Molecular biological access to the chemistry of unknown soil microbes: a new frontier for natural products. *Chem Biol* 1998; **5**: R245–9.

12 Rosenberg E, Zilber-Rosenberg I. Symbiosis and development: the hologenome concept. *Birth Defects Res C Embryo Today* 2011; **93**: 56–66.

13 Zoetendal EG, Akkermans AD, De Vos WM. Temperature gradient gel electrophoresis analysis of 16S rRNA from human fecal samples reveals stable and host-specific communities of active bacteria. *Applied and environmental microbiology* 1998; **64**: 3854–9.

14 Turnbaugh PJ, Hamady M, Yatsunenko T *et al.* A core gut microbiome in obese and lean twins. *Nature* 2009; **457**: 480–4.

15 Qin J, Li R, Raes J *et al.* A human gut microbial gene catalogue established by metagenomic sequencing. *Nature* 2010; **464**: 59–65.

16 Human Microbiome Project C. A framework for human microbiome research. *Nature* 2012; **486**: 215–21.

17 Eckburg PB, Bik EM, Bernstein CN *et al.* Diversity of the human intestinal microbial flora. *Science* 2005; **308**: 1635–8.

18 Turnbaugh PJ, Quince C, Faith JJ *et al.* Organismal, genetic, and transcriptional variation in the deeply sequenced gut microbiomes of identical twins. *Proceedings of the National Academy of Sciences of the USA* 2010; **107**: 7503–8.

19 Burke C, Steinberg P, Rusch D *et al.* Bacterial community assembly based on functional genes rather than species. *Proceedings of the National Academy of Sciences of the USA* 2011; **108**: 14288–93.

20 Tap J, Mondot S, Levenez F *et al.* Towards the human intestinal microbiota phylogenetic core. *Environ Microbiol* 2009; **11**: 2574–84.

21 Arumugam M, Raes J, Pelletier E *et al.* Enterotypes of the human gut microbiome. *Nature* 2011; **473**: 174–80.

22 Biagi E, Nylund L, Candela M *et al.* Through ageing, and beyond: gut microbiota and inflammatory status in seniors and centenarians. *PLoS One* 2010; **5**: e10667.

23 Ley RE, Lozupone CA, Hamady M *et al.* Worlds within worlds: evolution of the vertebrate gut microbiota. *Nat Rev Microbiol* 2008; **6**: 776–88.

24 Ochman H, Worobey M, Kuo CH *et al.* Evolutionary relationships of wild hominids recapitulated by gut microbial communities. *PLoS Biol* 2010; **8**: e1000546.

25 Dominguez-Bello MG, Costello EK, Contreras M *et al.* Delivery mode shapes the acquisition and structure of the initial microbiota across multiple body habitats in newborns. *Proceedings of the National Academy of Sciences of the USA* 2010; **107**: 11971–5.

26 Dominguez-Bello MG, Blaser MJ. Do you have a probiotic in your future? *Microbes and Infection* 2008; **10**: 1072–6.

27 Penders J, Thijs C, Vink C *et al.* Factors influencing the composition of the intestinal microbiota in early infancy. *Pediatrics* 2006; **118**: 511–21.

28 Isolauri E. Development of healthy gut microbiota early in life. *Journal of Paediatrics and Child Health* 2012; **48**: 1–6.

29 Claesson MJ, Cusack S, O'Sullivan O *et al.* Composition, variability, and temporal stability of the intestinal microbiota of the elderly. *Proceedings of the National Academy of Sciences of the USA* 2011; **108 Suppl 1**: 4586–91.

30 Dethlefsen L, Relman DA. Incomplete recovery and individualized responses of the human distal gut microbiota to repeated antibiotic perturbation. *Proceedings of the National Academy of Sciences of the USA* 2011; **108 Suppl 1**: 4554–61.

31 Jernberg C, Lofmark S, Edlund C *et al.* Long-term impacts of antibiotic exposure on the human intestinal microbiota. *Microbiology* 2010; **156**: 3216–23.

32 Turnbaugh PJ, Ley RE, Mahowald MA *et al.* An obesity-associated gut microbiome with increased capacity for energy harvest. *Nature* 2006; **444**: 1027–31.

33 Hehemann JH, Correc G, Barbeyron T *et al.* Transfer of carbohydrate-active enzymes from marine bacteria to Japanese gut microbiota. *Nature* 2010; **464**: 908–12.

34 Guarner F, Malagelada JR. Gut flora in health and disease. *Lancet* 2003; **361**: 512–9.

35 Backhed F, Ley RE, Sonnenburg JL *et al.* Host-bacterial mutualism in the human intestine. *Science* 2005; **307**: 1915–20.

36 LeBlanc JG, Milani C, de Giori GS *et al.* Bacteria as vitamin suppliers to their host: a gut microbiota perspective. *Current Opinion in Biotechnology* 2013; **24**: 160–8.

37 Topping DL, Clifton PM. Short-chain fatty acids and human colonic function: roles of resistant starch and nonstarch polysaccharides. *Physiol Rev* 2001; **81**: 1031–64.

38 Ottman N, Smidt H, de Vos WM *et al.* The function of our microbiota: who is out there and what do they do? *Frontiers in Cellular and Infection Microbiology* 2012; **2**.

39 De Filippo C, Cavalieri D, Di Paola M *et al.* Impact of diet in shaping gut microbiota revealed by a comparative study in children from Europe and rural Africa. *Proceedings of the National Academy of Sciences of the USA* 2010; **107**: 14691–6.

40 Ley RE, Backhed F, Turnbaugh P *et al.* Obesity alters gut microbial ecology. *Proceedings of the National Academy of Sciences of the USA* 2005; **102**: 11070–5.

41 Saulnier DM, Ringel Y, Heyman MB *et al.* The intestinal microbiome, probiotics and prebiotics in neurogastroenterology. *Gut Microbes* 2013; **4**: 17–27.

42 Grice EA, Segre JA. The skin microbiome (vol 9, pg 244, 2011). *Nat Rev Microbiol* 2011; **9**.

43 Gao Z, Tseng CH, Strober BE *et al.* Substantial alterations of the cutaneous bacterial biota in psoriatic lesions. *PLoS One* 2008; **3**: e2719.

44 McDowell A, Gao A, Barnard E *et al.* A novel multilocus sequence typing scheme for the opportunistic pathogen Propionibacterium acnes and characterization of type I cell surface-associated antigens. *Microbiology* 2011; **157**: 1990–2003.

45 Bek-Thomsen M, Lomholt HB, Kilian M. Acne is not associated with yet-uncultured bacteria. *J Clin Microbiol* 2008; **46**: 3355–60.

46 Grice EA, Segre JA. The skin microbiome. *Nat Rev Microbiol* 2011; **9**: 244–53.

47 Price LB, Liu CM, Melendez JH *et al.* Community Analysis of Chronic Wound Bacteria Using 16S rRNA Gene-Based Pyrosequencing: Impact of Diabetes and Antibiotics on Chronic Wound Microbiota. *Plos One* 2009; **4**.

48 Grice EA, Snitkin ES, Yockey LJ *et al.* Longitudinal shift in diabetic wound microbiota correlates with prolonged skin defense response (vol 107, pg 14799, 2010). *Proceedings of the National Academy of Sciences of the USA* 2010; **107**: 17851.

49 Charlson ES, Bittinger K, Haas AR *et al.* Topographical continuity of bacterial populations in the healthy human respiratory tract. *Am J Respir Crit Care Med* 2011; **184**: 957–63.

50 Garcia-Rodriguez JA, Fresnadillo Martinez MJ. Dynamics of nasopharyngeal colonization by potential respiratory pathogens. *J Antimicrob Chemother* 2002; **50 Suppl S2**: 59–73.

51 Bogaert D, Keijser B, Huse S *et al.* Variability and diversity of nasopharyngeal microbiota in children: a metagenomic analysis. *PLoS One* 2011; **6**: e17035.

52 Heikkinen T, Jarvinen A. The common cold. *Lancet* 2003; **361**: 51–9.

53 Lysholm F, Wetterbom A, Lindau C *et al.* Characterization of the viral microbiome in patients with severe lower respiratory tract infections, using metagenomic sequencing. *PLoS One* 2012; **7**: e30875.

54 Allander T, Tammi MT, Eriksson M *et al.* Cloning of a human parvovirus by molecular screening of respiratory tract samples. *Proceedings of the National Academy of Sciences of the USA* 2005; **102**: 12891–6.

55 Greninger AL, Chen EC, Sittler T *et al.* A metagenomic analysis of pandemic influenza A (2009 H1N1) infection in patients from North America. *PLoS One* 2010; **5**: e13381.

56 Beck JM, Young VB, Huffnagle GB. The microbiome of the lung. *Transl Res* 2012; **160**: 258–66.

57 Erb-Downward JR, Thompson DL, Han MK *et al.* Analysis of the lung microbiome in the "healthy" smoker and in COPD. *PLoS One* 2011; **6**: e16384.

58 Hilty M, Burke C, Pedro H *et al.* Disordered microbial communities in asthmatic airways. *PLoS One* 2010; **5**: e8578.

59 Huang YJ, Nelson CE, Brodie EL *et al.* Airway microbiota and bronchial hyperresponsiveness in patients with suboptimally controlled asthma. *The Journal of allergy and clinical immunology* 2011; **127**: 372–81 e1–3.

60 Kiley JP. Advancing respiratory research. *Chest* 2011; **140**: 497–501.

61 Martin R, Soberon N, Vazquez F *et al.* [Vaginal microbiota: composition, protective role, associated pathologies, and therapeutic perspectives]. *Enferm Infecc Microbiol Clin* 2008; **26**: 160–7.

62 Yamamoto T, Zhou X, Williams CJ *et al.* Bacterial populations in the vaginas of healthy adolescent women. *J Pediatr Adolesc Gynecol* 2009; **22**: 11–8.

63 Eschenbach DA, Davick PR, Williams BL *et al.* Prevalence of hydrogen peroxide-producing Lactobacillus species in normal women and women with bacterial vaginosis. *J Clin Microbiol* 1989; **27**: 251–6.

64 Redondo-Lopez V, Cook RL, Sobel JD. Emerging role of lactobacilli in the control and maintenance of the vaginal bacterial microflora. *Rev Infect Dis* 1990; **12**: 856–72.

65 Fettweis JM, Serrano MG, Girerd PH *et al.* A new era of the vaginal microbiome: advances using next-generation sequencing. *Chem Biodivers* 2012; **9**: 965–76.

66 Hyman RW, Fukushima M, Diamond L *et al.* Microbes on the human vaginal epithelium. *Proceedings of the National Academy of Sciences of the USA* 2005; **102**: 7952–7.

67 Zhou X, Bent SJ, Schneider MG *et al.* Characterization of vaginal microbial communities in adult healthy women using cultivation-independent methods. *Microbiology* 2004; **150**: 2565–73.

68 Zhou X, Brown CJ, Abdo Z *et al.* Differences in the composition of vaginal microbial communities found in healthy Caucasian and black women. *Isme Journal* 2007; **1**: 121–33.

69 Stevenssimon C, Jamison J, Mcgregor JA *et al.* Racial variation in vaginal ph among healthy sexually active adolescents. *Sexually Transmitted Diseases* 1994; **21**: 168–72.

70 Ravel J, Gajer P, Abdo Z *et al.* Vaginal microbiome of reproductive-age women. *Proceedings of the National Academy of Sciences of the USA* 2011; **108**: 4680–7.

71 Fiscella K, Klebanoff MA. Are racial differences in vaginal pH explained by vaginal flora? *American Journal of Obstetrics and Gynecology* 2004; **191**: 747–50.

72 Fredricks DN. Molecular methods to describe the spectrum and dynamics of the vaginal microbiota. *Anaerobe* 2011; **17**: 191–5.

73 McCook A. The vagina catalogues. *Nat Med* 2011; **17**: 765–7.

74 Boris S, Barbes C. Role played by lactobacilli in controlling the population of vaginal pathogens. *Microbes and infection/Institut Pasteur* 2000; **2**: 543–6.

75 Zarate G, Nader-Macias ME. Influence of probiotic vaginal lactobacilli on in vitro adhesion of urogenital pathogens to vaginal epithelial cells. *Lett Appl Microbiol* 2006; **43**: 174–80.

76 Boris S, Suarez JE, Vazquez F *et al.* Adherence of human vaginal lactobacilli to vaginal epithelial cells and interaction with uropathogens. *Infection and immunity* 1998; **66**: 1985–9.

77 Boskey ER, Cone RA, Whaley KJ *et al.* Origins of vaginal acidity: high D/L lactate ratio is consistent with bacteria being the primary source. *Hum Reprod* 2001; **16**: 1809–13.

78 Velraeds MM, van de Belt-Gritter B, van der Mei HC *et al.* Interference in initial adhesion of uropathogenic bacteria and yeasts to silicone rubber by a Lactobacillus acidophilus biosurfactant. *Journal of medical microbiology* 1998; **47**: 1081–5.

79 Velraeds MM, van de Belt-Gritter B, Busscher HJ *et al.* Inhibition of uropathogenic biofilm growth on silicone rubber in human urine by lactobacilli — a teleologic approach. *World J Urol* 2000; **18**: 422–6.

80 Martin R, Suarez JE. Biosynthesis and degradation of H_2O_2 by vaginal lactobacilli. *Applied and environmental microbiology* 2010; **76**: 400–5.

81 Pavlova SI, Kilic AO, Kilic SS *et al*. Genetic diversity of vaginal lactobacilli from women in different countries based on 16S rRNA gene sequences. *J Appl Microbiol* 2002; **92**: 451–9.

82 Fredricks DN, Fiedler TL, Marrazzo JM. Molecular identification of bacteria associated with bacterial vaginosis. *N Engl J Med* 2005; **353**: 1899–911.

83 Ley RE, Turnbaugh PJ, Klein S *et al*. Microbial ecology — human gut microbes associated with obesity. *Nature* 2006; **444**: 1022–3.

84 Martin R, Miquel S, Ulmer J *et al*. Role of commensal and probiotic bacteria in human health: a focus on inflammatory bowel disease. *Microbial cell factories* 2013; **12**.

85 Martin R, Miquel S, Langella P *et al*. The role of metagenomics in understanding the human microbiome in health and disease. *Virulence* 2014; **5**: 413–23.

86 Littman DR, Pamer EG. Role of the commensal microbiota in normal and pathogenic host immune responses. *Cell host & microbe* 2011; **10**: 311–23.

87 Turner JR. Intestinal mucosal barrier function in health and disease. *Nat Rev Immunol* 2009; **9**: 799–809.

88 Quigley EM. Probiotics in gastrointestinal disorders. *Hosp Pract (1995)* 2010; **38**: 122–9.

89 Coombes JL, Powrie F. Dendritic cells in intestinal immune regulation. *Nat Rev Immunol* 2008; **8**: 435–46.

90 Kelly D, Mulder IE. Microbiome and immunological interactions. *Nutr Rev* 2012; **70 Suppl 1**: S18–30.

91 Ivanov, II, Frutos Rde L, Manel N *et al*. Specific microbiota direct the differentiation of IL-17-producing T-helper cells in the mucosa of the small intestine. *Cell host & microbe* 2008; **4**: 337–49.

92 Round JL, Mazmanian SK. Inducible Foxp3+ regulatory T-cell development by a commensal bacterium of the intestinal microbiota. *Proceedings of the National Academy of Sciences of the USA* 2010; **107**: 12204–9.

93 Hwang JS, Im CR, Im SH. Immune disorders and its correlation with gut microbiome. *Immune Netw* 2012; **12**: 129–38.

94 Hojo K, Nagaoka S, Ohshima T *et al*. Bacterial interactions in dental biofilm development. *J Dent Res* 2009; **88**: 982–90.

95 Rogler G. The importance of gut microbiota in mediating the effect of NOD2 defects in inflammatory bowel disease. *Gut* 2010; **59**: 153–4.

96 de Moreno de LeBlanc A, Dogi CA, Galdeano CM *et al*. Effect of the administration of a fermented milk containing Lactobacillus casei DN-114001 on intestinal microbiota and gut associated immune cells of nursing mice and after weaning until immune maturity. *BMC Immunol* 2008; **9**: 27.

97 Hviid A, Svanstrom H, Frisch M. Antibiotic use and inflammatory bowel diseases in childhood. *Gut* 2011; **60**: 49–54.

98 Sokol H, Lay C, Seksik P *et al*. Analysis of bacterial bowel communities of IBD patients: What has it revealed? *Inflammatory bowel diseases* 2008; **14**: 858–67.

99 Sokol H, Pigneur B, Watterlot L *et al*. Faecalibacterium prausnitzii is an anti-inflammatory commensal bacterium identified by gut microbiota analysis of Crohn disease patients. *Proceedings of the National Academy of Sciences of the USA* 2008; **105**: 16731–6.

100 Willing BP, Dicksved J, Halfvarson J *et al*. A pyrosequencing study in twins shows that gastrointestinal microbial profiles vary with inflammatory bowel disease phenotypes. *Gastroenterology* 2010; **139**: 1844–U105.

101 Rajilic-Stojanovic M, Heilig HGHJ, Molenaar D *et al*. Development and application of the human intestinal tract chip, a phylogenetic microarray: analysis of universally conserved phylotypes in the abundant microbiota of young and elderly adults. *Environmental Microbiology* 2009; **11**: 1736–51.

102 Garrett WS, Gallini CA, Yatsunenko T *et al*. Enterobacteriaceae act in concert with the gut microbiota to induce spontaneous and maternally transmitted colitis. *Cell host & microbe* 2010; **8**: 292–300.

103 Lepage P, Hasler R, Spehlmann ME *et al*. twin study indicates loss of interaction between microbiota and mucosa of patients with ulcerative colitis. *Gastroenterology* 2011; **141**: 227–36.

104 Ohman L, Simren M. Intestinal microbiota and its role in irritable bowel syndrome (IBS). *Curr Gastroenterol Rep* 2013; **15**: 323.

105 Ford AC, Talley NJ. Irritable bowel syndrome. *BMJ* 2012; **345**: e5836.

106 Dupont HL. Review article: evidence for the role of gut microbiota in irritable bowel syndrome and its potential influence on therapeutic targets. *Aliment Pharmacol Ther* 2014; **39**: 1033–42.

107 Bercik P, Wang L, Verdu EF *et al*. Visceral hyperalgesia and intestinal dysmotility in a mouse model of postinfective gut dysfunction. *Gastroenterology* 2004; **127**: 179–87.

108 Lee BJ, Bak YT. Irritable bowel syndrome, gut microbiota and probiotics. *J Neurogastroenterol Motil* 2011; **17**: 252–66.

109 Collins S, Verdu E, Denou E *et al*. The role of pathogenic microbes and commensal bacteria in irritable bowel syndrome. *Dig Dis* 2009; **27 Suppl 1**: 85–9.

110 Ringel Y, Maharshak N. Intestinal microbiota and immune function in the pathogenesis of irritable bowel syndrome. *Am J Physiol Gastrointest Liver Physiol* 2013; **305**: G529–41.

111 Spiller R. Postinfectious functional dyspepsia and postinfectious irritable bowel syndrome: different symptoms but similar risk factors. *Gastroenterology* 2010; **138**: 1660–3.

112 Salonen A, de Vos WM, Palva A. Gastrointestinal microbiota in irritable bowel syndrome: present state and perspectives. *Microbiology* 2010; **156**: 3205–15.

113 Codling C, O'Mahony L, Shanahan F *et al*. A molecular analysis of fecal and mucosal bacterial communities in irritable bowel syndrome. *Digestive diseases and sciences* 2010; **55**: 392–7.

114 Claesson MJ, Jeffery IB, Conde S *et al*. Gut microbiota composition correlates with diet and health in the elderly. *Nature* 2012; **488**: 178–84.

115 Collins SM, Bercik P. The relationship between intestinal microbiota and the central nervous system in normal gastrointestinal function and disease. *Gastroenterology* 2009; **136**: 2003–14.

116 Zitvogel L, Galluzzi L, Viaud S *et al*. Cancer and the gut microbiota: An unexpected link. *Science translational medicine* 2015; **7**: 271ps1.

117 Plottel CS, Blaser MJ. Microbiome and malignancy. *Cell host & microbe* 2011; **10**: 324–35.

118 Lazarova DL, Bordonaro M, Carbone R *et al*. Linear relationship between Wnt activity levels and apoptosis in colorectal carcinoma cells exposed to butyrate. *Int J Cancer* 2004; **110**: 523–31.

119 Wu S, Rhee KJ, Albesiano E *et al*. A human colonic commensal promotes colon tumorigenesis via activation of T helper type 17 T cell responses. *Nat Med* 2009; **15**: 1016–22.

120 Castellarin M, Warren RL, Freeman JD *et al*. Fusobacterium nucleatum infection is prevalent in human colorectal carcinoma. *Genome Research* 2012; **22**: 299–306.

121 Kostic AD, Gevers D, Pedamallu CS *et al*. Genomic analysis identifies association of Fusobacterium with colorectal carcinoma. *Genome Research* 2012; **22**: 292–8.

122 Krisanaprakornkit S, Kimball JR, Weinberg A *et al*. Inducible expression of human beta-defensin 2 by Fusobacterium nucleatum in oral epithelial cells: Multiple signaling pathways and role of commensal bacteria in innate immunity and the epithelial barrier. *Infection and immunity* 2000; **68**: 2907–15.

123 Parekh PJ, Arusi E, Vinik AI *et al*. The role and influence of gut microbiota in pathogenesis and management of obesity and metabolic syndrome. *Front Endocrinol (Lausanne)* 2014; **5**: 47.

124 Ley RE, Turnbaugh PJ, Klein S *et al*. Microbial ecology: human gut microbes associated with obesity. *Nature* 2006; **444**: 1022–3.

125 Backhed F, Manchester JK, Semenkovich CF *et al*. Mechanisms underlying the resistance to diet-induced obesity in germ-free mice. *Proceedings of the National Academy of Sciences of the USA* 2007; **104**: 979–84.

126 Rabot S, Membrez M, Bruneau A *et al*. Germ-free C57BL/6J mice are resistant to high-fat-diet-induced insulin resistance and have altered cholesterol metabolism. *Faseb Journal* 2010; **24**: 4948–.

127 Backhed F, Ding H, Wang T *et al*. The gut microbiota as an environmental factor that regulates fat storage. *Proceedings of the National Academy of Sciences of the USA* 2004; **101**: 15718–23.

128 Ajslev TA, Andersen CS, Gamborg M *et al*. Childhood overweight after establishment of the gut microbiota: the role of delivery mode, pre-pregnancy weight and early administration of antibiotics. *Int J Obes (Lond)* 2011; **35**: 522–9.

129 Arora T, Singh S, Sharma RK. Probiotics: Interaction with gut microbiome and antiobesity potential. *Nutrition* 2013; **29**: 591–6.

130 Marco ML, Peters THF, Bongers RS *et al*. Lifestyle of Lactobacillus plantarum in the mouse caecum. *Environmental Microbiology* 2009; **11**: 2747–57.

131 Luoto R, Kalliomaki M, Laitinen K *et al*. The impact of perinatal probiotic intervention on the development of overweight and obesity: follow-up study from birth to 10 years. *Int J Obes (Lond)* 2010; **34**: 1531–7.

132 Sonnenburg JL, Chen CTL, Gordon JI. Genomic and metabolic studies of the impact of probiotics on a model gut symbiont and host. *Plos Biology* 2006; **4**: 2213–26.

133 Li JV, Ashrafian H, Bueter M *et al*. Metabolic surgery profoundly influences gut microbial-host metabolic cross-talk. *Gut* 2011; **60**: 1214–23.

134 Cho I, Yamanishi S, Cox L *et al*. Antibiotics in early life alter the murine colonic microbiome and adiposity. *Nature* 2012; **488**: 621–6.

135 Larsen N, Vogensen FK, van den Berg FWJ *et al*. Gut microbiota in human adults with type 2 diabetes differs from non-diabetic adults. *Plos One* 2010; **5**.

136 Qin JJ, Li YR, Cai ZM *et al*. A metagenome-wide association study of gut microbiota in type 2 diabetes. *Nature* 2012; **490**: 55–60.

137 Le Chatelier E, Nielsen T, Qin JJ *et al*. Richness of human gut microbiome correlates with metabolic markers. *Nature* 2013; **500**: 541–.

138 Cotillard A, Kennedy SP, Kong LC *et al*. Dietary intervention impact on gut microbial gene richness (vol 500, pg 585, 2013). *Nature* 2013; **502**.

Microbiota-microbiota and microbiota-host interactions in health and disease

CHAPTER 7

Systems biology of bacteria-host interactions

Almut Heinken, Dmitry A. Ravcheev and Ines Thiele
Luxembourg Centre for Systems Biomedicine, University of Luxembourg, Belval, Luxembourg

7.1 Introduction

The preceding chapter has highlighted how the field of human microbiome research is growing exponentially thanks to advances in genome sequencing and bioinformatics. Worldwide research initiatives, such as the Human Microbiome Project[1] and MetaHit[2], have yielded large amounts of high-throughput "omics" data. Metagenomic[3], metatranscriptomic[4], metaproteomic[5], and metabolomic[6] analyses have greatly improved our insight into the metabolic potential of the ecosystem residing in the mammalian intestine. A variety of publicly accessible resources including high-quality genome sequences[7], metagenomic datasets[8], tools for metabolic pathway prediction[9,10], and tools generating organism-specific networks[11–16] are available that can support computational analyses of the gut microbiota (Table 1).

For a complete understanding of the metabolic interactions between the mammalian microbiota and its host, an integrated view of the components that make up the whole host-microbe "superorganism" is necessary. The challenge to integrate analyses of metabolites, macromolecules, and cells on a "systems" level has motivated the field of systems biology[17]. The aim of systems biology is to use computational methods to gain a complete, systems-level understanding of a cell, organism, or ecosystem. To this end, computational networks models are commonly used, including kinetic models and regulatory models as well as stoichiometric networks where nodes represent metabolites and links represent reactions[17]. Two approaches can be distinguished: top-down and bottom-up systems biology. Top-down systems biology approaches take advantage of the increased availability of high-throughput "omics" data at affordable costs. From these experimental datasets, network structures are inferred. The aim of top-down modeling is to predict novel biological hypotheses, which must subsequently be validated experimentally[17]. In contrast, bottom-up modeling aims to be mechanism-based, well-structured, and accurate. Network structures are formulated manually at the molecular level based on detailed mechanistic knowledge[17]. Predictions derived from these models must subsequently be validated experimentally, with *in vitro* or *in vivo* results informing the model in an iterative

The Human Microbiota and Chronic Disease: Dysbiosis as a Cause of Human Pathology, First Edition.
Edited by Luigi Nibali and Brian Henderson.
© 2016 John Wiley & Sons, Inc. Published 2016 by John Wiley & Sons, Inc.

Table 1 Publicly available resources for computational analyses of the gut microbiota. GENRE = genome-scale reconstruction; FBA = flux balance analysis.

Resource	Description	URL	Ref.
Model generation and analysis			
CellNetAnalyzer	Matlab toolbox for the analysis of GENREs and signal-flow networks, as well as network visualization with interactive maps.	http://www2.mpi-magdeburg.mpg.de/projects/cna/cna.html	111,12
COBRA (constraint-based reconstruction and analysis) Toolbox	A toolbox for implementing biased and unbiased COBRA methods to analyze metabolic reconstructions.	http://opencobra.github.io/	113,14
FAME (Flux Analysis and Modeling Environment)	Online platform for building stoichiometric models, running flux balance analysis, and visualization of results.	http://f-a-m-e.fame-	15
MicrobesFlux	Online platform for GENRE draft reconstruction and constraint-based modeling with FBA and dynamic FBA.	http://www.microbesflux.org/	14
Model SEED	Automated reconstruction of analysis-ready bacterial and archaeal models from user genomes annotated by the RAST Server[115].	http://rast.nmpdr.org/	11
OptFlux	Toolbox designed for *in silico* metabolic engineering. Contains several GENRE model analysis and optimization methods.	http://www.optflux.org/	116
Pathway Tools	Software for the prediction of metabolic pathways in organisms based on their annotated genomes.	http://bioinformatics.ai.sri.com/ptools/	16
RAVEN (Reconstruction, Analysis, and Visualization of Metabolic Networks) Toolbox	Software that generates semi-automated reconstructions based on the KEGG database and published models.	http://biomet-toolbox.org/index.php?page=downtools-raven	12
rBioNet	COBRA Toolbox extension for the reconstruction of GENREs that enforces physicochemical constraints.	http://opencobra.github.io/	117
SuBliMinaL Toolbox	A toolbox providing independent modules for many common steps in the reconstruction of metabolic networks.	http://www.mcisb.org/resources/subliminal/	13
Omics databases and analysis			
EBI Metagenomics	Automated analysis and archiving of metagenomic data.	https://www.ebi.ac.uk/metagenomics/	118
EnsembleGenomes	Genome database that enables the annotation, analysis, and visualization of genomes from all domains of life.	http://ensemblgenomes.org/	119
Genome and protein databases			
COG (Clusters of Orthologous Groups of proteins)	Resource listing clusters of orthologous proteins that were constructed from sequenced prokaryotic and eukaryotic genomes.	http://www.ncbi.nlm.nih.gov/COG/	120
eggNOG (evolutionary genealogy of genes: Non-supervised Orthologous Groups)	Tools for orthology assignment to sequenced genomes. Also includes a pipeline for the reconstruction and visualization of phylogenetic relationships.	eggnogdb.embl.de/	121

Table 1 (*Continued*)

Resource	Description	URL	Ref.
GenBank	Collection of all publicly available DNA sequences.	http://www.ncbi.nlm.nih.gov/genbank/	122
HMP (Human Microbiome Project)	Repository of over 3,000 genome sequences isolated from the human body.	http://www.hmpdacc.org/	123,24
HUMAnN (HMP Unified Metabolic Analysis Network)	A pipeline that determines the pathways present or absent in metagenomic data from microbial communities.	http://huttenhower.sph.harvard.edu/humann	50
IMG (Integrated Microbial Genomes)	Database containing genome sequences from the three domains of life, plasmids, viruses, and genome fragments, also containing analysis tools for comparing the structural and functional annotations of genomes.	http://img.jgi.doe.gov/	63
MicrobesOnline	Database and tools for comparative and functional genomics of bacteria, archaea and fungi.	http://microbesonline.org/	125
MG-RAST (Metagenomics RAST)	A platform that provides automated analysis of prokaryotic metagenomic data.	http://metagenomics.anl.gov/	126
PATRIC (Pathosystems Resource Integration Center)	Contains genome annotations, genome metadata, and data integration of bacterial organisms. Offers automated annotations of bacterial genome sequences.	http://patricbrc.vbi.vt.edu/	127
Pfam	Database listing protein families, which are each defined by two alignments and a profile hidden Markov model (HMM).	http://pfam.xfam.org/	128
RAST (Rapid Annotations using Subsystems Technology)	Resource for annotation of bacterial and archaeal genomes based on SEED subsystems.	http://rast.nmpdr.org/	115
Pathway resources			
BioCyc	A collection of metabolic pathways and genomes of sequenced organisms. Contains tools to predict metabolic pathways, genes encoding pathway gaps, and operons.	http://biocyc.org/	129
KEGG (Kyoto Encyclopedia of Genes and Genomes) PATHWAY	Manually drawn metabolic pathway maps representing our knowledge on biochemical reactions and molecular interactions.	http://www.genome.jp/kegg/pathway.html	130,131
Reactome	Manually curated database of human pathways and reactions covering about 7,000 proteins. Mainly signaling pathways are covered.	http://www.reactome.org	75
Others			
BiomeNet	Bayesian modeling framework that constructs a set of metabolic subnetworks from metagenomic samples and infers differences in prevalence of these subnetworks between groups.	http://sourceforge.net/projects/biomenet/	56

(*Continued*)

Table 1 (*Continued*)

Resource	Description	URL	Ref.
BRENDA (Braunschweig Enzyme Database)	Enzyme information repository including EC numbers, reactions, kinetic enzyme parameters and organism-specific information.	http://www.brenda-enzymes.de	132
CAZy (Carbohydrate-Active enZYmes)	Database listing carbohydrate-degrading enzymes and binding sites, including many enzymes targeting dietary and host polysaccharides.	http://www.cazy.org/	133
KBase	Genome assembly and annotation, GENRE reconstruction and simulation, data integration, and data sharing for predictions of biological functions.	http://kbase.us/	134
HMDB (Human Metabolome Database)	Database of the small molecule metabolites that occur in human cells. Links chemical, clinical, and biochemical data.	http://www.hmdb.ca/	135–137
NetCooperate	Web-based tool that predicts the cooperative potential of two species based on their metabolic networks. Syntrophic or competitive interactions can be elucidated.	http://elbo.gs. washington.edu/ software_netcooperate.html	138
PathPred	Web-based platform that predicts reaction pathways of a query compound based on known xenobiotics biodegradation pathways or the biosynthesis of secondary metabolites.	http://www.genome.jp/tools/pathpred/	10
Virtual Metabolic Human	A comprehensive knowledge base on human metabolism and human-associated microbes. Includes the metabolic reactions and the corresponding gene associations that are known to occur in human cells.	https://vmh.uni.lu/	

process[18]. In this chapter, computational systems biology approaches and their applications to human gut microbiome research are described, with particular focus on constraint-based modeling.

7.2 Computational analysis of host-microbe interactions

7.2.1 Analysis of metagenomic data

Metagenomic methods access the genomic contents of microbial communities by means of 16S ribosomal RNA sequencing and whole metagenome shotgun sequencing[19]. These techniques have the advantage of being culture-independent and capturing thousands of species[19]. Metagenomic methods include taxonomic profiling, which describes the diversity of the community, and functional profiling, which captures its functional repertoire[20]. Since the output of metagenomic methods is too large for manual human inspection, bioinformatic analyses are required for their interpretation. A variety of computational studies have successfully elucidated novel biological knowledge from metagenomic datasets. For instance, the "enterotypes" concept, which stratifies human individuals according to their microbiome composition, has been proposed[21]. Another metagenomic study linked low microbial gene richness with a detrimental host phenotype, including adiposity, insulin resistance, and more pronounced inflammation[22]. Based on the metagenomes of 145 European women with normal, impaired, and diabetic glucose control, a mathematical model was developed that could predict type 2 diabetes from metagenomic data[23]. Yet another metagenomic study of diabetes patients identified gut microbial markers of type 2 diabetes, e.g., a decrease in butyrate producers[24]. The strain-level variation in the human microbiome was investigated based on metagenomic analysis. The analysis revealed that a variety of species displayed variations in gene copy-number with implications for obesity and inflammatory bowel disease[25]. For an in-depth discussion of metagenomic analysis of the human microbiome, refer to chapter 6.

7.2.2 Metabolic reconstruction through comparative genomics

Metagenome analysis provides large amounts of information about bacterial communities. However, it represents only an approximation since the genetic material is divided between hundreds of distinct microbial species in one individual. The reference genome collections available at the NCBI and IMG databases (Table 1) include around 2900 sequenced genomes of human microbes, of which around 25% belong to the gastrointestinal tract. Additionally, the reference set of genomes includes a large numbers of pan-genomes, i.e., unions of genome sets of multiple strains of a species. The most abundant pan-genome sets are represented for *Enterococcus faecalis* (80 strains), *Propionibacterium acnes* (74 strains), *Enterococcus faecium* (72 strains), *Escherichia coli* (71 strains), and *Helicobacter pylori* (64 strains). The presence of such huge amount of genomic data provides an exciting opportunity for genomic-based studies beyond metagenomic analysis.

Comparative genomics approaches are used for the genome-based reconstruction of bacterial metabolism and regulation. The most widely used annotation method predicts gene functions based on the protein sequence similarity between annotated genes and previously studied proteins. The prediction of gene functions can be substantially improved by increasing the number of genome context methods used, such as the analysis of co-occurrence and co-localization of genes[26]. Although experimental data for many gut microbes is sparse, comparative genomics techniques can be used to provide more confidence in functional annotation. Genome context-based comparative genomics techniques, such as analysis of conserved chromosomal clusters, phyletic patterns, and prediction of regulatory interactions (Figure 1), are used to predict genes that encode the missing functions, enabling gap-filling of the reconstructed pathways and thus the discovery of novel functions. The chromosomal clustering technique is based on the observation that genes involved in the same pathways are often co-clustered on the chromosome and that this clustering is conserved among related genomes[26].

Comparative genomics approaches that combine both sequence similarity searches and genome context methods are regularly applied for the reconstruction of so-called subsystems, i.e. functionally related genes, such as genes for a single metabolic pathway. For instance, comparative genomics was used to

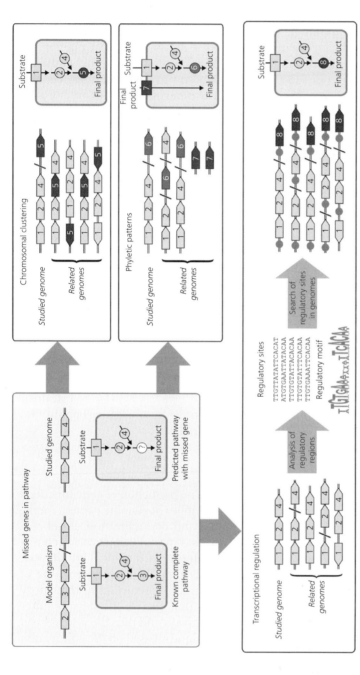

Figure 1 Overview of context-based methods in comparative genomics to identify missing genes in a pathway. This scheme describes the situation when a full pathway (1 to 4) was described in a model organism but in another organism the gene required for the last reaction (4) of the pathway was not found. In all techniques, the gap-filling approach is based on the comparison of the studied genomes with closely related genomes. In the *"Chromosomal clustering"* approach, an alternative form of the gene (5) for the final reaction is predicted by co-localization of genes in multiple genomes. In the *"Phyletic patterns"* approach, two possible scenarios are shown: prediction of an alternative form of the gene (6) for the final reaction by co-occurrence in multiple genomes, and prediction of a transporter (7) for the final product. The transporter was predicted by reciprocal presence of gene (7) with the pathway encoding genes. In the *"Transcriptional regulation"* approach, promoter regions of the genes are analyzed for the presence of conserved sequence motifs. Being identified, these motifs are used for prediction of candidate regulatory sites in multiple genomes. An alternative form of the gene (8) for the final reaction is predicted by the presence of candidate sites in promoter regions of the gene (8) in multiple genomes. Genomes are represented by double lines and individual genes are illustrated as pentagons according to the gene direction. Slashes separated distinct gene loci on the genome. Reactions are shown by arrows, metabolic reactions are marked by circles, while transport reactions are marked by rectangles. Previously known and novel predicted genes are shown by different colors.

reconstruct the metabolic and regulatory carbohydrate degradation networks of *Bacteroides thetaiotaomicron*[27], *Clostridium difficile*[28], *Shewanella* spp.[29,30], and *Thermotoga maritima*[31,32]. Comparative genomics approaches were also used for the reconstruction of numerous metabolic pathways, such as utilization of arabinose[33], N-acetylgalactosamine[34], and rhamnose[35], biosynthesis of methionine[36,37] and NAD[33,38], and metal homeostasis[39,40], and the prediction of a novel pathway for menaquinone biosynthesis[41]. In another application, the genomes of 256 common human gut bacteria were assessed for the presence of biosynthesis pathways for eight B-vitamins[42]. Each of the eight vitamins was produced by 40–65% of the analyzed human gut microbiota genomes. The distribution of synthetized B-vitamins suggests that human gut bacteria actively exchange B-vitamins among each other, thereby enabling the survival of organisms that cannot synthesize some of these essential cofactors. A systematic analysis of respiratory reductases of electron acceptors in 254 human gut microbes revealed oxygen as the most widespread electron acceptor as aerobic and microaerobic reductases were found in 140 genomes[43]. Among the electron acceptors for anaerobic respiration, fumarate was the most widespread as fumarate reductases were found in 93 genomes. Other widespread electron acceptors included nitrite (57 genomes), nitrate (46 genomes), molecular hydrogen (29 genomes), and dimethyl sulfoxide (23 genomes). The results also suggested that gut microbes may co-operate on the level of respiration as, for example, *Lactobacillus* spp. can only reduce nitrate to nitrite, which then is further reduced to ammonia by the other bacteria, such as *Bacteroides* spp. and *Parabacteroides* spp[43]. These examples show that refined, targeted annotation efforts are required to elucidate the complex interaction map between microbes commonly found in the human gut as well as to understand their co-evolution.

In summary, comparative genomics approaches yield in the reconstruction of pathways, the characterization of pathway gaps and the improvement of genome annotations. Such applications are of particular value for metabolic modeling approaches depending on metagenomic data or sequenced genomes (see below), which require high-quality genome annotations.

7.3 Network-based modeling

Moving beyond genomic analysis of the human gut microbiome, the next step is the construction of well-defined, predictive computational models for the gut microbial ecosystem and its host. Rather than merely categorizing the components of the gut ecosystem and describing measured phenotypes, such models can lead to novel, non-intuitive hypotheses. Common network-based modeling approaches used for the contextualization of metagenomic data include topological network modeling and constraint-based modeling.

7.3.1 Topological network modeling

Topological networks can be directly constructed from shotgun metagenomics data. Typically, such networks summarize the enzymatic potential of an ecosystem with nodes denoting metabolites and edges representing the reactions connecting them[44]. Reaction and metabolite information may be taken from databases

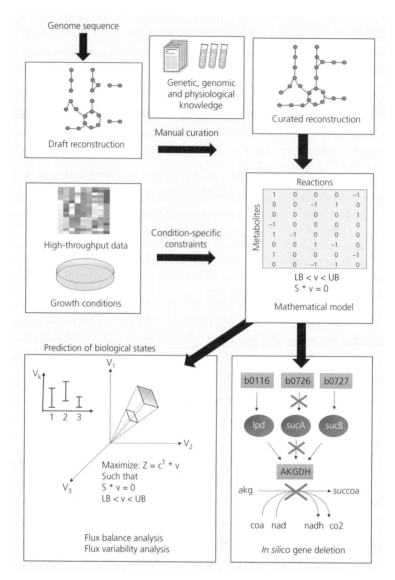

Figure 2 Overview of the constraint-based reconstruction and analysis (COBRA) approach. A target organism's metabolic reconstruction is generated from a draft reconstruction based on the genome sequence and is then manually validated and curated against the organism's bibliome and available experimental data. The curated reconstruction is converted into a mathematical model in the form of a stoichiometric matrix, with rows representing metabolites and columns representing reactions. The mathematical model is then converted into a predictive simulation platform by implementing condition-specific constraints, e.g., high-throughput data or growth conditions, such as nutrient availability. A variety of methods for biological predictions exist. For instance, flux balance analysis predicts the flux through an objective function Z, e.g., biomass production. The simulation yields the feasible solution space under the given constraints (implemented as lower and upper bounds on fluxes through reactions). Flux variability analysis predicts alternate flux distributions by computing the allowed flux spans for all reactions in the model. *In silico* gene deletion studies simulate gene knock-out phenotypes. If the deleted gene encodes an enzyme for which no isoenzymes exist,

such as KEGG[45] or MetaCyc[46]. Topological networks often use the "supra-organism" approach rather than separating enzyme content by species. Networks with this structure treat multi-species communities as a single organism, ignoring species-species boundaries[44]. The main advantage of supra-organismal networks is that they can be rapidly constructed from metagenomic data without the need for intensive manual curation[47]. Several studies have applied topological metabolic modeling to gut microbiome research[48–53] (reviewed in detail in[47,54,55]). Another modeling framework, called Bayesian inference of metabolic networks (BiomeNet), uses a Bayesian model to infer the prevalence of metabolic subnetworks in microbial communities[56]. Here, a metabolic network is constructed from metagenomic data and divided into a set of metabosystems, which themselves are made up of connected subnetworks. Metabosystems were found to discriminate the gut microbiomes of healthy individuals and IBD patients. The latter had a higher prevalence in metabosystems associated with host glycan degradation and ascorbate metabolism, suggesting that the IBD-associated microbiota exploits host-derived glycans and interferes with host antioxidant uptake[56].

7.3.2 Constraint-based modeling

Constraint-Based Modeling and Analysis (COBRA) aims to predict the behavior of a species by simulating fluxes through a network describing the target organism. The most common method is flux balance analysis, in which fluxes through the network are computed while optimizing a defined objective (e.g., biomass production)[57]. Limitations on fluxes through reactions represent constraints that living systems are forced to obey[58]. At the heart of the COBRA approach are accurate, well-structured metabolic reconstructions based on the target organisms' genome sequences. Such genome-scale reconstructions (GENREs) are constructed in a bottom-up manner and describe the target organism's metabolism[58]. The mathematical form of a GENRE is the so-called stoichiometric matrix (S) of size $m \times n$ where the rows represent compounds (m) and the columns represent reactions (n)[57] (Figure 2). A GENRE is unique for the target organism, but one can derive many different condition-specific models from a single reconstruction by the definition of condition-specific constraints (e.g., nutrient availability and gene knock-out).

More than 130 manually curated GENREs have been published to date[59–62]. The currently available manually curated human gut microbial GENREs account for the main bacterial and archaeal phyla found in the human intestine as well as commensal, probiotic, and pathogenic species ([60], see also Table 2). Further development in the field is slowed by the intensive curation effort needed to build a GENRE, as the reconstruction of a well-studied bacterial species can take

Figure 2 (*Continued*) the gene deletion results in a simulated enzyme defect. LB = lower bounds, UB = upper bounds, S = representation of the stoichiometric matrix, Z = objective function (e.g., biomass), b0116/b0726/b0727 = example for the gene annotation of an enzyme (in this case, 2-oxoglutarate dehydrogenase, encoded in *Escherichia coli* str K-12 substr. MG1655), lpd/sucA/sucB = example for multiple complexes that compose an enzyme (in this case, 2-oxoglutarate dehydrogenase, EC number 1.2.4.2), AKGDH = 2-oxoglutarate dehydrogenase, akg = 2-oxoglutarate, co2 = carbon dioxide, coa = Coenzyme A, nad = nicotinamide adenine dinucleotide, nadh = reduced nicotinamide adenine dinucleotide, succoa = succinyl-CoA.

up to six months[18]. Several resources have been developed that can generate analysis-ready draft reconstructions from genome sequences[11–16] (Table 1). These tools, together with the increasing number of high-quality genome sequences available in repositories, such as Integrated Microbial Genomes[63], will speed up the metabolic reconstruction of other health-relevant human microbes. However, these draft reconstructions require further refinement based on biochemical and phenotypic data, to ensure that the *in silico* model represents as accurately as possible to metabolic processes and phenotypic capabilities of the reconstructed organism[64,65].

7.3.3 Metabolic reconstructions of human metabolism

A prerequisite for modeling human-microbe interactions is a high-quality network of human metabolism. The first genome-scale reconstruction and model of human metabolism, Recon1, was published in 2007[66]. It was based on Build 35 of the human genome and integrated information from more than 1,500 primary and review publications. Recon1 captured the functions of 2,004 proteins, 2,766 metabolites, and 3,311 metabolic and transport reactions. Recon1 is a generic model that represents the union of the metabolism of cells throughout the human body. A variety of subsequent studies successfully applied Recon1 to the prediction of human metabolic states in health and disease[67–69]. Recon2[70] was generated by expanding and refining Recon1 in a joint effort of the systems biology research community. Information was added from other manually curated metabolic databases, namely the Edinburgh Human Metabolic Network (EHMN)[71], a liver-specific metabolic model[72], an Acylcarnitine/Fatty Acid Oxidation module[73], and a human small intestinal enterocyte reconstruction[74]. Further, various metabolite and gene-specific information was incorporated into Recon2, permitting the mapping of omics data and cross-reference with other resources. The gene information was updated to Build 37. Recon2 has twice as many reactions as Recon1, and the majority of the existing pathways were expanded through the combination of the different resources of human metabolism, thus closing many of the knowledge-gaps present in Recon1 as well as adding further metabolic information[70]. More than 100 studies have been published using Recon1 or Recon2[69].

Other databases of human metabolism include Reactome[75] and the Human Metabolic Reaction database (HMR)[76,77]. While Reactome[75] is a highly curated database covering many cellular processes including metabolism, it focuses on signaling pathways. HMR[77] is an integrated database of Recon1[66], EHMN[71], hepatocyte[72], and an adipocyte-specific metabolic model[78].

7.3.4 Constraint-based modeling of host-microbe interactions

Several constraint-based efforts have modeled microbe-microbe interactions of biotechnologically and medically important human bacteria (Table 2). The availability of high-quality reconstructions of human metabolism and of other host organisms (e.g., mouse[79–81]), enables the computational modeling of host-microbe interactions[54]. In a first effort, Bordbar *et al.* constructed a host-pathogen model simulating the infection of the human alveolar macrophage with *Mycobacterium tuberculosis*[82]. The model was constructed by placing a *M. tuberculosis* reconstruction into an *in silico* compartment simulating the intracellular phagosome. This approach enabled the simulation of the pathogen's lifestyle inside a host cell, resulting in the accurate prediction of gene deletion phenotypes[82]. In another

Table 2 Applications of constraint-based modeling to the simulation of multi-species interactions.

Community description	Modeling method	Main results	Ref.
A sulfate reducer and a methanogen found in the intestines of ruminants and other ecosystems	Flux balance analysis[57] (see Figure 2)	The syntrophic growth of the two species was modeled. Mutualistic cross-feeding depended on hydrogen transfer from the sulfate reducer *Desulfovibrio vulgaris* to the methanogen *Methanococcus maripaludis*.	139
Artificial community of seven microbes	Flux balance analysis[57]	Pairwise behavior (mutualism, neutralism, competition) was predicted in every possible combination while subjecting the pairs to random media and genetic perturbations.	140
Natural consortium of thermophilic, phototrophic organisms	Elementary mode analysis[141], which decomposes a metabolic network into a set of pathways each obeying physiologically reasonable constraints.	The syntrophic mass and energy flow between three groups, oxic phototrophs, filamentous anoxic phototrophs, and sulfate-reducing bacteria, was modeled for a daylight and a nighttime scenario.	142
Collection of *Escherichia coli* mutant strains auxotrophic for essential biomass precursors	Minimization of metabolic adjustment[143], which identifies a flux solution in a mutant strain that most closely resembles the wild-type strain.	Certain auxotrophic pairs were able to complement each other's growth by exchanging essential metabolites (e.g., amino acids). These findings were confirmed *in vitro*.	144
Fe(III)-reducing *Geobacter* and *Rhodoferax* species found in natural anoxic soil communities	Dynamic flux balance analysis[145], which enables the simulation of temporal dynamics.	The competition between two soil Fe(III) reducers, which has implications for uranium bioremediation, was simulated. A long-term bioremediation strategy was proposed based on the model.	146,147
Natural soil community consisting of a sulfate reducer, a methanogen, and a cellobiose degrader	OptCom (a multi-objective optimization problem representing total community growth and individual growth of each microbe)	The metabolite cross-feeding between the three species grown in association was simulated and found to agree with experimental measurements.	148
Collection of 118 species from various environments	Flux variability analysis[149] (see Figure 2)	The microbes were joined pairwise in all combinations and cooperative or competitive behavior was predicted. A network of give-take relationship loops was constructed.	150
Two *E. coli* strains adapted to laboratory evolution	Community flux balance analysis, which computes optimal community growth at balanced growth of the included species	The cross-feeding of two *E. coli* strains, which were adapted to certain carbon sources after long-term laboratory evolution, was predicted.	151

(Continued)

Table 2 (*Continued*)

Community description	Modeling method	Main results	Ref.
Two Fe(III)-reducing *Geobacter* species	Contextualization with meta-omic data, flux balance analysis[57]	Syntrophic growth of the two species through direct interspecies electron transfer was modeled. Their transcriptional adaptation to each other's presence was shown.	152
Methionine-auxotrophic *E. coli* mutant and a methionine-producing *Salmonella enterica* mutant	Grid-based modeling framework incorporating temporal dynamics and spatial distribution	The modeling approach captured the colony shape of *E. coli* and experimentally determined ratios between the two species. In a follow-up study, cooperative and competitive interactions were predicted while subjecting the two bacteria to systematic genetic perturbations.	153
Human gut commensals *Bacteroides thetaiotaomicron*, *Eubacterium rectale* and *Methanobreviibacter smithii*	Contextualization with transcriptomic data from gnotobiotic mice[100], identification of reporter metabolites and transcriptional regulation	The behavior of the three species in the mouse intestine was modeled. *B. thetaiotaomicron* produced acetate that was converted to butyrate by *E. rectale*. CO_2 and hydrogen produced by *E. rectale* were converted to methane by *M. smithii*.	101
Human gut commensals *Faecalibacterium prausnitzii* and *Bifidobacterium adolescentis*	Flux balance analysis[57], OptCom[148]	The cross-feeding between the two species on glucose was predicted. The growth-stimulating effect of acetate produced by *B. adolescentis* on the butyrate producer *F. prausnitzii* was shown.	154
Community of 11 human gut microbes including commensals, probiotics and pathogens	Flux balance analysis[57], flux variability analysis[149]	Pairwise interactions such as commensalism, mutualism, and competition were predicted in all combinations under varying environmental constraints. The probiotic *Lactobacillus plantarum* displayed mutualistic interactions with several species including *E. coli* under anoxic conditions, which were abolished in the presence of oxygen.	155
~1,500 microbes from various natural environments	Flux balance analysis[57]	Species co-occurrences were identified in metagenomic samples from different environments (e.g., soil, water, human gut). Metabolic exchanges in the co-occurring pairs (e.g., amino acids, sugars) were predicted.	156
Methanogenic community consisting of the bacterium *Syntrophobacter fumaroxidans* and the archaeon *Methanospirillum hungatei*	Thermodynamics-based metabolic flux analysis (TMFA)[157,158], which extends flux balance analysis by thermodynamic constraints	The syntrophic growth of the two species on propionate was modeled. *M. hungatei* converted formate and H_2 produced by *S. fumaroxidans* to methane. Conditions in which the products of *S. fumaroxidans* are fully consumed by *M. hungatei* were proposed.	159

model representing the symbiotic host-microbe relationship between mouse[80] and the commensal gut microbe, *B. thetaiotaomicron*[65], the metabolic dependencies have been systematically analyzed. An *in silico* global gene-deletion analysis revealed that the presence of *B. thetaiotaomicron* could rescue otherwise lethal mouse knockout phenotypes, including three known inborn errors of metabolism[65]. In more extensive follow-up work, the global human reconstruction Recon2[70] was joined with 11 manually curated gut microbe reconstructions[83] to predict microbially induced health-relevant metabolites. The prediction included many of the known co-metabolites (e.g., phenylacetylglutamine, 4-hydroxyphenylacetate) as well as amino acids, hormone precursors, hormones, and vitamins. A simplified model of microbial dysbiosis predicted a significantly lower potential to positively influence host metabolism, e.g., through the production of detrimental metabolites, such as phenolic compounds, sulfide, and ammonia. Moreover, cadaverine, which is elevated in the fecal extracts of ulcerative colitis (UC) patients[84], was predicted to be synthesized only by gamma-Proteobacteria[83]. Higher abundances of Proteobacteria are commonly found in inflammatory bowel disease microbiomes[85]. On a larger scale, such computational analyses could identify valuable links between potentially detrimental metabolites and microbial taxa associated with dysbiosis.

7.4 Other computational modeling approaches

Computational model approaches that are not network-based are typically based on solving a set of differential equations. For instance a computational model of the dynamics between antibiotic-tolerant and -sensitive gut microbes during antibiotic therapy was constructed by defining a set of mathematical equations[86]. A singular value decomposition of metagenomic samples was then performed that resulted in the classification of the samples' microbiota according to antibiotic resistance[86]. Another approach, called iDynoMiCS (individual-based Dynamics of Microbial Communities Simulator), uses mathematical modeling to simulate the dynamics of biofilm formation[87]. The model includes spatiotemporal parameters and can predict outcomes such as biofilm thickness and structure[87]. An interesting ongoing effort is the tool eGut, which is based on iDynoMiCS[88]. This computational framework will also include representations of the host mucosa and lumen and allow simulation of the influence of mucosal biofilm on host immunity and inflammation[88]. Common approaches briefly described below include ODE (ordinary differential equation) modeling and kinetic modeling.

7.4.1 Ordinary differential equation (ODE) models

ODE models may be based on Lotka-Volterra equations, which describe the dynamics of a predator-prey relationship. The generalized Lotka-Volterra form enables one to model general species-species relationships in an ecosystem[89]. For example, an ODE model of the primary colonization in the murine gut has been constructed[90]. Fecal samples from germfree mice inoculated with an adult microbiota were analyzed at time points from zero to 21 days. A mathematical model accounting for time-dependent dynamics revealed that microbe-microbe interactions during colonization were mostly parasitic among Bacteroidetes and competitive among Firmicutes[90]. Another recent application of the generalized Lotka-Volterra equation, Learning Interactions from MIcrobial Time Series (LIMITS), modeled the dynamics of the gut microbial ecosystem[91]. The time-dependent dynamics of the gut microbiota were first modeled with a training dataset and subsequently with the species abundance data from two individuals' microbiomes. The analysis revealed a network of mutualistic and competitive interactions between 14 abundant species. *Bacteroides fragilis* and *Bacteroides stercosis* had the highest number of outgoing interactions and could thus be considered keystone species[91]. Stein et al. constructed an ecological model based on the Lotka-Volterra equation and simulated the dynamics of *Clostridium difficile* infection in clindamycin-treated mice[92]. Model parameters were inferred from experimental data collected from mice infected with *C. difficile*. Positive and negative genus-genus interactions were predicted. For example, *Akkermansia* and *Blautia* had a negative effect on the growth of *C. difficile*, while the *Enterococccus* genus interacted positively with the latter. A subnetwork of taxa with a combined protective effect against *C. difficile* was proposed[92]. Computational modeling can thus provide valuable insight into the dynamics of antibiotic-induced dysbiosis and perturbations by opportunistic pathogens. Following up on this work, the Lotka-Volterra model was combined with time-resolved gut microbiota data from patients with *C. difficile* infection and *C. difficile* carriers without infection[93]. The mathematical model supplemented with experimental data revealed that certain gut inhabitants were associated with *C. difficile* infection resistance. The bacterium *Clostridium scindens* mediated *C. difficile* resistance via the production of secondary bile acids, which inhibit the growth of the latter[93]. This mathematical modeling approach thus elucidated a natural mechanism protecting against antibiotic-induced dysbiosis. Recently, the time-dependent metagenomic data of Stein *et al.*[92] was also integrated into a dynamic Boolean network model and subnetworks of clindamycin treatment and of *C. difficile* infection were constructed[94]. A topological genome-scale metabolic network was constructed at the genus level and the competitive and cooperative potential of the genera was determined via seed set analysis[52,95]. The analysis combined with subsequent experimental validation revealed that the genus *Barnesiella* can slow down the growth of *C. difficile*[94].

7.4.2 Kinetic modeling

Kinetic models of metabolism represent enzymatic reactions occurring in an organism as a set of differential equations based on experimentally determined parameters. One such mathematical model of cross-feeding in the gut described the interaction between lactate-producing bacteria and two species that can convert lactate to butyrate, *Eubacterium hallii* and *Anaerostipes coli* SS2/1[96]. A more comprehensive kinetic model has recently been published that accounts for the major metabolic functions in the gut ecosystem[97]. Ten functional groups accounting for group-specific pathways, substrate preferences, and preferred pH range were used to represent the major representatives in the human microbiota, e.g. propionate and butyrate producers. This setup allowed the prediction of metabolic cross-feeding in the presence and absence of certain functional groups. For example, omitting the acetate-propionate-succinate producer group representing *Bacteroides* ssp. caused non-butyrate producing starch degraders representing, e.g., *Ruminococcus bromii* to dominate the remaining groups[97]. Another kinetic modeling approach investigated the relationship between pH value and microbial diversity[98]. The model defined "specialists" and "generalists" that varied in their tolerance to fluctuations in pH value. The analysis suggested that pH fluctuations due to microbial activity may promote microbial diversity[98]. Compared with constraint-based genome-scale models, kinetic models have the advantage of incorporating experimentally validated kinetic variables, such as pH value. However, unlike genome-scale models, kinetic models do not include the genomic information of target organisms. A promising application would be the integration of kinetic models into genome-scale gut microbe models. Methods for the implementation of predicted or experimentally validated kinetic parameters into constraint-based models are available[99].

7.5 Conclusion

A variety of computational methods exist that can aid the interpretation of *in vitro* and *in vivo* data as well as generate novel biological hypotheses. These methods can valuably complement *in vitro* and *in vivo* approaches. For instance, constraint-based models have successfully predicted microbe-microbe interactions (Table 2). Expanding these community models to a size that captures the taxonomic or functional gut microbiota diversity will enable us systematically to explore microbial interactions *in silico*. The potential of *in vitro* cultivation approaches to elucidate novel metabolic microbe-microbe interactions is hampered by the complexity of the microbiome and the difficult cultivation requirements of most gut inhabitants. Computational models provide excellent complementary tools that may predict unexpected, non-intuitive cross-feeding between microbes. Such predictions could subsequently be validated *in vitro*. Species-resolved community models, such as constraint-based models, are particularly useful in this context as they can elucidate the individual microbial "players". Computational models have the advantage of predicting the outcomes of any number of hypothetical scenarios, thereby reducing time, risk, and cost of experiments. Simulating host-microbe interactions is particularly valuable in this context since it could be used to minimize the number of animal experiments. For instance, computational methods make possible the simulation of a germfree host colonized with specific microbes[83], which is impossible to perform experimentally for humans and labor- and cost-intensive for animal models.

Another useful application of computational modeling is the contextualization of high-throughput data. One method is the construction of topological networks from metagenomic data (e.g.,[49,53]). Another approach is overlaying constraint-based reconstructions with metagenomic, metatranscriptomic or metabolomic data, thus generating condition-specific models (Figure 2). The integration of transcriptomic data from the intestines of gnotobiotic mice[100] into a constraint-based framework has been performed for the human commensal bacteria *B. thetaiotaomicron* and *Eubacterium rectale*[101] and could be readily be applied to a more extensive gut microbial community. For instance, constraint-based models could be overlaid with high-throughput metagenomic, metatranscriptomic, and metabolomic data derived from patients suffering from inflammatory bowel diseases and other conditions associated with an imbalanced microbiota. A number of studies have derived human disease- and cell-type-specific models from a global human reconstruction[66,70,102–106]. A human cell model could be tailored condition-specific, e.g., to intestinal inflammation, and combined with a disease-specific gut microbiota community model. This host-microbiota model could serve the simulation of altered metabolic states due to microbial dysbiosis, thus furthering our insight into the complex ecosystem residing in the gut and its interactions with the human host.

Moreover, dietary or pharmacological interventions that can restore a dysbiotic to a healthy microbiota could be systematically predicted. For instance, drugs that can selectively suppress gut microbial residents associated with dysbiosis, e.g., *E. coli* in inflammatory bowel disease patients[85] and *C. difficile* in relapsing infections following antibiotic treatment[107] would be valuable. Constraint-based reconstructions are useful tools for the prediction of antimicrobial drug targets[108]. For instance, a flux balance analysis-based framework identified existing FDA-approved drugs that are effective against the parasite *Leishmania manor*[109]. The available genome-scale models for *E. coli*[110] and *C. difficile*[62] could be applied for the prediction of metabolic drug targets that could be subsequently validated *in vitro*. Another possibility is the simulations of probiotic interventions that suppress undesirable gut microbiota inhabitants. Several constraint-based reconstructions of probiotics are available[60]. Finally, the simulation of different diets and their gut microbiota-mediated effects on host metabolism has already been performed for a small-scale gut microbe community[83] and could readily be extended to a larger, more representative scale. In summary, the computational modeling approaches discussed in this chapter will be valuable tools for studying microbial dysbiosis and its impact on host metabolism.

TAKE-HOME MESSAGE

- Systems biology is a valuable tool complementing and contextualizing experimental approaches. Established computational methods include network models (topological network modeling, constraint-based modeling), kinetic models, and ordinary differential equation (ODE) models.

- Genome-scale metabolic reconstructions provide predictive, mechanistic modeling frameworks. Their predictive potential depends on accurate, species-specific genome annotations. Comparative genomics approaches are useful tools for the improvement of genome annotations and the discovery of novel pathways.

- Constraint-based models have successfully predicted the metabolic interactions in microbial communities and between a host and its commensal microbiota in a mechanism-based manner. Expanding these models to a scale representative of the human gut microbiota will yield novel insight on the impact of gut microbial dysbiosis on host health and disease.

Acknowledgments

The authors are grateful to Mrs S. Magnusdottir for assistance in the preparation of the tables. This work has been funded by an ATTRACT programme grant to I.T. (FNR/A12/01) from the Luxembourg National Research Fund (FNR). DAR was supported by the National Research Fund (#6847110), Luxembourg, and cofunded under the Marie Curie Actions of the European Commission (FP7-COFUND).

References

1 Human Microbiome Project C. Structure, function and diversity of the healthy human microbiome. *Nature* 2012; **486**: 207–14.

2 Qin J, Li R, Raes J, Arumugam M, Burgdorf KS, Manichanh C *et al*. A human gut microbial gene catalogue established by metagenomic sequencing. *Nature* 2010; **464**: 59–65.

3 Gill SR, Pop M, Deboy RT, Eckburg PB, Turnbaugh PJ, Samuel BS *et al*. Metagenomic analysis of the human distal gut microbiome. *Science* 2006; **312**: 1355–9.

4 Maurice CF, Haiser HJ, Turnbaugh PJ. Xenobiotics shape the physiology and gene expression of the active human gut microbiome. *Cell* 2013; **152**: 39–50.

5 Verberkmoes NC, Russell AL, Shah M, Godzik A, Rosenquist M, Halfvarson J *et al*. Shotgun metaproteomics of the human distal gut microbiota. *Isme J* 2009; **3**: 179–89.

6 Wikoff WR, Anfora AT, Liu J, Schultz PG, Lesley SA, Peters EC *et al*. Metabolomics analysis reveals large effects of gut microflora on mammalian blood metabolites. *Proc Natl Acad Sci USA* 2009; **106**: 3698–703.

7 Markowitz VM, Chen IM, Palaniappan K, Chu K, Szeto E, Grechkin Y *et al*. IMG: the Integrated Microbial Genomes database and comparative analysis system. *Nucleic Acids Research* 2012; **40**: D115–22.

8 Integrative HMPRNC. The Integrative Human Microbiome Project: dynamic analysis of microbiome-host omics profiles during periods of human health and disease. *Cell Host & Microbe* 2014; **16**: 276–89.

9 Hatzimanikatis V, Li C, Ionita JA, Henry CS, Jankowski MD, Broadbelt LJ. Exploring the diversity of complex metabolic networks. *Bioinformatics* 2005; **21**: 1603–9.

10 Moriya Y, Shigemizu D, Hattori M, Tokimatsu T, Kotera M, Goto S *et al*. PathPred: an enzyme-catalyzed metabolic pathway prediction server. *Nucleic Acids Research* 2010; **38**: W138–43.

11 Henry CS, DeJongh M, Best AA, Frybarger PM, Linsay B, Stevens RL. High-throughput generation, optimization and analysis of genome-scale metabolic models. *Nat Biotechnol* 2010; **28**: 977–82.

12 Agren R, Liu L, Shoaie S, Vongsangnak W, Nookaew I, Nielsen J. The RAVEN toolbox and its use for generating a genome-scale metabolic model for Penicillium chrysogenum. *PLoS Comput Biol* 2013; **9**: e1002980.

13 Swainston N, Smallbone K, Mendes P, Kell D, Paton N. The SuBliMinaL Toolbox: automating steps in the reconstruction of metabolic networks. *Journal of Integrative Bioinformatics* 2011; **8**: 186.

14 Feng X, Xu Y, Chen Y, Tang YJ. MicrobesFlux: a web platform for drafting metabolic models from the KEGG database. *BMC Syst Biol* 2012; **6**: 94.

15 Boele J, Olivier BG, Teusink B. FAME, the Flux Analysis and Modeling Environment. *BMC Syst Biol* 2012; **6**: 8.

16 Karp PD, Paley SM, Krummenacker M, Latendresse M, Dale JM, Lee TJ *et al*. Pathway Tools version 13.0: integrated software for pathway/genome informatics and systems biology. *Briefings in bBoinformatics* 2010; **11**: 40–79.

17 Bruggeman FJ, Westerhoff HV. The nature of systems biology. *Trends in Microbiology* 2007; **15**: 45–50.

18 Thiele I, Palsson BO. A protocol for generating a high-quality genome-scale metabolic reconstruction. *Nat Protoc* 2010; **5**: 93–121.

19 Segata N, Boernigen D, Tickle TL, Morgan XC, Garrett WS, Huttenhower C. Computational meta'omics for microbial community studies. *Molecular Systems Biology* 2013; **9**: 666.

20 Franzosa EA, Hsu T, Sirota-Madi A, Shafquat A, Abu-Ali G, Morgan XC *et al*. Sequencing and beyond: integrating molecular 'omics' for microbial community profiling. *Nature Reviews Microbiology* 2015; **13**: 360–72.

21 Arumugam M, Raes J, Pelletier E, Le Paslier D, Yamada T, Mende DR *et al*. Enterotypes of the human gut microbiome. *Nature* 2011; **473**: 174–80.

22 Le Chatelier E, Nielsen T, Qin J, Prifti E, Hildebrand F, Falony G *et al*. Richness of human gut microbiome correlates with metabolic markers. *Nature* 2013; **500**: 541–6.

23 Karlsson FH, Tremaroli V, Nookaew I, Bergstrom G, Behre CJ, Fagerberg B *et al*. Gut metagenome in European women with normal, impaired and diabetic glucose control. *Nature* 2013; **498**: 99–103.

24 Qin J, Li Y, Cai Z, Li S, Zhu J, Zhang F *et al*. A metagenome-wide association study of gut microbiota in type 2 diabetes. *Nature* 2012; **490**: 55–60.

25 Greenblum S, Carr R, Borenstein E. Extensive strain-level copy-number variation across human gut microbiome species. *Cell* 2015; **160**: 583–94.

26 Osterman A, Overbeek R. Missing genes in metabolic pathways: a comparative genomics approach. *Current Opinion in Chemical Biology* 2003; **7**: 238–51.

27 Ravcheev DA, Godzik A, Osterman AL, Rodionov DA. Polysaccharides utilization in human gut bacterium Bacteroides thetaiotaomicron: comparative genomics reconstruction of metabolic and regulatory networks. *BMC Genomics* 2013; **14**: 873.

28 Antunes A, Camiade E, Monot M, Courtois E, Barbut F, Sernova NV *et al.* Global transcriptional control by glucose and carbon regulator CcpA in *Clostridium difficile*. *Nucleic Acids Res* 2012; **40**: 10701–18.

29 Rodionov DA, Yang C, Li X, Rodionova IA, Wang Y, Obraztsova A, *et al.* Genomic encyclopedia of sugar utilization pathways in the *Shewanella* genus. *BMC Genomics* 2010; **11**: 494.

30 Leyn SA, Li X, Zheng Q, Novichkov PS, Reed S, Romine MF *et al.* Control of proteobacterial central carbon metabolism by the HexR transcriptional regulator. A case study in *Shewanella oneidensis*. *J Biol Chem* 2011; **286**: 35782–94.

31 Rodionov DA, Rodionova IA, Li X, Ravcheev DA, Tarasova Y, Portnoy VA *et al.* Transcriptional regulation of the carbohydrate utilization network in *Thermotoga maritima*. *Front Microbiol* 2013; **4**: 244.

32 Rodionova IA, Yang C, Li X, Kurnasov OV, Best AA, Osterman AL *et al.* Diversity and versatility of the Thermotoga maritima sugar kinome. *J Bacteriol* 2012; **194**: 5552–63.

33 Zhang L, Leyn SA, Gu Y, Jiang W, Rodionov DA, Yang C. Ribulokinase and transcriptional regulation of arabinose metabolism in *Clostridium acetobutylicum*. *J Bacteriol* 2012; **194**: 1055–64.

34 Leyn SA, Gao F, Yang C, Rodionov DA. N-Acetylgalactosamine utilization pathway and regulon in proteobacteria: Genomic reconstruction and experimental characterization in *Shewanella*. *J Biol Chem* 2012; **287**: 28047–56.

35 Rodionova IA, Li X, Thiel V, Stolyar S, Stanton K, Fredrickson JK *et al.* Comparative genomics and functional analysis of rhamnose catabolic pathways and regulons in bacteria. *Front Microbiol* 2013; **4**: 407.

36 Leyn SA, Suvorova IA, Kholina TD, Sherstneva SS, Novichkov PS, Gelfand MS *et al.* Comparative genomics of transcriptional regulation of methionine metabolism in proteobacteria. *PLoS One* 2014; **9**: e113714.

37 Novichkov PS, Li X, Kuehl JV, Deutschbauer AM, Arkin AP, Price MN *et al.* Control of methionine metabolism by the SahR transcriptional regulator in Proteobacteria. *Environ Microbiol* 2014; **16**: 1–8.

38 Rodionov DA, Li X, Rodionova IA, Yang C, Sorci L, Dervyn E *et al.* Transcriptional regulation of NAD metabolism in bacteria: genomic reconstruction of NiaR (YrxA) regulon. *Nucleic Acids Res* 2008; **36**: 2032–46.

39 Kazakov AE, Rajeev L, Luning EG, Zane GM, Siddartha K, Rodionov DA *et al.* New family of tungstate-responsive transcriptional regulators in sulfate-reducing bacteria. *J Bacteriol* 2013; **195**: 4466–75.

40 Leyn SA, Rodionov DA. Comparative genomics of DtxR-family regulons for metal homeostasis in Archaea. *J Bacteriol* 2015; **197**: 451–458;: e113714.

41 Hiratsuka T, Furihata K, Ishikawa J, Yamashita H, Itoh N, Seto H, *et al.* An alternative menaquinone biosynthetic pathway operating in microorganisms. *Science* 2008; **321**: 1670–3.

42 Magnusdottir S, Ravcheev D, de Crecy-Lagard V, Thiele I. Systematic genome assessment of B-vitamin biosynthesis suggests co-operation among gut microbes. *Front Genetics* 2015; **6**: 148.

43 Ravcheev DA, Thiele I. Systematic genomic analysis reveals the complementary aerobic and anaerobic respiration capacities of the human gut microbiota. *Front Nicrobiology* 2014; **5**: 674.

44 Borenstein E. Computational systems biology and in silico modeling of the human microbiome. *Briefings in Bioinformatics* 2012; **13**: 769–80.

45 Kanehisa M, Araki M, Goto S, Hattori M, Hirakawa M, Itoh M *et al.* KEGG for linking genomes to life and the environment. *Nucleic Acids Research* 2008; **36**: D480–4.

46 Caspi R, Altman T, Dale JM, Dreher K, Fulcher CA, Gilham F *et al.* The MetaCyc database of metabolic pathways and enzymes and the BioCyc collection of pathway/genome databases. *Nucleic Acids Research* 2010; **38**: D473–9.

47 Greenblum S, Chiu HC, Levy R, Carr R, Borenstein E. Towards a predictive systems-level model of the human microbiome: progress, challenges, and opportunities. *Curr Opin Biotechnol* 2013; **24**: 810–20.

48 Faust K, Sathirapongsasuti JF, Izard J, Segata N, Gevers D, Raes J *et al.* Microbial co-occurrence relationships in the human microbiome. *PLoS Comput Biol* 2012; **8**: e1002606.

49 Jacobsen UP, Nielsen HB, Hildebrand F, Raes J, Sicheritz-Ponten T, Kouskoumvekaki I *et al.* The chemical interactome space between the human host and the genetically defined gut metabotypes. *Isme J* 2013; **7**: 730–42.

50 Abubucker S, Segata N, Goll J, Schubert AM, Izard J, Cantarel BL *et al.* Metabolic reconstruction for metagenomic data and its application to the human microbiome. *PLoS Comput Biol* 2012; **8**: e1002358.

51 Sridharan GV, Choi K, Klemashevich C, Wu C, Prabakaran D, Pan LB *et al.* Prediction and quantification of bioactive microbiota metabolites in the mouse gut. *Nat Commun* 2014; **5**: 5492.

52 Levy R, Borenstein E. Metabolic modeling of species interaction in the human microbiome elucidates community-level assembly rules. *Proc Natl Acad Sci USA* 2013; **110**: 12804–9.

53 Greenblum S, Turnbaugh PJ, Borenstein E. Metagenomic systems biology of the human gut microbiome reveals topological shifts associated with obesity and inflammatory bowel disease. *Proc Natl Acad Sci USA* 2012; **109**: 594–9.

54 Heinken A, Thiele I. Systems biology of host-microbe metabolomics. *Wiley Interdisciplinary Reviews Systems Biologya And Medicine* 2015; **7**: 195–219.

55 Manor O, Levy R, Borenstein E. Mapping the inner workings of the microbiome: genomic- and metagenomic-based study of metabolism and metabolic interactions in the human microbiome. *Cell Metabolism* 2014; **20**: 742–752.

56 Shafiei M, Dunn KA, Chipman H, Gu H, Bielawski JP. BiomeNet: a Bayesian model for inference of metabolic divergence among microbial communities. *PLoS Comput Biol* 2014; **10**: e1003918.

57 Orth JD, Thiele I, Palsson BO. What is flux balance analysis? *Nat Biotechnol* 2010; **28**: 245–8.

58 Palsson B. *Systems Biology: Properties of Reconstructed Networks.* Cambridge: Cambridge University Press, 2006.

59 Monk J, Nogales J, Palsson BO. Optimizing genome-scale network reconstructions. *Nat Biotechnol* 2014; **32**: 447–52.

60 Thiele I, Heinken A, Fleming RM. A systems biology approach to studying the role of microbes in human health. *Curr Opin Biotechnol* 2013; **24**: 4–12.

61 Monk JM, Charusanti P, Aziz RK, Lerman JA, Premyodhin N, Orth JD *et al.* Genome-scale metabolic reconstructions of multiple Escherichia coli strains highlight strain-specific adaptations to nutritional environments. *Proc Natl Acad Sci USA* 2013; **110**: 20338–43.

62 Larocque M, Chenard T, Najmanovich R. A curated C. difficile strain 630 metabolic network: prediction of essential targets and inhibitors. *BMC Syst Biol* 2014; **8**: 117.

63 Markowitz VM, Chen IM, Palaniappan K, Chu K, Szeto E, Pillay M *et al.* IMG 4 version of the integrated microbial genomes comparative analysis system. *Nucleic Acids Research* 2014; **42**: D560–7.

64 Heinken A, Khan MT, Paglia G, Rodionov DA, Harmsen HJ, Thiele I. A functional metabolic map of Faecalibacterium prausnitzii, a beneficial human gut microbe. *J Bacteriol* 2014; **196**: 3289–302.

65 Heinken A, Sahoo S, Fleming RM, Thiele I. Systems-level characterization of a host-microbe metabolic symbiosis in the mammalian gut. *Gut Microbes* 2013; **4**: 28–40.

66 Duarte NC, Becker SA, Jamshidi N, Thiele I, Mo ML, Vo TD *et al.* Global reconstruction of the human metabolic network based on genomic and bibliomic data. *Proc Natl Acad Sci USA* 2007; **104**: 1777–82.

67 Bordbar A, Palsson BO. Using the reconstructed genome-scale human metabolic network to study physiology and pathology. *Journal of Internal Medicine* 2012; **271**: 131–41.

68 Mardinoglu A, Nielsen J. Systems medicine and metabolic modelling. *Journal of Internal Medicine* 2012; **271**: 142–54.

69 Aurich MK, Thiele I. Computational modeling of human metabolism and its application to systems biomedicine. *Methods Mol Biol* 2016; **1386**: 253–281.

70 Thiele I, Swainston N, Fleming RM, Hoppe A, Sahoo S, Aurich MK *et al.* A community-driven global reconstruction of human metabolism. *Nat Biotechnol* 2013; **31**: 419–25.

71 Ma H, Sorokin A, Mazein A, Selkov A, Selkov E, Demin O *et al.* The Edinburgh human metabolic network reconstruction and its functional analysis. *Molecular Systems Biology* 2007; **3**: 135.

72 Gille C, Bolling C, Hoppe A, Bulik S, Hoffmann S, Hubner K *et al.* HepatoNet1: a comprehensive metabolic reconstruction of the human hepatocyte for the analysis of liver physiology. *Molecular Systems Biology* 2010; **6**: 411.

73 Sahoo S, Franzson L, Jonsson JJ, Thiele I. A compendium of inborn errors of metabolism mapped onto the human metabolic network. *Mol Biosyst* 2012; **8**: 2545–58.

74 Sahoo S, Thiele I. Predicting the impact of diet and enzymopathies on human small intestinal epithelial cells. *Human Molecular Genetics* 2013; **22**: 2705–22.

75 Croft D, Mundo AF, Haw R, Milacic M, Weiser J, Wu G *et al.* The Reactome pathway knowledgebase. *Nucleic Acids Research* 2014; **42**: D472–7.

76 Mardinoglu A, Nielsen J. New paradigms for metabolic modeling of human cells. *Curr Opin Biotechnol* 2015; **34C**: 91–7.

77 Agren R, Bordel S, Mardinoglu A, Pornputtapong N, Nookaew I, Nielsen J. Reconstruction of genome-scale active metabolic networks for 69 human cell types and 16 cancer types using INIT. *PLoS Comput Biol* 2012; **8**: e1002518.

78 Mardinoglu A, Agren R, Kampf C, Asplund A, Nookaew I, Jacobson P *et al.* Integration of clinical data with a genome-scale metabolic model of the human adipocyte. *Molecular Systems Biology* 2013; **9**: 649.

79 Selvarasu S, Karimi IA, Ghim GH, Lee DY. Genome-scale modeling and in silico analysis of mouse cell metabolic network. *Mol Biosyst* 2010; **6**: 152–61.

80 Sigurdsson MI, Jamshidi N, Steingrimsson E, Thiele I, Palsson BO. A detailed genome-wide reconstruction of mouse metabolism based on human Recon 1. *BMC Syst Biol* 2010; **4**: 140.

81 Quek LE, Nielsen LK. On the reconstruction of the Mus musculus genome-scale metabolic network model. *Genome Inform* 2008; **21**: 89–100.

82 Bordbar A, Lewis NE, Schellenberger J, Palsson BO, Jamshidi N. Insight into human alveolar macrophage and M. tuberculosis interactions via metabolic reconstructions. *Molecular Systems Biology* 2010; **6**: 422.

83 Heinken A, Thiele I. Systematic prediction of health-relevant human-microbial co-metabolism through a computational framework. *Gut Microbes* 2015.

84 Le Gall G, Noor SO, Ridgway K, Scovell L, Jamieson C, Johnson IT *et al.* Metabolomics of fecal extracts detects altered metabolic activity of gut microbiota in ulcerative colitis and irritable bowel syndrome. *Journal of Proteome Research* 2011; **10**: 4208–18.

85 Loh G, Blaut M. Role of commensal gut bacteria in inflammatory bowel diseases. *Gut Microbes* 2012; **3**: 544–55.

86 Bucci V, Bradde S, Biroli G, Xavier JB. Social interaction, noise and antibiotic-mediated switches in the intestinal microbiota. *PLoS Comput Biol* 2012; **8**: e1002497.

87 Lardon LA, Merkey BV, Martins S, Dotsch A, Picioreanu C, Kreft JU *et al.* iDynoMiCS: next-generation individual-based modelling of biofilms. *Environ Microbiol* 2011; **13**: 2416–34.

88 Bucci V, Xavier JB. Towards predictive models of the human gut microbiome. *Journal of Molecular Biology* 2014; **426**: 3907–16.

89 Faust K, Raes J. Microbial interactions: from networks to models. *Nature Reviews Microbiology* 2012; **10**: 538–50.

90 Marino S, Baxter NT, Huffnagle GB, Petrosino JF, Schloss PD. Mathematical modeling of primary succession of murine intestinal microbiota. *Proc Natl Acad Sci USA* 2014; **111**: 439–44.

91 Fisher CK, Mehta P. Identifying keystone species in the human gut microbiome from metagenomic timeseries using sparse linear regression. *PloS One* 2014; **9**: e102451.

92 Stein RR, Bucci V, Toussaint NC, Buffie CG, Ratsch G, Pamer EG *et al.* Ecological modeling from time-series inference: insight into dynamics and stability of intestinal microbiota. *PLoS Comput Biol* 2013; **9**: e1003388.

93 Buffie CG, Bucci V, Stein RR, McKenney PT, Ling L, Gobourne A *et al.* Precision microbiome reconstitution restores bile acid mediated resistance to Clostridium difficile. *Nature* 2015; **517**: 205–8.

94 Steinway SN, Biggs MB, Loughran TP, Jr., Papin JA, Albert R. Inference of network dynamics and metabolic interactions in the gut microbiome. *PLoS Comput Biol* 2015; **11**: e1004338.

95 Borenstein E, Kupiec M, Feldman MW, Ruppin E. Large-scale reconstruction and phylogenetic analysis of metabolic environments. *Proc Natl Acad Sci USA* 2008; **105**: 14482–7.

96 Munoz-Tamayo R, Laroche B, Walter E, Dore J, Duncan SH, Flint HJ *et al.* Kinetic modelling of lactate utilization and butyrate production by key human colonic bacterial species. *FEMS Microbiology Ecology* 2011; **76**: 615–24.

97 Kettle H, Louis P, Holtrop G, Duncan SH, Flint HJ. Modelling the emergent dynamics and major metabolites of the human colonic microbiota. *Environ Microbiol* 2014; **17**: 1615–30.

98 Kettle H, Donnelly R, Flint HJ, Marion G. pH feedback and phenotypic diversity within bacterial functional groups of the human gut. *J Theoretical Biology* 2014; **342**: 62–9.

99 Reed JL. Shrinking the metabolic solution space using experimental datasets. *PLoS Comput Biol* 2012; **8**: e1002662.

100 Mahowald MA, Rey FE, Seedorf H, Turnbaugh PJ, Fulton RS, Wollam A *et al.* Characterizing a model human gut microbiota composed of members of its two dominant bacterial phyla. *P Natl Acad Sci USA* 2009; **106**: 5859–64.

101 Shoaie S, Karlsson F, Mardinoglu A, Nookaew I, Bordel S, Nielsen J. Understanding the interactions between bacteria in the human gut through metabolic modeling. *Scientific Reports* 2013; **3**: 2532.

102 Aurich MK, Paglia G, Rolfsson O, Hrafnsdóttir S, Magnúsdóttir M, Stefaniak MM *et al.* Prediction of intracellular metabolic states from extracellular metabolomic data. *Metabolomics* 2014: 1–17.

103 Zelezniak A, Pers TH, Soares S, Patti ME, Patil KR. Metabolic network topology reveals transcriptional regulatory signatures of type 2 diabetes. *PLoS Comput Biol* 2010; **6**: e1000729.

104 Li L, Zhou X, Ching WK, Wang P. Predicting enzyme targets for cancer drugs by profiling human metabolic reactions in NCI-60 cell lines. *BMC Bioinformatics* 2010; **11**: 501.

105 Agren R, Mardinoglu A, Asplund A, Kampf C, Uhlen M, Nielsen J. Identification of anticancer drugs for hepatocellular carcinoma through personalized genome-scale metabolic modeling. *Molecular Systems Biology* 2014; **10**: 721.

106 Yizhak K, Gaude E, Le Devedec S, Waldman YY, Stein GY, van de Water B *et al.* Phenotype-based cell-specific metabolic modeling reveals metabolic liabilities of cancer. *eLife* 2014; **3**.

107 Bartlett JG, Gerding DN. Clinical recognition and diagnosis of Clostridium difficile infection. *Clinical infectious diseases : an official publication of the Infectious Diseases Society of America* 2008; **46** Suppl 1: S12–8.

108 Chavali AK, D'Auria KM, Hewlett EL, Pearson RD, Papin JA. A metabolic network approach for the identification and prioritization of antimicrobial drug targets. *Trends in Microbiology* 2012; **20**: 113–23.

109 Chavali AK, Blazier AS, Tlaxca JL, Jensen PA, Pearson RD, Papin JA. Metabolic network analysis predicts efficacy of FDA-approved drugs targeting the causative agent of a neglected tropical disease. *BMC Syst Biol* 2012; **6**: 27.

110 Orth JD, Conrad TM, Na J, Lerman JA, Nam H, Feist AM *et al.* A comprehensive genome-scale reconstruction of Escherichia coli metabolism--2011. *Molecular Systems Biology* 2011; **7**: 535.

111 Klamt S, Saez-Rodriguez J, Gilles ED. Structural and functional analysis of cellular networks with CellNetAnalyzer. *BMC Systems Biology* 2007; **1**: 2.

112 Klamt S, von Kamp A. An application programming interface for CellNetAnalyzer. *Bio Systems* 2011; **105**: 162–8.

113 Becker SA, Feist AM, Mo ML, Hannum G, Palsson BO, Herrgard MJ. Quantitative prediction of cellular metabolism with constraint-based models: the COBRA Toolbox. *Nat Protoc* 2007; **2**: 727–38.

114 Schellenberger J, Que R, Fleming RM, Thiele I, Orth JD, Feist AM *et al.* Quantitative prediction of cellular metabolism with constraint-based models: the COBRA Toolbox v2.0. *Nat Protoc* 2011; **6**: 1290–307.

115 Aziz RK, Bartels D, Best AA, DeJongh M, Disz T, Edwards RA *et al.* The RAST Server: rapid annotations using subsystems technology. *BMC Genomics* 2008; **9**: 75.

116 Rocha I, Maia P, Evangelista P, Vilaca P, Soares S, Pinto JP *et al.* OptFlux: an open-source software platform for in silico metabolic engineering. *BMC Systems Biology* 2010; **4**: 45.

117 Thorleifsson SG, Thiele I. rBioNet: A COBRA toolbox extension for reconstructing high-quality biochemical networks. *Bioinformatics* 2011; **27**: 2009–10.

118 Hunter S, Corbett M, Denise H, Fraser M, Gonzalez-Beltran A, Hunter C *et al.* EBI metagenomics — a new resource for the analysis and archiving of metagenomic data. *Nucleic Acids Research* 2014; **42**: D600–6.

119 Kersey PJ, Allen JE, Christensen M, Davis P, Falin LJ, Grabmueller C *et al.* Ensembl Genomes 2013: scaling up access to genome-wide data. *Nucleic Acids Research* 2014; **42**: D546–52.

120 Tatusov RL, Fedorova ND, Jackson JD, Jacobs AR, Kiryutin B, Koonin EV *et al.* The COG database: an updated version includes eukaryotes. *BMC Bioinformatics* 2003; **4**: 41.

121 Powell S, Forslund K, Szklarczyk D, Trachana K, Roth A, Huerta-Cepas J *et al.* eggNOG v4.0: nested orthology inference across 3686 organisms. *Nucleic Acids Research* 2014; **42**: D231–9.

122 Benson DA, Cavanaugh M, Clark K, Karsch-Mizrachi I, Lipman DJ, Ostell J *et al.* GenBank. *Nucleic Acids Research* 2013; **41**: D36–42.

123 Human Microbiome Project Consortium. A framework for human microbiome research. *Nature* 2012; **486**: 215–21.

124 Structure, function and diversity of the healthy human microbiome. *Nature* 2012; **486**: 207–14.

125 Dehal PS, Joachimiak MP, Price MN, Bates JT, Baumohl JK, Chivian D *et al*. MicrobesOnline: an integrated portal for comparative and functional genomics. *Nucleic Acids Research* 2010; **38**: D396–400.

126 Meyer F, Paarmann D, D'Souza M, Olson R, Glass EM, Kubal M *et al*. The metagenomics RAST server — a public resource for the automatic phylogenetic and functional analysis of metagenomes. *BMC Bioinformatics* 2008; **9**: 386.

127 Wattam AR, Abraham D, Dalay O, Disz TL, Driscoll T, Gabbard JL *et al*. PATRIC, the bacterial bioinformatics database and analysis resource. *Nucleic Acids Research* 2014; **42**: D581–91.

128 Finn RD, Bateman A, Clements J, Coggill P, Eberhardt RY, Eddy SR *et al*. Pfam: the protein families database. *Nucleic Acids Research* 2014; **42**: D222–30.

129 Caspi R, Altman T, Billington R, Dreher K, Foerster H, Fulcher CA *et al*. The MetaCyc database of metabolic pathways and enzymes and the BioCyc collection of Pathway/Genome Databases. *Nucleic Acids Research* 2014; **42**: D459–71.

130 Kanehisa M, Goto S, Sato Y, Kawashima M, Furumichi M, Tanabe M. Data, information, knowledge and principle: back to metabolism in KEGG. *Nucleic Acids Research* 2014; **42**: D199–205.

131 Kanehisa M, Goto S. KEGG: kyoto encyclopedia of genes and genomes. *Nucleic Acids Research* 2000; **28**: 27–30.

132 Schomburg I, Chang A, Placzek S, Sohngen C, Rother M, Lang M *et al*. BRENDA in 2013: integrated reactions, kinetic data, enzyme function data, improved disease classification: new options and contents in BRENDA. *Nucleic Acids Research* 2013; **41**: D764–72.

133 Lombard V, Golaconda Ramulu H, Drula E, Coutinho PM, Henrissat B. The carbohydrate-active enzymes database (CAZy) in 2013. *Nucleic Acids Research* 2014; **42**: D490–5.

134 Png CW, Linden SK, Gilshenan KS, Zoetendal EG, McSweeney CS, Sly LI *et al*. Mucolytic bacteria with increased prevalence in IBD mucosa augment in vitro utilization of mucin by other bacteria. *American Journal of Gastroenterology* 2010; **105**: 2420–8.

135 Wishart DS, Tzur D, Knox C, Eisner R, Guo AC, Young N, *et al*. HMDB: the Human Metabolome Database. *Nucleic Acids Research* 2007; **35**: D521–.

136 Wishart DS, Knox C, Guo AC, Eisner R, Young N, Gautam B *et al*. HMDB: a knowledgebase for the human metabolome. *Nucleic Acids Research* 2009; **37**: D603–10.

137 Wishart DS, Jewison T, Guo AC, Wilson M, Knox C, Liu Y *et al*. HMDB 3.0 — The Human Metabolome Database in 2013. *Nucleic Acids Research* 2013; **41**: D801–7.

138 Levy R, Carr R, Kreimer A, Freilich S, Borenstein E. NetCooperate: a network-based tool for inferring host-microbe and microbe-microbe cooperation. *BMC Bioinformatics* 2015; **16**: 164.

139 Stolyar S, Van Dien S, Hillesland KL, Pinel N, Lie TJ, Leigh JA *et al*. Metabolic modeling of a mutualistic microbial community. *Molecular Systems Biology* 2007; **3**: 92.

140 Klitgord N, Segre D. Environments that induce synthetic microbial ecosystems. *PLoS Comput Biol* 2010; **6**.

141 Gagneur J, Klamt S. Computation of elementary modes: a unifying framework and the new binary approach. *BMC Bioinformatics* 2004; **5**: 175.

142 Taffs R, Aston JE, Brileya K, Jay Z, Klatt CG, McGlynn S *et al*. In silico approaches to study mass and energy flows in microbial consortia: a syntrophic case study. *BMC Syst Biol* 2009; **3**: 114.

143 Segre D, Vitkup D, Church GM. Analysis of optimality in natural and perturbed metabolic networks. *Proc Natl Acad Sci USA* 2002; **99**: 15112–7.

144 Wintermute EH, Silver PA. Emergent cooperation in microbial metabolism. *Molecular Systems Biology* 2010; **6**: 407.

145 Mahadevan R, Edwards JS, Doyle FJ, 3rd. Dynamic flux balance analysis of diauxic growth in Escherichia coli. *Biophysical Journal* 2002; **83**: 1331–40.

146 Zhuang K, Izallalen M, Mouser P, Richter H, Risso C, Mahadevan R, *et al*. Genome-scale dynamic modeling of the competition between Rhodoferax and Geobacter in anoxic subsurface environments. *Isme J* 2011; **5**: 305–16.

147 Zhuang K, Ma E, Lovley DR, Mahadevan R. The design of long-term effective uranium bioremediation strategy using a community metabolic model. *Biotechnology and Bioengineering* 2012; **109**: 2475–83.

148 Zomorrodi AR, Maranas CD. OptCom: a multi-level optimization framework for the metabolic modeling and analysis of microbial communities. *PLoS Comput Biol* 2012; **8**: e1002363.

149 Thiele I, Fleming RM, Bordbar A, Schellenberger J, Palsson BO. Functional characterization of alternate optimal solutions of Escherichia coli's transcriptional and translational machinery. *Biophysical Journal* 2010; **98**: 2072–81.

150 Freilich S, Zarecki R, Eilam O, Segal ES, Henry CS, Kupiec M *et al.* Competitive and cooperative metabolic interactions in bacterial communities. *Nat Commun* 2011; **2**: 589.

151 Khandelwal RA, Olivier BG, Roling WF, Teusink B, Bruggeman FJ. Community flux balance analysis for microbial consortia at balanced growth. *PLoS One* 2013; **8**: e64567.

152 Nagarajan H, Embree M, Rotaru AE, Shrestha PM, Feist AM, Palsson BO *et al.* Characterization and modelling of interspecies electron transfer mechanisms and microbial community dynamics of a syntrophic association. *Nat Commun* 2013; **4**: 2809.

153 Harcombe WR, Riehl WJ, Dukovski I, Granger BR, Betts A, Lang AH *et al.* Metabolic resource allocation in individual microbes determines ecosystem interactions and spatial dynamics. *Cell Reports* 2014; **7**: 1104–15.

154 El-Semman IE, Karlsson FH, Shoaie S, Nookaew I, Soliman TH, Nielsen J. Genome-scale metabolic reconstructions of Bifidobacterium adolescentis L2-32 and Faecalibacterium prausnitzii A2-165 and their interaction. *BMC Syst Biol* 2014; **8**: 41.

155 Heinken A, Thiele I. Anoxic Conditions promote species-specific mutualism between gut microbes in silico. *Appl Environ Microbiol* 2015; **81**: 4049–61.

156 Zelezniak A, Andrejev S, Ponomarova O, Mende DR, Bork P, Patil KR. Metabolic dependencies drive species co-occurrence in diverse microbial communities. *Proc Natl Acad Sci USA* 2015; **112**: 6449–54.

157 Henry CS, Broadbelt LJ, Hatzimanikatis V. Thermodynamics-based metabolic flux analysis. *Biophysical Journal* 2007; **92**: 1792–805.

158 Hamilton JJ, Dwivedi V, Reed JL. Quantitative assessment of thermodynamic constraints on the solution space of genome-scale metabolic models. *Biophysical Journal* 2013; **105**: 512–22.

159 Hamilton JJ, Calixto Contreras M, Reed JL. Thermodynamics and H2 Transfer in a Methanogenic, Syntrophic Community. *PLoS Comput Biol* 2015; **11**: e1004364.

CHAPTER 8

Bacterial biofilm formation and immune evasion mechanisms

Jessica Snowden

University of Nebraska Medical Center, Omaha, Nebraska, United States

8.1 Introduction

Bacterial biofilms are increasingly recognized as an important part of many infectious disease processes. Biofilms are organized communities of bacteria, characterized by changes in cell metabolism and growth, that reside within a self-produced matrix that protects the organisms from the host immune response and from antibiotic treatment[1,2]. In this chapter, we introduce key concepts in biofilm formation and the current understanding of the complex and impaired interactions between the immune system and these bacterial communities.

8.2 Biofilms in human disease

Bacterial biofilms are now known to play an important role in the pathophysiology of many infectious diseases, including device-associated infections, chronic otitis media and sinusitis, dental disease and colonization in the lungs of cystic fibrosis patients[1,3,4]. These infections are typically chronic and difficult to diagnose and treat[5]. Many types of bacteria are capable of surviving in a relatively sessile lifestyle, termed a biofilm, in which the organisms cluster in a self-produced matrix attached to a surface[1,6,7]. Biofilm formation has been described in many different strains of bacteria, such as *Staphylococcus aureus, Staphylococcus epidermidis, Pseudomonas aeruginosa, Streptococcus mutans, Porphyromonas gingivalis* and others[1,5,8,9]. These communities may consist of a single or multiple strains or species of bacteria and can also occur in certain fungi, such as *Candida*[5,9]. As our ability to culture organisms within biofilms improves through new culture techniques and better sampling through biopsy or device sonication, our knowledge of the bacterial composition of biofilms in human disease is improving (Table 1)[5].

Device-associated biofilm infections include vascular devices, such as central venous catheters, intracranial devices, such as ventricular shunts, and orthopedic and dental implants among others. These devices are all at risk of infection from

The Human Microbiota and Chronic Disease: Dysbiosis as a Cause of Human Pathology, First Edition.
Edited by Luigi Nibali and Brian Henderson.
© 2016 John Wiley & Sons, Inc. Published 2016 by John Wiley & Sons, Inc.

Table 1 Biofilm-mediated clinical disease and associated organisms.

Clinical Disease	Typical biofilm-forming organisms

Device-associated infection
  -Vascular devices (e.g. central venous line, port)
  -CNS devices (e.g. shunt)
  -Orthopedic implant (e.g. artificial joint)
	Staphylococcus epidermidis
	Staphylococcus aureus
	Gram-negative bacilli
	Candida
	Enterococcus faecalis/faecium

Periodontal disease	*Porphyromonas gingivalis* *Treponema denticola* *Tannerella forsythia* *Aggregatibacter actinomycetemcomitans* *Filifactor alocis* *Staphylococcus aureus* Streptococci Genus *Desulfobulbus* *Candida albicans*
Catheter-associated urinary tract infection	*Escherichia coli* *Enterococcus* species *Acinetobacter* species *Klebsiella* species *Serratia* species *Pseudomonas aeruginosa*
Ventilator-associated pneumonia	*Pseudomonas aeruginosa* *Acinetobacter* species *Escherichia coli* *Klebsiella* species *Staphylococcus aureus*
Chronic wound infection	*Staphylococcus aureus* *Pseudomonas aeruginosa*
Chronic lung infections (e.g. cystic fibrosis)	*Pseudomonas aeruginosa* *Staphylococcus aureus*
Endocarditis	*Staphylococcus aureus* *Streptococcus* species (including *Enterococcus*, viridans group, etc.)
Chronic sinusitis, otitis	*Pseudomonas aeruginosa* *Staphylococcus aureus* *Staphylococcus epidermidis* *Haemophilus influenza*

Sources: American Thoracic Society and Infectious Diseases Society of America 2005, 388–416, Archer *et al.* 2011, 445–459, Baddour *et al.* 2005, e394–434, Donlan and Costerton 2002, 167–193, Hamilos 2014, 640–53.e4, Hoiby *et al.* 2015, Hooton *et al.* 2010, 625–663, Osmon *et al.* 2013, e1–e25, van Gennip *et al.* 2012, 2601–2607.

the introduction of skin flora at the time the device is placed or accessed, as well as infection via hematogenous seeding[1,10]. Frequent organisms identified in these patients are potent biofilm-formers such as *S. epidermidis*, *S. aureus* and *Enterococcus*[1]. Other device-associated biofilm infections include catheter-associated urinary tract infections, secondary to *Escherichia coli* or other Gram negatives, and ventilator-associated pneumonia, with overgrowth of pathogens such as *Pseudomonas* species and normal respiratory flora in biofilms adhering to the endotracheal tube[11–13].

Biofilms have also been implicated in tissue-based diseases, such as endocarditis, pulmonary colonization in cystic fibrosis and chronic upper respiratory infections such as otitis and sinusitis. Endocarditis occurs when bacteria adhere to either damaged native valvular tissue or prosthetic valves to form biofilms[1,14]. These chronic and potentially lethal infections are caused most often by *S. aureus* and streptococcal species, including *Enterococcus* and viridans group streptococci[1,14]. Chronic pulmonary infections in cystic fibrosis patients are also now known to involve biofilm formation, particularly with *Pseudomonas aeruginosa*[4]. Chronic rhinosinusitis is also now known to involve biofilm formation, with polymicrobial biofilms or those containing *S. aureus* being associated with more severe disease[15].

Chapter 1 discussed the initial description of biofilms as "animalcules" in the plaque of Van Leeuwenhoek's teeth in the seventeenth century[1,8]. Hence, it is not surprising that biofilms also play a key role in periodontal disease, where key biofilm-forming pathogens include *Porphyromonas gingivalis*, *Treponema denticola*, *Tannerella forsythia*, *Aggregatibacter actinomycetemcomitans*, *Streptococcus mitis*, *Streptococcus salivarius*, *Streptococcus mutans*, *Filifactor alocis*, *Staphylococcus aureus* and the genus *Desulfobulbus*[8,16]. The role of the microbial biofilms and dysbiosis in inducing periodontal disease is discussed by Hajishengallis and Lamont in chapter 14.

Antibiotic resistance in biofilm infections is thought to be caused by a variety of factors, including metabolic alterations in bacteria within the biofilm, decreased penetration of antibiotics due to the extracellular matrix, inactivation of the antibiotic by compounds within the extracellular matrix, inoculum effects related to the very large number of bacteria in the biofilm relative to the available antibiotic molecules and increased exchange of bacterial resistance mechanisms as bacteria reside in close proximity to each other[17–19]. Many studies have demonstrated that bacterial biofilms are more resistant to antibiotic killing than are identical bacterial strains growing planktonically as individual, free-floating cells[1,5,19,20]. For example, a recent study by Claessens et al. demonstrated low killing of *S. epidermidis* within a biofilm by vancomycin and teicoplanin, two glycopeptide antibiotics frequently used for treatment of planktonic *S. epidermidis* infections[19]. Additionally, studies in *Streptococcus mutans*, an important component of dental biofilms, demonstrated that chemical treatment of biofilms with chlorhexidine may kill the bacteria but leave residual biofilm matrix structures that appear to increase the ease with which the surface becomes re-infected[21]. Because of this, removal or debridement of the infected surface is frequently required for adequate treatment[12,13,22,23]. In situations in which removal of the infected surface is not a viable option, such as chronic sinus disease, prolonged treatment with multiple and higher-dose antibiotics may be needed. This can significantly add to the expense and morbidity associated with these infections. Given the inherent limitations of antibiotic treatment in biofilm-mediated infections, better understanding of the interactions between the immune system and the biofilm is key to developing adjunctive treatment strategies for these infections.

8.3 Biofilm formation

There are three stages that characterize biofilm formation: 1) attachment, 2) proliferation and 3) dispersal. By adapting to this lifestyle, through changes in metabolic pathway and other key genes, the bacteria are able to survive, despite

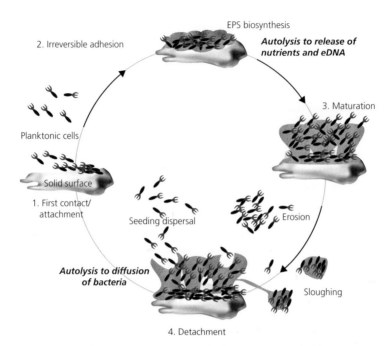

Figure 1 Biofilm formation and development. Initially, planktonic cells adhere to a solid surface (1), and production of extracellular polymeric substances (EPS) stabilizes the adhered colony (2). Some of the cells undergo autolysis, releasing nutrients and eDNA that promote growth and maturation of the biofilm (3). Cells are dispersed from the biofilm and can colonize other sites through three mechanisms: erosion, sloughing, and seeding dispersal (4). Seeding dispersal implicates an active process of autolysis resulting in release of single bacterial cells and cavity formation. Used under Creative Commons License from "Die for the community: an overview of programmed cell death in bacteria" by N Allocati, M Masulli, C Di Ilio and V De Laurenzi in *Cell Death and Disease*, January 2015 (Figure 3 original text).

pressure from antibiotics and the immune system, before spreading to other sites of infection with dispersal of the biofilm (Figure 1)[9]. These stages have been extensively studied in staphylococci as well as *Pseudomonas*, with on-going studies in the polymicrobial biofilms seen in dental and other disease sites.

Initial attachment to the host surface or device can be triggered by environmental stresses like high temperature, high osmolarity or the presence of ethanol or antibiotics, according to *in vitro* studies[24,25]. Attachment is partially aided by the host response to the placement of a medical device, with the adherence of host proteins like collagen and fibronectin to the device, which can serve as microbial surface components recognizing adhesive matrix molecules (MSCRAMMS) that facilitate bacterial attachment[7]. Similarly, a host-derived proteinaceous matrix coats teeth very shortly following tooth brushing or other cleaning and forms the basis for initial bacterial attachment[1,26]. Several genes and molecules have been identified as significant virulence factors that facilitate this attachment, which have been extensively explored as they are attractive therapeutic or prophylactic targets in managing these infections. In staphylococci, the best described of these is polysaccharide intercellular adhesion (PIA), which renders *S. aureus* and *S. epidermidis* strains better able to form biofilms and more resistant to phagocytosis by neutrophils and antibacterial peptides[27–29]. It is particularly important in *S. epidermidis*, where approximately 85% of strains from patients with vascular-device associated bacteremia were positive for PIA[30]. Other key molecules in staphylococci include Embp (fibronectin-binding protein) and accumulation associated protein (Aap)[27,31–33]. In *Pseudomonas aeruginosa*, similar genes such as algC encoding the synthesis of an exopolysaccharide, alginate, are upregulated shortly after adhesion to a solid surface[34].

Once the bacteria attach to start biofilm formation, proliferation, maturation and detachment occur to establish and spread biofilm-related infections[35]. The proliferation phase is characterized by accumulation of the bacteria on the device or host surface as well as physiologic and metabolic changes in the cells living within the biofilm[35]. In staphylococci, these changes are controlled by the master regulator *agr*, which significantly influences biofilm formation as well as production of virulence factors[36]. Many factors participate in these changes in various biofilm forming bacterial strains. Programmed cell death is an important component of biofilm formation, in which certain organisms autolyse to provide nutrients for the surrounding colony, provide key extracellular matrix components such as extracellular DNA (eDNA) and aid in dispersal of the bacteria to other sites[37–39]. Through quorum sensing, a process in which increased cell density results in the accumulation of certain molecules that can effect changes in gene expression, bacteria can communicate with each other within multicellular communities like biofilms[9]. Quorum sensing is of particular importance in *Pseudomonas*, where biofilm formation and production of other virulence factors is regulated by a complex network of quorum signaling molecules[40,41].

Biofilm formation in periodontal disease is particularly interesting as it involves interaction between many different types of bacteria. Bacteria adhere to a proteinaceous film on the surface of the teeth, with the gram-positive cocci and other bacteria of the normal oral flora forming the initial attachments to this film through the production of extracellular glucans[1]. Gram-positive organisms such as *Streptococcus* species and actinomycetes predominate after several days, leading to the production of the extracellular matrix and local changes in oxygen and pH typical of biofilm formation[26]. Adhesins then provide the ability of multiple other bacterial types to adhere to this biofilm community by a process termed co-aggregation or co-adhesion to form a plaque within 2–3 weeks (see also chapter 14)[1,26].

Studies are ongoing to explore further the signals triggering release of bacteria from the biofilm structure, which leads to spread of infection and worsened morbidity and mortality. By understanding the broad mechanisms regulating biofilm formation across multiple strains, as well as the species-specific factors in bacterial strains of key clinical significance, we may be able to develop new therapeutic and preventative strategies.

8.4 Immune responses to biofilms

In the decades since the description of biofilms as key players in human disease, much work has been done to elucidate the bacterial and host factors that participate in this process. In this section, we will review the current understanding of innate and adaptive immune responses to infection, as well as other cell types with immune and non-immune functions, such as fibroblasts and epithelial cells, that may play a role. These immune responses are distinct and somewhat attenuated in comparison with planktonic infections, and thus may play a role in persistence of these infections.

8.4.1 Innate immune responses

Innate immune cells such as macrophages and neutrophils are important first responders to bacterial infection. These cells detect conserved molecular patterns in bacteria through membrane or cytoplasmic pattern recognition receptors such as Toll-like receptors (TLRs) and NOD-like receptors (NLRs)[42–45]. Myeloid differentiation factor 88 (MyD88) is an important adaptor that mediates many of these immune signaling pathways and has been shown to be important in *in vivo* responses to staphylococcal biofilm infection. MyD88 KO mice display significant increases in *S. aureus* burdens on catheters as well as in surrounding tissues[46]. Interestingly, however, there is a disconnect between *in vitro* and *in vivo* phenotypes of several other immune signaling pathways, demonstrating the multiple evasion mechanisms available to bacteria during *in vivo* infection. For example, TLR2 recognizes peptidoglycan, teichoic acids (TA) and lipoproteins (Lpp) associated with Gram-positive bacteria such as *S. aureus*[44,47,48]. Phenol-soluble modulins and PIA associated with *S. epidermidis* are also recognized by TLR2 *in vitro*[2,49]. While biofilm matrix components such as PIA, peptidoglycan and e-DNA can stimulate TLR2 and TLR9 *in vitro*, no role for these receptors has been demonstrated in staphylococcal biofilms *in vivo*, suggesting that redundant immune evasion mechanisms are available to the staphylococcal biofilm[49,50]. Furthermore, *in vivo* studies evaluating the immune responses to staphylococcal biofilms in the periphery and the central nervous system have demonstrated that levels of inflammatory mediators such as IL-1β, TNF-α, CXCL2, and CCL2 were reduced in biofilm-infected mice compared to uninfected controls or parenchymal infection[50,51]. This suggests that there are multiple avenues of immune stimulation and evasion available to staphylococci. In *Pseudomonas* biofilms, *in vitro* studies have shown that the eDNA in the biofilm matrix is a major pro-inflammatory stimulus via TLR9 stimulation, although the *in vivo* significance of this is unknown at this point[52]. Multiple TLRs, including TLR9, have been shown to play a role in planktonic responses to infection with *Pseudomonas in vivo*, but given the differences observed between planktonic and biofilm immune responses in other bacteria, it is reasonable to assume that this may be distinct in *Pseudomonas* as well[51,53]. The mechanism for this attenuated immune response to biofilms is still being defined. While innate immune cells such as macrophages and neutrophils are recruited to the site of biofilm infection, they are not as effective nor recruited at the same levels as observed in planktonic or parenchymal infections[50,51,54,55]. Macrophages are likely to be important in coordinating the ongoing response to chronic infections like biofilms, as they reside in all tissues of the body, are long-lived and are capable of producing a broad array of chemokines, cytokines and other mediators to coordinate the surrounding immune response[46]. *In vitro* experiments have demonstrated that monocytes are capable of responding to *Pseudomonas* biofilm matrix molecule alginate via TLR2- and TLR4-dependent mechanisms, but the role of monocytes/macrophages *in vivo* with *Pseudomonas* biofilms has not been well defined[56,57]. However, this has been an area of much study in staphylococcal biofilms over the last several years, with new insights into the mechanisms of the attenuated immune responses to biofilms. Macrophages have been shown in multiple *in vitro* studies to be impaired in their pro-inflammatory mediator production, phagocytic ability in response to *S. aureus* and *S. epidermidis* biofilms[50,55]. In *S. aureus,* this is at least partially mediated by a factor secreted in the biofilm, as biofilm supernatants are capable of triggering this impaired macrophage performance[2]. Studies in *S. epidermidis* biofilms found that biofilm formation mediated through either Aap, PIA or Embp were equally resistant to phagocytosis by macrophages, suggesting this disruption is at least partially due to a common biofilm structural rather than a specific matrix molecule[31]. The macrophages in these cultures also produced significantly lower amounts of pro-inflammatory IL-1β than similar co-cultures with planktonic *S. epidermidis*[31]. In animal models of *S. aureus* biofilm infection, the macrophages recruited to peripheral sites of infection have been shown to have an alternatively activated, or M2, phenotype that is associated with anti-inflammatory mediator production[50,58]. These studies have been confirmed with *in vitro* experiments showing inhibition of NF-κB activation and IL-1β production, as well as decreased activation of peritoneal macrophages in response to biofilm infection[55,59,60]. This skewed macrophage response appears to be crucial for biofilm immune evasion as adoptive transfer of classically activated, or M1, macrophages resulted in restoration of a pro-inflammatory immune milieu and clearance of the biofilm[58].

Neutrophils also play an important role in the containment of biofilm infections and production of reactive oxygen species as a means of combatting these infections[54,61]. Both IL-1β and the presence of neutrophils at the site of infection appear to have a protective role in a post-arthroplasty model of staphylococcal biofilm infection[54,62]. However, the function of neutrophils in response to staphylococcal biofilms is limited by the production of poly-N-acetylglucosamine and poly-y-acid (PGA), which prevent bacterial engulfment by neutrophil and antibody- and complement-mediated phagocytosis. Antibody binding and complement deposition are inhibited by the large amounts of extracellular matrix present, through mechanical and other means, further inhibiting effective neutrophil function through a lack of effective opsonization[25,29,54,63,64]. In contrast to the previously cited utilization of classically activated macrophages, adoptive transfer of neutrophils does not appear to aid in clearance of biofilms, confirming their limitations in interacting effectively with these bacteria[58]. This corresponds with other studies showing a shift in *S. aureus* biofilm gene expression on microarray when exposed to macrophages, but not neutrophils[65]. In other infection models, such as periodontal disease, neutrophils are effectively recruited to the site of biofilm infection by inflammatory signals from epithelial cells as well as small quorum sensing molecules in the biofilm itself[61]. However, as the neutrophils are limited in their ability to engulf the organism, the inflammatory activity of the frustrated neutrophils through production of ROS and other cytotoxic and proteolytic factors may contribute to overall pathology by damaging nearby host tissue[2,66–68]. Certain oral flora, such as *Porphyromonas gingivalis,* are able to dampen production of chemokines and other inflammatory molecules to limit more effectively the immune response[68]. Similarly, in *Pseudomonas* neutrophils are recruited to the site of infection but undergo degranulation and are unable to phagocytose the bacteria effectively[69,70]. *Pseudomonas* is actually capable of utilizing the neutrophil degradation byproducts in biofilm matrix assembly, truly optimizing its subversion of the immune response[71]. Cumulatively, these studies suggest that the role of neutrophils in the response to staphylococcal biofilm is significantly limited and may actually contribute to pathology in biofilm disease states.

Myeloid-derived suppressor cells (MDSCs) have recently been shown to play a crucial role in response to peripheral biofilm infections. MDSCs are a heterogeneous population of immature myeloid cells that are potent inhibitors of T cell activation[72]. An animal model of *S. aureus* biofilm infection demonstrated a significant influx of these cells to the tissue surrounding the infection, with clearance of the biofilm and restoration of classical macrophage inflammatory mediators and phagocytosis following depletion of these cells[72]. This suggests that these cells play an important role in dampening the immune response to allow the development and persistence of the biofilm *in vivo*. These cells are under ongoing study in other infection models to determine the applicability of this mechanism across bacterial species and sites of infection.

8.4.2 Adaptive immune responses

A better understanding of adaptive immune responses may aid in the development of vaccines or other preventive strategies against these common human infections. Unfortunately, little information regarding adaptive immune responses to biofilm infections as opposed to planktonic infection is currently available. Given the key role for functional innate immune responses in shaping later, effective adaptive immune responses, it would be expected that adaptive immune responses to biofilms would be impaired, similar to the innate immune responses. Dendritic cells are essential antigen presenting cells, but have not been studied well in regards to biofilm infection. Regional T lymphocytes responses to a staphylococcal biofilm infection in the bone appear to be skewed to an early, ineffective Th1/Th17 early and a later immunosuppressive Th2/Treg phenotype in one study utilizing *S. aureus*[73]. In this study, Th1-associated antibodies IgG2a and IgG2b were elevated early in infection, with production of Th2-associated antibody IgG1 later in infection. T lymphocytes have also been implicated in that Th1/Th17 inhibition by anti-IL12p40 antibodies early in the course of infection has the ability to prevent chronic biofilm infection in this bone infection model[74]. Th17 cells, important in autoimmunity and immune responses to extracellular infection, have also been implicated in the response to dental biofilms, particularly *Porphyromonas gingivalis*[75]. The role of T cells appears to be limited by the presence of MDSC which inhibit T cell activation and further recruitment, which may explain the lack of T cell influx or responses in many biofilm models of infection[72].

B cell responses to staphylococci are limited by Protein A (SpA), which binds to IgM molecules exposed on the surface of B lymphocytes, leading to apoptosis of these cells[76]. Importantly, mice infected with SpA mutants demonstrated improved bacterial clearance as well as humoral immune responses[77,78]. Cystic fibrosis patients with pseudomonal biofilms have been shown to produce antibodies to bacterial elastase and proteases that may aid in clearance of infection *in vitro*, but as in the staphylococcal studies, it is difficult to attribute antibody development and efficacy to biofilm versus planktonic bacteria[79]. Interestingly, secretory IgA is produced in the mouth in response to bacteria that make up the biofilm in that setting, such as *Streptococcus mutans* and *salivarius,* but it is not clear what role these antibodies may play in infection[16]. The biofilm matrix itself may inhibit the efficacy of antibody in response to these infections, as antibody binding is inhibited by the large amounts of extracellular matrix present, through mechanical and other means[25,29,54,63,64].

8.4.3 Fibroblasts, epithelial cells and other immune responses

While most is known about the role of immune cells such as neutrophils, macrophages and monocytes in response to biofilm infection, there is emerging data suggesting that other non-immune cell types may participate in the immune response to biofilms as well. For example, epithelial cells in the gingiva have been shown to express intercellular adhesion molecule-1 (ICAM-1) and IL-8, important factors in the migration of neutrophils to the site of infection, in a gradient that directs these cells to the interface with the dental biofilm and away from deeper cells that may be sensitive to inflammatory damage[80]. Additionally, *in vitro* experiments of co-culture with fibroblasts and polymicrobial biofilms typical of dental infection revealed up-regulation of TLR, IL-1 and IL-17 pathways by microarray[81]. However, fibroblast responses may be attenuated by specific pathogens, such as *P. gingivalis*, which decreases inflammasome activation and subsequent IL-1β production, as well as degrading chemokines produced by these fibroblasts[82,83]. Sinonasal epithelial cells are also important components of the immune response to chronic sinusitis, as these cells express multiple TLRs and are capable of producing antimicrobial peptides as well as chemokines and cytokines in response to infection[15,84–86]. While these cells have not yet been studied specifically in relation to biofilm formation, patients with chronic rhinosinusitis, presumably with biofilm-mediated infection, have been shown to have significant alterations in TLR expression in some studies[15].

Antimicrobial peptides are also important in the innate immune response to biofilms, by destroying pathogens directly or by targeting them for destruction by other immune cells. These antimicrobial peptides have been studied in staphylococcal and pseudomonal biofilms, with some promise for future therapeutic development. RNAIII-inhibiting peptide (RIP) has been shown to down-regulate expression of biofilm genes in *S. aureus*[87]. A synthetic antimicrobial peptide, DASamP1, has also shown promise at specifically inhibiting *S. aureus* biofilms *in vitro* and *in vivo*[88]. Cathelicidin LL-37 is an important human cationic host defense peptide found at mucosal surfaces, in the granules of phagocytes and in bodily fluids[4]. It has been shown to inhibit biofilm formation in staphylococci and in *Pseudomonas*, making it a particularly attractive target for adjunctive therapeutic strategies in cystic fibrosis patients who experience pulmonary biofilm infections with both of these groups of bacteria[3,4,89]. It has shown similar efficacy with regards to the oral pathogen *Aggregatibacter actinomyetemcomitans*[90]. However, the therapeutic application of this compound must be carefully evaluated as LL-37 has also been shown to be pro-inflammatory and ciliotoxic, which may limit its use clinically[3].

8.5 Biofilm immune evasion strategies

Biofilm structures also produce molecules that directly lyse immune cells and inhibit their function, such as phenol soluble modulins (PSMs) in staphylococci. PSMs are synthesized by staphylococcal strains and, in addition to their role in biofilm adhesion and formation, have lytic activity against leukocytes and red blood cells[91,92]. PSMs also trigger chemotaxis and priming of human neutrophils and cytokine expression, so their role in promoting or inhibiting an inflammatory response is still under investigation and may vary depending on the type of infection[91,93,94]. Additionally, leukocidins produced by *S. aureus*, such as leukocidin A/B (LukA/B), have been shown to kill macrophages, dendritic cells and neutrophils[95,96]. Community-acquired methicillin-resistant *S. aureus* strains also form alpha-toxin, leukotoxin LukGH and Panton-Valentine leucocidin (PVL), pore-forming toxins capable of destroying many host cells[29,97]. Nuclease produced by staphylococci can also aid in escape from neutrophil responses by degrading neutrophil extracellular traps (NETs)[98]. Quorum-sensing molecules important in pseudomonal biofilm formation, such as acyl-homoserine lactones, have been shown to have complex interactions with the immune response, stimulating neutrophil chemotaxis via increased IL-8 production and NK-κB activation but conversely decreasing production of other pro-inflammatory mediators and lymphocyte proliferation[40,41]. Rhamnolipid, a quorum-sensing regulated virulence factor in *Pseudomonas*, has also been shown to lyse macrophages and neutrophils, likely contributing to the impaired immune response to this infection[99]. As previously noted, *P. gingivalis* also produces factors that can inhibit immune cell function, although the mechanisms of this are not yet fully defined[82,83]. These examples highlight an added layer of immune evasion utilized by bacteria within biofilms to aid in persistence of infection, in addition to hiding from innate immune receptors and skewing immune responses as discussed previously.

8.6 Vaccines and biofilm therapeutics

Several of the molecules discussed above as participants in biofilm formation and the immune responses to biofilm are currently being explored as vaccine candidates to prevent biofilm-associated infections. For example, in staphylococci, PIA-independent biofilm forming strains demonstrate anti-phagocytic properties attributable to a 20-kDa polysaccharide currently under study as a possible vaccine component[100]. While antibodies have been produced to some staphylococcal antigens, an optimal cocktail for efficient clearance and prevention of these infections has not yet been developed[2,101–104]. This may be due to the interference provided by the extracellular matrix of biofilms, limiting the ability of the antibody effectively to opsonize the bacteria. A successful vaccine approach will likely require a combination of immunostimulatory and bacterial factors specific to biofilm growth and thus requires more in-depth understanding of this interplay. In addition to the defensins and antimicrobial peptides such as LL-37 discussed above, other biologic approaches to adjuvant therapy are being explored. Bacteriophages are bacterial viruses that replicate within the bacteria, but not eukaryotic cells, resulting in lysis of the cell[105]. Given the close proximity of bacteria within the biofilm, this "infectious" approach is particularly attractive as an alternative or adjunct to antibiotic therapy. Preliminary data suggests that this may be an effective option, with rapid spread of the phage colony and destabilization of the extracellular matrix[105]. However, this remains under further evaluation to monitor effects on resistance and host responses.

Other anti-biofilm therapeutics currently under investigation for pseudomonal biofilms include anthranilate, a degradation product of tryptophan that can disrupt biofilm structures at later stages in biofilm maturation[106]. Additionally, quorum-sensing inhibitors have shown some promise in increasing the efficacy of the innate immune response to pseudomonal biofilm infections[107]. Innate immune cells themselves have also been proposed as a therapeutic option, with adoptive transfer of classically activated (M1) macrophages at the time of orthopedic implant showing great promise in reducing surgical device infection in animal models[58]. As our understanding of the bacterial and immunologic factors involved in these infections continues to increase, we expect to continue to identify therapeutic targets with greater frequency in the future.

8.7 Conclusions

Our understanding of the complex interactions between biofilms and the host immune response continues to advance, but much work is still needed. In particular, interactions with polymicrobial biofilms and infections in privileged spaces, such as the central nervous system, remain unclear. The role of age in the immune response to biofilms is an emerging avenue of investigation, with work within our own lab exploring the decreased inflammatory response that occurs in neonates in response to typical biofilm-forming organisms. On the other end of the spectrum, immunosenescence is used to describe the declining immune function seen in the elderly. This has been proposed as a possible mechanism for the increase of biofilm-associated dental disease observed in this population[108]. In adults and the aging, this is of particular interest as data emerges linking the chronic inflammation associated with periodontal biofilms with insulin resistance, heart disease and other diseases of aging[108,109]. Given the increasing complexity of medical care and medical devices available, biofilm-associated infections can only be expected to increase in frequency. Ideally, immunomodulatory adjunctive therapies could be developed to add to our current antibiotics to speed resolution of these infections while ameliorating the tissue damage associated with the host response. New preventive and therapeutic strategies are desperately needed to avoid the significant cost, in terms of patient morbidity as well as health care costs associated with prolonged therapy and added surgical procedures, associated with these infections.

TAKE-HOME MESSAGE

- Biofilms are an important part of many chronic human diseases. The unique growth structure and metabolic adaptations of this bacterial growth phenotype allow resistance to antibiotics as well as immune clearance.

- The mechanisms of the attenuated immune response to biofilms are multiple and still under investigation, including skewing of macrophage responses away from effective anti-bacterial function, lysis of immune cells by bacterially derived factors, masking of bacterial antigens by biofilm structures and absorption and deactivation of opsonizing factors such as complement and antibodies.

- Improving our understanding of the complex interplay between biofilm and the immune response *in vivo* is needed to allow new therapeutic and preventive modalities for these infections.

References

1 Donlan RM, Costerton JW. Biofilms: survival mechanisms of clinically relevant microorganisms. *Clin Microbiol Rev* 2002; **15**: 167–193.

2 Hanke ML, Kielian T. Deciphering mechanisms of staphylococcal biofilm evasion of host immunity. *Front Cell Infect Microbiol* 2012; **2**: 62.

3 Chennupati SK, Chiu AG, Tamashiro E *et al.* Effects of an LL-37-derived antimicrobial peptide in an animal model of biofilm Pseudomonas sinusitis. *Am J Rhinol Allergy* 2009; **23**: 46–51.

4 Overhage J, Campisano A, Bains M *et al.* Human host defense peptide LL-37 prevents bacterial biofilm formation. *Infect Immun* 2008; **76**: 4176–4182.

5 Hoiby N, Bjarnsholt T, Moser C *et al.* ESCMID guideline for the diagnosis and treatment of biofilm infections 2014. *Clin Microbiol Infect* 2015; **21** Suppl 1: S1–25.

6 Costerton JW, Stewart PS, Greenberg EP. Bacterial biofilms: a common cause of persistent infections. *Science* 1999; **284**: 1318–1322.

7 Scherr TD, Heim CE, Morrison JM, Kielian T. Hiding in plain sight: interplay between staphylococcal biofilms and host immunity. *Front Immunol* 2014; **5**: 37.

8 Pollanen MT, Paino A, Ihalin R. Environmental stimuli shape biofilm formation and the virulence of periodontal pathogens. *Int J Mol Sci* 2013; **14**: 17221–17237.

9 Allocati N, Masulli M, Di Ilio C, De Laurenzi V. Die for the community: an overview of programmed cell death in bacteria. *Cell Death Dis* 2015; **6**: e1609.

10 Gutierrez-Murgas Y, Snowden JN. Ventricular shunt infections: Immunopathogenesis and clinical management. *J Neuroimmunol* 2014; **276**: 1–8.

11 American Thoracic Society, Infectious Diseases Society of America. Guidelines for the management of adults with hospital-acquired, ventilator-associated, and healthcare-associated pneumonia. *Am J Respir Crit Care Med* 2005; **171**: 388–416.

12 Hooton TM, Bradley SF, Cardenas DD *et al.* Diagnosis, prevention, and treatment of catheter-associated urinary tract infection in adults: 2009 International Clinical Practice Guidelines from the Infectious Diseases Society of America. *Clin Infect Dis* 2010; **50**: 625–663.

13 Osmon DR, Berbari EF, Berendt AR *et al.* Diagnosis and management of prosthetic joint infection: clinical practice guidelines by the Infectious Diseases Society of America. *Clin Infect Dis* 2013; **56**: e1–e25.

14 Baddour LM, Wilson WR, Bayer AS *et al.* Infective endocarditis: diagnosis, antimicrobial therapy, and management of complications: a statement for healthcare professionals from the Committee on Rheumatic Fever, Endocarditis, and Kawasaki Disease, Council on Cardiovascular Disease in the Young, and the Councils on Clinical Cardiology, Stroke, and Cardiovascular Surgery and Anesthesia, American Heart Association: endorsed by the Infectious Diseases Society of America. *Circulation* 2005; **111**: e394–434.

15 Hamilos DL. Host-microbial interactions in patients with chronic rhinosinusitis. *J Allergy Clin Immunol* 2014; **133**: 640–53.e4.

16 Nogueira RD, King WF, Gunda G, *et al.* Mutans streptococcal infection induces salivary antibody to virulence proteins and associated functional domains. *Infect Immun* 2008; **76**: 3606–3613.

17 Mathur T, Singhal S, Khan S, *et al.* Adverse effect of staphylococci slime on in vitro activity of glycopeptides. *Jpn J Infect Dis* 2005; **58**: 353–357.

18 Belfield K, Bayston R, Birchall JP, Daniel M. Do orally administered antibiotics reach concentrations in the middle ear sufficient to eradicate planktonic and biofilm bacteria? A review. *Int J Pediatr Otorhinolaryngol* 2015; **79**: 296–300.

19 Claessens J, Roriz M, Merckx R, *et al.* Inefficacy of vancomycin and teicoplanin in eradicating and killing Staphylococcus epidermidis biofilms in vitro. *Int J Antimicrob Agents* 2015; **45**: 368–75.

20 Hess DJ, Henry-Stanley MJ, Wells CL. Interplay of antibiotics and bacterial inoculum on suture-associated biofilms. *J Surg Res* 2012; **177**: 334–340.

21 Ohsumi T, Takenaka S, Wakamatsu R *et al.* Residual structure of Streptococcus mutans biofilm following complete disinfection favors secondary bacterial adhesion and biofilm re-development. *PLoS One* 2015; **10**: e0116647.

22 Baddour LM, Epstein AE, Erickson CC *et al.* Update on cardiovascular implantable electronic device infections and their management: a scientific statement from the American Heart Association. *Circulation* 2010; **121**: 458–477.

23 Tunkel AR, Hartman BJ, Kaplan SL *et al*. Practice guidelines for the management of bacterial meningitis. *Clin Infect Dis* 2004; **39**: 1267–1284.

24 Laverty G, Gorman SP, Gilmore BF. Biomolecular mechanisms of staphylococcal biofilm formation. *Future Microbiol* 2013; **8**: 509–524.

25 Kristian SA, Birkenstock TA, Sauder U *et al*. Biofilm formation induces C3a release and protects Staphylococcus epidermidis from IgG and complement deposition and from neutrophil-dependent killing. *J Infect Dis* 2008; **197**: 1028–1035.

26 Marsh PD. Microbiology of dental plaque biofilms and their role in oral health and caries. *Dent Clin North Am* 2010; **54**: 441–454.

27 Rohde H, Burdelski C, Bartscht K *et al*. Induction of Staphylococcus epidermidis biofilm formation via proteolytic processing of the accumulation-associated protein by staphylococcal and host proteases. *Mol Microbiol* 2005; **55**: 1883–1895.

28 Vuong C, Voyich JM, Fischer ER *et al*. Polysaccharide intercellular adhesin (PIA) protects Staphylococcus epidermidis against major components of the human innate immune system. *Cell Microbiol* 2004; **6**: 269–275.

29 Cheung GY, Rigby K, Wang R *et al*. Staphylococcus epidermidis strategies to avoid killing by human neutrophils. *PLoS Pathog* 2010; **6**: e1001133.

30 Mack D, Haeder M, Siemssen N, Laufs R. Association of biofilm production of coagulase-negative staphylococci with expression of a specific polysaccharide intercellular adhesin. *J Infect Dis* 1996; **174**: 881–884.

31 Schommer NN, Christner M, Hentschke M *et al*. Staphylococcus epidermidis uses distinct mechanisms of biofilm formation to interfere with phagocytosis and activation of mouse macrophage-like cells 774A.1. *Infect Immun* 2011; **79**: 2267–2276.

32 Rohde H, Frankenberger S, Zähringer U, Mack D. Structure, function and contribution of polysaccharide intercellular adhesin (PIA) to Staphylococcus epidermidis biofilm formation and pathogenesis of biomaterial-associated infections. *Eur J Cell Biol* 2010; **89**: 103–111.

33 Christner M, Franke GC, Schommer NN *et al*. The giant extracellular matrix-binding protein of Staphylococcus epidermidis mediates biofilm accumulation and attachment to fibronectin. *Mol Microbiol* 2010; **75**: 187–207.

34 Davies DG, Geesey GG. Regulation of the alginate biosynthesis gene algC in Pseudomonas aeruginosa during biofilm development in continuous culture. *Appl Environ Microbiol* 1995; **61**: 860–867.

35 Fey PD, Olson ME. Current concepts in biofilm formation of Staphylococcus epidermidis. *Future Microbiol* 2010; **5**: 917–933.

36 Yao Y, Vuong C, Kocianova S *et al*. Characterization of the Staphylococcus epidermidis accessory-gene regulator response: quorum-sensing regulation of resistance to human innate host defense. *J Infect Dis* 2006; **193**: 841–848.

37 Bayles KW. The biological role of death and lysis in biofilm development. *Nat Rev Microbiol* 2007; **5**: 721–726.

38 Rice KC, Mann EE, Endres JL *et al*. The cidA murein hydrolase regulator contributes to DNA release and biofilm development in Staphylococcus aureus. *Proc Natl Acad Sci U S A* 2007; **104**: 8113–8118.

39 Wagner VE, Iglewski BH. P. aeruginosa Biofilms in CF Infection. *Clin Rev Allergy Immunol* 2008; **35**: 124–134.

40 Hughes DT, Sperandio V. Inter-kingdom signalling: communication between bacteria and their hosts. *Nat Rev Microbiol* 2008; **6**: 111–120.

41 Williams P, Camara M. Quorum sensing and environmental adaptation in Pseudomonas aeruginosa: a tale of regulatory networks and multifunctional signal molecules. *Curr Opin Microbiol* 2009; **12**: 182–191.

42 Savva A, Roger R. Targeting Toll-like receptors: promising therapeutic strategies for the management of sepsis-associated pathology and infectious diseases. *Front Immunol* 2013; **4**: 387.

43 Suresh R, Mosser DM. Pattern recognition receptors in innate immunity, host defense, and immunopathology. *Adv Physiol Educ* 2013; **37**: 284–291.

44 Akira S, Uematsu S, Takeuchi O. Pathogen recognition and innate immunity. *Cell* 2006; **124**: 783–801.

45 Hanke ML, Kielian T. Toll-like receptors in health and disease in the brain: mechanisms and therapeutic potential. *Clin Sci (Lond)* 2011; **121**: 367–387.

46 Hanke ML, Angle A, Kielian T. MyD88-dependent signaling influences fibrosis and alternative macrophage activation during Staphylococcus aureus biofilm infection. *PLoS One* 2012; **7**: e42476.

47 Hussain M, Hastings JGM, White PJ. Comparison of cell-wall teichoic acid with high-molecular-weight extracellular slime material from Staphylococcus epidermidis. *Journal of Medical Microbiology* 1992; **37**: 368–375.

48 Jabbouri S, Sadovskaya I. Characteristics of the biofilm matrix and its role as a possible target for the detection and eradication of Staphylococcus epidermidis associated with medical implant infections. *FEMS Immunol Med Microbiol* 2010; **59**: 280–291.

49 Stevens NT, Sadovskaya I, Jabbouri S *et al.* Staphylococcus epidermidis polysaccharide intercellular adhesin induces IL-8 expression in human astrocytes via a mechanism involving TLR2. *Cell Microbiol* 2009; **11**: 421–432.

50 Thurlow LR, Hanke ML, Fritz T *et al.* Staphylococcus aureus biofilms prevent macrophage phagocytosis and attenuate inflammation in vivo. *J Immunol* 2011; **186**: 6585–6596.

51 Snowden JN, Beaver M, Beenken K *et al.* Staphylococcus aureus sarA Regulates Inflammation and Colonization during Central Nervous System Biofilm Formation. *PLoS One* 2013; **8**: e84089.

52 Fuxman Bass JI, Russo DM, Gabelloni ML *et al.* Extracellular DNA: a major proinflammatory component of Pseudomonas aeruginosa biofilms. *J Immunol* 2010; **184**: 6386–6395.

53 Benmohamed F, Medina M, Wu YZ *et al.* Toll-like receptor 9 deficiency protects mice against Pseudomonas aeruginosa lung infection. *PLoS One* 2014; **9**: e90466.

54 Meyle E, Brenner-Weiss G, Obst U *et al.* Immune defense against S. epidermidis biofilms: components of the extracellular polymeric substance activate distinct bactericidal mechanisms of phagocytic cells. *Int J Artif Organs* 2012; **35**: 700–712.

55 Cerca F, Andrade F, Franca A *et al.* Staphylococcus epidermidis biofilms with higher proportions of dormant bacteria induce a lower activation of murine macrophages. *J Med Microbiol* 2011; **60**: 1717–1724.

56 Otterlei M, Sundan A, Skjak-Braek G *et al.* Similar mechanisms of action of defined polysaccharides and lipopolysaccharides: characterization of binding and tumor necrosis factor alpha induction. *Infect Immun* 1993; **61**: 1917–1925.

57 Flo TH, Ryan L, Latz E *et al.* Involvement of toll-like receptor (TLR) 2 and TLR4 in cell activation by mannuronic acid polymers. *J Biol Chem* 2002; **277**: 35489–35495.

58 Hanke ML, Heim CE, Angle A *et al.* targeting macrophage activation for the prevention and treatment of Staphylococcus aureus biofilm infections. *J Immunol* 2013; **190**: 2159–2168.

59 Park KR, Bryers JD. Effect of macrophage classical (M1) activation on implant-adherent macrophage interactions with Staphylococcus epidermidis: A murine in vitro model system. *J Biomed Mater Res A* 2012; **100**: 2045–2053.

60 Prosser A, Hibbert J, Strunk T *et al.* Phagocytosis of neonatal pathogens by peripheral blood neutrophils and monocytes from newborn preterm and term infants. *Pediatr Res* 2013; **74**: 503–510.

61 Hirschfeld J. Dynamic interactions of neutrophils and biofilms. *J Oral Microbiol* 2014; **6**: 26102.

62 Bernthal NM, Pribaz JR, Stavrakis AI *et al.* Protective role of IL-1beta against post-arthroplasty Staphylococcus aureus infection. *J Orthop Res* 2011; **29**: 1621–1626.

63 Cerca N, Jefferson KK, Oliveira R *et al.* Comparative antibody-mediated phagocytosis of Staphylococcus epidermidis cells grown in a biofilm or in the planktonic state. *Infect Immun* 2006; **74**: 4849–4855.

64 Guenther F, Stroh P, Wagner C *et al.* Phagocytosis of staphylococci biofilms by polymorphonuclear neutrophils: S. aureus and S. epidermidis differ with regard to their susceptibility towards the host defense. *Int J Artif Organs* 2009; **32**: 565–573.

65 Scherr TD, Roux CM, Hanke ML *et al.* Global transcriptome analysis of Staphylococcus aureus biofilms in response to innate immune cells. *Infect Immun* 2013; **81**: 4363–4376.

66 Wagner C, Obst U, Hansch GM. Implant-associated posttraumatic osteomyelitis: collateral damage by local host defense? *Int J Artif Organs* 2005; **28**: 1172–1180.

67 Wagner C, Kondella K, Bernschneider T *et al.* Post-traumatic osteomyelitis: analysis of inflammatory cells recruited into the site of infection. *Shock* 2003; **20**: 503–510.

68 Bostanci N, Belibasakis GN. Porphyromonas gingivalis: an invasive and evasive opportunistic oral pathogen. *FEMS Microbiol Lett* 2012; **333**: 1–9.

69 van Gennip M, Christensen LD, Alhede M *et al.* Interactions between polymorphonuclear leukocytes and Pseudomonas aeruginosa biofilms on silicone implants in vivo. *Infect Immun* 2012; **80**: 2601–2607.

70 Jesaitis AJ, Franklin MJ, Berglund D, *et al.* Compromised host defense on Pseudomonas aeruginosa biofilms: characterization of neutrophil and biofilm interactions. *J Immunol* 2003; **171**: 4329–4339.

71 Walker TS, Tomlin KL, Worthen GS, *et al.* Enhanced Pseudomonas aeruginosa biofilm development mediated by human neutrophils. *Infect Immun* 2005; **73**: 3693–3701.

72 Heim CE, Vidlak D, Scherr TD *et al.* Myeloid-derived suppressor cells contribute to staphylococcus aureus orthopedic biofilm infection. *The Journal of Immunology* 2014; **192**: 3778–3792.

73 Prabhakara R, Harro JM, Leid JG *et al.* Murine immune response to a chronic Staphylococcus aureus biofilm infection. *Infect Immun* 2011; **79**: 1789–1796.

74 Prabhakara R, Harro JM, Leid JG *et al.* Suppression of the inflammatory immune response prevents the development of chronic biofilm infection due to methicillin-resistant Staphylococcus aureus. *Infect Immun* 2011; **79**: 5010–5018.

75 Awang RA, Lappin DF, MacPherson A *et al.* Clinical associations between IL-17 family cytokines and periodontitis and potential differential roles for IL-17A and IL-17E in periodontal immunity. *Inflamm Res* 2014; **63**: 1001–1012.

76 Goodyear CS, Silverman GJ. Staphylococcal toxin induced preferential and prolonged in vivo deletion of innate-like B lymphocytes. *Proc Natl Acad Sci U S A* 2004; **101**: 11392–11397.

77 Kim HK, Emolo C, Missiakas D, Schneewind O. A monoclonal antibody that recognizes the E domain of staphylococcal protein A. *Vaccine* 2014; **32**: 464–469.

78 Kim HK, Kim HY, Schneewind O, Missiakas D. Identifying protective antigens of Staphylococcus aureus, a pathogen that suppresses host immune responses. *FASEB J* 2011; **25**: 3605–3612.

79 Doring G, Goldstein W, Roll A *et al.* Role of Pseudomonas aeruginosa exoenzymes in lung infections of patients with cystic fibrosis. *Infect Immun* 1985; **49**: 557–562.

80 Tonetti MS, Imboden MA, Lang NP. Neutrophil migration into the gingival sulcus is associated with transepithelial gradients of interleukin-8 and ICAM-1. *J Periodontol* 1998; **69**: 1139–1147.

81 Belibasakis GN, Bao K, Bostanci N. Transcriptional profiling of human gingival fibroblasts in response to multi-species in vitro subgingival biofilms. *Mol Oral Microbiol* 2014; **29**: 174–183.

82 Khalaf H, Lonn J, Bengtsson T. Cytokines and chemokines are differentially expressed in patients with periodontitis: possible role for TGF-beta1 as a marker for disease progression. *Cytokine* 2014; **67**: 29–35.

83 Belibasakis GN, Guggenheim B, Bostanci N. Down-regulation of NLRP3 inflammasome in gingival fibroblasts by subgingival biofilms: involvement of Porphyromonas gingivalis. *Innate Immun* 2013; **19**: 3–9.

84 Yamin M, Holbrook EH, Gray ST *et al.* Cigarette smoke combined with Toll-like receptor 3 signaling triggers exaggerated epithelial regulated upon activation, normal T-cell expressed and secreted/CCL5 expression in chronic rhinosinusitis. *J Allergy Clin Immunol* 2008; **122**: 1145–1153 .e3.

85 Ramanathan M, Jr, Lee WK, Dubin MG *et al.* Sinonasal epithelial cell expression of Toll-like receptor 9 is decreased in chronic rhinosinusitis with polyps. *Am J Rhinol* 2007; **21**: 110–116.

86 Dong Z, Yang Z, Wang C. Expression of TLR2 and TLR4 messenger RNA in the epithelial cells of the nasal airway. *Am J Rhinol* 2005; **19**: 236–239.

87 Lopez-Leban F, Kiran MD, Wolcott R, Balaban N. Molecular mechanisms of RIP, an effective inhibitor of chronic infections. *Int J Artif Organs* 2010; **33**: 582–589.

88 Menousek J, Mishra B, Hanke ML *et al.* Database screening and in vivo efficacy of antimicrobial peptides against methicillin-resistant Staphylococcus aureus USA300. *Int J Antimicrob Agents* 2012; **39**: 402–406.

89 Dean SN, Bishop BM, van Hoek ML. Natural and synthetic cathelicidin peptides with anti-microbial and anti-biofilm activity against Staphylococcus aureus. *BMC Microbiol* 2011; **11**: 114-2180-11-114.

90 Sol A, Ginesin O, Chaushu S *et al.* LL-37 opsonizes and inhibits biofilm formation of Aggregatibacter actinomycetemcomitans at subbactericidal concentrations. *Infect Immun* 2013; **81**: 3577–3585.

91 Wang R, Braughton KR, Kretschmer D *et al.* Identification of novel cytolytic peptides as key virulence determinants for community-associated MRSA. *Nat Med* 2007; **13**: 1510–1514.

92 Otto M. Molecular basis of Staphylococcus epidermidis infections. *Semin Immunopathol* 2012; **34**: 201–214.

93 Queck SY, Khan BA, Wang R *et al.* Mobile genetic element-encoded cytolysin connects virulence to methicillin resistance in MRSA. *PLoS Pathog* 2009; **5**: e1000533.

94 Vuong C, Durr M, Carmody AB *et al.* Regulated expression of pathogen-associated molecular pattern molecules in Staphylococcus epidermidis: quorum-sensing determines pro-inflammatory capacity and production of phenol-soluble modulins. *Cell Microbiol* 2004; **6**: 753–759.

95 Dumont AL, Nygaard TK, Watkins RL *et al.* Characterization of a new cytotoxin that contributes to Staphylococcus aureus pathogenesis. *Mol Microbiol* 2011; **79**: 814–825.

96 DuMont AL, Yoong P, Day CJ, *et al.* Staphylococcus aureus LukAB cytotoxin kills human neutrophils by targeting the CD11b subunit of the integrin Mac-1. *Proc Natl Acad Sci U S A* 2013; **110**: 10794–10799.

97 Hanamsagar R, Torres V, Kielian T. Inflammasome activation and IL-1beta/IL-18 processing are influenced by distinct pathways in microglia. *J Neurochem* 2011; **119**: 736–748.

98 Berends ET, Horswill AR, Haste N, *et al.* Nuclease expression by Staphylococcus aureus facilitates escape from neutrophil extracellular traps. *J Innate Immun* 2010; **2**: 576–586.

99 Alhede M, Bjarnsholt T, Givskov M, Alhede M. Pseudomonas aeruginosa biofilms: mechanisms of immune evasion. *Adv Appl Microbiol* 2014; **86**: 1–40.

100 Spiliopoulou AI, Kolonitsiou F, Krevvata MI *et al.* Bacterial adhesion, intracellular survival and cytokine induction upon stimulation of mononuclear cells with planktonic or biofilm phase Staphylococcus epidermidis. *FEMS Microbiol Lett* 2012; **330**: 56–65.

101 Anderson AS, Miller AA, Donald RG *et al.* Development of a multicomponent Staphylococcus aureus vaccine designed to counter multiple bacterial virulence factors. *Hum Vaccin Immunother* 2012; **8**: 1585–1594.

102 Harro JM, Peters BM, O'May GA *et al.* Vaccine development in Staphylococcus aureus: taking the biofilm phenotype into consideration. *FEMS Immunol Med Microbiol* 2010; **59**: 306–323.

103 Brady RA, O'May GA, Leid JG *et al.* Resolution of Staphylococcus aureus biofilm infection using vaccination and antibiotic treatment. *Infect Immun* 2011; **79**: 1797–1803.

104 Van Mellaert L, Shahrooei M, Hofmans D, Eldere JV. Immunoprophylaxis and immunotherapy of Staphylococcus epidermidis infections: challenges and prospects. *Expert Rev Vaccines* 2012; **11**: 319–334.

105 Taylor PK, Yeung AT, Hancock RE. Antibiotic resistance in Pseudomonas aeruginosa biofilms: towards the development of novel anti-biofilm therapies. *J Biotechnol* 2014; **191**: 121–130.

106 Kim SK, Park HY, Lee JH. Anthranilate deteriorates biofilm structure of Pseudomonas aeruginosa and antagonizes the biofilm-enhancing indole effect. *Appl Environ Microbiol* 2015; **81**: 2328–38.

107 Bjarnsholt T, Jensen PO, Burmolle M *et al.* Pseudomonas aeruginosa tolerance to tobramycin, hydrogen peroxide and polymorphonuclear leukocytes is quorum-sensing dependent. *Microbiology* 2005; **151**: 373–383.

108 Rajendran M, Priyadharshini V, Arora G. Is immunesenescence a contributing factor for periodontal diseases? *J Indian Soc Periodontol* 2013; **17**: 169–174.

109 Kuo LC, Polson AM, Kang T. Associations between periodontal diseases and systemic diseases: a review of the inter-relationships and interactions with diabetes, respiratory diseases, cardiovascular diseases and osteoporosis. *Public Health* 2008; **122**: 417–433.

CHAPTER 9

Co-evolution of microbes and immunity and its consequences for modern-day life

Markus B. Geuking

Mucosal Immunology Lab, University of Bern, Switzerland

9.1 Introduction

We live in a world full of microbes and consequently all of our body surfaces are colonized with a plethora of mostly beneficial symbiotic microbes. How our immune system handles these microbes and how these microbes influence our immune system is currently a very active research field. It is now becoming evident that few, if any, body systems remain unaffected by the presence of our microbial inhabitants. The dramatic increase of immune-mediated disorders such as allergic, atopic, autoimmune, and auto-inflammatory diseases over the last few decades has fuelled research aiming to understand host-microbial interactions. Several hypotheses such as the "hygiene hypothesis"[1], "old friends hypothesis"[2], or "microflora hypothesis"[3] have been put forward to postulate that an altered environment, be it through hygiene, sanitation, antibiotics, vaccines, diet, or absence of parasites, has changed microbial and/or infectious exposure in such a way that the immune system no longer receives appropriate or sufficient immuno-regulatory signals. This dysregulated immune system is therefore more prone to act inappropriately, and hence the observed increase in immune-mediated disorders. Importantly, on an evolutionary scale, the environmental (not genetic) changes that occurred over the last few decades happened in the blink of an eye. It is therefore possible that the adaptive immune system, although evolutionarily flexible to adapt to environmental changes such as pathogen load and parasites, has not yet caught up with the observed changes in our environment, which includes changes in our microbiota. In this chapter we put the co-evolution of microbes and the immune system into the context of host-microbial immune interactions and assess how this co-evolution might continue.

The Human Microbiota and Chronic Disease: Dysbiosis as a Cause of Human Pathology, First Edition.
Edited by Luigi Nibali and Brian Henderson.

9.2 Symbiosis in eukaryotic evolution

Bacteria evolved more than 3 billion years ago and formed a dominant part of the environment (Figure 1). They probably filled every available ecological niche and played a key role in one of the most important evolutionary events, namely the origin of eukaryotes[4] and later the evolution of animals[5,6]. Animals therefore evolved in a milieu filled with bacteria, and this close association with bacteria, including predation, commensals, and infection with pathogenic bacteria, has consequently been a constant force during animal evolution.

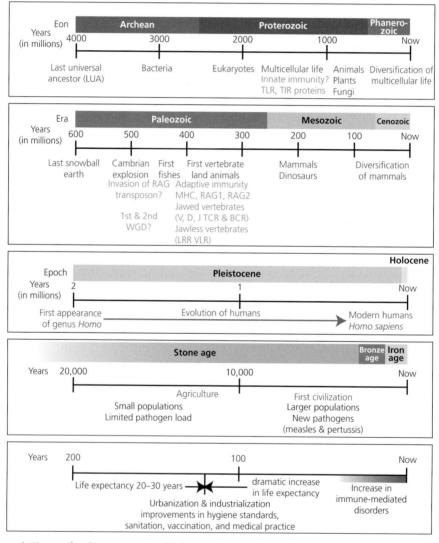

Figure 1 Time scale of immune-microbial co-evolution. The evolution of life is shown and important evolutionary (red), immunological (green), and modernization events (black) are indicated. The different eons, eras, epochs, and ages are indicated in the colored boxes. (*see color plate section for color details*).

Besides the presence of bacteria in the same environment, animals have also always lived with a plethora of microorganisms on their external (e.g. skin) as well as internal (e.g. oral cavity or intestines) body surfaces. This tendency of eukaryotes, and mammals in particular, to form close microbial associations is an evolutionary route that is markedly different from the bacterial counterpart, which associate among themselves in a variety of consortia, but with only a few members associating with eukaryotic hosts. Only specific members from 11 of the 55 bacterial phyla described so far have been shown to associate with eukaryotes[7]. In the case of the human (or mammalian) gut this number reduces to bacterial members from only eight phyla[8]. The two most dominant phyla in the human intestine are Bacteroidetes and Firmicutes. This already indicates that colonization of the intestine is not a random event that is easily achieved by most bacterial species, but rather that this is an evolutionarily established process[9]. The intestine provides a very specific niche that, in a competitive real-life situation, can only be colonized by a select group of bacteria that eventually form a consortium that, in its entirety, is beneficial to both the host and the microbiota. Interestingly, although the large intestine seems to be a very selective niche, bacterial colonization can reach densities up to 10^{12} microbes per gram. This is the highest bacterial density observed so far in bacterial ecosystems[10]. The advantages for the microbes are therefore the presence of an almost perfect niche with excellent nutrient availability. In addition, the metabolic contribution of the microbiota to the overall metabolic capacity of the host-microbial super-organism is tremendous[10,11]. Host-microbial symbiosis provides the host with additional metabolic capacities and colonization resistance to pathogenic bacteria[12]. Mammals developed, in addition to innate immune defenses, a powerful adaptive immune system as well. Therefore, the co-evolution of bacterial symbiosis and the immune system required that careful regulatory mechanisms be in place to maintain health and avoid unnecessary immune and inflammatory responses[13].

9.3 Evolution of the (innate and adaptive) immune system

As mentioned above, bacteria occupied every single possible niche at the time eukaryotes and animals evolved. Many of these bacteria would now be called pathogens since they were capable of killing other organisms or using them to generate copies of themselves. Therefore eukaryotes have constantly evolved defense mechanisms to be able to survive these attacks. Of course, in return microorganisms constantly applied new attack mechanisms, generating a host-versus-pathogen arms race[14].

9.3.1 Immune proteins

Proteins and protein domains involved in immune defense against pathogens are universal in animals and therefore most likely evolved before the origin of animals[15]. Such proteins include the Toll/Interleukin 1 receptor/resistance domain (TIR) and the interleukin-1 receptor-like domain. Toll-like receptors (TLR) also predate the divergence of plants and animals (Figure 1). TLRs belong to germ-line encoded pattern-recognition receptors (PRR) and form a crucial part of the innate immune system[16]. They have evolved to recognize highly conserved pathogen-associated patterns (PAMP) such as the bacterial endotoxin, lipopolysaccharide (LPS) (recognized by TLR4), peptidoglycan (recognized by TLR2), and bacterial DNA (recognized by TLR9). It is likely that these receptors already fulfilled a defensive function before the divergence of plants and animals.

9.3.2 Evolution of adaptive immunity

In addition to the innate immune system with its germ-line encoded components, jawed vertebrates (gnathostomes) have evolved a lymphocyte-based adaptive immune system that generates a vast repertoire of antigen-specificity by rearranging immunoglobulin (Ig) V, D, J gene segments to form T and B cell receptors[17,18]. Immunoglobulins first appeared in cartilaginous fish. To cover the vast possible repertoire and allow for positive and negative clonal selection, the adaptive immune system required a large number of lymphocytes in each individual. Large-bodied vertebrates were the first animals to fulfil this requirement. It is believed that two important events, the invasion of a recombination-activating gene (RAG) transposon[19] and two whole-genome duplications (WGD)[20], provided the key for the rapid appearance of the adaptive immune system (Figure 1).

Rag-1 and *Rag-2* are crucial genes for rearrangement of the V, D, and J regions. They have no homologous regions and no introns but are closely linked to each other. Therefore, vertebrates might have acquired them as part of a transposon[21,22]. However, when exactly the RAG transposon invaded the genome to allow the appearance of an adaptive immune system remains controversial[19,23]. It is generally accepted that the vertebrate genome underwent two rounds of WGD before the radiation of jawed vertebrates[20] (Figure 1). These WGD likely played a crucial role in the evolution of the adaptive immune system since many essential components of the adaptive immune system, including MHC genes, form paralogy groups consisting of paralogues resulting from WGD[24].

9.3.3 Two separate adaptive immune systems evolved

Interestingly, it looks as if two very different recombinatorial systems for the generation of antigen-receptors appeared about 500 million years ago at the dawn of vertebrate evolution in jawless vs jawed vertebrates (Figure 1). In contrast to the recombinatorial VDJ system of jawed vertebrates, jawless vertebrates assemble their antigen receptors by genomic rearrangement of leucine-rich repeat (LRR)-encoding modules to form variable lymphocyte receptors (VLR)[25–27]. This almost simultaneous appearance of two types of recombinatorial immune systems within such a short evolutionary time frame, the immense energy investment to generate a repertoire of lymphocytes of which most are negatively selected, and the high risk of generating autoimmunity, raise the question of what was the evolutionary pressure that caused the appearance of these two independent adaptive immune systems.

LRR-containing proteins are ancient mediators of antimicrobial responses and it is therefore likely that the last common ancestors of plants and animals used LRR-containing proteins for microbial detection[28]. Indeed, it is speculated that the LRR-containing proteins and other immune receptors may have evolved because they play a pivotal role in the maintenance and surveillance of endosymbiotic bacterial communities[14]. It is possible that this complex long-term coexistence between animals and commensal microbes may have favored the evolution of vast repertoires of very specific antigen receptors because the strategy of the innate immune system to recognize PAMPs indiscriminatively targets all microorganisms. Therefore, the evolutionary force provided by the microbiota may have driven the evolution of an adaptive immune system not only to detect and clear microbial pathogens, but also to detect, control, and appropriately respond to the non-pathogenic commensal microbes present. The adaptive immune system is exquisitely equipped to sense and respond to foreign non-self antigens. The microbiota therefore presents a particular challenge due to its enormous load of foreign antigens. One way to deal with this is by immunological ignorance through spatial separation of the bacteria from the immune system or by down-modulation of innate and adaptive immunity to prevent overt inflammation[29]. Another way to deal with this situation is through the induction of immunological tolerance, which is a likely scenario since it has been demonstrated that some gut bacteria are excellent inducers of regulatory T cells[30,31]. This, of course, raises the possibility that the main contribution of the microbiota to the evolution of the adaptive immune system involved the establishment of regulatory mechanisms to suppress unwanted inflammation toward the microbiota[32–34].

9.4 Hygiene hypothesis

How does the important role of the microbiota in the evolution of our immune system affect our understanding of immune-mediated disorders in modern life? The Hygiene Hypothesis, originally proposed by Strachan[1], is based on the idea that our immune system is no longer exposed to the same environmental triggers that it encountered when it evolved. Although this hypothesis was originally postulated to explain allergic disorders, it was extended and modified several times to incorporate the observed increase in not only allergic, but also a range of chronic inflammatory and auto-immune disorders (see the "old friends hypothesis"[35] and the "microflora hypothesis"[3]). It is clear that changes in host genetics alone cannot explain the marked increase in disease susceptibility observed over a time-span of only 40 years and therefore, these hypotheses focused on what might have changed in the environment.

Initially, epidemiological studies revealed an inverse correlation between hygiene levels and the susceptibility to disease. For example, children with older siblings, children who attended daycare centers within their first year, and children who grew up on a farm with livestock all had a reduced prevalence of asthma[1,36]. The term "old friends" refers to the idea that the immune system evolved to mainly respond to the microbiota and organisms that modify the microbiota and to "old" infections that had the capacity to persist in smaller populations because they were rather harmless or subclinical or could produce a carrier state (Figure 1). These organisms had to be tolerated by the immune system and were therefore likely involved in the evolution of immune regulatory mechanisms[2]. Such "old infections" also included parasites such as helminths. Later, when populations got bigger during urbanization, a different kind of infections emerged that either killed the host or induced strong immunity.

These infections included childhood virus infections and naturally, these agents were not involved in the development or evolution of immune-regulatory mechanisms.

Improvements in hygiene standards, sanitation, and medical practice during industrialization (Figure 1) probably contributed to the disappearance of non-lethal parasites, in particular helminths. Importantly, parasitic helminth infections are rarely lethal and induce immunoregulatory mechanisms[37]. Interestingly, evolutionarily, helminths appeared at the same time as the adaptive immune system[38]. It can be imagined that due to the disappearance of these parasites the immune system no longer received appropriate immune regulatory signals, which may contribute to an increase in immune-mediated disorders.

Lastly, the microflora hypothesis of allergic diseases proposed by Noverr and Huffnagel[3] argues that changes in the microbiota composition, caused by, for example, antibiotic use early in life and dietary changes in industrialized countries, led to altered immunological tolerance and regulation.

9.5 What drives the composition of the microbiota?

As mentioned above, the selective advantage of acquiring and maintaining a very diverse microbiota that provides colonization resistance, metabolic pathways, and immune education is likely to have been involved in evolution of the adaptive immune system, whereas invertebrates that solely rely on innate immunity show much simpler microbiotas. So what determines the composition of the microbiota? The composition of the intestinal microbiota is very dynamic during the first three years of life until it establishes an adult-like microbial community, which appears to be relatively stable. Immediately following birth, the newborn intestine is not completely anaerobic and therefore allows for colonization by facultative aerobes like *Enterobacteriaceae*. This initial colonization causes the intestinal lumen to become extremely anaerobic within a few days, thus allowing other obligate anaerobic species, such as *Bifidobacteria, Clostridia* spp. and *Bacteriodetes,* to populate the intestine. These initial colonization events are heavily influenced by the mode of delivery (also covered in chapters 3 and 6 of this book)[39,40]. The microbiota of children delivered by Cesarean section (resembling the skin microbiota) largely differs from children born vaginally (vaginal microbiota)[40]. Correlative studies suggest that babies delivered by Cesarean section are more likely to develop obesity[41], inflammatory bowel disease (IBD)[42], and asthma or atopic diseases[43] before adulthood.

Diet is also a critical determination of microbial composition and the use of breast milk versus formula feeding during the first month of life also impacts on the composition of the microbiota[40]. The introduction of solid food also leads to shifts in the composition and complexity of the microbiota. The presence of new substrates in the intestine naturally results in compositional shifts towards species that can utilize these new compounds. By three years of age, the microbiota becomes more "adult-like" in terms of stability, diversity and composition. The two most dominant bacterial phyla within the adult microbiota are *Firmicutes* and *Bacteroidetes*. Firmicutes are comprised mainly of the genera Clostridium, Faecalibacterium, Blautia, Ruminococcus, and Lactobacillus, while

Bacteroidetes are mainly represented by Bacteroides and Prevotella. Less abundant are other phyla such as Actinobacteria (Bifidobacteria) or Proteobacteria (Enterobacteriaceae).

Although the gut microbiota is relatively stable in healthy adults, it is influenced by a variety of environmental factors. Antibiotics are obvious candidates for modulating the microbiota. Excessive use of antibiotics in neonates has also been associated with higher risk to develop asthma[44,45], cow's-milk allergy[46], irritable bowel syndrome[47], IBD[48], and obesity[49]. The "western" life style is also characterized by a particular diet, which can have profound effects on the microbiota[50].

It is now widely accepted that early-life exposure to microbes is particularly important for the development of the immune system and for tuning the immune system reactivity later in life[51–53]. This early-life window of opportunity is even likely to extend into the prenatal phase during pregnancy.

9.6 The pace of evolution

As a species, humans went through different phases that are thought to differ in terms of population density and pathogen load. Very early human societies had rather small population sizes (Figure 1) and therefore carried a relatively limited pathogen load with high transmission rates but conferring little immunity. The rise of agriculture around 10,000 years ago (Figure 1) led to the development of much larger human communities with concomitant appearance of new pathogens such as measles and pertussis[54]. Besides hygiene, environment, and pathogenic or infectious burden, changes in life expectancy must also be considered. During most of human history, until the 19th century, the average life expectancy was between 20 and 30 years (Figure 1). Most casualties were likely due to infections[55]. Within only 150 years, the average life expectancy in developed industrialized countries has risen dramatically, which is obviously not the result of genetic adaptation to pathogens but is caused by massive improvements in hygiene, sanitation, and development of vaccines and antibiotics. As a consequence, technological advances and adaptations have hugely outpaced genetic adaptation.

Genomic studies at the (coding and non-coding) regulatory level of immune response genes indicate that these genes are generally less evolutionary constrained and seem to be more frequently targeted by positive selection than those genes involved in other biological processes[56–59]. Despite this relative genetic flexibility of the immune system to adapt to changes in pathogen presence, in evolutionary terms, it must be kept in mind that it was likely impossible for the immune system to adapt to the recent changes that have occurred in such a short time frame.

The most prominent example of genes that maintain variability due to selection pressure is the major histocompatibility (MHC) genes. MHC genes display a high degree of polymorphism, which is at least in part due to pathogen-driven selective pressure[60]. It could be envisioned that the immune system will eventually catch up and adapt to the dramatic changes that occurred in our environment, and in particular within our microbiota, over the last several decades, but this is pure speculation.

TAKE-HOME MESSAGE

- The interaction between the microbiota and the immune system is very complex and delicate and was established by co-evolution over millions of years.

- It is possible that careful management of the host-microbial mutualistic relationship was one of the main driving forces for the evolution of the adaptive immune system.

- The environmental changes (hygiene, sanitation, vaccination, diet, etc....) over the last few decades have in evolutionary terms happened in the blink of an eye.

References

1 Strachan DP. Hay fever, hygiene, and household size. *BMJ* 1989; **299**: 1259–60.

2 Rook GA, Raison CL, Lowry CA. Microbial 'old friends', immunoregulation and socioeconomic status. *Clinical and Experimental Immunology* 2014; **177**: 1–12.

3 Noverr MC, Huffnagle GB. The 'microflora hypothesis' of allergic diseases. *Clinical and Experimental Allergy: Journal of the British Society for Allergy and Clinical Immunology* 2005; **35**: 1511–20.

4 Alegado RA, King N. Bacterial influences on animal origins. *Cold Spring Harb Perspect Biol* 2014; **6**.

5 Knoll AH. The multiple origins of complex multicellularity. *Annual Review of Earth and Planetary Sciences* 2011; **39**: 217–39.

6 Narbonne GM. The Ediacara biota: neoproterozoic origin of animals and their ecosystems. *Annual Review of Earth and Planetary Sciences* 2005; **33**: 421–42.

7 Sachs JL, Skophammer RG, Regus JU. Evolutionary transitions in bacterial symbiosis. *Proc Natl Acad Sci U S A* 2011; **108 Suppl 2**: 10800–7.

8 Turnbaugh PJ, Ley RE, Hamady M *et al.* The human microbiome project. *Nature* 2007; **449**: 804–10.

9 Seedorf H, Griffin NW, Ridaura VK *et al.* Bacteria from diverse habitats colonize and compete in the mouse gut. *Cell* 2014; **159**: 253–66.

10 O'Hara AM, Shanahan F. The gut flora as a forgotten organ. *EMBO Reports* 2006; **7**: 688–93.

11 Wikoff WR, Anfora AT, Liu J *et al.* Metabolomics analysis reveals large effects of gut microflora on mammalian blood metabolites. *Proc Natl Acad Sci U S A* 2009; **106**: 3698–703.

12 Geuking MB, Koller Y, Rupp S *et al.* The interplay between the gut microbiota and the immune system. *Gut Microbes* 2014; **5**: 411–8.

13 Lee YK, Mazmanian SK. Has the microbiota played a critical role in the evolution of the adaptive immune system? *Science* 2010; **330**: 1768–73.

14 Pancer Z, Cooper MD. The evolution of adaptive immunity. *Annu Rev Immunol* 2006; **24**: 497–518.

15 Srivastava M, Larroux C, Lu DR *et al.* Early evolution of the LIM homeobox gene family. *BMC Biology* 2010; **8**: 4.

16 Beutler B, Poltorak A. Sepsis and evolution of the innate immune response. *Critical Care Medicine* 2001; **29**: S2–6; discussion S-7.

17 Cannon JP, Haire RN, Rast JP *et al.* The phylogenetic origins of the antigen-binding receptors and somatic diversification mechanisms. *Immunol Rev* 2004; **200**: 12–22.

18 Flajnik MF, Du Pasquier L. Evolution of innate and adaptive immunity: can we draw a line? *Trends in Immunology* 2004; **25**: 640–4.

19 Fugmann SD, Messier C, Novack LA *et al.* An ancient evolutionary origin of the Rag1/2 gene locus. *Proc Natl Acad Sci U S A* 2006; **103**: 3728–33.

20 Panopoulou G, Poustka AJ. Timing and mechanism of ancient vertebrate genome duplications — the adventure of a hypothesis. *Trends in Genetics: TIG* 2005; **21**: 559–67.

21 Oettinger MA, Schatz DG, Gorka C *et al.* RAG-1 and RAG-2, adjacent genes that synergistically activate V(D)J recombination. *Science* 1990; **248**: 1517–23.

22 Thompson CB. New insights into V(D)J recombination and its role in the evolution of the immune system. *Immunity* 1995; **3**: 531–9.

23 Crisp A, Boschetti C, Perry M *et al.* Expression of multiple horizontally acquired genes is a hallmark of both vertebrate and invertebrate genomes. *Genome Biology* 2015; **16**: 50.

24 Kasahara M. What do the paralogous regions in the genome tell us about the origin of the adaptive immune system? *Immunol Rev* 1998; **166**: 159–75.

25 Kim HM, Oh SC, Lim KJ *et al.* Structural diversity of the hagfish variable lymphocyte receptors. *J Biol Chem* 2007; **282**: 6726–32.

26 Pancer Z, Amemiya CT, Ehrhardt GR *et al.* Somatic diversification of variable lymphocyte receptors in the agnathan sea lamprey. *Nature* 2004; **430**: 174–80.

27 Pancer Z, Saha NR, Kasamatsu J *et al.* Variable lymphocyte receptors in hagfish. *Proc Natl Acad Sci U S A* 2005; **102**: 9224–9.

28 Meyerowitz EM. Plants compared to animals: the broadest comparative study of development. *Science* 2002; **295**: 1482–5.

29 Hooper LV. Do symbiotic bacteria subvert host immunity? *Nature reviews. Microbiology* 2009; **7**: 367–74.

30 Atarashi K, Tanoue T, Shima T *et al.* Induction of colonic regulatory T cells by indigenous Clostridium species. *Science* 2011; **331**: 337–41.

31 Geuking MB, Cahenzli J, Lawson MA *et al.* Intestinal bacterial colonization induces mutualistic regulatory T cell responses. *Immunity* 2011; **34**: 794–806.

32 O'Mahony C, Scully P, O'Mahony D *et al.* Commensal-induced regulatory T cells mediate protection against pathogen-stimulated NF-kappaB activation. *PLoS Pathog* 2008; **4**: e1000112.

33 Round JL, Mazmanian SK. Inducible Foxp3+ regulatory T-cell development by a commensal bacterium of the intestinal microbiota. *Proc Natl Acad Sci U S A* 2010; **107**: 12204–9.

34 Sokol H, Pigneur B, Watterlot L *et al.* Faecalibacterium prausnitzii is an anti-inflammatory commensal bacterium identified by gut microbiota analysis of Crohn disease patients. *Proc Natl Acad Sci U S A* 2008; **105**: 16731–6.

35 Rook GA, Adams V, Hunt J *et al.* Mycobacteria and other environmental organisms as immunomodulators for immunoregulatory disorders. *Springer seminars in immunopathology* 2004; **25**: 237–55.

36 Strachan DP. Allergy and family size: a riddle worth solving. *Clinical and Experimental Allergy: Journal of the British Society for Allergy and Clinical Immunology* 1997; **27**: 235–6.

37 Maizels RM, McSorley HJ, Smyth DJ. Helminths in the hygiene hypothesis: sooner or later? *Clinical and Experimental Immunology* 2014; **177**: 38–46.

38 Jackson JA, Friberg IM, Little S *et al.* Review series on helminths, immune modulation and the hygiene hypothesis: immunity against helminths and immunological phenomena in modern human populations: coevolutionary legacies? *Immunology* 2009; **126**: 18–27.

39 Penders J, Gerhold K, Thijs C *et al.* New insights into the hygiene hypothesis in allergic diseases: mediation of sibling and birth mode effects by the gut microbiota. *Gut Microbes* 2014; **5**: 239–44.

40 Penders J, Thijs C, Vink C *et al.* Factors influencing the composition of the intestinal microbiota in early infancy. *Pediatrics* 2006; **118**: 511–21.

41 Kalliomaki M, Collado MC, Salminen S *et al.* Early differences in fecal microbiota composition in children may predict overweight. *The American Journal of Clinical Nutrition* 2008; **87**: 534–8.

42 Bager P, Simonsen J, Nielsen NM *et al.* Cesarean section and offspring's risk of inflammatory bowel disease: a national cohort study. *Inflamm Bowel Dis* 2012; **18**: 857–62.

43 Kolokotroni O, Middleton N, Gavatha M *et al.* Asthma and atopy in children born by caesarean section: effect modification by family history of allergies — a population based cross-sectional study. *BMC Pediatrics* 2012; **12**: 179.

44 Russell SL, Gold MJ, Hartmann M *et al.* Early life antibiotic-driven changes in microbiota enhance susceptibility to allergic asthma. *EMBO Reports* 2012; **13**: 440–7.

45 Stensballe LG, Simonsen J, Jensen SM *et al.* Use of antibiotics during pregnancy increases the risk of asthma in early childhood. *The Journal of Pediatrics* 2013; **162**: 832–8 e3.

46 Metsala J, Lundqvist A, Virta LJ *et al.* Mother's and offspring's use of antibiotics and infant allergy to cow's milk. *Epidemiology* 2013; **24**: 303–9.

47 Villarreal AA, Aberger FJ, Benrud R *et al.* Use of broad-spectrum antibiotics and the development of irritable bowel syndrome. *WMJ: official publication of the State Medical Society of Wisconsin* 2012; **111**: 17–20.

48 Shaw SY, Blanchard JF, Bernstein CN. Association between the use of antibiotics in the first year of life and pediatric inflammatory bowel disease. *American Journal of Gastroenterology* 2010; **105**: 2687–92.

49 Trasande L, Blustein J, Liu M *et al.* Infant antibiotic exposures and early-life body mass. *International Journal of Obesity* 2013; **37**: 16–23.

50 Maslowski KM, Mackay CR. Diet, gut microbiota and immune responses. *Nat Immunol* 2011; **12**: 5–9.

51 Cahenzli J, Koller Y, Wyss M *et al.* Intestinal microbial diversity during early-life colonization shapes long-term IgE levels. *Cell Host Microbe* 2013; **14**: 559–70.

52 Djuardi Y, Wammes LJ, Supali T *et al.* Immunological footprint: the development of a child's immune system in environments rich in microorganisms and parasites. *Parasitology* 2011; **138**: 1508–18.

53 Olszak T, An D, Zeissig S *et al.* Microbial exposure during early life has persistent effects on natural killer T cell function. *Science* 2012; **336**: 489–93.

54 Wolfe ND, Dunavan CP, Diamond J. Origins of major human infectious diseases. *Nature* 2007; **447**: 279–83.

55 Casanova JL, Abel L. Inborn errors of immunity to infection: the rule rather than the exception. *J Exp Med* 2005; **202**: 197–201.

56 Blekhman R, Man O, Herrmann L *et al.* Natural selection on genes that underlie human disease susceptibility. *Current Biology: CB* 2008; **18**: 883–9.

57 Castillo-Davis CI, Kondrashov FA, Hartl DL *et al.* The functional genomic distribution of protein divergence in two animal phyla: coevolution, genomic conflict, and constraint. *Genome Research* 2004; **14**: 802–11.

58 Kosiol C, Vinar T, da Fonseca RR *et al.* Patterns of positive selection in six Mammalian genomes. *PLoS Genetics* 2008; **4**: e1000144.

59 Sironi M, Menozzi G, Comi GP *et al.* Analysis of intronic conserved elements indicates that functional complexity might represent a major source of negative selection on non-coding sequences. *Human Molecular Genetics* 2005; **14**: 2533–46.

60 Prugnolle F, Manica A, Charpentier M *et al.* Pathogen-driven selection and worldwide HLA class I diversity. *Current Biology: CB* 2005; **15**: 1022–7.

CHAPTER 10

How viruses and bacteria have shaped the human genome: the implications for disease

Frank Ryan

The Academic Unit of Medical Education, University of Sheffield, United Kingdom

10.1 Genetic symbiosis

In 1966, while studying *Amoeba proteus* at the State University of New York at Buffalo, biologist Kwang Jeon experienced an unexpected calamity: his cultures had been struck down by a lethal plague. When he investigated, he found that the plague was caused by an unknown strain of a bacterium he would subsequently call the x-bacterium. Jeon was interested in cytoplasmic inheritance so, rather than sacrifice his cultures and begin again, he allowed the plague to take its natural course, observing its progress. The plague exterminated the majority of his amoebae, but a tiny minority survived despite the persistence of large numbers of x-bacteria within vacuoles in their cytoplasm (Figures 1 and 2). Then he observed a remarkable change. Roughly a year after the plague had appeared — the equivalent of 40 host generations — the *A. proteus* and the x-bacterium became mutually interdependent[1]. When Jeon removed the cytoplasmic bacteria by the application of antibiotics, the amoebae died. If the amoebae were cut open to release the x-bacteria, the bacteria were no longer capable of independent growth on culture plates. It is possible to transfer nuclei from one amoeba to another within the same species. But when Jeon transplanted the nucleus of the Amoeba-x-bacterium into an amoeba of the parental strain, it proved lethal to the parental strain. It appeared that the union of the parental strain and the x-bacterium had given rise to a new species of amoeba-x-bacterium.

Over many more years of study, Jeon confirmed that there had been genetic exchange between the *A. proteus* and the x-bacterium. They had entered into what symbiologists term a "genetic symbiosis." What then is genetic symbiosis and how does it fit with the classical ideas of Darwinian evolution?

In 1878 Anton de Bary defined symbiosis as "the living together of differently-named organisms". Today we might interpret symbiosis as an interaction between two or more species in which one or more of the interacting "partners" benefits. The partners are called symbionts and the partnership as a whole is called the holobiont. Since only one of the partners needs to benefit, symbiosis includes

The Human Microbiota and Chronic Disease: Dysbiosis as a Cause of Human Pathology, First Edition.
Edited by Luigi Nibali and Brian Henderson.

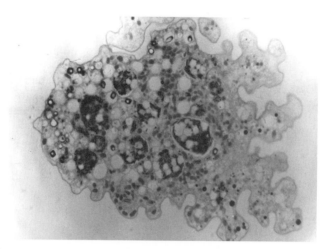

Figure 1 *Amoeba proteus* (courtesy of Professor Kwang Jeon).

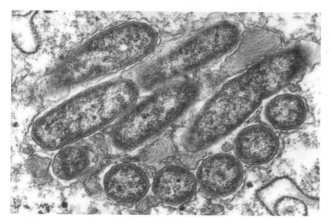

Figure 2 Close-up of vesicle containing x-bacteria (courtesy of Professor Kwang Jeon).

parasitism and commensalism as well as mutualism. Indeed mutualism, in which two or more partners benefit, often evolves from parasitism, with many observed examples of symbiosis lying somewhere in between the two extremes of outright parasitism and mutualism. Symbioses are extremely common in nature, with a wide range of different interactions.

Symbioses are defined by the levels of interactive exchange, including metabolic, behavioral and genetic[2,3]. Historically the earliest examples of symbiosis to be studied were the lichens, which are metabolic partnerships of algae and fungi, and the metabolic symbioses between fungi and the roots of plants, known as mycorrhizae. Other common examples of symbioses include the partnerships between microbes and their hosts. These include bacterial and archaeal symbioses with other microbes, as well as microbial symbioses with animals and plants, including for example the tubeworms of the oceanic vents, marine bioluminescence, *Wolbachia* in its remarkable symbioses with arthropods and parasitic nematodes, and the many other symbionts of insect digestive systems, and the gut flora of humans[4-6]. How intriguing that study of genetic transfer in *Wolbachia* reveals a key role for bacteriophage, a group of viruses that are increasingly seen to play a very important role in many different examples of viral symbiosis[7].

Symbiosis also plays a role in some of the elemental cycles essential for life. For example, the nitrogen cycle is enabled by metabolic symbioses in legumes, such as peas and clover, which form symbiotic unions with soil-based nitrogen-fixing proteobacteria known as rhizobia[8]. The rhizobia stimulate the formation of nodules within the plant roots, where they fix atmospheric nitrogen into organic nitrogenous products needed by the metabolic processes of the plants; meanwhile the plants supply the rhizobia with high-energy compounds from photosynthesis to feed the rhizobia. However, there is an additional layer of complexity to the symbiosis. In soil many natural strains of rhizobia lack the ability to fix nitrogen. But they can acquire it through the transfer of a "symbiotic island" of six genes from a donor strain of rhizobia, *Mesorhysobium loti*[9,10]. Since this involves the sharing of genes between different strains of rhizobia it is an example of genetic symbiosis. But there is a further wrinkle still. A key gene in the transfer package is a bacteriophage integrase, essential for the symbiotic island to be integrated into the receptor bacterium's genome. The integrase is the "smoking gun" of a preceding virus-phage symbiosis that has provided an essential requirement for the bacterium to bacterium genetic symbiosis.

When symbiosis gives rise to evolutionary change, it is termed "symbiogenesis"[11]. Genetic symbioses are a source of rapid evolutionary innovation involving the transfer of pre-evolved genes or genetic sequences from one evolutionary lineage to another. At its most powerful, genetic symbiosis involves the union of entire disparate genomes to create a novel "holobiontic" genome — a genome made up of two or more disparate evolutionary lineages brought into union by genetic symbiosis.

10.2 Mitochondria: symbiogenesis in the human

During the Proterozoic eon, microbes evolved the ability to convert the energy of sunlight to storable compounds through photosynthesis. A by-product of photosynthesis was highly reactive oxygen, which would have been poisonous to many of the co-existing microbes at that time. This in turn led to the evolution of another strain of proteobacterium capable of breathing oxygen. The stage was now set for two of the most important of all the evolutionary transitions in the history of biodiversity: the symbiogenetic incorporation of photosynthetic microbes into archaic eukaryotic ancestors — the latter may have evolved from a clade of archaea[12] — that would result in the evolution of plastids, including chloroplasts; and the symbiogenetic incorporation of the oxygen-breathing bacterium into an archaic eukaryote that would evolve into mitochondria[13]. These two holobiontic genomic unions were crucial to the evolution of the animals, plants, fungi and other photosynthetic and oxygen-breathing life forms we see today.

Genetic sequencing of plastids reveals that eukaryotic photosynthesis evolved again and again, involving different photosynthesizing microbial symbionts. But it would appear that all mitochondria have derived from a single holobiontic evolutionary event. Thus the genome of our human mitochondria derives from that archaic oxygen-breathing proteobacterium. It also implies that our human genome is holobiontic: it combines two distinct evolutionary lineages, that of the oxygen-breathing bacterium and that of its original eukaryotic host, brought into union through symbiogenesis. This has important implications for the genetics of mitochondrial disease.

The original bacterial symbiont would have possessed perhaps 1,500 to 2,000 genes. Over the two billion years of genomic union, the genome of the human mitochondrial organelle has been whittled down to a residuum of 37 genes, encoding 22 transfer RNAs, two types of ribosomal RNA and 13 proteins involved in cellular oxidative phosphorylation[14]. However, during the vast time period of holobiontic evolution, an additional 300 or so of the original bacterial genes have been transferred to the nucleus, where some are still involved in a functional nucleus-mitochondrial genetic linkage[15]. Since the mitochondria are part of the cytoplasmic inheritance of the ovum, and since they reproduce by a bacterial-style budding, independently of the nucleus, mitochondrial inheritance is non-Mendelian and exclusively maternal.

With just 37 genes, when compared to the nuclear protein-coding genome of approximately 20,400, we might anticipate numerically fewer inherited disorders of metabolism arising from mitochondrial mutations when compared to those arising from mutations of nuclear genes. However, where most of our nuclear DNA is non-protein-coding, so that mutations are less likely to prove deleterious in terms of physiological function, the bulk of mitochondrial DNA is protein-coding and thus any mutation is more likely to cause disease. Bacterial genes are also more error-prone than eukaryotic nuclear genes, so that mutations in mitochondrial genes are ten to twenty times commoner than we find in nuclear genes.

Disease-causing mutations have been identified in more than 30 of the 37 mitochondrial genes and in more than 30 of the related nuclear genes[16]. Mitochondrial diseases usually present at birth or in early childhood, causing a progressive neuro-degenerative disorder, with oxygenation problems also affecting the heart, liver and skeletal muscle. Clinical categories include Leigh's syndrome, a subacute sclerosing encephalopathy that affects about 1 in 40,000 neonates, and NARP, a syndrome of neurogenic muscular weakness, ataxia and retinitis pigmentosa[17]. Leber's hereditary optic neuropathy, one of the commonest inherited forms of eye disease, is another example of disease caused by mitochondrial gene mutations. Added to this, mitochondrial dysfunction can also arise from mutations in the linked genes that were transposed to the nucleus from the original mitochondrial genome. Friedreich's ataxia is an illustrative example, where the affected nuclear gene, known as *FXN*, codes for the mitochondrial protein, frataxin. Frataxin enables the removal of iron from the mitochondria, so that when it is defective or absent, the mitochondria accumulate excess iron and suffer free radical damage, with resulting oxidation dysfunction[18].

Only through understanding the symbiogenetic evolutionary origins of mitochondria can we understand the genetics involved in normal mitochondrial function as well as the metabolic disorders arising from mutations arising in the mitochondrial, and linked, genes.

10.3 Viral symbiogenesis

Symbioses involving bacteria, archaea and algae have been extensively studied within the world of biology. But until recently the contribution of viruses to symbiogenesis has received little attention. Yet as long ago as 1926 Felix d'Herelle, in his landmark book *The Bacteriophage*, devoted a chapter to the symbiotic behavior of the bacteriophage, comparing this to the root-fungus symbiosis, known as mycorrhizae, that feeds the orchid[19]. During the ensuing decades, entomologists recognized the symbiotic nature of polydnaviruses in their intimate partnerships with the Ichneumonid and Braconid parasitic wasps, which is one of the most successful survival strategies in the world of entomology.There are at least 60,000 species of these wasps in nature, which have entered into symbiotic partnerships with approximately 20,000 species of polydnaviruses. Many of the viral genomes are integrated into the wasp genome. Whether integrated into the wasp genome or multiplying in the wasp's reproductive system, the viruses emerge to coat the eggs of the wasp and are injected along with the eggs into the Lepidopteran caterpillar prey. Here the viruses block the cellular (hemocyte) immune rejection of the eggs and disrupt the caterpillar endocrine system, suppressing the thoracic gland and thus the metamorphic development into the pupa and adult. Within the caterpillar the viruses also induce the production of sugars to feed wasp larvae. A conserved gene family has been found in viruses of different wasp subfamilies, which would fit with a common phylogenetic origin[20]. Estimations of the first establishment of the symbiosis, based on mitochondrial 165 rRNA and cytochrome oxidase I together with nuclear 28SrRNA of the polydnavirus-bearing clade of braconid wasps, suggest a unique evolutionary event 73.7 ± 10 million years ago[21,22]. Despite such observations the concept of viral symbiosis remained largely unexplored, with little in the way of systematic study.

As recently as 2009, Moreira and López-Gardia suggested that viruses be viewed as "pickpockets" of host genomes[23]. While some biologists might agree with this viewpoint, others see this as a somewhat blinkered perspective. Exchange of genetic material between viruses and hosts has been amply confirmed, but there is no escaping the fact that the majority of viral genes, and particularly so the genes of phage and retrovirus lineages, are not found in the genomes of their hosts. Indeed, the greater part of the exchange would appear to be from virus to host[24]. Recent observations of oceanic viruses reveal them to be "the largest reservoir of genetic diversity on the planet," playing key roles in oceanic biomass as well as global geochemical cycles[25,26]. In 2014, a new strain of bacteriophage, labelled crAssphage, was identified by metagenomic analysis of human fecal microbial communities[27]. Its 97kbp genome appears to be six times more abundant in publicly available metagenomes than all other phages put together. Yet the majority of its proteins have no known match in the standard databases, perhaps helping to explain why it was not detected earlier. Its host may be a *Bacteroides*, but as yet its role in fecal microbiology remains obscure.

By the early 1990s observation of the behavior of plague viruses, and in particular plague viruses that entered into persistent co-evolutionary relationships with their hosts, suggested that viruses might fulfil symbiotic roles in evolution. In a multicentre study of the interaction between HIV-1 and the human MHC, Kiepiela and colleagues found that the rate of disease progression was strongly associated with a particular HLA-B but not HLA-A allele expression[28]. They also found that substantially greater selection pressure is imposed on HIV-1 by HLA-B alleles than by HLA-A, and that HLA-B gene frequencies in the population are those likely to be most influenced by HIV disease. In other words, virus and host are influencing one another's evolution, a symbiogenetic pattern of interaction. There was a pressing need for clarity of definition of what viral symbiosis might involve, including the evolutionary mechanisms and a systematic approach to investigative methodology. Concepts such as "aggressive symbiosis," "species gene pool culling" and co-evolution (symbiosis) between the survivors and persisting virus were introduced to explain what virologists were actually witnessing[29,30]. Botanical virologist Marilyn Roossinck tested these definitions and methodologies and confirmed their validity in a series of innovative experiments in nature[31–33]. Further definitional and methodological development took the concept of viral symbiosis to the World Congress of the International Symbiosis Society, in 2006, leading to two defining papers[34,35]. A more recent review, aimed at a microbiological readership, showed how viral symbiosis can be seen to work in a very similar range of mechanisms to what has been established with bacteria[36]. Today there is overwhelming evidence that retroviruses have played an important symbiotic role in the evolution of the mammals, and in particular, in the evolution of the human genome.

Retroviruses reproduce by means of a viral enzyme, reverse transcriptase, that converts the viral RNA to the complementary DNA, after which viral integrase inserts the viral genome into the chromosomes of the host target cell's genome to create "proviruses" that act as templates for viral reproduction. An extension of this viral reproductive strategy gives retroviruses a remarkable evolutionary potential. Epidemic retroviruses can insert their genomes into the germ cells of their hosts, a process known as "endogenization," with the viral inserts now classed as "endogenous retroviruses," or ERVs. Such endogenous retroviruses have made a major symbiogenetic contribution to the evolution of the mammalian, primate and in particular the human genome.

Successive plague retroviral colonizations of the ancestral human genome have given rise to some 30 to 50 endogenous viral "families," and these are further subdivided into more than 200 groups and subgroups. Families in this sense do not refer to exogenous viral families but rather independent colonizing lineages, and the endogenous viruses are now known as human endogenous retroviruses, or HERVs. The many different lineages suggest that the ancestral pre-hominin and hominin populations will have been subjected to the typical "aggressive symbiosis" pattern of host-viral interaction, with species gene pool culling and subsequent co-evolution of the culled host population and virus, with accompanying endogenization of the host genome. We are observing this pattern of retroviral colonizations in the prevailing Koala retrovirus epidemic in Australia, or KoRV, where, over the course of an estimated hundred or so years, the majority of koalas have already been infected. The epidemic has caused a high level of plague culling, with deaths from immunodeficiency and blood and bone marrow dyscrasias similar to what was seen in the early years of the HIV-1 epidemic in humans. The koala retrovirus is also invading the koala germ line, with some examples already showing 46 viral loci scattered throughout the chromosomes[37,38]. The ancestral stage of primate evolution appears to have coincided with repeated colonizations involving the

Figure 3 Schematic of the genome of a retrovirus.

endogenous "family" known as the HERV-Ks. At least ten of these HERV-Ks are unique to humans, suggesting they entered the ancestral genome after the evolutionary divergence of hominins from chimpanzees[39,40]. A small number of HERV-Ks are such recent arrivals that they appear to have entered the human genome after the presumed migration of modern humans out of Africa, so they are only found in a subset of humans in Africa and the south Asian coastal areas[41].

As we saw with the pruning of the mitochondrial genome by selection over time, many of the HERV loci have also been pruned through mutations, including stop codons, insertions and deletions. This led an earlier generation of geneticists to assume the viral and other repeat' sequences were "junk DNA."[42,43] In fact a significant number have become incorporated into the functional holobiontic nuclear genome. And just as the mitochondrial genome retained key aspects of its bacterial evolutionary inheritance, so also do the HERV loci retain key aspects that are quintessentially viral in character.

A typical retrovirus genome is shown in Figure 3, comprising three pathognomonic "genes" labelled *gag*, *pol* and *env*. These are essentially genetic domains with multiple properties, with *gag* coding for matrix and core shell proteins, *pol* coding for enzymes such as reverse transcriptase, protease, ribonuclease and integrase, and *env* coding for the surface and transmembrane glycoproteins. These are flanked by the viral regulatory regions, called long terminal repeats, or LTRs. The colonization of the human genome by a wide variety of different retroviral genomes, often in many copies, opened up the possibility of novel holobiontic evolutionary potentials. Although the viral genomes derived from very different evolutionary lineages to that of the pre-human and hominin hosts, we also need to consider that these colonizing viruses also evolved as obligate genetic parasites of the same hosts and so the viral genomes were pre-evolved to interact with the hosts at genetic, biochemical and physiological levels. This has important implications for human evolution, embryology, physiology and biochemistry.

Natural selection plays a critical role in the evolution of such virus-host holobiontic genomic unions. The invading viral genome will initially be silenced by epigenetic means, particularly methylation. But over the course of evolutionary time the viral loci will have opportunities to express their genes and regulatory sections. And this in turn will offer the potential of novel holobiontic evolution at every stage in the host life cycle, including reproduction, placentation, embryonic development, and in the development and function of every different human cell, tissue and organ. Given the quintessentially viral nature of HERVs and the fact they were formerly genetic parasites of the same host, we might anticipate a highly "interactive" pattern of genomic coadaptation. We can also anticipate that selection will now preserve viral or former host sequences that enhance survival of the holobiontic organism, and it will select against former host or viral sequences that impair survival.

Today we recognize that many viral genes and related genetic sequences have been co-opted in this way to become involved in a wide range of structural and genomic functions as well as regulatory and transcriptional activity[44–47]. Related "transposable elements," including long-interspersed elements (LINEs), short-interspersed elements (SINEs) and HERV LTRs, are frequently found within protein-coding regions of the genome[48]. Once integrated into the holobiontic genome, the HERV LTRs in particular contain promoter and enhancer elements as well as polyadenylation signals capable of controlling not only the expression of viral genes but also those derived from the former host[49,50]. Indeed, at least 50% of the human specific HERV-K(HML-2) LTRs serve as active promoters for host DNA transcription[51].

10.4 HERV proteins

HERV proteins, translated from endogenous viral loci, have increasingly come under scrutiny in recent years. The first to be discovered was "syncytin-1", coded by the *env* gene of a HERV-W, known as the ERVWE1 locus on chromosome 7, and promoted by the viral 5'LTR[52,53]. As can be seen in Figure 4, a schematic of the ERVWE1 locus after Gimenez and Mallet[54], the HERV-W genomic structure is essentially intact. The intronic insert between the 5'LTR and the *gag* and *pol* domains silences these two domains, so that the desired domain, the *env*, can be translated using the viral promoter in the 5'LTR. The promoter is itself regulated by a complex array of upstream elements, including the 5'LTR of another viral locus, the MaLR and a medley of host elements. Thus, even in a single active viral locus, we witness a very similar action of selection to what we saw with the mitochondria, with selective pruning of unwanted portions of the viral genome and tight conservation and regulatory control of those elements that have been incorporated into holobiontic function. Moreover, the complexity of upstream regulatory control might also allow syncytin-1 to be expressed in a range of tissues other than placenta.

What function does this quintessentially viral gene, promoted by the viral promoter within a conserved viral genome, serve in the human placenta? It changes the fate of the cells at the placental interface, converting what were formally trophoblast cells into syncytiotrophoblasts, and thus fusing the interface into a syncytium. Three more HERVs, HERV-FRD, ERV-3, and HERV-K, have been identified as playing roles in cytotrophoblast fusogenesis and there is evidence to support the HERV-FRD and the HERV-K playing immunosuppressive roles in the protection of the foetus from maternal immunological rejection[55–58]. Both the

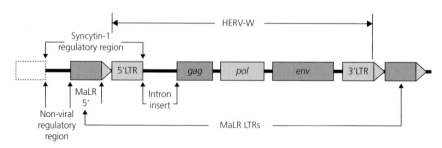

Figure 4 Schematic of the ERVWE1 locus (after Gimenez J and Mallet F).[54]

fusogenic and the immunosuppressive roles are of course pre-evolved viral traits, deriving from the genetic parasitic inheritance. At least eight more viral loci appear to be involved with human reproduction, though their roles, in health and disease, are still under investigation[36].

Given the highly interactive symbiogenetic potential of HERVs in relation to the host genome and the growing evidence for their roles in reproduction, it seems likely that they will have evolved roles in human embryogenesis. But until recently it was difficult, from both an ethical and a practical standpoint, to investigate this. When Spadafora and colleagues blocked the action of reverse transcriptase in mouse embryos it caused an irreversible arrest of development beyond the four-cell stage, which was accompanied by substantial reprogramming of developmental and translational genes[59]. In pioneering work on human embryos recovered from tubal pregnancies, Larsson and colleagues found that one of the placentally expressed HERVs, known as ERV3, was highly expressed as mRNA in foetal adrenal cortex, kidney tubules, tongue, heart, liver and central nervous system[60]. More recently, work on human embryos has revealed profound influences of HERVs on human embryogenesis. For example, HERV-H, a virus with a great many loci in the human chromosomes, is abundantly expressed in human embryonic stem cells, where it appears to play a key role in human embryonic stem cell identity. The viral LTRs appear to be functioning as enhancers. Meanwhile, multiple viral loci of the same endogenous retrovirus appear to be acting epigenetically as long non-coding RNAs that maintain human embryonic cells in a state of pluripotency[61,62]. As we saw above, the HERV-Ks are the most recently acquired group of HERVs, which are particularly associated with the primate stage of evolution, both before and after the split with a common ancestor with the chimpanzees. Many largely intact HERV-K loci remain scattered throughout the human chromosomes. Working with preimplantation embryos and pluripotent cells, Grow and colleagues have shown that HERV-K is widely transcribed during normal human embryogenesis, beginning with the eight-cell stage of embryogenesis, continuing through to the emergence of epiblast cells in preimplantation blastocysts and ceasing during human embryonic stem cell derivation from blastocyt outgrowths[63]. They concluded that "complex interactions between retroviral proteins and host factors can fine tune pathways of early human development."

Given their viral origins and highly interactive nature within the genome, HERVs have obvious potential for genetic disorders. Micro-deletions of the Y chromosome can cause male infertility through non-allelic homologous recombination between HERV elements on the Y chromosome[64]. The expression of HERVs and related genetic sequences has been associated with in a wide variety of human diseases, including psoriasis, hemophilia and muscular dystrophy, schizophrenia, and bipolar disorders[65-73]. Elsewhere I have extensively reviewed the putative role of HERVs in a wide variety of auto-immune diseases, notably multiple sclerosis and systemic lupius erythematosus (SLE), but also including rheumatoid arthritis, alopecia areata, Sjögren's syndrome, congenital heart block, type 1 diabetes, multiple sclerosis and primary biliary cirrhosis[74]. I have also extensively reviewed the role of HERVs in a wide variety of hematological malignancies, as well as cancers of the lung, oesophagus, stomach, intestine, bone marrow, bladder, prostate, breast, cervix, melanoma, seminomas and teratocarcinomas[75]. In all such cases the evidence for a role for HERVs in the pathogenetic processes is suggestive though as yet inconclusive. For example, the expression of a HERV protein in diseased tissue might signify a role in normal immunological reaction, or it might signify a dysregulation of normal expression such as is currently being suggested for syncytin-1 in MS[74], or a recruitment of a HERV protein to the actual process of carcinogenesis.

A key problem in the further exploration of the links to auto-immunity and carcinogenesis has been the fact that we still lack sufficient knowledge of the evolutionary and the normal physiological role of the HERVs and related sequences in the human genome. In particular, we need new methodologies for studying the quantitative expression of HERV proteins in normal cells, tissues and organs. Previous studies of the role of HERVs in embryogenesis and normal physiology relied on mRNA detection, which, through suggestive, is insufficiently diagnostic since mRNAs can be inactivated by epigenetic mechanisms. One such new methodology for deep sequencing of HERV proteins has been introduced by Larsson and colleagues at Uppsala, Sweden, who have generated dual affinity-purified polyclonal antibodies to HERV *env* encoded protein, allowing a more accurate detection of HERV ERV3 protein expression pattern in a wide range of normal human cells, tissues and organs[76]. This would suggest that a variety of HERV proteins may indeed contribute to the normal physiology of many different tissues and organs. The logical progression is to plot the entire HERV contribution to holobiontic function, and to evaluate the entire HERV transcriptome in human cells, tissues and organs. This is likely to open a new and important window of understanding to the HERV contribution to human evolution and disease.

TAKE-HOME MESSAGE

- Genetic symbioses are a source of evolution involving the transfer of pre-evolved genes or genetic sequences from one evolutionary lineage to another, including humans.

- Human endogenous retroviruses have played an important symbiogenetic role in the evolution of the human genome, with implications for reproduction, placentation, embryological development, and the normal biochemistry and physiology of human cells, tissues and organs.

- There is also growing evidence to suggest that HERVs are also playing an important role in a wide spectrum of disease, including hereditary disorders of metabolism, many cancers and autoimmune diseases.

- We need to plot the contribution of the entire HERV genome to the holobiontic human genome, extrapolating to the role of the entire HERV transcriptome in human physiology before we can fully evaluate the nature and extent of the HERV contribution to evolution, physiology and disease.

References

1 Margulis L, Fester R eds, 1991. *Symbiosis as a Source of Evolutionary Innovation*. Cambridge, Mass: MIT Press. Chapter 9. Jeon K. Amoeba and x-bacteria: symbiont acquisition and possible species change.
2 Douglas AE, 1994. *Symbiotic Interactions*. Oxford: Oxford University Press.
3 Margulis L, 1981. *Symbiosis in Cell Evolution*. New York, WH Freeman & Co.
4 Paracer S, Ahmadjian V, 2000. *Symbiosis: An Introduction to Biological Associations*. Oxford: Oxford University Press.

5 Werren JH, Guo L, Windsor DW. Distribution of Wolbachia among neotropical arthropods. *Proc Roy Soc B* 1995; **262** (1364): 197–204.

6 Kozek WJ, Rao RU. The discovery of Wolbachia in arthropods and nematodes — a historical perspective. *Issues Infect Dis* 2007; **5** (Wolbachia: a bug's life in another bug): 1–14. doi:10.1159/000104228.

7 Kent BN, Salichos L, Gibbons JG *et al*. Complete bacteriophage transfer in a bacterial endosymbiont (*Wolbachia*) determined by targeted genome capture. *Genome Biol Evol* 2011; **3**:209–18.

8 Young JPW, Johnston AWB. The evolution of specificity in the legume-rhizobium symbiosis. *Trends in Ecol and Evol* 1989; **4**: 341–9.

9 Sullivan JT, Ronson CW. Evolution of rhizobia by acquisition of a 500-kb symbiosis island that integrates into a phe-tRNA gene. *Proc Natl Acad Sci USA* 1998; **95**: 5145–9.

10 Long SR. Genes and signals in the rhizobium-legume symbiosis. *Plant Physiol* 2001; **125**: 69–72.

11 Sapp J, 1994. *Evolution by Association: A History of Symbiosis*. New York: Oxford University Press.

12 Spang A, Saw JH, Jorgensen SL *et al*. Complex archaea that bridge the gap between prokaryotes and eukaryotes. *Nature* 2015; doi:10.1039/nature14447.

13 Margulis L, Sagan D, 1997. *Microcosmos: Four Billion Years of Microbial Evolution*. Berkeley, Los Angeles, London: University of California Press.

14 http://www.biochemistry.org/education/pdfs/basc/10/BASC10_full.pdf.

15 Timmis JN, Ayliffe MA, Huang CY *et al*. Endosymbiotic gene transfer: organelle genomes forge eukaryotic chromosomes. *Nature Rev Genet* 2004; **5**: 123–35.

16 Thorburn DR. Mitochondrial disorders: prevalence, myths and advances. *J Inherit Metab Dis* 2004; **27**: 349–62.

17 http://www.ncbi.nlm.nih.gov/bookshelf/picrender.fcgi?book=gene&&partid=1173&blobtype=pdf.

18 Boddaert N, Sang KHLQ, Rötig A *et al*. Selective iron chelation in Friedreich ataxia: biologic and clinical implications. *Blood* 2007; **110**: 401–8.

19 D'Herelle F, 1926. *Bacteriophage and its Behaviour*. London: Ballière, Tindall and Cox. Trans George H. Smith.

20 Provost B, Varricchio P, Arana E *et al*. Bracoviruses contain a large multigene family coding for protein tyrosine phosphatases. *J Virol* 2004; **78**: 13090–103.

21 Whitfield JB. Estimating the age of the polydnavirus/braconid wasp symbiosis. *Proc Natl Acad Sci USA* 2002; **99**: 7508–13.

22 Belle E, Beckage NE, Rousselet J *et al*. Visualization of polydnavirus sequences in a parasitoid wasp chromosome. *J Virol* 2002; **76**: 5793–6.

23 Moreira D, López-Garcia P. Ten reasons to exclude viruses from the tree of life. *Nature Rev Microbiol*, 2009; **7**: 306–11.

24 Villarreal L, 2005. *Viruses and the Evolution of Life*. Washington DC: ASM Press.

25 Hambly E, Suttle CA. The viriosphere, diversity, and genetic exchange within phage communities. *Curr Opin Microbiol* 2005; **8**: 444–50.

26 Edwards RA, Rohwer F. Viral metagonomics. *Nature* 2005; **3**: 504–10.

27 Dutilh BE, Cassman N, McNair K *et al*. A highly abundant bacteriophage discovered in the unknown sequences of human faecal metagenomes. *Nature Comm*. Doi: 10.1038/ncomms5498.

28 Kiepiela P, Leslie AJ, Honeyborne I *et al*. Dominant influence of HLA-B in mediating the potential co-evolution of HIV and HLA. *Nature* 2004; **432**: 769–74.

29 Ryan F, 1996. *Virus X*. London, HarperCollins Publishers. In the US, Boston, New York, 1997; Little, Brown.

30 Ryan F, 2002. *Darwin's Blind Spot*. Boston, New York; Houghton Mifflin.

31 Roossinck MJ. Symbiosis versus competition in plant virus evolution. *Nature* 2005; **3**: 917–24.

32 Márquez LM, Redman RS, Rodriguez RJ, Roossinck MJ. A virus in a fungus in a plant — three-way symbiosis required for thermal tolerance. *Science* 2007; **315**: 513–515.

33 Roossinck MJ, Saha PI, Wiley HB. Ecogenomics: using massively parallel pyrosequencing to understand virus ecology. *Mol Ecol* 2010; **19**(1): 1–8.

34 Villarreal LP. Viral persistence and symbiosis: are they related? *Symbiosis* 2007; **43**: 1–9.

35 Ryan FP. Viruses as symbionts. *Symbiosis* 2007; **44:** 11–21.

36 Villarreal LP, Ryan FP. Viruses in host evolution: general principles and future extrapolations. *Curr Top Virol* 2011; **9**: 79–90.

37 Tarlinton R, Meers J, Young PR. Retroviral invasion of the koala genome. *Nature* 2006; **442**: 79–81.

38 Tarlinton R, Meers J, Young P. Biology and evolution of the endogenous koala retrovirus. *Cell Mol Life Sci* 2008. doi 10.1007/s00018-008—8499-y.

39 Barbulescu M, Turner G, Seaman MI. Many human endogenous retrovirus K (HERV-K) proviruses are unique to humans. *Curr Biol* 1999; **9**: 861–8.

40 Turner G, Barbulescu M, Su M *et al.* Insertional polymorphisms of full-length endogenous retroviruses in humans. *Curr Biol* 2001; **11**: 1531–35.

41 Jha AR, Nixon DF, Rosenberg MG *et al.* Human endogenous retrovirus K106 (HERV-K106) was infectious after the emergence of anatomically modern humans. *PLoS One* 2011; **6**: e20234.

42 Orgel LE, Crick FHC. Selfish DNA: the ultimate parasite. *Nature* 1980; **284**: 604–7.

43 Doolittle WF, Sapienza C. Selfish genes, the phenotype paradigm and genomic evolution. *Nature* 1980; **284**: 601–3.

44 Shapiro JA, von Sternberg R. Why repetitive DNA is essential to genome function. *Biol Rev* 2005; **80**: 1–24.

45 Hughes JF, Coffin JM. Evidence for genomic rearrangements mediated by human endogenous retroviruses during primate evolution. *Nature Genet* 2001; **29**: 487–9.

46 Sverdlov ED. Retroviruses and human evolution. *BioEssays* 2000; **22**: 161–71.

47 Pérot P, Mugnier N, Montgiraud C *et al.* Microarray-based sketches of the HERV transcriptome landscape. *PLoS ONE* 2012; **7**(6): e40194.

48 Nekrutenko A, Li W-H. Transposable elements are found in a large number of human protein-coding genes. *Trends in Genet* 2001; **17**: 619–21.

49 Liu M, Eiden MV. Role of human endogenous retroviral long terminal repeats (LTRs) in maintaining the integrity of the human germ line. *Viruses* 2011; **3**: 901–5.

50 Gogvadze E, Stukacheva E, Buzdin A, Sverdlov E. Human specific modulation of transcriptional activity provided by endogenous retroviral inserts. *J Virol* 2009; doi:10.1128/JVI.00123-09.

51 Buzdin A, Kovalskaya-Alexandrova E, Gogvadze E *et al.* At least 50% of human-specific HERV-K (HML-2) long terminal repeats serve in vivo as active promoters for host nonrepetitive DNA transcription. *J Virol* 2006; **80**: 10752–62.

52 Mi S, Lee X, Li X *et al.* Syncytin is a captive retroviral envelope protein involved in human placental morphogenesis. *Nature* 2000; **403**: 785–9.

53 Blond J-L, Lavillette D, Cheynet V *et al.* An envelope glycoprotein of the human endogenous retrovirus HERV-W is expressed in the human placenta and fuses cells expressing the type D mammalian retrovirus receptor. *J Virol* 2000; **74**: 3321–9.

54 Gimenez J, Mallet F. 2007 (September), *Atlas Genet.Cytogenet. Oncol. Haematol.* See also http://AtlasGeneticsOncology.org/Genes/ERVWE1ID40497ch7q21.html.

55 Blaise S, de Parseval N, Bénit L *et al.* Genomewide screening for fusogenic human endogenous retrovirus envelopes identifies syncytin 2, a gene conserved on primate evolution. *Proc Natl Acad Sci USA* 2003; **100**: 13013–8.

56 Rote NS, Chakrabarti S, Stetzer BP. The role of human endogenous retroviruses in trophoblast differentiation and placental development. *Placenta* 2004; **25**: 673–83.

57 Kämmerer U, Germeyer A, Stengel S *et al.* Human endogenous retrovirus K (HERV-K) is expressed in villous and extravillous cytotrophoblast cells of the human placenta. *J Reprod Immunol* 2011; **91**: 1–8.

58 Lokossou AG, Toudic C, Barbeau B. Implication of human endogenous retrovirus envelope proteins in placental functions. *Viruses* 2014; **6**: 4609–27.

59 Spadafora C. A reverse transcriptase-dependent mechanism plays central roles in fundamental biological processes. *Syst Biol Reprod Med* 2008; **54**: 11–21.

60 Andersson A-C, Venables PJW, Tönjes RR *et al.* Developmental expression of HERV-R (ERV-3) and HERV-K in human tissue. *Virology* 2002; **297**: 220–5.

61 Santoni FA, Guerra J, Luban J. HERV-H is abundant in human embryonic stem cells and a precise marker for pluripotency. *Retrovirology* 2012; **9**: 111. doi:10.1186/1742-4690-9-111.

62 Lu X, Sachs F, Ramsay LA *et al.* The retrovirus HERVH is a long noncoding RNA required for human embryonic stem cell identity. *Nature Struct Mol Biol* 2014; **21**: 423–5.

63 Grow EJ, Flynn RA, Chavez SL *et al.* Intrinsic retroviral reactivation in human preimplantation embryos and pluripotent cells. *Nature* 2015; doi:10.1038/nature14308.

64 Blanco P, Shlumukova M, Sargent CA *et al.* Divergent outcomes of intra-chromosomal recombination on the human Y chromosome: male infertility and recurrent polymorphism. *J Med Genet* 2000; **37**: 752–758.

65 Deininger PL, Batzer MA. Alu repeats and human disease. *Mol Genet Metab* 1999; **67**: 183–93.

66 Nelson PN, Hooley P, Roden D *et al*. Human endogenous retroviruses: transposable elements with potential? *Clin Exper Immun* 2004;**138**: 1–9.

67 Foerster J, Nolte I, Junge J *et al*. Haplotype sharing analysis identifies a retroviral dUTPase as candidate susceptibility gene for psoriasis. *J Invest Derm* 2005; **124**: 99–102.

68 Kazazian HH, Wong C, Youssoufian H *et al*. Haemophilia A resulting from de novo insertion of L1 sequences represents a novel mechanism for mutation in man. *Nature* 1988; **332**: 164–66.

69 Narita N, Nishio H, Kitoh Y *et al*. Insertion of a 5′ truncated L1 element into the 3′ end of exon 44 in the dystrophin gene resulted in skip of the exon during splicing in a case of Duchenne muscular dystrophy. *J Clin Invest* 1993; **91**(5): 1862–7.

70 Frank O, Giehl M, Zheng C *et al*. Human endogenous retrovirus expression profiles in samples from brains of patients with schizophrenia and bipolar disorders. *J Virol* 2005; **79**(17): 10890–901.

71 Yolken RH, Karlsson, Yee F *et al*. Endogenous retroviruses and schizophrenia. *Brain Res Rev* 2000; **31**(2-3): 193–9.

72 Karlsson H, Bachmann S, Schroder J *et al*. Retroviral RNA identified in the cerebrospinal fluids and brains of individuals with schizophrenia. *Proc Nat Acad Sci USA* 2001; **98**(8): 4634–9.

73 Otowa T, Tochigi M, Rogers M *et al*. Insertional polymorphisms of endogenous retrovirus HERV-K115 in schizophrenia. *Neurosci Lett* 2006; **408**(3): 226–9.

74 Ryan FP. An alternative approach to medical genetics based on modern evolutionary biology. Part 3: HERVs in diseases. *J Roy Soc Med* 2009; **102**: 415–24.

75 Ryan FP. An alternative approach to medical genetics based on modern evolutionary biology. Part 4: HERVs in cancer. *J Roy Soc Med* 2009; **102**: 474–80.

76 Fei Chen, Christina Atterby, Per-Henrik Edqvist *et al*. Expression of HERV-R ERV3 encoded Env-protein in human tissues: introducing a novel protein-antibody-based proteomics. *J Roy Soc Med* 2013; **107**(1) 22–29.

CHAPTER 11

The microbiota as an epigenetic control mechanism

Boris A. Shenderov

Laboratory of Biology of Bifidobacteria, Head of Research Group Probiotics and Functional Foods,
Gabrichevsky Research Institute of Epidemiology and Microbiology, Moscow, Russia

11.1 Introduction

The global incidence of chronic metabolic diseases (metabolic syndrome, type 2 diabetes, autoimmune diseases, cancer, neurodegenerative diseases, some infections, etc.) is rapidly increasing and has become a major problem in developed countries. Previously, it was thought that these human conditions are predominantly driven by acquired genetic mutations. Now, it is becoming clear that any phenotype is the result of complex interactions between genotype, epigenome and environment. Unlike the stable codon structure of the DNA molecule, the chemical changes in DNA that constitute the "epigenome" are much more labile and responsive to environmental and endogenous signals. Epigenetic mechanisms are fundamentally important because human health, well-being, and the vast majority of chronic metabolic diseases are closely connected with the functions of these molecular mechanisms[1-4]. Among the influences on the epigenome, it has recently been proposed that exogenous and endogenous microorganisms can have actions relevant to human health and disease[5]. The adult human body contains trillions of microorganisms, representing hundreds of species and thousands of subspecies and strain variants (collectively known as the microbiome) and having a predominantly symbiotic relationship with their host. Adult humans are more prokaryotic than eukaryotic, with 90% of our cells estimated to be of microbial and only 10% of human origin[6-9]. In the healthy adult gut, 80% of identified microbiota belongs to three dominant phyla: *Bacteroidetes, Firmicutes,* and *Actinobacteria*. We have previously reported (chapter 6) that the composition of microbiota populating a wide range of niches both on the surface of and inside all multicellular organisms is not stable over the whole lifetime of an individual and is influenced by our environment and lifestyle, including medicines, disease status, seasonal changes, age, and especially diet[10]. This means that a balanced or imbalanced diet will always lead to a structural or functional variation of microbiome composition, with significant consequence for the health of the individual. Diverse microbial populations, some beneficial, others not, play a strategic role

The Human Microbiota and Chronic Disease: Dysbiosis as a Cause of Human Pathology, First Edition.
Edited by Luigi Nibali and Brian Henderson.
© 2016 John Wiley & Sons, Inc. Published 2016 by John Wiley & Sons, Inc.

in controlling human health, well-being, metabolism and diseases[11,12]. Owing to low-molecular-weight (LMW) structural, metabolic and signaling molecules, microorganisms are able to sense environmental factors, interact with corresponding cell surface, membrane, cytoplasm and nucleic acid receptors, supporting or changing human homeostasis and health in the specific environmental conditions. These microbial molecules produce modulation of epigenomic regulation of gene expression and/or alteration of the information exchange in numerous bacterial and bacteria-host systems[13–15]. Therefore, various microbiota-associated functions could be the novel original biomarkers of the physiological and pathological status of each person[6–8,11,14–16]. To obtain objective and complete information regarding the host/microbe cross-talk in various physiological and extreme environment conditions, a set of novel molecular, genetic and biochemical "omic" technologies have been developed and introduced into biomedical science and practice (genomics, epigenomics, transcriptomics, metabolomics, phenomics)[17,18]. The multidisciplinary omic approach permits a better understanding the molecular bases of the health, and diseases of humans and animals. Advances in the omic fields, especially coupled with current and novel germ-free and gnotobiotic technologies, open a new era in characterizing the role of the microbiota in the host genomic and epigenomic stability, metabolomics and phenotypic alterations and have far-reaching implications relevant to health and metabolic diseases.

The intestinal epithelium and gut microbiota sense each other in order to trigger various cellular and molecular responses when required, by releasing host cell and microbial signaling molecules regulating the expression of appropriate genes and or post-translation modification of their final products. Understanding how small molecules of microbial origin interact with the human host through metagenome and metaepigenome, physiological functions, metabolism and multiple signal transduction pathways will permit the design of a new generation of genetic- and epigenetic-based drugs and functional foods targeting complex chronic metabolic diseases. This chapter concentrates on the potential role of the microbiota as a source of epigenetic factors in the regulation of mammalian development, health, chronic metabolic diseases, and the contribution of micro-ecological approaches to the improvement of our understanding in these fields.

11.2 Background on epigenetics and epigenomic programming/reprograming

Epigenetics is a rapidly expanding field of medico-biological science focusing on the processes that regulate how and when certain genes are turned on, off, up-regulated or down-regulated. The epigenome is a dynamic organization of the DNA as a result of the interaction of the genome with its cellular and extracellular environment. Epigenetic modifications (the spectrum of "working and non-working" genes) can arise in the first stages of ontogenesis and persist from one cell division to the next during several cell generations. Epigenetics does not involve the nucleotide sequence of the DNA; epigenomic effects are connected with the covalent attachment of different chemical groups to DNA, chromatin, histones and to different other associated proteins during the post-translation period. Human epigenetics helps to explain the relationship between individual genotype and the environment as well as some features and mechanisms of various multifactorial chronic metabolic diseases including onset, gender effects and individual fluctuation of symptoms. The epigenetic code can be individual as well as tissue- and cell-specific. Many types of epigenetic processes have been identified (DNA methylation, chromatin modifications, histone methylation, acetylation, phosphorylation, ubiquitination, ADP-ribosylation, biotinylation, sumoyalation, hydroxylation, repeat-induced gene silencing, microRNA interference, etc.). It has been demonstrated that these and other reactions (glycosylation, glucoranidation, sulfatation, etc.) can occur with proteins that are not connected with chromatin[1-3,19,20].

Methylation is an enzymatic process in which DNA nucleotides and/or amino acid residues in the N-termini of histones are methylated by various types of methyltransferases; a number of demethylases mediate the removal of methyl groups (CH_3). 5-adenosyl-methionine (SAM) is used as the methyl donor compound in this biochemical reaction. DNA methylation usually is involved in the long-term silencing of gene expression, whereas histone methylation has a short-term and flexible effect[4,21,22]. The transcriptional consequences of histone methylation/demethylation depend on the specific site being methylated/demethylated and on the number of methyl groups fixed on the amino acid residue. There are two types of methyltransferases: site-specific, which can be found only in eukaryotic cells, and methyltransferases used by both prokaryotic and eukaryotic cells[20,23,24]. It should be noted that histone amino acid residues may be mono-, di-, and/or tri-modified. For example, methylation can occur several times on one lysine residue and such modification can result in different biological outcomes[25]. In lysine acetylation the acetyl moiety from acetyl-CoA transfers to the ε-group of a lysine residue and is dynamically governed by two groups of counteracting enzymes known as histone acetyltransferases (HATs) catalyzing lysine acetylation and protein deacetylases (HDACs) deacetylating a lysine residue, respectively. Acetylation of specific lysine residues on histone is generally associated with transcriptional activation, while histone deacetylation results in transcriptional repression. Eighteen HDAC enzymes are known in mammals and are divided into four classes (I-IV). Among them there is a family of proteins acting as nicotinamide adenine dinucleotide (NAD)-dependent deacetylases (sirtuins); class I and II HDACs are zinc-dependent[20,24,26-28].

Biotinylation is an epigenomic process characterized by the addition of the water-soluble vitamin, biotin, to histone lysine residues. Mammalian cells cannot synthesize biotin and absolutely depend on a constant supply of this vitamin via food or from the intestinal microbiota. All main types of histones in the human cell chromatin may be biotinylated. The expression activity of about 10% structural genes in human chromosomes is connected with histone biotinylation[29,30]. Reversible protein phosphorylation (addition or elimination of the phosphate group from adenosine triphosphate (ATP) to histones) connected with epigenetic control for gene expression occurs in both eukaryotic and prokaryotic cells. Approximately half of human proteins are phosphorylated. Protein kinases, together with their related phosphatases, are responsible for these processes. The transfer of phosphate groups from ATP to reactive side chains of amino acids (serine, threonine, tyrosine, histidine, and aspartate) in proteins can control the target protein activity either directly by inducing different conformation modifications in the active site of protein structure or indirectly, by regulating protein-protein interactions, resulting in the activation or inactivation of many genes and correspondingly enzymes. Phosphorylated histones usually stimulate gene activity; on the other hand, histone de-phosphorylation as rule suppresses its activity[31,32]. ADP- Ribosylation includes non-covalent specific adjusting ADP-ribose group to various mostly chromosomal proteins, resulting in reversible destruction of chromatin structure with gene expression modification. This process is mediated by the NAD+-depending ADP-ribosyltransferase that transfers ADP-Ribose groups from NAD+ to amino acid residues (for instance, arginine) of corresponding proteins. There are other enzymes (specific hydrolases) that can remove ADP-Ribose group from amino acid. Poly- ADP-Ribosylation is characterized by transfer of multiple ADP-Ribose groups with the formation of modified proteins carrying long (up to 200 units) branched chains ADP-Ribose units. It occurs in eukaryotic as well as in prokaryotic organisms. The most common Poly-ADP-ribosylation affects nuclear and chromosomal proteins (histones, endonucleases, DNA-ligases II)[33,34].

Ubiquitination is the post-translational modification of a histone lysine residue by the covalent attachment of one or more ubiquitin (Ub) monomers, targeted for proteasomal degradation. Ub is a 76-amino-acid protein. Ubiquitination/deubiquitination is linked to both activation and/or inhibition of transcription regulation, cell differentiation, apoptosis, proteasome protein degradation, autophagy, lysosome systems, immune responses, endocytosis, DNA repair, and the transmission of molecular signals from a cell's exterior to its interior. Ubiquitination plays also important role in the onset and progression of many metabolic diseases, infection and muscle dystrophies[35]. MicroRNAs (miRNAs) are a class of widely distributed non-coding RNAs, many of which are involved in cell growth, differentiation, and other functions. MicroRNA-interference is an important epigenomic process in which specific genes can be turned off, or silenced, via mechanisms mediated by a class of endogenous non-coding RNA molecules of the same size in plants, worms, flies, animals (including human), and microbes. These small RNAs (short — about 20, and long — typically 200 nucleotides) are implicated in regulating gene expression on the different post-transcriptional levels acting as inhibitors of transcription, modulators of DNA and histone methylation, chromatin reconstruction, protein activity, modulating mature miRNA abundance, localization, and its stability[36]. Currently more than 5000 mammalian miRNAs have been identified that can potentially target up to 30% of all protein-coding genes. The miRNAs affect gene expression by interacting with RNA and other cellular factors, resulting in either the degradation or suppression of RNA processing or modification at genomic sites, clusters, and regions[18,36–38]. Determination in the body tissues and fluids of methylated DNA and proteins, acetylated, phosphorylated and other modified proteins, miRNA, specific substrate enzymes, co-substrates and co-factors participating in the epigenomic processes could represent a biomarker for detecting epigenomic programing/reprograming condition in some metabolic diseases[3,14,18,20,24,37,39].

All cells have identical genomes. However, each cell has unique assembling of epigenetic instructions for developing and supporting its own gene-specific expression profiles. In recent years, it has become increasingly clear that epigenetic processes play a critical role in the epigenomic programming/reprogramming of human health and risk of metabolic and other diseases during all stages of life, from fertilized egg to old age. The term "epigenomic programming" describes the net sum of interactions between the host metabolism and microbiota through which exposure of lifestyle of pregnant women as well as various effectors of epigenetic changes, including environment, exogenous (medicines, pathogenic microbes, various pollutants) and endogenous (hormones, indigenous host microbiota) factors and agents during pregnancy and nursing period can alter fetal growth and development, resulting in permanent functional and metabolic changes *in utero* and increased susceptibility to disease later in life. During early embryogenesis the mammalian genome is "cleaned" of most epigenetic mutations/modifications, the latter then being progressively re-established during embryonic development. Developing organisms appear to be particularly susceptible to epigenetic changes[3,4,19,40].

Epigenomic reprogramming of cell genome and post-translation modification of gene expression are essential mechanisms for the development, regeneration and postnatal life of higher eukaryotic organisms. Many chronic metabolic and degenerative disorders and diseases are initiated and/or influenced by non-optimal epigenomic programming/reprogramming, often taking place during the first 1,000 days of life — from conception into early infancy[3,4,39]. Food and gut microbiota are the most important life-long environmental factors, often playing the critical role in the epigenomic programming/reprogramming, developing and progression of chronic metabolic diseases. It is well known that foodstuffs contain plenty of various macro- and micronutrients that may interact with the epigenetic machinery. Indigenous gut, vaginal and/or placenta microbiota of pregnant women also produce multiple LMW structural, metabolic and signal molecules that can penetrate via the placenta into the fetus, resulting in permanent effects on its development programming, cognitive function, metabolism, and body composition in the natal and postnatal periods of life, through epigenomic variations of gene expression[3,19]. Pregnant women, prenatal and early postnatal infants, children in the pubertal period, and elderly people have very distinct and specific dietary and microbial LMW bioactive requirements[3,15,19]. The formation and alteration of epigenetic programming/reprograming of growth and development of human organism, health and diseases first of all depend on such exogenous and endogenous factors as:

- which genes are under epigenomic influence;
- environment and lifestyle factors (nutrition ration, water-salt imbalance, chronic deficiency and/or surplus of certain essential bioactives of food or microbial origin, smoking, alcohol, chemical pollutants, medicines)
- mitochondrial and bacterial energy metabolism and anti-oxidative protective state;
- trans-generational maternal-fetal flow of nutrients and other compounds closely related with placental function;
- physiological status of the individual;
- number and composition of microbial bioactive molecules serving as additional sources of different chemical groups, amino and other organic acids, modified bases, miRNA, other metabolic and signaling molecules, enzymes or their co-factors;
- intra- and inter-species signaling adequacy in eukaryotic and prokaryotic cells.

The above-mentioned information leads to the hypothesis that both nutritional imbalance and long-lasting significant alterations in maternal indigenous microbiota may induce long-term metabolic consequences in offspring as a result of the modulation in epigenetic gene control and disorders of development in programming/reprogramming during embryonic, fetal, and infant development. Supplementation to pregnant women of a diet with food bioactives (enzymes, relevant co-substrate, co-factors, or their precursors) or microbial-origin bioactives and/or restoration of women gut microbiota with probiotics, prebiotics, or metabiotics may replenish the intracellular concentration of metabolic or signal molecules necessary for an epigenomic program favorable for health and well-being[3,4,39,41].

11.3 Epigenomics and link with energy metabolism

Epigenomic DNA and chromatin remodeling are closely linked with the levels of total caloric intake in mammalian cells. The synthesis of energy in the form of ATP involves more than 100 proteins organized into five respiratory chain complexes on the inner mitochondrial membrane. Mammalian cells contain from a few hundred to some thousands of copies of mtDNA. The synthesis of thirteen key proteins is determined by genes localized in the circular double-stranded mitochondrion DNA; all the remaining respiratory chain components and proteins are encoded by the cell nucleus. It is known that mitochondria play a central role in how eukaryotic cells respond to environmental stressors, via energy and increased oxygen species production, participation in apoptosis, lipid homeostasis, steroid synthesis, innate immune response, oxidative stress and inflammation, control of redox signaling pathways, and stem cell functions. However, mitochondria have also important roles in the formation of numerous metabolites needed in the tricarboxylic acid cycle participating in epigenetic gene regulation. They provide key metabolites (NAD+, ATP, 2-oxoglutarate, acetyl CoA and some others) that are co-factors required for numerous transcriptional and epigenetic reactions. Now about 1,600 compounds of various origin have been identified as able to alter mitochondrial functions[42–46]. Recent evidence suggests that mitochondria can participate in epigenetic modification of both nuclear and mitochondrion genomes, in alteration of DNA methylation, chromatin remodeling, acetylation, ADP-ribosylation, oxidative phosphorylation, and in miRNA formation in response to environmental stress[42–44,47]. Despite the mitochondrion's bacterial origins (see chapter 10 by Ryan) and its unique chromosome, replication and metabolism, many functions of this organelle are controlled by proteins encoded by nuclear genomic DNA. Mitochondrial dysfunctions are associated with accelerating aging and a variety of human disorders (cardiovascular and neurodegenerative diseases, cancer, kidney and liver diseases, infertility, etc.)[46]. Mitochondria and indigenous microbiota may share common features due to the prokaryotic origin of mitochondria. Mitochondria in the human body cells resemble bacteria not only in the size and shape. They also have their own kind of DNA that is very similar to bacterial DNA. In addition, there are other biochemical, molecular and cell biological data bearing out the prokaryotic origin of mitochondria[45,48]. Mitochondria and bacteria have also some similar pathways for energy production, especially in the glycolytic formation of pyruvate and ATP[49–51]. Despite the mitochondria producing 60-80% energy necessity for human cells, the microbial fermentation of 50–60 g of carbohydrates daily produces about 600 mmol of short-chain fatty acids (SCFA), with a total energy value to 750 kJ[52] (see chapter 1 for further discussion on the role of SCFAs in microbiota/host interactions). Representatives of the *Bacteroidetes* usually are linked with carbohydrate metabolism, while the members of the *Firmicutes* are associated with carbohydrate transport systems. In humans, microbial gut fermentation of indigestible carbohydrates provides 5–30% of total energy requirement via the production and metabolism of SCFA. 59–80% of energy needs in intestinal epithelial cells is provided by butyrate produced by members of the gut microbiota; acetate and propionate are the important source of energy for brain, heart and muscle cells[49,52,53].

Therefore, from an evolutionary point of view, mitochondria and the host microbiota should functionally be considered to be both a collective metabolically active internal "organ" affecting the host's energy metabolism, and the regulator of the other energy-connected functions including gene expression in the mitochondrial and nuclear genomes of mammalian organisms[39,42,47]. Cell energy metabolism needs about one hundred proteins, fifty various enzyme co-substrates and co-factors of food and/or microbial origin: vitamins (B1, B2, B5, B6, B7, B9, B12, C, phyloquinone-K1, menaquinone-K2), non-vitamins (NAD, NADH, NADP+, NADPH), ATP, cytidine triphosphate (CTP), SAM, 3-"phosphoadenosine-5"-phosphosulfate (PAPS), glutathione (GSH), Coezyme B, Coenzyme M, Cofactor F-430, Coenzyme Q10, haem, alpha lipoic acid, methanofuran, molybdopterin/molybdenum cofactor, pyrroloquinoline quinine (PQQ), tetrahydrobiopterin (THB/BH4), tetrahydromethanopterin (THMPT/H4MPT), minerals (Ca, Cu, Fe^{++}, Fe^{+++}, Mg, Mn, Mo, Ni, Se, Zn), amino acids (arginine, lysine, methionine, cysteine, beta-alanine, serine, threonine, histidine, tryptophan, aspartic acid), Krebs cycle organic acids and some nucleotides (pyrimidine), miRNA[5,14,42,46,50,54–57]. Any deficiency or excess of the above-mentioned compounds may produce mitochondria structural and/or functional imbalance, quite often resulting in a wide spectrum of metabolic diseases[42,46,55,56]. Molecular investigations of epigenomic-associated processes have demonstrated that most major participants in the epigenomic machinery are formed during energy metabolism in the eukaryotic cell mitochondria and in the prokaryotic cell membranes[3,5,20,24,55].

11.4 The microbiota as a potential epigenetic modifier

Exactly how the human microbiota changes human gene expression at the cell and molecular level is being explored. It is well known that microorganisms produce multiple LMW substances that can quickly distribute within the human host and interact with different targets in cells, tissue and, organs[13–15]. The human gut microbiota contributes 36% of the small molecules in human blood[17]. These LMW compounds are considered to be potential auto-inducers regulating prokaryotic and eukaryotic cells, bacteria-bacteria and host-bacteria relationships. Recent OMIC-based studies have provided insights into how the microbiota (including pathogenic, indigenous and probiotics) sense and adapt to the gastrointestinal tract environment and interfere with host gene expression processes. It should be remembered that according to recent data all human proteins possess motifs present also in bacterial proteins. Besides, there are bacteria-human similarities for pathogenic, opportunistic and symbiotic species[58,59]. Microbial epigenetics investigate how these microbial-derived compounds participate in the control of stability of prokaryotic and eukaryotic DNA, its transcription and translation, as well as switch on or switch off corresponding genes via direct interaction with specific DNA, RNA or chromatin receptors by using various kinases or other epigenetic-associated microbial enzymes and their co-factors[5,23,39,40,60–63]. It has been revealed that some LMW substances of microbial origin might be really endogenous environmental modifying factors of the genomic and epigenomic processes both in the host and prokaryotic cells[5,15,39,62]. It is thought that epigenetic effectors can act through direct or indirect mechanisms[3,5,13]. The direct effects may be associated with the ability of microbial compounds to affect DNA methylation, chromatin and histone architecture or miRNA interference. For example, the production of bacterial methyl or acetyl groups, biotin, eukaryote-like serine/threonine kinases, phosphatases, methyltransferases, acetylases/deacetylases, or BirA ligase may directly affect chromatin and histone architecture or DNA methylation, or other epigenetic processes in eukaryotes. DNA from most prokaryotes and eukaryotes contains the methylated bases (4-methylcytosine, 5-methylcytosine, and 6-methyladenine)[5]. About 100 different methyltransferases have been identified in animal, plant, and microbial cells[64], including more than 20 lysine and 10 arginine histone methyltransferases[65]. Methylation/demethylation[22,23], acetylation/deacetylation[26,27], phosphorylation/dephosphorylating[31,32,66] are reversible and can be involved in both gene activation and silencing in a wide variety of prokaryotes and eukaryotes. There are data that some representatives of these groups of enzymes, functionally (and partly structurally), are very similar in eukaryotic and prokaryotic cells[23,27,31,67]. Mammalian DNA methyltransferases change the methylation profile for specific compartments of the genome in a tissue-specific manner. In contrast, prokaryotic methyltranferases modify all of their recognition sites[5]. It has been found that lysine methyltransferase activity may be connected with the SET (**S**u(var)3-9, **E**nhancer of Zeste, **T**rithorax) domains composed of approximately 130 amino acids. Genes coding for SET-domain-containing proteins are present in all eukaryotic genomes[71]. Prokaryotes do not have histones or highly organized chromatin. During the last years, through the use of next-generation sequencing (NGS), it has been demonstrated that the vast majority of bacterial and archaea genomes also contains genes coding for SET-domain proteins that functionally are similar DNA-binding proteins of eukaryotic histone methyltransferases[25,68]. Stimulation of host HDAC activity is a common feature of microbial infection, resulting in the epigenetic repression of genes responsible for immunity[69]. In eukaryotic organisms a phosphate-group usually adjoins serine, threonine and rarely tyrosine, in prokaryotes — mostly to histidine and aspartic acid residues by specific protein kinases. Bacterial protein-tyrosine kinases occur in most bacterial phyla and can regulate polysaccharide biosynthesis, lysogenization, heat-shock response, DNA replication, cell cycle, virulence, resistance to cationic antimicrobial peptides and cell shape, including spore germination[31,32]. The eukaryotic type serine-threonine protein kinases (eSTPKs) found in the normal human microbiota and in pathogenic bacteria carry catalytic domains that share structural and functional homology with eukaryotic STPKs. Such enzymes are now considered ubiquitous. Eukaryote-type serine-threonine protein phosphatases (eSTPPs) participating in the dephosphorylation processes are present in both Gram-negative and Gram-positive bacteria and Archaea. They have conserved catalytic amino acids domains that are structurally nearly identical to that of eukaryotes. Eukaryotic- like eSTPKs and eSTPPs are usually encoded in the same operon and can regulate normal cell growth, division and translation in various bacteria[31]. It has been demonstrated that proteasomes are present in all eukaryotes and many Archaea, and prokaryotes. The discovery of an Ub-modifier in prokaryotes has shown bacteria may also possess such post-translational protein-tagging systems affecting various cellular processes[70]. For example, *Bacteroides fragilis* is one of the predominant members of the symbiotic human gut microbiota but may also be causative reason of some autoimmune, bowel inflammatory diseases and opportunistic infections. The sequence of *B. fragilis* genome has shown that it contains a gene (ubb), the product of which has 63% identity to human Ub. Because Ub is involved in the innate and adaptive immune responses, the discovery of Ub homologue in a key member of normal resident gut microbiota suggests that the presence in the intestinal tract of an aberrant bacterial Ub could inappropriately activate the host immune system via interference with eukaryotic epigenetic Ub activity[72]. Biotin deficiency causes abnormally low biotinylation of histones and results in an aberrant gene regulation (e.g. derepression of retrotransposons leading to chromosomal instability)[73]. Biotinylation of histones in mammalian cells is mediated by eukaryotic holocarboxylase synthetase (HCS), biotinidase and microbial non-selective enzyme (BirA ligase). Bir A ligase plays a key role in cell signaling and chromatin remodeling during biotin biosynthesis in prokaryotes. Similar mechanisms of gene regulation for HCS have been reported in eukaryotes[74]. More than half known prokaryotic cell RNA can be refer to miRNAs[37,75]. They belong to three special groups: small RNAs involving in trans-regulation activities via the binding of target mRNAs, long UTRs RNAs and other classes of translation or transcription attenuators. These non-coding RNA mostly fulfill functional, regulatory roles. In addition, they can redefine the transcription units of coding genes, which is important for transcript quantification and for the RNA-RNA interactions[75]. Various epigenetics mechanisms affecting gene expression without involving mutations operate not only in prokaryotes and Archaea but in eukaryotic microorganisms (yeast, protozoa) as well. These epigenetic changes enable such organisms to adapt more rapidly to environmental conditions including host immunity, medical intervention, etc.[76]. The indirect effects to epigenetic homeostasis are linked with microbial bioactive influences to different signaling pathways (e.g., altered expression of growth factors, receptors, other co-substrates and co-factors that in turn modify transcription regulatory net at gene promoters). An example of indirectly action of microbial LMW molecules to chromatin remodeling is the deficiency of some co-substrates (methionine, betaine, and choline) and/or cofactors (folate, vitamins B2, B6 and B12) produced by indigenous intestinal microbiota[5,14,63,77–79]. Being a methyl donor (or co-factors), all the above-mentioned substances participate in the one-carbon pathway. Gut bacteria also affect the bioavailability of many dietary sources of methyl groups. Inadequate dietary and/or microbe methyl group and co-factor provision alters carbon metabolism and leads to hypomethylation in many important epigenomic-associated pathways.

This alteration can impair DNA methylation, resulting in elevated plasma homocysteine concentrations, decreased SAM content, and increased risk of various coronary, cerebral, hepatic, vascular diseases, and malignancy[41,63,64]. Another example of an indirect effect of indigenous microorganisms to human epigenome could be gut microbiota contributions to the transformation, bio-viability, absorption and/or excretion of some biochemical elements (zinc, iodine, selenium, cobalt, and others) that are co-factors of various enzymes participating in the work of different epigenomic processes[77,80]. Some LMW molecules of microbial origin can directly or indirectly activate or inhibit epigenomic regulation via interference of enzymatic activity[5,13,40,81] or through modifications of cell receptors (e.g., methylation and histone deacetylation of TLR4-like receptors in intestinal cells[81]) that take part in epigenomic reprogramming and/or post-translation modification of histones and other proteins. Bacteria can also produce metabolic compounds acting as inhibitors or activators of chromatin-modifying enzymes[68]. For example, SCFA modulation of epigenomic effects may be connected with their involvement into energy metabolism[14,39,53] and in such epigenomic processes as chromatin acetylation/deacetylation[68,82,83], phosphorylation/dephosphorylation, DNA methylation[82,84,85], ubiquitination[86] and miRNA synthesis[62,86,87]. The treatment of epithelial cells with butyric acid results in hyperacetylation through the inhibition of sirtuin activity[88]. Microbe-derived butyrate, at physiological concentrations, can trigger cell cycle arrest and apoptosis in colon cell line through interference in the regulation of host gene expression via colonic epithelial HDAC inhibition[89] or decreased expression of the miRNAs (miR-106b)[62]. SCFA, and especially butyrate, are capable of inducing the formation of ROS in enterocytes, resulting in the blockade of IκB-α ubiquitination in them; ubiquitin lygase degradation breaks the function of signal NF-κB pathway[86]. The miRNA responsible for expression or inhibition of p21gene may also be the target for butyrate[62]. During microbial metabolism of cruciferous vegetables or garlic a corresponding number of sulforaphane cysteine/sulforaphane N-acetyl-cysteine and allyl mercaptan/diallyl disulfide are formed that can inhibit activity of histone deacethytransferases enzymes[26, 90].

11.5 Epigenetic control of the host genes by pathogenic and opportunistic microorganisms

It is known that the vast majority of pathogenic and opportunistic microorganisms can modify global host responses during infection to optimize their infectious cycles. Various pathogens can either inhibit or mimic host functional or biochemical reactions related to epigenetic alterations[25,34,59]. Pathogen-induced epigenetic deregulations may affect host cell function either to suppress host defense or to permit pathogen persistence. The virulence potential of such microorganisms includes the different sensor/signaling networks allowing them to directly interfere with transcription, chromatin remodeling, RNA slicing or DNA replication and repair[23,40,66,68]. *Epstein-Barr* virus, hepatitis viruses B and C, Human papilloma virus, *Streptococcus bovis, Chlamydia pneumoniae, C. trachomatis, Campylobacter rectus, Helicobacter pylori, Bacillus anthracis, Listeria monocytogenes, Moraxella catarrhalis* as well as toxins of *Cholera vibrio, Corynebacterium diphtheria,* and *Clostridium* toxins can contribute to the host epigenetic changes via the inhibition of chromatin-silencing complexes or on similarity of bacterial effector molecules to eukaryotic histone-modifying enzymes, resulting in the onset and progression of some autoimmune, metabolic, inflammation, and infectious connected diseases[40,68,69]. For example, many Chlamydia strains have a SET-domain coding for genes with histone H3 methytransferase activity. The functionally similar SET domains can be found in *Legionella pneumophila*, *Bordetella bronchiseptica*, *Burkholderia thailandensis*, and *B. anthracis*. Some representatives of Archaea (*Methanosarcina mazei*, *Sulfolobus islandicus*) also contain a set of DNA-binding proteins that are able to produce post-translational modifications, similar to the histone alterations in the chromatin of eukaryotic cells. Various LMW bacterial components including virulence factors (flagellins, lipopolysaccharides, lysins) modify histone acetylation/deacetylation and/or phosphorylation/dephosphorylation by direct or indirect activating host signal cascades (e.g., NF-κB, MAPKs, P13K, MSK1/2, IKKα, AKT)[40]. Recent discovery of the existence of eukaryote-like serine-threonine protein kinases and phosphatases in a number of prokaryotes (*M. tuberculosis*, *Y. pseudotuberculosis*, *L. monocytogenes*, *S. pneumonia*, *E.faecalis*, *P. aeruginosa*, and *S.enterica* serovar *Typhi*, *S. aureus*) has revealed the participation of these enzymes in the modulation of host/pathogen interactions via alteration of phosphorylation/dephosphorylation. Several bacteria eSTPKs are involved in virulence of *Yersinia* spp. and *Salmonella* spp. through disruption of the eukaryotic actin cytoskeleton and increasing microbial resistance to phagocytosis by macrophages[31,66]. Certain pathogenic bacteria (e.g., *L.pneumophila*, *Shigella* spp.) possessing eSTPKs can interact with host defense factors through disruption of the NF-κB signaling pathways connected with elimination of the pathogens[66]. Eukaryotic pathogens possess an endogenous ubiquitination system allowing them to modify and use its components as effector molecules. On the other hand, bacterial pathogens can produce effector proteins with structural and functional similarities to eukaryotic E3-ligase or secrete molecules inhibiting Ub-mediated degradation; effector mimicry of components of the Ub machinery occurs via gene flow. The ability to interfere with the processes of ubiquitination/deubiquitination by "mimicking" host-cell proteins has been found in yersinia, chlamidia, Escherichia, listeria, herpesvirus, cytomegalovirus, and adenovirus[25,91]. This mechanism plays an important role in their pathogenic characteristics through suppression of host immune responses. Bacterial infection (e.g. *Campylobacter rectus*) can induce epigenetic changes in the genes of placenta cells; such epigenetically affected genes may play an important role in the disorders of fetal development[68]. Rapid elimination of microbe-linked epigenetic alterations may restore defense mechanisms, protect host organism from chronic or latent infections, and decrease a risk of cancer, autoimmune, and inflammatory diseases[68,83]. Some pathogenic bacteria can produce long-term epigenetic memory associated with immune memory in vertebrates[40].

11.6 Epigenetic control of the host genes by indigenous (probiotic) microorganisms

Host/microbe dialogue occurs between indigenous (probiotic) bacteria and the host. It is well known that host epithelial cells are continuously exposed to large numbers of symbiotic microorganisms. Using different experimental animal models it was shown that during passages through the gastrointestinal tract some silent genes of probiotic bacteria may be induced by host cell signals[62,92–94]. In these conditions the induction of 72 probiotic *L. plantarum* WSFS1 genes was observed[92]. On the other hand, probiotic (e.g. *Lactobacillus* GG) inside the intestinal tract can additionally induce expression of over 400 genes involved in immune response, inflammation, cell growth and differentiation, apoptosis, and cell to cell signaling[93]. This means that microbial symbionts (both prokaryotic and eukaryotic) and eukaryotic cells inside the small intestine and colon may be involved in the mutual bacteria-host cell cross-talk, possibly via various epigenomic post-translational modifications[5,14,39,40,81]. For example, it is established that host epithelial cells sense intestinal microorganisms by Toll- and NOD-like receptors, recognizing such microbe compounds as LPS, lipoproteins, and CpG DNA. Excessive response of TLR to commensal bacteria produces down-regulation of the NF-κB pathway by inducing or maintaining DNA methylation of the gene, which prevents excessive inflammation in the gut[40,81]. Though the implication of the indigenous gut microorganisms in epigenetic variations has been discussed[5,25,40,95], the importance of probiotics on the base of living symbiotic bacteria or their LMW substances (metabiotics), as epigenetic modifying factors in the treatment of chronic metabolic diseases remains unclear; this is further discussed in chapter 30 of this book. Investigation of the nine bifidobacterial genomes revealed that some of them had genes coding for the methyltransferases. The strain of *B.longum* GT15 possessed two different DNA methyltransferases (BLGT_01880 and BLGT_03390)[96]. There are data that sodium butyrate (SB) administration in animal models may have beneficial effects on Alzheimer's disease (AD)[97], on restoration of dendritic spine density in hippocampal neurons[98], on associative memory[99], on Parkinson's disease[100] and on protection of neurons against MPP+ toxicity[101]. Production of butyrate by gut symbiotic (probiotic) bacteria and their anti-inflammatory action may be connected with the ability of SB to inhibit HDAC enzymes in eukaryotic cells[40, 83]. Butyrate acts as an inhibitor of HDACs and inflammation, suggesting the use of butyrate-producing probiotic bacteria (e.g. *Faecalibacterium prausnitzii*) as a novel type of immunosuppresor in preventing and treatment of chronic immune-mediated diseases[83]. Targeting the epigenome, mainly with small drugs such as inhibitor HDAC capable of crossing the blood brain barrier, delayed the onset and progression of the symptoms in animal models of neurodegenerative disease[102]. The various inhibitors of HDAC used in clinics for treatment of different forms of cancer (valpronic acid-C8, phenyl-butyrate, sodium butyrate) produce their effects on these enzymes in mM concentrations[83].

Bifidobacteria dominate fecal communities throughout the first year of life. Children are significantly more different from one another than are adults in terms of their fecal bacterial community phylogenetic structure; they are also more different in terms of their repertoires of microbiome-encoded functions. A prominent example of these shared age-related changes involves the metabolism of vitamins B12 and folic acid[7]. In contrast to folate, which is synthesized by microbes and plants, B12 is primarily produced by microbes[103,104]. The gut microbiomes of babies are enriched in genes involved in the *de novo* biosynthesis of folate, whereas microbiomes of adult persons possess significantly higher representation of genes that metabolize dietary folate and its reduced form tetrahydrofolate (THF). Unlike *de novo* folate biosynthetic pathway components, which decrease with age, the procobalamin biosynthesis increases with age. The folate and cobalamin pathways are linked functionally by methionine synthase, which catalyses the formation of THF from 5-methyl-TNF and L-homocysteine, requiring cobalamin as a cofactor; the representation of this enzyme also increases with age. The low relative abundance of functional gene groups involved in cobalamin biosynthesis in the fecal microbiomes of babies correlates with the lower representation of members of *Bacteroidetes, Firmicutes* and *Archaea* in their microbiota. Although the biosynthetic pathway for cobalamin is well represented in the genomes of these organisms, *Bifidobacterium, Streptococcus, Lactobacillus and Lactococcus*, which dominate the baby gut microbiota, are deficient in these genes. By contrast, several of these early gut colonizers contain genes involved in folate biosynthesis and metabolism. These changes in the microbe vitamin biosynthetic pathways correlate with published reports indicating that blood levels of folate decrease and B12 increase with age. Genes involved in fermentation, methagenesis and the metabolism of arginine, glutamate, aspartate and lysine were higher in the adult microbiome, whereas genes involved in the metabolism of cysteine and fermentation pathways found in lactic acid bacteria of babies[105]. The ability of microbiota members to access host-derived glycans and other compounds plays a key part in establishing a gut microbial community. These findings suggest that indigenes microbiota (probiotics) should be considered when assessing the nutritional needs of human at various stages of the epigenomic program of development.

11.7 Concluding remarks and future directions

Taken together, the data presented above convincingly affirm that many microbial LMW molecules could be the significant exogenous and endogenous effectors causing persistent changes in energy-connected epigenomic functions in human beings. These epigenetic changes may be beneficial for the host or may have non-desirable side effects. The existence of an energy-connected mitochondria/microbiota/epigenetics axis in mammalian organisms may be illustrated by the following examples. Many epigenetic processes are susceptible to changes in the levels of SAM, ATP, acetyl-CoA, NAD[24,39]. A NAD+ dependent histone deacetylase (sirtuin) is a key modifier of chromatin structure. Acetylation and deacetylalation of histone proteins cause changes in the cell gene expression. The main donor of acetyl groups for formation of acetyl-CoA participating in epigenomic acetylation reactions is the gut microbiota. Bacteria/eukaryotes share a common pathway for acetyl-CoA biosynthesis from pantothenate, cysteine, and β-alanine. These essential co-substrates and co-factors may be found in the foodstuffs in small quantities; they are also generated endogenously and/or by the various gut microorganisms. Deficiencies in the above-mentioned LMW bioactives of dietary or microbial origin may result in disorders of acetylation, ribosylation, and other epigenetic processes participating in chromatin remodeling and regulating mitochondrion and nucleus gene expression.

Hence, the disruption of the close cooperation between eukaryotic mitochondria and symbiotic prokaryote membranes in the host energy synthesis may become the key factor predisposing to a wide spectrum of both dynamic energy imbalances and multiple epigenomic alterations. Such alterations can in turn result in onset and progression of many human chronic diseases. It has been shown that combination of such dietary supplements as L-carnitine, α-lipoic acid, coenzyme Q10, reduced nicotinamide adenine dinucleotide, membrane phospholipids and some others can reduce mitochondrial dysfunction and improve the common complaints of persons suffering from chronic metabolic diseases linked with energy imbalance[46,56]. The impact of "epigenetic diets" (caloric/dietary restriction, diet enriched with nutrients involved in one-carbon metabolism and/or other bioactive food components) in the anti-ageing program has been described[41]. The regulation of the main transcription factors involved in host immunity may be mediated by probiotics via complex epigenetic alterations[83]. For these reasons, the search for natural dietary and indigenous microbial modifiers of energy-connected epigenomic machines is still emerging, especially with regard to personal medicine[65,90,106–108]. However, the vast majority of reports concerning host/probiotic bacteria cross-talk have focused on single bacterial strains, while in real life mammalian gut bacteria never act on the host cells in isolation. The structural and functional variety of microbial molecular effectors, their interactions with other chemical components, competitive inhibitors for specific transport proteins or absorption site, differing study designs and molecular diagnostic tests used are additional reasons for the apparently inconsistent results of studies on modifiers of energy-connected epigenomic processes[39,77]. Therefore, probiotics and their bioactives (metabiotics) should be considered for the control of epigenetic phenomena in different individuals in specific environmental conditions. However, epigenetically "sensitive" regions in the human genome, epigenome, metabolome and signaling pathways and potential side effects of such specific probiotics need to be identified first[13,39,77]. Special attention has been paid to the design of epigenetic-based probiotics for pregnant women and infants. Animal models have established that probiotics can modify the epigenomic reprograming[109]. Currently accumulated data have already led us to suppose that known enzymes and other participants of mitochondrial and epigenomic machines could represent potential targets as microbial "epigenetic-based" preventive agents in chronic metabolic diseases[5,39,102,110]. The typical example of a beneficial modifier of a cell epigenome might be the microbial SCFA[53,62,83]. Further investigations will clarify what known food, microbial or artificial pan- or selective inhibitors and activators of enzymes involved in epigenetic DNA and chromatin modifications might be really introduced in clinical practice[20,97,102,110,111]. In preclinical investigations of potential epigenomic modifiers, various gnotobiological models and the modern OMIC-technologies should be included[18,39,112]. All known and novel potential probiotics and metabiotics should be investigated for their ability to interfere with gene expression in eukaryotic and prokaryotic cells in detail to exclude their possible side effects. Such approach, which has been termed "pro-bioepigenomics," allows a better evaluation of the efficacy and safety of probiotics prepared on the base of living bacteria, their structural compounds, metabolites and/or signal molecules early in their development program.

TAKE-HOME MESSAGE

- Many substances of microbial origin with chemical and functional similarity to dietary nutrients and/or endogenous substances can modify gene expression profiles in eukaryotic cells.

- Mitochondria and bacterial membranes functionally should be considered as a collective metabolically active internal "organ" affecting the host's energy metabolism and energy-connected epigenomic functions.

- A host micro-ecological imbalance caused by various factors and agents produces different epigenetic abnormalities that can result in the onset and progression of metabolic diseases.

- The design of microbial epigenetic-based drugs and functional foods could be a very important scientific-applied approach for the treatment and prophylaxis of epigenetic-associated human diseases. Pro-bioepigenetics employing complex OMIC-technologies and gnotobiological models should be used for better assessment of efficacy and safety of known and potential probiotics.

References

1 Feinberg AP. Epigenetics at the epicenter of modern medicine. *JAMA* 2008; **299**: 1345–50.
2 Furrow RE, Christiansen B, Feldman MW. Environment-sensitive epigenetics and the heritability of complex diseases. *Genetics* 2011; **189**: 1377–87.
3 Kanherkar RR, Bhatia-Dey N, Csoka AB. Epigenetics across the human lifespan. *Front Cell Develop Biol* 2014; **2**. DOI: 10.3389/fcell.2014.00049.
4 Gluckman PD, Hanson MA, Buklijas T, Low FM, Beedle AS. Epigenetic mechanisms that underpin metabolic and cardiovascular diseases. *Nature* 2009; **5**: 401–408.
5 Shenderov BA. Gut indigenous microbiota and epigenetics. *Microb Ecol Health Dis* 2012; **23**: 17461. DOI: 10.3402/mehd.v23i0.17461.
6 Grice EA, Segre JA. The human microbiome: our second genome. *Annu Rev Genomics Hum Genet* 2012; **13**: 151–70.
7 Yatsunenko T, Rey FE, Minary MJ, Tehan I, Dominguez-Bello MG, Contreras M *et al.* Human gut microbiome viewed across age and geography. *Nature* 2012; **486**: 222–227.
8 Bengmark S. Gut microbiota, immune development and function. *Pharmacol Res* 2013; **69(1)**: 87–113.
9 Moloney ED, Desbonnet L, Clarke G, Dinan TG, Cryan JF. The microbiome: stress, health and disease. *Mamm Genome* 2014; **25**: 49–74.
10 David LA, Maurice CF, Carmody RN, Gootenberg DB, Button JE, Wolfe BE, *et al.* Diet rapidly and reproducibly alters the human gut microbiome. *Nature* 2014; **505(7484)**: 559–63.
11 Suvorov A. Gut microbiota, probiotics, and human health. *Bioscience of Microbiota, Food and Health* 2013; **32(3)**: 81–91.
12 World Digestive Health Day, WDHD, May 29, 2014. *WGO Handbook on Gut Microbes — Importance in Health & Disease*: 1–64.
13 Shenderov BA. Probiotic (symbiotic) bacterial languages. *Anaerobe* 2011; **17**: 490–5.
14 Jeffery IB, O'Toole PW. Diet-microbiota interactions and their implications for healthy living. *Nutrients* 2013; **5**: 234–252.
15 Soen Y. Environmental disruption of host-microbe co-adaptation as a potential driving force in evolution. *Front genet* 2014; **5**. DOI: 10.3389/fgenr.2014.00168.
16 Nibali L, Henderson B, Sadiq ST, Donos N. Genetic dysbiosis: the role of microbial insults in chronic inflammatory diseases. *J Oral Microbiology* 2014; **6**: 22962. DOI: 10.3402/jom.v6.22962.
17 Hood L. Tackling the microbiome. *Science* 2012; **336**: 1209.
18 Shenderov BA. "OMIC"-technologies and their importance in current prophylactic and regenerative medicine. *Regenerative Med J* 2012; **3(49)**: 70–83 (in Russian).
19 Kussmann M, Van Bladeren PJ. The extended nutrigenomics-understanding the interplay between the genomes of food, gut microbes, and human host. *Front Genet* 2011; **2**: 21. DOI: 10.3389/fgene.2011.00021.

20 Finley A, Copeland RA. Small molecule control of chromatin remodeling. *Chemistry & Biology* 2014; **21**. DOI: 10/1016/j.chembiol.2014.07.024.

21 Ziller MJ, Gu H, Muller F, Donaghey J, Tsai LT, Kohlbacher O *et al*. Charting a dynamic DNA methylation landscape of the human genome. *Nature* 2013; **500**: 477–481.

22 Ehrlich M, Lacey M. DNA methylation and differentiation: silencing, upregulation and modulation of gene expression. *Epigenomics* 2013; **5(5)**: 553–68.

23 Casadesus L, Low D. Epigenetic gene regulation in the bacterial world. *Microbiol Mol Biol Rev* 2006; **70**: 830–56.

24 Kaelin WG, McKnight SL. Influence of metabolism on epigenetics and disease. *Cell* 2013; **153**. DOI: 10.1016/j.cell.2013.03.004.

25 Alvares-Venegas R. Bacterial SET domain proteins and their role in eukaryotic chromatin modification. *Frontiers in genetics epigenomics and epigenetics* 2014; doi: 10.3389/fgene. 2014. 00065.

26 Kim GW, Gocevski G, Wu CJ, Yang XJ. Dietary, metabolic, and potentially environmental modulation of the lysine acetylation machinery. *Int J Cell Biol* 2010; 632739. DOI: 10.1155/2010/632739.

27 Wang Q, Zhang Y, Yang C, Xiong H, Lin Y, Yao J *et al*. Acetylation of metabolic enzymes coordinates carbon source utilization and metabolic flux. *Science* 2010; **327**: 1004–7.

28 Lazo-Gomez R, Ramirez-Jarquin UN, Tovar-y-Romo LB, Tapia R. Histone deacetylases and their role in motor neuron degeneration. *Front Cell Neurosci* 2013; **7**: 243. DOI: 10.3389/fncel.2013.00243.

29 Zempleni J, Wijeratne SS, Hassan YI. Biotin. *Biofactors* 2009; **35**: 36–46.

30 Hassan YI, Moriyama H, Zempleni J. The polypeptide Syn67 interacts physically with human holocarboxylase synthetase, but is not a target for biotinylation. *Arch Biochem Biophys* 2010; **495**: 35–41.

31 Pereira SFF, Goss L, Dworkin J. Eukaryote-like serine/threonine kinases and phosphatases in bacteria. *Microbiol Mol Biol Rev* 2011; **75(1)**: 192–212.

32 Shi L, Ji B, Kolar-Znika L, Boskovic A, Jadeau F, Combet C *et al*. Evolution of bacterial protein-tyrosine kinases and their relaxed specificity toward substrates. *Genome Biol Evol* 2014; **6(4)**: 800–17.

33 Sharan RN. Poly-ADP-Ribosylation in cancer. In: *Cancer epigenetics* (Tollefsbol T. ed). Florida. CRS Press 2009; 265–76.

34 Simon NC, Aktories K, Barbieri JT. Novel bacterial ADP-ribosylating toxins: structure and function. *Nature Rev Microbiol* 2014; **12**: 599–611.

35 Popovic D, Vucic D, Dikic I. Ubiquitination in disease pathogenesis and treatment. *Nature Medicine* 2014; **20**: 1242–53.

36 Castel SE, Martienssen RA. RNA interference (RNAi) in the nucleus: roles for small RNA in transcription, epigenetics and beyond. *Nat Rev Genet* 2013; **14(2)**: 100–12.

37 O'Flaherty S, Klaenhammer TR. The impact of omic technologies on the study of food microbes. *Annu Rev Food Sci Technol* 2011; **2**: 353–71.

38 Yang L, Froberg JE, Lee JT. Long noncoding RNAs: fresh perspectives into the RNA world. *Trends Biochem Sci* 2014; **39(1)**: 35–43.

39 Shenderov BA, Midtvedt T. Epigenomic programing: a future way to health? *Microb Ecol Health Dis* 2014; **25**: 24145. DOI: 10.3402/mehd.v25.24145.

40 Takahashi K. Influence of bacteria on epigenetic gene control. *Cell Mol Life Sciences* 2014; **71(6)**: 1045–54.

41 Bacalini MG, Friso S, Olivieri F, Pirazzini C, Giuliani C, Capri M *et al*. Present and future of anti-ageing epigenetic diets. *Mech Ageing Dev* 2014; DOI: 10.1016/j.mad.2013.12.006.

42 Wallace DC. The epigenome and the mitochondrion: bioenergetics and the environment. *Gen Dev* 2010; **24**: 1571–3.

43 Chinnery PF, Elliott HR, Hudson G, Samuels DC, Relton CL. Epigenetics, epidemiology and mitochondrial DNA diseases. *Int J Epidemiol* 2012; **41**: 177–87.

44 Shaughnessy DT, McAllister K, Worth L, Haugen AC, Meyer JN, Domann FE, *et al*. Mitochondria, energetics, epigenetics, and cellular responses to stress. *Environ Health Perspect* 2014. DOI: 10.1289/ehp.1408418.

45 Friedman JR, Nunnari J. Mitochondrial form and function. *Nature* 2014; **505**: 335–43.

46 Nicolson GL. Mitochondrial dysfunction and chronic disease: treatment with natural supplements. *Altern Ther Health Med* 2014; **20** (suppl 1): 18–25.

47 Bellizzi D, D'Aquila P, Giordano M, Montesanto A, Passarino G. Global DNA methylation levels are modulated by mitochondrial DNA variants. *Epigenomics* 2012; **4**: 17–27.

48 Gray MW. Mitochondrial evolution. *Cold Spring Harb Perspect Biol* 2012; **4**: a011403.

49 Arora T, Sharma R. Fermentation potential of the gut microbiome: implications for energy homeostasis and weight management. *Nutr Rev* 2011; **69**: 99–106.

50 Soballe B, Poole RK. Microbial ubiquinones: multiple roles in respiration, gene regulation and oxidative stress management. *Microbiology* 1999; **145**: 1817–30.

51 Herrman G, Jayamani E, Mai G, Buckel W. Energy conservation via electron-transferring flavoprotein in anaerobic bacteria. *J Bacteriol* 2008; **190**: 784–91.

52 McNeil NI. The contribution of the large intestine to energy supplies in man. *Am J Clin Nutr* 1984; **39**: 338–342.

53 Shenderov BA. Targets and effects of short chain fatty acids. *Current Medical Science* 2013; **N1-2**: 21–50 (in Russian).

54 Barrey E, Saint-Auret G, Bonnamy B, Damas D, Boyer O, Gidrol X. Pre-micro RNA and mature micro RNA in human mitochondria. *PLoS One* 2011; **6**: e20220. DOI: 10.1371/journal.pone.0020220.

55 Wallace DC, Fan W. Energetics, epigenetics, mitochondrial genetics. *Mitochondrion* 2010; **10**: 12–31.

56 Wallace DC, Fan W, Procaccio V. Mitochondrial energetics and therapeutics. *Ann Rev Pathol* 2010; **5**: 297–348.

57 Jimenez-Chillaron JC, Duaz R, Martinez D, Pentinat T, Ramon-Krauel M, Ribo S, Plosch T. The role of nutrition on epigenetic modifications and their implications on health. *Biochimie* 2012; **94**: 2242–63.

58 Trost B, Lucchese G, Stufano A, Bickis M, Kusalik A., Kanduc D. No human protein is exempt from bacterial motifs, not even one. *Self Nonself* 2010; **1(4)**: 328–334.

59 Doxey AC, McConkey BJ. Prediction of molecular mimicry candidates in human pathogenic bacteria. *Virulence* 2013; **4(6)**: 453–66.

60 Holmes E, Li JV, Athanasiou T, Ashrafian H, Nicholson JK. Understanding the role of gut microbiome-host metabolic signal disruption in health and disease. *Trends Microbiol* 2011; **19**: 349–59.

61 Kinross JM, Darzi AW, Nicholson JK. Gut microbiome-host interactions in health and disease. *Genome Medicine* 2011; **19**: 349–59.

62 Hu S, Dong TS, Dalai SR, Wu F, Bissonnette M, Rwon JH *et al.* The microbe-derived short chain fatty acid butyrate targets miRNA-dependent p21 gene expression in human colon cancer. *PLoS ONE* 2011; **6**: e16221. DOI: 10.1371/journal.pone. 0016221.

63 Martin F-PJ, Sprenger N, Montoliu I, Rezzi S, Kochhar S, Nicholson JK. Dietary modulation of gut functional ecology studied by fecal metabonomics. *J Proteome Res* 2010; **9**: 5284–95.

64 Stead LM, Brosnan JT, Brosnan ME, Vance DE, Jacobs RL. Is it time to reevaluate methyl balance in humans? *Am J Clin Nutr* 2006; **83**: 5–10.

65 Bissinger E-M, Heinke R, Sippl W, Jung M. Targeting epigenetic modifiers: Inhibitors of histone methyltransferases. *Med Chem Commun* 2010; **1**: 114–124.

66 Canova MJ, Molle V. Bacterial serine/threonine protein kinases in host-pathogen interactions. *J Biol Chem* 2014; **289(14)**: 9473–79.

67 Ishikawa K, Handa N, Sears L, Raleigh EA, Kobayashi I. Cleavage of a model DNA replication fork by a methyl-specific endonuclease. *Nucl Acids Res* 2011; **39**: 5489–98.

68 Bierne H, Hamon M, Cossart P. Epigenetic and bacterial infections. *Cold Spring Harb Perspect Med* 2012; **2(12)**: a010272. DOI: 10.1101/cshperspect.a010272.

69 Paschos K, Allday MJ. Epigenetic reprogramming of host genes in viral and microbial pathogenesis. *Trends Microbiol* 2010; **18(10)**: 439–47.

70 Darwin KH Prokaryotic ubiquitin-like protein, proteasomes, and pathogenesis. *Nat Rev Microbiol* 2009; **7(7)**: 485–91.

71 Wood A, Shilatifard A. Posttranslational modifications of histones by methylation. *Advances in Protein Chemistry* 2004; **67**: 201–222.

72 Patrick S, Jobling KL, O'Connor D, Thacher Z, Dryden DTF, Blakely GW. A unique homologue of the eukaryotic protein-modifier ubiquitin present in the bacterium Bacteroides fragilis, a predominant resident of the human gastrointestinal tract. *Microbiology* 2011; **157**: 3071–78.

73 Chew YC, West JT, Kratzer SJ, Ilvarsonn AM, Eissenberg JC, Dave BJ *et al.* Biotinylation of histones repress transposable elements in human and mouse cells and cell lines, and in Drosophila melanogaster. *J Nutrition* 2008; **138**: 2316–22.

74 Kobza K, Sarath G, Zempleni J. Prokaryotic BirA ligase biotinylates K4, K9, K18 and K23 in histone H3. *BMB reports* 2008; **41**: 1–6.

75 Toffano-Nioche C, Luo Y, Kuchly C, Wallon C, Steinbach D, Zytnicki M, Jacq A, Gautheret D. Detection of non-coding RNA in bacteria and archaea using the DETR'PROK Galaxy pipeline. *Methods* 2013; **63**: 60–65.

76 Verstrepen KJ, Fink GR. Genetic and epigenetic mechanisms underlying cell-surface variability in protozoa and fungi. *Annu Rev Genet* 2009; **43**: 1–24.

77 Shenderov BA. Metabiotics: novel idea or natural development of probiotic conception. *Microb Ecol Health Dis* 2013; **24**: 20399. DOI: 10.3402/mehd.v24i0.20399.

78 Craig SAS. Betaine in human nutrition. *Am J Clin Nutr* 2004; **89**: 539–49.

79 Rossi M, Amaretti A, Raimondi S. Folate production by probiotic bacteria. *Nutrients* 2011; **3**: 118–34.

80 Shenderov BA. *Functional foods in metabolic syndrome prophylaxis.* Moscow: DeLi Print; 2008. 319 pages (in Russian).

81 Takahashi K, Sugi Y, Nakano K, Tsuda M, Kurihara K, Hosono A, Kaminogawa S. Epigenetic control of the host gene by commensal bacteria in large intestinal epithelial cells. *J Biol Chem* 2011; **286(41)**: 35755–62.

82 Canani RB, Di Costanzo M, Leone L, Pedata M, Meli R, Calignano A. Potential beneficial effects of butyrate in intestinal and extra-intestinal diseases. *World J Gastroenterol* 2011; **17(12)**: 1519–28.

83 Licciardi PV, Ververis K, Karagiannis TC. Histone deacetylase inhibition and dietary short-chain fatty acids. *ISRN Allergy* 2011; 869647. Doi:10.5402/2011/869647.

84 MacFabe DF. Short-chain fatty acid fermentation products of the gut microbiome: implications in autism spectrum disorders. *Microb Ecol Health Dis.* 2012; **23**: 19260. DOI: 10.3402/mehd.v23i0.19260.

85 Daly K, Shirazi-Beechey SP. Microarray analysis of butyrate regulated genes in colonic epithelial cells. *DNACell Biol* 2006; **25(1)**: 49–62.

86 Kumar A, Wu H, Coiller-Hyams LS, Young-Man Kwon, Hanson JM, Neish AS. The bacterial fermentation product butyrate influences epithelial signaling via reactive oxygen species-mediated changes in cullin-1 neddylation. *J Immunol* 2009; **182(1)**: 538–46.

87 Aoyama M, Kotani J, Usami M. Butyrate and propionate induced activated or non-activated neutrophil apoptosis via HDAC inhibitor activity but without activating GPR-41/GPR-43 pathways. *Nutrition* 2010; **26**: 653–61.

88 Soret R, Chevalier J, De Coppet P, Poupeau G, Derkinderen P, Segain JP, Neunlist M. Short-chain fatty acids regulate the enteric neurons and control gastrointestinal motility in rats. *Gastroenterology* 2010; **138(5)**: 1772–82.

89 Rooks MG, Garrett S. Bacteria, food, and cancer. *F1000 Biology Reports* 2011; **3**: 12. DOI:103410/B3-12

90 Gerhauser C. Cancer chemoprevention and nutrigenetics: state of the art and future challenges. *Top Curr Chem* 2014; **329**: 73–132.

91 Spallek T, Robatzek S, Gohre V. How microbes utilize host ubiquitination. *Cellular Microbiology* 2009; **11(10)**: 1425–34.

92 Bron PA, Grangette C, Mercenier A, de Vos WM, Kleerebezem M. Identification of Lactobacillus plantarum genes that are induced in the gastrointestinal tract of mice. *J Bacteriol* 2004; **186**: 5721–9.

93 Di Caro S, Tao H, Grillo A, Elia C, Gasbarrini G, Sepulveda AR, Gasbarrini A. Effects of Lactobacillus GG on gene expression pattern in small bowel mucosa. *Dig Liver Dis* 2005; **37**: 320–9.

94 Yuan J, Wang B, Sun Z, Bo X, Yuan X, He X et al. Analysis of host-inducing proteome changes in Bifidobacterium longum NCC2705 grown in vivo. *J Proteome Res* 2008; **7(1)**: 375–85.

95 Dumas ME, Barton RH, Toye A, Cloarec O, Blancher C, Rothwell A et al. Metabolic profiling reveals a contribution of gut microbiota to fatty liver phenotype in insulin-resistant mice. *Proc Natl Acad Sci USA* 2006; **103**: 12511–6.

96 Zakharevich NV, Averina OV, Klimina KM, Kudryavtseva AV, Kasianov AS, Makeev VJ, Danilenko VN. Complete genome sequence of Bifidobacterium longum GT15: identification and characterization of unique and global regulatory genes. *Microbial Ecology* 2015; **70(3)**: 819–834.

97 Fisher A. Targeting histone-modifications in Alzheimer's disease: What is the evidence that this is a promising therapeutic avenue? *Neuropharmacology* 2014; **80**: 95–102.

98 Ricobaraza A, Cuadrado-Tejedor M, Marco S, Perez-Otano I, Garcia-Osta A. Phenylbutyrate rescues dendritic spine loss associated with memory deficits in a mouse model of Alzheimer's disease. *Hippocampus* 2010; **22**: 1040–1050.

99 Govindarajan N, Agis-balboa RC, Walter J, Sananbenesi F, Fischer A. Sodium butyrate improves memory function in an Alzheimer's disease mouse model when administered at an advanced stage of disease progression. *J Alzheimer's Dis* 2011; **26**: 187–197.

100 Rane P, Shields J, Heffernan M, Guo Y, Akbarian S, King JA. The histone deacetylase inhibitor, sodium butyrate, alleviates cognitive deficits in pre-motor stage PD. *Neuropharmacology* 2012; **62**: 2409–2412.

101 Zhou W, Bercury K, Cummiskey J, Luong N, Lebin J, Freed CR. Phenylbutyrate up-regulates the DJ-1 protein and protects neurons in cell culture and in animal models of Parkinson disease. *J Biol Chem* 2011; **286**: 14941–14951. DOI: 10.1074/jbc.M110.211029.

102 Coppede F. The potential of epigenetic therapies in neurodegenerative diseases. *Front Genet* 2014; **5**: 220. DOI: 10.3389/fgene.2014.00220.

103 Krautler B. Vitamin B12: chemistry and biochemistry. *Biochem Soc Trans* 2005; **33**: 806–810.

104 Pompei A, Cordisco L, Amaretti A, Zanoni S, Raimondi S, Matteuzzi D, Rossi M. Administration of folate-producing bifidobacteria enhances folate status in Wistar rats. *J Nutr* 2007; **137**: 2742–46.

105 Monsen AL, Refsum H, Markestad T, Ueland FM. Cobalamin status and its biochemical methyl-malonic acid and homocysteine in different age groups from 4 days to 19 years. *Clin Chem* 2003; **49**: 2067–75.

106 Lawless MW, Norris S, O'Byme KJ, Gray SG. Targeting histone deacetylases for the treatment of disease. *J Cell Mol Medicine* 2009; **13**: 826–52.

107 Cheng X, Horton JR, Yang Z, Kalman D, Zhang Z, Hattman S, *et al.* 2010. Small Molecule inhibitors of Bacterial DAM DNA Methyltransferases. USA Patent 11/720971.A1. Available from http//:www.freepatentsonline.com/y2010/0035945.html.

108 Thakur VS, Deb G, Babcook MA, Gupta S. Plant phytochemicals as epigenetic modulators: role in cancer chemoprevention. *American Association of Pharmaceutical Scientists Journal (AAPS J)* 2014; **16(1)**: 151–63.

109 Luoto R, Ruuskanen O, Waris M, Kalliomaki M, Salminen S, Isolauri E. Prebiotics and probiotics supplementation prevents rhinovirus infections in preterm infants: a randomized placebo-controlled trial. *J Allergy Clin Immunol* 2014; **133**: 405–13.

110 Coppede F. Advances in the genetics and epigenetics of neurodegenerative diseases. *Epigenetics Neurodegener Dis* 2014; **1**: 3–31.

111 Harrison IF, Dexter DT. Epigenetic targeting of histone deacetylase: Therapeutic potential in Parkinson's disease? *Pharmacol Ther* 2013; **140**: 34–52.

112 Serino M, Chabo C, Burcelin R. Intestinal microOMICS to define health and disease in human and mice. *Curr Pharm Biotechnol* 2012; **13(5)**: 746–58.

CHAPTER 12

The emerging role of propionibacteria in human health and disease

Holger Brüggemann

Department of Biomedicine, Aarhus University, Aarhus, Denmark

12.1 Introduction

Propionibacteria can be found as part of the skin microbiota on almost every human being. In particular, *Propionibacterium acnes* (*P. acnes*) successfully and often exclusively colonizes sebaceous follicles of the face and the back. For many decades propionibacteria have been regarded as harmless skin commensals, with the exception of the association with the highly prevalent skin disorder acne vulgaris. In the last decade, genome sequencing, typing efforts, host-bacterium interaction studies and improved microbial cultivation approaches have revealed that these bacteria can also act as opportunistic pathogens, and new disease associations have been uncovered. At the same time new findings regarding the possible beneficial role of cutaneous propionibacteria for the host have been made, highlighting the microbial interaction with the host immune system. This chapter summarizes current knowledge on propionibacteria with special emphasis on *P. acnes*. Aspects of its phylogenetic diversity, host-interacting properties, host immune responses and evidence for its association with human pathologies will be presented and discussed.

12.2 Microbiological features of propionibacteria

The Gram-positive genus *Propionibacterium* belongs to the phylum Actinobacteria and contains classical (or dairy) and cutaneous species. Propionibacteria are ubiquitous in nature. They can be found in the gastrointestinal tract of cows and sheep, being responsible for the conversion of lactate into propionic acid. Dairy propionibacteria such as *Propionibacterium freudenreichii* and *Propionibacterium acidipropionicii* can be found in cheese, where they add to the maturation and flavor[1]. Such dairy propionibacteria are considered to have probiotic effects[2].

The three most important cutaneous species are *P. acnes*, *Propionibacterium avidum* (*P. avidum*) and *Propionibacterium granulosum* (*P. granulosum*); these species

The Human Microbiota and Chronic Disease: Dysbiosis as a Cause of Human Pathology, First Edition.
Edited by Luigi Nibali and Brian Henderson.
© 2016 John Wiley & Sons, Inc. Published 2016 by John Wiley & Sons, Inc.

can be found on the skin of virtually every human being, where *P. acnes* dominates in sebaceous-rich areas of the scalp, forehead, ear, back, and alae nasi[3]. *P. avidum* prefers moist rather than oily areas; it is found mainly in the anterior nares, axilla, and rectum[4]. The skin microbiota of other animals has not been studied extensively, and thus it is not clear if cutaneous propionibacteria are human-specific. Propionibacteria have usually a coryneform, irregular shape, and the species differ morphologically (Figure 1). The length of the small rods can vary depending on the environment, with smaller rods being formed under unfavorable conditions. Surface appendices are produced by some propionibacteria. Most strains of *P. avidum* produce exopolysaccharide and *P. granulosum* can produce pili-like structures[5]. *P. acnes* seems not to produce distinct cell appendices. Within the population of *P. acnes*, strain variations exist. Type III strains of *P. acnes* form long filamentous-like cells, whereas cells of type I *P. acnes* are much smaller[6]. Propionibacteria are non-motile, although some species can employ a twitching motility. They do not form spores, but they are able to survive for prolonged times under unfavorable growth conditions. The relation to oxygen is species-dependent and not clearly resolved. They are aerotolerant and possess the enzyme catalase[1]. Most species prefer anaerobic conditions, but growth with oxygen is possible, for instance in the case of *P. acnes*. This species also possesses a respiratory chain[7]. They can

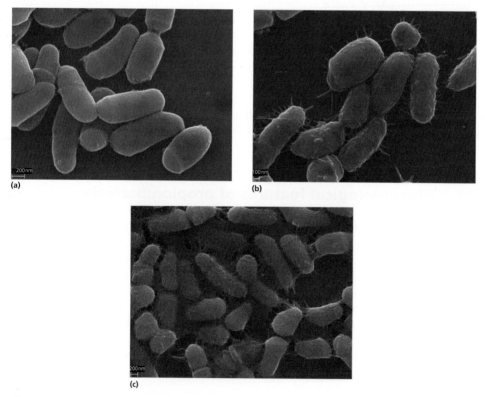

Figure 1 Electron microscope pictures of three prominent cutaneous propionibacteria. (a) *P. acnes* (type IA); (b) *P. granulosum*; (c) *P. avidum*. Pictures by V. Brinkmann, Max-Planck-Institute for Infection Biology, Berlin, Germany.

actually grow better under very low oxygen pressures compared to strict anaerobic conditions, and thus they are sometimes referred to as "nanoaerophilic." Energy is conserved primarily by the methylmalonyl-CoA pathway[1]. Here, pyruvate is first converted, via oxaloacetate, to succinate, employing the biotin-dependent enzyme methylmalonyl-CoA carboxytransferase. Energy is conserved via anaerobic respiration with fumarate reductase. Succinate is further converted into the end product propionate, using the B12-dependent enzyme methylmalonyl-CoA mutase. Transcriptome data indicate that *P. acnes* can also carry out aerobic respiration under low oxygen pressure, and anaerobic respiration using nitrate as terminal electron acceptor[8]. Another highly expressed system is the arginine deiminase (ADI) pathway that catalyzes the conversion of arginine to ornithine via citrulline, thereby generating NH_3 and ATP. This energy-conserving system might counteract the acidification in the course of propionate formation, akin to *Streptococcus suis* and sourdough lactic acid bacteria, which use the ADI pathway to facilitate survival in acidic conditions[9]. The metabolic flexibility of *P. acnes* might be an advantage when residing in sebaceous follicles of the human skin where sebum amount and composition vary.

12.3 Population structure of *P. acnes*

Every bacterial species can be phylogenetically divided into types and subtypes that can have different properties. At present, a detailed map of the population structure exists only for *P. acnes*[10,11]. This map is based on typing schemes such as Multi-Locus-Sequence Typing (MLST) and recently on whole genome comparison[12,13]. The main phylogenetically distinct lineages of *P. acnes* are called type I, II and III. Type I strains are further subdivided into IA, IB and IC, and IA strains can be further distinguished into subtypes and clonal complexes.

The co-evolutionary history of *P. acnes* with the human host is unknown. Compared with other species, the population of *P. acnes* shows limited diversity. The pan-genome of the species is small, with a high genomic synteny[13,14]. This suggests that *P. acnes* has associated with humans relatively recently, in contrast to other members of the human microbiota, such as *Helicobacter pylori*, that had already colonized humans before they migrated out of Africa 125,000 years ago[15]. It has been shown that the human skin of each individual is usually colonized with heterogeneous and multiphyletic communities of *P. acnes* strains, with type IA strains being the most predominant[12,13]. A recent study that analyzed the skin microbiome of 14 healthy humans showed that the heterogeneity of *P. acnes* strains was more individual- than site-specific, i.e. similar *P. acnes* communities colonized different sites of the person's skin, and each person had a typical assemblage of *P. acnes* types[16].

A few studies have shown that *P. acnes* communities of acne-affected skin differ from communities found on healthy skin in terms of the predominant *P. acnes* subtypes[17–19]. For instance, clonal complexes (CCs) such as CC18 (within type IA) are more often associated with acne-affected skin, whereas other CCs are preferentially associated with healthy skin. Currently, it is not known if acne-associated strains such as CC18 strains have elevated pathogenic properties that could be involved in disease initiation. Alternatively, such strains might have a growth advantage as a consequence of a changed sebaceous follicular environment during acne. The studies mentioned have an important drawback: *P. acnes* was cultivated from skin swabs and typing was done for isolated colonies. Such typing results do not necessarily represent the actual assemblages on the skin. A direct detection approach of all types of *P. acnes* in a given sample is necessary to assess assemblage differences between normal and acne-affected skin. A recently developed single-locus sequence typing scheme could be employed for such a task[13].

Typing of *P. acnes* has been applied not only to human skin isolates, but also for other clinical isolates that were suspected to be responsible for infections. These efforts showed, for instance, that *P. acnes* isolates from deep tissue infections more often belonged to types IB and II than to the predominant skin type IA[12,17,18]. The types IB and II might have an elevated potential to colonize and survive in deep tissue sites. Thus, typing helps differentiate the "true" presence of *P. acnes* in such tissue sites from skin-derived contaminations that could occur during sample taking or sample processing. Currently, more efforts are being made to study the different properties of *P. acnes* types. So far, comparative genome analyses have highlighted the type-specific presence of a few genomic regions that encode a variety of gene clusters, including a bacteriocin synthesis locus, a tight adherence locus, a thiopeptide synthesis cluster, a clustered regularly interspaced short palindromic repeats (CRISPR)/cas locus, a non-ribosomal peptide synthetase locus, and others[14,20]. However, future studies need to show if the encoded functions are actually produced *in vivo* and add to the phenotype of the respective *P. acnes* types.

12.4 Propionibacteria as indigenous probiotics of the skin

Cutaneous propionibacteria are ubiquitously present on human skin. Several possible explanations exist for their predominance: (I) they have special and unique metabolic capabilities that allow them to survive and propagate on human skin; (II) they have traits that enable them to successfully out-compete other bacterial or fungal species and; (III) the human host actively supports their presence, e.g. by providing nutrients or by producing antimicrobial substances that preferentially kill competing species. It is sensible to assume that all three explanations play a role. A few host beneficial properties of propionibacteria have often been mentioned in this context:

1 Propionibacteria secrete propionate (and acetate) as a metabolic end product. Such short-chain fatty acids (SCFAs) could inhibit the growth of other bacteria[21]. In addition, the host beneficial effects of propionate and acetate, together with butyrate, have lately been studied in more detail, since these SCFAs are also produced by the gut microbiota (discussed more extensively in chapter 23 on gut microbiota). Besides being important as energy sources for the host cell metabolism, they have immunomodulatory effects[22]. Butyrate and propionate are able to inhibit histone deacetylases in colonic and immune cells (see chapter 11 by Shenderov). This leads to histone hyperacetylation and impacts gene expression with diverse consequences for cellular differentiation. In addition, SCFAs have been shown to modulate host cell signaling via G protein-coupled receptors (GPR41, GPR43, GPR109A)[23]. Taking these effects together, SCFAs

modulate the differentiation and activity of colonic regulatory T cells that control intestinal inflammation[24]. The skin microbiota also affects T cell differentiation (see section 12.7 below), but it remains to be shown whether propionate produced by propionibacteria on the skin can exert such effects, as previously suggested[25-27]. Besides SCFAs, other fatty acids are generated by *P. acnes* due to the strong lipolytic activity that degrades sebum lipids. Such fatty acids might also impact the host tissue and/or the proximate microbiota.

2 Propionibacteria produce bacteriocins that potentially kill competing microorganisms in close vicinity. Non-cutaneous propionibacteria, such as *Propionibacterium thoenii* and *Propionibacterium jensenii* can produce antimicrobial substances, designated thoeniicin and jenseniin, respectively[2,28]. These bacteriocins usually act against closely related species but some, such as propionicin PLG-1 produced by *P. thoenii* P-127, have a more widespread activity, inhibiting even some Gram-negative pathogens and fungi[29]. For cutaneous propionibacteria, it has been shown that some strains of *P. avidum* produce bacteriocins that can inhibit *P. acnes* strains[30]. It was also reported that some strains of *P. acnes* produce a bacteriocin-like substance, designated acnecin, that acts against other *P. acnes* strains[31]. Another substance with bacteriocin-like properties, isolated from *P. acnes* strains recovered from dental plaque, is bacteriostatic and active against both Gram-positive and Gram-negative anaerobes[32]. A comparative study including the bacteriocin activity of all *P. acnes* types is so far lacking. Some type IA strains (CC3) harbor a gene cluster that is homologous to the *sag* gene cluster for streptolysin S synthesis in *Streptococcus pyogenes*[20]. Another subtype (IB, CC36) harbor genes for the biosynthesis of a thiazolylpeptide[8]. A precursor peptide is encoded in this genomic island that is identical to the precursor of Berninamycin A from *Streptomyces bernensis*, which is a potent inhibitor of protein synthesis in Gram-positive bacteria. But so far no study has reported the isolation of this Berninamycin-like substance from *P. acnes*.

3 *P. acnes* might exert anti-inflammatory effects directly or indirectly. Besides the activity of *P. acnes* on the T cell differentiation (see section 12.7), a possible effect could be mediated by the *P. acnes* enzyme linoleic acid isomerase that catalyzes the formation of conjugated linoleic acid (CLAs)[33]. CLAs have been reported to modulate immune responses. For instance, they can modulate dendritic cell cytokine production, which plays a role in inflammation by directing T helper (Th) cell differentiation[34]. The structure of the linoleic acid isomerase of *P. acnes* has been resolved but its physiological role remains to be studied.

12.5 Propionibacteria as opportunistic pathogens

Many articles and reviews have discussed the role of *P. acnes* in the pathogenesis of acne vulgaris. Clear-cut evidence for a true etiological role is still lacking, but there are convincing arguments supporting the assumption that *P. acnes* plays an important part in the course of acne[35,36]. One key argument is the beneficial effect of antimicrobial treatment for acne patients: antibiotics such as erythromycin, clindamycin and tetracycline usually ameliorate the disease, and treatment failure has been associated with antibiotic-resistant *P. acnes* strains[37,38]. Another key argument is the impact of *P. acnes* on the innate and adaptive host immunity (see Section 12.7). In addition, recent discoveries indicating that only certain types of *P. acnes* could account for disease formation while others represent harmless or beneficial commensals of normal human skin, further strengthen an etiological role of *P. acnes* in acne. However, since an appropriate animal model for human skin is lacking, it is very difficult to determine further the precise role of *P. acnes*. Therefore, doubts about the importance of *P. acnes* in acne formation and/or progression are often expressed[39].

Other disease associations of *P. acnes* at non-skin sites have been summarized in the review of Perry and Lambert[40]. The evidence of an etiological role of *P. acnes* in the conditions mentioned varies a lot. Accumulating evidence exists for a role of *P. acnes* in postoperative and device-related infections, including those of joint prostheses, shunts and prosthetic heart valves. Circumstantial evidence exists for a role in cases of endocarditis, post-neurosurgical CNS infections, dental infections, bacterial endophthalmitis, postoperative spine infections, and the SAPHO (synovitis, acne, pustulosis, hyperostosis, osteitis) syndrome. In addition, recent studies have shown evidence that *P. acnes* can cause cases of sarcoidosis. These findings are summarized by Eishi[41]. Another association concerns prostate pathologies: several studies report the isolation and cultivation of *P. acnes* from radical prostatectomy specimens and its visualization in cancerous tissue[42–44]. Furthermore, animal experiments have shown that the bacterium can induce a persistent infection of the prostate gland[45,46]. Despite this circumstantial evidence, the role of *P. acnes* in prostate pathologies is a matter of controversial debate. The organism might actually be introduced during surgery or during previous surgical intervention, since patients that undergo radical prostatectomy usually have been previously biopsied. Thus, for most of these disease associations of *P. acnes*, if not for all, it is difficult to differentiate a true infection from skin-derived contamination that might have occurred during surgery or specimen handling. Moreover, it cannot be ruled out that *P. acnes* rather represents a secondary bystander with little to no impact on disease formation. Each clinical study needs a careful design that guarantees highly sterile conditions during surgery and specimen processing. Convincing evidence for a disease association includes multiple approaches: the detection of *P. acnes* by DNA/RNA-based methods, the cultivation from clinical samples and the visualization of bacteria in histological tissue samples, e.g. by immunohistochemistry or fluorescence in situ hybridization. Moreover, the presence of the microorganism in deep tissues sites should go along with an inflammatory signature in the vicinity. Nevertheless, several lines of evidence, including animal experimentation, suggest that *P. acnes* can persist for extended times in deep tissue sites. This might be related to its ability to evade the degradation pathway within macrophages and/or invade epithelial host cells. A persistent *P. acnes* infection could in theory cause low-grade inflammation that, with time, might lead to more serious conditions.

Little is known about the association of *P. avidum* and *P. granulosum* with human diseases. *P. avidum* has been found to cause abscess formation, in particular after surgical intervention; it has been described as the cause of abdominal wall and intra-peritoneal, perianal, psoas, splenic, and breast abscesses[47–49]. The disease association of *P. granulosum* is less clear, though it has been found in a few cases of endocarditis and endophthalmitis, and has been associated, like *P. acnes*, with sarcoidosis[41,50].

12.6 Host interacting traits and factors of propionibacteria

Propionibacteria live in close contact with human skin tissue, where they encounter keratinocytes and, possibly, sebocytes. Different strategies to adhere to human tissues have been developed (Figure 1): *P. avidum* can produce an exopolysaccharide, a meshwork that connects cells and might support a biofilm-like structure[5]. In contrast, *P. granulosum* expresses pili that potentially support the attachment to surfaces. For *P. acnes*, no such obvious surface-attached structures are visible. However, *P. acnes* biofilms have been seen in normal human sebaceous follicles[51]. Surface molecules such as polysaccharides and adhesins are likely to be involved.

To date only very few gene products of *P. acnes* have been studied at the molecular level. This is due to difficulties in genetically manipulating the organism. So far, knockout mutants can only be generated in one strain of *P. acnes* (type IB, KPA171202)[52]. Most often discussed factors of *P. acnes* are CAMP (Christie-Atkins-Munch-Peterson) factors[7,53] and surface-exposed proteins designated dermatan sulphate adhesins. Interestingly, these factors of *P. acnes* have recently been detected in human sebaceous follicles *in vivo*[54]. The genome of *P. acnes* encodes five CAMP factors (CAMP1 to CAMP5). At least one of the CAMP factors is responsible for the co-hemolytic reaction of *P. acnes* with both sheep and human erythrocytes[55], since a knockout mutant of CAMP2 showed reduced co-hemolytic activity in the CAMP reaction[52]. The CAMP reaction is a synergistic hemolysis of sheep erythrocytes by the CAMP factor and the β-toxin (sphingomyelinase C) from *S. aureus*. In addition, inhibition of CAMP2 by neutralising antibodies efficiently attenuated *P. acnes*-induced inflammation in the mouse ear model[56]. It was also shown that CAMP2 can act as an exotoxin, exhibiting cytotoxic activity on host cells[57]. Although their exact *in vivo* role remains to be unraveled, the importance of CAMP factors is supported by the conservation of all five CAMP paralogs across the different *P. acnes* types[12]. In contrast, only two and one CAMP factors are encoded in the genomes of *P. avidum* and *P. granulosum*, respectively[5]. Both latter organisms do not possess CAMP1, CAMP2 and CAMP4. It is tempting to suggest that the presence of five CAMP factors in *P. acnes* is a key to its success as skin colonizer. A possibility is that the pore-forming activity of the CAMP factor supports the organism in obtaining certain nutrients from the host tissue. Whether the hemolytic activity of many strains of *P. acnes*, as seen on blood agar plates, is also related to the CAMP factors remains unclear.

The two dermatan sulphate adhesins DsA1 and DsA2 are cell surface-exposed, weakly similar to the M-like protein of *Streptococcus equi* and display dermatan sulphate binding activity[58]. They have immunoreactive properties[59]. Dermatan sulphate is expressed in many mammalian tissues and is the predominant glycan present in skin. Interestingly, DsA1 and DsA2 are associated with phase and antigenic variation; the latter is due to a proline-threonine rich repeat in the C-terminus of the protein, which can result in a heterogenic phenotype of a clonal population[58–60]. This could represent a strategy of *P. acnes* for immune evasion. Apart from that, the exact role of DsA1 and DsA2 is unknown. Expression of DsA1 and DsA2 is type-dependent and mainly restricted to type IA *P. acnes*.

P. acnes has saprophytic properties; it acquires nutrients for growth from within sebaceous follicles, such as lipids of the sebum, which is mainly composed of triglycerides (~40%). A lipolytic activity of *P. acnes* has been shown, and a secreted lipase, the triacylglycerol lipase GehA, has been investigated[61]. However, a recent study using a proteomic approach showed that the predominate *P. acnes* lipase produced in sebaceous follicles is GehB, which is a paralog of GehA[54]. It is not known if GehA and GehB use different substrates. In addition, secreted or surface-exposed enzymes with hydrolytic activity exist, such as two putative endoglycoceramidases that might catalyse the hydrolysis of the glycosidic linkage between oligosaccharides and ceramides of glycosphingolipids[60].

A recent study identified the most abundant surface-attached factors of propionibacteria, determined by a surfome-approach that is based on trypsin cleavage of surface-exposed protein moieties[5]. In all three *Propionibacterium* species the most abundantly detected surface-attached proteins were RlpA (rare lipoprotein A) -domain containing lipoproteins, containing bacterial SH3 domains and a peptidoglycan-binding domain. Bacterial lipoproteins are heterogeneous; some can also function as ligands of the host cell receptor Toll-like receptor 2 (TLR2), thus triggering an innate immune reaction. It remains to be shown in the future whether the RlpA-like lipoproteins of *P. acnes* are TLR2 ligands.

12.7 Host responses to *P. acnes*

12.7.1 Innate immune responses

Cell culture experiments with *P. acnes* have been carried out in the last two decades in order to study the possible inflammatory role of this organism. Table 1 summarizes prominent original findings[62-77]. Many studies have been carried out with keratinocytes. The expression and/or secretion of several cytokines and chemokines is triggered by *P. acnes* when co-incubated with keratinocytes. Similar results have been obtained using sebocytes (Table 1). TLR2-dependent NF-kappaB activation has been identified as an underlying mechanism of cytokine activation. It has not been clarified which TLR2-ligand of *P. acnes* is involved. At present it is not well understood if different *P. acnes* types have distinct pro-inflammatory activities, but a few studies suggest such differences[63,78]. More recently, *P. acnes* has been shown to activate the NLRP3 inflammasome, a system responsible for the activation of inflammatory processes via IL-1beta maturatio[70,73,74]. In acne lesions, mature caspase-1 and NLRP3 were detected around the pilosebaceous follicles. The studies suggested that the encounter of *P. acnes* with macrophages could locally result in the release of IL-1beta and therefore exacerbate inflammation. Based on the above, it is clear that *P. acnes* has the ability to activate the innate immune system in several different cell lines. However, the *in vivo* meaning of these findings remains puzzling. A key question remains: how does normal skin that is also colonized with *P. acnes* control the innate immune response to *P. acnes*? Bacterial proliferation might be tightly controlled in healthy sebaceous follicles and also strain types with differing inflammatory potential seem to exist. In addition, the impact of *P. acnes* on T cell proliferation and differentiation might play an important role in guaranteeing skin homeostasis (see below).

Table 1 Selected studies on the pro-inflammatory activity of *P. acnes*.

Cell system	Effect	Comments/mechanistic aspects	Reference
Keratinocytes	IL-1alpha, TNFalpha, GM-CSF	triggered by viable *P. acnes* and protein GroEL	Graham et al., 2004[62]
	hBD2, IL-8	TLR2 and TLR4 dependent; hBD2 activation is *P. acnes* strain-dependent	Nagy et al., 2005[63]
	IL-1beta, GM-CSF, IL-8	possible involvement of coproporphyrin III	Schaller et al., 2005[64]
	TLR2, TLR4, MMP9	TLR2, TLR4, MMP9 also present in acne lesions	Jugeau et al., 2005[65]
	ROS (super-oxide anion), IL-8	ROS→IL-8 → apoptosis	Grange et al., 2009[66]
	IL-8	mediated by ERK, p38 and JNK kinases	Mak et al., 2012[67]
Sebocytes	hBD2, TNF-alpha, CXCL8	sebocyte viability and differentiation altered	Nagy et al., 2006[68]
	IL-8	anti-*P. acnes* antibodies attenuates effect	Nakatsuji et al., 2008[69]
	caspase-1, IL-1beta	NLRP3-dependent; triggered by ROS/protease activity	Li et al., 2014[70]
Monocytes	IL-1beta, TNFalpha, IL-8	soluble factor of *P. acnes*, possibly TLR4 ligand	Vowels et al., 1995[71]
	IL-12, IL-8	TLR2 → NF-kappaB activation	Kim et al., 2002[72]
	IL-1beta maturation	NLRP3 inflammasome activation depends on phagocytosis, lysosomal destabilization, ROS production	Kistowska et al., 2014[73]
	caspase-1, caspase-5, IL-1beta maturation	NLPR3-dependent; involvement of potassium efflux	Qin et al., 2014[74]
Dermal fibroblasts	MMP2, TNF-alpha	NF-kappaB-dependent	Choi et al., 2008[75]
Macrophages	iNOS/NO, COX-2/PGE2	via ROS-dependent stimulation of ERK, JNK, NF-kappaB, AP-1	Tsai et al., 2013[76]
Prostate epithelial cells	IL-6, IL-8, GM-CSF	TLR2; NF-kappaB	Drott et al., 2010[77]
	IL-6, IL-8, GM-CSF, MMP9	NF-kappaB and STAT3	Fassi Fehri et al., 2011[43]; Mak et al. 2012[67]

12.7.2 Adaptive immune responses

The interaction of *P. acnes* with T cells has been studied in several waves. In the '70s, the anti-tumor activity of *P. acnes* (then called *Corynebacterium parvum*) was investigated, since intratumoral administration of *P. acnes* successfully inhibited tumor progression in the mouse model. It was suggested that this activity was due to *P. acnes*-activated macrophages that collaborated with T cells, leading to an "enhanced T cell reactivity"[79,80]. However, the exact mechanism of the antitumor activity of *P. acnes* remained unsolved. In later years, it has been clarified that *P. acnes* can trigger a strong Th (T helper cell) 1 type response; induced Th1-type cytokines include IL-12, IFN-γ and TNF-α[81]. It is likely that this Th1-type response plays an important role in the anti-tumor activity of *P. acnes*. The impact of *P. acnes* on T helper cells might also be relevant in the immunopathogenesis of acne. Very recent studies have shown another layer of T cell interaction: *P. acnes* is not only a potent inducer of a Th1 type response but also triggers a Th17 response, at least in human peripheral blood mononuclear cells (PBMCs)[82]. *P. acnes*-induced IL-17 production was observed, and both Th1 and Th17 effector cytokines are strongly upregulated in acne lesions. Therefore, it was suggested that acne might be a Th17-mediated disease. Another study showed that *P. acnes* can promote Th17 responses as well as mixed Th17/Th1 responses by inducing the concomitant secretion of IL-17A and IFN-gamma from specific CD4+ T cells[83]. *P. acnes*-specific Th17 and Th17/Th1 cells can also be found in the peripheral blood of patients suffering from acne. Other data point towards a host beneficial role of a Th17 response[26,27]. The outcome might depend on the balance between Th1 and Th17 responses. Taken together, the immune responses to *P. acnes* are manifold. In acne, early and late stages must be differentiated. It seems reasonable to assume that an early stage is characterized by the innate immune activation due to a temporary massive proliferation of *P. acnes*. The second phase might include macrophages that infiltrate affected follicles. This in turn leads to inflammasome activation and activated macrophages collaborate with immune cells to induce a Th1/Th17 type-driven adaptive response.

12.7.3 Host cell tropism of *P. acnes*

Several studies have investigated the effect of *P. acnes* on different cell types. One study compared the effect in two different cell lines, in keratinocytes (HaCaT cell line) and in a prostatic epithelial cell line (RWPE1)[67]. Here, clear-cut differences have been observed in both the magnitude of the transcriptional responses and their temporal regulation between the two cell lines. The data showed that *P. acnes* can trigger an acute but transient inflammatory response in keratinocytes; in contrast, the bacterium triggers a delayed but sustained inflammatory response in prostate cells that is associated with prolonged NF-kappaB activation. This goes along with the observation that *P. acnes* efficiently invades the prostatic cell line, but does this only poorly in keratinocytes. These data suggest that host cell tropism exists. Therefore, the encounter of *P. acnes* in different tissue sites can have largely different consequences for the nature and duration of inflammation. In particular, the ability of *P. acnes* to invade certain cell types might be relevant to its persistence. Several studies have addressed the invasion capacity of *P. acnes*. In a study attempting to find an etiological link between sarcoidosis and *P. acnes* infection, invasion of HEK293T cells by *P. acnes* was correlated with bacterial serotype and genotype[84]. In another sarcoidosis study, the invasion of A549 cells by clinical *P. acnes* isolates was observed[85]. In studies examining the interaction of *P. acnes* with prostatic cells, intracellular location of *P. acnes* was observed[43,44]. The intracellular fate of *P. acnes* has been also studied in professional phagocytes, where *P. acnes* can resist or delay its intracellular degradation *in vitro*[86]. However, whether or not these findings are relevant *in vivo* remains to be investigated.

12.8 *Propionibacterium*-specific bacteriophages

Recently, the virome, i.e. the entire gene pool of viral/bacteriophage populations, and its interaction with the microbiome has become a new focus of attention[87–89]. Similar to the gut, the skin also harbors many different viruses; a large portion is bacteriophages that are specific to certain members of the microbiota[16]. Bacteriophages that specifically infect propionibacteria have also been identified[90–92]. They are abundant on certain areas of human skin. *P. acnes*-infecting phages are members of the family of siphoviruses. These are doubled-stranded DNA viruses that contain an icosahedral head and have a flexible long non-contractile tail. Phages that lyse *P. acnes* can be easily isolated from healthy human skin by using a phage plaque assay. Sequencing of their genomes revealed their highly conserved nature[92,93]. The genomes of *P. acnes*-specific siphoviruses consist of about 45 putative genes; half of the genome is strongly conserved (gp1-gp19). This conserved part encodes proteins important for DNA packing, structure, and assembly of viron particles[92], including terminases, portal proteins and tail proteins. The 3′end of the genome (gp19-gp45) is less well conserved and encodes, among others, a putative amidase and holin, which play an essential role in the release of phage progeny from the host. A common feature of the phages that infect *P. acnes* is the "pseudolysogenic" lifecycle. No lysogeny that involves insertion of the phage genome into the host genome has been observed. It is assumed that the phage is maintained in the bacterial cytoplasm as an unstable linear or circular plasmidal prophage, and stays this way until induced by either stressful or favorable conditions. In this sense, the pseudolysogenic state can be regarded as a "pause" in the lytic cycle[94]. When induced, the phage enters the lytic pathway, where replication, packing, assembly of viron particles and release of phage progeny take place.

Nothing is known about the role of these *P. acnes*-specific phages. They display a broad host range against most types of *P. acnes*, but phage-resistant *P. acnes* have been found[91,92]. At least some of these strains harbor the CRISPR/cas immunity system[92,95]. It remains to be investigated in the future whether certain phages can actually alter/shape the *P. acnes* community of the skin. Such alterations could directly or indirectly impact skin homeostasis.

12.9 Concluding remarks

Several remarkable discoveries have been made regarding our skin microbiota. New technologies such as next-generation sequencing have revealed the extent of microbial diversity and the community composition of our abundant microbial species on the skin (also discussed in chapter 5). The dominance of propionibacteria in certain, usually lipid-rich areas of the skin has initiated research regarding their beneficial role. In this respect, new data suggest that propionibacteria (and others, most importantly Staphylococci) are able to shape a T-cell population with skin-protective consequences. The role of secreted propionic acid and other metabolites in shaping such responses is currently being examined. However, the interaction of propionibacteria and especially *P. acnes* with the skin appears to be a fragile balance: changes in the microenvironment of the skin (e.g. increased sebaceous gland activity during puberty) might lead to altered colonization patterns, such as a different *P. acnes* type composition (dysbiosis) and/or altered bacterial proliferation. Such changes could impact skin homeostasis, e.g. by triggering an uncontrolled innate immune response and/or by affecting the balance between Th1 and Th17 responses. Such an out-of-balance scenario could explain the role of *P. acnes* in acne vulgaris. In addition, several lines of evidence suggest a pathogenic scenario if *P. acnes* "accidentally" reaches other, non-skin sites, e.g. via wounds or during surgery. In such deep tissue sites, the bacteria encounter cell types that react differently than skin cells. In particular, in immunocompromised hosts, the bacterium might lead to a low-grade persistent infection due to certain properties, such as the capability to invade host cells and the partial resistance to phagosomal degradation. However, research in this field is challenging: the quest for causality is hindered by the high risk of skin-derived bacterial contamination. New innovative research approaches are needed to unravel the full significance of *P. acnes* in health and disease.

TAKE-HOME MESSAGE

- The propionibacterial species *P. acnes*, *P. avidum* and *P. granulosum* (and *Propionibacterium*-specific bacteriophages) are ubiquitously found on human skin, preferentially in lipid-rich areas of the face and back.

- The population of *P. acnes* can be phylogenetically differentiated into lineages and clonal complexes. Strains belonging to different phylotypes show a few genomic and phenotypic differences, including the pro-inflammatory activity.

- The presence of propionibacteria, in particular of *P. acnes*, is associated with acne vulgaris as well as infections at non-skin sites such as postoperative and device-related infections.

- The *P. acnes* population of acne-affected skin differs from healthy skin. If this dysbiosis is causing acne or is the consequence of acne formation is currently not known.

- Highly secreted and surface-exposed factors of *P. acnes* are CAMP factors and dermatan-sulphate adhesins, with co-hemolytic and immunogenic properties, respectively.

- Several properties of propionibacteria, e.g. the production of short-chain and free fatty acids, the secretion of antimicrobial substances, and the impact on T-cell differentiation and stimulation might have host beneficial consequences.

References

1 Cummins CS, Johnson JL. The genus *Propionibacterium*. In: *The prokaryotes* (Balows E, Truper HG, Dworkin M, Harder W, Schleifer KH, eds), 2nd edn. New York: Springer-Verlag, 1992; 834–49.

2 Poonam, Pophaly SD, Tomar SK *et al*. Multifaceted attributes of dairy propionibacteria: a review. *World J Microbiol Biotechnol* 2012; **28**: 3081–95.

3 Grice EA, Segre JA. The skin microbiome. *Nat Rev Microbiol* 2011; **9**: 244–53.

4 McGinley KJ, Webster GF, Leyden JJ. Regional variations of cutaneous propionibacteria. *Appl Environ Microbiol* 1978; **35**: 62–66.

5 Mak TN, Schmid M, Brzuszkiewicz E *et al*. Comparative genomics reveals distinct host-interacting traits of three major human-associated propionibacteria. *BMC Genomics* 2013; **14**: 640.

6 McDowell A, Perry AL, Lambert PA, Patrick S. A new phylogenetic group of *Propionibacterium acnes*. *J Med Microbiol* 2008; **57**: 218–24.

7 Brüggemann H, Henne A, Hoster F *et al*. The complete genome sequence of *Propionibacterium acnes*, a commensal of human skin. *Science* 2004; **305**: 671–3.

8 Brzuszkiewicz E, Weiner J, Wollherr A *et al*. Comparative genomics and transcriptomics of *Propionibacterium acnes*. *PLoS One* 2011; **6**: e21581.

9 Gruening P, Fulde M, Valentin-Weigand P, Goethe R. Structure, regulation, and putative function of the arginine deiminase system of *Streptococcus suis*. *J Bacteriol* 2006; **188**: 361–9.

10 Kilian M, Scholz CF, Lomholt HB. Multilocus sequence typing and phylogenetic analysis of *Propionibacterium acnes*. *J Clin Microbiol* 2012; **50**: 1158–65.

11 Yu Y, Champer J, Garbán H, Kim J. Typing of *Propionibacterium acnes*: a review of methods and comparative analysis. *Br J Dermatol* 2015; Jan 20.

12 McDowell A, Nagy I, Magyari M *et al.* The opportunistic pathogen *Propionibacterium acnes*: insights into typing, human disease, clonal diversification and CAMP factor evolution. *PLoS One* 2013; **8**: e70897.

13 Scholz CF, Jensen A, Lomholt HB *et al.* A novel high-resolution single locus sequence typing scheme for mixed populations of *Propionibacterium acnes in vivo. PLoS One* 2014; **9**: e104199.

14 Brüggemann H, Lomholt HB, Kilian M. The flexible gene pool of *Propionibacterium acnes. Mob Genet Elements* 2012; **2**: 145–8.

15 Linz B, Balloux F, Moodley Y *et al.* An African origin for the intimate association between humans and *Helicobacter pylori. Nature* 2007; **445**: 915–8.

16 Oh J, Byrd AL, Deming C *et al.* Biogeography and individuality shape function in the human skin metagenome. *Nature* 2014; **514**: 59–64.

17 Lomholt HB, Kilian M. Population genetic analysis of *Propionibacterium acnes* identifies a subpopulation and epidemic clones associated with acne. *PLoS One* 2010; **5**: e12277.

18 McDowell A, Barnard E, Nagy I *et al.* An expanded multilocus sequence typing scheme for *Propionibacterium acnes*: investigation of 'pathogenic', 'commensal' and antibiotic resistant strains. *PLoS One* 2012; **7**: e41480.

19 Fitz-Gibbon S, Tomida S, Chiu BH *et al. Propionibacterium acnes* strain populations in the human skin microbiome associated with acne. *J Invest Dermatol* 2013; **133**: 2152–60.

20 Brüggemann H, Lomholt HB, Tettelin H, Kilian M. CRISPR/cas loci of type II *Propionibacterium acnes* confer immunity against acquisition of mobile elements present in type I *P. acnes. PLoS One* 2012; **7**: e34171.

21 Shu M, Wang Y, Yu J *et al.* Fermentation of *Propionibacterium acnes*, a commensal bacterium in the human skin microbiome, as skin probiotics against methicillin-resistant *Staphylococcus aureus. PLoS One* 2013; **8**: e55380.

22 Louis P, Hold GL, Flint HJ. The gut microbiota, bacterial metabolites and colorectal cancer. *Nat Rev Microbiol* 2014; **12**: 661–72.

23 Kim MH, Kang SG, Park JH *et al.* Short-chain fatty acids activate GPR41 and GPR43 on intestinal epithelial cells to promote inflammatory responses in mice. *Gastroenterology* 2013; **145**: 396–406.

24 Arpaia N, Campbell C, Fan X *et al.* Metabolites produced by commensal bacteria promote peripheral regulatory T-cell generation. *Nature* 2013; **504**: 451–5.

25 Andresen L, Hansen KA, Jensen H *et al.* Propionic acid secreted from propionibacteria induces NKG2D ligand expression on human-activated T lymphocytes and cancer cells. *J Immunol* 2009; **183**: 897–906.

26 Belkaid Y, Segre JA. Dialogue between skin microbiota and immunity. *Science* 2014; **346**: 954–9.

27 Naik S, Bouladoux N, Linehan JL *et al.* Commensal-dendritic-cell interaction specifies a unique protective skin immune signature. *Nature* 2015; Jan 5.

28 Faye T, Holo H, Langsrud T *et al.* The unconventional antimicrobial peptides of the classical propionibacteria. *Appl Microbiol Biotechnol* 2011; **89**: 549–54.

29 Lyon WJ, Glatz BA. Isolation and purification of propionicin PLG-1, a bacteriocin produced by a strain of *Propionibacterium thoenii. Appl Environ Microbiol* 1993; **59**: 83–8.

30 Ko HL, Pulverer G, Jeljaszewicz J. Propionicins, bacteriocins produced by *Propionibacterium avidum. Zentralbl Bakteriol Orig* 1978; **241**: 325–8.

31 Fujimura S, Nakamura T. Purification and properties of a bacteriocin-like substance (acnecin) of oral *Propionibacterium acnes. Antimicrob Agents Chemother* 1978; **14**: 893–8.

32 Paul GE, Booth SJ. Properties and characteristics of a bacteriocin-like substance produced by *Propionibacterium acnes* isolated from dental plaque. *Can J Microbiol* 1988; **34**: 1344–7.

33 Liavonchanka A, Hornung E, Feussner I, Rudolph MG. Structure and mechanism of the *Propionibacterium acnes* polyunsaturated fatty acid isomerase. *Proc Natl Acad Sci U S A* 2006; **103**: 2576–81.

34 Draper E, DeCourcey J, Higgins SC *et al.* Conjugated linoleic acid suppresses dendritic cell activation and subsequent Th17 responses. *J Nutr Biochem* 2014; **25**: 741–9.

35 Dessinioti C, Katsambas AD. The role of *Propionibacterium acnes* in acne pathogenesis: facts and controversies. *Clin Dermatol* 2010; **28**: 2–7.

36 Williams HC, Dellavalle RP, Garner S. Acne vulgaris. *Lancet* 2012; **379**: 361–72.

37 Eady EA, Cove JH, Holland KT, Cunliffe WJ. Erythromycin resistant propionibacteria in antibiotic treated acne patients: association with therapeutic failure. *Br J Dermatol* 1989; **121**: 51–7.

38 Ozolins M, Eady EA, Avery AJ *et al.* Comparison of five antimicrobial regimens for treatment of mild to moderate inflammatory facial acne vulgaris in the community: randomised controlled trial. *Lancet* 2004; **364**: 2188–95.

39 Shaheen B, Gonzalez M. Acne sans *P. acnes. J Eur Acad Dermatol Venereol* 2013; **27**: 1–10.

40 Perry A, Lambert P. *Propionibacterium acnes*: infection beyond the skin. *Expert Rev Anti Infect Ther* 2011; **9**: 1149–56.

41 Eishi Y. Etiologic link between sarcoidosis and *Propionibacterium acnes. Respir Investig* 2013; **51**: 56–68.

42 Alexeyev OA, Marklund I, Shannon B *et al.* Direct visualization of *Propionibacterium acnes* in prostate tissue by multicolor fluorescent *in situ* hybridization assay. *J Clin Microbiol* 2007; **45**: 3721–8.

43 Fassi Fehri L, Mak TN, Laube B *et al.* Prevalence of *Propionibacterium acnes* in diseased prostates and its inflammatory and transforming activity on prostate epithelial cells. *Int J Med Microbiol* 2011; **301**: 69–78.

44 Bae Y, Ito T, Iida T *et al.* Intracellular *Propionibacterium acnes* infection in glandular epithelium and stromal macrophages of the prostate with or without cancer. *PLoS One* 2014; **9**: e90324.

45 Olsson J, Drott JB, Laurantzon L *et al.* Chronic prostatic infection and inflammation by *Propionibacterium acnes* in a rat prostate infection model. *PLoS One* 2012; **7**: e51434.

46 Shinohara DB, Vaghasia AM, Yu SH *et al.* A mouse model of chronic prostatic inflammation using a human prostate cancer-derived isolate of *Propionibacterium acnes. Prostate* 2013; **73**: 1007–15.

47 Panagea S, Corkill JE, Hershman MJ, Parry CM. Breast abscess caused by *Propionibacterium avidum* following breast reduction surgery: case report and review of the literature. *J Infect* 2005; **51**: e253–5.

48 Vohra A, Saiz E, Chan J *et al.* Splenic abscess caused by *Propionibacterium avidum* as a complication of cardiac catheterization. *Clin Infect Dis* 1998; **26**: 770–1.

49 Janvier F, Delacour H, Larréché S *et al.* Abdominal wall and intra-peritoneal abscess by *Propionibacterium avidum* as a complication of abdominal parietoplasty. *Pathol Biol* 2013; **61**: 223–5.

50 Armstrong RW, Wuerflein RD. Endocarditis due to *Propionibacterium granulosum*. *Clin Infect Dis* 1996; **23**: 1178–9.

51 Jahns AC, Alexeyev OA. Three-dimensional distribution of *Propionibacterium acnes* biofilms in human skin. *Exp Dermatol* 2014; **23**: 687–9.

52 Sörensen M, Mak TN, Hurwitz R *et al.* Mutagenesis of *Propionibacterium acnes* and analysis of two CAMP factor knock-out mutants. *J Microbiol Methods* 2010; **83**: 211–6.

53 Valanne S, McDowell A, Ramage G *et al.* CAMP factor homologues in *Propionibacterium acnes*: a new protein family differentially expressed by types I and II. *Microbiology* 2005; **151**: 1369–79.

54 Bek-Thomsen M, Lomholt HB, Scavenius C *et al.* Proteome analysis of human sebaceous follicle infundibula extracted from healthy and acne-affected skin. *PLoS One* 2014; **9**: e107908.

55 Choudhury TK. Synergistic lysis of erythrocytes by *Propionibacterium acnes. J Clin Microbiol* 1978; **8**: 238–41.

56 Liu PF, Nakatsuji T, Zhu W *et al.* Passive immunoprotection targeting a secreted CAMP factor of *Propionibacterium acnes* as a novel immunotherapeutic for acne vulgaris. *Vaccine* 2011; **29**: 3230–8.

57 Nakatsuji T, Tang DC, Zhang L *et al. Propionibacterium acnes* CAMP factor and host acid sphingomyelinase contribute to bacterial virulence: potential targets for inflammatory acne treatment. *PLoS One* 2011; **6**: e14797.

58 McDowell A, Gao A, Barnard E *et al.* A novel multilocus sequence typing scheme for the opportunistic pathogen *Propionibacterium acnes* and characterization of type I cell surface-associated antigens. *Microbiology* 2011; **157**: 1990–2003.

59 Lodes MJ, Secrist H, Benson DR *et al.* Variable expression of immunoreactive surface proteins of *Propionibacterium acnes. Microbiology* 2006; **152**: 3667–81.

60 Holland C, Mak TN, Zimny-Arndt U *et al.* Proteomic identification of secreted proteins of *Propionibacterium acnes. BMC Microbiol* 2010; **10**: 230.

61 Miskin JE, Farrell AM, Cunliffe WJ, Holland KT. *Propionibacterium acnes*, a resident of lipid-rich human skin, produces a 33 kDa extracellular lipase encoded by *gehA. Microbiology* 1997; **143**: 1745–55.

62 Graham GM, Farrar MD, Cruse-Sawyer JE *et al.* Proinflammatory cytokine production by human keratinocytes stimulated with *Propionibacterium acnes* and *P. acnes* GroEL. *Br J Dermatol* 2004; **150**: 421–8.

63 Nagy I, Pivarcsi A, Koreck A *et al.* Distinct strains of *Propionibacterium acnes* induce selective human beta-defensin-2 and interleukin-8 expression in human keratinocytes through toll-like receptors. *J Invest Dermatol* 2005; **124**: 931–8.

64 Schaller M, Loewenstein M, Borelli C *et al.* Induction of a chemoattractive proinflammatory cytokine response after stimulation of keratinocytes with *Propionibacterium acnes* and coproporphyrin III. *Br J Dermatol* 2005; **153**: 66–71.

65 Jugeau S, Tenaud I, Knol AC *et al.* Induction of toll-like receptors by *Propionibacterium acnes*. *Br J Dermatol* 2005; **153**: 1105–13.

66 Grange PA, Chéreau C, Raingeaud J *et al.* Production of superoxide anions by keratinocytes initiates *P. acnes*-induced inflammation of the skin. *PLoS Pathog* 2009; **5**: e1000527.

67 Mak TN, Fischer N, Laube B *et al.* *Propionibacterium acnes* host cell tropism contributes to vimentin-mediated invasion and induction of inflammation. *Cell Microbiol* 2012; **14**: 1720–33.

68 Nagy I, Pivarcsi A, Kis K *et al.* *Propionibacterium acnes* and lipopolysaccharide induce the expression of antimicrobial peptides and proinflammatory cytokines/chemokines in human sebocytes. *Microbes Infect* 2006; **8**: 2195–205.

69 Nakatsuji T, Liu YT, Huang CP *et al.* Antibodies elicited by inactivated *Propionibacterium acnes*-based vaccines exert protective immunity and attenuate the IL-8 production in human sebocytes: relevance to therapy for acne vulgaris. *J Invest Dermatol* 2008; **128**: 2451–7.

70 Li ZJ, Choi DK, Sohn KC *et al.* *Propionibacterium acnes* activates the NLRP3 inflammasome in human sebocytes. *J Invest Dermatol* 2014; **134**: 2747–56.

71 Vowels BR, Yang S, Leyden JJ. Induction of proinflammatory cytokines by a soluble factor of *Propionibacterium acnes*: implications for chronic inflammatory acne. *Infect Immun* 1995; **63**: 3158–65.

72 Kim J, Ochoa MT, Krutzik SR *et al.* Activation of toll-like receptor 2 in acne triggers inflammatory cytokine responses. *J Immunol* 2002; **169**: 1535–41.

73 Kistowska M, Gehrke S, Jankovic D *et al.* IL-1beta drives inflammatory responses to *Propionibacterium acnes in vitro* and *in vivo*. *J Invest Dermatol* 2014; **134**: 677–85.

74 Qin M, Pirouz A, Kim MH *et al.* *Propionibacterium acnes* induces IL-1beta secretion via the NLRP3 inflammasome in human monocytes. *J Invest Dermatol* 2014; **134**: 381–8.

75 Choi JY, Piao MS, Lee JB *et al.* *Propionibacterium acnes* stimulates pro-matrix metalloproteinase-2 expression through tumor necrosis factor-alpha in human dermal fibroblasts. *J Invest Dermatol* 2008; **128**: 846–54.

76 Tsai HH, Lee WR, Wang PH *et al.* *Propionibacterium acnes*-induced iNOS and COX-2 protein expression via ROS-dependent NF-kappaB and AP-1 activation in macrophages. *J Dermatol Sci* 2013; **69**: 122–31.

77 Drott JB, Alexeyev O, Bergström P *et al.* *Propionibacterium acnes* infection induces upregulation of inflammatory genes and cytokine secretion in prostate epithelial cells. *BMC Microbiol* 2010; **10**: 126.

78 Jasson F, Nagy I, Knol AC *et al.* Different strains of *Propionibacterium acnes* modulate differently the cutaneous innate immunity. *Exp Dermatol* 2013; **22**: 587–92.

79 Baum M, Breese M. Antitumour effect of *Corynebacterium parvum*. Possible mode of action. *Br J Cancer* 1976; **33**: 468–73.

80 McBride WH, Peters LJ, Mason KA, Milas L. A role for T lymphocytes in the antitumour action of systemic *C. parvum*. *Dev Biol Stand* 1977; **38**: 253–7.

81 Tsuda K, Yamanaka K, Linan W *et al.* Intratumoral injection of *Propionibacterium acnes* suppresses malignant melanoma by enhancing Th1 immune responses. *PLoS One* 2011; **6**: e29020.

82 Agak GW, Qin M, Nobe J *et al.* *Propionibacterium acnes* induces an IL-17 response in acne vulgaris that is regulated by vitamin A and vitamin D. *J Invest Dermatol* 2014; **134**: 366–73.

83 Kistowska M, Meier B, Proust T *et al.* *Propionibacterium acnes* promotes Th17 and Th17/Th1 responses in acne patients. *J Invest Dermatol* 2015; **135**: 110–8.

84 Furukawa A, Uchida K, Ishige Y *et al.* Characterization of *Propionibacterium acnes* isolates from sarcoid and non-sarcoid tissues with special reference to cell invasiveness, serotype, and trigger factor gene polymorphism. *Microb Pathog* 2009; **46**: 80–7.

85 Tanabe T, Ishige I, Suzuki Y *et al.* Sarcoidosis and NOD1 variation with impaired recognition of intracellular *Propionibacterium acnes*. *Biochim Biophys Acta* 2006; **1762**: 794–801.

86 Fischer N, Mak TN, Shinohara DB *et al.* Deciphering the intracellular fate of *Propionibacterium acnes* in macrophages. *Biomed Res Int* 2013: 603046.

87 Reyes A, Semenkovich NP, Whiteson K *et al.* Going viral: next-generation sequencing applied to phage populations in the human gut. *Nat Rev Microbiol* 2012; **10**: 607–17.

88 Abeles SR, Pride DT. Molecular bases and role of viruses in the human microbiome. *J Mol Biol* 2014; **426**: 3892–906.

89 Saey TH. Beyond the microbiome: The vast virome: scientists are just beginning to get a handle on the many roles of viruses in the human ecosystem. *Science News* 2014; **185**: 18–21.

90 Farrar MD, Howson KM, Bojar RA *et al.* Genome sequence and analysis of a *Propionibacterium acnes* bacteriophage. *J Bacteriol* 2007; **189**: 4161–7.

91 Lood R, Mörgelin M, Holmberg A *et al.* Inducible siphoviruses in superficial and deep tissue isolates of *Propionibacterium acnes*. *BMC Microbiol* 2008; **8**: 139.

92 Marinelli LJ, Fitz-Gibbon S, Hayes C *et al.* *Propionibacterium acnes* bacteriophages display limited genetic diversity and broad killing activity against bacterial skin isolates. *MBio* 2012; **3**: e00279–12.

93 Lood R, Collin M. Characterization and genome sequencing of two *Propionibacterium acnes* phages displaying pseudolysogeny. *BMC Genomics* 2011; **12**: 198.

94 Los M, Wegrzyn G. Pseudolysogeny. *Adv Virus Res* 2012; **82**: 339–49.

95 Brüggemann H, Lood R. Bacteriophages infecting *Propionibacterium acnes*. *Biomed Res Int* 2013: 705741.

Dysbioses and bacterial diseases: Metchnikoff's legacy

CHAPTER 13

The periodontal diseases: microbial diseases or diseases of the host response?

Luigi Nibali

Centre for Oral Clinical Research, Queen Mary University of London, London, United Kingdom

13.1 The tooth: a potential breach in the mucosal barrier

The teeth have unique features in the human body, being non-shedding structures connected with the outside environment (through the oral cavity) at one end and at the other end connected to the jaw through the alveolar bone. Hence, the teeth can be seen as a potential breach in the body's mucosal barrier. In healthy conditions, such a gap is "sealed" by the presence of a tight dento-gingival junction. More precisely, the tooth is surrounded by an attachment apparatus, named the "periodontium" (from ancient Greek "around the tooth"), which provides its support and function. The periodontium includes the alveolar bone (the part of the jaw bone where the teeth are embedded), the periodontal ligament, which connects the tooth to the alveolar bone, the cementum (the part of the tooth where the periodontal ligament fibres are inserted) and the gingiva, which provides protection to the deeper structures by forming a connective tissue-epithelial attachment to the tooth (see Figure 1). Part of the epithelium of the junction between tooth and gingiva is named the "junctional epithelium" and forms a continuum with the oral epithelium.

13.2 The periodontium from health to disease

Bacterial colonization starts on the tooth immediately upon tooth eruption[1]. Indeed, sub-clinical inflammatory signs, with presence of a limited inflammatory infiltrate, exist in the junctional epithelium even in healthy clinical conditions[2].

The first line of periodontal defense is provided by the innate immune response, in the form of neutrophils and macrophages, fibroblasts, epithelial and dendritic cells, which are normally continuously engaged in responses to bacterial plaque in a state of pre-clinical "physiological" inflammation. Landmark studies in the 1960s proved that upon prolonged microbial colonization of the tooth-gingiva interface, gingival inflammation inevitably develops[3,4]. This process is characterized by redness and swelling of the marginal gingiva around the teeth and is known as gingivitis.

The Human Microbiota and Chronic Disease: Dysbiosis as a Cause of Human Pathology, First Edition.
Edited by Luigi Nibali and Brian Henderson.
© 2016 John Wiley & Sons, Inc. Published 2016 by John Wiley & Sons, Inc.

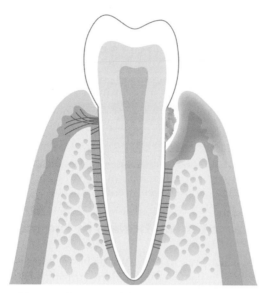

Figure 1 A tooth affected by periodontitis. While the left side of the tooth presents alveolar bone and periodontal ligament fibres in their correct location, the right side presents deposits attached to the root surface, associated with apical migration of the periodontal attachment apparatus and bone loss and gingival inflammation and swelling.

The events leading to the onset of gingivitis are triggered by the presence of bacteria and their by-products sub-gingivally, which stimulate the host release of inflammatory mediators[5]. This process initially occurs by vascular endothelial responses including neutrophil migration into the gingival crevice, regulated by cytokine and chemokines and accompanied by proliferation of the junctional epithelium ("early gingivitis"). If the accumulation of bacteria within dental plaque in the gingiva-tooth interface continues, the inflammatory infiltrate shifts towards a higher number of plasma cells together with T lymphocytes, macrophages and neutrophils, gradually moving to "established gingivitis" and, in genetically predisposed individuals, eventually to the "periodontitis" lesion, with predominant plasma cells and the occurrence of tissue destruction.

Already during tooth eruption, the junctional epithelium has a high content of mononuclear cells, providing a favorable environment for the start of local immune responses[6]. When this response cannot control microbial accumulation, complex inflammatory cascades are activated and an adaptive immune response is called upon by antigen-presenting cells, with a progressive shift from a predominantly T-cell lesion to a B-cell dominated lesion. The balance between T-helper subtypes may be important for the development of the periodontitis lesion. Among T-helper cells, Th-1 is associated with cell-mediated immunity and Th-2 with humoral immunity[7]. A predominant Th2-type response in the periodontium was proposed to be associated with disease progression[8], although this has been challenged[9] and the role of Th-17 and T regulatory (Treg) cells, respectively as effector and suppressive cells may modulate this process[10,11]. Tregs are thought to be important in host-microbe homeostasis and to have a protective role in the periodontium by guaranteeing a balanced host response following microbial challenge[12]. In the advanced periodontitis lesion, B cells are transformed into antibody-producing plasma cells and become the dominating cell subtype[5]. In periodontitis cases, upon chronic subgingival bacterial colonization, the junctional epithelium converts into "pocket epithelium," with loss of cellular continuity in its coronal portion[13]. This is likely to be due to the effect of bacteria able to invade the epithelial tissues and to the effect of cells involved in the local host response[14]. In particular, the microbial challenge is thought to indirectly determine T-cell responses through Toll-like receptors (TLR) and antigen-presenting cells[15]. The inflammatory response mediated by TLR-2 may represent the main pathway to elicit a host response in periodontitis cases[16]. A pathogenic shift in the microbial biofilm associated with a complex network of antibodies, cytokines, growth factors, prostanoids, reactive oxygen species, and other mediators leads to slow, continuous apical migration of the dental plaque front in the pocket. Evidence points to the crucial role of immune cells in periodontal tissue destruction[17]. The apical migration of the plaque front in turn leads to the creation of a chronic inflammatory lesion in the gingival crevice that tends to progress apically with concomitant apical migration of the attachment apparatus at the expenses of the alveolar bone, which undergoes a process of bone resorption typical of the advanced periodontal lesion. This ultimately results in tooth mobility, migration and drifting of the teeth, and eventually tooth loss. In addition to leukocytes, resident fibroblasts and junctional epithelial cells can participate in the tissue-destruction process, with the activation of metalloproteinases leading to connective tissue destruction, apical migration of the epithelial attachment and the formation of a pocket, typical of the established periodontal lesion[18]. A multilevel hierarchical model has been proposed to understand the pathogenesis of periodontitis, including clinical parameters at the top level, and tissue, cellular, and subcellular layers divided into biologic networks below. In this system, microbial factors and combinations of genetic and environmental factors determine biologic expression of the immune-inflammatory network and bone and connective tissue network[5]. Further knowledge of gene expression, proteomics and metabolomics would shed light into patterns of disease associated with specific genetic, microbial and environmental factors.

13.3 Periodontitis: one of the most common human diseases

A recent survey[19] showed that among dentate adults in England, 54% experienced gum bleeding and 45% had periodontitis. These figures are higher than in a previous survey[20], suggesting an upward trend, with increased prevalence over the last 10 years. In the U.S. in 2009-2010, data from the NHANES study estimated a prevalence of 47% for periodontitis in adults[21]. In 2010, severe periodontitis was estimated to be the sixth most prevalent condition in the world, affecting 743 million people worldwide, with a prevalence of 11.2% and an age-standardized incidence of 701 cases per 100,000 person-years. The prevalence increased gradually with age, with a greater increase between the third and fourth decades of life[22].

Furthermore, the importance of periodontitis is compounded by its association and potential co-morbidity with other chronic inflammatory diseases such as diabetes, cardiovascular disease, metabolic syndrome and rheumatoid arthritis and by its association with pregnancy outcomes[23]. The periodontitis-systemic diseases axis is explained by shared risk factors (genetics, environmental, behavioral) as well as by a potential direct effect of periodontal bacteria in inducing systemic inflammation and oxidative stress. In particular, periodontopathogenic bacteria are thought be able to enter the bloodstream through infected periodontia[24], with a possible final destiny including atheromatous plaques[25] or amniotic fluid of pregnant women[26]. A possible interaction between oral bacteria and gut bacteria is discussed by Yamazaki in chapter 15. However, a clear causative effect of periodontitis on systemic diseases has not yet been established.

13.4 Periodontal treatment: a non-specific biofilm disruption

Destructive forms of periodontal disease are sub-classified in the more common chronic periodontitis (CP) and aggressive periodontitis (AgP), less common and usually affecting younger individuals with a rapid rate of progression[27]. The treatment of both aggressive and chronic periodontitis involves a non-specific disruption of the microbial biofilm above and below the gingival margin[28], achieved by oral hygiene instructions and non-surgical periodontal therapy (NSPT). In the case of aggressive periodontitis, the adjunctive use of local or systemic antibiotics has been advocated in order to improve clinical gains[29]. Host-modulating therapies have been suggested, but without strong backing from research evidence at present[30,31]. More advanced cases need surgical treatment or extraction of teeth. After initial and/or subsequent surgical therapy, subjects susceptible to periodontitis are then generally enrolled in long-term maintenance programmes in order to reduce the risk of relapse and tooth loss. Periodontal therapy can be very successful in treating periodontal disease and strict maintenance therapy has been shown to be successful and vital for tooth retention in the great majority of individuals[32]. However, there seems to be a subgroup of individuals who tend to have progression and tooth loss despite treatment[33-35], presumably because of increased susceptibility determined by a combination of microbial, genetic and environmental factors.

13.5 Microbial etiology

Studies in the 1970s investigated a possible microbiological cause of periodontitis and identified bacteria associated with chronic[36-39] and aggressive periodontitis[40]. In his pioneering work on the microbial etiology of periodontal disease, Sigmund Socransky tailored Koch's criteria to define causative microbes for periodontitis[41] and proposed the clustering of bacteria in "color complexes" to facilitate their assignment to a health-associated or disease-associated category. In this model, the "red complex" characterized by *Porhyromonas gingivalis*, *Treponema denticola* and *Tannerella forsythia* was associated with disease progression[42]. The concept of "finding the microbial cause of periodontitis" was underpinned by the belief that periodontitis was caused by specific bacteria[43], in contrast to the earlier belief that just the sheer volume of plaque could cause it, irrespective of the species involved[44]. Clearly, both are simplistic views, highlighted by the fact that even bacteria recognized as "periodontopathogenic" by the American Academy of Periodontology in 1996 (namely *Porhyromonas gingivalis*, *Tannerella forsythia* and *Actinobacillus actinomycetemcomitans*) may not be present in diseased individuals[45] and may also be detected in healthy sites and healthy subjects with no periodontitis[46-48]. Even the association between aggressive periodontitis and *A. actinomycetemcomitans*, based on its high prevalence[49], on the elevated serum antibody response to this bacterium[50,51], association with disease progression[52] and correlation with treatment response[53], has recently been called into question[54,55]. One of the most compelling examples of association between microbes and periodontitis is represented by the highly leukotoxic JP2 strain of *A. actinomycetemcomitans*, associated with onset and progression of aggressive periodontitis in some North African populations[56-59].

Hence, the bacteria now recognized as periodontopathogenic are certainly not the only possible pathogens predisposing to periodontitis. Recent studies, some involving metagenomic analysis of subgingival plaque samples in periodontitis and healthy individuals, suggest the possible role of non-cultivable bacteria in its pathogenesis, including among others *Filifactor alocis*[60], *Selenomonas genus*[61], *Actinobacter baumannii*[62], *Treponema lecithinolyticum*[63] and other microbes including the *Archaea* domain[64] and viruses such as Human cytomegalovirus (HCMV) and Epstein Barr virus (EBV)[65,66]. New theories on the microbial etiology of periodontitis have recently emerged[67] and are discussed exhaustively in the next chapter (chapter 14). However, the search for the role of bacteria in the pathogenesis of periodontitis is complicated by the nature of the subgingival biofilm (discussed in chapter 4) and by the fundamental role of the host response in reacting to the microbial challenge and in determining tissue damage.

13.6 The host response in periodontitis

Seventy years ago, Orban and Weinmann proposed the concept that "periodontosis" (what we now call aggressive periodontitis) was caused by a metabolic imbalance[68]. Thirty years later, Baer hinted at the possible role of genetically inherited defects in the host response in disease onset[69]. A substantial body of evidence has since then accumulated to suggest large inter-individual variations in the response to dental plaque accumulation, leading to different degrees of gingival inflammation in different subjects[70]. In other words, subjects respond differently to the same degree of plaque accumulation, and highly susceptible individuals exhibit an enhanced response, with increased and more rapid inflammatory reactions[71]. It is also striking that studies on Sri Lankan tea laborers showed that, despite absence of routine dental prophylaxis and appropriate dental care, three patterns of susceptibility are identifiable, namely very susceptible, marginally susceptible and resistant individuals[72]. Equally, among patients initially diagnosed with periodontitis and treated in a private dental practice, three "progression" groups were identified over time, with well-maintained individuals, "downhill" and "extreme downhill" subjects (based on the number of teeth lost over more than 20 years of maintenance therapy)[73]. Potential host response defects were sought among neutrophils, suggested by evidence of their presence in gingival lesions and in root surfaces of periodontitis cases[52,74]. An array of possible malfunctions were suggested, including increased adhesion, reduced chemotaxis, increased superoxide and nitric oxide production and reduced phagocytosis in AgP[75–77] and CP cases[78]. These findings back the theory that, if chronically activated by certain microbial triggers, the host response could contribute to the periodontal tissue damage. Other disease-predisposing pathways have been suspected in over-amplifications of inflammatory responses mediated by cytokine production[5], by defects in mounting an appropriate antibody response to periodontopathogenic bacteria[79,80] or indeed in the effect of cytokines and other inflammatory mediators in inducing increased antibody responses[81]. A compelling example of high predisposition to periodontal destruction is represented by localized aggressive periodontitis (LAgP), which usually has a circumpubertal onset and affects first molars and incisors, the first teeth to erupt in the mouth at age 6–8. The disease progresses very rapidly, with periodontal attachment and alveolar bone loss, but then it often "burns out," with stability sometimes observed even in the absence of therapy[82], suggesting a possible host response adaptation or the mounting of an effective and "mature" antibody response to pathogenic bacteria such as *A. actinomycetemcomitans*[55].

Landmark twin studies attribute to genetically inherited factors a strong role in the susceptibility to periodontitis[83]. The search for an inherited genetic mutation or variant predisposing to periodontitis has encountered some difficulties owing to the small sample size and poor methodological quality of most studies and often to the lack of hypothesis-driven research. Although specific gene defects causing early-onset periodontitis have been discovered, they are usually responsible for periodontitis as part of syndromes or for the rare pre-pubertal periodontitis[84,85]. Such defects involve genes affecting neutrophil function or metabolic, structural or immune protein defects. This is the case, for example, in the COL3A and COL5A genes affecting collagen production and responsible for Ehlers-Danlos syndrome and of the CTSC gene, coding for a neutrophil protease (cathepsin C) and responsible for Papillon-Lefevre syndrome and pre-pubertal periodontitis[85]. It is now accepted that most cases of periodontitis, including AgP and CP, have a complex susceptibility profile attributed by genes with a minor, modifying effect on periodontal homeostasis through inflammatory, immune, metabolic, structural or nervous system pathways or a combination of these. Promising putative single-nucleotide polymorphisms (SNPs) recently identified in genome-wide studies in the GLT6D1 and vitamin C transporter SVCT1 genes may increase the susceptibility to AgP[86,87], while gene variants in the ANRIL, NPY, VAMP3/CAMTA1, NIN and WNT5A genes have shown trends for association with CP[88–90]. The exact function of most of these putative genes and hence their role in periodontal pathogenesis remains largely obscure at this stage. However, it is becoming clearer that the interaction between the complex subgingival biofilms generated in the interface between teeth and gum and common genetic variations in *Homo sapiens* is responsible for aberrant responses that lead to periodontal pathology. Emerging evidence on how this host response to specific periodontopathogenic bacteria such as *P. gingivalis* might cause protein citrullination and have systemic effects is discussed in chapter 18.

13.7 Conclusions

The overall microbial load in the tooth-gingiva interface, especially in the subgingival biofilm, the type of microbes and strains present and their interactions, the host response and environmental and behavioral factors all act together to predispose to periodontitis onset and progression. A complex network of interactions between all these factors can define a specific expression pattern that determines the shift from health to periodontal disease[5]. Potential keystone periodontal microbes may have an important role in disrupting tissue homeostasis and in driving a dysbiotic change in the subgingival biofilm[91] (see chapter 14).

TAKE-HOME MESSAGE

- The teeth could be considered a unique example of non-shedding structures connecting the outside environment (oral cavity) with the bone. Therefore, they can be a means for a microbial "breach" of the mucosal barrier.

- Periodontitis is a very common chronic destructive inflammatory disease affecting the supporting apparatus of the teeth, resulting from the interaction between a predisposing host genetic profile and the presence of subgingival microbes.

- Microbial dysbiosis is emerging as a likely pathogenic mechanisms causing periodontitis.

References

1 Gudino S, Rojas N, Castro C *et al*. Colonization of mutans streptococci in Costa Rican children from a high-risk population. *J Dent Child (Chic)* 2007; **74**: 36–40.
2 Brecx MC, Schlegel K, Gehr P *et al*. Comparison between histological and clinical parameters during human experimental gingivitis. *J Periodontal Res* 1987; **22**: 50–7.
3 Loe H, Theilade E, Jensen SB. Experimental gingivitis in man. *J Periodontol* 1965; **36**: 177–87.
4 Theilade E, Wright WH, Jensen SB, Loe H. Experimental gingivitis in man. II. A longitudinal clinical and bacteriological investigation. *J Periodontal Res*. 1966; **1**: 1–13.
5 Kornman KS. Mapping the pathogenesis of periodontitis: a new look. *J Periodontol* 2008; **79**: 1560–8.
6 Schroeder HE, Listgarten MA. The gingival tissues: the architecture of periodontal protection. *Periodontol 2000* 1997; **13**: 91–120.
7 Murphy KM, Reiner SL. The lineage decisions of helper T cells. *Nat Rev Immunol* 2002; **2**: 933–44.
8 Gemmell E, Yamazaki K, Seymour GJ. Destructive periodontitis lesions are determined by the nature of the lymphocytic response. *Crit Rev Oral Biol Med* 2002; **13**: 17–34.
9 Berglundh T, Liljenberg B, Lindhe J. Some cytokine profiles of T-helper cells in lesions of advanced periodontitis. *J Clin Periodontol* 2002; **29**: 705–9.
10 Gaffen SL, Hajishengallis G. A new inflammatory cytokine on the block: re-thinking periodontal disease and the Th1/Th2 paradigm in the context of Th17 cells and IL-17. *J Dent Res* 2008; **87**: 817–28.
11 Di Benedetto A, Gigante I, Colucci S, Grano M. Periodontal disease: linking the primary inflammation to bone loss. *Clin Dev Immunol*. 2013; 503754.
12 Garlet GP, Sfeir CS, Little SR. Restoring host-microbe homeostasis via selective chemoattraction of Tregs. *J Dent Res* 2014; **93**: 834–9.
13 Schluger S. Periodontics today: dentistry tomorrow. *J Dist Columbia Dent Soc* 1977; 6–8.
14 Bosshardt DD, Lang NP. The junctional epithelium: from health to disease. *J Dent Res* 2005; **84**: 9–20.
15 Iwasaki A, Medzhitov R. Toll-like receptor control of the adaptive immune responses. *Nat Immunol* 2004; **5**: 987–95.
16 Myneni SR, Settem RP, Sharma A. Bacteria take control of tolls and T cells to destruct jaw bone. *Immunol Invest* 2013; **42**: 519–31.

17 Hernandez M, Dutzan N, Garcia-Sesnich J *et al*. Host-pathogen interactions in progressive chronic periodontitis. *J Dent Res* 2011; **90**: 1164–70.

18 Page RC, Kornman KS. The pathogenesis of human periodontitis: an introduction. *Periodontol 2000* 1997; **14**: 9–11.

19 White D, Pitts N, Steele J, Sadler K, Chadwick B. *Disease and related disorders: a report from the Adult Dental Health Survey 2009*. The NHS Health and Social Care Information Centre, 2011.

20 Kelly M, Steele J, Nuttall N, Bradnock G, Morris J, Nunn J, Pine C, Pitts N, Treasure E, White D (1999) *Adult Dental Health Survey. Oral Health in the United Kingdom*. London. National Statistics.

21 Eke PI, Page RC, Wei L *et al*. Update of the case definitions for population-based surveillance of periodontitis. *J Periodontol* 2012; **83**: 1449–54.

22 Kassebaum NJ, Bernabe E, Dahiya M *et al*. Global burden of severe periodontitis in 1990-2010: a systematic review and meta-regression. *J Dent Res* 2014; **93**: 1045–53.

23 Linden GJ, Herzberg MC. Periodontitis and systemic diseases: a record of discussions of working group 4 of the Joint EFP/AAP Workshop on Periodontitis and Systemic Diseases. *J Clin Periodontol* 2013; **40** Suppl 143: S20–23.

24 Forner L, Larsen T, Kilian M *et al*. Incidence of bacteremia after chewing, tooth brushing and scaling in individuals with periodontal inflammation. *J Clin Periodontol* 2006; **33**: 401–7.

25 Haraszthy VI, Zambon JJ, Trevisan M *et al*. Identification of periodontal pathogens in atheromatous plaques. *J Periodontol* 2000; **71**: 1554–60.

26 Leon R, Silva N, Ovalle A *et al*. Detection of Porphyromonas gingivalis in the amniotic fluid in pregnant women with a diagnosis of threatened premature labor. *J Periodontol* 2007; **78**: 1249–55.

27 Armitage GC. Development of a classification system for periodontal diseases and conditions. *Northwest Dent* 2000; **79**: 31–5.

28 Heitz-Mayfield LJ, Trombelli L, Heitz F *et al*. A systematic review of the effect of surgical debridement vs non-surgical Chambrone L, Chambrone D, Lima LA *et al*. Predictors of tooth loss during long-term periodontal maintenance: a systematic review of observational studies. *J Clin Periodontol* 2010; **37**: 675–84 debridement for the treatment of chronic periodontitis. *J Clin Periodontol* 2002; **29 Suppl 3**: 92–102.

29 Sgolastra F, Gatto R, Petrucci A *et al*. Effectiveness of systemic amoxicillin/metronidazole as adjunctive therapy to scaling and root planing in the treatment of chronic periodontitis: a systematic review and meta-analysis. *J Periodontol* 2012; **83**: 1257–69.

30 Reddy MS, Geurs NC, Gunsolley JC. Periodontal host modulation with antiproteinase, anti-inflammatory, and bone-sparing agents. A systematic review. *Ann Periodontol* 2003; **8**: 12–37.

31 Gulati M, Anand V, Govila V *et al*. Host modulation therapy: An indispensable part of perioceutics. *J Indian Soc Periodontol* 2014; **18**: 282–8.

32 Axelsson P, Nystrom B, Lindhe J. The long-term effect of a plaque control program on tooth mortality, caries and periodontal disease in adults. Results after 30 years of maintenance. *J Clin Periodontol* 2004; **31**: 749–57.

33 Hirschfeld L, Wasserman B. A long-term survey of tooth loss in 600 treated periodontal patients. *J Periodontol* 1978; **49**: 225–37.

34 Eke PI, Dye BA, Wei L *et al*. Prevalence of periodontitis in adults in the United States: 2009 and 2010. *J Dent Res* 2012; **91**: 914–20.

35 Eke PI, Dye BA, Wei L *et al*. Prevalence of periodontitis in adults in the United States: 2009 and 2010. *J Dent Res* 2012; **91**: 914–20.

36 Slots J. The predominant cultivable microflora of advanced periodontitis. *Scand J Dent Res* 1977; **85**: 114–21.

37 Socransky SS. Microbiology of periodontal disease — present status and future considerations. *J Periodontol* 1977; **48**: 497–504.

38 Tanner AC, Haffer C, Bratthall GT *et al*. A study of the bacteria associated with advancing periodontitis in man. *J Clin Periodontol* 1979; **6**: 278–307.

39 Moore WE, Holdeman LV, Smibert RM *et al*. Bacteriology of severe periodontitis in young adult humans. *Infect Immun* 1982; **38**: 1137–48.

40 Newman HN. The apical border of plaque in chronic inflammatory periodontal disease. *Br Dent J* 1976; **141**: 105–13.

41 Socransky SS. Criteria for the infectious agents in dental caries and periodontal disease. *J Clin Periodontol* 1979; **6**: 16–21.

42 Socransky SS, Haffajee AD, Cugini MA et al. Microbial complexes in subgingival plaque. *J Clin Periodontol* 1998; **25**: 134–144.

43 Loesche WJ. Clinical and microbiological aspects of chemotherapeutic agents used according to the specific plaque hypothesis. *J Dent Res* 1979; **58**: 2404–12.

44 Macdonald JB, Sutton RM, Knoll ML, *et al.* (1956) The pathogenic components of an experimental fusospirochetal infection. *J Infect Dis* **98**: 15.

45 Moore WE, Holdeman LV, Cato EP *et al.* Comparative bacteriology of juvenile periodontitis. *Infect Immun* 1985; **48**: 507–19.

46 Albandar JM, Brown LJ, Loe H. Putative periodontal pathogens in subgingival plaque of young adults with and without early-onset periodontitis. *J Periodontol* 1997; **68**: 973–81.

47 Gafan GP, Lucas VS, Roberts GJ *et al.* Prevalence of periodontal pathogens in dental plaque of children. *J Clin Microbiol* 2004; **42**: 4141–6.

48 Diaz R, Ghofaily LA, Patel J *et al.* Characterization of leukotoxin from a clinical strain of Actinobacillus actinomycetemcomitans. *Microb Pathog* 2006; **40**: 48–55.

49 Zambon JJ, Christersson LA, Slots J. Actinobacillus actinomycetemcomitans in human periodontal disease. Prevalence in patient groups and distribution of biotypes and serotypes within families. *J Periodontol* 1983; **54**: 707–11.

50 Ebersole JL, Cappelli D, Sandoval MN. Subgingival distribution of A. actinomycetemcomitans in periodontitis. *J Clin Periodontol* 1994; **21**: 65–75.

51 Albandar JM, DeNardin AM, Adesanya MR *et al.* Associations between serum antibody levels to periodontal pathogens and early-onset periodontitis. *J Periodontol* 2001; **72**: 1463–9.

52 Fine DH, Markowitz K, Furgang D *et al.* Aggregatibacter actinomycetemcomitans and its relationship to initiation of localized aggressive periodontitis: longitudinal cohort study of initially healthy adolescents. *J Clin Microbiol* 2007; **45**: 3859–69.

53 Mandell RL, Ebersole JL, Socransky SS. Clinical immunologic and microbiologic features of active disease sites in juvenile periodontitis. *J Clin Periodontol* 1987; **14**: 534–40.

54 Faveri M, Figueiredo LC, Duarte PM *et al.* Microbiological profile of untreated subjects with localized aggressive periodontitis. *J Clin Periodontol* 2009; **36**: 739–49.

55 Kulkarni C, Kinane DF. Host response in aggressive periodontitis. *Periodontol 2000* 2014; **65**: 79–91.

56 Brogan JM, Lally ET, Poulsen K *et al.* Regulation of Actinobacillus actinomycetemcomitans leukotoxin expression: analysis of the promoter regions of leukotoxic and minimally leukotoxic strains. *Infect Immun* 1994; **62**: 501–8.

57 Ennibi OK, Benrachadi L, Bouziane A *et al.* The highly leukotoxic JP2 clone of Aggregatibacter actinomycetemcomitans in localized and generalized forms of aggressive periodontitis. *Acta Odontol Scand* 2012; **70**: 318–22.

58 Bandhaya P, Saraithong P, Likittanasombat K *et al.* Aggregatibacter actinomycetemcomitans serotypes, the JP2 clone and cytolethal distending toxin genes in a Thai population. *J Clin Periodontol* 2012; **39**: 519–25.

59 Haubek D, Ennibi OK, Poulsen K *et al.* Risk of aggressive periodontitis in adolescent carriers of the JP2 clone of Aggregat Yapar M, Saygun I, Ozdemir A *et al.* Prevalence of human herpesviruses in patients with aggressive periodontitis. *J Periodontol* 2003; **74**: 1634–40 ibacter (Actinobacillus) actinomycetemcomitans in Morocco: a prospective longitudinal cohort study. *Lancet* 2008; **371**: 237–42.

60 Schlafer S, Riep B, Griffen AL *et al.* Filifactor alocis--involvement in periodontal biofilms. *BMC Microbiol* 2010; **10**: 66.

61 Drescher J, Schlafer S, Schaudinn C *et al.* Molecular epidemiology and spatial distribution of Selenomonas spp. in subgingival biofilms. *Eur J Oral Sci* 2010; **118**: 466–74.

62 da Silva-Boghossian CM, do Souto RM, Luiz RR *et al.* Association of red complex, A. actinomycetemcomitans and non-oral bacteria with periodontal diseases. *Arch Oral Biol* 2011; **56**: 899–906.

63 Riep B, Edesi-Neuss L, Claessen F *et al.* Are putative periodontal pathogens reliable diagnostic markers? *J Clin Microbiol* 2009; **47**: 1705–11.

64 Matarazzo F, Ribeiro AC, Feres M *et al.* Diversity and quantitative analysis of Archaea in aggressive periodontitis and periodontally healthy subjects. *J Clin Periodontol* 2011; **38**: 621–7.

65 Contreras A, Slots J. Mammalian viruses in human periodontitis. *Oral Microbiol Immunol* 1996; **11**: 381–6.

66 Yapar M, Saygun I, Ozdemir A *et al.* Prevalence of human herpesviruses in patients with aggressive periodontitis. *J Periodontol* 2003; **74**: 1634–40.

67 Hajishengallis G, Lamont RJ. Beyond the red complex and into more complexity: the polymicro-bial synergy and dysbiosis (PSD) model of periodontal disease etiology. *Mol Oral Microbiol* 2012; **27**: 409–19.

68 Orban B, Weinmann JP. Diffuse atrophy of the alveolar bone (periodontosis). *J Periodontol* 1942; **13**: 31.

69 Baer PN. The case for periodontosis as a clinical entity. *J Periodontol* 1971; **42**: 516–20.

70 Trombelli L, Tatakis DN, Scapoli C *et al*. Modulation of clinical expression of plaque-induced gingi-vitis. II. Identification of "high-responder" and "low-responder" subjects. *J Clin Periodontol* 2004; **31**: 239–52.

71 Trombelli L, Scapoli C, Tatakis DN *et al*. Modulation of clinical expression of plaque-induced gingi-vitis: response in aggressive periodontitis subjects. *J Clin Periodontol* 2006; **33**: 79–85.

72 Loe H, Anerud A, Boysen H, Morrison E. Natural history of periodontal disease in man. Rapid, moderate and no loss of attachment in Sri Lankan Laborers 14 to 46 years of age. *J. Clin Periodontol.* 1986; **13**(5): 431–45.

73 Hirschfeld L, Wasserman B. A long-term survey of tooth loss in 600 treated periodontal patients. *J Periodontol* 1978; **49**: 225–37.

74 Rams TE, Keyes PH. Direct microscopic features of subgingival plaque in localized and generalized juvenile periodontitis. *Pediatr Dent* 1984; **6**: 23–7.

75 Shapira L, Borinski R, Sela MN *et al*. Superoxide formation and chemiluminescence of peripheral polymorphonuclear leukocytes in rapidly progressive periodontitis patients. *J Clin Periodontol* 1991; **18**: 44–8.

76 Shibata K, Warbington ML, Gordon BJ *et al*. Nitric oxide synthase activity in neutrophils from patients with localized aggressive periodontitis. *J Periodontol* 2001; **72**: 1052–8.

77 Van Dyke TE, Serhan CN. Resolution of inflammation: a new paradigm for the pathogenesis of periodontal diseases. *J Dent Res* 2003; **82**: 82–90.

78 Fredriksson MI, Gustafsson AK, Bergstrom KG *et al*. Constitutionally hyperreactive neutrophils in periodontitis. *J Periodontol* 2003; **74**: 219–24.

79 Schenkein HA, Berry CR, Burmeister JA *et al*. Anti-cardiolipin antibodies in sera from patients with periodontitis. *J Dent Res* 2003; **82**: 919–22.

80 Hwang AM, Stoupel J, Celenti R *et al*. Serum antibody responses to periodontal microbiota in chronic and aggressive periodontitis: a postulate revisited. *J Periodontol* 2014; **85**: 592–600.

81 Schenkein HA, Barbour SE, Tew JG. Cytokines and inflammatory factors regulating immuno-globulin production in aggressive periodontitis. *Periodontol 2000* 2007; **45**: 113–27.

82 Mros ST, Berglundh T. Aggressive periodontitis in children: a 14-19-year follow-up. *J Clin Periodontol* 2010; **37**: 283–7.

83 Michalowicz BS, Diehl SR, Gunsolley JC *et al*. Evidence of a substantial genetic basis for risk of adult periodontitis. *J Periodontol* 2000; **71**: 1699–707.

84 Hewitt C, McCormick D, Linden G *et al*. The role of cathepsin C in Papillon-Lefevre syndrome, prepubertal periodontitis, and aggressive periodontitis. *Hum Mutat* 2004; **23**: 222–8.

85 Hart TC, Atkinson JC. Mendelian forms of periodontitis. *Periodontol 2000* 2007; **45**: 95–112.

86 Schaefer AS, Richter GM, Nothnagel M *et al*. A genome-wide association study identifies GLT6D1 as a susceptibility locus for periodontitis. *Hum Mol Genet* 2010; **19**: 553–62.

87 de Jong TM, Jochens A, Jockel-Schneider Y et al. SLC23A1 polymorphysm rs6596473 in the vita-min C transporter SVCT1 is associated with aggressive periodontitis. *J Clin Periodontol* 2014; **41**: 531–540.

88 Schaefer AS, Bochenek G, manke T et al. Validation of reported genetic risk factors for periodontitis in a large-scale replication study. *J Clin Periodontol* 2013; **40**: 563–572.

89 Bochenek G, Hasler R, El Mokhtari NE *et al*. The large non-coding RNA ANRIL, which is associated with atherosclerosis, periodontitis and several forms of cancer, regulates ADIPOR1, VAMP3 and C11ORF10. *Hum Mol Genet* 2013; **22**: 4516–27.

90 Divaris K, Monda KL, North KE *et al*. Exploring the genetic basis of chronic periodontitis: a genome-wide association study. *Hum Mol Genet* 2013; **22**: 2312–24.

91 Hajishengallis G, Lamont RJ. Beyond the red complex and into more complexity: the polymicro-bial synergy and dysbiosis (PSD) model of periodontal disease etiology. *Mol Oral Microbiol* 2012; **27**: 409–19.

CHAPTER 14

The polymicrobial synergy and dysbiosis model of periodontal disease pathogenesis

George Hajishengallis[1] and Richard J. Lamont[2]

[1] School of Dental Medicine, University of Pennsylvania, Philadelphia, United States
[2] School of Dentistry, University of Louisville, Louisville, KY, United States

14.1 Introduction

The indigenous oral microbiota is a complex and diverse assemblage of microorganisms that usually exists in balance with the host. However, despite millennia of co-evolution, the oral microbiota also commonly initiates diseases of the mouth. Periodontal diseases are ancient and prevalent afflictions characterized by destruction of the soft and hard tissues supporting the teeth. An overview of periodontal diseases was given in chapter 13. For almost as long as a microbiological component to periodontal diseases has been recognized, debate has raged over the relative contributions of bacteria and of destructive host responses, and over the necessity for colonization by specific organisms as opposed to a non-specific bacterial threshold. Recent advances in microbiome analyses, coupled with increasingly precise molecular biology tools and the availability of mice with specific defects in innate and adaptive immunity, have begun to resolve these issues. A current model of periodontal disease etiology and pathogenesis, that accommodates distinct properties of oral bacterial communities, along with protective and destructive aspects of host responses, is the polymicrobial synergy and dysbiosis (PSD) model[1] (Figure 1a). In this model, interactive bacterial communities comprised of physiologically compatible organisms are, in most individuals, most of the time, controlled by the host inflammatory response. However, colonization by immune-subversive bacteria, such as *Porphyromonas gingivalis*, synergistically elevates the pathogenic potential of the entire community[2] and leads to disease in susceptible individuals. Central to the disease process is initiation, by the now dysbiotic community, of an immune response that is ineffective, uncontrolled and destructive and that, moreover, can sustain "inflammophilic" members of the microbial community through the provision of nutrients derived from tissue breakdown[3].

At the heart of microbial community dependent diseases such as periodontal disease lies the functional specialization that arises among the participating

The Human Microbiota and Chronic Disease: Dysbiosis as a Cause of Human Pathology, First Edition.
Edited by Luigi Nibali and Brian Henderson.
© 2016 John Wiley & Sons, Inc. Published 2016 by John Wiley & Sons, Inc.

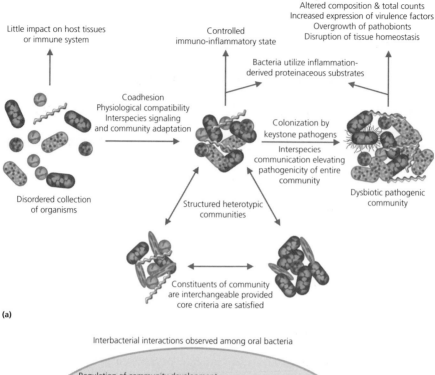

(a)

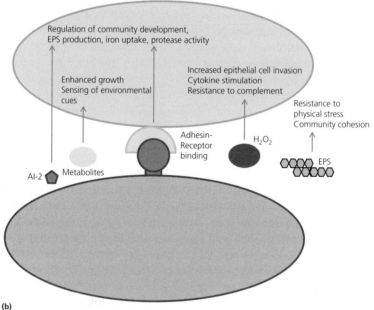

(b)

Figure 1 The polymicrobial synergy and dysbiosis (PSD) model of periodontal disease etiology.
(a) Model overview. In health, communities assembled through co-adhesion and physiological
compatibility participate in balanced interactions with the host in a controlled immuno-
inflammatory state. In disease, there is colonization by keystone pathogens such as *P. gingivalis*
that enhance community virulence through interactive communication with accessory pathogens
such as *S. gordonii*, and disruption of immune surveillance. The dysbiotic community proliferates,
pathobionts (green) overgrow and become more active, and tissue destruction ensues. The
participation of individual species is less important than the presence of the appropriate suite of
genes. (*see color plate section for color details*). (b) Summary of synergistic interbacterial interactions
that have been documented among oral bacteria. Signaling can occur through direct contact,
primarily adhesin-receptor binding, and sensing of compounds in solution. Interspecies
communication can modulate the phenotypic properties of partner organisms, and ultimately
affect virulence potential. From Lamont and Hajishengallis, 2014 (Ref. 77) with permission.

organisms. In the past it has been convenient to think of bacteria as being either "commensals" or "pathogens." Progress in a number of areas of bacteriology, however, has shown that such concepts are restrictive and do not reflect the more nuanced roles that microorganisms can play. Many members of the oral microbiota that do not cause disease themselves (and hence had been considered commensals) nonetheless support or enhance the virulence of other organisms, and are thus more accurately designated accessory pathogens[4]. Other species, the pathobionts, also common in healthy microenvironments, can promote pathology under conditions of disrupted homeostasis[5]. Organisms that are essential for a loss of homeostatic balance are the keystone pathogens that exert their influence even in low numbers through manipulation of community structure and host immune responses[3,5]. In order to understand the pathogenesis of periodontal disease, we must therefore consider synergistic interactions among contributing bacteria, how these give rise to a dysbiotic community, and the means by which host immune responses are incapacitated and redirected to allow the dysbiotic bacterial community to flourish.

14.2 A (very) polymicrobial etiology of periodontitis

The "classical" bacterial diseases, such as tetanus and diphtheria, that were brought under control during what is called the golden era of microbiology are examples where a single organism causes one clinically well-defined disease. In contrast, many of the diseases that are poorly controlled in the present day are caused by heterotypic communities of organisms that display polymicrobial synergy, whereby one organism enhances the colonization and/or virulence of another[6]. These include skin and wound infections, respiratory tract infections, vaginosis, rhinosinusitis, and, *par excellence*, periodontal disease[7]. A significant body of evidence based on cultural analyses of periodontal lesions combined with *in vitro* studies of infection along with animal models of disease had implicated the pathogenic triumvirate of *P. gingivalis*, *Tannerella forsythia*, and *Treponema denticola* as the etiological agents of chronic periodontitis[8–10]. However, more recent molecular microbiome studies show that the periodontal microbiota is more heterogeneous and diverse than previously thought, with a number of newly recognized species showing a good correlation with disease status[11–13]. Novel disease-associated species include the Gram-positives: *Filifactor alocis* and *Peptostreptococcus stomatis* and other species from the genera *Prevotella*, *Megasphaera*, *Selenomonas*, *Desulfobulbus*, *Dialister*, and *Synergistetes*[11–15]. Moreover, microbiome studies often show that the organisms of the classical triad are not numerically dominant at disease sites. Recognition that *P. gingivalis* can be a low-abundance member of the disease-associated microbiota in fact dates back to at least the early 1980s[16]. More recent studies in experimental animals show that while *P. gingivalis* failed to cause periodontitis in the absence of commensal bacteria, i.e., in germ-free mice, at low colonization levels *P. gingivalis* could remodel the indigenous symbiotic community into a dysbiotic state that triggered inflammatory bone loss[17]. Mechanistically, this involved manipulation of the host response to create a permissive environment allowing the conversion from a symbiotic commensal community to a dysbiotic community structure capable of causing destructive inflammation[2]. Examples by which

P. gingivalis can subvert host defenses — often in synergy with other organisms — and enhance the fitness of the entire microbial community are given below. The capacity of *P. gingivalis* to exert a community-wide impact that tips the balance towards dysbiosis, while being a quantitatively minor constituent, established the organism as a keystone pathogen. Other more recently appreciated organisms may also function as keystone pathogens or as pathobionts[18], their categorization requiring further study.

14.3 Synergism among periodontal bacteria

The synergistic pathogenicity of periodontal organisms in animal models has been recognized for some time[19] and nutrient transfer among *P. gingivalis* and other species such as *T. denticola* is well defined[20,21]. Indeed, co-culture of *P. gingivalis* with *T. denticola* induces an alteration in *P. gingivalis* hemin uptake strategies and changes in the abundance of enzymes involved in glutamate and glycine catabolism[21]. *P. gingivalis* can provide a source of free glycine and isobutyric acid for *T. denticola* growth, while *T. denticola* produces succinic acid which enhances growth of *P. gingivalis*[20,21]. Polymicrobial synergy extends well beyond cross-feeding, however, and through sophisticated signaling mechanisms microbial communities are tightly integrated and capable of collectively regulating activities including expression of virulence factors such as proteases[21–23] (Figure 1b).

The initial encounter between physiologically compatible organisms on the supra- and sub-gingival tooth surfaces frequently results in interbacterial binding or coadhesion[24]. The cohesive accretion of organisms, an early stage of biofilm development, confers resistance to removal by mechanical shearing forces in the mouth. Periodontal organisms often possess multiple adhesins with differing specificities that will broaden the range of potential binding partners and increase the affinity of binding[25]. *P. gingivalis* and the antecedent colonizer *S. gordonii* bind to each other through two sets of adhesion-receptor pairs. The FimA and Mfa1 component fimbriae of *P. gingivalis* engage the streptococcal GAPDH and SspA/B surface proteins respectively[26]. In addition to attachment, the close association between bacteria can be a means that allows the organisms to sense their surroundings and modify their phenotype accordingly. In the case of *P. gingivalis*, binding of Mfa1 to SspB modulates tyrosine (de)phosphorylation signal transduction that converges on the transcriptional regulator CdhR[27,28] (Figure 2). Active CdhR is a repressor of *mfa1* and of *luxS* that encodes the synthetic enzyme producing the autoinducer (AI)-2 signalling molecule necessary for optimal *P. gingivalis-S. gordonii* community development. The end result, therefore, is a restriction of further *P. gingivalis* accumulation. Conversely, *P. gingivalis* cells become primed for community development through detection of metabolites of *S. gordonii*[29]. Initial interaction of *P. gingivalis* with *S. gordonii* extracellular components induces autophosphorylation and activation of the tyrosine kinase Ptk1, which acts to limit production of the CdhR transcriptional repressor and thus increase production of the community effectors Mfa1 and LuxS[30]. Moreover, Ptk1 is also involved in the secretion of extracellular polysaccharide by *P. gingivalis*, thereby influencing pathogenic potential, as demonstrated in murine models of alveolar bone loss in which dual infection with *P. gingivalis* and *S. gordonii* results in greater periodontal

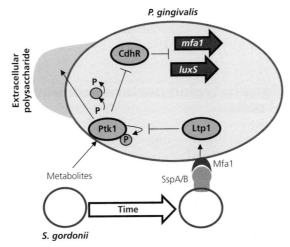

Figure 2 Interactions between *S. gordonii* and *P. gingivalis*. Metabolites produced by *S. gordonii* induce autophosphorylation of the Ptk1 tyrosine kinase of *P. gingivalis*. Ptk1 initiates a signaling cascade that results in inactivation of the transcriptional repressor CdhR. Consequently, expression of the minor fimbrial adhesin subunit Mfa1 and the AI-2 biosynthetic enzyme LuxS is elevated, and *P. gingivalis* is "primed" for attachment to *S. gordonii*. Ptk1 is also a participant in the machinery of secretion of extracellular polysaccharide, and active Ptk1 increases the level of polysaccharide material on the surface of *P. gingivalis*. Over time, direct contact mediated by Mfa1-SspA/B binding elevates expression of the tyrosine phosphatase Ltp1, which dephosphorylates Ptk1, ultimately relieving repression of ChdR. Expression of Mfa1 and LuxS is reduced, and community development is constrained. From Lamont and Hajishengallis, 2014 (Ref. 77) with permission.

bone loss than infection with either organism alone[31]. *S. gordonii* thus fulfills the criteria for an accessory pathogen, a role also established with *Aggregatibacter actinomycetemcomitans*[32].

Localized Aggressive Periodontitis (LAP) is a clinically distinct manifestation of periodontal disease that is primarily associated with *A. actinomycetemcomitans* infection (discussed by Nibali in chapter 13). *S. gordonii and A. actinomycetemcomitans* interact on a number of levels. *S. gordonii* produce lactate as an end product of metabolism, and *A. actinomycetemcomitans* displays resource partitioning to favor lactate as a carbon source even in the presence of alternative carbon sources such as glucose or fructose[33]. Another metabolic by-product of *S. gordonii* is H_2O_2, and *A. actinomycetemcomitans* possesses two mechanisms, known as fight or flight responses, to avoid H_2O_2 toxicity. First, detection of H_2O_2 induces upregulation of catalase (KatA), which degrades H_2O_2 (the fight response), and second there is increased production of Dispersin B (DspB), an enzyme that releases *A. actinomycetemcomitans* cells from biofilms (the flight response)[34]. In vivo, these opposing responses are delicately balanced, and in the murine dual species abscess infection model, spatial analyses show that *A. actinomycetemcomitans* is optimally positioned at around 4 µm from *S. gordonii*, which allows cross-feeding but reduces exposure to inhibitory levels of H_2O_2[34]. Both responses are also necessary for virulence of *A. actinomycetemcomitans* in this model, and additionally growth of *S. gordonii* in abscesses is enhanced in the presence of *A. actinomycetemcomitans*[34]. Similar to *P. gingivalis*, therefore, *A. actinomycetemcomitans* has keystone pathogen properties in that it can impact the fitness of the community as a whole[34].

14.4 Interactions between bacterial communities and epithelial cells

Among the first host cells that encounter periodontal organisms are the gingival epithelial cells (GECs) that line the gingival crevice. GECs both form a physical barrier to microbial intrusion and constitute an interactive interface that senses organisms and signals their presence to the underlying cells of the immune system. The outcomes of the interaction between bacteria and GECs thus impact the early stages of both innate and acquired immunity. Many periodontal bacteria can actively internalize and survive within GECs, a location that provides a nutrient-rich, generally reducing environment that is partially protected from immune effector molecules[35]. Synergistic interactions among periodontal bacteria include co-operation to promote invasion. For example, consortia of *P. gingivalis* and *Fusobacterium nucleatum* invade GECs in higher number than either organism alone[36]. *F. alocis* and *P. gingivalis* also exhibit synergistic infection of epithelial cells, and dual species invasion elicits a distinct pattern of host cell responses[37].

The production of cytokines and chemokines by GECs, and consequent ordered recruitment of immune effector cells, in response to bacterial challenge is a major contributor to the maintenance of periodontal health. Disruption of the production, secretion or function of cytokines by one organism will provide blanket protection to the community as a whole, and is a significant feature of dysbiotic communities. Proteolytic organisms such as *P. gingivalis* or *T. denticola* can degrade cytokines and chemokines, along with other immune effector molecules, directly[38]. However, the dysbiotic impact of periodontal community members on the immune status of epithelial cells occurs on multiple levels (Figure 3). *P. gingivalis* induces what is known as a localized chemokine paralysis by preventing the production of neutrophil and T-cell chemokines, even in the presence of otherwise stimulatory organisms[39–41]. Inhibition of production of the neutrophil chemokine CXCL8 (IL-8) by *P. gingivalis* is effectuated by a serine phosphatase SerB, which specifically dephosphorylates the S536 residue on the p65 subunit of NF-kB. Dephosphorylation prevents nuclear translocation of p65 and transcription of the *IL8* gene which is primarily under the control of NF-kB comprised of p65 homodimers[42]. Similarly, endothelial cells are targeted by *P. gingivalis* whereby upregulation of E-selectin by other periodontal bacteria is diminished, thus impacting the leukocyte adhesion and transmigration aspects of neutrophil recruitment[43]. While this dysbiosis may be transient, it can have sustained consequences for the composition and virulence of the microbial community[2] and for host tissue damage, as a SerB deficient mutant of *P. gingivalis* induces less bone loss in experimental animals than the parental strain[44]. *P. gingivalis* also inhibits production of the T-cell chemokines CXCL10 (IP-10), CXCL9 (Mig) and CXCL11 (ITAC) from GECs, by diminishing signal transduction through Stat1 and IRF-1[45]. As with CXCL8, *P. gingivalis* also prevents production of these cytokines by pro-inflammatory organisms such as *F. nucleatum*. The potential biological consequences are that *P. gingivalis*

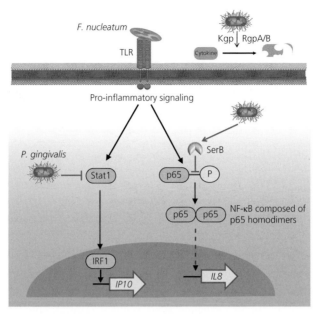

Figure 3 Cytokine disruption by *P. gingivalis*. Oral bacteria such as *F. nucleatum* engage TLRs on epithelial cell surfaces and activate pro-inflammatory signaling pathways. *P. gingivalis,* however, suppresses production of the neutrophil chemokine IL-8 (CXCL8) and the T-cell chemokine IP-10 (CXCL10) from epithelial cells. Mechanistically, invasive *P. gingivalis* inactivates Stat1 which in turn reduces expression of IP10 promoted by the IRF1 transcription factor. Intracellular *P. gingivalis* also secrete the serine phosphatase SerB, which dephosphorylates the serine 536 residue of the p65 NF-κB subunit and prevents the formation of p65 homodimers and consequently translocation of NF-κB into the nucleus. Transcription of the *IL8* gene, which is under the control of NF-κB, is reduced. The gingipain proteases Rgp A/B and Kgp, which are secreted by *P. gingivalis*, can degrade a number of cytokines and chemokines including IL-8. From Lamont and Hajishengallis, 2014 (Ref. 77) with permission.

may reduce the level Th1 development and allow Th17-mediated inflammation to flourish, consistent with the organism's ability to induce Th17-promoting cytokines (e.g., IL-6 and IL-23, but not the Th1-related IL-12) in antigen-presenting cells[46]. Th17 cells, which are abundant in periodontitis lesions, can function as an osteoclastogenic subset that links T-cell activation to inflammatory bone loss.

14.5 Manipulation of host immunity

In periodontitis, the diseased periodontal pockets become a "breeding ground" for periodontal bacteria despite the presence of viable leukocytes capable of eliciting immune responses, and of other defense mechanisms such as complement[47,48]. It can thus be inferred that the periodontitis-associated community can successfully escape host defenses. However, evasion of immune-mediated killing is not sufficient by itself to ensure the persistence of periodontal inflammophilic communities, the growth of which depend crucially on the presence of inflammatory tissue breakdown products, e.g., degraded collagen peptides and haem-containing

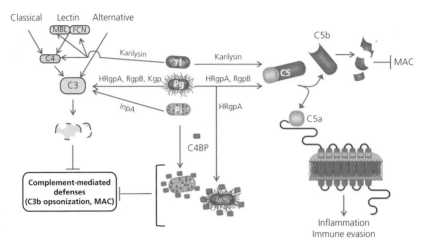

Figure 4 Synergistic inhibition of complement-dependent antimicrobial activities. *P. gingivalis, P. intermedia,* and *T. forsythia* can inhibit the classical, lectin, and alternative pathways of complement activation by degrading C3, C4, mannose-binding lectin (MBL), or ficolins (FCN) through the action of proteases, as indicated. Such activities are synergistic and prevent the deposition of C3b opsonin or the membrane attack complex (MAC) on the surface of these pathogens as well as bystander bacterial species. Moreover, *P. gingivalis* and *P. intermedia* also protect themselves against complement also by using surface molecules (HRgpA gingipain for *P. gingivalis,* an undefined molecule for *P. intermedia*) to capture the circulating C4b-binding protein (C4BP), a physiological negative regulator of the classical and lectin pathways. Furthermore, *P. gingivalis* (via its Arg-specific gingipains HRgpA and RgpB) and *T. forsythia* (via its karilysin) can release biologically active C5a from C5, thereby stimulating inflammation through the activation of the C5a receptor (C5aR). This receptor is also exploited by the bacteria for immune evasion of leukocyte killing. InpA, interpain A; Kgp, Lys-specific gingipain. From Lamont and Hajishengallis, 2014 (Ref. 77) with permission.

compounds[5]. In other words, if the periodontal bacteria resorted to immunosuppression, a common evasion strategy of many pathogens[49], this would inhibit bacterial killing, but — at the same time — would deprive the bacteria of the "inflammatory spoils" required for their survival[5]. As outlined below, periodontal bacteria can subvert the host response in ways that interfere with bacterial killing while promoting inflammation. Whereas several bacteria are capable of manipulating the host response *in vitro, in vivo* supporting evidence is perhaps strongest in the case of *P. gingivalis*[3], which as discussed above is a well-documented keystone pathogen.

  P. gingivalis and certain other periodontitis-associated bacteria can suppress complement activation through the action of specific proteolytic enzymes. The gingipains of *P. gingivalis*[50], interpain A of *Prevotella intermedia*[51], and karilysin of *Tannerella forsythia*[52] can all block complement activation by degrading the central complement component C3, or upstream components of the complement cascade such as the pattern-recognition molecules mannose-binding lectin and ficolins (Figure 4**)**. These inhibitory mechanisms block the deposition of opsonins or the membrane attack complex on surfaces of the bacteria. Importantly, the gingipains act synergistically with karilysin or interpain A and can thus inhibit complement even at low concentrations[50–52]. This suggests that the proteases could interfere with complement activation even after their release and diffusion within the microbial community,

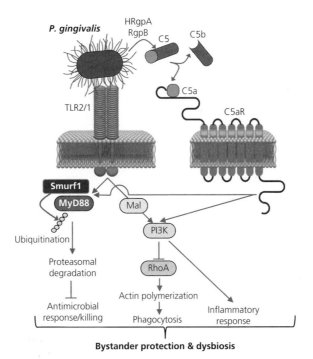

Figure 5 Subversion of neutrophil function and dysbiosis. *P. gingivalis* co-activates TLR2 and C5aR in neutrophils. and the resulting crosstalk leads to E3 ubiquitin ligase Smurf1-dependent ubiquitination and proteasomal degradation of MyD88, thereby inhibiting a host-protective antimicrobial response. Moreover, the C5aR-TLR2 crosstalk activates PI3K, which prevents phagocytosis through inhibition of RhoA activation and actin polymerization, while stimulating an inflammatory response. In contrast to MyD88, another TLR2 adaptor, Mal, is involved in the subversive pathway and acts upstream of PI3K. The integrated mechanism provides "bystander" protection to otherwise susceptible bacterial species and promotes polymicrobial dysbiotic inflammation *in vivo*. From Maekawa *et al.*, 2014 (Ref. 59) with permission. (*see color plate section for color details*).

thereby promoting the survival of otherwise susceptible bystander bacterial species. Moreover, *P. gingivalis* and *P. intermedia* can hijack physiological soluble inhibitors of the complement cascade, such as the C4b-binding protein[53,54] (Figure 4), whereas *T. denticola* expresses an 11.4-kDa cell surface lipoprotein which binds factor H, also a major soluble physiological inhibitor of complement[55]. These mechanisms provide an additional layer of protection against complement.

Notably, even under conditions that suppress the canonical activation of the complement cascade, *P. gingivalis* Arg-specific gingipains (HRgpA, RgpB) can act directly on C5 to release biologically active C5a[50,56]. This has been a puzzling finding, since C5a is a potent activator of phagocytes[57]. However, more recent studies have shown that *P. gingivalis* exploits the C5a receptor (C5aR; CD88) to subvert the function of leukocytes[58,59]. This subversive action requires a cross-talk with Toll-like receptor 2 (TLR2), which is involved in innate immune recognition of *P. gingivalis*[60]. The key experiments underpinning the molecular strategy by which *P. gingivalis* instigates a subversive C5aR-TLR2 crosstalk in neutrophils to disengage bacterial clearance from inflammation (Figure 5) are briefly outlined below.

In the murine chamber model that allows the quantitative assessment of bacterium–neutrophil interactions, mice deficient in either C5aR (*C5ar*[-/-]) or TLR2 (*Tlr2*[-/-]) could effectively kill *P. gingivalis*, which however persisted in wild-type mice[59]. Inhibition of either C5aR or TLR2 in wild-type mice resulted in effective killing of *P. gingivalis* by neutrophils, suggesting that these receptors engage in cooperative crosstalk to prevent bacterial killing. In contrast to *P. gingivalis*, *F. nucleatum* was susceptible to neutrophil killing in wild-type mice. Intriguingly, however, the survival of *F. nucleatum* was dramatically enhanced in the presence of *P. gingivalis*, unless C5aR or TLR2 signaling was blocked[59]. Consistent with these findings, pharmacological inhibition of either C5aR or TLR2 in *P. gingivalis*-colonized mice eliminates *P. gingivalis* from the periodontal tissue, reverses the increase in total microbiota counts induced by *P. gingivalis* colonization, and blocks periodontal inflammation. In signalling experiments using both human and mouse neutrophils, it was shown that *P. gingivalis*-induced C5aR-TLR2 crosstalk causes ubiquitination and proteasomal degradation of the TLR2 signalling adaptor MyD88, thereby suppressing its antimicrobial effects[59] (Figure 5). Moreover, the C5aR-TLR2 crosstalk activates an alternative TLR2 pathway in which the adaptor Mal induces phosphoinositide 3-kinase (PI3K) signalling, which in turn blocks GTPase RhoA-dependent actin polymerization and hence the phagocytosis of *P. gingivalis* and bystander bacteria[59]. In summary, *P. gingivalis* manipulates neutrophils through two distinct mechanisms that together ensure bacterial survival and the perpetuation of inflammation.

Similarly to its subversion of neutrophils, *P. gingivalis* requires intact C5aR and TLR2 to manipulate the function of macrophages. However, the signaling mechanisms underlying the exploitation of C5aR and TLR2 in macrophages are entirely different[58]. In macrophages, the *P. gingivalis*-induced C5aR-TLR2 crosstalk leads to synergistic activation of cAMP-dependent protein kinase A, which in turn inhibits nitric oxide-dependent intracellular killing[58]. Maximal activation of the cAMP-dependent protein kinase A requires the participation of the chemokine receptor CXCR4, which is engaged by the pathogen's FimA fimbriae. In contrast, the inhibition of this cAMP pathway in neutrophils has no effect on *P. gingivalis* survival[59]. Consistent with the notion discussed earlier that *P. gingivalis* blocks innate defenses but not inflammation, the *P. gingivalis*-induced C5aR-TLR2-CXCR4 crosstalk impairs the killing capacity of macrophages without significantly affecting their inflammatory responses (TNF, IL-1β, and IL-6)[56].

Besides TLR2, *P. gingivalis* can interact with TLR4 by means of its LPS, although in a rather unusual way as compared to most other Gram-negative bacteria in the oral cavity or elsewhere. *P. gingivalis* can enzymatically modify the lipid A moiety of its LPS to either evade or antagonize TLR4 activation, in contrast to the classical LPS which is a potent TLR4 agonist[43,61]. The shifting of the lipid A activity from TLR4-evasive to TLR4-suppressive is dependent upon lipid A phosphatase activity[62], which was shown to be critical for the capacity of *P. gingivalis* to colonize and increase the total periodontal bacterial load in a rabbit model of ligature-induced periodontitis[63]. Interestingly, *P. gingivalis* deficient in lipid A phosphatase activity not only fails to avoid recognition by TLR4 but is rendered more susceptible to cationic antimicrobial peptide killing, due to associated changes in its outer surface that promote the binding of the cationic antimicrobial peptides[62,64]. In macrophage experiments using wild-type *P. gingivalis* and isogenic phosphatase mutants with a "locked" lipid A profile, the expression of evasive or antagonistic lipid A was associated with suppression of the non-canonical inflammasome and increased bacterial survival as compared to the expression of TLR4 agonist lipid A[65]. In this regard, proper activation of the non-canonical inflammasome leads to pyroptosis, a proinflammatory mechanism of lytic cell death that protects the host against bacterial infection[66]. The non-canonical mode of inflammasome activation, which requires the participation of caspase-11 and can be triggered by the LPS of gram-negative bacteria, is thought to discriminate cytosolic from vacuolar bacteria that can be detected by the canonical inflammasome, which comprises the NLRP3 protein (NOD-like receptor family pyrin domain-containing-3), the adaptor molecule ASC (Apoptosis-associated speck-like protein containing CARD), and caspase-1[66]. In the non-canonical mechanism, cytosolic LPS directly binds and activates caspase 11[67]. LPS with antagonistic lipid A can also bind caspase 11 but fails to induce oligomerization which is required for activation[67]. In addition to blocking the non-canonical pathway, *P. gingivalis* can inhibit canonical inflammasome activation by other bacteria. Indeed, *P. gingivalis* was shown to suppress endocytic events required for *F. nucleatum*-induced NLRP3 inflammasome activation in macrophages[68]. This mechanism may therefore potentially promote the virulence of the entire microbial community.

The opsonization of bacteria with complement fragments (*e.g.*, with C3b opsonin) or specific antibodies can enhance the phagocytic killing of bacteria. This mechanism might be counteracted by *P. gingivalis* which, at least *in vitro*, can effectively degrade both C3 and IgG[50,69–72]. Conversely, hyperimmune rabbit sera to *P. gingivalis* proteases reverses their capacity to degrade C3 and IgG and compromises the resistance of *P. gingivalis* to neutrophil killing[69]. In principle, therefore, specific antibodies induced in the course of periodontitis against *P. gingivalis* could promote its killing, although periodontitis progresses despite the presence of specific antibodies to *P. gingivalis* and other periodontal bacteria[73]. This is probably because naturally induced antibodies to periodontal bacteria are of low affinity and poor functionality[73,74], perhaps attributed to the capacity of periodontal bacteria to interfere also with adaptive immunity. In this regard, the recruitment and development of T helper subsets (which can regulate the function of B and plasma cells) appears to be subverted by periodontal bacteria[40,46,75]. For instance, *P. gingivalis* causes suboptimal activation of dendritic cells, leading to T cell responses that are thought unlikely to provide host protection[76].

14.6 Conclusions

The literature discussed in this chapter provides compelling evidence that no single microbial species (nor even a select few species) is sufficiently pathogenic on its own to cause periodontitis. The orchestration of inflammatory bone loss involves the participation of keystone pathogens, which subvert the host response to disrupt homeostasis, and pathobionts that over-activate the dysfunctional host response, leading to dysbiotic inflammation. If a keystone pathogen is a conductor in this orchestra, the pathobionts are its musicians. According to the PSD model, moreover, dysbiosis is crucially dependent not on the particular microbial roster but rather on the specific gene combinations or collective virulence activity within the dysbiotic microbial community[77]. The situation has been likened to that of a rowing crew. "All the oars need to be manned but the identities of individual crew members, provided they are capable of rowing, are not important for forward progression. Some functions, however, such as cox, are more important for coordinating characteristics such as direction and speed[1]." In support of this notion, a recent metatranscriptomic analysis has shown that periodontitis-associated microbial communities display conserved metabolic and virulence gene expression profiles, although the composition of these communities is highly variable among patients[78].

Periodontitis can also be seen as a side-effect of survival and nutritional strategies of dysbiotic communities. From a microbial standpoint, the proactive induction of periodontal inflammation is not a potentially destructive event but a vital force. Indeed, inflammation can drive the selection and enrichment of "inflammophilic" communities by providing proteinaceous and haem-containing substrates derived from inflammatory tissue breakdown. In contrast, those species that cannot benefit from the altered ecological conditions of the inflammatory environment, or for which host inflammation is detrimental, are likely to be outcompeted and perhaps represent species associated with periodontal health[3]. Therefore, inflammation and dysbiosis engage in a positive feedback loop in which each reinforces each other.

The PSD model is also consistent with a novel approach for the treatment of periodontitis. Specifically, it predicts that the most promising strategies should be those aiming to restore homeostasis rather than to perform "infection control" through antimicrobials. One such option is community manipulation, e.g., to favor the growth of organisms that are antagonistic to *P. gingivalis*[79,80] and/or to reduce the levels or activity of its accessory pathogens, such as *S. gordonii*. Both approaches would be expected to prevent or reverse the transition to a dysbiotic community. Another potentially effective strategy involves targeted modulation of the host response to block destructive inflammation and counteract the microbial subversive strategies that promote dysbiosis. Community manipulation and host modulation strategies have been successful in preclinical models of periodontitis[31,81–83], thereby fueling reasonable optimism for promising clinical trials in the near future.

TAKE-HOME MESSAGE

- Polymicrobial synergy and dysbiosis drive periodontitis in a susceptible host.
- Dysbiosis involves specialized accessory and keystone pathogens and pathobionts.
- Microbial immune subversion is central to the persistence of dysbiotic communities.
- The dysbiotic microbiota sustains itself by feasting on the "inflammatory spoils."

References

1 Hajishengallis G, Lamont RJ. Beyond the red complex and into more complexity: The Polymicrobial Synergy and Dysbiosis (PSD) model of periodontal disease etiology. *Mol Oral Microbiol* 2012; **27**: 409–19.

2 Hajishengallis G, Liang S, Payne MA *et al.* Low-abundance biofilm species orchestrates inflammatory periodontal disease through the commensal microbiota and complement. *Cell Host Microbe* 2011; **10**: 497–506.

3 Hajishengallis G, Lamont RJ. Breaking bad: Manipulation of the host response by *Porphyromonas gingivalis*. *Eur J Immunol* 2014; **44**: 328–38.

4 Whitmore SE, Lamont RJ. The pathogenic persona of community-associated oral streptococci. *Mol Microbiol* 2011; **81**: 305–1.

5 Hajishengallis G. Immunomicrobial pathogenesis of periodontitis: keystones, pathobionts, and host response. *Trends Immunol* 2014; **35**: 3–11.

6 Petersen C, Round JL. Defining dysbiosis and its influence on host immunity and disease. *Cell Microbiol* 2014; **16**: 1024–33.

7 Short FL, Murdoch SL, Ryan RP. Polybacterial human disease: the ills of social networking. *Trends Microbiol* 2014; **22**: 508–16.

8 Byrne SJ, Dashper SG, Darby IB *et al.* Progression of chronic periodontitis can be predicted by the levels of Porphyromonas gingivalis and Treponema denticola in subgingival plaque. *Oral Microbiol Immunol* 2009; **24**: 469–77.

9 Socransky SS, Haffajee AD. Periodontal microbial ecology. *Periodontol 2000* 2005; **38**: 135–87.

10 Socransky SS, Haffajee AD, Cugini MA *et al.* Microbial complexes in subgingival plaque. *J Clin Periodontol* 1998; **25**: 134–44.

11 Kumar PS, Leys EJ, Bryk JM *et al.* Changes in periodontal health status are associated with bacterial community shifts as assessed by quantitative 16S cloning and sequencing. *J Clin Microbiol* 2006; **44**: 3665–73.

12 Dewhirst FE, Chen T, Izard J *et al.* The human oral microbiome. *J Bacteriol* 2010; **192**: 5002–17.

13 Griffen AL, Beall CJ, Campbell JH *et al.* Distinct and complex bacterial profiles in human periodontitis and health revealed by 16S pyrosequencing. *ISME J* 2012; **6**: 1176–85.

14 Abusleme L, Dupuy AK, Dutzan N *et al.* The subgingival microbiome in health and periodontitis and its relationship with community biomass and inflammation. *ISME J* 2013; **7**: 1016–25.

15 Kumar PS, Griffen AL, Moeschberger ML *et al.* Identification of candidate periodontal pathogens and beneficial species by quantitative 16S clonal analysis. *J Clin Microbiol* 2005; **43**: 3944–55.

16 Moore WE, Holdeman LV, Smibert RM *et al.* Bacteriology of severe periodontitis in young adult humans. *Infect Immun* 1982; **38**: 1137–48.

17 Hajishengallis G, Liang S, Payne MA *et al.* Low-abundance biofilm species orchestrates inflammatory periodontal disease through the commensal microbiota and complement. *Cell Host Microbe* 2011; **10**: 497–506.

18 Jiao Y, Darzi Y, Tawaratsumida K *et al.* Induction of bone loss by pathobiont-mediated Nod1 signaling in the oral cavity. *Cell Host Microbe* 2013; **13**: 595–601.

19 Kesavalu L, Sathishkumar S, Bakthavatchalu V *et al.* Rat model of polymicrobial infection, immunity, and alveolar bone resorption in periodontal disease. *Infect Immun* 2007; **75**: 1704–12.

20 Grenier D. Nutritional interactions between two suspected periodontopathogens, Treponema denticola and Porphyromonas gingivalis. *Infect Immun* 1992; **60**: 5298–301.

21 Tan KH, Seers CA, Dashper SG *et al.* Porphyromonas gingivalis and Treponema denticola exhibit metabolic symbioses. *PLoS Pathog* 2014; **10**: e1003955.

22 Kuboniwa M, Hendrickson EL, Xia Q *et al.* Proteomics of Porphyromonas gingivalis within a model oral microbial community. *BMC Microbiol* 2009; **9**: 98.

23 Frias-Lopez J, Duran-Pinedo A. Effect of periodontal pathogens on the metatranscriptome of a healthy multispecies biofilm model. *J Bacteriol* 2012; **194**: 2082–95.

24 Wright CJ, Burns LH, Jack AA *et al.* Microbial interactions in building of communities. *Mol Oral Microbiol* 2013; **28**: 83–101.

25 Jenkinson HF, Lamont RJ. Oral microbial communities in sickness and in health. *Trends Microbiol* 2005; **13**: 589–95.

26 Kuboniwa M, Lamont RJ. Subgingival biofilm formation. *Periodontol 2000* 2010; **52**: 38–52.

27 Maeda K, Tribble GD, Tucker CM *et al.* A Porphyromonas gingivalis tyrosine phosphatase is a multifunctional regulator of virulence attributes. *Mol Microbiol* 2008; **69**: 1153–64.

28 Chawla A, Hirano T, Bainbridge BW *et al.* Community signalling between Streptococcus gordonii and Porphyromonas gingivalis is controlled by the transcriptional regulator CdhR. *Mol Microbiol* 2010; **78**: 1510–22.

29 Kuboniwa M, Tribble GD, James CE *et al.* Streptococcus gordonii utilizes several distinct gene functions to recruit Porphyromonas gingivalis into a mixed community. *Mol Microbiol* 2006; **60**: 121–39.

30 Wright CJ, Xue P, Hirano T *et al.* Characterization of a bacterial tyrosine kinase in Porphyromonas gingivalis involved in polymicrobial synergy. *Microbiologyopen* 2014.

31 Daep CA, Novak EA, Lamont RJ *et al.* Structural dissection and in vivo effectiveness of a peptide inhibitor of Porphyromonas gingivalis adherence to Streptococcus gordonii. *Infect Immun* 2011; **79**: 67–74.

32 Ramsey MM, Rumbaugh KP, Whiteley M. Metabolite cross-feeding enhances virulence in a model polymicrobial infection. *PLoS Pathog* 2011; **7**: e1002012.

33 Ramsey MM, Whiteley M. Polymicrobial interactions stimulate resistance to host innate immunity through metabolite perception. *Proc Natl Acad Sci U S A* 2009; **106**: 1578–83.

34 Stacy A, Everett J, Jorth P *et al.* Bacterial fight-and-flight responses enhance virulence in a polymicrobial infection. *Proc Natl Acad Sci U S A* 2014; **111**: 7819–24.

35 Tribble GD, Lamont RJ. Bacterial invasion of epithelial cells and spreading in periodontal tissue. *Periodontol 2000* 2010; **52**: 68–83.

36 Saito A, Kokubu E, Inagaki S *et al.* Porphyromonas gingivalis entry into gingival epithelial cells modulated by Fusobacterium nucleatum is dependent on lipid rafts. *Microb Pathog* 2012; **53**: 234–42.

37 Aruni W, Chioma O, Fletcher HM. Filifactor alocis: The newly discovered kid on the block with special talents. *J Dent Res* 2014; **93**: 725–32.

38 Potempa J, Banbula A, Travis J. Role of bacterial proteinases in matrix destruction and modulation of host responses. *Periodontol 2000* 2000; **24**: 153–92.

39 Darveau RP, Belton CM, Reife RA *et al.* Local chemokine paralysis, a novel pathogenic mechanism for *Porphyromonas gingivalis*. *Infect Immun* 1998; **66**: 1660–5.

40 Jauregui CE, Wang Q, Wright CJ *et al.* Suppression of T-cell chemokines by *Porphyromonas gingivalis*. *Infect Immun* 2013; **81**: 2288–95.

41 Bostanci N, Belibasakis GN. Porphyromonas gingivalis: an invasive and evasive opportunistic oral pathogen. *FEMS Microbiol Lett* 2012; **333**: 1–9.

42 Takeuchi H, Hirano T, Whitmore SE *et al.* The serine phosphatase SerB of Porphyromonas gingivalis suppresses IL-8 production by dephosphorylation of NF-kappaB RelA/p65. *PLoS Pathog* 2013; **9**: e1003326.

43 Darveau RP. Periodontitis: a polymicrobial disruption of host homeostasis. *Nat Rev Microbiol* 2010; **8**: 481–90.

44 Bainbridge B, Verma RK, Eastman C *et al.* Role of Porphyromonas gingivalis phosphoserine phosphatase enzyme SerB in inflammation, immune response, and induction of alveolar bone resorption in rats. *Infect Immun* 2010; **78**: 4560–9.

45 Jauregui CE, Wang Q, Wright CJ *et al.* Suppression of T-cell chemokines by Porphyromonas gingivalis. *Infect Immun* 2013; **81**: 2288–95.

46 Moutsopoulos NM, Kling HM, Angelov N *et al.* *Porphyromonas gingivalis* promotes Th17 inducing pathways in chronic periodontitis. *J Autoimmun* 2012; **39**: 294–303.

47 Ryder MI. Comparison of neutrophil functions in aggressive and chronic periodontitis. *Periodontol 2000* 2010; **53**: 124–37.

48 Hajishengallis G. Complement and periodontitis. *Biochem Pharmacol* 2010; **80**: 1992–2001.

49 Cyktor JC, Turner J. Interleukin-10 and immunity against prokaryotic and eukaryotic intracellular pathogens. *Infect Immun* 2011; **79**: 2964–73.

50 Popadiak K, Potempa J, Riesbeck K *et al.* Biphasic effect of gingipains from *Porphyromonas gingivalis* on the human complement system. *J Immunol* 2007; **178**: 7242–50.

51 Potempa M, Potempa J, Kantyka T *et al.* Interpain A, a cysteine proteinase from *Prevotella intermedia*, inhibits complement by degrading complement factor C3. *PLoS Pathog* 2009; **5**: e1000316.

52 Jusko M, Potempa J, Karim AY *et al.* A metalloproteinase karilysin present in the majority of *Tannerella forsythia* isolates inhibits all pathways of the complement system. *J Immunol* 2012; **188**: 2338–49.

53 Potempa M, Potempa J, Okroj M *et al.* Binding of complement inhibitor C4b-binding protein contributes to serum resistance of *Porphyromonas gingivalis. J Immunol* 2008; **181**: 5537–44.

54 Malm S, Jusko M, Eick S *et al.* Acquisition of complement inhibitor serine protease factor I and its cofactors C4b-binding protein and factor H by *Prevotella intermedia. PLoS One* 2012; **7**: e34852.

55 McDowell JV, Huang B, Fenno JC *et al.* Analysis of a unique interaction between the complement regulatory protein factor H and the periodontal pathogen *Treponema denticola. Infect Immun* 2009; **77**: 1417–25.

56 Liang S, Krauss JL, Domon H *et al.* The C5a receptor impairs IL-12-dependent clearance of *Porphyromonas gingivalis* and is required for induction of periodontal bone loss. *J Immunol* 2011; **186**: 869–77.

57 Ward PA. Functions of C5a receptors. *J Mol Med* 2009; **87**: 375–8.

58 Wang M, Krauss JL, Domon H *et al.* Microbial hijacking of complement-Toll-like receptor crosstalk. *Sci Signal* 2010; **3**: ra11.

59 Maekawa T, Krauss JL, Abe T *et al. Porphyromonas gingivalis* manipulates complement and TLR signaling to uncouple bacterial clearance from inflammation and promote dysbiosis. *Cell Host Microbe* 2014; **15**: 768–78.

60 Burns E, Bachrach G, Shapira L *et al.* Cutting Edge: TLR2 is required for the innate response to *Porphyromonas gingivalis*: Activation leads to bacterial persistence and TLR2 deficiency attenuates induced alveolar bone resorption. *J Immunol* 2006; **177**: 8296–300.

61 Darveau RP, Hajishengallis G, Curtis MA. *Porphyromonas gingivalis* as a potential community activist for disease. *J Dent Res* 2012; **91**: 816–20.

62 Coats SR, Jones JW, Do CT *et al.* Human Toll-like receptor 4 responses to *P. gingivalis* are regulated by lipid A 1- and 4′- phosphatase activities. *Cell Microbiol* 2009; **11**: 1587–99.

63 Zenobia C, Hasturk H, Nguyen D *et al. Porphyromonas gingivalis* lipid A phosphatase activity is critical for colonization and increasing the commensal load in the rabbit ligature model. *Infect Immun* 2014; **82**: 650–9.

64 Curtis MA, Percival RS, Devine D *et al.* Temperature dependent modulation of *Porphyromonas gingivalis* lipid A structure and interaction with the innate host defences. *Infect Immun* 2011; **79**: 1187–93.

65 Slocum C, Coats SR, Hua N *et al.* Distinct lipid A moieties contribute to pathogen-induced site-specific vascular inflammation. *PLoS Pathog* 2014; **10**: e1004215.

66 Lamkanfi M, Dixit VM. Mechanisms and functions of inflammasomes. *Cell* 2014; **157**: 1013–22.

67 Shi J, Zhao Y, Wang Y *et al.* Inflammatory caspases are innate immune receptors for intracellular LPS. *Nature* 2014; **514**: 187–92.

68 Taxman DJ, Swanson KV, Broglie PM *et al. Porphyromonas gingivalis* mediates inflammasome repression in polymicrobial cultures through a novel mechanism involving reduced endocytosis. *J Biol Chem* 2012; **287**: 32791–9.

69 Cutler CW, Arnold RR, Schenkein HA. Inhibition of C3 and IgG proteolysis enhances phagocytosis of Porphyromonas gingivalis. *J Immunol* 1993; **151**: 7016–29.

70 Sundqvist G, Carlsson J, Herrmann B *et al.* Degradation of human immunoglobulins G and M and complement factors C3 and C5 by black-pigmented Bacteroides. *J Med Microbiol* 1985; **19**: 85–94.

71 Grenier D, Mayrand D, McBride BC. Further studies on the degradation of immunoglobulins by black-pigmented Bacteroides. *Oral Microbiol Immunol* 1989; **4**: 12–8.

72 Vincents B, Guentsch A, Kostolowska D *et al.* Cleavage of IgG1 and IgG3 by gingipain K from Porphyromonas gingivalis may compromise host defense in progressive periodontitis. *FASEB J* 2011; **25**: 3741–50.

73 Gemmell E, Yamazaki K, Seymour GJ. The role of T cells in periodontal disease: homeostasis and autoimmunity. *Periodontol 2000* 2007; **43**: 14–40.

74 Schenkein HA. Host responses in maintaining periodontal health and determining periodontal disease. *Periodontol 2000* 2006; **40**: 77–93.

75 Gaddis DE, Maynard CL, Weaver CT *et al.* Role of TLR2 dependent-IL-10 production in the inhibition of the initial IFN-γ T cell response to *Porphyromonas gingivalis. J Leukoc Biol* 2012; **93**: 21–31.

76 Zeituni AE, Jotwani R, Carrion J *et al.* Targeting of DC-SIGN on human dendritic cells by minor fimbriated *Porphyromonas gingivalis* strains elicits a distinct effector T cell response. *J Immunol* 2009; **183**: 5694–704.

77 Lamont RJ, Hajishengallis G. Polymicrobial synergy and dysbiosis in inflammatory disease. *Trends Mol Med* 2015; **21**: 172–183.

78 Jorth P, Turner KH, Gumus P *et al.* Metatranscriptomics of the human oral microbiome during health and disease. *MBio* 2014; **5**: e01012–14.

79 Wang BY, Wu J, Lamont RJ *et al.* Negative correlation of distributions of Streptococcus cristatus and Porphyromonas gingivalis in subgingival plaque. *J Clin Microbiol* 2009; **47**: 3902–6.

80 Xie H, Hong J, Sharma A *et al.* Streptococcus cristatus ArcA interferes with Porphyromonas gingivalis pathogenicity in mice. *J Periodontal Res* 2012; **47**: 578–83.

81 Maekawa T, Abe T, Hajishengallis E *et al.* Genetic and intervention studies implicating complement C3 as a major target for the treatment of periodontitis. *J Immunol* 2014; **192** 6020–7.

82 Glowacki AJ, Yoshizawa S, Jhunjhunwala S *et al.* Prevention of inflammation-mediated bone loss in murine and canine periodontal disease via recruitment of regulatory lymphocytes. *Proc Natl Acad Sci U S A* 2013; **110**: 18525–30.

83 Hasturk H, Kantarci A, Van Dyke TE. Paradigm shift in the pharmacological management of periodontal diseases. *Front Oral Biol* 2012; **15**: 160–76.

CHAPTER 15

New paradigm in the relationship between periodontal disease and systemic diseases: effects of oral bacteria on the gut microbiota and metabolism

Kazuhisa Yamazaki

Division of Oral Science for Health Promotion, Niigata University Graduate School of Medical and Dental Sciences, Niigata, Japan

15.1 Introduction

A large variety of microbes colonize our body surfaces, and our oral cavity and gut lumen are no exceptions[1]. These microbes are called commensals since the interactions between bacteria and epithelial surfaces are important for homeostasis. Accordingly, disruption in the balance of commensal microbes (dysbiosis) can result in a variety of metabolic and autoimmune diseases[2].

Periodontal disease is a chronic inflammatory disease likely resulting from dysbiosis of the oral microbiota. Epidemiological studies indicate its association with increased risk of various diseases such as diabetes, atherosclerotic vascular diseases and rheumatoid arthritis[3] (Figure 1). It is possible that common disease susceptibilities and risk factors could explain the association between these diseases. Although direct evidence is still lacking, causal mechanisms proposed include endotoxemia, proinflammatory cytokines and molecular mimicry[4].

Interestingly, the diseases reported as affected by periodontal disease are often described in association with dysbiosis of the gut microbiota. Under physiological conditions, bacteria in the intestine are commensal and mediate food digestion, strengthen the immune system, and prevent pathogens from invading tissues and organs. However, once dysbiosis occurs and detrimental bacteria become predominant, noxious agents such as bacterial toxins and metabolites damage the gut epithelial wall. These are then absorbed into the systemic circulation through the disrupted epithelium, resulting in impairment of various tissues and organs such as the liver, heart, kidney, pancreas and blood vessels[5].

Orally administered live bacteria have been shown to affect the composition of the gut microbiota, as evidenced by oral probiotics[6]. The compositions of the oral microbiota and gut microbiota are taxonomically distinct[7]. This suggests that if sufficient numbers of oral bacteria reach the intestine, these bacteria could affect

The Human Microbiota and Chronic Disease: Dysbiosis as a Cause of Human Pathology, First Edition.
Edited by Luigi Nibali and Brian Henderson.
© 2016 John Wiley & Sons, Inc. Published 2016 by John Wiley & Sons, Inc.

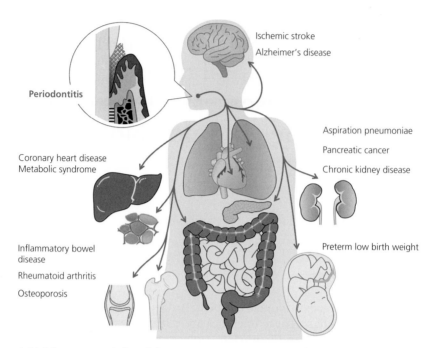

Figure 1 Link between periodontal disease and systemic diseases. Clinical and epidemiological studies strongly suggest that periodontal disease increases the risk of the various diseases in the figure.

the composition of the gut microbiota. The alteration of the gut microbiota by oral bacteria is an interesting supposition that could partially explain causal mechanisms of systemic diseases resulting from periodontal diseases. In this chapter, the author considers our recent observations and traditional mechanistic hypotheses to inform a discussion on systemic diseases potentially associated with both periodontal disease and gut microbiota dysbiosis.

15.2 Association between periodontal and systemic diseases

15.2.1 Periodontal disease and diabetes

A small body of evidence supports significant adverse effects of periodontal disease on glycemic control, diabetes complications, and development of type 2 diabetes[8]. A recent meta-analysis of five studies reported a significant weighted mean reduction in HbA1c of 0.4% over a follow-up period of 3–9 months after periodontal therapy, despite heterogeneous data[9].

The mechanisms by which periodontal disease induces insulin resistance have yet to be determined. TNF-α is a well-known cytokine that inhibits insulin signalling[10,11]. However, inconsistent data have been presented regarding the levels of TNF-α in gingival crevicular fluids. Several studies reported increased levels of TNF-α[12], whereas others found no difference between periodontitis patients and healthy subjects[13]. Few reports have demonstrated elevated levels of TNF-α in the blood of patients with periodontal disease[14]. In contrast, increased levels of serum IL-6 in periodontitis patients are well documented[15]. IL-6 induces CRP synthesis in the liver, and both IL-6 and CRP are known to induce insulin resistance[16]. Elevated levels of both may account for insulin resistance in periodontitis patients.

In an animal model, experimental periodontitis induced by ligature placement has been demonstrated to induce glucose intolerance in a Zucker rat strain exhibiting obesity, hyperlipidemia, and hyperinsulinemia due to a leptin gene mutation[17]. However, no difference in blood level TNF-α, a well-known cytokine inhibiting insulin signalling, was seen dependent on experimental periodontitis. In contrast, oral administration of *P. gingivalis* (strain W50) in Tallyho/JngJ mice (type 2 diabetes) and streptozotocin-induced diabetic C57BL/6J mice (type 1 diabetes) demonstrated significantly higher serum TNF-α levels compared with sham administration, although no difference in glucose intolerance was observed[18]. As such, the role of TNF-α in insulin resistance during periodontal infection remains to be determined.

15.2.2 Periodontal disease and atherosclerotic vascular diseases

A number of epidemiological studies have demonstrated that periodontal disease is associated with increased incidence and fatality rates of atherosclerotic vascular diseases[19]. Subsequent systematic reviews and meta-analyses have revealed a weak but positive correlation between periodontal diseases and coronary heart disease even after adjusting for risk factors such as smoking, diabetes, alcohol consumption, obesity, and blood pressure[19–22]. A comprehensive review by a working group from the American Heart Association also concluded that periodontal disease is associated with atherosclerotic vascular disease independent of known confounders. However, the working group further concluded that there was no evidence for a causal link[23].

Endothelial dysfunction is the earliest indicator of cardiovascular disease, and its severity may be modulated by the intensity of systemic inflammation. This inflamed state can be evaluated by measuring different markers such as high-sensitivity C-reactive protein, tumor necrosis factor-α (TNF-α), and interleukin-6 (IL-6), which have also been reported at elevated levels in patients with periodontitis. An association between infection and atherosclerotic vascular disease has long been suggested. A number of studies have shown the presence of periodontopathic bacterial DNA in arteries, suggesting the involvement of periodontal disease in atherosclerotic vascular diseases[24–31]. However, a recent study demonstrated that a higher load and mean diversity of bacteria were detected in vascular biopsies from patients with vascular diseases with chronic periodontitis (CP) compared with those without CP. It is of note that gut microbes were frequently identified in patients with CP in addition to some known oral bacterial taxa. The authors speculated that low-grade systemic inflammation caused by periodontal disease may affect intestinal permeability and thereby translocation of enteric bacteria[32]. Several murine animal studies aimed at clarifying the effects of periodontopathic bacterial infection on atherogenesis have documented atheromatous plaque formation and systemic inflammatory marker elevation[33–35].

15.2.3 Periodontal disease and rheumatoid arthritis

An association between periodontal disease and rheumatoid arthritis has recently been suggested. Epidemiological studies have demonstrated a higher prevalence and severity of periodontitis in patients with rheumatoid arthritis (RA)[36]. Periodontal disease and RA share many aspects in terms of their etiology and pathogenesis: genetic predispositions, such as polymorphisms in the IL-1 gene and HLA-DRB1 allele, and environmental factors, such as smoking, are associated with increased risk of both diseases[37–42]. Therefore, rather than a causal relation between the two diseases, it could be that common risk factors manifest through different diseases. While a few reports contrarily assert that periodontal disease could indeed be a risk factor for RA[43,44] mechanisms underlying the relationship between the two diseases have not been clarified.

It is now clear that the majority of RA cases are triggered by an autoimmune response to citrullinated proteins (this is discussed further in chapter 18 by Detert). These proteins are generated by endogenous peptidyl arginine deiminase (PAD) under physiological conditions, but the loss of tolerance in genetically susceptible individuals drives the generation of autoantibodies against citrullinated proteins in the synovia[45–47]. *Porphyromonas gingivalis* is the only known periodontopathic bacterium that expresses a bacterial PAD[48]. Highlighting the importance of this connection, in one study a subset of early RA patients demonstrated positive *P. gingivalis* antibody responses, and these responses correlated with anti-cyclic citrullinated peptide antibody reactivity[43]. However, another study that analysed new-onset RA patients showed no correlation with the presence of *P. gingivalis* in periodontal pockets and anti-citrullinated protein antibody titres, despite the fact that more advanced forms of periodontitis were already present in the patients[44]. Furthermore, the suggestion that PAD is the link between RA and periodontal disease has been disputed[49]. Thus, the role of *P. gingivalis* PAD in the pathogenesis of human RA remains elusive.

An animal model study demonstrated that oral infection with *P. gingivalis* exacerbated the development and severity of collagen-induced arthritis by favoring a Th17 response[50]. Another study has shown that pre-existing periodontitis augments the development of collagen antibody-induced arthritis in mice[51]. A comparison of wild-type *P. gingivalis* with PAD-deficient *P. gingivalis* or *P. intermedia* (which is naturally PAD-deficient) has implicated bacterial PAD as the mechanistic link between periodontal infection with *P. gingivalis* and RA[52]. However, this study used chamber inoculation of bacteria, which is quite different from the natural course of infection in periodontal disease and cannot be extrapolated to human periodontitis. Thus, the mechanisms by which periodontal disease elevates auto-immune responses are still unclear.

15.2.4 Periodontal disease and non-alcoholic fatty liver disease

Non-alcoholic fatty liver disease (NAFLD) is one of the most common liver diseases in the western world and is characterized by steatosis, inflammatory cytokines, and insulin resistance. NAFLD ranges from benign simple steatosis to steatohepatitis (NASH). Clinical studies have suggested that microbial factors are the driving force for hepatic steatosis and inflammation[53]. Brandl and Schnabl discuss in chapter 22 how intestinal dysbiosis could predispose to inflammation followed by increased gut permeability, bacterial translocation and liver inflammation leading to liver pathology. These mechanistic insights led researchers to investigate the association between periodontal disease, especially *P. gingivalis* infection, and NAFLD. The frequency of *P. gingivalis* detection in the saliva of NAFLD patients was significantly higher than in healthy subjects and was further elevated in patients with NASH. In NAFLD patients positive for *P. gingivalis*, non-surgical periodontal treatment combined with local antibiotic therapy not only improved periodontal conditions but also ameliorated liver function parameters, such as aspartate aminotransferase and alanine aminotransaminase serum levels[53].

In an experiment with animals fed a high-fat diet, *P. gingivalis* administration from the cervical vein resulted in a marked increase not only in body and liver weight compared with sham-administered mice, but also in lipid accumulation, triglycerides, and ALT levels. The notion that the effects were specific to *P. gingivalis* was confirmed by the finding that administration of non-periodontopathic oral bacteria, such as *Streptococcus sanguinis* and *Streptococcus salivalius*, had no such effects[54].

Another experiment with high-fat-diet-fed animals in which *P. gingivalis* was administered through exposed dental pulp tissue demonstrated translocation of *P. gingivalis* to the liver and exacerbated steatohepatitis, possibly due to increased expression of TLR2 in hepatocytes and subsequent elevated production of proinflammatory cytokines[55]. However, it should be noted that intravenous inoculation and dental pulp-mediated infection are different from the natural route of periodontal bacterial infection.

A two-hit model has been proposed to drive the pathogenesis of NASH from NAFLD[56]. The first hit is hepatic steatosis, mainly induced by an increase in free fatty acids (FFA) and a decrease in the oxidation of FFA in the liver, which sensitizes the liver to additional inflammatory insults. The second hits include increased generation of reactive oxygen species and TLR agonists.

Inflammasome-dependent processing of IL-1β and IL-18 is considered to be important in the progression of NAFLD because inflammasomes are the sensors for pathogen-associated molecular patterns or damage-associated molecular patterns[57]. In fact, activation of the NLRP3 inflammasome has been reported to be involved in insulin resistance and atheromatous plaque formation[58]. In contrast, it has also been reported that a lack of inflammasomes exacerbated NAFLD induced by a methionine-choline-deficient diet in mice. This phenotype was transferable to wild-type mice when co-housed with inflammasome-deficient mice, suggesting the involvement of gut microbiota alterations[59].

Interestingly, as described by Snowden in chapter 8, *P. gingivalis* is known to suppress inflammasome activity that is induced by other bacteria or bacterial products[60]. Precise mechanisms of reduced inflammasome activity remain to be elucidated, although suppression of endocytosis by *P. gingivalis* has been proposed[61]. This may be a mechanism linking periodontal disease and NAFLD.

15.2.5 Periodontal disease and pre-term birth

The association of bacterial infections with preterm low birth weight has been reported since the 1980s[62]. An early small case-control study demonstrated an association between preterm low-birthweight deliveries and maternal periodontal diseases[63]. Several subsequent studies supported these findings[64–66]. There is increased detection frequency of *F. nucleatum* DNA in the amniotic fluid in cases of preterm birth, whereas no bacterial DNA is detectable in the amniotic fluid for normal births[67]. *F. nucleatum* can colonize the placenta with FadA adhesin, proliferate[68], and induce an inflammatory response through TLR-dependent signalling[69]. Other biological mechanisms involved in preterm birth include induction of early uterus contractions by increased levels of PGE2 and TNF-α in the blood from periodontal tissue inflammation[69–71]. Despite the theoretical viability of possible mechanisms by which periodontal disease promotes preterm birth, intervention studies analysing the effects of periodontal therapy on preterm births have produced conflicting results[72]. In mouse experiments, *P. gingivalis* and *Campylobacter rectus* have been shown to restrict foetal growth when infected in a subcutaneous chamber[73,74]. However, again, these models do not reflect the natural route of infection in periodontal disease.

15.2.6 Periodontal disease and obesity

It has been reported that the prevalence and severity of periodontal disease are higher in obese individuals than in lean individuals. Based on these results, obesity can be considered a risk factor for periodontal disease[74–78]. Obesity in association with periodontal tissue destruction was also reported in a mouse experiment: *P. gingivalis*-containing ligature placements induced significantly greater alveolar bone resorption in high-fat diet-fed obese mice than in normal-diet-fed lean mice[79]. However, the potential effects of periodontal disease on obesity have been investigated neither in periodontitis patients nor in mice with experimental periodontal infections.

15.2.7 Periodontal disease and cancer

The incidence of cancer development is higher in the presence of chronic inflammatory conditions[80]. Therefore, it is reasonable to consider that periodontal disease may potentially be a risk factor for cancer. In the Health Professionals Follow-up Study in the United States, the likelihood of suffering from cancer was analysed prospectively for individuals self-reporting alveolar bone resorption (considered as a marker of periodontal disease). Results showed there was a 64% increase in the relative risk of pancreatic cancer in those classified with periodontal disease after 16 years of follow-up[81]. A further study in the same cohort demonstrated that men who had never smoked who reported periodontal disease had a slight but significant increased total cancer incidence of 14%. After adjustment, there were significantly increased risks of lung, kidney, pancreatic, and hematological cancers[82].

A causal association between periodontal infection and incidence of cancer or promotion of malignant alteration has not been demonstrated in an experimental periodontitis model. An overabundance of *Fusobacterium nucleatum*, a periodontopathic bacterium, has been reported in colorectal cancer tissue[83,84]; however, it is unclear whether the presence of *F. nucleatum* had a pathogenic role or was rather a consequence of the cancer. Recently, it has become evident that *F. nucleatum* adheres to, invades, and induces oncogenic and inflammatory responses to stimulate cancer cell growth through its unique FadA adhesion[85]. However, where *F. nucleatum* comes from and how it is overrepresented in the colon have not been clarified.

15.2.8 Periodontal disease and inflammatory bowel disease

Inflammatory bowel disease (IBD) is characterized by refractory and persistent inflammation of the intestinal tract and primarily includes ulcerative colitis and Crohn's disease, comprehensively discussed in chapter 19 of this book. The pathophysiology of IBD is multifactorial and based upon a complex interaction of genetic, immunological, and environmental factors leading to substantial defects of the intestinal mucosal barrier[86,87]. A genome-wide disease association study clearly demonstrated that a frameshift mutation in the nucleotide-binding oligomerization domain-containing protein 2 gene (*NOD2*) is associated with susceptibility to Crohn's disease[88]. *NOD2* is expressed in monocytes and macrophages and plays an important role in the recognition of muramyl dipeptide. The association of this gene and periodontal disease was explored since the mutation leads to a significant change in NF-κB activation and subsequent proinflammatory cytokine production; however, no positive relationship was observed[89,90].

Case-control studies have demonstrated a significantly higher incidence of gingivitis and periodontitis in IBD patients[91]. Prevalence, severity, and extent of periodontitis were also greater in IBD patients[92]. Another study showed that IBD patients harbor higher levels of bacteria in the oral cavity that are related to opportunistic infections in inflamed subgingival sites even when matched for other clinical periodontal parameters[93]. Interestingly, an analysis of salivary microbiota observed dysbiosis characterized by a dominance of *Streptococcus*, *Prevotella*, *Neisseria*, *Haemophilus*, *Veillonella*, and *Gemella* in IBD patients. Furthermore, a strong correlation was shown between IL1-β levels and relative abundance of some of these bacteria, including *Prevotella*. These data suggest that dysbiosis of salivary microbiota is associated with inflammatory responses in IBD patients, suggesting a possible link to gut microbiota dysbiosis[94].

15.3 Issues in causal mechanisms of periodontal disease for systemic disease

Although the possible significance of common susceptibility cannot be discounted, there are several hypothetical causal mechanisms linking periodontal disease and systemic diseases. First, bacteria from dental plaque invade gingival tissue through ulcerated sulcular epithelial linings of periodontal pockets and then disseminate into systemic circulation. Second, various proinflammatory cytokines are produced in inflamed periodontal tissue, which can also enter systemic circulation. Third, autoantigens are generated from the immune response to periodontopathic bacteria cross-reacting with self-antigens through molecular mimicry or *P. gingivalis*-derived PAD (Figure 2). However, each hypothesis suffers from contradictory points, preventing comprehensive acceptance.

15.3.1 Endotoxemia (bacteremia)

The surface area of periodontal pockets exposed to bacterial biofilms ranges from $15\,cm^2$ to $20\,cm^2$[95]. Periodontopathic bacteria likely gain access to the systemic circulation through ulcerated pocket epithelia. Bacteremia and bacterial invasion by periodontopathic bacteria have been observed in humans[96]; however, the detection of bacteria in blood or affected tissues has not been consistent. In atheromatous

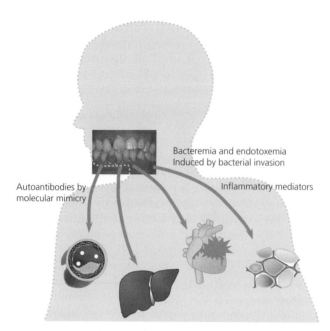

Figure 2 Hypothetical mechanisms linking periodontitis and systemic diseases. Bacteria and/or products invade into systemic circulation via the disrupted epithelial barrier of periodontal pockets. Inflammatory mediators produced in the inflamed periodontal tissues gain access to the systemic circulation. Antibodies against bacterial products cross-react with host molecules, resulting in tissue inflammation.

plaques, average detection rates of periodontopathic bacteria, such as *P. gingivalis,* is 30% in Western populations, which is similar to the rates in Japanese populations[4]. Viable *A. actinomycetemcomitans* and *P. gingivalis* have also been detected in human atherosclerotic plaques[97,98]. Although some periodontopathic bacteria have been reported to be able to invade cultured cells *in vitro*[99,100], little evidence exists concerning alteration of cell function due to invasion. Collectively, there has been no evidence that invading bacteria play a direct role in atherogenesis in humans.

Without inducing bacteremia, we demonstrated that oral infection with *P. gingivalis* enhanced atherogenesis in apolipoprotein E (ApoE)-mutant mice, notably accompanied by an increase in LDL cholesterol and a decrease in HDL cholesterol, with altered gene expression profiles related to cholesterol transport[101,102]. Although the occurrence of transient bacteremia from manipulation of the teeth and periodontal tissues is common, the extent of the bacteremia resulting from a dental procedure is relatively low (less than 10^4 colony-forming units (CFUs) of bacteria per millilitre)[103]. Moreover, there is no information available on the relationship between the severity and extent of disease and the extent of bacteremia. Therefore, the findings observed after direct inoculation of bacteria may not represent potential findings from naturally occurring periodontitis. In addition, most studies used ApoE-deficient mice that phenotypically develop hyperlipidemia and atherosclerosis. Whether or not infection is a trigger for the development of atherosclerosis in healthy subjects cannot be addressed using such a model.

15.3.2 Inflammatory mediators

A number of studies in periodontitis patients have demonstrated that serum levels of high-sensitivity C-reactive protein (hs-CRP) and IL-6 are elevated with increasing severity of the disease, which decreased after successful periodontal treatment[104,105]. These findings are evidence of the elevated state of systemic inflammation in periodontitis patients[15]. Hs-CRP is synthesized mainly in the liver in response to IL-6. Therefore, IL-6 is a key cytokine in systemic inflammation in periodontitis patients. However, there is no evidence of elevated systemic IL-6 being derived from inflamed periodontal tissues.

Histological analysis has demonstrated that oral administration of *P. gingivalis* induced little inflammation of gingival tissues in mice[101]. Taken together, there is no direct evidence that bacteremia from gingival tissue and local inflammatory responses are likely to be involved in systemic inflammation, at least in a mouse model.

15.3.3 Autoimmune response from molecular mimicry

Autoantibodies can often be detected in the sera of periodontitis patients. These include anti-heat-shock protein 60 (HSP60) antibody and β-2-glycoprotein I (β2GPI)-dependent anti-cardiolipin antibody (anti-CL). HSP60 belongs to a family of related proteins that have been highly conserved during evolution. Its expression can be elevated by various stimulants such as heat, bacterial infection, and cytokiness[106]. We revealed the presence of T-cells reactive to HSP60 and *P. gingivalis* GroEL, a homologue of HSP60[107], and cross-reactive anti-HSP60 and anti-*P. gingivalis* GroEL antibodies[108] in the tissue of periodontitis patients. Later, it was demonstrated that these T-cells also exist in the peripheral circulation of atherosclerosis patients[31]. These results suggest that T-cell clones with the same specificity may be involved in the pathogenesis of the two diseases.

It has been reported that anti-CL levels are elevated in a fraction of individuals with generalized periodontitis[109]. Anti-CL is associated with clinical sequelae of thrombosis, stroke, myocardial infarction, early atherosclerosis, and adverse pregnancy outcomes, all of which have been associated with periodontal infections[110].

Involvement of molecular mimicry between GroEL and HSP60 in atherogenesis has been demonstrated by intraperitoneal immunization with *P. gingivalis* in ApoE-deficient mice. The increased pathogen burden of *P. gingivalis* worsened atherosclerosis severity. HSP60 was detected in lesions, and in *P. gingivalis*-immunized mice, lesion development was correlated with anti-GroEL antibody levels[111]. Recent studies have implied that bacterial and viral infections can induce cross-reactive anti-CL autoantibodies that recognize the TLRVYK sequence of β2GPI[112]. Mice immunized with microbial pathogens with homologous sequences related to TLRVYK produced cross-reactive anti-β2GPI[113,114]. Furthermore, it has been demonstrated that the anti-cardiolipin antibody fraction of anti-*P. gingivalis* antibodies causes foetal loss in mice. This effect was not observed when an arg-gingipain-deficient *P. gingivalis* strain was used for immunization, suggesting that arg-gingipain is important for the induction of the anti-cardiolipin antibody fraction of the anti-*P. gingivalis* antibody[115]. The systemic effects of cross-reactive antibodies induced by infection with periodontopathic bacteria might explain the link between periodontal disease and atherosclerosis and adverse pregnancy outcomes. However, this mechanism is not necessarily applicable to other diseases.

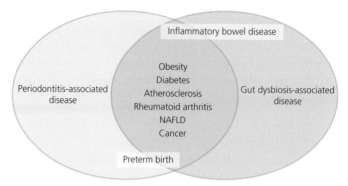

Figure 3 Association between periodontal disease and gut dysbiosis. The diseases associated with periodontal disease are also reported to be affected by gut dysbiosis.

15.4 New insights into the mechanisms linking periodontal disease and systemic disease

Most diseases linked with periodontal disease also seem to be characterized by dysbiosis of the gut microbiota[2,5,116,117]. Potential mechanisms linking intestinal dysbiosis and other systemic diseases are specifically discussed in chapters 22 to 27. Decreased gut barrier function and subsequent endotoxemia, imbalance of gut immune function and adverse effects of bacterial metabolites are considered underlying mechanisms, although details of these remain to be elucidated (Figure 3).

It is well known that large numbers of bacteria live in the saliva of healthy individuals. In patients with severe periodontitis, the concentration of *P. gingivalis* can reach $10^6/ml$[118–120]. Since the proportion of *P. gingivalis* in the oral flora is estimated to be 0.8%[121] and humans produce 1–1.5 L of saliva a day, patients with severe periodontitis could be swallowing 10^{12} to 10^{13} (10^9 to 10^{10} as *P. gingivalis*) bacteria per day. Given that the bacterial flora of the oral cavity is quite different from that of the gut, it is possible that a large amount of swallowed oral bacteria could alter the gut flora. In support of this idea, the effect of oral bacteria on the gut microbiota has recently been shown in patients with liver cirrhosis[122]. The study revealed a major change in the gut microbiota in patients with liver cirrhosis occurring because of a massive invasion of the gut by oral bacterial species. The correlation in disease severity with the abundance of the invading species suggests that they may play an active role in pathology.

15.5 Effect of oral administration of *P. gingivalis* on metabolic change and gut microbiota

To investigate this hypothesis, changes in the gut microbiota, insulin and glucose intolerance, and levels of tissue inflammation were analysed in mice after oral administration of *P. gingivalis* strain W83 twice a week for five weeks[123]. Pyrosequencing of ileum contents revealed that populations of Bacteroidales were significantly elevated in *P. gingivalis*-administered mice, which coincided with increases in insulin resistance and systemic inflammation (Figure 4). In *P. gingivalis*-administered mice,

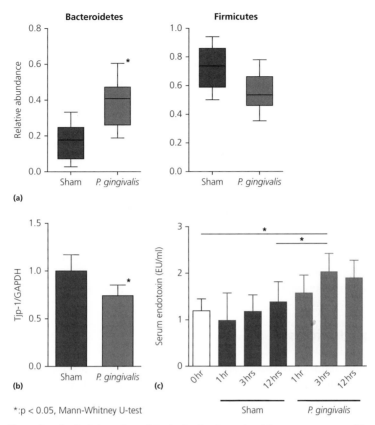

(a)

(b)

(c)

*:p < 0.05, Mann-Whitney U-test

Figure 4 Effect of oral administration of *P. gingivalis*. Gut microbiota was compared between *P. gingivalis*-administered and sham-administered mice by 16S rRNA sequencing analysis. Relative abundances of each bacterial group in the phylum are shown. The proportion of Bacteroidetes was significantly higher in *P. gingivalis*-administered mice than in sham-administered mice. The proportion of Firmicutes tended to be lower in *P. gingivalis*-administered mice than in sham-administered mice (a). The gene expression of tight junction protein (*tjp1*) was significantly lower in *P. gingivalis*-administered mice than in sham-administered mice (b). Serum endotoxin levels increased after *P. gingivalis* administration (c).

blood endotoxin levels tended to be higher, whereas gene expression of tight junction proteins in the ileum decreased significantly. Interestingly, the expression of genes coding for intestinal alkaline phosphatase (Akp3) in the small intestine were both downregulated in *P. gingivalis*-administered mice compared with sham-administered mice. It was demonstrated that a defect in intestinal alkaline phosphatase was associated with high-fat-diet-induced metabolic syndrome, and endogenous and orally supplemented IAP inhibited endotoxin absorption as well as reversed metabolic syndrome in mice[124].

Gene expression profiles revealed that expression of proinflammatory genes was upregulated, whereas the genes that improve insulin sensitivity were downregulated in the adipose tissue of *P. gingivalis*-administered mice compared with sham-administered mice. In the liver, oral administration of *P. gingivalis* also led to increased mRNA expression of proinflammatory cytokines and decreased mRNA expression of molecules having potentially anti-inflammatory properties.

In the other direction, the possibility that dysbiosis of gut microbiota affects periodontal tissue destruction has also been investigated. Dextran sodium sulphate (DSS)-colitis is a well-known mouse model of IBD with changes in gut microbiota[125,126]. Inflammatory bowel disease is considered to be a result of an imbalance of the gut microbiota, epithelium, and immune system[127]. Although DSS-colitis may not have a direct relationship with periodontal tissue homeostasis, enhanced alveolar bone resorption in mice with DSS-colitis was shown irrespective of oral bacteria manipulations[128]. These results imply a change in the gut microbiota or its effects on inflammation-induced bone resorption. Feeding mice with a diet containing n-3 polyunsaturated fatty acid (n-3 PUFA)-rich fish oil prevented *P. gingivalis*-induced alveolar bone resorption compared with feeding a diet containing saturated fatty acid-rich corn oil[129]. Since n-3 PUFA is shown to have a favorable effect on gut microbiota[130], it is suggested that the suppressive effect of n-3 PUFA on alveolar bone resorption could be attributed to the suppression of systemic inflammation through gut microbiota modulation. These results provide a new paradigm on the interrelationship between periodontitis and systemic diseases.

15.6 Conclusions

It has become evident that oral administration of *P. gingivalis,* a periodontopathic bacterium, induces an alteration of the gut microbiota and an elevation of blood endotoxin levels in a mouse model. These results are similar to findings observed in either diet-induced obese mice or genetically modified obese mice and diabetic model mice. Accumulating evidence suggests that dysbiosis of the gut microbiota is associated with an increased risk of diabetes, atherosclerosis, NAFLD, obesity, and rheumatoid arthritis. Our study showing that swallowing large amounts of oral bacteria may alter the gut microbiota[123] provides a rationale for the biological basis of a causal association between periodontal disease and systemic diseases that cannot be explained by existing hypotheses (Figure 5). Although the compositions of the microbiota of each body site such as oral cavity and gut are taxonomically distinctive, classification of the patterns of predominant phyla designated as community types in each location of the body do have a close association with each other[131]. This indicates that knowing the oral community types could allow us to speculate gut community types, and vice versa. Thus, analysis of the composition of the microbiota of periodontitis patients will make it possible to identify subjects prone to particular systemic diseases. In addition, it may be useful for developing oral probiotic therapies targeted towards the gut microbiota.

Further studies are needed to determine whether oral bacteria other than *P. gingivalis* can induce similar changes in gut microbiota and metabolism, to identify bacterial components responsible for gut microbiota changes and to determine gut bacteria that become pathogenic due to the influx of oral bacteria. Most importantly of all, research is needed to clarify whether the gut microbiota of patients with periodontal disease is different from those of periodontally healthy subjects. Exploration of the gut microbiota of these patients could elucidate a potential association with those of patients with systemic diseases.

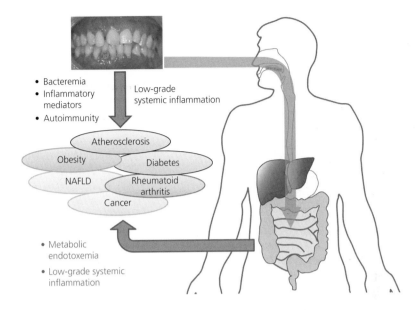

Figure 5 Proposed mechanism for the effect of periodontal infection on various diseases. A large number of swallowed bacteria induce change in gut microbiota as well as gut barrier function. This results in the endotoxemia that is similar to the metabolic endotoxemia seen in obesity; it causes low-grade systemic inflammation in various tissues and organs and increases the risk of various systemic diseases.

TAKE-HOME MESSAGE

- Periodontitis is associated with a number of other chronic diseases, including diabetes, cardiovascular disease and rheumatoid arthritis.

- Oral administration of periodontopathic bacterium *P. gingivalis* induces an alteration of the gut microbiota and an elevation of blood endotoxin levels in a mouse model, similar to that induced by diet or genetic modifications.

- Swallowing large amounts of oral bacteria may induce dysbiosis of the gut microbiota.

- Pathogenic and/or therapeutic alteration of the subgingival biofilm may have implications for other diseases linked with microbial-host response homeostasis.

References

1 Human Microbiome Project Consortium. Structure, function and diversity of the healthy human microbiome. *Nature* 2012; **486**: 207–14.
2 Cho I, Blaser MJ. The human microbiome: at the interface of health and disease. *Nat Rev Genet* 2012; **13**: 260–70.
3 Kaur S, White S, Bartold PM. Periodontal disease and rheumatoid arthritis: a systematic review. *J Dent Res* 2013; **92**: 399–408.
4 Cullinan MP, Seymour GJ. Periodontal disease and systemic illness: will the evidence ever be enough? *Periodontol 2000* 2013; **62**: 271–86.
5 Clemente JC, Ursell LK, Parfrey LW, Knight R. The impact of the gut microbiota on human health: an integrative view. *Cell* 2012; **148**: 1258–70.

6 Delzenne NM, Neyrinck AM, Backhed F, Cani PD. Targeting gut microbiota in obesity: effects of prebiotics and probiotics. *Nat Rev Endocrinol* 2011; **7**: 639–46.

7 Koren O, Spor A, Felin J, Fak F, Stombaugh J, Tremaroli V, Behre CJ, Knight R, Fagerberg B, Ley RE, Backhed F. Human oral, gut, and plaque microbiota in patients with atherosclerosis. *Proc Natl Acad Sci U S A* 2011; **108 Suppl 1**: 4592–8.

8 Borgnakke WS, Ylostalo PV, Taylor GW, Genco RJ. Effect of periodontal disease on diabetes: systematic review of epidemiologic observational evidence. *J Clin Periodontol* 2013; **40 Suppl 14**: S135–52.

9 Teeuw WJ, Gerdes VE, Loos BG. Effect of periodontal treatment on glycemic control of diabetic patients: a systematic review and meta-analysis. *Diabetes Care* 2010; **33**: 421–7.

10 Hotamisligil GS, Peraldi P, Budavari A, Ellis R, White MF, Spiegelman BM. IRS-1-mediated inhibition of insulin receptor tyrosine kinase activity in TNF-α- and obesity-induced insulin resistance. *Science* 1996; **271**: 665–8.

11 Hotamisligil GS, Shargill NS, Spiegelman BM. Adipose expression of tumor necrosis factor-α: direct role in obesity-linked insulin resistance. *Science* 1993; **259**: 87–91.

12 Ikezawa I, Tai H, Shimada Y, Komatsu Y, Galicia JC, Yoshie H. Imbalance between soluble tumour necrosis factor receptors type 1 and 2 in chronic periodontitis. *J Clin Periodontol* 2005; **32**: 1047–54.

13 Ribeiro FV, de Mendonca AC, Santos VR, Bastos MF, Figueiredo LC, Duarte PM. Cytokines and bone-related factors in systemically healthy patients with chronic periodontitis and patients with type 2 diabetes and chronic periodontitis. *J Periodontol* 2011; **82**: 1187–96.

14 Andrukhov O, Ulm C, Reischl H, Nguyen PQ, Matejka M, Rausch-Fan X. Serum cytokine levels in periodontitis patients in relation to the bacterial load. *J Periodontol* 2011; **82**: 885–92.

15 Schenkein HA, Loos BG. Inflammatory mechanisms linking periodontal diseases to cardiovascular diseases. *J Clin Periodontol* 2013; **40 Suppl 14**: S51–69.

16 Dandona P, Aljada A, Bandyopadhyay A. Inflammation: the link between insulin resistance, obesity and diabetes. *Trends Immunol* 2004; **25**: 4–7.

17 Pontes Andersen CC, Flyvbjerg A, Buschard K, Holmstrup P. Periodontitis is associated with aggravation of prediabetes in Zucker fatty rats. *J Periodontol* 2007; **78**: 559–65.

18 Li H, Yang H, Ding Y, Aprecio R, Zhang W, Wang Q, Li Y. Experimental periodontitis induced by *Porphyromonas gingivalis* does not alter the onset or severity of diabetes in mice. *J Periodontal Res* 2013; **48**: 582–90.

19 Humphrey LL, Fu R, Buckley DI, Freeman M, Helfand M. Periodontal disease and coronary heart disease incidence: a systematic review and meta-analysis. *J Gen Intern Med* 2008; **23**: 2079–86.

20 Bahekar AA, Singh S, Saha S, Molnar J, Arora R. The prevalence and incidence of coronary heart disease is significantly increased in periodontitis: a meta-analysis. *Am Heart J* 2007; **154**: 830–7.

21 Janket SJ, Baird AE, Chuang SK, Jones JA. Meta-analysis of periodontal disease and risk of coronary heart disease and stroke. *Oral Surg Oral Med Oral Pathol Oral Radiol Endod* 2003; **95**: 559–69.

22 Khader YS, Albashaireh ZS, Alomari MA. Periodontal diseases and the risk of coronary heart and cerebrovascular diseases: a meta-analysis. *J Periodontol* [Meta-Analysis Review]. 2004; **75**: 1046–53.

23 Lockhart PB, Bolger AF, Papapanou PN, Osinbowale O, Trevisan M, Levison ME, Taubert KA, Newburger JW, Gornik HL, Gewitz MH, Wilson WR, Smith SC, Jr., Baddour LM. Periodontal disease and atherosclerotic vascular disease: does the evidence support an independent association? a scientific statement from the American Heart Association. *Circulation* 2012; **125**: 2520–44.

24 Haraszthy VI, Zambon JJ, Trevisan M, Zeid M, Genco RJ. Identification of periodontal pathogens in atheromatous plaques. *J Periodontol* 2000; **71**: 1554–60.

25 Ishihara K, Nabuchi A, Ito R, Miyachi K, Kuramitsu HK, Okuda K. Correlation between detection rates of periodontopathic bacterial DNA in coronary stenotic artery plaque and in dental plaque samples. *J Clin Microbiol* 2004; **42**: 1313–5.

26 Kurihara N, Inoue Y, Iwai T, Umeda M, Huang Y, Ishikawa I. Detection and localization of periodontopathic bacteria in abdominal aortic aneurysms. *Eur J Vasc Endovasc Surg* 2004; **28**: 553–8.

27 Nakano K, Nemoto H, Nomura R, Inaba H, Yoshioka H, Taniguchi K, Amano A, Ooshima T. Detection of oral bacteria in cardiovascular specimens. *Oral Microbiol Immunol* 2009; **24**: 64–8.

28 Ohki T, Itabashi Y, Kohno T, Yoshizawa A, Nishikubo S, Watanabe S, Yamane G, Ishihara K. Detection of periodontal bacteria in thrombi of patients with acute myocardial infarction by polymerase chain reaction. *Am Heart J* 2012; **163**: 164–7.

29 Pucar A, Milasin J, Lekovic V, Vukadinovic M, Ristic M, Putnik S, Kenney EB. Correlation between atherosclerosis and periodontal putative pathogenic bacterial infections in coronary and internal mammary arteries. *J Periodontol* 2007; **78**: 677–82.

30 Stelzel M, Conrads G, Pankuweit S, Maisch B, Vogt S, Moosdorf R, Flores-de-Jacoby L. Detection of *Porphyromonas gingivalis* DNA in aortic tissue by PCR. *J Periodontol* 2002; **73**: 868–70.

31 Yamazaki K, Ohsawa Y, Itoh H, Ueki K, Tabeta K, Oda T, Nakajima T, Yoshie H, Saito S, Oguma F, Kodama M, Aizawa Y, Seymour GJ. T-cell clonality to *Porphyromonas gingivalis* and human heat shock protein 60s in patients with atherosclerosis and periodontitis. *Oral Microbiol Immunol* 2004; **19**: 160–7.

32 Armingohar Z, Jorgensen JJ, Kristoffersen AK, Abesha-Belay E, Olsen I. Bacteria and bacterial DNA in atherosclerotic plaque and aneurysmal wall biopsies from patients with and without periodontitis. *J Oral Microbiol* 2014; **6**.

33 Gibson FC, 3rd, Hong C, Chou HH, Yumoto H, Chen J, Lien E, Wong J, Genco CA. Innate immune recognition of invasive bacteria accelerates atherosclerosis in apolipoprotein E-deficient mice. *Circulation* 2004; **109**: 2801–6.

34 Lalla E, Lamster IB, Hofmann MA, Bucciarelli L, Jerud AP, Tucker S, Lu Y, Papapanou PN, Schmidt AM. Oral infection with a periodontal pathogen accelerates early atherosclerosis in apolipoprotein e-null mice. *Arterioscler Thromb Vasc Biol* 2003; **23**: 1405–11.

35 Li L, Messas E, Batista EL, Jr., Levine RA, Amar S. *Porphyromonas gingivalis* infection accelerates the progression of atherosclerosis in a heterozygous apolipoprotein E-deficient murine model. *Circulation* 2002; **105**: 861–7.

36 Rutger Persson G. Rheumatoid arthritis and periodontitis — inflammatory and infectious connections: r eview of the literature. *J Oral Microbiol* 2012; **4**.

37 Bonfil JJ, Dillier FL, Mercier P, Reviron D, Foti B, Sambuc R, Brodeur JM, Sedarat C. A "case control" study on the role of HLA DR4 in severe periodontitis and rapidly progressive periodontitis: identification of types and subtypes using molecular biology (PCR.SSO). *J Clin Periodontol* 1999; **26**: 77–84.

38 de Pablo P, Dietrich T, McAlindon TE. Association of periodontal disease and tooth loss with rheumatoid arthritis in the US population. *J Rheumatol* 2008; **35**: 70–6.

39 Firatli E, Kantarci A, Cebeci I, Tanyeri H, Sonmez G, Carin M, Tuncer O. Association between HLA antigens and early onset periodontitis. *J Clin Periodontol* 1996; **23**: 563–6.

40 Katz J, Goultschin J, Benoliel R, Brautbar C. Human leukocyte antigen (HLA) DR4. Positive association with rapidly progressing periodontitis. *J Periodontol* 1987; **58**: 607–10.

41 Marotte H, Farge P, Gaudin P, Alexandre C, Mougin B, Miossec P. The association between periodontal disease and joint destruction in rheumatoid arthritis extends the link between the HLA-DR shared epitope and severity of bone destruction. *Ann Rheum Dis* 2006; **65**: 905–9.

42 Kobayashi T, Ito S, Kuroda T, Yamamoto K, Sugita N, Narita I, Sumida T, Gejyo F, Yoshie H. The interleukin-1 and Fcγ receptor gene polymorphisms in Japanese patients with rheumatoid arthritis and periodontitis. *J Periodontol* 2007; **78**: 2311–8.

43 Arvikar SL, Collier DS, Fisher MC, Unizony S, Cohen GL, McHugh G, Kawai T, Strle K, Steere AC. Clinical correlations with *Porphyromonas gingivalis* antibody responses in patients with early rheumatoid arthritis. *Arthritis Res Ther* 2013; **15**: R109.

44 Scher JU, Ubeda C, Equinda M, Khanin R, Buischi Y, Viale A, Lipuma L, Attur M, Pillinger MH, Weissmann G, Littman DR, Pamer EG, Bretz WA, Abramson SB. Periodontal disease and the oral microbiota in new-onset rheumatoid arthritis. *Arthritis Rheum* 2012; **64**: 3083–94.

45 Klareskog L, Catrina AI, Paget S. Rheumatoid arthritis. *Lancet* 2009; **373**: 659–72.

46 Klareskog L, Ronnelid J, Lundberg K, Padyukov L, Alfredsson L. Immunity to citrullinated proteins in rheumatoid arthritis. *Annu Rev Immunol* 2008; **26**: 651–75.

47 Wegner N, Lundberg K, Kinloch A, Fisher B, Malmstrom V, Feldmann M, Venables PJ. Autoimmunity to specific citrullinated proteins gives the first clues to the etiology of rheumatoid arthritis. *Immunol Rev* 2010; **233**: 34–54.

48 McGraw WT, Potempa J, Farley D, Travis J. Purification, characterization, and sequence analysis of a potential virulence factor from *Porphyromonas gingivalis*, peptidylarginine deiminase. *Infect Immun* 1999; **67**: 3248–56.

49 Konig MF, Paracha AS, Moni M, Bingham CO 3rd, Andrade F. Defining the role of *Porphyromonas gingivalis* peptidylarginine deiminase (PPAD) in rheumatoid arthritis through the study of PPAD biology. *Ann Rheum Dis* 2015; **74**: 2054–61.

50 Marchesan JT, Gerow EA, Schaff R, Taut AD, Shin SY, Sugai J, Brand D, Burberry A, Jorns J, Lundy SK, Nunez G, Fox DA, Giannobile WV. *Porphyromonas gingivalis* oral infection exacerbates the development and severity of collagen-induced arthritis. *Arthritis Res Ther* 2013; **15**: R186.

51 Cantley MD, Haynes DR, Marino V, Bartold PM. Pre-existing periodontitis exacerbates experimental arthritis in a mouse model. *J Clin Periodontol* 2011; **38**: 532–41.

52 Maresz KJ, Hellvard A, Sroka A, Adamowicz K, Bielecka E, Koziel J, Gawron K, Mizgalska D, Marcinska KA, Benedyk M, Pyrc K, Quirke AM, Jonsson R, Alzabin S, Venables PJ, Nguyen KA, Mydel P, Potempa J. *Porphyromonas gingivalis* facilitates the development and progression of destructive arthritis through its unique bacterial peptidylarginine deiminase (PAD). *PLoS Pathog* 2013; **9**: e1003627.

53 Moschen AR, Kaser S, Tilg H. Non-alcoholic steatohepatitis: a microbiota-driven disease. *Trends Endocrinol Metab* 2013; **24**: 537–45.

54 Yoneda M, Naka S, Nakano K, Wada K, Endo H, Mawatari H, Imajo K, Nomura R, Hokamura K, Ono M, Murata S, Tohnai I, Sumida Y, Shima T, Kuboniwa M, Umemura K, Kamisaki Y, Amano A, Okanoue T, Ooshima T, Nakajima A. Involvement of a periodontal pathogen, *Porphyromonas gingivalis on the pathogenesis of non-alcoholic fatty liver disease. BMC Gastroenterol* 2012; **12**: 16.

55 Furusho H, Miyauchi M, Hyogo H, Inubushi T, Ao M, Ouhara K, Hisatune J, Kurihara H, Sugai M, Hayes CN, Nakahara T, Aikata H, Takahashi S, Chayama K, Takata T. Dental infection of *Porphyromonas gingivalis* exacerbates high-fat-diet-induced steatohepatitis in mice. *J Gastroenterol* 2013; **48**: 1259–70.

56 Day CP, James OF. Steatohepatitis: a tale of two "hits"? *Gastroenterology* 1998; **114**: 842–5.

57 Strowig T, Henao-Mejia J, Elinav E, Flavell R. Inflammasomes in health and disease. *Nature* 2012; **481**: 278–86.

58 Wen H, Ting JP, O'Neill LA. A role for the NLRP3 inflammasome in metabolic diseases — did Warburg miss inflammation? *Nature Immunology*. [Research Support, N.I.H., Extramural Research Support, Non-U.S. Gov't Review]. 2012; **13**: 352–7.

59 Henao-Mejia J, Elinav E, Jin C, Hao L, Mehal WZ, Strowig T, Thaiss CA, Kau AL, Eisenbarth SC, Jurczak MJ, Camporez JP, Shulman GI, Gordon JI, Hoffman HM, Flavell RA. Inflammasome-mediated dysbiosis regulates progression of NAFLD and obesity. *Nature*. [Research Support, N.I.H., Extramural Research Support, Non-U.S. Gov't Research Support, U.S. Gov't, Non-P.H.S.]. 2012; **482**: 179–85.

60 Belibasakis GN, Guggenheim B, Bostanci N. Down-regulation of NLRP3 inflammasome in gingival fibroblasts by subgingival biofilms: involvement of *Porphyromonas gingivalis. Innate Immun* 2013; **19**:3–9.

61 Taxman DJ, Swanson KV, Broglie PM, Wen H, Holley-Guthrie E, Huang MT, Callaway JB, Eitas TK, Duncan JA, Ting JP. *Porphyromonas gingivalis* mediates inflammasome repression in polymicrobial cultures through a novel mechanism involving reduced endocytosis. *J Biol Chem* 2012; **287**: 32791–9.

62 McGregor JA, French JI, Lawellin D, Todd JK. Preterm birth and infection: pathogenic possibilities. *Am J Reprod Immunol Microbiol* 1988; **16**: 123–32.

63 Offenbacher S, Katz V, Fertik G, Collins J, Boyd D, Maynor G, McKaig R, Beck J. Periodontal infection as a possible risk factor for preterm low birth weight. *J Periodontol* 1996; **67**: 1103–13.

64 Dasanayake AP, Boyd D, Madianos PN, Offenbacher S, Hills E. The association between *Porphyromonas gingivalis*-specific maternal serum IgG and low birth weight. *J Periodontol* 2001; **72**: 1491–7.

65 Jeffcoat MK, Geurs NC, Reddy MS, Cliver SP, Goldenberg RL, Hauth JC. Periodontal infection and preterm birth: results of a prospective study. *J Am Dent Assoc* 2001; **132**: 875–80.

66 Offenbacher S, Lieff S, Boggess KA, Murtha AP, Madianos PN, Champagne CM, McKaig RG, Jared HL, Mauriello SM, Auten RL, Jr., Herbert WN, Beck JD. Maternal periodontitis and prematurity. Part I: Obstetric outcome of prematurity and growth restriction. *Ann Periodontol* 2001; **6**: 164–74.

67 Han YW, Shen T, Chung P, Buhimschi IA, Buhimschi CS. Uncultivated bacteria as etiologic agents of intra-amniotic inflammation leading to preterm birth. *J Clin Microbiol* 2009; **47**: 38–47.

68 Han YW, Redline RW, Li M, Yin L, Hill GB, McCormick TS. *Fusobacterium nucleatum* induces premature and term stillbirths in pregnant mice: implication of oral bacteria in preterm birth. *Infect Immun* 2004; **72**: 2272–9.

69 Liu H, Redline RW, Han YW. *Fusobacterium nucleatum* induces fetal death in mice via stimulation of TLR4-mediated placental inflammatory response. *J Immunol* 2007; **179**: 2501–8.

70 Jared H, Boggess KA, Moss K, Bose C, Auten R, Beck J, Offenbacher S. Fetal exposure to oral pathogens and subsequent risk for neonatal intensive care admission. *J Periodontol* 2009; **80**: 878–83.
71 Madianos PN, Lieff S, Murtha AP, Boggess KA, Auten RL, Jr., Beck JD, Offenbacher S. Maternal periodontitis and prematurity. Part II: Maternal infection and fetal exposure. *Ann Periodontol* 2001; **6**: 175–82.
72 Wimmer G, Pihlstrom BL. A critical assessment of adverse pregnancy outcome and periodontal disease. *J Clin Periodontol* 2008; **35**: 380–97.
73 Lin D, Smith MA, Elter J, Champagne C, Downey CL, Beck J, Offenbacher S. *Porphyromonas gingivalis* infection in pregnant mice is associated with placental dissemination, an increase in the placental Th1/Th2 cytokine ratio, and fetal growth restriction. *Infect Immun* 2003; **71**: 5163–8.
74 Yeo A, Smith MA, Lin D, Riche EL, Moore A, Elter J, Offenbacher S. *Campylobacter rectus* mediates growth restriction in pregnant mice. *J Periodontol* 2005; **76**: 551–7.
75 Alabdulkarim M, Bissada N, Al-Zahrani M, Ficara A, Siegel B. Alveolar bone loss in obese subjects. *J Int Acad Periodontol* 2005; **7**: 34–8.
76 Al-Zahrani MS, Bissada NF, Borawskit EA. Obesity and periodontal disease in young, middle-aged, and older adults. *Journal of Periodontology* 2003; **74**: 610–5.
77 Buhlin K, Gustafsson A, Pockley AG, Frostegard J, Klinge B. Risk factors for cardiovascular disease in patients with periodontitis. *Eur Heart J.* [Comparative Study Research Support, Non-U.S. Gov't] 2003; **24**: 2099–107.
78 Saito T, Shimazaki Y, Sakamoto M. Obesity and periodontitis. *New England Journal of Medicine.* [Letter]. 1998; **339**: 482–3.
79 Amar S, Zhou Q, Shaik-Dasthagirisaheb Y, Leeman S. Diet-induced obesity in mice causes changes in immune responses and bone loss manifested by bacterial challenge. *Proc Natl Acad Sci U S A* 2007; **104**: 20466–71.
80 Coussens LM, Werb Z. Inflammation and cancer. *Nature* 2002; **420**:860–7.
81 Michaud DS, Joshipura K, Giovannucci E, Fuchs CS. A prospective study of periodontal disease and pancreatic cancer in US male health professionals. *J Natl Cancer Inst* 2007; **99**:171–5.
82 Michaud DS, Liu Y, Meyer M, Giovannucci E, Joshipura K. Periodontal disease, tooth loss, and cancer risk in male health professionals: a prospective cohort study. *Lancet Oncol* 2008; **9**: 550–8.
83 Castellarin M, Warren RL, Freeman JD, Dreolini L, Krzywinski M, Strauss J, Barnes R, Watson P, Allen-Vercoe E, Moore RA, Holt RA. *Fusobacterium nucleatum* infection is prevalent in human colorectal carcinoma. *Genome Res* 2012; **22**: 299–306.
84 Kostic AD, Gevers D, Pedamallu CS, Michaud M, Duke F, Earl AM, Ojesina AI, Jung J, Bass AJ, Tabernero J, Baselga J, Liu C, Shivdasani RA, Ogino S, Birren BW, Huttenhower C, Garrett WS, Meyerson M. Genomic analysis identifies association of Fusobacterium with colorectal carcinoma. *Genome Res* 2012; **22**: 292–8.
85 Rubinstein MR, Wang X, Liu W, Hao Y, Cai G, Han YW. *Fusobacterium nucleatum* promotes colorectal carcinogenesis by modulating E-cadherin/β-catenin signaling via its FadA adhesin. *Cell Host Microbe* 2013; **14**: 195–206.
86 Eckburg PB, Relman DA. The role of microbes in Crohn's disease. *Clin Infect Dis* 2007; **44**: 256–62.
87 Schreiber S, Rosenstiel P, Albrecht M, Hampe J, Krawczak M. Genetics of Crohn disease, an archetypal inflammatory barrier disease. *Nat Rev Genet* 2005; **6**: 376–88.
88 Ogura Y, Bonen DK, Inohara N, Nicolae DL, Chen FF, Ramos R, Britton H, Moran T, Karaliuskas R, Duerr RH, Achkar JP, Brant SR, Bayless TM, Kirschner BS, Hanauer SB, Nunez G, Cho JH. A frameshift mutation in NOD2 associated with susceptibility to Crohn's disease. *Nature* 2001; **411**: 603–6.
89 Folwaczny M, Glas J, Torok HP, Mauermann D, Folwaczny C. The 3020insC mutation of the NOD2/CARD15 gene in patients with periodontal disease. *Eur J Oral Sci* 2004; **112**: 316–9.
90 Laine ML, Murillo LS, Morre SA, Winkel EG, Pena AS, van Winkelhoff AJ. CARD15 gene mutations in periodontitis. *J Clin Periodontol* 2004; **31**: 890–3.
91 Vavricka SR, Manser CN, Hediger S, Vogelin M, Scharl M, Biedermann L, Rogler S, Seibold F, Sanderink R, Attin T, Schoepfer A, Fried M, Rogler G, Frei P. Periodontitis and gingivitis in inflammatory bowel disease: a case-control study. *Inflamm Bowel Dis* 2013; **19**: 2768–77.
92 Habashneh RA, Khader YS, Alhumouz MK, Jadallah K, Ajlouni Y. The association between inflammatory bowel disease and periodontitis among Jordanians: a case-control study. *J Periodontal Res* 2012; **47**: 293–8.

93 Brito F, Zaltman C, Carvalho AT, Fischer RG, Persson R, Gustafsson A, Figueredo CM. Subgingival microflora in inflammatory bowel disease patients with untreated periodontitis. *Eur J Gastroenterol Hepatol* 2013; **25**: 239–45.

94 Said HS, Suda W, Nakagome S, Chinen H, Oshima K, Kim S, Kimura R, Iraha A, Ishida H, Fujita J, Mano S, Morita H, Dohi T, Oota H, Hattori M. Dysbiosis of salivary microbiota in inflammatory bowel disease and its association with oral immunological biomarkers. *DNA Res* 2014; **21**: 15–25.

95 Hujoel PP, White BA, Garcia RI, Listgarten MA. The dentogingival epithelial surface area revisited. *J Periodontal Res* 2001; **36**: 48–55.

96 Tomas I, Diz P, Tobias A, Scully C, Donos N. Periodontal health status and bacteraemia from daily oral activities: systematic review/meta-analysis. *J Clin Periodontol* 2012; **39**: 213–28.

97 Kozarov EV, Dorn BR, Shelburne CE, Dunn WA, Jr., Progulske-Fox A. Human atherosclerotic plaque contains viable invasive *Actinobacillus actinomycetemcomitans* and *Porphyromonas gingivalis*. *Arterioscler Thromb Vasc Biol* 2005; **25**: e17–8.

98 Rafferty B, Jonsson D, Kalachikov S, Demmer RT, Nowygrod R, Elkind MS, Bush H, Jr., Kozarov E. Impact of monocytic cells on recovery of uncultivable bacteria from atherosclerotic lesions. *J Intern Med* 2011; **270**: 273–80.

99 Dorn BR, Dunn WA, Jr., Progulske-Fox A. Invasion of human coronary artery cells by periodontal pathogens. *Infect Immun* 1999; **67**: 5792–8.

100 Fardini Y, Wang X, Temoin S, Nithianantham S, Lee D, Shoham M, Han YW. *Fusobacterium nucleatum* adhesin FadA binds vascular endothelial cadherin and alters endothelial integrity. *Mol Microbiol* 2011; **82**: 1468–80.

101 Maekawa T, Takahashi N, Tabeta K, Aoki Y, Miyashita H, Miyauchi S, Miyazawa H, Nakajima T, Yamazaki K. Chronic oral infection with *Porphyromonas gingivalis* accelerates atheroma formation by shifting the lipid profile. *PloS One* 2011; **6**: e20240.

102 Miyauchi S, Maekawa T, Aoki Y, Miyazawa H, Tabeta K, Nakajima T, Yamazaki K. Oral infection with *Porphyromonas gingivalis* and systemic cytokine profile in C57BL/6.KOR-ApoE[shl] mice. *J Periodontal Res* 2012; **47**: 402–8.

103 Wilson W, Taubert KA, Gewitz M, Lockhart PB, Baddour LM, Levison M, Bolger A, Cabell CH, Takahashi M, Baltimore RS, Newburger JW, Strom BL, Tani LY, Gerber M, Bonow RO, Pallasch T, Shulman ST, Rowley AH, Burns JC, Ferrieri P, Gardner T, Goff D, Durack DT. Prevention of infective endocarditis: guidelines from the American Heart Association: a guideline from the American Heart Association Rheumatic Fever, Endocarditis, and Kawasaki Disease Committee, Council on Cardiovascular Disease in the Young, Council on Clinical Cardiology, Council on Cardiovascular Surgery and Anesthesia, and the Quality of Care and Outcomes Research Interdisciplinary Working Group. *Circulation* 2007; **116**: 1736–54.

104 Nakajima T, Honda T, Domon H, Okui T, Kajita K, Ito H, Takahashi N, Maekawa T, Tabeta K, Yamazaki K. Periodontitis-associated up-regulation of systemic inflammatory mediator level may increase the risk of coronary heart disease. *J Periodontal Res* 2009.

105 Vidal F, Figueredo CM, Cordovil I, Fischer RG. Periodontal therapy reduces plasma levels of interleukin-6, C-reactive protein, and fibrinogen in patients with severe periodontitis and refractory arterial hypertension. *J Periodontol* 2009; **80**: 786–91.

106 Ueki K, Tabeta K, Yoshie H, Yamazaki K. Self-heat shock protein 60 induces tumour necrosis factor-α in monocyte-derived macrophage: possible role in chronic inflammatory periodontal disease. *Clin Exp Immunol* 2002; **127**: 72–7.

107 Yamazaki K, Ohsawa Y, Tabeta K, Ito H, Ueki K, Oda T, Yoshie H, Seymour GJ. Accumulation of human heat shock protein 60-reactive T cells in the gingival tissues of periodontitis patients. *Infect Immun* 2002; **70**: 2492–501.

108 Tabeta K, Yamazaki K, Hotokezaka H, Yoshie H, Hara K. Elevated humoral immune response to heat shock protein 60 (hsp60) family in periodontitis patients. *Clin Exp Immunol* 2000; **120**: 285–93.

109 Schenkein HA, Berry CR, Burmeister JA, Brooks CN, Barbour SE, Best AM, Tew JG. Anti-cardiolipin antibodies in sera from patients with periodontitis. *J Dent Res* 2003; **82**: 919–22.

110 Schenkein HA, Best AM, Brooks CN, Burmeister JA, Arrowood JA, Kontos MC, Tew JG. Anti-cardiolipin and increased serum adhesion molecule levels in patients with aggressive periodontitis. *J Periodontol* 2007; **78**: 459–66.

111 Ford PJ, Gemmell E, Timms P, Chan A, Preston FM, Seymour GJ. Anti-*P. gingivalis* response correlates with atherosclerosis. *J Dent Res* 2007; **86**: 35–40.

112 Blank M, Krause I, Fridkin M, Keller N, Kopolovic J, Goldberg I, Tobar A, Shoenfeld Y. Bacterial induction of autoantibodies to β2-glycoprotein-I accounts for the infectious etiology of antiphospholipid syndrome. *J Clin Invest* 2002; **109**: 797–804.

113 Wang D, Nagasawa T, Chen Y, Ushida Y, Kobayashi H, Takeuchi Y, Umeda M, Izumi Y. Molecular mimicry of *Aggregatibacter actinomycetemcomitans* with β2 glycoprotein I. *Oral Microbiol Immunol* 2008; **23**: 401–5.

114 Chen YW, Nagasawa T, Wara-Aswapati N, Ushida Y, Wang D, Takeuchi Y, Kobayashi H, Umeda M, Inoue Y, Iwai T, Ishikawa I, Izumi Y. Association between periodontitis and anti-cardiolipin antibodies in Buerger disease. *J Clin Periodontol* 2009; **36**: 830–5.

115 Schenkein HA, Bradley JL, Purkall DB. Anticardiolipin in *Porphyromonas gingivalis* antisera causes fetal loss in mice. *J Dent Res* 2013; **92**: 814–8.

116 Belkaid Y, Hand TW. Role of the microbiota in immunity and inflammation. *Cell* 2014; **157**: 121–41.

117 Fukuda S, Ohno H. Gut microbiome and metabolic diseases. *Semin Immunopathol* 2014; **36**: 103–14.

118 Boutaga K, Savelkoul PH, Winkel EG, van Winkelhoff AJ. Comparison of subgingival bacterial sampling with oral lavage for detection and quantification of periodontal pathogens by real-time polymerase chain reaction. *J Periodontol* 2007; **78**: 79–86.

119 Saygun I, Nizam N, Keskiner I, Bal V, Kubar A, Acikel C, Serdar M, Slots J. Salivary infectious agents and periodontal disease status. *J Periodontal Res* 2011; **46**: 235–9.

120 von Troil-Linden B, Torkko H, Alaluusua S, Jousimies-Somer H, Asikainen S. Salivary levels of suspected periodontal pathogens in relation to periodontal status and treatment. *J Dent Res* 1995; **74**: 1789–95.

121 Kumar PS, Leys EJ, Bryk JM, Martinez FJ, Moeschberger ML, Griffen AL. Changes in periodontal health status are associated with bacterial community shifts as assessed by quantitative 16S cloning and sequencing. *J Clin Microbiol* 2006; **44**: 3665–73.

122 Qin N, Yang F, Li A, Prifti E, Chen Y, Shao L, Guo J, Le Chatelier E, Yao J, Wu L, Zhou J, Ni S, Liu L, Pons N, Batto JM, Kennedy SP, Leonard P, Yuan C, Ding W, Hu X, Zheng B, Qian G, Xu W, Ehrlich SD, Zheng S, Li L. Alterations of the human gut microbiome in liver cirrhosis. *Nature* 2014; **513**: 59–64.

123 Arimatsu K, Yamada H, Miyazawa H, Minagawa T, Nakajima M, Ryder MI, Gotoh K, Motooka D, Nakamura S, Iida T, Yamazaki K. Oral pathobiont induces systemic inflammation and metabolic changes associated with alteration of gut microbiota. *Sci Rep* 2014; **4**: 4828.

124 Kaliannan K, Hamarneh SR, Economopoulos KP, Nasrin Alam S, Moaven O, Patel P, Malo NS, Ray M, Abtahi SM, Muhammad N, Raychowdhury A, Teshager A, Mohamed MM, Moss AK, Ahmed R, Hakimian S, Narisawa S, Millan JL, Hohmann E, Warren HS, Bhan AK, Malo MS, Hodin RA. Intestinal alkaline phosphatase prevents metabolic syndrome in mice. *Proc Natl Acad Sci U S A* 2013; **110**: 7003–8.

125 Kim HS, Berstad A. Experimental colitis in animal models. *Scand J Gastroenterol* 1992; **27**: 529–37.

126 Ohkawara T, Nishihira J, Takeda H, Hige S, Kato M, Sugiyama T, Iwanaga T, Nakamura H, Mizue Y, Asaka M. Amelioration of dextran sulfate sodium-induced colitis by anti-macrophage migration inhibitory factor antibody in mice. *Gastroenterology* 2002; **123**: 256–70.

127 Braun J, Wei B. Body traffic: ecology, genetics, and immunity in inflammatory bowel disease. *Annu Rev Pathol* 2007; **2**: 401–29.

128 Oz HS, Ebersole JL. A novel murine model for chronic inflammatory alveolar bone loss. *J Periodontal Res* 2010; **45**: 94–9.

129 Kesavalu L, Bakthavatchalu V, Rahman MM, Su J, Raghu B, Dawson D, Fernandes G, Ebersole JL. Omega-3 fatty acid regulates inflammatory cytokine/mediator messenger RNA expression in *Porphyromonas gingivalis*-induced experimental periodontal disease. *Oral Microbiol Immunol* 2007; **22**: 232–9.

130 Yu HN, Zhu J, Pan WS, Shen SR, Shan WG, Das UN. Effects of fish oil with a high content of n-3 polyunsaturated fatty acids on mouse gut microbiota. *Arch Med Res* 2014; **45**: 195–202.

131 Ding T, Schloss PD. Dynamics and associations of microbial community types across the human body. *Nature* 2014; **509**: 357–60.

CHAPTER 16

The vaginal microbiota in health and disease

S. Tariq Sadiq and Phillip Hay

St. George's, University of London, United Kingdom

16.1 What makes a healthy microbiota

Clinically, a healthy vagina is defined as the absence of vaginal discharge and symptoms of vaginitis and a low pH of vaginal secretions. For many years the study of microbial associations of the healthy vaginal state has relied on traditional culture techniques, which have demonstrated a protective role of some microbes, particularly various *Lactobacillus* species (described by Wilson in chapter 1). More recently, the use of molecular approaches including targeted PCR, conventional Sanger and high throughput parallel sequencing has allowed more detailed descriptions of the microbial constituents of the vagina.

Some of these studies have attempted to categorize the different states of the healthy vaginal microbiota and one such recent investigation, which was adjusted for ethnicity, suggested that microbial composition can be divided in to five cluster-types, four of which are dominated by lactobacilli, *L. crispatus, L. iners, L. jensenii and L. gasserii*[1]. A fifth cluster-type, perhaps comprising up to a quarter of healthy women, has a more diverse microbiotic composition in which lactobacilli may be present, but are depleted in relation to others organisms, many of which are traditionally associated with bacterial vaginosis (BV), the commonest vaginal disease syndrome worldwide. Interestingly, this study raised caution over interpreting vaginal microbiome findings in studies that defined vaginal health exclusively using the Nugent score[2], which is a gold-standard semi-quantitative microscopic morphological scoring system designed to diagnose BV, as many women carrying lactobacilli-dominant cluster-types would have been diagnosed with BV using Nugent.

The differences in vaginal microbiota in health and disease appear to be influenced by ethnicity. Black and Hispanic women, for example, have a higher preponderance towards lactobacilli-depleted vaginal microbiota compared to Asian and Caucasian women[3,4]. This in itself may go some way to explain observed differences in apparent susceptibility to sexually transmitted infections (STIs) and HIV in different populations (see below). These may also reflect a host-microbiota evolutionary trade-off between the need to protect against the threat of ascending

The Human Microbiota and Chronic Disease: Dysbiosis as a Cause of Human Pathology, First Edition.
Edited by Luigi Nibali and Brian Henderson.
© 2016 John Wiley & Sons, Inc. Published 2016 by John Wiley & Sons, Inc.

pathogenic infection of the reproductive tract with its impact on fertility and the nutritional and environmental constraints faced by different human populations, although much more research is required to demonstrate this.

16.1.1 How does the vaginal microbiota mediate healthiness?

Lactobacillus spp. are thought to maintain vaginal health through competitive exclusion of other microbes, for example by expressing proteins that affect adherence of harmful microbes such as *Gardnarella vaginalis, Staphylococcal aureus*, group B streptococcus and *E. coli* to vaginal mucosa[5,6]. Lactobacilli may also produce microbicidal and virucidal compounds such as hydrogen peroxide (H_2O_2)[7] and lactic acid, the latter of which is particularly active when the overall pH of vaginal secretions is low[8], as well as other compounds[9] that have activity against urogenital tract pathogens, including STIs such as *N. gonorrhoea* and HIV.

In turn, a healthy microbiota is able to induce host tolerance in vaginal tissue, to the microbes comprising the microbiota itself, through mechanisms dependent on bacterial compounds such as lactic acid[10] or by recognition of microbial associated molecular patterns (MAMPs) by the host innate immune system[11]. In contrast, unhealthy microbiota associated with BV and STIs are able to induce host pro-inflammatory responses and have been associated with recruitment of cellular immune cells that may act as targets for other STIs and HIV infection[12,13].

Finally, an abnormal vaginal microbiota may inhibit the host's own anti-microbial factors released at mucosal surfaces, increasing vulnerability to both HIV infection and STIs. A good example of this effect is secretory leukocyte protease inhibitor (SLPI), which has natural anti-HIV activity. BV and other STIs have been shown to reduce vaginal concentrations of SLPI[14].

16.1.2 Establishment of the vaginal microbiota

How vaginal microbial communities are established in early life is as yet unclear. Maternal microbiota may be important, potentially inferred from evidence from the development of adult gut microbiota, where exposure of sterile fetal gut to maternal vaginal, gastro-intestinal and fecal microbes during delivery may play a role[15]. Work from different geographical regions has found post-natal fecal microbiota differences in babies born by elective cesarian and vaginal delivery (discussed by Bermudez in chapter 6)[16,17]. It is tempting to speculate that pre-pubertal vaginal microbiota environments, which generally lack lactobacilli, might be influenced in a similar way. Additional sources of vaginal microbiota in adulthood may include ascending colonization from perineal skin and gut or even blood-borne translocation from other anatomical microbiomes such as the oral cavity[18]. One interesting area for which there is some evidence is the link between periodontitis and BV both epidemiologically and microbiologically[19]. The vaginal microbiome once established may extend into the uterus without any evidence of inflammation, suggesting stable carriage[20].

16.1.3 The role of host genetic variation on vaginal health

The variable susceptibility to BV, its predominance in certain ethnic populations and the interaction between microbiome and innate immunity suggest that host genetics may play a role in the maintenance of a healthy vagina. However, the fact that many common causes and practices strongly influence the development of BV makes this association challenging to demonstrate. A recent study looking at Toll-like receptor (TLR) variants in TLR1, TLR2, TLR4, TLR6 and TLR9 among African American women infected with HIV infection found associations with BV for two allelic variants in TLR4 and TLR9[21]. TLR4 activation triggers responses to Gram-negatives through induction of pro-inflammatory cytokines, particularly tumor necrosis factor (TNF) and interleukin-1 beta (IL1-β) and interestingly, the same TLR4 non-synonomous SNP (896 A > G, resulting in replacement of aspartic acid with glycine in the final TLR protein), has been associated with differences in susceptibility to various infections[22] and specifically in relation to vaginal health, reduced IL1-β, a tenfold higher load of *G. vaginalis,* a higher pH and with presence of other anaerobic Gram-negative rods in pregnant women[23,24]. More work will be needed to establish the role of host genetic predisposition to vaginal ill-health, but a clearer understanding may allow for a more targeted approach to manage patients.

16.1.4 Impact of age, menstrual cycle and environmental factors on vaginal health

The onset of puberty and the menstrual cycle, perhaps through fluctuations of hormonal oestrogen concentrations and consequent glycogen production, impacts the lactobacilli load in the reproductive tract. During menses, low glycogen levels may result in depletion of lactobacilli, perhaps particularly for *L. crispatus*[25], and high vaginal iron concentrations cause growth of *G. vaginalis*[26]. A recent finding, perhaps suggestive of adaption during menses, is that some strains of *L. jensenii* appear to co-opt iron, which may enable better adhesion of the lactobacillus to vaginal epithelium[27].

Studies have demonstrated that menopause is associated with decreased expression of glycogen from vaginal epithelia,[28] that in turn reduces the bacterial load of lactobacilli. There is perhaps mixed evidence for the role of hormonal contraception on reducing the likelihood of BV[29,30], although the same studies demonstrate reduction in abundance of some microbial species. An independent association with cigarette smoking and "unhealthy vaginal microbiota" has been previously demonstrated[31].

16.2 The vaginal microbiota in disease

16.2.1 Bacterial vaginosis

BV is the most common cause of abnormal vaginal discharge in women of child-bearing age. It is a syndrome of unknown cause characterized by depletion of the normal Lactobacillus population and an overgrowth or dysbiosis of vaginal anaerobes, accompanied by loss of the usual vaginal acidity. In 1983 the term "bacterial vaginosis" replaced the older term "Gardnerella vaginitis." This recognized the fact that many anaerobic or facultative anaerobic bacteria are present and that classical signs of inflammation are absent[32]. Women with symptomatic BV report an offensive, fishy-smelling discharge that is most noticeable after unprotected intercourse or at the time of menstruation. The diagnosis can be confirmed by microscopy ± additional tests. About 50% of cases are asymptomatic. BV is associated with infective complications in pregnancy and following gynecological surgery, and is a risk factor for the acquisition of STIs including HIV (see below). The global prevalence of microscopically defined BV varies considerably but may be as high as 50% in sub-Saharan Africa[33] compared to between 10–20%, in the UK in unselected populations. The prevalence may regress in pregnancy[34] and is predictably higher in women presenting to sexual health clinics[35].

16.2.1.1 Etiology and pathogenesis of BV

The triggers for BV are probably multiple but may include perhaps over-zealous vaginal cleaning[36], the onset of menses, a recent change of sex partner[37], exposure to semen[38], cigarette smoking[31] and having an STI, particularly *T. vaginalis* infection and *herpes simplex* type 2[39,40]. Recent work has suggested that HIV infection, whether stable or progressive, plays little part in the makeup of the vaginal microbiota[31], although this cohort was studied in a developed world setting.

When BV develops, lactobacilli reduce in concentration and may disappear, whilst there is an increased concentration of anaerobic and facultative anaerobic organisms. These changes reduce the inhibitory mediators described above such as lactic acid, H_2O_2, and host factors, resulting in an increase in vaginal pH from the normal 3.5–4.5 to as high as 7.0. More alkaline vaginal secretions, whether derived from bioactive amines of BV microbes or from semen, reduce the inhibitory effect of H_2O_2 on anaerobic growth further[38,41]. Hormonal changes and inoculation with organisms from a partner may also be important. The organisms classically associated with BV using culture include *Gardnerella vaginalis*, Bacteroides (Prevotella), *Mycoplasma hominis* and *Mobiluncus* spp. Those more recently identified using molecular techniques include *Atopobium vaginae*, *Prevotella* spp., *Leptotrichia* spp., *Megasphaera* spp., *Eggerthella* spp., *Dialister* spp., *Bifidobacterium* spp., *Slackia* spp. and BV associated bacteria type 1 (BVAB1), BVAB2 and BVAB3 (reviewed in[42]). An interesting description of BV is that of a development of a biofilm in which *G. vaginalis* is at the centre of BV pathogenesis[43] and accounts for 90% of bacteria seen in the biofilm. In this description, *A. vaginae* was the only other numerically important organism. Lactobacilli predominate in women with normal flora, but do not form a biofilm. Since the pathogenesis of BV is likely to be at the epithelial surface, this raises the possibility that all the other organisms found in vaginal fluid are epiphenomena, exploiting an altered vaginal environment, whilst Gardnerella is the critical factor.

Although BV often arises spontaneously around the time of menstruation, it may resolve spontaneously in mid-cycle. Post-menopause, despite lactobacillus load being reduced, it is not known how often BV occurs. BV is a clinical diagnosis and should be suspected in any woman presenting with an offensive, typically fishy-smelling vaginal discharge. Speculum examination shows a thin, homogeneous, white or yellow discharge adherent to the walls of the vagina. Amsel criteria[44] have been the mainstay of diagnosis in settings such as sexual health clinics where Gram stain microscopy can be performed. Epithelial cells covered with so many small bacteria that the border is fuzzy are termed "clue cells" because their presence is a clue to the diagnosis[45].

Recent studies have suggested that there is a continuum from normal Lactobacillus dominated microbiota to "severe BV." This is recognized in Gram-stain scoring systems such as the previously mentioned Nugent score but also the Hay/Ison[46] score, but not with Amsel criteria. For a comparison of these criteria for diagnosing BV, see Table 1. More recently, attempts to exploit knowledge of

Table 1 Comparison of diagnostic criteria for bacterial vaginosis.

Composite (Amsel)	Nugent Score	Hay-Ison criteria
Criteria	Gram stain of vaginal smear. Gram stain of vaginal smear. 10–20 averaged	Gram stain of vaginal smear. 1000 × oil immersion fields
At least three of the four criteria must be fulfilled to make a diagnosis of Bacterial Vaginosis	Total scores for increasing presence of following morphotypes: Lactobacillus-like - 0–4; Gardnerella-like - 0–4; Mobiluncus-like - 0–2.	Comparison of the mixture of different bacterial morphotypes
Vaginal pH >4.5	0–3: Normal	Grade 1 (Normal) Lactobacillus dominant
Release of a fishy smell on addition of alkali (10% potassium hydroxide)	4–6: Intermediate (if clue cells are found, diagnosed as BV)	Grade 2 (Intermediate) Mixed Lactobacillus with Gardnerella/Mobiluncus
Characteristic discharge on examination	7–10: Bacterial Vaginosis	Grade 3 (Bacterial vaginosis) Few lactobacilli outnumbered by Gardnerella/Mobiluncus
Presence of "clue cells" on microscopy of vaginal fluid mixed with normal saline		

the relationship between relative abundance of key bacterial species associated with BV and the BV phylotype have led to use of molecular methods to diagnose BV. One such recently proposed method includes a quantitative bacterial specific rRNA PCR. Molecular methods such as these will need to be validated in proper diagnostic and clinical evaluations before being used to make definitive diagnoses[47].

The microbiota in BV responds rapidly to antibiotic treatment within hours. BVABs are cleared more quickly following treatment with metronidazole compared to other facultative anaerobes such as *G. vaginalis* and the environmental niche repopulated with lactobacilli species relatively quickly. However, both BVAB and Gardnerella can quickly repopulate the vaginal epithelium when antibiotics are stopped[48]. Maintaining healthy microbiota in the face of BV may be achieved by various strategies including regular screening and treatment for BV[31], the use of directly applied probiotics containing lactobacilli[49], prebiotics containing iron chelators such as human lactoferrin[50] and oestrogens. All these strategies are undergoing investigation for the prevention of pre-term birth[51].

It has been a matter of debate as to whether BV is sexually transmissible. A meta-analysis concluded that BV has the characteristics of an STI, being associated with partner change and other STIs[37], and further support comes from a study that reported no BV in women denying any oral or digital genital contact[52]. An older study suggested a clinical link between BV in women and urethritis in their male partners[53] and more recent studies have suggested BV associated organisms as urethral pathogens[54]. There is a stronger case for sexual transmission of BV organisms between women who have sex with women[55]. The strongest evidence against BV being an STI has come from studies reporting similar rates in self-reported virgin and non-virgin women[56,57].

16.2.2 Clinical consequences of altered vaginal microbiota (see Figure 1)

BV is associated with second-trimester miscarriage and preterm birth. The reported odds ratio is 1.4–7.0. It is thought that women with BV are at increased risk of chorio-amnionitis, which can stimulate preterm birth through the release of pro-inflammatory cytokines[58]. Several studies have assessed the value of screening for and treatment of BV in preventing adverse outcomes in pregnancy. The results have been variable; some studies show a benefit with treatment in terms of reducing preterm birth rates, but the largest study to date showed no benefit from treatment with short courses of metronidazole[59]. On the basis of these studies, it cannot be concluded that antibiotic treatment of BV in pregnancy will universally reduce the incidence of preterm birth. This was confirmed by the most recent Cochrane Review[60].

Women infected with *Chlamydia trachomatis* who undergo elective termination of pregnancy are at high risk of endometritis and pelvic inflammatory disease. BV also confers an increased risk, and may be present in almost 30% of such women. A double-blind, placebo-controlled trial in Sweden showed that the risk of endometritis in women without Chlamydia was 12.2% in a placebo-treated group and 3.8% in those prescribed oral metronidazole before termination[61]. A more recent randomized controlled trial in Sweden found a fourfold reduction in infective complications with clindamycin cream compared with placebo[62].

BV has been associated with vaginal cuff cellulitis, wound infection and abscess formation after hysterectomy. No randomized trials have been performed to investigate the value of screening and treatment before such surgery. The potential role of BV in infections following IUD insertion, hysteroscopy, and dilatation and curettage has not been systematically studied.

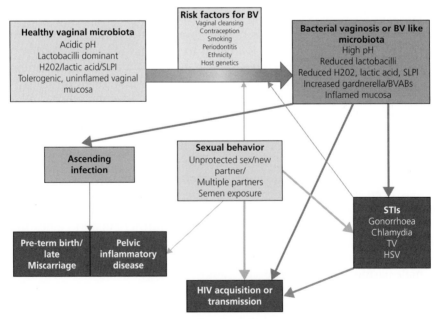

Figure 1 The complex relationship among the transition from healthy to unhealthy vaginal microbiota, sexual behavior and poor clinical outcomes. The transition is associated with severe clinical and global health consequences. SLPI: secretory leukocyte protease inhibitor.

16.2.3 Vaginal microbiota and transmission and susceptibility to HIV infection

There is considerable evidence that abnormal vaginal microbiota or BV is associated with an increased risk of acquiring HIV infection following exposure. An excellent review of the literature recently described the most up-to-date evidence and the multiple, interacting and cyclical ways in which disturbed vaginal microbiota might influence HIV transmission[42]. Diagnoses of abnormal flora in the reviewed studies was made by Nugent score, Amsel criteria or both and in some studies the risk of acquiring HIV was associated for both cases of BV and in intermediate states using the Nugent score. All studies reviewed were conducted in high-HIV-prevalence sub-Saharan African countries. Adjusted hazard ratios (adjusted for concurrent STIs, sexual behavior and other known risk factors) for HIV acquisition ranged from 1.15 to 2.02. In settings where both HIV and bacterial vaginosis/intermediate flora are high, these perhaps seemingly low risk ratios translate into a considerable number of actual HIV acquisitions attributable to HIV. For example, it has been estimated that if BV prevalence was 30 percent and the adjusted hazards ratio 1.6, the proportion of HIV infections attributable to BV would be approximately 15 percent. This is similar to the attributable risk associated with some STIs where hazards ratios are generally higher and in the order of 2–5[63]. There is also growing evidence for male partners of women infected with HIV who also have BV to be at increased risk of HIV infection. However, the attributable risks for such transmission are unknown but may also be considerable[64], with hazards ratios of acquiring HIV in uninfected male partners of greater than 3 observed[65]. It is likely, given the potential mechanisms of mucosal protection described above, that future studies will delineate more clearly the precise mechanisms through which this happens.

16.3 Conclusions

The recent advances in high-throughput genome sequencing have allowed greater understanding of the compositions of microbiota that inhabit the human vagina, including more in-depth descriptions of shifts in microbiota from health to disease, vulnerability to HIV infection and STIs, and the risks of pre-term birth and late miscarriage. Even greater understanding is likely to be gained as gene, protein and metabolic functional descriptions are integrated with knowledge of microbial composition, perhaps allowing future personalized treatment and prevention interventions tailored to specific vaginal micro-ecologies.

TAKE-HOME MESSAGE

- A healthy vaginal microbiota is vital to the maintenance of female reproductive health and to global health.
- Improved understanding of mechanism of altered vaginal microbiota should lead to better prevention and therapeutic strategies to prevent serious clinical outcomes such as HIV infection, pre-term birth and late miscarriage.

References

1 Ravel J, Gajer P, Abdo Z *et al.* Vaginal microbiome of reproductive-age women. *Proceedings of the National Academy of Sciences of the United States of America* 2011; **108 Suppl 1**: 4680–7.

2 Nugent RP, Krohn MA, Hillier SL. Reliability of diagnosing bacterial vaginosis is improved by a standardized method of gram stain interpretation. *Journal of Clinical Microbiology* 1991; **29**: 297–301.

3 Zhou X, Brown CJ, Abdo Z *et al.* Differences in the composition of vaginal microbial communities found in healthy Caucasian and black women. *The ISME Journal* 2007; **1**: 121–33.

4 Jespers V, Menten J, Smet H *et al.* Quantification of bacterial species of the vaginal microbiome in different groups of women, using nucleic acid amplification tests. *BMC Microbiology* 2012; **12**: 83.

5 Ojala T, Kankainen M, Castro J *et al.* Comparative genomics of Lactobacillus crispatus suggests novel mechanisms for the competitive exclusion of Gardnerella vaginalis. *BMC genomics* 2014; **15**: 1070.

6 Cadieux PA, Burton J, Devillard E *et al.* Lactobacillus by-products inhibit the growth and virulence of uropathogenic Escherichia coli. *Journal of Physiology and Pharmacology: an Official Journal of the Polish Physiological Society* 2009; **60 Suppl 6**: 13–8.

7 Schwebke JR. Role of vaginal flora as a barrier to HIV acquisition. *Current Infectious Disease Reports* 2001; **3**: 152–5.

8 O'Hanlon DE, Moench TR, Cone RA. Vaginal pH and microbicidal lactic acid when lactobacilli dominate the microbiota. *PLoS one* 2013; **8**: e80074.

9 Witkin SS, Linhares IM, Giraldo P. Bacterial flora of the female genital tract: function and immune regulation. *Best Practice & Research. Clinical Obstetrics & Gynaecology* 2007; **21**: 347–54.

10 Mossop H, Linhares IM, Bongiovanni AM *et al.* Influence of lactic acid on endogenous and viral RNA-induced immune mediator production by vaginal epithelial cells. *Obstetrics and Gynecology* 2011; **118**: 840–6.

11 Hedges SR, Barrientes F, Desmond RA *et al.* Local and systemic cytokine levels in relation to changes in vaginal flora. *The Journal of Infectious Diseases* 2006; **193**: 556–62.

12 Fichorova RN, Buck OR, Yamamoto HS *et al.* The villain team-up or how Trichomonas vaginalis and bacterial vaginosis alter innate immunity in concert. *Sexually Transmitted Infections* 2013; **89**: 460–6.

13 Shattock RJ, Haynes BF, Pulendran B *et al.* Improving defences at the portal of HIV entry: mucosal and innate immunity. *PLoS Medicine* 2008; **5**: e81.

14 Huppert JS, Huang B, Chen C *et al.* Clinical evidence for the role of Trichomonas vaginalis in regulation of secretory leukocyte protease inhibitor in the female genital tract. *The Journal of infectious Diseases* 2013; **207**: 1462–70.

15 Tomasello G, Tralongo P, Damiani P *et al.* Dismicrobism in inflammatory bowel disease and colorectal cancer: changes in response of colocytes. *World Journal of Gastroenterology : WJG* 2014; **20**: 18121–30.

16 Liu D, Yu J, Li L *et al.* Bacterial community structure associated with elective cesarean section versus vaginal delivery in chinese newborns. *Journal of Pediatric Gastroenterology and Nutrition* 2015; **60**: 240–6.

17 Musilova S, Rada V, Vlkova E *et al.* Colonisation of the gut by bifidobacteria is much more common in vaginal deliveries than Caesarean sections. *Acta Paediatrica* 2015.

18 Payne MS, Bayatibojakhi S. Exploring preterm birth as a polymicrobial disease: an overview of the uterine microbiome. *Frontiers in Immunology* 2014; **5**: 595.

19 Nibali L, Henderson B, Sadiq ST *et al.* Genetic dysbiosis: the role of microbial insults in chronic inflammatory diseases. *Journal of Oral Microbiology* 2014; **6**.

20 Mitchell CM, Haick A, Nkwopara E *et al.* Colonization of the upper genital tract by vaginal bacterial species in nonpregnant women. *American Journal of Obstetrics And Gynecology* 2014.

21 Royse KE, Kempf MC, McGwin G, Jr. *et al.* Toll-like receptor gene variants associated with bacterial vaginosis among HIV-1 infected adolescents. *Journal of Reproductive Immunology* 2012; **96**: 84–9.

22 Ziakas PD, Prodromou ML, El Khoury J *et al.* The role of TLR4 896 A>G and 1196 C>T in susceptibility to infections: a review and meta-analysis of genetic association studies. *PLoS One* 2013; **8**: e81047.

23 Genc MR, Vardhana S, Delaney ML *et al.* Relationship between a toll-like receptor-4 gene polymorphism, bacterial vaginosis-related flora and vaginal cytokine responses in pregnant women. *European Journal Of Obstetrics, Gynecology, and Reproductive Biology* 2004; **116**: 152–6.

24 Ryckman KK, Williams SM, Krohn MA *et al.* Interaction between interleukin-1 receptor 2 and Toll-like receptor 4, and cervical cytokines. *Journal of Reproductive Immunology* 2011; **90**: 220–6.

25 Santiago GL, Cools P, Verstraelen H *et al.* Longitudinal study of the dynamics of vaginal microflora during two consecutive menstrual cycles. *PloS One* 2011; **6**: e28180.

26 Srinivasan S, Liu C, Mitchell CM *et al.* Temporal variability of human vaginal bacteria and relationship with bacterial vaginosis. *PloS One* 2010; **5**: e10197.

27 Martin R, Sanchez B, Urdaci MC *et al.* Effect of iron on the probiotic properties of the vaginal isolate Lactobacillus jensenii CECT 4306. *Microbiology* 2015.

28 Mirmonsef P, Modur S, Burgad D *et al.* Exploratory comparison of vaginal glycogen and Lactobacillus levels in premenopausal and postmenopausal women. *Menopause* 2014.

29 Vodstrcil LA, Hocking JS, Law M *et al.* Hormonal contraception is associated with a reduced risk of bacterial vaginosis: a systematic review and meta-analysis. *PloS One* 2013; **8**: e73055.

30 Borgdorff H, Verwijs MC, Wit FW *et al.* The impact of hormonal contraception and pregnancy on sexually transmitted infections and on cervicovaginal microbiota in African sex workers. *Sexually Transmitted Diseases* 2015; **42**: 143–52.

31 Mehta SD, Donovan B, Weber KM *et al.* The vaginal microbiota over an 8- to 10-year period in a cohort of HIV-infected and HIV-uninfected women. *PloS One* 2015; **10**: e0116894.

32 Easmon CS, Hay PE, Ison CA. Bacterial vaginosis: a diagnostic approach. *Genitourinary Medicine* 1992; **68**: 134–8.

33 Wawer MJ, Gray RH, Sewankambo NK *et al.* A randomized, community trial of intensive sexually transmitted disease control for AIDS prevention, Rakai, Uganda. *AIDS* 1998; **12**: 1211–25.

34 Hay PE, Morgan DJ, Ison CA *et al.* A longitudinal study of bacterial vaginosis during pregnancy. *British Journal of Obstetrics and Gynaecology* 1994; **101**: 1048–53.

35 Blackwell AL, Thomas PD, Wareham K *et al.* Health gains from screening for infection of the lower genital tract in women attending for termination of pregnancy. *Lancet* 1993; **342**: 206–10.

36 Fashemi B, Delaney ML, Onderdonk AB *et al.* Effects of feminine hygiene products on the vaginal mucosal biome. *Microbial Ecology in Health and Disease* 2013; **24**.

37 Fethers KA, Fairley CK, Hocking JS *et al.* Sexual risk factors and bacterial vaginosis: a systematic review and meta-analysis. *Clinical infectious Diseases: an Official Publication of the Infectious Diseases Society of America* 2008; **47**: 1426–35.

38 O'Hanlon DE, Lanier BR, Moench TR *et al.* Cervicovaginal fluid and semen block the microbicidal activity of hydrogen peroxide produced by vaginal lactobacilli. *BMC Infectious Diseases* 2010; **10**: 120.

39 Martin DH, Zozaya M, Lillis RA *et al.* Unique vaginal microbiota that includes an unknown Mycoplasma-like organism is associated with Trichomonas vaginalis infection. *The Journal of Infectious Diseases* 2013; **207**: 1922–31.

40 Masese L, Baeten JM, Richardson BA *et al.* Incident herpes simplex virus type 2 infection increases the risk of subsequent episodes of bacterial vaginosis. *The Journal of Infectious Diseases* 2014; **209**: 1023–7.

41 O'Hanlon DE, Moench TR, Cone RA. In vaginal fluid, bacteria associated with bacterial vaginosis can be suppressed with lactic acid but not hydrogen peroxide. *BMC Infectious Diseases* 2011; **11**: 200.

42 Buve A, Jespers V, Crucitti T *et al.* The vaginal microbiota and susceptibility to HIV. *AIDS* 2014; **28**: 2333–44.

43 Swidsinski A, Mendling W, Loening-Baucke V *et al.* Adherent biofilms in bacterial vaginosis. *Obstetrics and Gynecology* 2005; **106**: 1013–23.

44 Amsel R, Totten PA, Spiegel CA *et al.* Nonspecific vaginitis. Diagnostic criteria and microbial and epidemiologic associations. *The American Journal of Medicine* 1983; **74**: 14–22.

45 Gardner HL, Dukes CD. Haemophilus vaginalis vaginitis: a newly defined specific infection previously classified non-specific vaginitis. *American journal of Obstetrics And Gynecology* 1955; **69**: 962–76.

46 Chawla R, Bhalla P, Chadha S *et al.* Comparison of Hay's criteria with Nugent's scoring system for diagnosis of bacterial vaginosis. *BioMed Research International* 2013; **2013**: 365194.

47 Kurakawa T, Ogata K, Tsuji H *et al.* Establishment of a sensitive system for analysis of human vaginal microbiota on the basis of rRNA-targeted reverse transcription-quantitative PCR. *Journal of Microbiological Methods* 2015; **111C**: 93–104.

48 Mayer BT, Srinivasan S, Fiedler TL *et al.* Rapid and profound shifts in the vaginal microbiota following antibiotic treatment for bacterial vaginosis. *The Journal of Infectious Diseases* 2015.

49 Bodean O, Munteanu O, Cirstoiu C *et al.* Probiotics – a helpful additional therapy for bacterial vaginosis. *Journal of Medicine and Life* 2013; **6**: 434–6.

50 Jarosik GP, Land CB. Identification of a human lactoferrin-binding protein in Gardnerella vaginalis. *Infection and Immunity* 2000; **68**: 3443–7.

51 Otsuki K, Tokunaka M, Oba T *et al.* Administration of oral and vaginal prebiotic lactoferrin for a woman with a refractory vaginitis recurring preterm delivery: appearance of lactobacillus in vaginal flora followed by term delivery. *The Journal of Obstetrics and Gynaecology Research* 2014; **40**: 583–5.

52 Fethers KA, Fairley CK, Morton A *et al.* Early sexual experiences and risk factors for bacterial vaginosis. *The Journal of Infectious Diseases* 2009; **200**: 1662–70.

53 Keane FE, Thomas BJ, Whitaker L *et al.* An association between non-gonococcal urethritis and bacterial vaginosis and the implications for patients and their sexual partners. *Genitourinary Medicine* 1997; **73**: 373–7.

54 Manhart LE, Khosropour CM, Liu C *et al.* Bacterial vaginosis-associated bacteria in men: association of Leptotrichia/Sneathia spp. with nongonococcal urethritis. *Sexually Transmitted Diseases* 2013; **40**: 944–9.

55 Vodstrcil LA, Walker SM, Hocking JS *et al.* Incident Bacterial vaginosis (BV) in women who have sex with women is associated with behaviors that suggest sexual transmission of BV. *Clinical infectious Diseases: an Official Publication of the Infectious Diseases Society of America* 2014.

56 Bump RC, Buesching WJ, 3rd. Bacterial vaginosis in virginal and sexually active adolescent females: evidence against exclusive sexual transmission. *American Journal of Obstetrics and Gynecology* 1988; **158**: 935–9.

57 Vaca M, Guadalupe I, Erazo S *et al.* High prevalence of bacterial vaginosis in adolescent girls in a tropical area of Ecuador. *BJOG: an International Journal of Obstetrics And Gynaecology* 2010; **117**: 225–8.

58 Goldenberg RL, Hauth JC, Andrews WW. Intrauterine infection and preterm delivery. *The New England Journal of Medicine* 2000; **342**: 1500–7.

59 Carey JC, Klebanoff MA, Hauth JC *et al.* Metronidazole to prevent preterm delivery in pregnant women with asymptomatic bacterial vaginosis. National Institute of Child Health and Human Development Network of Maternal-Fetal Medicine Units. *The New England Journal of Medicine* 2000; **342**: 534–40.

60 McDonald HM, Brocklehurst P, Gordon A. Antibiotics for treating bacterial vaginosis in pregnancy. *The Cochrane Database of Systematic Reviews* 2007: CD000262.

61 Larsson PG, Platz-Christensen JJ, Thejls H *et al.* Incidence of pelvic inflammatory disease after first-trimester legal abortion in women with bacterial vaginosis after treatment with metronidazole: a double-blind, randomized study. *American Journal of Obstetrics And Gynecology* 1992; **166**: 100–3.

62 Larsson PG, Platz-Christensen JJ, Dalaker K *et al.* Treatment with 2% clindamycin vaginal cream prior to first trimester surgical abortion to reduce signs of postoperative infection: a prospective, double-blinded, placebo-controlled, multicenter study. *Acta Obstetricia et Gynecologica Scandinavica* 2000; **79**: 390–6.

63 Atashili J, Poole C, Ndumbe PM *et al.* Bacterial vaginosis and HIV acquisition: a meta-analysis of published studies. *AIDS* 2008; **22**: 1493–50.

64 Cone RA. Vaginal microbiota and sexually transmitted infections that may influence transmission of cell-associated HIV. *The Journal of Infectious Diseases* 2014; **210 Suppl 3**: S616–21.

65 Cohen CR, Lingappa JR, Baeten JM *et al.* Bacterial vaginosis associated with increased risk of female-to-male HIV-1 transmission: a prospective cohort analysis among African couples. *PLoS Medicine* 2012; **9**: e1001251.

SECTION 4

Dysbioses and chronic diseases: is there a connection?

CHAPTER 17

Reactive arthritis: the hidden bacterial connection

John D. Carter

University of South Florida Morsani College of Medicine, Tampa, FL, United States

17.1 Introduction

The spondyloarthritides are a group or arthritides that share clinical, radiographic, genetic and laboratory features. These include ankylosing spondylitis, psoriatic arthritis, inflammatory bowel disease-related spondyloarthritis, reactive arthritis, and undifferentiated spondyloarthritis. The articular features include an inflammatory axial arthritis with predilection for the sacroiliac joints, inflammatory arthritis of the peripheral joints that typically occurs in an oligoarticular pattern with predilection for the large joints of the lower extremities, enthesitis and dactylitis. Enthesitis represents inflammation at the bony tendinous insertions, such as Achilles tendonitis, plantar fasciitis, or epicondylitis. Dactylitis is diffuse swelling of a finger or toe, known as "sausage digit." At first glance, it is obvious that the joint-based inflammation seen in spondyloarthritis (SpA) is broad, including the synovium, tendons, and surrounding tissues. Further, the inflammation that accompanies SpA is not just articular based. Other organs that are frequently involved in the inflammatory process of SpA include the skin, eyes, and mucous membranes. Occasionally, internal organs such as the heart or lungs are also involved. Therefore, the disease process of SpA is not strictly an inflammatory arthritis; it is a systemic inflammatory state.

Much recent evidence suggests that the microbiota can contribute, or potentially be etiologic, to several different types of inflammatory arthritis[1–3]. These data are emerging, specifically, in relation to ankylosing spondylitis[4] and psoriatic arthritis[5]. There are also abundant data implicating the microbiota as causative agent for inflammatory bowel disease[6]: as stated, patients with inflammatory bowel disease can develop SpA. Interestingly, the gut microbiota seen in patients with psoriatic arthritis is similar to that of patients with inflammatory bowel disease[5]. Further links between the microbiome and the various types of SpA exist. It is well documented that many patients with ankylosing spondylitis will have histological evidence consistent with inflammatory bowel disease, even in the absence of gastrointestinal symptoms[7,8].

The Human Microbiota and Chronic Disease: Dysbiosis as a Cause of Human Pathology, First Edition.
Edited by Luigi Nibali and Brian Henderson.

As discussed extensively in this book, changes in the composition of the microbiota can lead to dysbiosis (defined by Curtis in chapter 2); in turn, this can cause human disease. Dysbiosis reflects the pathologic state of a microbial imbalance in or on the human body. This imbalance is commensal in nature, but clearly individual bacteria contribute to this imbalance, leading to human disease. These individual bacteria can vary between disease states potentially leading to variation in phenotypic disease expression. A group of researchers have argued for years that *Klebsiella* is the causative of ankylosing spondylitis[9]. Subsequent research could not replicate these findings, which questioned the true etiologic role of this bacterium in this specific SpA phenotype[10]. Ebringer and Wilson review the evidence for the interaction between Klebsiella and HLA-B27 in chapter 20. More recently, *Bacteroides* has been implicated as the cause of this same disease[11]. Perhaps a truer picture is that certain commensal groups of gastrointestinal bacteria may predispose to disease.

17.2 Reactive arthritis

There is perhaps no clearer example of a single bacterium causing secondary disease than reactive arthritis (ReA). In general, there are two types of ReA, post-venereal and post-dysentery ReA. ReA is an inflammatory disease that occurs 1–6 weeks after exposure to certain genitourinary and gastrointestinal infections. *Chlamydia, Salmonella, Shigella, Campylobacter*, and *Yersinia* are unequivocally responsible for the genesis of ReA in persons with the appropriate genetic constitution[12]; other organisms have been implicated as possible causes of ReA, but this chapter will focus on the definite bacterial triggers. The "classic triad" of symptoms of ReA includes synovitis, urethritis, and conjunctivitis, but the majority of patients do not present with the "classic triad" of symptoms and several other types of clinical features are frequently encountered[12]. These include a couple of specific inflammatory skin lesions: keratoderma blennorrhagicum and circinate balanitis. The former is a pustular and/or plaque-like rash that typically occurs on the palms and soles and the latter is a similar rash that occurs on the penis. Interestingly, both are histologically indistinct from pustular psoriasis[13]. Although the initial description of the "classic triad" of symptoms included urethritis and conjunctivitis[14], patients with ReA can also develop chronic gastrointestinal symptoms and histologic changes consistent with inflammatory bowel disease. Anterior uveitis is another frequent complication of ReA[12]. Because these clinical manifestations of ReA represent pathologic sequelae resulting from a known bacterial trigger and these same clinical manifestations are also clinical features of other types of SpA, this suggest further links between the different types of SpA.

Chlamydial infections and certain enteric infections are known etiologic agents for ReA. It is interesting to note that these different bacterial agents culminate in a set of clinical symptoms that are indistinct regardless of the original infecting organism. Further, important differences in the pathophysiology of post-chlamydial ReA compared to the post-enteric variety exist, yet the phenotypic features are the same[12]. This suggests a more complex role of the triggering bacteria than originally thought.

17.3 Pathophysiology of ReA

The pathophysiology of ReA shows the classic interplay between environment and genetics. The causative organisms are Gram-negative with lipopolysaccharide (LPS) as a key component in their cell wall. While we are beginning to understand aspects of the bacterial products and their effects during infection, we have very little understanding of the host's response to those products. For years it was apparent that these causative organisms had the capability of triggering ReA; yet there was a disconnect. Patients would experience an acute venereal or enteric infection and in the ensuing weeks develop systemic inflammatory disease, i.e. ReA. Investigations were performed on these patients to demonstrate the presence of these causative organisms in the synovial fluid of these ReA patients. However, cultures were routinely sterile. It was questioned if these organisms made it to the synovium at all; the thought was that these triggering organisms would initiate an autoimmune response that resulted in ReA. However, subsequent studies utilizing polymerase chain reaction (PCR) technology documented the presence of all the definite bacterial triggers of ReA, or their bacterial products, in the synovial tissue or fluid of patients with ReA[15-22]. This has been demonstrated in multiple studies involving many different laboratories. While patients with ReA do not have systemic infection in the traditional sense, these microbes disseminate with particular tropism for certain organs. In spite of the fact that all of these triggering microbes disseminate from their initial site of infection, there are important differences. The most important distinction involves viable bacterial persistence.

Chlamydia trachomatis is the most common cause of ReA[12]. When a patient acquires *C. trachomatis* as part of an acute genital infection, the epithelial cells are the most frequent targets for this primary infection. When these cells become infected the organisms proceed through a transcriptionally-governed developmental cycle. Differentiation from the extracellular, metabolically-inactive elementary body (EB) form to the intracellular, metabolically active reticulate body (RB) form occurs within the first several hours of intracellular growth[23]. Approximately 48 hours post-infection, most RB dedifferentiate back to the EB form and are released from the host cell by exocytosis or host cell lysis[24]. In those patients who experience an acute *C. trachomatis* infection, about 5% will develop ReA[25]. In patients with ReA, much data suggest that the monocytes are the vehicles of dissemination of the microbe from its site of primary infection. However, the journey of disseminated chlamydiae unfolds along a quite different pathway. Infection of monocytes appears relatively normal for the first 24 to 48 hours, but by 72 hours post-infection all intracellular chlamydiae have entered the unusual infection state designated "persistence"; transition to this state also is transcriptionally governed. During persistence, RB exhibit aberrant morphology, growing much larger than during active infections[26,27]. Persistence also is characterized by a low metabolic rate[17,18], ultrastructural changes[29], and resistance to host defenses[30]. The persistent state of *C. trachomatis* is characterized by the down-regulation of most genes required for cell division and differential upregulation of three paralog heat shock protein (Hsp) 60 genes (Ct110, Ct604, and Ct755)[28]. While much progress has been made in recent years in understanding the molecular and cellular biology of this unusual pathogen in its normal infection state, many aspects critical to understanding the genetic and other bases for persistent infection remain to be elucidated.

Like chlamydiae, the triggering microbes responsible for post-enteric ReA all have LPS components of the bacterial cell walls. All four of these organisms (*Salmonella, Shigella, Campylobacter,* and *Yersinia*) are also motile to varying degrees. *Salmonella* is one of the most common enteric infections in the U.S., and is the most frequently studied enteric bacteria associated with ReA. After salmonellosis, individuals of Caucasian descent may be more likely than those of Asian descent, to develop ReA[31]; and children are less susceptible than adults[32]. This further highlights host interaction and suggests that repetitive exposures might be more likely to cause disease. The attack rate of all four of the bacterial triggers for ReA ranges from about 1–30% of people exposed. Evidence of all the post-enteric organisms has been demonstrated in the synovial tissue of patients with ReA, and one of these studies suggested the organism (*Yersinia*) was viable[16], as is the case with chlamydiae.

In *Chlamydia*-induced ReA, specifically, data exist suggesting that previous infections with similar organisms can work in concert to elicit a more robust inflammatory response. The different chlamydiae have demonstrated an additive or synergistic effect in determining ReA attack rate or incidence. Given that *Chlamydia pneumoniae* is a common infection, previous exposure to *C. pneumoniae* could have an effect on a subsequent response to a *C. trachomatis* infection or *vice versa*. Reports have demonstrated that prior *C. pneumoniae* infection primes a Th1 T-cell response to *C. trachomatis* antigens[33].

Although recent data suggest the role that HLA-B27 plays in disease susceptibility has traditionally been overstated[12], it does play a role. HLA-B27 has multiple alleles that could influence host response and disease susceptibility. Few studies have analyzed the specific HLA-B27 alleles in the setting of ReA. One study suggests that HLA-B*2705 is the most common allele observed in B27-positive ReA patients[34]. This allele is seen less frequently than in the other SpA's and in B27 healthy controls. Another study suggests that HLA-B*5703 increases the risk for the classic triad of symptoms of ReA[35]. Data also suggest that HLA-B27 could lend towards bacterial persistence, specifically with *C. trachomatis* and *Salmonella*[36,37]. Further analysis of the role of HLA-B27 is presented in chapter 20.

Because ReA represents the classic interplay of host and environment and the bacterial triggers of disease are defined, much research has centered on the host in order to determine disease susceptibility. Although there is a link with HLA-B27, this is certainly not the sole determinant; variations in the microbe itself might be uniquely etiologic. In the case of *C. trachomatis*, this bacterial species is comprised of 14 strains/serovars (A-C are ocular/trachoma serovars, D-K are genital serovars), plus three biovars of lymphogranuloma venereum (LGV, L1-L3). Surprising recent data have demonstrated that these synovial-based persistent chlamydiae uniquely belong to the ocular, not genital, serovars[38]. It is important to note that acute chlamydial genital infections are not usually clonal and that a small percentage of these acute infections, usually in the order of 1–5%, include ocular strains[39,40]. Interestingly, this percentage mirrors the attack rate of *Chlamydia*-induced arthritis. Therefore, the triggering microbes of ReA might have uniquely arthritogenic features.

17.4 Questions remain

The question remains, however, whether these positive PCR findings demonstrating bacterial persistence in the synovial materials of ReA patients indicate that these patients experience a disseminated persistent infection that is the driving force of their ReA, or whether the disseminated presence of these causative organisms simply lays the foundation for an autoimmune phenomenon through a mechanism not completely understood. The causative bacteria of ReA have occasionally been demonstrated in the synovial tissue of patients with various types of arthritis and even some asymptomatic patients, so the importance of this finding has been questioned[41–43]. Further, bacterial DNA from various pathogens not associated with ReA has been discovered in synovial tissue[44,45]. These findings have led some to conclude that the triggering organisms are just that: they trigger an inflammatory response that becomes systemic, and an auto-regulated inflammatory condition ensues. The fact remains, however, that the synovial-based causative organisms are routinely documented at a statistically higher rate in patients with disease (ReA) compared to the control group[46]. Further, a randomized controlled trial assessing the effect of combination antibiotics in patients with PCR-documented *Chlamydia*-induced ReA demonstrated significant PCR clearance and resultant clinical benefit compared to placebo-treated patients[47].

Even more novel is the notion that the persistent bacteria or bacterial products that play a role in ReA could incorporate themselves in, and perhaps alter, an existing microbiome of the synovial tissue and other organs. A very interesting study was performed more than 10 years ago that assessed synovial samples of over 200 patients with various types of arthritis using a "pan-bacterial" PCR primer. The researchers discovered that ~10% of the 237 subjects were PCR positive[44]. This might represent a low number because any samples from patients that were known to be positive for *Chlamydia, Borrelia*, and *Mycoplasma* were excluded and many of the samples studied were synovial fluid (tissue is more likely to give positive results in such a study). It was not clear if such a finding had any clinical significance, so this and similar data led to much of the medical community downplaying the significance. Perhaps this study revealed a truth that was hidden in plain sight? Perhaps a synovial-based microbiome exists and when new or foreign bacterial incorporate into this microbiome a dysbiosis ensues.

Emerging data indicate that dysbiosis in the gut can exacerbate or even be causative of many chronic diseases including certain types of SpA, namely psoriatic arthritis and ankylosing spondylitis. Because of the clinical overlap of these types of SpA with ReA, a case can be made that dysbiosis could play a similar role in this inflammatory arthritis. In the case of post-enteric ReA, the initial gastrointestinal infection could alter the gut microbiome leading to dysbiosis; with postchlamydial ReA, the dysbiosis might occur at the level of the synovia. Shenderov introduced the concept of the microbiota as an epigenetic control mechanism in chapter 11. It is also very likely that epigenetics plays a role in the establishment of disease. It has been demonstrated (and described above) that the gene production of persistent chlamydiae is transcriptionally altered. It remains to be elucidated if this transcriptional regulation occurs independently with *chlamydia*, or as a result of some commensal effect. It is well documented that in ReA, whether the triggering infection is gastrointestinal or genitourinary, the result is the same phenotypic disease. Along these same lines, it is entirely possible that dysbiosis, and subsequent epigenetics, in the gut versus the joint could also lead to similar phenotypic manifestations.

17.5 Conclusion

The quest to prove that infectious agents are etiologic for chronic human diseases dates back to antiquity. Perhaps there is no better example of a bacterial infection leading to a secondary sub-acute or chronic disease as with the case of ReA. This condition represents a classic example of host-pathogen interaction leading to human disease. Modern laboratory techniques have proven that the triggering bacteria of ReA travel from the initial site of the infection to the synovial tissue. In the case of *Chlamydia trachomatis*, these bacteria are transcriptionally down-regulated, but persistently viable. In the case of the post-enteric organisms, these often lead to chronic gastrointestinal symptoms suggesting a change in the gut microbiota. Both types of infections could lead to dysbiosis. The effect that treatment, especially antibiotics, has on this dysbiosis and resulting human disease remains largely unknown. More studies need to be done assessing the possible role of dysbiosis in ReA.

TAKE-HOME MESSAGE

- Reactive arthritis represents the classic interplay of host with environment.
- Certain bacteria have unequivocally been shown to be responsible for the genesis of reactive arthritis.
- These same bacteria have been documented to travel from the initial infecting site to the synovium.
- In the case of *chlamydiae*, these synovial-based organisms are persistently viable.
- The possibility remains that these same triggering microbes could create a dysbiosis in the gut or synovium resulting in phenotypic disease.

References

1 Scher JU, Abramson SB. The microbiome and rheumatoid arthritis. *Nature Reviews Rheumatology.* 2011; **7**(10): 569–78.
2 Brusca SB, Abramson SB, Scher JU. Microbiome and mucosal inflammation as extra-articular triggers for rheumatoid arthritis and autoimmunity. *Curr Opin Rheumatol.* 2014 Jan; **26**(1): 101–7.
3 McLean MH, Dieguez D Jr, Miller LM, Young HA. Does the microbiota play a role in the pathogenesis of autoimmune diseases? *Gut.* 2015; **64**(2): 332–341.
4 Costello ME, Ciccia F, Willner D, Warrington N, Robinson PC, Gardiner B, Marshall M, Kenna TJ, Triolo G, Brown MA. Intestinal dysbiosis in ankylosing spondylitis. *Arthritis Rheumatol* 2014 Nov 21. doi: 10.1002/art.38967. [Epub ahead of print].
5 Scher JU, Ubeda C, Artacho A, Attur M, Isaac S, Reddy SM, Marmon S, Neimann A, Brusca S, Patel T, Manasson J, Pamer EG, Littman DR, Abramson SB. Decreased bacterial diversity characterizes the altered gut microbiota in patients with psoriatic arthritis, resembling dysbiosis in inflammatory bowel disease. *Arthritis Rheumatol* 2015 **67**(1): 128–39.

6 Matsuoka K, Kanai T. The gut microbiota and inflammatory bowel disease. *Semin Immunopathol* 2015; **37**(1): 47–55.

7 Mielants H, Veys EM, Goemaere S. Gut inflammation in the spondyloarthropathies: clinical, radiologic, biologic and genetic features in relation to the type of histology: a prospective study. *J Rheumatol* 1991; **18**(10): 1542–51.

8 Leirisalo-Repo M, Turunen U, Stenman S. High frequency of silent inflammatory bowel disease in spondylarthropathy. *Arthritis Rheum* 1994; **37**(1): 23–31.

9 Ebringer R, Cooke D, Cawdell DR, Cowling P, Ebringer A. Ankylosing spondylitis: klebsiella and HL-A B27. *Rheumatol Rehabil* 1977; **16**(3): 190–6.

10 Stone MA, Payne U, Schentag C, Rahman P, Pacheco-Tena C, Inman RD. Comparative immune responses to candidate arthritogenic bacteria do not confirm a dominant role for Klebsiella pneumonia in the pathogenesis of familial ankylosing spondylitis. *Rheumatology* (Oxford) 2004; **43**(2): 148–55.

11 Stebbings SM, Taylor C, Tannock GW, Baird MA, Highton J. The immune response to autologous bacteroides in ankylosing spondylitis is characterized by reduced interleukin 10 production. *J Rheumatol* 2009; **36**(4): 797–800.

12 Carter JD, Hudson AP. Reactive arthritis: clinical aspects and medical management. *Rheum Dis Clin North Am* 2009; **35**(1): 21–44.

13 Schneider JM, Matthews JH, Graham BS. Reiter's syndrome. *Cutis* 2003; **71**(3): 198–200.

14 Bauer W, Engelmann EP. Syndrome of unknown aetiology characterized by urethritis, conjunctivitis, and arthritis (so-called Reiter's Disease). *Trans Ass Amer Physics* 1942; **57**: 307–8.

15 Braun J, Tuszewski M, Eggens U, Mertz A, Schauer-Petrowskaja C, Döring E, Laitko S, Distler A, Sieper J, Ehlers S. Nested polymerase chain reaction strategy simultaneously targeting DNA sequences of multiple bacterial species in inflammatory joint diseases. I. Screening of synovial fluid samples of patients with spondyloarthropathies and other arthritides. *J Rheumatol* 1997; **24**(6): 1092–100.

16 Gaston JS, Cox C, Granfors K. Clinical and experimental evidence for persistent Yersinia infection in reactive arthritis. *Arthritis Rheum* 1999; **42**(10): 2239–42.

17 Gerard HC, Branigan PJ, Schumacher HR Jr, Hudson AP. Synovial chlamydia trachomatis in patients with reactive arthritis/Reiter's syndrome are viable but show aberrant gene expression. *J Rheumatol* 1998; **25**(4): 734–42.

18 Granfors K, Jalkanen S, Toivanen P, Koski J, Lindberg AA. Bacterial lipopolysaccharide in synovial fluid cells in Shigella-triggered reactive arthritis. *J. Rheumatology* 1992: **19**(3): 500.

19 Nikkari S, Rantakokko K, Ekman P, *et al*. Salmonella-triggered reactive arthritis: use of polymerase chain reaction, immunocytochemical staining, and gas-chromatography-mass spectrometry in the detection of bacterial components from synovial fluid. *Arthritis Rheum* 1999; **42**(1): 84–9.

20 Taylor-Robinson D, Gilroy CB, Thomas BJ, Keat AC. Detection of chlamydia trachomatis DNA in joints of reactive arthritis patients by polymerase chain reaction. *Lancet* 1992; **340**(8811): 81–2.

21 Gerard HC, Wang Z, Whittum-Hudson JA *et al*. Cytokine and chemokine mRNA produced in synovial tissue chronically infected with Chlamydia trachomatis and C. pneumoniae. *J Rheumatol* 2002; **29**(9): 1827–35.

22 Nikkari S, Merilahti-Palo R, Saario R, Soderstom KO, Granfors K, Skurnik M, Toivanen P. Yersinia-triggered reactive arthritis. Use of polymerase chain reaction and immunocytochemical in the detection of bacterial components from synovial specimens. *Arthritis Rheum* 1992; **35**(6): 682–7.

23 Abdelrahman YM, Belland RJ. The chlamydial developmental cycle. *FEMS Microbiol Rev* 2005; **29**: 949–959.

24 Hatch TP. Developmental biology. In R.S. Stephens (ed.), *Chlamydia – Intracellular Biology, Pathogenesis, and Immunity.* American Society for Microbiology, Washington DC. 1999; 29–67.

25 Carter JD, Rehman A, Guthrie JP, Gerard HC, Stanich J, Hudson AP. Attack rate of Chlamydia-induced reactive arthritis and effect of the CCR5-Delta-32 mutation: a prospective analysis. *J Rheumatol* 2013; **40**(9): 1578–82.

26 Beatty WL, Byrne GI, Morrison RP. Morphological and antigenic characterization of interferon γ-mediated persistent Chlamydia trachomatis infection in vitro. *Proc Natl Acad Sci USA* 1993; **90**: 3998–4002.

27 Beatty WL, Morrison RP, Byrne GI. Reactivation of persistent C. trachomatis infection in cell culture. *Infect Immun* 1995; **63**: 199–205.

28 Gérard HC, Freise J, Wang Z, Roberts G, Rudy D, Krauß-Opatz B, Köhler L, Zeidler H, Schumacher HR, Whittum-Hudson JA, Hudson AP. Chlamydia trachomatis genes whose products are related to

energy metabolism are expressed differentially in active vs. persistent infection. *Microb Infect* 2002; **4**: 13–22.

29 Koehler L, Nettelnbreker E, Hudson AP, Ott N, Gérard HC, Branigan PJ, Schumacher HR, Drommer W, Zeidler H. Ultrastructural and molecular analyses of the persistence of Chlamydia trachomatis (serovar K) in human monocytes. *Microb Pathog* 1997; **22**: 133–142.

30 Mpiga P, Ravaoarinoro M. Chlamydia trachomatis persistence: an update. *Microbiological Research* 2006; **161**: 9–19.

31 McColl GJ, Diviney MB, Holdsworth RF *et al*. HLA-B27 expression and reactive arthritis suscepti-bility in two patient cohorts infected with Salmonella Typhimurium. *Int Med J* 2000; **30**(1): 28–32.

32 Rudwaleit M, Richter S, Braun J, Sieper J. Low incidence of reactive arthritis in children following a salmonella outbreak. *Ann Rheum Dis* 2001; **60**(11): 1055–7.

33 Telyatnikova N, Hill Gaston JS. Prior exposure to infection with Chlamydia pneumoniae can influ-ence the T-cell-mediated response to Chlamydia trachomatis. *FEMS Immunol Med Microbiol* 2006; **47**(2): 190–8.

34 Sampaio-Barros PD, Conde RA, Donadi EA, Bonfiglioli R, Costallat LT, Samara AM, Bértolo MB. Frequency of HLA-B27 and its alleles in patients with Reiter syndrome: comparison with the fre-quency in other spondyloarthropathies and a healthy control population. *Rheumatol Int* 2008; **28**(5): 483–6. Epub 2007 Aug 24.

35 Díaz-Peña R, Blanco-Gelaz MA, Njobvu P, López-Vazquez A, Suárez-Alvarez B, López-Larrea C. Influence of HLA-B*5703 and HLA-B*1403 on susceptibility to spondyloarthropathies in the Zambian population. *J Rheumatol* 2008; **35**(11): 2236–40.

36 Kuipers JG, Bialowons A, Dollmann P *et al*. The modulation of chlamydial replication by HLA-B27 depends on the cytoplasmic domain of HLA-B27. *Clin Exp Rheumatol* 2001; **19**(1): 47–52.

37 Ge S, He Q, Granfors K. HLA-B27 modulates intracellular growth of Salmonella pathogenicity island 2 mutants and production of cytokines in infected monocytic U937 cells. *PLoS One* 2012; **7**(3):e34093. doi: 10.1371/journal.pone.0034093. Epub 2012 Mar 28.

38 Gerard HC, Stanich JA, Whittum-Hudson JA, Schumacher HR, Carter JD, Hudson AP. Patients with Chlamydia-associated arthritis have ocular (trachoma), not genital, serovars of C. trachomatis in synovial tissue. *Microb Pathog* 2010; **48**(2): 62–8.

39 Mittal A. Serovar distribution of Chlamydia trachomatis isolates collected from the cervix: use of the polymerase chain reaction and restriction endonuclease digestion. *Br J Biomed Sci* 1998; **55**(3): 179–183.

40 Workowski KA, Stevens CE, Suchland RJ, Holmes KK, Eschenback DA, Pettinger MB *et al*. Clinical manifestations of genital infection due to Chlamydia trachomatis in women: differences related to serovar. *Clin Infect Dis* 1994; **19**(4): 756–760.

41 Cox CJ, Kempsell KE, Gaston JS. Investigation of infectious agents associated with arthritis by reverse transcription PCR of bacterial rRNA. *Arthritis Res Ther* 2003; **5**(1): R1–8.

42 Cuchacovich R, Japa S, Huang WQ *et al*. Detection of bacterial DNA in Latin American patients with reactive arthritis by polymerase chain reaction and sequencing analysis. *J Rheumatol* 2002; **29**(7): 1426–9.

43 Wilkinson NZ, Kingsley GH, Jones HW *et al*. The detection of DNA from a range of bacterial species in the joints of patients with a variety of arthritidies using a nested, broad-range polymerase chain reaction. *Rheumatology* 1999; **38**(3): 260–6.

44 Gerard HC, Wang Z, Wang GF *et al*. Chromosomal DNA from a variety of bacterial species is present in synovial tissue from patients with various forms of arthritis. *Arthritis Rheu*. 2001; **44**(7): 1689–97.

45 Wilkinson NZ, Kingsley GH, Jones HW *et al*. The detection of DNA from a range of bacterial species in the joints of patients with a variety of arthritidies using a nested, broad-range polymerase chain reaction. *Rheumatology* 1999; **38**(3): 260–6.

46 Carter JD, Gérard HC, Espinoza LR, Ricca LR, Valeriano J, Snelgrove J, Oszust C, Vasey FB, Hudson AP. Chlamydiae as etiologic agents in chronic undifferentiated spondylarthritis. *Arthritis Rheum* 2009; **60**(5): 1311–6.

47 Carter JD, Espinoza LR, Inman RD, Sneed KB, Ricca LR, Vasey FB, Valeriano J, Stanich JA, Oszust C, Gerard HC, Hudson AP. Combination antibiotics as a treatment for chronic Chlamydia-induced reactive arthritis: a double-blind, placebo-controlled, prospective trial. *Arthritis Rheum* 2010; **62**(5): 1298–307.

CHAPTER 18

Rheumatoid arthritis: the bacterial connection

Jacqueline Detert

Charité-Universitätsmedizin Berlin, Berlin, Germany

18.1 Preclinical rheumatoid arthritis

Rheumatoid arthritis (RA) is a multifactorial inflammatory autoimmune disease affecting the synovial cells of joints. Many internal organs (e.g. lung, heart, vessels) may also be affected as part of an inflammatory systemic disease. Despite a great deal of research, the very cause of autoimmune diseases is still not well understood. Hypotheses currently assume that they develop from genetic predisposition and a combination of various external influences such as stress, infection and pregnancy.

Data from healthy blood donors show that serological autoimmune changes (e.g., rheumatoid factor (RF) IgM, anti-cyclic citrullinated peptide [anti-CCP]) are at a very early stage significantly detectable while giving no clinical symptoms. It was assumed that this early stage might continue for at least seven years prior to any clinical manifestation of RA[1], Early changes may thereby involve arthralgia and positive serological antibody detection[2]. Accumulation of various autoantibodies was described in the preclinical phase. This reflects the process of the epitopes spreading and the increased development of pro-inflammatory and inflammatory cytokines/chemokines and autoantibodies, which can predict the onset of clinical arthritis[3]. Titer analyses during the early phase of the disease are important for clinical diagnosis and they seem to be dependent on age. Corresponding titers of >40 year-old patients can be much higher than those of younger patients[4]. This early manifestation is also referred to as pre-clinical RA[5,6]. Due to this long interval between the onset of autoimmune phenomena and clinical detectable changes, the actual main etiologic factor is difficult to find. Thus, it is also possible that the causative agent acted during childhood[7]. The current findings and hypotheses lead to the conclusion that several requirements must be met before the development of an autoimmune disease such as RA[7,8].

The Human Microbiota and Chronic Disease: Dysbiosis as a Cause of Human Pathology, First Edition.
Edited by Luigi Nibali and Brian Henderson.

18.2 Predisposition to RA

Although the inheritance of RA cannot be traced according to the Mendelian laws, the results of twin studies suggest a hereditary predisposition[9-11]. For direct descendants of RA patients, the risk ratio varies between 4 and 8%[12]. The concordance rates of genetic involvement vary from 15 to 32%[13,14].

18.3 MCH-HLA and genetic predisposition to RA

The MHC (major histocompatibility complex) genes are now very well studied and represent the most likely candidate genes for RA predisposition[15]. A group of alleles in the HLA-DRB1 locus on the short arm of chromosome 6 (6p21.3) — which share a common amino-acid sequence at residues 70–74 — is associated with RA and called the shared epitope (SE)[15-17]. HLA-DRB1 alleles and non-HLA alleles such as the PTPN22 allele and PADI4 alleles have been noted in different populations. Apparently there is an ethnic heterogeneity as to genetic variation and how genes are associated with the disease[18]. This heterogeneity may explain differences in the disease process and its adaptation to this process[19]. Nonetheless, further intensive research is necessary to clarify the genetic risk and physiological mechanisms in disease development[18]. The SE encodes the MHC class II β-chain. This epitope is expressed on antigen-presenting cells such as B cells and dendritic cells[20]. The RA risk alleles of the SE differ among various ethnic groups. Within Europeans, HLA-DRB1*0401, 0404, and *0101 have often been noted[16,21]. In contrast, among East Asians, HLA-DRB1*0405 and *0101 are detectable[22,23].

Additional important clinical RF antibodies used in RA diagnosis include ACPA (anti-citrullinated protein antibodies). The positive ACPA titers are predictive of disease severity[24,25]. HLA-DRB1 SE alleles associated with ACPA-positive RA[19,25-27]. However, their role is still unclear.

In the context of MHC in autoimmune pathogenesis, various bacteria and viruses have been proposed as potential microbial triggers[28-31]. Often the viral-associated arthralgias are associated with rashes that can be easily overlooked. The occurrence of rheumatological antibodies, such as the RF or anti-nuclear antibodies (ANA), is usually temporarily observed in the acute phase or after the infection[32-36]. In some cases, follow-up studies showed that autoantibodies can be detected even three years after the infection with Sindbis virus (SINV) and seroconversion to positive RF[31]. Interestingly, the SINV-infection and RA show similarities in their cytokine profiles, such as up-regulation of macrophage migration inhibitory factor expression[37-39] Evidence that alpha viruses (e.g. Ross River virus, chikungunya virus, SINV) can lead to the development of autoimmune diseases of the joints is apparently lacking in other studies, especially long-term observations[29].

Other studies support a role for some viruses like human herpesvirus 6 (HHV-6), Epstein-Barr virus (EBV), human cytomegalovirus (CMV), parvovirus B19, rubella virus, human T-cell leukemia virus, and hepatitis B virus) in the initiation of RA, mostly on the basis of epidemiological data and abnormal immune responses to the virus[40-43]. However, the role of viruses in RA also remains uncertain.

18.4 Molecular mimicry in RA

Chapters 9 and 10 of this book have described how the human host and its colonizing microbes have co-evolved. Thus, microbes can have similar, almost homologous molecules that are also used by humans and may use the same cellular pathways. The main function of the immune system, i.e. the recognition between self and non-self, can be prevented by this feature of the microbes[44]. The ability of microorganisms with this similarity of some molecules to prevent the immune response of the host was first described in 1964 by Damian as molecular mimicry. Damian referred to a thesis of Rowley and Jenkin, who had a few years previously found an antigenic cross-relationship between parasite and host[45,46]. They later changed the meaning of the term "molecular mimicry." Thereafter, the mimicry refers to the antigenic determinants of microorganisms that can cause a damaging autoimmune response in the host[47]. If a pathogen cannot be tolerated despite this described strategy of the host's mimicry, can serious consequences for the host be expected? Although an effective immune reaction against the pathogen may be seen as beneficial, a strong overreaction against the body's own similar structures was deemed possible[48]. Thus, a molecular mimicry certainly triggers autoimmune diseases or affects their progression[47–49].

Thus, the hypothesis was developed that while a single microorganism might be capable of inducing an autoimmune disease, its course could then be exacerbated by many other microorganisms ("burden of infections")[7,48,49]. On the other hand, it was in fact postulated by the hygiene hypothesis that infections can protect people from autoimmune diseases[7].

18.5 Innate immune system and RA

Dysregulation of the adaptive immune system is central to the discussion of the pathogenesis of RA[50]. The RA as joint disease is systemic and may involve different organs over its course. Women are affected more often than men, suggesting hormonal influences. A typical feature is the formation of autoantibodies against endogenous structures. For example, the RF is formed, which is directed against the Fc portion of the immunoglobulin[51]. The RF is not specific for RA but is found in other diseases (bacterial infections, hepatic disease, lymphoproliferative disease, osteoarthritis, etc.)[52–54]. Activated lymphocytes, monocytes and macrophages infiltrate the synovia of joints. In addition to the T-cells of type T_H1, predominantly proinflammatory and inflammatory cytokines are formed (interleukins [ILs] -1, -2, -6, -8, -10, -17; tumor necrosis factor-α [TNF-α]; platelet-derived growth factor; insulin-like growth factor; and transforming growth factor β)[55]. Anti-inflammatory cytokines like IL-4 and IL-10 exhibit immunosuppressive activities. The expansion of Th17 cells of the type is supported by the proinflammatory acting IL-23 and inhibited by IL-27 as an anti-inflammatory cytokine[56–58] In particular, the proinflammatory cytokines involved in the differentiation and activation of pathogenic cells, the migration of these cells in the target organ in the process of neovascularization, and the development and activation of osteoclasts are thus involved in the process of bone destruction[59,60]. The drug inhibition of these cytokines is now the central approach of much drug development[61]. It is interesting that this has already been realized, inasmuch as the actual cause of RA is not due to any single specific autoantigen.

In recent years, a two-step model for the development of autoimmune diseases, as well as RA, has been adopted[50,62]. Taking into account the subsequent immunological response to the development of chronic disease, we can designate three phases[63]. Thus, not only the adaptive immune system, but also the innate immune system is involved in disease initiation and development. The first phase involves the interaction between a foreign antigen (e.g. microorganisms, microbial products) and the innate immune system in the presence of predisposing factors, such as a certain HLA-DR type. In the second stage, initial joint inflammation appears, followed by a third stage of joint destruction and reconstruction[50,62–64].

18.6 Bystander activation and pattern recognition receptors

The discovery of Toll-like receptors (TLRs) brought a change in the scientific view of the innate immune system. TLRs specifically recognize bacteria and virus particles, in the so-called PAMP (pathogen-associated molecular pattern). They are found in many cells of the adaptive immune system and have a modulatory influence on the immune system[65–67]. TLRs are also found in antigen-presenting cells, especially dendritic cells. The activation of dendritic cells by PAMP results in a clonal expansion of antigen-specific T and B cells[67].

Dendritic cells after phagocytosis are subject to different antigens genes going on and off. They utilize the antigenic stimulus as a form of encoding. The resulting gene products (including cytokines) thereby provide, e.g. for T-helper cells, additional signals to modulate the T-cell response against the particular antigen. Cytokines are synthesized only by activated cells and only act on cells expressing the corresponding cytokine receptors[68,69]. Thus, only clones that recognize an antigen with multiple epitopes will expand. Cytokines themselves act non-specifically. After the activation of a specific cell and its cytokines, all activated cells respond to these cytokines in the same place and at the same time (irrespective of which antigen has been activated)[68,69]. Time-conformist responses to various antigens may therefore also influence each other (bystander activation). Autoimmunity, such as that towards RA, could likely be a bystander effect of the cytokine storm that causes immune dysregulation[70,71].

During the initial induction of inflammation, TLRs appear to play important roles; this holds true especially for the pattern recognition receptors TLR2 and TLR9, which are connected with RA in particular. The intra-articular administration of CpG-DNA leads to an increased production of TNF and arthritis in which TLR9 was identified as a ligand[72,73]. DNA from different pathogens has been found in the joints of RA patients[74–76] Moreover, TLR may be involved in the RA-tripping. This leads to an up-regulation of co-stimulatory molecules on antigen-presenting cells, which are dormant until then, prior to a microbial infection. This mechanism leads to a loss of T-cell anergy. Otherwise, in the immune system, hidden antigens are presented by the inflammatory response and support the "epitope spreading[77]."

18.7 Antibodies and neoepitopes

It is undisputed that the detection of RF or ACPA is associated with HLA-DRB1[15]. Smoking increases the risk of developing ACPA[78,79]. It was postulated that stress may activate the mucosal surface (of lung, oral cavity, and gut) and thus promote post-translational modification by citrullination of proteins in the membrane[79-81]. This tolerance-associated loss of neoepitopes can be found in ACPA-positive RA patients[79-81]. In particular, products of infectious agents (so-called virulence factors) can induce an immune response by citrullination (e.g. lipopolysaccharide, peptidoglycans, superantigens, bacterial DNA, and Hsp). A number of such citrullinated autoantigens (e.g., ACPA, keratin, fibrinogen, fibronectin, collagen, vimentin and α-enolase) have been identified in recent years in various locations (e.g., by extracellular matrix of joints, lung, skin, and mucosal tissue)[82-84]. During and after infection, positive RF are often observed and associated with the induction of RA[51,85]. RA-specific antibodies may be found to exist 10 to 15 years before the joint disease is established[86].

18.8 Superantigens

Superantigens have a molecular weight of 24-30 kDa and are hydrophilic and heat-resistant. They are distributed as exotoxins of bacteria[87]. They interact with T-cell receptors (TCR) and MHC-II molecules and trigger a permanent signal in T-cells[88,89]. Thus, they lead to a permanent over-activation of the cells. Superantigens appear to have nothing in common with antigens. For Vβ the region is identified as a binding site for superantigens and is localized within the variable part of the beta chain of the TCR[88,89], demonstrated in RA patients more frequently than in control groups[87]. Most superantigens studied are formed by Staphylococcus *aureus* and Streptococcus *pyogenes*[90].

18.9 LPS

Lipopolysaccharide (LPS) is an endotoxin from the cell wall of gram-negative bacteria. Through activation of immune cells of the host, pro-inflammatory factors are formed, and these can lead to severe inflammation[91]. Pro-inflammatory signaling pathways are induced by binding to specific surface receptors (e.g., TLR4)[92,93]. This will stimulate multiple signaling pathways (such as the phosphatidylinositol 3-kinase (PI-3K)/protein kinase B (Akt), mitogen-activated protein kinase (MAPK) and nuclear factor-κB (NF-κB)[93,94]. The NF-kB is activated in synovial fluid, which plays an essential role in cartilage destruction[95,96]. NF-kB is involved in the regulation of numerous genes that are generally activated during infection, adhesion, cell cycle, apoptosis, survival and in the inflammation process and therefore in the production of IL-1β, TNF-α, IL-6, cyclooxygenase-2 (COX-2) and matrix metalloproteinases (MMPs)[97,98].

18.10 Bacterial DNA and peptidoglycans

Detection of bacterial DNA, peptidoglycans and antibody titers against these bacteria in serum and synovial fluid of RA patients support the hypothesis that infections play a role in RA pathogenesis[75,99]. A chronic bacteremia of highly pathogenic bacteria can damage organs that are not located close to each other. Through a long-term continuous sowing of these bacterial products, synovial inflammation is followed by erosions of antigens present in cartilage and then by autoimmunity[100,101], although this is still controversial[75,99–101].

18.11 Heat-shock proteins

Heat-shock proteins (HSPs) are molecules produced in higher quantities of cells under stress[102]. Because of their role as helpers in protein folding, they are also referred to as molecular chaperones. HSPs play a critical role in the immune system[103], are capable of inducing a humoral, as well as a cellular immune response and may activate the complement[104]. Some HSPs are known to play an important role in autoimmune diseases[104] Starting from the fact that HPSs and ultimately the immunoglobulin heavy-chain binding protein (BiP) itself are involved in the assembly of immunoglobulins and MHC molecules, an immune-modulatory role seems obvious. In addition, some HSPs have been identified as ligands of TLR[104,105]. It has been argued whether or not there is a cross-reactivity between bacterial and human proteins due to the high conservation status of HSPs. Autoantibodies to HSPs, e.g. hsp70 and hsp90 were found in patients with RA[106] and BiP has been described as a RA-specific autoantigen by several working groups[106].

18.12 Toll-like and bacterial infections

A variety of pro-inflammatory cytokines (e.g. TNFα, IL-1, and IL-6, matrix metalloproteinases [MMPs]) are induced by TLR[107,108]. Furthermore, a number of interactions between antigen-presenting cells and T cells have been described alongside costimulatory molecules as upregulated. Moreover, it is known that TLR can activate a self-limiting arthritis (Lyme disease, Chlamydia-induced arthritis) and thus contribute to bacterial clearance[109–111].

18.13 Proteus mirabilis

Proteus mirabilis is a bacterium that is not pathogenic in most individuals and occurs ubiquitously as a component of the microbial flora of soil and water, and also in the human gastrointestinal tract. As it enters in the urinary tract or upon interaction with the bloodstream, *P. mirabilis* produces large amounts of hemolysin and urease, contributing to the hydrolysis of urea to ammonia. This causes an increase in the pH value of urine and thus the displacement of the solubility product with the formation of crystals. Furthermore, it explains why *P. mirabilis* can be found in connection with kidney stones. It has been proposed that the enzymes (urease, hemolysin) lead to changes in the immune system and also in genetically predisposed humans work as a RA trigger[43,112,113]. Proteus bacteria enhanced the bacterial immune response in RA patients. In these patients, increased antibody titers (ESRAAL and anti-IRRET antibodies) against Proteus antigenic epitopes were observed. These antibodies are cross-reactive with tissues containing EQ/KRRA or LRREI peptide sequences, particularly hyaline cartilage in the joints[114]. These antibodies lead to an induction of the inflammatory cascade, including the formation of pro-inflammatory and inflammatory cytokines, as well as harmful cytotoxic effects on joint tissues[50,115].

18.14 Porphyromonas gingivalis and RA

Each milliliter of human saliva usually contains hundreds of millions of bacteria belonging to more than 600 different types. Chapter 14 discussed the role of subgingival biofilms in the developments of periodontal diseases. Periodontopathogenic bacteria include *Porphyromonas gingivalis*, *Prevotella intermedia*, *Tannerella forsythia*, *Treponema denticola* and *Aggregatibacter actinomycetemcomitans*[116–119]. In recent years, interesting epidemiological and pathogenetic relationships between periodontitis and RA have emerged, mainly associated with the bacterium *P. gingivalis*. Epidemiological evidence showed that, compared to healthy subjects, RA patients have an increased incidence of periodontitis, while patients with periodontitis have a higher prevalence of RA than patients without periodontitis[116,120–124]. RA patients also seem to have a higher risk of alveolar bone loss and subsequent tooth loss[125–127].

Detection of bacterial DNA and high antibody titers of anaerobes of the oral cavity in serum and synovial fluid of RA patients in different stages of the disease support the hypothesis that oral infectious agents play a role in the pathogenesis of RA[100,128,129]. Such pathogens, like *P. gingivalis*, can affect the integrity of the epithelium. This allows them to penetrate into the human endothelial cells and affect both the transcription and protein synthesis, while being able to enter the bloodstream[130]. With the help of extracellular cysteine proteases (gingipains), this bacterium manages to use the innate immune response of the host to its own benefit[131–133]. Gingipains affect the pro-inflammatory signaling by stimulation of the proteinase-activated receptor-2 (PAR-2) on the neutrophils[134,135].

Chronic oral infection prior to the start of arthritis stimulates the formation and mobilization of neutrophils Th17, which play an important role in the defense against extracellular bacteria[136]. With the increase of Th17 cell immune response, a number of cytokines such as IL-1β, IL-6, IL-22, TNF-α, transforming growth factor-β, and IL-23 are secreted[137]. It could be demonstrated that *P. gingivalis* penetrates into primary human chondrocytes, where it is able to induce cellular effects (Pischon *et al.* 2009) such as a delay in the progression of the cell cycle and increase in the apoptosis of chondrocytes[120,123,138,139].

The enzymatic deimination of arginine to citrulline residues by the enzyme peptidyl arginine deiminase (PAD) is a form of post-translational protein modification (Schellekens *et al.* 2000). With this modification, the structure of the protein's biochemical and antigenic properties is changed. *P. gingivalis* is currently the only known bacterium with the expression of peptidyl arginine deiminase (PAD)[140–142], even if its PAD is not completely homologous to the human PAD. It leads inter alia to an irreversible, post-translational conversion of arginine to citrulline[120,143] The process of citrullination was mainly found for proteins of the cytoskeleton in the course of apoptosis (e.g., cytokeratin, vimentin, filaggrin). Diseases such as RA are the result of a (patho) physiological citrullinati of the structural proteins as well as a result of an increased accumulation of citrullinated proteins (e.g. citrullinated vimentin [MCV])[101]. The PAD of *P. gingivalis* (PPAD) is activated at a higher pH and therefore requires no calcium as opposed to human PAD[140]. The PPAD modifies the C-terminal arginine using the gingipains[101,144]. In addition, citrullinated peptides are produced by the PPAD via the degradation of fibrinogen, α-enolase, fibrin, and vimentin[101,130]. The PPAD thus seems to have an important influence on the formation of neo-epitopes associated with disease progression[130,145]. The potential citrullinated autoantigens include fibrinogen, enolase, vimentin and collagen II[146–149] Since the PPAD can be found on the bacterial surface, they can citrullinate on bacterial adhesins' host proteins. Fibrinogen-derived peptides can be generated quickly with the carboxy-terminal citrulline residues[150,151]. Thus, citrullinated enolase peptide-1 (CEP-1) — identified as an RA autoantigen and a substrate of the PPAD and B-cell epitopes of the α-enolase — dominates. CEP-1 antibodies may thus also cross-react with the enolase of *P. gingivalis* derived through the PPAD[142,152]. The pathophysiological function of auto-citrullinated PPAD in *P. gingivalis* and its relevance in autoimmune diseases such RA needs to be explored more in the future[101,140,144,153,154]. The reduced immune tolerance to citrullinated proteins seems to lead to an increased formation of autoantibodies, thus representing a key feature of this disease[144]. The practical relevance of citrullination can be seen in ACPA titer studies. Recently, patients with an aggressive periodontitis showed increased ACPA titers[155]. A genetic predisposition has also been suggested as modulating the role of *P. gingivalis* in autoimmune diseases[120,127,154,156].

18.15 Gastrointestinal flora and RA

The hypothesis that elements of the gastrointestinal flora might cause joint-associated diseases was formulated back in the previous century[157–159]. Joint disease-related microflora and their analyses were, however, not pursued intensively for many years[159]. Scientific debate on this in the literature has increased only in recent years. Study of fecal microflora using the 16S rRNA hybridization technique in patients with newly diagnosed RA (NORA)[160] showed fewer counts of bifido bacteria and bacteroides (both of the Prophyromonas-Prevotella group), *Bacteroides fragilis* subgroup, and the *Eubacterium rectal-Clostdridium coccoides* group compared with fibromyalgia patients[160–162].) In addition, a high level of *Prevotella copri* was detected in NORA patients[162]. Moreover, higher levels of Lactobacillus *salivarius*, L. *iners* and L. *ruminis* were observed in the fecal community of patients with early RA compared with controls[161]. One conclusion from this evidence is that the genetic predisposition to an altered bacterial spectrum leads to an arthritogenic potential[160,163].

18.16 Smoking, lung infection and RA

Cigarette smoking leads to an increased expression and protein citrullination PAD in bronchoalveolar lavage cells of smokers without arthritis[164]. A subgroup of patients with lung diseases, however, showed positive anti-CCP antibodies in plasma[165]. Overall, smoking seems to result in a greater risk for the development of RA. (Gan *et al.* 2013) and signs of inflammation in the bronchi and lung tissue were detected in patients with early RA[166,167]. Furthermore, it was demonstrated that the accumulation of supraglottic pathogens (e.g. Prophyromonas, Prevotella *sp*) in the lungs was associated with airway inflammation[168]. Perhaps the lung is thus involved in the process of citrullination and thus autoimmunity[166–168].

18.17 Where to go from here?

In the EIRA study, data were collected from 6,401 persons between 1996 and 2009 to find social and environmental factors that contribute to the risk of RA[169]. Among 2,831 patients in whom chronic arthritis occurred during the study period, the influence of infections was evaluated[170]. A further 3,570 persons served as matched controls (in regard to age at diagnosis, calendar year, gender and place of residence). All study participants (mean age 52 years; 72% women) had indicated the appearance of gastric infection with diarrhea, urinary tract infections and genital infections in the last two years. In addition, they were asked (when relevant) about inflammation of the prostate, antibiotic-treated sinusitis, tonsillitis, pharyngitis or pneumonia over the same period of time[170]. For participants who reported gastroenteritis, a reduced RA risk of about 29% was detected; for participants with past urinary or genital infections, this risk reduction was 22% and 20% respectively. For participants diagnosed in the previous two years with all three of those infections, the risk for chronic joint disease was reduced by 50%. In contrast, paranasal sinuses, tonsillitis and pneumonia appeared to bear absolutely no relation to the risk of RA[170]. Nevertheless, many research data were able to show a relation between infectious agents and RA disease, so that such results in epidemiological research of course remain a matter of controversy. The background is first that autoimmune diseases are often multifactorial and thus all conditions must be met in order to develop such a disease. Secondly, social and environmental influences — very diverse and, above all, important parameters for disease induction — must also be explored in long-term studies.

TAKE-HOME MESSAGE

- Detection of bacterial DNA, peptidoglycans and antibody titers against specific bacteria in serum and synovial fluid of rheumatoid arthritis patients support the hypothesis that infections play a role in RA pathogenesis.
- Dysbiosis at a distant location such as the gut or oral cavity could predispose to RA via citrullination and autoimmune reactions.

References

1 Nielen, MMJ, van Schaardenburg D, Reesink HW *et al.* Specific autoantibodies precede the symptoms of rheumatoid arthritis: a study of serial measurements in blood donors. *Arthritis and Rheumatism* 2004; **50**(2): 380–86. doi:10.1002/art.20018.

2 Kung TN, Bykerk VP. Detecting the earliest signs of rheumatoid arthritis: symptoms and examination. *Rheumatic Diseases Clinics of North America* 2014; **40**(4): 669–83. doi:10.1016/j.rdc.2014.07.009.

3 Sokolove J, Bromberg R, Deane KD *et al.* Autoantibody epitope spreading in the pre-clinical phase predicts progression to rheumatoid arthritis." *PloS One* 2012; **7**(5): e35296. doi:10.1371/journal.pone.0035296.

4 Deane KD, O'Donnell CI, Hueber W, Majka DS, Lazar AA, Derber LA, Gilliland WR *et al.* The number of elevated cytokines and chemokines in preclinical seropositive rheumatoid arthritis predicts time to diagnosis in an age-dependent manner. *Arthritis and Rheumatism* 2010; **62**(11): 3161–72. doi:10.1002/art.27638.

5 Raza K, Gerlag DM. Preclinical inflammatory rheumatic diseases: an overview and relevant nomenclature. *Rheumatic Diseases Clinics of North America* 2014; **40**(4): 569–80. doi:10.1016/j.rdc.2014.07.001.

6 Deane D. Preclinical rheumatoid arthritis (autoantibodies): an updated review. *Current Rheumatology Reports* 2014; **16**(5): 419. doi:10.1007/s11926-014-0419-6.

7 Kivity S, Agmon-Levin N, Blank M, Shoenfeld Y. Infections and autoimmunity – friends or foes? *Trends in Immunology* 2009; **30**(8): 409–14. doi:10.1016/j.it.2009.05.005.

8 von Herrath MG, Fujinami RS, Whitton JL. Microorganisms and autoimmunity: making the barren field fertile? *Nature Reviews. Microbiology* 2003; **1**(2): 151–57. doi:10.1038/nrmicro754.

9 Seldin MF, Amos CI, Ward R, Gregersen PK. The genetics revolution and the assault on rheumatoid arthritis. *Arthritis and Rheumatism* 1999; **42**(6): 1071–79. doi:10.1002/1529-0131(199906)42:6<1071:AID-ANR1>3.0.CO;2-8.

10 MacGregor AJ, Snieder H, Rigby AS *et al.* Characterizing the quantitative genetic contribution to rheumatoid arthritis using data from twins. *Arthritis Rheum* 2000; **43**(1): 30–7.

11 Lettre G, Rioux JD. Autoimmune diseases: insights from genome-wide association studies. *Human Molecular Genetics* 2008; **17**(R2): R116–21. doi:10.1093/hmg/ddn246.

12 Grant SF, Thorleifsson G, Frigge ML, Thorsteinsson J, Gunnlaugsdóttir B, Geirsson AJ, Gudmundsson M *et al.* The inheritance of rheumatoid arthritis in Iceland. *Arthritis and Rheumatism* 2001; **44**(10): 2247–54.

13 Silman AJ, MacGregor AJ, Thomson W, Holligan S, Carthy D, Farhan A, Ollier WE. Twin concordance rates for rheumatoid arthritis: results from a nationwide study. *British Journal of Rheumatology* 1993; **32**(10): 903–7.

14 Sawada S Takei M. Epstein-Barr virus etiology in rheumatoid synovitis. *Autoimmunity Reviews* 2005; **4**(2): 106–10. doi:10.1016/j.autrev.2004.08.034.

15 Gregersen PK, Silver J, Winchester RJ. The shared epitope hypothesis. An approach to understanding the molecular genetics of susceptibility to rheumatoid arthritis. *Arthritis and Rheumatism* 1987; **30** (11): 1205–13.

16 du Montcel, ST, Michou L, Petit-Teixeira E, Osorio J, Lemaire I, Lasbleiz S, Pierlot C *et al.* New classification of HLA-DRB1 alleles supports the shared epitope hypothesis of rheumatoid arthritis susceptibility. *Arthritis and Rheumatism* 2005; **52**(4): 1063–68. doi:10.1002/art.20989.

17 Michou L, Croiseau P, Petit-Teixeira E, du Montcel ST, Lemaire I, Pierlot C, Osorio J *et al.* Validation of the reshaped shared epitope HLA-DRB1 classification in rheumatoid arthritis. *Arthritis Research & Therapy* 2006; **8**(3): R79. doi:10.1186/ar1949.

18 Suzuki A, Kochi Y, Okada Y Yamamoto K. Insight from genome-wide association studies in rheumatoid arthritis and multiple sclerosis. *FEBS Letters* 2011; **585**(23): 3627–32. doi:10.1016/j.febslet.2011.05.025.

19 Bossini-Castillo L. de Kovel C, Kallberg H, van 't Slot R, Italiaander A, Coene M, Tak PP *et al.* A genome-wide association study of rheumatoid arthritis without antibodies against citrullinated peptides. *Annals of the rheumatic diseases* 2014. doi:10.1136/annrheumdis-2013-204591.

20 Weyand CM, Goronzy JJ. Association of MHC and rheumatoid arthritis. HLA polymorphisms in phenotypic variants of rheumatoid arthritis. *Arthritis Research* 2000; **2**(3): 212–16. doi:10.1186/ar90.

21 Kochi Y, Suzuki A, Yamada R, Yamamoto K. Ethnogenetic heterogeneity of rheumatoid arthritis-implications for pathogenesis. *Nature Reviews. Rheumatology* 2010; **6**(5): 290–95. doi:10.1038/nrrheum.2010.23.

22 Lee H-S, Lee KW, Song GG, Kim H-A, Kim S-Y, Bae C-S. Increased susceptibility to rheumatoid arthritis in Koreans heterozygous for HLA-DRB1*0405 and *0901. *Arthritis and Rheumatism* 2004; **50**(11): 3468–75. doi:10.1002/art.20608.

23 Ahn S, Choi H-B, Kim T-G. 2011. HLA and disease associations in Koreans. *Immune Network* 2011; **11** (6): 324–35. doi:10.4110/in.2011.11.6.324.

24 Kastbom, A., Strandberg G, Lindroos A, Skogh T. Anti-CCP antibody test predicts the disease course during 3 years in early rheumatoid arthritis (the Swedish TIRA project). *Annals of the rheumatic diseases* 2004; **63**(9): 1085–89. doi:10.1136/ard.2003.016808.

25 van der Linden, MPM, van der Woude D, Ioan-Facsinay Levarht A, Nivine EW, Stoeken-Rijsbergen G, Huizinga TWJ, Toes REM, van der Helm-van Mil AHM. Value of anti-modified citrullinated vimentin and third-generation anti-cyclic citrullinated peptide compared with second-generation anti-cyclic citrullinated peptide and rheumatoid factor in predicting disease outcome in undifferentiated arthritis and rheumatoid arthritis. *Arthritis and Rheumatism* 2009; **60**(8): 2232–41. doi:10.1002/art.24716.

26 van Gaalen, FA, van Aken J, Huizinga TWJ, Schreuder GMT, Breedveld FC, Zanelli E, van Venrooij WJ, Verweij CL, Toes, REM, de Vries, RRP. Association between HLA class II genes and autoantibodies to cyclic citrullinated peptides (CCPs) influences the severity of rheumatoid arthritis. *Arthritis and Rheumatism* 2004; **50**(7): 2113–21. doi:10.1002/art.20316.

27 Nordang, GBN, Flåm ST, Maehlen MT, Kvien TK, Viken MK, Lie BA. HLA-C alleles confer risk for anti-citrullinated peptide antibody-positive rheumatoid arthritis independent of HLA-DRB1 alleles. *Rheumatology (Oxford, England)* 2013; **52**(11): 1973–82. doi:10.1093/rheumatology/ket252.

28 Kerr J R, Mattey DL, Thomson W, Poulton KV, Ollier, WER. Association of symptomatic acute human parvovirus B19 infection with human leukocyte antigen class I and II alleles. *The Journal of Infectious Diseases* 2002; **186**(4): 447–52. doi:10.1086/341947.

29 Kerr, JR, Mattey DL. Preexisting psychological stress predicts acute and chronic fatigue and arthritis following symptomatic parvovirus B19 infection. *Clinical Infectious Diseases: An Official Publication of the Infectious Diseases Society of America* 2008; **46**(9): e83–7. doi:10.1086/533471.

30 Lunardi C, Tinazzi E, Bason C, Dolcino M, Corrocher R, Puccetti A. Human parvovirus B19 infection and autoimmunity. *Autoimmunity Reviews* 2008; **8**(2): 116–20. doi:10.1016/j.autrev.2008.07.005.

31 Sane J, Guedes S, Ollgren J, Kurkela S, Klemets P, Vapalahti O, Kela E, Lyytikäinen O,Nuorti JP. Epidemic sindbis virus infection in Finland: a population-based case-control study of risk factors. *The Journal of Infectious Diseases* 2011; **204**(3): 459–66. doi:10.1093/infdis/jir267.

32 Salonen EM, Vaheri A, Suni J, Wager O. Rheumatoid factor in acute viral infections: interference with determination of IgM, IgG, and IgA antibodies in an enzyme immunoassay. *The Journal of Infectious Diseases* 1980; **142**(2): 250–55.

33 Schattner A, Rager-Zisman B. Virus-induced autoimmunity. *Reviews of Infectious Diseases* 1990; **12**(2): 204–22.

34 Tan EM, Feltkamp TE, Smolen JS, Butcher B, Dawkins R, Fritzler MJ, Gordon T *et al*. Range of antinuclear antibodies in "healthy" individuals. *Arthritis and Rheumatism* 1997; **40**(9): 1601–11. doi:10.1002/1529-0131(199709)40:9.

35 Franssila R Hedman K. Infection and musculoskeletal conditions: Viral causes of arthritis. *Best Practice & Research. Clinical Rheumatology* 2006; **20**(6): 1139–57. doi:10.1016/j.berh.2006.08.007.

36 Louthrenoo W. Rheumatic manifestations of human immunodeficiency virus infection. *Current Opinion in Rheumatology* **20**(1): 92–99. doi:10.1097/BOR.0b013e3282f1fea7.

37 Assunção-Miranda I, Bozza MT, Da Poian AT. Pro-inflammatory response resulting from sindbis virus infection of human macrophages: implications for the pathogenesis of viral arthritis. *Journal of Medical Virology* 2010; **82** (1): 164–74. doi:10.1002/jmv.21649.

38 Herrero LJ, Nelson M, Srikiatkhachorn A, Gu R, Anantapreecha S, Fingerle-Rowson G, Bucala R, Morand E, Santos LL Mahalingam S. Critical role for macrophage migration inhibitory factor (MIF) in Ross River virus-induced arthritis and myositis. *Proceedings of the National Academy of Sciences USA* 2011; **108**(29): 12048–53. doi:10.1073/pnas.1101089108.

39 Herrero LJ, Sheng K-C, Jian P, Taylor A, Her Z, Herring BL, Chow A *et al*. Macrophage migration inhibitory factor receptor CD74 mediates alphavirus-induced arthritis and myositis in murine models of alphavirus infection. *Arthritis and Rheumatism* 2013; **65**(10): 2724–36. doi:10.1002/art.38090.

40 Olson JK, Croxford JL, Miller SD. Virus-induced autoimmunity: potential role of viruses in initiation, perpetuation, and progression of T-cell-mediated autoimmune disease. *Viral Immunology* 2001; **14**(3): 227–50.

41 Toussirot E, Roudier J. Pathophysiological links between rheumatoid arthritis and the Epstein-Barr virus: an update. *Joint, Bone, Spine: Revue du Rheumatisme* 2007; **74**(5): 418–26. doi:10.1016/j.jbspin.2007.05.001.

42 Alvarez-Lafuente R, Fernández-Gutiérrez B, de Miguel S, Jover JA, Rollin R, Loza E, Clemente D, Lamas JR. Potential relationship between herpes viruses and rheumatoid arthritis: analysis with quantitative real time polymerase chain reaction. *Annals of the Rheumatic Diseases* 2005; **64**(9): 1357–59. doi:10.1136/ard.2004.033514.

43 Newkirk MM, Watanabe Duffy KN, Leclerc J, Lambert N, Shiroky JB. Detection of cytomegalovirus, Epstein-Barr virus and herpes virus-6 in patients with rheumatoid arthritis with or without Sjögren's syndrome. *British Journal of Rheumatology* 1994; **33**(4): 317–22.

44 Sigal LH Molecular biology and immunology for clinicians, 9 pathogenesis of autoimmunity: molecular mimicry. *Journal of Clinical Rheumatology: Practical Reports on Rheumatic & Musculoskeletal Diseases* 1999; **5**(5): 293–96.

45 Rowley D, Jenkin CR. Antigenic cross-reaction between host and parasite as a possible cause of pathogenicity. *Nature* 1962; **193**: 151–54.

46 Damian RT. Molecular mimicry: antigen sharing by parasite and host and its consequences. *American Naturalist* 1964; **98**: 129–49.

47 Blank M, Ori Barzilai O, Shoenfeld Y. Molecular mimicry and auto-immunity. *Clinical Reviews in Allergy & Immunology* 2007; **32**(1): 111–18.

48 van Heemst J, van der Woude D, Huizinga TW, Toes RE. HLA and rheumatoid arthritis: how do they connect? *Annals of Medicine* 2014; **46**(5): 304–10. doi:10.3109/07853890.2014.907097.

49 von Herrath MG, Fujinami RS, Whitton JL. Microorganisms and autoimmunity: making the barren field fertile?. *Nature Reviews. Microbiology* 2003; **1**(2): 151–57. doi:10.1038/nrmicro754.

50 Firestein GS. 2003. Evolving concepts of rheumatoid arthritis. *Nature* 2003; **423** (6937): 356–61. doi:10.1038/nature01661.

51 Rycke L de, Peene I, Hoffman IEA, Kruithof E, Union A, Meheus L, Lebeer K *et al.* Rheumatoid factor and anticitrullinated protein antibodies in rheumatoid arthritis: diagnostic value, associations with radiological progression rate, and extra-articular manifestations. *Annals of the Rheumatic Diseases* 2004; **63**(12): 1587–93. doi:10.1136/ard.2003.017574.

52 Rose HM, Ragan C. Differential agglutination of normal and sensitized sheep erythrocytes by sera of patients with rheumatoid arthritis. *Proceedings of the Society for Experimental Biology and Medicine (New York, N.Y.)* 1948; **68**(1): 1–6.

53 Pike RM, Sulkin SE, Coggeshall HC. Serological reactions in rheumatoid arthritis; factors affecting the agglutination of sensitized sheep erythrocytes in rheumatid-arthritis serum. *Journal of Immunology* 1950; **63**(4): 441–46.

54 Dörner T, Egerer K, Feist E, Burmester GR. Rheumatoid factor revisited. *Current Opinion in Rheumatology* 12004; **6**(3): 246–53.

55 Kirkham BW, Lassere MN, Edmonds JP, Juhasz KM,. Bird PA, Lee CS, Shnier R, Portek IJ. Synovial membrane cytokine expression is predictive of joint damage progression in rheumatoid arthritis: a two-year prospective study (the DAMAGE study cohort). *Arthritis and Rheumatism* 2006; **54**(4): 1122–31. doi:10.1002/art.21749.

56 Tang C, Chen S, Qian H, Huang W. 2012. Interleukin-23: as a drug target for autoimmune inflammatory diseases. *Immunology* 2012; **135**(2): 112–24. doi:10.1111/j.1365-2567.2011.03522.x.

57 Benedetti G, Miossec P. Interleukin 17 contributes to the chronicity of inflammatory diseases such as rheumatoid arthritis. *European Journal of Immunology* 2014; **44**(2): 339–47. doi:10.1002/eji.201344184.

58 Adamopoulos IE, Stefan Pflanz. The emerging role of Interleukin 27 in inflammatory arthritis and bone destruction. *Cytokine & Growth Factor Reviews* 2013; **24**(2): 115–21. doi:10.1016/j.cytogfr.2012.10.001.

59 Azizi G, Jadidi-Niaragh F, Mirshafiey A. 2013. Th17 cells in immunopathogenesis and treatment of rheumatoid arthritis. *International Journal of Rheumatic Diseases* 2013; **16** (3): 243–53. doi:10.1111/1756-185X.12132.

60 Gaffen SL, Jain R, Garg AV, Cua DJ. The IL-23-IL-17 immune axis: from mechanisms to therapeutic testing. *Nature Reviews. Immunology* 2014; **14**(9): 585–600. doi:10.1038/nri3707.

61 Venkatesha SH, Dudics S, Acharya B, Moudgil KD. Cytokine-modulating strategies and newer cytokine targets for arthritis therapy. *International Journal of Molecular Sciences* 2014; **16**(1): 887–906. doi:10.3390/ijms16010887.

62 Smolen JS, Günter Steiner G. 2003. Therapeutic strategies for rheumatoid arthritis. *Nature Reviews. Drug Discovery 2003;* **2**(6): 473–88. doi:10.1038/nrd1109.

63 Holmdahl R, Malmström V, Burkhardt H. Autoimmune priming, tissue attack and chronic inflammation — the three stages of rheumatoid arthritis. *European Journal of Immunology* 2014; **44**(6): 1593–99. doi:10.1002/eji.201444486.

64 Ebringer A, Wilson C. HLA molecules, bacteria and autoimmunity. *Journal of Medical Microbiology* 2000' **49**(4): 305–11.

65 Janeway A, Medzhitov R Innate immune recognition. *Annual Review of Immunology* 2002; **20**: 197–216.

66 Heine H, Ulmer AJ 2005. Recognition of bacterial products by toll-like receptors. *Chemical Immunology and Allergy* 2005; **86**: 99–119.

67 Takeda K, Akira S.. Toll-like receptors. *Current Protocols in Immunology,* ed. Coligan JE *et al.* Chapter 14: Unit 14.12.

68 McGuirk P, Mills, KHG. Pathogen-specific regulatory T cells provoke a shift in the Th1/Th2 paradigm in immunity to infectious diseases. *Trends in Immunology* 2002; **23**(9): 450–55.

69 Roumier T, Capron M, Dombrowicz D Faveeuw C. Pathogen induced regulatory cell populations preventing allergy through the Th1/Th2 paradigm point of view. *Immunologic Research* 2008; **40**(1): 1–17. doi:10.1007/s12026-007-0058-3.

70 Luckey D, Behrens M, Smart M, Luthra H, David CS, Taneja V DRB1*0402 may influence arthritis by promoting naive CD4+ T-cell differentiation in to regulatory T cells. *European Journal of Immunology* 2014; **44**(11): 3429–38. doi:10.1002/eji.201344424.

71 Stelekati E, Shin H, Doering TA, Dolfi DV, Ziegler CG, Beiting DP, Dawson L *et al.* Bystander chronic infection negatively impacts development of CD8(+) T cell memory. *Immunity* 2014; **40**(5): 801–13. doi:10.1016/j.immuni.2014.04.010.

72 Deng GM., Nilsson IM, Verdrengh M, Collins LV, Tarkowski A. Intra-articularly localized bacterial DNA containing CpG motifs induces arthritis. *Nature Medicine* 1999; **5**(6): 702–5. doi:10.1038/9554.

73 Zeuner RA, Ishii KJ, Lizak MJ, Gursel I, Yamada H, Klinman DM, Verthelyi D. 2002. Reduction of CpG-induced arthritis by suppressive oligodeoxynucleotides. *Arthritis and Rheumatism* 2002; **46**(8): 2219–24. doi:10.1002/art.10423.

74 Schaeverbeke T, Renaudin H, Clerc M, Lequen L,Vernhes JP, de Barbeyrac B, Bannwarth B, Bébéar C, Dehais J. Systematic detection of mycoplasmas by culture and polymerase chain reaction (PCR) procedures in 209 synovial fluid samples. *British Journal of Rheumatology* 1997; **36**(3): 310–14.

75 van der Heijden IM, Wilbrink B, Tchetverikov I, Schrijver IA, Schouls LM, Hazenberg MP, Breedveld FC, Tak PP. Presence of bacterial DNA and bacterial peptidoglycans in joints of patients with rheumatoid arthritis and other arthritides. *Arthritis and Rheumatis*.

76 Reichert S, Maximilian Haffner, Gernot Keyßer, Christoph Schäfer, Jamal M. Stein, Hans-Schaller G, Wienke A, Strauss H, Heide S. Schulz S. Detection of oral bacterial DNA in synovial fluid. *Journal of Clinical Periodontology* 2013; **40**(6): 591–98. doi:10.1111/jcpe.12102.

77 Craft J, Fatenejad S. Self antigens and epitope spreading in systemic autoimmunity. *Arthritis and Rheumatism* 1997; **40**(8): 1374–82. doi:10.1002/1529-0131(199708)40:8<1374:AID-ART3> 3.0.CO;2-7.

78 Li Y, Xu L, Olsen BR. Lessons from genetic forms of osteoarthritis for the pathogenesis of the disease. *Osteoarthritis and cartilage/OARS, Osteoarthritis Research Society* 2007; **15**(10): 1101–5. doi:10.1016/j.joca.2007.04.013.

79 Ytterberg AJ. Joshua V, Reynisdottir G, Tarasova NK, Rutishauser D, Ossipova E, Hensvold AH *et al.* Shared immunological targets in the lungs and joints of patients with rheumatoid arthritis: identification and validation. *Annals of the Rheumatic Diseases* 2014. doi:10.1136/annrheumdis-2013-204912.

80 Kobayashi T, Yoshie H. Host responses in the link between periodontitis and rheumatoid arthritis. *Current Oral Health Reports* 2015; **2**: 1.

81 McLean MH, Dieguez D, Miller LM, Young HA. Does the microbiota play a role in the pathogenesis of autoimmune diseases? *Gut* 2015; **64**(2): 332–41. doi:10.1136/gutjnl-2014-308514.

82 van der Woude D, Rantapää-Dahlqvist S, Ioan-Facsinay A, Onnekink C, Schwarte CM, Verpoort KN,Drijfhout JW, Huizinga TWJ, Toes REM, Pruijn GJM. Epitope spreading of the anti-citrullinated protein antibody response occurs before disease onset and is associated with the disease course of early arthritis. *Annals of the Rheumatic Diseases* 2010; **69**(8): 1554–61. doi:10.1136/ard.2009.124537.

83 okolove J, Bromberg R, Deane KD, Lahey LL, Derber LA, Piyanka E. Chandra, Edison JD *et al.* Autoantibody epitope spreading in the pre-clinical phase predicts progression to rheumatoid arthritis. *PloS One* 2012; **7**(5): e35296. doi:10.1371/journal.pone.0035296.

84 Pablo P de, Dietrich T, Chapple ILC, Milward M, Chowdhury M, Charles PJ, Buckley CD, Venables PJ. The autoantibody repertoire in periodontitis: a role in the induction of autoimmunity to citrullinated proteins in rheumatoid arthritis? *Annals of the Rheumatic Diseases* 2014; **73**(3): 580–86. doi:10.1136/annrheumdis-2012-202701.

85 Sokolove J, Johnson DS, Lahey LJ, Wagner CA, Cheng D, Thiele GM, Michaud K *et al.* Rheumatoid factor as a potentiator of anti-citrullinated protein antibody-mediated inflammation in rheumatoid arthritis. *Arthritis & Rheumatology (Hoboken, N.J.)* 2014; **66**(4): 813–21. doi:10.1002/art.38307.

86 de Hair MJH, van de Sande, MGH, Ramwadhdoebe TH, Hansson M, Landewé R, van der Leij C, Maas M *et al.* Features of the synovium of individuals at risk of developing rheumatoid arthritis: implications for understanding preclinical rheumatoid arthritis. *Arthritis & Rheumatology (Hoboken, N.J.)* 2014; **66**(3): 513–22. doi:10.1002/art.38273.

87 Irwin, MJ. Hudson KR. Fraser JD, Gascoigne.NR.Enterotoxin residues determining T-cell receptor V beta binding specificity. *Nature* 1992; **359**(6398): 841–43. doi:10.1038/359841a0.

88 Gascoigne NR, Ames KT. 1991. Direct binding of secreted T-cell receptor beta chain to superantigen associated with class II major histocompatibility complex protein. *Proceedings of the National Academy of Sciences USA 1991*; **88**(2): 613–16.

89 Hayball JD, Robinson JH, O'Hehir RE, Verhoef A, Lamb JR, Lake RA. Identification of two binding sites in staphylococcal enterotoxin B that confer specificity for TCR V beta gene products. *International Immunology* 1994; **6**(2): 199–211.

90 Petersson K, Forsberg G, Walse B. Interplay between superantigens and immunoreceptors. *Scandinavian Journal of Immunology* 2004; **59**(4): 345–55. doi:10.1111/j.0300-9475.2004.01404.x.

91 Gutierrez-Ramos JC, Bluethmann H. Molecules and mechanisms operating in septic shock: lessons from knockout mice. *Immunology Today* 1997; **18**(7): 329–34.

92 Ulevitch RJ, Tobias PS. Receptor-dependent mechanisms of cell stimulation by bacterial endotoxin. *Annual Review of Immunology* 1995; **13**: 437–57.

93 Ni M, MacFarlane AW, Toft M, Lowell CA, Campbell KS, Hamerman JA. B-cell adaptor for PI3K (BCAP) negatively regulates Toll-like receptor signaling through activation of PI3K. *Proceedings of the National Academy of Sciences USA* 2012; **109**(1): 267–72. doi:10.1073/pnas.1111957108.

94 Sethi G, Ahn KS, und Aggarwal BB. Targeting nuclear factor-kappa B activation pathway by thymoquinone: role in suppression of antiapoptotic gene products and enhancement of apoptosis. *Molecular Cancer Research: MCR* 2008; **6**(6): 1059–70. doi:10.1158/1541-7786.MCR-07-2088.

95 Grall F, Gu X, Tan L, Cho J-Y, Inan MS, Pettit AR Thamrongsak U *et al.* Responses to the proinflammatory cytokines interleukin-1 and tumor necrosis factor alpha in cells derived from rheumatoid synovium and other joint tissues involve nuclear factor kappaB-mediated induction of the Ets transcription factor ESE-1. *Arthritis and Rheumatism* 2003; **48**(5): 1249–60. doi:10.1002/art.10942.

96 Benito MJ, Murphy E, Murphy EP, van den Berg WB, FitzGerald O, Bresnihan B. Increased synovial tissue NF-kappa B1 expression at sites adjacent to the cartilage-pannus junction in rheumatoid arthritis. *Arthritis and Rheumatism* 2004; **50**(6): 1781–87. doi:10.1002/art.20260.

97 Kumar A, Takada Y, Boriek AM, Aggarwal BB. Nuclear factor-kappaB: its role in health and disease. *Journal of Molecular Medicine (Berlin, Germany)* 2004; **82**(7): 434–48. doi:10.1007/s00109-004-0555-y.

98 Shakibaei M, John T, Schulze-Tanzil G, Lehmann I, Mobasheri A. Suppression of NF-kappaB activation by curcumin leads to inhibition of expression of cyclo-oxygenase-2 and matrix metalloproteinase-9 in human articular chondrocytes: Implications for the treatment of osteoarthritis. *Biochemical Pharmacology* 2007; **73**(9): 1434–45. doi:10.1016/j.bcp.2007.01.005.

99 Chen T, Rimpiläinen M, Luukkainen R, Möttönen T, Yli-Jama T, Jalava J, Olli Vainio Toivanen P. Bacterial components in the synovial tissue of patients with advanced rheumatoid arthritis or osteoarthritis: analysis with gas chromatography-mass spectrometry and pan-bacterial polymerase chain reaction. *Arthritis and Rheumatism* 2003; **49**(3): 328–34. doi:10.1002/art.11119.

100 Moen K. Brun JG, Valen M, Skartveit L, Eribe EK, Olsen RI Jonsson RJ. Synovial inflammation in active rheumatoid arthritis and psoriatic arthritis facilitates trapping of a variety of oral bacterial DNAs. *Clinical and Experimental Rheumatology* 2006; **24**(6): 656–63.

101 Wegner N, Lundberg K, Kinloch A, Fisher B, Malmström V, Feldmann M, Venables PJ. Autoimmunity to specific citrullinated proteins gives the first clues to the etiology of rheumatoid arthritis. *Immunological Reviews* 2010; **233**(1): 34–54. doi:10.1111/j.0105-2896.2009.00850.x.

102 van Eden W, 1991. Heat-shock proteins as immunogenic bacterial antigens with the potential to induce and regulate autoimmune arthritis. *Immunological Reviews* 1991; **121**: 5–28.

103 Lambrecht S, Juchtmans N, Elewaut D. Heat-shock proteins in stromal joint tissues: innocent bystanders or disease-initiating proteins?. *Rheumatology (Oxford, England)* 2014; **53**(2): 223–32. doi:10.1093/rheumatology/ket277.

104 Multhoff G. Heat shock proteins in immunity. *Handbook of Experimental Pharmacology* 2006 (**172**): 279–304.

105 Vabulas RM, Wagner H, Schild H. Heat shock proteins as ligands of toll-like receptors. *Current Topics in Microbiology and Immunology* 2002; **270**: 169–84.

106 Panayi GS, Corrigall VM. BiP regulates autoimmune inflammation and tissue damage. *Autoimmunity Reviews* 2006; **5**(2): 140–42. doi:10.1016/j.autrev.2005.08.006.

107 Re F, Strominger JL. Toll-like receptor 2 (TLR2) and TLR4 differentially activate human dendritic cells. *The Journal of Biological Chemistry* 2001; **276**(40): 37692–99. doi:10.1074/jbc.M105927200.

108 Akira S, Takeda K. Toll-like receptor signalling. *Nature Reviews. Immunology* 2004 **4**(7): 499–511. doi:10.1038/nri1391.

109 Guerau-de-Arellano M, Huber BT. Chemokines and Toll-like receptors in Lyme disease pathogenesis. *Trends in Molecular Medicine* 2005; **11**(3): 114–20. doi:10.1016/j.molmed.2005.01.003.

110 Strle K, Shin JJ, Glickstein LJ, Steere AC. Association of a Toll-like receptor 1 polymorphism with heightened Th1 inflammatory responses and antibiotic-refractory Lyme arthritis. *Arthritis and Rheumatism* 2012; **64**(5): 1497–1507. doi:10.1002/art.34383.

111 Zhang X, Glogauer M, Zhu F, Kim T-H, Chiu B, Inman RD. Innate immunity and arthritis: neutrophil Rac and toll-like receptor 4 expression define outcomes in infection-triggered arthritis. *Arthritis and Rheumatism* 2005; **52**(4): 1297–1304. doi:10.1002/art.20984.

112 Coker C, Poore CA, Li X, Mobley HL. Pathogenesis of Proteus mirabilis urinary tract infection. *Microbes and Infection/Institut Pasteur* 2000; **2**(12): 1497–1505.

113 Ebringer A, Rashid T. Rheumatoid arthritis is caused by a Proteus urinary tract infection. *APMIS: Acta Pathologica, Microbiologica, et Immunologica Scandinavica* 2014; **122**(5): 363–68. doi:10.1111/apm.12154.

114 Rashid T, Jayakumar KS, A. Binder A, Ellis S, Cunningham P, Ebringer A. Rheumatoid arthritis patients have elevated antibodies to cross-reactive and non cross-reactive antigens from Proteus microbes. *Clinical and Experimental Rheumatology* 2007; **25**(2): 259–67.

115 Choy E Understanding the dynamics: pathways involved in the pathogenesis of rheumatoid arthritis. *Rheumatology (Oxford, England)* 2012; **51** Suppl 5: v3-11.

116 Mercado FB, Marshall RI, Klestov AC, Bartold PM. Relationship between rheumatoid arthritis and periodontitis. *Journal of Periodontology* 2001; **72**(6): 779–87. doi:10.1902/jop.2001.72.6.779.

117 Socransky SS, Haffajee AD Dental biofilms: difficult therapeutic targets. *Periodontology 2000* 2002; **28**: 12–55.

118 Lovegrove JM. Dental plaque revisited: bacteria associated with periodontal disease. *Journal of the New Zealand Society of Periodontology* 2004; (**87**): 7–21.

119 Hajishengallis G, Liang S, Payne MA, Hashim A, Jotwani R, Eskan MA, McIntosh ML *et al.* Low-abundance biofilm species orchestrates inflammatory periodontal disease through the commensal microbiota and complement. *Cell Host & Microbe* 2011; **10**(5): 497–506. doi:10.1016/j.chom.2011.10.006.

120 Rosenstein ED, Greenwald RA, Kushner LJ, Weissmann G. Hypothesis: the humoral immune response to oral bacteria provides a stimulus for the development of rheumatoid arthritis. *Inflammation* 2004; **28**(6): 311–18. doi:10.1007/s10753-004-6641-z.

121 Pischon N. Pischon T, Kröger J, Gülmez E, Kleber B-M, Bernimoulin J-P, Landau H *et al.* Association among rheumatoid arthritis, oral hygiene, and periodontitis. *Journal of Periodontology* 2008; **79** (6): 979–86. doi:10.1902/jop.2008.070501.

122 Georgiou TO, Marshall RI Bartold PM. Prevalence of systemic diseases in Brisbane general and periodontal practice patients. *Australian Dental Journal* 2004; **49**(4): 177–84.

123 Pischon N, Röhner E, Hocke A, N'Guessan P, Müller HC, Matziolis G. Kanitz V *et al.* Effects of Porphyromonas gingivalis on cell cycle progression and apoptosis of primary human chondrocytes. *Annals of the Rheumatic Diseases* 2009; **68**(12): 1902–7. doi:10.1136/ard.2008.102392.

124 Pablo P de, Dietrich T, McAlindon TE. Association of periodontal disease and tooth loss with rheumatoid arthritis in the US population. *The Journal of Rheumatology* 2008; **35**(1): 70–76.

125 Lagervall M, Jansson L, Bergström J. Systemic disorders in patients with periodontal disease. *Journal of Clinical Periodontology* 2003; **30**(4): 293–99.

126 Monsarrat P, Vergnes J-N, Blaizot A, Constantin de Grado A, Fernandez G, Ramambazafy H, Sixou M, Cantagrel A, Nabet C. Oral health status in outpatients with rheumatoid arthritis: the OSARA study. *Oral Health and Dental Management* 2014; **13**(1): 113–19.

127 Demmer RT, Molitor JA, Jacobs DR, Michalowicz BS. Periodontal disease, tooth loss and incident rheumatoid arthritis: results from the First National Health and Nutrition Examination Survey and its epidemiological follow-up study. *Journal of Clinical Periodontology* 2011; **38**(11): 998–1006. doi:10.1111/j.1600-051X.2011.01776.x.

128 Mikuls TR, Payne JB, Reinhardt RA, Thiele GM, Maziarz, E Cannella AC, Holers VM, Kuhn KA, O'Dell J. Antibody responses to Porphyromonas gingivalis (P. gingivalis) in subjects with rheumatoid arthritis and periodontitis. *International Immunopharmacology* 2009; **9**(1): 38–42. doi:10.1016/j.intimp.2008.09.008.

129 Martinez-Martinez RE. Abud-Mendoza C, Patiño-Marin N, Rizo-Rodríguez JC, Little JW, Loyola-Rodríguez JP. Detection of periodontal bacterial DNA in serum and synovial fluid in refractory rheumatoid arthritis patients. *Journal of Clinical Periodontology* 2009; **36**(12): 1004–10. doi:10.1111/j.1600-051X.2009.01496.x.

130 Lundberg K, Kinloch A, Fisher BA, Wegner N, Wait R Charles P,. Mikuls TR, Venables PJ. Antibodies to citrullinated alpha-enolase peptide 1 are specific for rheumatoid arthritis and cross-react with bacterial enolase. *Arthritis and Rheumatism* 2008; **58**(10): 3009–19. doi:10.1002/art.23936.

131 Guo A, Nguyen K-A, Potempa J. Dichotomy of gingipains action as virulence factors: from cleaving substrates with the precision of a surgeon's knife to a meat chopper-like brutal degradation of proteins. *Periodontology 2000* 2010; **54**(1): 15–44. doi:10.1111/j.1600-0757.2010.00377.x.

132 Krauss JL, Potempa J, Lambris JD, Hajishengallis G. Complementary Tolls in the periodontium: how periodontal bacteria modify complement and Toll-like receptor responses to prevail in the host. *Periodontology 2000* 2010; **52**(1): 141–62. doi:10.1111/j.1600-0757.2009.00324.x.

133 Potempa M, Potempa J. Protease-dependent mechanisms of complement evasion by bacterial pathogens. *Biological Chemistry* 2012; **393**(9): 873–88. doi:10.1515/hsz-2012-0174.

134 Liu Y-CG, Lerner UH, Teng Y-TA. Cytokine responses against periodontal infection: protective and destructive roles. *Periodontology 2000* 2010; **52**(1): 163–206. doi:10.1111/j.1600-0757.2009.00321.x.

135 Trindade F, Oppenheim FG, Helmerhorst EJ, Amado F, Gomes PS, Vitorino R.. Uncovering the molecular networks in periodontitis. *Proteomics. Clinical Applications* 2014; **8**(9-10): 748–61. doi:10.1002/prca.201400028.

136 Cardoso CR, Garlet GP, Crippa GE, Rosa AL, Júnior WM, Rossi MA, Silva.JS. Evidence of the presence of T helper type 17 cells in chronic lesions of human periodontal disease. *Oral Microbiology and Immunology* 2009; **24**(1): 1–6. doi:10.1111/j.1399-302X.2008.00463.x.

137 Marchesan JT, Gerow EA, Schaff R, Taut AD, Shin S-Y, Sugai J, Brand D *et al.*. Porphyromonas gingivalis oral infection exacerbates the development and severity of collagen-induced arthritis. *Arthritis Research & Therapy* 2013; **15**(6): R186. doi:10.1186/ar4376.

138 Röhner E, Detert J, Kolar P, Hocke A, N'Guessan P, Matziolis G, Kanitz V *et al.* Induced apoptosis of chondrocytes by Porphyromonas gingivalis as a possible pathway for cartilage loss in rheumatoid arthritis. *Calcified Tissue International* 2010; **87**(4): 333–40. doi:10.1007/s00223-010-9389-5.

139 Röhner E, Hoff P, Matziolis G, Perka C, Riep B, Nichols FC, Kielbassa AM *et al.* 2012. The impact of Porphyromonas gingivalis lipids on apoptosis of primary human chondrocytes. *Connective Tissue Research* 2012; **53**(4): 327–33. doi:10.3109/03008207.2012.657308.

140 McGraw WT, J. Potempa J, Farley D, Travis J. Purification, characterization, and sequence analysis of a potential virulence factor from Porphyromonas gingivalis, peptidylarginine deiminase. *Infection and Immunity* 1999; **67**(7): 3248–56.

141 Abdullah S-N, Farmer E-A, Spargo L, Logan R, Gully N. Porphyromonas gingivalis peptidylarginine deiminase substrate specificity. *Anaerobe* 2013; **23**:102–8.

142 Maresz KJ, Hellvard A, Sroka A, Adamowicz K, Bielecka E, Koziel J, Gawron K *et al*. Porphyromonas gingivalis facilitates the development and progression of destructive arthritis through its unique bacterial peptidylarginine deiminase (PAD). *PLoS Pathogens* 2013; **9**(9): e1003627. doi:10.1371/journal.ppat.1003627.

143 Schellekens GA, Visser H, de Jong, BA, van den Hoogen FH, Hazes JM, Breedveld FC, van Venrooij WJ. The diagnostic properties of rheumatoid arthritis antibodies recognizing a cyclic citrullinated peptide. *Arthritis and Rheumatism* 2000; **43**(1): 155–63. doi:10.1002/1529-0131 (200001)43:1<155:AID-ANR20>3.0.CO;2-3.

144 Quirke A-M, Lugli EB Wegner N, Hamilton BC, Charles P, Chowdhury M, Ytterberg AJ *et al.* Heightened immune response to autocitrullinated Porphyromonas gingivalis peptidylarginine deiminase: a potential mechanism for breaching immunologic tolerance in rheumatoid arthritis. *Annals of the Rheumatic Diseases* 2014; **73**(1): 263–69. doi:10.1136/annrheumdis-2012-202726.

145 Mangat P, Wegner N, Venables PJ, Potempa J. Bacterial and human peptidylarginine deiminases: targets for inhibiting the autoimmune response in rheumatoid arthritis? *Arthritis Research & Therapy* 2010; **12**(3): 209. doi:10.1186/ar3000.

146 Kinloch AJ, Alzabin S, Brintnell W, Wilson E, Barra L, Wegner N, Bell DA, Ewa Cairns Venables PJ. Immunization with Porphyromonas gingivalis enolase induces autoimmunity to mammalian α-enolase and arthritis in DR4-IE-transgenic mice. *Arthritis and Rheumatism* 2011; **63**(12): 3818–23. doi:10.1002/art.30639.

147 Gilliam BE, Reed MR, Chauhan AK, Dehlendorf AB, Moore TL Evidence of fibrinogen as a target of citrullination in IgM rheumatoid factor-positive polyarticular juvenile idiopathic arthritis. *Pediatric Rheumatology Online Journal* 2011; **9**: 8.

148 Harre U, Georgess D, Bang H, Bozec A, Axmann R, Ossipova E, Jakobsson P-J *et al.* Induction of osteoclastogenesis and bone loss by human autoantibodies against citrullinated vimentin. *The Journal of Clinical Investigation* 2012; **122**(5): 1791–1802. doi:10.1172/JCI60975.

149 Gilliam BE, Chauhan AK, Moore.TL. Evaluation of anti-citrullinated type II collagen and anti-citrullinated vimentin antibodies in patients with juvenile idiopathic arthritis. *Pediatric Rheumatology Online Journal* 2013; **11**(1): 31. doi:10.1186/1546-0096-11-31.

150 Pablo P de, Dietrich T, Chapple ILC, Milward M, Chowdhury M, Charles PJ, Buckley CD, Venables PJ. The autoantibody repertoire in periodontitis: a role in the induction of autoimmunity to citrullinated proteins in rheumatoid arthritis?. *Annals of the Rheumatic Diseases* 2014; **73**(3): 580–86. doi:10.1136/annrheumdis-2012-202701.

151 Brink M, Hansson M, Rönnelid J, Klareskog L Dahlqvist SR. The autoantibody repertoire in periodontitis: a role in the induction of autoimmunity to citrullinated proteins in rheumatoid arthritis? Antibodies against uncitrullinated peptides seem to occur prior to the antibodies to the corresponding citrullinated peptides. *Annals of the Rheumatic Diseases* 2014; **73**(7): e46. doi:10.1136/annrheumdis-2014-205498.

152 Montes A, Dieguez-Gonzalez R, Perez-Pampin E, Calaza M, Mera-Varela A, Gomez-Reino JJ, Gonzalez A. Particular association of clinical and genetic features with autoimmunity to citrullinated α-enolase in rheumatoid arthritis. *Arthritis and Rheumatism* 2011; **63**(3): 654–61. doi:10.1002/art.30186.

153 Rodríguez SB, Stitt BL, Ash DE. 2009. Expression of peptidylarginine deiminase from Porphyromonas gingivalis in Escherichia coli: enzyme purification and characterization. *Archives of Biochemistry and Biophysics* 2009; **488**(1): 14–22. doi:10.1016/j.abb.2009.06.010.

154 Koziel J, Mydel P, Potempa J. The link between periodontal disease and rheumatoid arthritis: an updated review. *Current Rheumatology Reports* 2014; **16**(3): 408. doi:10.1007/s11926-014-0408-9.

155 Hendler A, Mulli TK, Hughes FJ, Perrett D, Bombardieri M, Houri-Haddad Y, Weiss EI, Nissim A. Involvement of autoimmunity in the pathogenesis of aggressive periodontitis. *Journal of Dental Research* 2010; **89**(12): 1389–94. doi:10.1177/0022034510381903.

156 Berthelot J-M, Le Goff B. Rheumatoid arthritis and periodontal disease. *Joint, Bone, Spine: Revue du Rheumatisme* 2010; **77**(6): 537–41. doi:10.1016/j.jbspin.2010.04.015.

157 Hunter W. *The Role of Sepsis and of Antisepsis in Medicine.* 1918: Philadelphia; S.S. White Dental Manufacturing Co., pp. 585–602.

158 Lyons R. An appraisal of the focal infection theory; with special reference to arthritis. *Journal of Endodontia* 1946; **1**(4): 39–45.

159 Hughes R.A. 1994. Focal infection revisited. *British Journal of Rheumatology* 1994; **33** (4): 370–77.

160 Vaahtovuo J, Munukka E, Korkeamäki M, Luukkainen L, Toivanen P. Fecal microbiota in early rheumatoid arthritis. *The Journal of Rheumatology* 2008; **35**(8): 1500–1505.

161 Liu X, Zou Q, Zeng B, Fang Y, Wei H. Analysis of fecal Lactobacillus community structure in patients with early rheumatoid arthritis. *Current Microbiology* 2013; **67**(2): 170–76. doi:10.1007/s00284-013-0338-1.

162 Scher JU, Sczesnak A, Longman RS, Segata N, Ubeda C, Bielski C, Rostron T *et al*. Expansion of intestinal Prevotella copri correlates with enhanced susceptibility to arthritis. *eLife* 2013; **2**:e01202.

163 Brusca SB, Abramson SB, Scher JU. Microbiome and mucosal inflammation as extra-articular triggers for rheumatoid arthritis and autoimmunity *Current Opinion in Rheumatology* 2014; **26**(1): 101–7. doi:10.1097/BOR.0000000000000008.

164 Makrygiannakis D, Hermansson M, Ulfgren A-K, Nicholas AP, Zendman, AJW,.Eklund A, Grunewald J, Skold CM, Klareskog L, Catrina AI. Smoking increases peptidylarginine deiminase 2 enzyme expression in human lungs and increases citrullination in BAL cells. *Annals of the Rheumatic Diseases* 2008; **67**(10): 1488–92. doi:10.1136/ard.2007.075192.

165 Kochi Y, Thabet MM, Suzuki A, Okada Y, Daha NA, Toes REM, Huizinga,TWJ *et al*. PADI4 polymorphism predisposes male smokers to rheumatoid arthritis. *Annals of the Rheumatic Diseases* 2011; **70**(3): 512–15. doi:10.1136/ard.2010.130526.

166 Demoruelle,MK., Weisman MH, Simonian PL, Lynch DA, Sachs PB, Pedraza IF, Annie R. Harrington *et al*. Brief report: airways abnormalities and rheumatoid arthritis-related autoantibodies in subjects without arthritis: early injury or initiating site of autoimmunity? *Arthritis and Rheumatism* 2012; **64**(6): 1756–61. doi:10.1002/art.34344.

167 Perry E, Kelly C, Eggleton P, de Soyza A, Hutchinson D. The lung in ACPA-positive rheumatoid arthritis: an initiating site of injury? *Rheumatology (Oxford, England)* 2014; **53**(11): 1940–50. doi:10.1093/rheumatology/keu195.

168 van Willis C, Demoruelle MK, Derber LA, Chartier-Logan CJ, Parish MC, Pedraza IF, Weisman MH, Norris JM, Holers VM, Deane KD. Sputum autoantibodies in patients with established rheumatoid arthritis and subjects at risk of future clinically apparent disease *Arthritis and Rheumatism* 2013 **65**(10): 2545–54. doi:10.1002/art.38066.

169 Källberg H, Vieira V, Holmqvist M, Hart JE, Costenbader KH, Bengtsson C, Klareskog L, Karlson EW, Alfredsson L. Regional differences regarding risk of developing rheumatoid arthritis in Stockholm County, Sweden: results from the Swedish Epidemiological Investigation of Rheumatoid Arthritis (EIRA) study. *Scandinavian Journal of rheumatology* 2013; **42**(5): 337–43. doi:10.3109/03 009742.2013.769062.

170 Sandberg MEC, Bengtsson C, Klareskog L, Alfredsson L, Saevarsdottir S. Recent infections are associated with decreased risk of rheumatoid arthritis: a population-based case-control study. *Annals of the Rheumatic Diseases*. 2015; doi:10.1136/annrheumdis-2014-206493.

CHAPTER 19

Inflammatory bowel disease and the gut microbiota

Nik Ding and Ailsa Hart

St. Mark's Hospital, London, United Kingdom

19.1 The microbiota in inflammatory bowel disease

The normal gut microbiota contains 100 trillion microbes, primarily bacteria, with >1100 prevalent species and a minimum of 160 species in each human individual[1]. Chapter 2 of this book introduced the concept of dysbiosis as a shift in the normal microbiota that can occur in different tissues and can result in a host-microbial relationship breakdown. Dysbiosis of the gut microbiota likely instigates inflammatory bowel disease (IBD) development. The shift from predominantly "symbiont" microbes to potentially harmful "pathobiont" microbes is thought to be pivotal in this proces[2]. There has been significant interest regarding the role of dysbiosis in chronic gastrointestinal disease, particularly IBD, because it is essential to mucosal integrity or dysfunction.

The human microbiota has been characterized using genome-wide association studies and various genetic analyses, through which links between intestinal microbiota profiles and altered immune responses have been discovered[3]. IBD pathogenesis can be explained by changes in these profiles, which indicate microbiota population shifts. This chapter outlines the major IBD patient studies in relation to dysbiosis mechanisms, recent research progress from 16S-based studies linking specific organisms with disease, and studies of the microbiota and its future applications with regard to metagenomics and metabolomics.

19.2 Dysbiosis and IBD pathogenesis

The relationship between the gut commensal microbiota and IBD is complex and can be explained by four broad mechanisms: (i) conventional microbiota dysbiosis, (ii) intestinal inflammation induction by pathogens and/or functionally altered commensal bacteria, (iii) genetic defects in the host resulting in the loss of the commensal microbiota, and (iv) defective immunoregulation in the host. Large-scale human genetics studies, totalling 75,000 cases and controls, revealed 163 host

The Human Microbiota and Chronic Disease: Dysbiosis as a Cause of Human Pathology, First Edition.
Edited by Luigi Nibali and Brian Henderson.
© 2016 John Wiley & Sons, Inc. Published 2016 by John Wiley & Sons, Inc.

susceptibility loci[4]. These loci are enriched for pathways involved in environmental modulation of intestinal homeostasis[5]. IBD incidence in North America and Europe rose dramatically towards the end of the 20th century, frequently doubling between decades[6]. The increased prevalence of IBD has expanded into developing countries over the last twenty years, with increased occurrence of ulcerative colitis (UC) compared with Crohn's disease (CD)[6]. Additionally, several studies of monozygotic twin pairs showed that their IBD concordance rate is significantly less than 50%, and CD had the least concordance[7]. Thus, IBD is a multifaceted disorder where germline genetics, the immune system, and environmental factors all play important roles[8]. The gut microbial community is increasingly gaining attention for its general health influences[9], particularly regarding IBD.

The largest microbial reservoir in the body is the gut. The gut microbiota coexists with the host. The numbers of bacteria in the gut microbiota vary throughout the gastrointestinal tract, reaching up to 10^{11} or 10^{12} cells/g luminal contents in the colon[10]. They perform several valuable functions for the host, including digesting host enzyme-inaccessible substrates, educating the immune system, and repressing harmful microorganism growth[11] (see chapter 3 for the physiological role of gut microbiota).

Advances in next-generation DNA sequencing have improved accuracy in documenting gut microbiota (see also chapter 6 by Bermudez). An exhaustive analysis of global gut bacterial communities from normal individuals suggests the existence of distinct enterotypes: *Bacteroides*, *Prevotella*, and *Ruminococcus*. These enterotypes seem to be influenced predominantly by dietary intake and are age- and body-mass-index (BMI) independent[12,13]. This will be discussed in further detail later. Studies suggest that the *Bacteroides* enterotype correlates to a "Western" protein-rich diet, whereas the *Prevotella* enterotype is associated with carbohydrate-rich diets[14]. Whether this Western enterotype is a distinct IBD risk factor is unclear at this time due to the large variations in what constitutes a particular diet.

The dysbiosis shift from beneficial "symbiont" microbes to potentially harmful "pathobiont" microbes is well documented[15]. While some of these gut microbiota changes were detected in a common IBD patient subset, some were clearly delineated between CD and UC patients. The best-defined change noted in IBD patients is a reduced phylum *Firmicutes* population[16]. Among *Firmicutes*, reduced *Faecalibacterium prausnitzii* in CD patients compared with controls has been well documented[17,18,19]. This is in contrast to a pediatric CD patient cohort that showed increased *F. prausnitzii* levels, suggesting that this bacterium may have a putative protective effect at the point of IBD onset[20]. In addition, *Firmicutes* diversity definitively decreased, and fewer constituent species were detected in IBD patients[21]. Unlike *Firmicutes*, some studies show increased phylum *Bacteroidetes* bacteria in IBD patients[22,23].

19.3 Environmental factors affecting microbiome composition

19.3.1 Diet

Dietary preference is a critical environmental factor affecting microbial composition (discussed by Gibson in chapter 29). Its possible role in determining microbiota composition throughout mammalian evolution has been demonstrated previously[24].

While no specific diet has been identified to directly cause, prevent, or treat IBD, the interactions between nutrients and microbiota must be taken into account when studying the microbiota's role in this disease. Thus far, limited information has been gathered in humans, due primarily to the difficulty of designing a large-scale controlled diet study. Wu *et al.* showed that long-term dietary patterns affected *Bacteroides*, *Prevotella*, and *Firmicutes* ratios, and that short-term dietary changes may not significantly influence it[25]. In addition, Zimmer *et al.* studied the impact of vegan or vegetarian diets on the gut microbiota[26]. They found that *Bacteroides* species, *Bifidobacterium* species, and *Enterobacteriaceae* were significantly reduced, whereas the total bacterial load was unaltered. Since *Enterobacteriaceae* are consistently increased in IBD patients, future studies should look at the role of short-term and long-term dietary patterns on the microbiome in IBD. Given the complexity of dietary effects, obtaining this information may only be feasible in large cohort studies[27].

19.3.2 Age

IBD exhibits age-related variations with three distinct onset stages. The peak onset age is typically 15 to 30 years old, with late onset occurring at approximately 60 years and early onset occurring at less than 10-years. Noticeably, the incidence of early-onset disease has significantly increased over the past decade[28]. These stages correspond to different phases of altered gut microbiota diversity and stability[29]. The microbiota usually has low complexity and is relatively unstable early in life. This is affected by birth route (vaginal or Cesarean section) and fluctuates as a result of dietary changes (switch from breastfeeding to solid foods), illness, and puberty[30]. Maximal microbial assembly stability and complexity is reached in adulthood, and then any variations are few and minor[31]. This stability decreases in elderly subjects (60 years of age or older)[32]. Given the different microbiota characteristics over the three distinct disease-onset stages, the microbiota may play different roles in disease initiation and progression.

19.4 Genetics and application to the immune system and dysbiosis in IBD

The interaction between host genetics and microbiome is one potential cause in the development of IBD. Host factors can contribute to dysbiosis in IBD patients. Polymorphisms have been observed in over 75 genes in CD patients, the majority of which can be broadly categorized as mediating abnormalities in innate immune responses, immunoregulation, or mucosal barrier function[33]. A number of these genetic alterations are expected to impact both the quantity and composition of the intestinal microflora. Nucleotide-binding oligomerization domain-containing protein 2 (NOD2), an intracellular pattern recognition receptor involved in innate immunity, was the first CD susceptibility gene identified[34], and the specific polymorphism results in defective α-defensin production and reduced clearance of intracellular bacteria[35]. NOD2 stimulates an immune reaction upon recognition of muramyl dipeptide, a peptidoglycan constituent of Gram-positive and Gram-negative bacterial cell walls. It is expressed in Paneth cells, which are predominantly located in the terminal ileum at the base of

intestinal crypts and produce antimicrobial defensins[36]. Therefore, *NOD2* mutations can significantly affect microbial composition. Indeed, IBD patients with
NOD2 mutations have increased numbers of mucosa-adherent bacteria and
decreased transcription of the anti-inflammatory cytokine, interleukin (IL) 10[37].
IBD patients carrying risk alleles for *NOD2* and autophagy-related 16-like 1 protein (ATG16L1), an IBD susceptibility gene involved in autophagy, have significant gut microbiota structure alterations, including decreased *Faecalibacterium*
levels and increased numbers of *Escherichia*[38].

Subjects homozygous for fucosyltransferase 2 (FUT2) loss-of-function alleles
are "non-secretors" and do not express ABO antigen in their gastrointestinal
mucosa or bodily secretions. Non-secretors have an increased risk for CD[39] and
exhibit substantial mucosa-associated microbiota alterations[40]. Thus, host genetics
play a strong role in establishing and shaping the gut microbiota. Indeed, monozygotic
twins frequently have more similar microbiomes than non-twin siblings[41]. Future
genome-wide studies examining interactions between common human genetic
variations and microbial ecosystem composition would be valuable. This is discussed
in more detail in chapter 29.

The microbiota is also involved in microbial selection: with the preservation
and selection of certain bacterial populations. Defective microbial killing may
result in increased gut bacteria exposure and pathogenic T cell activation. Paneth
cells in the small intestine are key antimicrobial peptide producers, generating the
α-defensins, human defensin 5 (HD5) and HD6. A study of ileal CD patient specimens found decreased Paneth cell HD5 expression and a concomitant reduction
in antimicrobial activity against *E. coli* and *Staphylococcus aureus*[42]. However, a similar
decrease in α-defensin production was not seen in colon CD, UC, or pouchitis
patients. α-defensin production decreases probably result from reduced expression of wingless-type MMTV integration site family member (Wnt) and transcription factor 4 (Tcf4), a transcription factor that controls the expression of several
downstream target genes that are critical in Paneth cell positioning, differentiation,
and maturation in intestinal crypts[43]. The functional consequences of reduced α-
defensin expression were subsequently confirmed in a mouse transgenic model.
HD5 expression changes had profound effects on intestinal microbiota, changing
from predominantly small bacilli and cocci in wild-type mice to mixed bacilli and
fusiform bacterial species in heterozygous mice and finally to predominantly fusiform bacteria in homozygous transgenic mice[44]. A second study using peptide
extracts taken from patients with active or inactive ileocolonic or colonic CD, UC,
or controls compared their antimicrobial activities against clinical *Bacteroides vulgatus*, *Enterococcus faecalis*, *E. coli*, and *S. aureus* isolates. A pronounced *B. vulgatus*, *E.
coli*, and *E. faecalis* antimicrobial activity reduction was observed in CD extracts,
but not in those from UC or controls[45].

Genetic polymorphism studies of ATG16L1 provide additional evidence that a
defect in the innate immune response exists in a subset of CD patients. ATG16L1
was initially identified in a genome-wide association scan as a candidate CD susceptibility gene[46]. Certain ATG16L1 polymorphisms affect the capacity of innate
immune cells to induce muramyl dipeptide-triggered autophagy, which is specifically recognized by NOD2. Ultimately, this deficiency results in antigen presentation
defects in the major histocompatibility complex, decreased immune activation,
and increased persistence of gut mucosa bacteria[47].

The role of the intestinal microbiota in the pathogenesis of IBD is also illustrated by the fact that most CD patients show substantial antibody responses to bacterial and fungal antigens. In a study conducted by Mow et al.[48], sera from 303 CD patients were tested for antibodies to I2, a CD-related bacterial sequence, E. coli outer membrane porin C (OmpC), and Saccharomyces cerevisiae (ASCA) and for perinuclear antineutrophil cytoplasmic antibodies. Overall, nearly 80% of patients with CD showed a serological response to I2, OmpC, or ASCA with the majority having serological responses to more than two microbial antigens. The antigens detected were significantly correlated with specific disease phenotypes. Patients with antibodies to I2 were more likely to have fibrostenosing CD, whereas patients with OmpC antibodies were more likely to have internal perforating disease. Patients positive for all three antigens were more likely to have undergone small-bowel surgery compared with non-reactive patients (72% vs. 23%; $P<0.001$).

19.5 An overview of gut microbiota studies in IBD

Many IBD susceptibility loci suggest an impaired microbe response in disease, but the causality of this relationship is unclear. IBD pathogenesis may result from dysregulation of the mucosal immune system, driving a pathogenic immune response against commensal gut flora. Some studies showed that the gut microbiota is an essential factor driving inflammation in IBD[49]. Indeed, short-term treatment with enterically coated antibiotics dramatically reduces intestinal inflammation[50] and has some efficacy in IBD, particularly in pouchitis[51]. Specifically, rifaximin has shown efficacy in recent CD trials[52]. Additionally, IBD patients show mucosal immunoglobulin G antibody secretion and mucosal T cell responses against commensal microbiota[53].

Significant improvements in DNA sequencing technology and analysis over the past decade have set the stage for IBD microbiome investigations. Many studies identified dysbiosis occurring in IBD, and a broad pattern has begun to emerge including biodiversity reduction, decreased representation of several Firmicutes phylum taxa, and increased Gammaproteobacteria[54]. Many studies consistently report a decrease in biodiversity, known as diversity or species richness in ecological terms and a measure of the total species number in a community. The colonic mucosa-associated microbiome in patients with UC shows a restriction in diversity at baseline (in remission) along with temporal instability, demonstrating that those with decreased bacterial richness were more likely to relapse within the next 12 months[55]. A fecal microbiome diversity reduction is seen in CD patients compared with healthy controls[56]. This was also found in monozygotic twin pairs that were discordant for CD[57]. This decreased diversity has been attributed to specific reduction within the Firmicutes phylum[58] and has also been associated with dominant taxa temporal instability in both UC and CD[59].

Reduced diversity has been observed in inflamed versus non-inflamed tissues, even within the same patient, and patients with CD have lower overall bacterial loads at inflamed regions[60]. To date, the largest IBD-related microbiome study is on early-onset CD in a multicenter pediatric cohort. More than 1000 treatment-naïve samples collected from multiple concurrent gastrointestinal locations from patients representing multiple disease phenotypes with respect to location, severity, and behavior were analysed. In addition to detailed characterization of the specific organisms lost or associated with the disease, this study indicated that rectal mucosa-associated microbiome assessment offers unique potential for convenient, early CD diagnosis.

Other nonbacterial microbiota members, namely fungi, viruses, archaea, and phages, may have a significant role in gastrointestinal disease[61] ; however, the vast majority of recent microbiota studies are based on 16S sequencing, thus largely ignoring these organisms. For example, in the context of intact gut microflora and mutated Atg16l1, norovirus infection is required for CD development in a mouse model[62]. Several studies noted a relationship between fungi and IBD[63], including an overall increase in fungal diversity in UC and CD[64]. The relationship between these organisms and IBD will no doubt be explored in more detail in the coming years, largely because microbiome studies are increasingly being performed using unbiased shotgun sequencing.

19.6 Specific bacterial changes in IBD

19.6.1 Potentiators

Specific taxonomic shifts have been reported in IBD. *Enterobacteriaceae* are increased in relative abundance, both in IBD patients and in mouse models[65]. *E. coli*, particularly adherent-invasive *E. coli* strains, have been isolated from ileal CD (iCD) biopsy specimens in culture-based studies[66] and are enriched in UC patients[67]. This enrichment is more pronounced in mucosal samples than in fecal samples[68]. The *Enterobacteriaceae* increase may indicate this clade's preference for an inflammatory environment. In fact, treatment with mesalamine, an anti-inflammatory drug used in IBD patients, decreases intestinal inflammation and is associated with a decrease in *Escherichia/Shigella*[69]. In addition to the luminal trends, several studies observed a shift in the microbes attached to the intestinal mucus layer. The small intestine has a single mucus layer, whereas the colon has two: a firmly attached, essentially sterile inner mucus layer and a variable- thickness outer mucus layer[70]. Each mucus layer consists of mucins, trefoil peptides, and secretory immunoglobulin A[71]. Although host-microbiota interactions are bidirectional, direct epithelium contact is limited by the mucus, and the production of antimicrobial factors such as defensins and regenerating islet-derived protein 3 gamma (RegIII-g)[72]. As long as the mucus layer is relatively healthy and intact, microbes attach to the mucus and do not have direct access to epithelial cells. The mechanisms of intestinal bacteria-mucus interaction are further discussed by Gibson in chapter 29. Patients with UC exhibit a greater overall attached bacterial density on the colonic mucus layer than healthy controls. Adherent-invasive *E. coli*, in particular, is more abundant in mucosal biopsy specimens from CD patients than in those from healthy subjects, and is particularly high in ileal specimens[73]. Adherent-invasive *E. coli* invades epithelial cells, replicates within macrophages[74], and induces granuloma formation *in vitro*[75]. In fact, *E. coli* has been found at higher levels in granulomas from CD patients relative to those from non-CD patients[76]. A second group of adherent invasive bacteria is the *Fusobacteria*.

The genus *Fusobacterium* is composed of Gram-negative anaerobes that principally colonize the oral cavity but also inhabit the gut. *Fusobacterium* species have been found in higher abundance in the colonic mucosa of UC patients relative to control subjects[77], and human isolates of *Fusobacterium varium* induce colonic mucosal erosion in mice by rectal enema[78]. Human *Fusobacterium* isolate invasiveness is positively correlated with the host's IBD status[79], suggesting that invasive *Fusobacterium* species may influence IBD pathology. Interestingly, *Fusobacterium* species were recently shown to be enriched in tumor vs. non-involved adjacent tissue in colorectal cancer[80], and human *Fusobacterium* isolates directly promote tumorigenesis in a mouse model[81]. Because IBD is among the highest risk factors for colorectal cancer development, *Fusobacterium* species may represent a potential link between these diseases.

19.6.2 Protectors

Evidence suggests that specific gut bacteria groups may have protective effects against IBD. For example, the colitis phenotype following dextran sulphate sodium treatment is more severe in mice that are reared germ-free compared with conventionally reared mice[82]. One mechanism for host protection by commensal microbiota is colonization resistance, in which commensals occupy niches within the host to prevent pathogen colonization[83] and outcompete pathogenic bacteria[84]. Interestingly, the microbiota can sometimes take the opposite role and facilitate viral infection. The commensal microbiota also has direct functional effects on potential pathogens (e.g., dampening virulence-related gene expression)[85]. In addition, the gut microbiota plays a role in shaping the mucosal immune system. *Bacteroides* and *Clostridium* species can induce regulatory T-cell expansion, reducing intestinal inflammation[86]. Other members of the microbiota can attenuate mucosal inflammation by regulating nuclear factor-κB activation[87]. Several bacterial species, most notably the *Bifidobacterium*, *Lactobacillus*, and *Faecalibacterium* genera, may protect the host from mucosal inflammation by several mechanisms including inflammatory cytokine down-regulation[88] or interleukin 10 stimulation[89]. *Faecalibacterium prausnitzii*, a proposed anti-inflammatory microbiota, is underrepresented in IBD[90]. *F. prausnitzii* is depleted in ileal CD biopsy samples, concomitant with increased *E. coli* abundance[91], and low mucosa-associated *F. prausnitzii* levels are associated with a higher risk of recurrent CD following surgery. Conversely, *F. prausnitzii* recovery following relapse is associated with maintaining clinical remission in UC[92].

Earlier in this book (chapters 1 and 6), we described how several gut microbiota constituents ferment dietary fibre, a prebiotic, to produce short-chain fatty acids (SCFAs), including acetate, propionate, and butyrate. SCFAs are the primary energy source for colonic epithelial cells[93] and have recently been shown to induce colonic regulatory T cell expansion[94]. *Ruminococcaceae*, particularly the butyrate-producing genus *Faecalibacterium*[95], are decreased in IBD, especially iCD. Other SCFA-producing bacteria, including *Odoribacter* and *Leuconostocaceae*, are reduced in UC, and *Phascolarctobacterium* and *Roseburia* are reduced in CD. Interestingly, *Ruminococcaceae* consume hydrogen and produce acetate, which can be used by *Roseburia* to produce butyrate. Therefore, it is consistent that both groups are reduced concomitantly in IBD.

19.6.3 Anti-inflammatory effects of microbiota (functional dysbiosis)

Chapter 14 of this book introduced the polymicrobial synergy and dysbiosis (PSD), according to which dysbiosis of the subgingival microbiota is not dependent on the particular microbial roster but rather on the specific gene combinations or collective virulence activity within the dysbiotic microbial community. A similar concept might be applied to the gut microbiota.

The intestinal microbiota's anti-inflammatory properties involve host–microbiota interactions using various metabolic pathways and the immune system. An early study showed that spontaneous colitis development and immune system activation requires the presence of enteric bacteria[96]. Metagenomic computational assessment of the intestinal microbiota identified an overall shift associated with IBD seen at both the microbial gene level and metabolic network level. These studies demonstrate the importance of evaluating microbial functionality changes to understand their effects on IBD pathogenesis.

CD microbiota functional signatures, obtained from stool samples, include alterations in bacterial carbohydrate metabolism, bacteria–host interactions, and host-secreted enzymes compared with healthy controls[97]. These signatures could be used as potential biomarker and therapeutic intervention targets. We have an increased appreciation for the potential role of specific bacteria in the etiopathogenesis of IBD by identification of these signatures and linkage to dysbiosis.

19.7 Functional composition of microbiota in IBD

At the phylogenetic level, there is generally high variability over time in the human microbiota, both between and within individual subjects. However, the functional composition (i.e., the functional potential of the metagenome's content) of the gut microbiota is remarkably stable[98]. Therefore, metagenomic approaches may provide greater insight into the gut microbiota's function in disease than taxonomic profiling[99].

Indeed, one IBD microbiome metagenomics study found that 12% of metabolic pathways differed significantly between IBD patients and healthy controls compared with just 2% variation between genus-level clades alone. Metagenomic and metaproteomic studies confirmed the decrease in butanoate and propanoate metabolism genes[100] and lower overall butyrate and other SCFA levels in iCD[101], consistent with the decreased SCFA-producing *Firmicutes* clades observed in taxonomic profiling studies. Another metagenomic trend identified in the IBD microbiome was altered pathobiont bacteria functional characteristics such as decreased amino acid biosynthesis and increased amino acid transport. These bacteria generally have a reduced capacity for nutrient production. Instead, the bacteria transport nutrients from the environment, primarily because they are readily available at inflammation and tissue-destruction sites[98].

Several studies noted an increase in sulphate-reducing bacteria in IBD[102]. Mesalamine, a common UC treatment, inhibits fecal sulphide production. Intriguingly, stool samples from patients not treated with mesalamine showed higher sulphide levels. Genes involved in metabolizing the sulphur-containing amino acid cysteine are increased in IBD, particularly in ileal CD, and there is increased sulphate transport in both UC and CD. In addition, saturated fat–derived taurine conjugates to bile acids, increasing free sulphur availability and causing a population expansion of the sulphate-reducing pathobiont *Bilophila wadsworthia*, which drives colitis in genetically susceptible Il10$^{-/-}$ but not wild-type mice[103]. The IBD metagenome has an increased capacity for managing oxidative stress, an inflammatory environment hallmark, as indicated by increased glutathione transport and riboflavin metabolism in UC. Patients with CD also exhibited an increase in type II secretion systems, which are involved in toxin secretion, and in virulence-related bacterial genes. These trends are indicative of a shift towards an inflammation-promoting microbiome.

The intestinal barrier is composed of a single epithelial cell layer linked through tight junctions and a mucus shield with strong surface hydrophobicity[104]. This cell layer is composed of various cell types including goblet and Paneth cells, which produce mucin and antimicrobial peptides (e.g., defensins, cathelicidins, and lysozymes), respectively. Additionally, goblet and Paneth cells provide strong barrier functions to control the equilibrium between microbe and non-self-antigen tolerance[105]. Consequently, epithelial barrier integrity disruption has been implicated in IBD pathogenesis via a wide range of specific mechanisms encompassing structural, metabolic, and innate immune pathways. Barrier defects include increased mucus permeability to water and various small or large molecules that elicit immune responses, either as a cause or a consequence of IBD.

Various bacteria produce hemolysins that damage host cell membranes. *E. coli*–secreted hemolysin increases the number of initial colon mucosa lesions, referred to as "focal leaks." These, in turn, reduce transepithelial electrical resistance and increase macromolecular uptake[106]. Mucosal samples from IBD patients and healthy controls show similar α-hemolysin-producing *E. coli* levels. However, patients with active UC display 10-fold higher α-hemolysin levels, suggesting a role for α-hemolysin in disrupting the epithelial barrier, and further highlighting the importance of measuring the microbiota's metabolic capacity rather than simply its composition[107].

Bacterial proteases are also known to affect intestinal integrity. Gelatinase, a metalloproteinase produced by *Enterococcus faecalis*, disrupts the epithelial barrier and increases inflammation in mice that are genetically susceptible to IBD[108]. Enterococcal IBD isolates show higher frequencies of virulence genes including gelE, which encodes for gelatinase[109]. Similar to other pathogenic species such as *Campylobacter*, *E. coli*, and *Shigella* species, certain *Helicobacter* species produce cytolethal distending toxins,[110] causing cell cycle arrest, chromatin fragmentation, and apoptosis, and are speculated to adversely influence IBD development[111]. Indeed, patients with CD and UC show increased enterohepatic *Helicobacter* species prevalence, although there are conflicting reports regarding whether *Helicobacter* species (specifically *Helicobacter pylori*) contribute to IBD progression[112]. As discussed above, these discrepancies may arise from different sampling methodologies. Adherent *Lactobacillus* spp. can upregulate extracellular mucin secretion and thus protect the gut mucosal surface against enteropathogen adherence. Frequent non-steroidal anti-inflammatory agent use is also a potential risk factor for IBD, because these agents disrupt intestinal tract mucosal integrity and may impact the microbiota composition.

19.8 Challenges

The challenges in all microbiome studies relate to the linkage of bacteria to function, but whether the changes can be used to indicate causality remains unanswered. Issues such as patient variability and study heterogeneity all serve to confound matters. Much of the data published so far refer to cross-sectional design. Validation in large-scale prospective study cohorts is needed via sequencing and targeted methods. Beyond the identification of bacterial species and functions in well-phenotyped cohorts, the future lies in the validation of novel marker species preceding disease onset or facilitating prediction of treatment outcomes.

19.9 Conclusion

Through improved technology and high-throughput techniques, the identification of specific bacteria that have been altered in the microbiota of IBD patients has increased knowledge of intestinal dysbiosis. This allows us to gain an understanding of bacterial composition and function in the potentiation and protection against IBD.

TAKE-HOME MESSAGE

- Host genetics play a strong role in establishing and shaping the gut microbiota resulting in dysbiosis.
- Environmental factors such as exposure to antibiotics and diet result in the potentiation of certain bacteria that may result in mucosal destruction and ulceration, as shown in various microbiota studies.

References

1 Qin J, Li R, Raes J, Arumugam M, Burgdorf KS, Manichanh C, Nielsen T, Pons N, Levenez F, Yamada T, Mende DR, Li J, Xu J, Li S, Li D, Cao J, Wang B, Liang H, Zheng H, Xie Y, Tap J, Lepage P, Bertalan M, Batto JM, Hansen T, Le Paslier D, Linneberg A, Nielsen HB, Pelletier E, Renault P, Sicheritz-Ponten T, Turner K, Zhu H, Yu C, Li S, Jian M, Zhou Y, Li Y, Zhang X, Li S, Qin N, Yang H, Wang J, Brunak S, Doré J, Guarner F, Kristiansen K, Pedersen O, Parkhill J, Weissenbach J, Bork P, Ehrlich SD, Wang J. A human gut microbial gene catalogue established by metagenomic sequencing. *Nature* 2010; **464**: 59–65 [PMID: 20203603 DOI: 10.1038/nature08821].

2 Kaur N, Chen CC, Luther J, Kao JY. Intestinal dysbiosis in inflammatory bowel disease. *Gut Microbes* 2011; **2**: 211–216 [PMID: 21983063 DOI: 10.4161/gmic.2.4.17863].

3 Hold GL, Smith W, Grange C, Watt ER, El-Omar EM, Mukhopadhya I. Role of the gut microbiota in inflammatory bowel disease pathogenesis: what have we learnt in the past 10 years? *World J Gastroenterol* 2014; **20**(5): 1192–210. doi:10.3748/wjg.v20.i5.1192.

4 Jostins L, Ripke S, Weersma RK *et al.* Host-microbe interactions have shaped the genetic architecture of inflammatory bowel disease. *Nature* 2012; **491**: 119–124.

5 Khor B, Gardet A, Xavier RJ. Genetics and pathogenesis of inflammatory bowel disease. *Nature* 2011; **474**: 307–317.

6 Molodecky NA, Soon IS, Rabi DM *et al.* Increasing incidence and prevalence of the inflammatory bowel diseases with time, based on systematic review. *Gastroenterology* 2012; **142**: 46–54.e42; quiz e30.

7 Halme L, Paavola-Sakki P, Turunen U *et al.* Family and twin studies in inflammatory bowel disease. *World J Gastroenterol* 2006; **12**: 3668–72.

8 Pillai S. Rethinking mechanisms of autoimmune pathogenesis. *J Autoimmun* 2013; **45**: 97–103.

9 Flint HJ, Scott KP, Louis P *et al.* The role of the gut microbiota in nutrition and health. *Nat Rev Gastroenterol Hepatol* 2012; **9**: 577–589.

10 Dave M, Higgins PD, Middha S *et al.* The human gut microbiome: current knowledge, challenges, and future directions. *Transl Res* 2012; **160**: 246–257.

11 O'Hara AM, Shanahan F. The gut flora as a forgotten organ. *EMBO Rep* 2006; **7**: 688–693.

12 Arumugam M, Raes J, Pelletier E, Le Paslier D, Yamada T, Mende DR, Fernandes GR, Tap J, Bruls T, Batto JM, Bertalan M, Borruel N, Casellas F, Fernandez L, Gautier L, Hansen T, Hattori M, Hayashi T, Kleerebezem M, Kurokawa K, Leclerc M, Levenez F, Manichanh C, Nielsen HB, Nielsen T, Pons N, Poulain J, Qin J, Sicheritz-Ponten T, Tims S, Torrents D, Ugarte E, Zoetendal EG, Wang J, Guarner F, Pedersen O, de Vos WM, Brunak S, Doré J, Antolín M, Artiguenave F, Blottiere HM, Almeida M, Brechot C, Cara C, Chervaux C, Cultrone A, Delorme C, Denariaz G, Dervyn R, Foerstner KU, Friss C, van de Guchte M, Guedon E, Haimet F, Huber W, van Hylckama-Vlieg J, Jamet A, Juste C, Kaci G, Knol J, Lakhdari O, Layec S, Le Roux K, Maguin E, Mérieux A, Melo Minardi R, M'rini C, Muller J, Oozeer R, Parkhill J, Renault P, Rescigno M, Sanchez N, Sunagawa S, Torrejon A, Turner K, Vandemeulebrouck G, Varela E, Winogradsky Y, Zeller G, Weissenbach J, Ehrlich SD, Bork P. Enterotypes of the human gut microbiome. *Nature* 2011; **473**: 174–180 [PMID: 21508958 DOI: 10.1038/nature09944].

13 Wu GD, Chen J, Hoffmann C, Bittinger K, Chen YY, Keilbaugh SA, Bewtra M, Knights D, Walters WA, Knight R, Sinha R, Gilroy E, Gupta K, Baldassano R, Nessel L, Li H, Bushman FD, Lewis JD. Linking long-term dietary patterns with gut microbial enterotypes. *Science* 2011; **334**: 105–108 [PMID: 21885731 DOI: 10.1126/science.1208344].

14 De Filippo C, Cavalieri D, Di Paola M, Ramazzotti M, Poullet JB, Massart S, Collini S, Pieraccini G, Lionetti P. Impact of diet in shaping gut microbiota revealed by a comparative study in children from Europe and rural Africa. *Proc Natl Acad Sci USA* 2010; **107**: 14691–14696 [PMID: 20679230 DOI: 10.1073/pnas.1005963107].

15 Kaur N, Chen CC, Luther J, Kao JY. Intestinal dysbiosis in inflammatory bowel disease. *Gut Microbes* 2011; **2**: 211–216 [PMID: 21983063 DOI: 10.4161/gmic.2.4.17863].

16 Frank DN, St Amand AL, Feldman RA, Boedeker EC, Harpaz N, Pace NR. Molecular-phylogenetic characterization of microbial community imbalances in human inflammatory bowel diseases. *Proc Natl Acad Sci USA* 2007; **104**: 13780–13785 [PMID: 17699621 DOI: 10.1073/.

17 Sokol H, Seksik P, Furet JP, Firmesse O, Nion-Larmurier I, Beaugerie L, Cosnes J, Corthier G, Marteau P, Doré J. Low counts of Faecalibacterium prausnitzii in colitis microbiota. *Inflamm Bowel Dis* 2009; **15**: 1183–1189 [PMID: 19235886 DOI: 10.1002/ibd.20903].

18 Swidsinski A, Ladhoff A, Pernthaler A, Swidsinski S, Loening-Baucke V, Ortner M, Weber J, Hoffmann U, Schreiber S, Dietel M, Lochs H. Mucosal flora in inflammatory bowel disease. *Gastroenterology* 2002; **122**: 44–54 [PMID: 11781279 DOI: 10.1053/gast.2002.30294].

19 Willing B, Halfvarson J, Dicksved J, Rosenquist M, Järnerot G, Engstrand L, Tysk C, Jansson JK. Twin studies reveal specific imbalances in the mucosa-associated microbiota of patients with ileal Crohn's disease. *Inflamm Bowel Dis* 2009; **15**: 653–660 [PMID: 19023901 DOI: 10.1002/ibd.20783].

20 Hansen R, Russell RK, Reiff C, Louis P, McIntosh F, Berry SH, Mukhopadhya I, Bisset WM, Barclay AR, Bishop J, Flynn DM, McGrogan P, Loganathan S, Mahdi G, Flint HJ, El-Omar EM, Hold GL. Microbiota of de-novo pediatric IBD: increased Faecalibacterium prausnitzii and reduced bacterial diversity in Crohn's but not in ulcerative colitis. *Am J Gastroenterol* 2012; **107**: 1913–1922 [PMID: 23 044767 DOI: 10.1038/ajg.2012.335].

21 Ott SJ, Musfeldt M, Wenderoth DF, Hampe J, Brant O, Fölsch UR, Timmis KN, Schreiber S. Reduction in diversity of the colonic mucosa associated bacterial microflora in patients with active inflammatory bowel disease. *Gut* 2004; **53**: 685–693 [PMID: 15082587 DOI: 10.1136/gut.2003.025403].

22 Neut C, Bulois P, Desreumaux P, Membré JM, Lederman E, Gambiez L, Cortot A, Quandalle P, van Kruiningen H, Colombel JF. Changes in the bacterial flora of the neoterminal ileum after ileocolonic resection for Crohn's disease. *Am J Gastroenterol* 2002; **97**: 939–946 [PMID: 12003430 DOI: 10.1111/j.1572-0241.2002.05613.x].

23 Andoh A, Kuzuoka H, Tsujikawa T, Nakamura S, Hirai F, Suzuki Y, Matsui T, Fujiyama Y, Matsumoto T. Multicenter analysis of fecal microbiota profiles in Japanese patients with Crohn's disease. *J Gastroenterol* 2012; **47**: 1298–1307 [PMID: 22576027 DOI: 10.1007/s00535-012-0605-0].

24 Ley RE, Hamady M, Lozupone C *et al.* Evolution of mammals and their gut microbes. *Science* 2008; **320**: 1647–1651.

25 Wu GD, Chen J, Hoffmann C *et al.* Linking long-term dietary patterns with gut microbial entero-types. *Science* 2011; **334**: 105–108.

26 Zimmer J, Lange B, Frick JS *et al.* A vegan or vegetarian diet substantially alters the human colonic faecal microbiota. *Eur J Clin Nutr* 2012; **66**: 53–60.

27 Moschen AR, Wieser V, Tilg H. Dietary factors: major regulators of the gut's microbiota. *Gut Liver* 2012; **6**: 411–416.

28 Martin-de-Carpi J, Rodriguez A, Ramos E *et al.* Increasing incidence of pediatric inflammatory bowel disease in Spain (1996-2009): the SPIRIT Registry. *Inflamm Bowel Dis* 2013; **19**: 73–80.

29 Spor A, Koren O, Ley R. Unravelling the effects of the environment and host genotype on the gut microbiome. *Nat Rev Microbiol* 2011; **9**: 279–290.

30 Dominguez-Bello MG, Blaser MJ, Ley RE *et al.* Development of the human gastrointestinal micro-biota and insights from high-throughput sequencing. *Gastroenterology* 2011; **140**: 1713–1719.

31 Lozupone CA, Stombaugh JI, Gordon JI *et al.* Diversity, stability and resilience of the human gut microbiota. *Nature* 2012; **489**: 220–230.

32 Claesson MJ, Cusack S, O'Sullivan O *et al.* Composition, variability, and temporal stability of the intestinal microbiota of the elderly. *Proc Natl Acad Sci USA* 2011; **108** (suppl 1): 4586–4591.

33 Waterman M, Xu W, Stempak JM *et al.* Distinct and overlapping genetic loci in Crohn's disease and ulcerative colitis: correlations with pathogenesis. *Inflamm Bowel Dis* 2011; **17**: 1936–42.

34 Ogura Y, Bonen DK, Inohara N *et al.* A frameshift mutation in NOD2 associated with susceptibility to Crohn's disease. *Nature* 2001; **411**: 603–606.

35 Hisamatsu T, Suzuki M, Reinecker HC *et al.* CARD15/NOD2 functions as an antibacterial factor in human intestinal epithelial cells. *Gastroenterology* 2003; **124**: 993–1000.

36 Stappenbeck TS, Hooper LV, Gordon JI. Developmental regulation of intestinal angiogenesis by indigenous microbes via Paneth cells. *Proc Natl Acad Sci USA* 2002; **99**: 15451–15455.

37 Philpott DJ, Girardin SE. Crohn's disease-associated Nod2 mutants reduce IL10 transcription. *Nat Immunol* 2009; **10**: 455–457.

38 Frank DN, Robertson CE, Hamm CM *et al.* Disease phenotype and genotype are associated with shifts in intestinal-associated microbiota in inflammatory bowel diseases. *Inflamm Bowel Dis* 2011; **17**: 179–184.

39 McGovern DP, Jones MR, Taylor KD *et al.* Fucosyltransferase 2 (FUT2) non-secretor status is associ-ated with Crohn's disease. *Hum Mol Genet* 2010; **19**: 3468–3476.

40 Rausch P, Rehman A, Kunzel S *et al.* Colonic mucosa- associated microbiota is influenced by an interaction of Crohn disease and FUT2 (Secretor) genotype. *Proc Natl Acad Sci USA* 2011; **108**: 19030–19035.

41 Turnbaugh PJ, Hamady M, Yatsunenko T *et al.* A core gut microbiome in obese and lean twins. *Nature* 2009; **457**: 480–484.

42 Wehkamp J, Salzman NH, Porter E *et al.* Reduced Paneth cell alpha-defensins in ileal Crohn's disease. *Proc Natl Acad Sci USA* 2005; **102**: 18129–34.

43 Wehkamp J, Wang G, Kubler I *et al.* The Paneth cell alpha-defensin deficiency of ileal Crohn's disease is linked to Wnt/Tcf-4. *J Immunol* 2007; **179**: 3109–18.

44 Salzman NH, Hung K, Haribhai D *et al.* Enteric defensins are essential regulators of intestinal micro-bial ecology. *Nat Immunol* 2010; **11**: 76–83.

45 Nuding S, Fellermann K, Wehkamp J *et al.* Reduced mucosal antimicrobial activity in Crohn's disease of the colon. *Gut* 2007; **56**: 1240–7.

46 Mow WS, Vasiliauskas EA, Lin YC *et al.* Association of antibody responses to microbial antigens and complications of small bowel Crohn's disease. *Gastroenterology* 2004; **126**: 414–24.

47 Sartor RB. Microbial influences in inflammatory bowel diseases. *Gastroenterology* 2008; **134**: 577–594.

48 Casellas F, Borruel N, Papo M *et al.* Antiinflammatory effects of enterically coated amoxicillin-clavulanic acid in active ulcerative colitis. *Inflamm Bowel Dis* 1998; **4**: 1–5.

49 Sartor RB. Therapeutic manipulation of the enteric microflora in inflammatory bowel diseases: antibiotics, probiotics, and prebiotics. *Gastroenterology* 2004; **126**: 1620–1633.

50 Rietdijk ST, D'Haens GR. Recent developments in the treatment of inflammatory bowel disease. *J Dig Dis* 2013; **14**: 282–287.

51 Pirzer U, Schonhaar A, Fleischer B *et al.* Reactivity of infiltrating T lymphocytes with microbial antigens in Crohn's disease. *Lancet* 1991; **338**: 1238–1239.

52 Morgan XC, Tickle TL, Sokol H *et al.* Dysfunction of the intestinal microbiome in inflammatory bowel disease and treatment. *Genome Biol* 2012; **13**: R79.

53 Ott SJ, Plamondon S, Hart A *et al*. Dynamics of the mucosa-associated flora in ulcerative colitis patients during remission and clinical relapse. *Journal of Clinical Microbiology*. 2008; **46**(10): 3510–3513. doi:10.1128/JCM.01512-08.

54 Manichanh C, Rigottier-Gois L, Bonnaud E *et al*. Reduced diversity of faecal microbiota in Crohn's disease revealed by a metagenomic approach. *Gut* 2006; **55**: 205–211.

55 Dicksved J, Halfvarson J, Rosenquist M *et al*. Molecular analysis of the gut microbiota of identical twins with Crohn's disease. *ISME J* 2008; **2**: 716–727.

56 Kang S, Denman SE, Morrison M *et al*. Dysbiosis of fecal microbiota in Crohn's disease patients as revealed by a custom phylogenetic microarray. *Inflamm Bowel Dis* 2010; **16**: 2034–2042.

57 Martinez C, Antolin M, Santos J *et al*. Unstable composition of the fecal microbiota in ulcerative colitis during clinical remission. *Am J Gastroenterol* 2008; **103**: 643–648.

58 Sepehri S, Kotlowski R, Bernstein CN *et al*. Microbial diversity of inflamed and noninflamed gut biopsy tissues in inflammatory bowel disease. *Inflamm Bowel Dis* 2007; **13**: 675–683.

59 Hunter P. The secret garden's gardeners: research increasingly appreciates the crucial role of gut viruses for human health and disease. *EMBO Rep* 2013; **14**: 683–685.

60 Cadwell K, Patel KK, Maloney NS *et al*. Virus-plus-susceptibility gene interaction determines Crohn's disease gene Atg16L1 phenotypes in intestine. *Cell* 2010; **141**: 1135–1145.

61 Trojanowska D, Zwolinska-Wcislo M, Tokarczyk M *et al*. The role of Candida in inflammatory bowel disease. Estimation of transmission of C. albicans fungi in gastrointestinal tract based on genetic affinity between strains. *Med Sci Monit* 2010; **16**: CR451–CR457.

62 Ott SJ, Kuhbacher T, Musfeldt M *et al*. Fungi and inflammatory bowel diseases: alterations of composition and diversity. *Scand J Gastroenterol* 2008; **43**: 831–841.

63 Lupp C, Robertson ML, Wickham ME *et al*. Host-mediated inflammation disrupts the intestinal microbiota and promotes the overgrowth of Enterobacteriaceae. *Cell Host Microbe* 2007; **2**: 119–129.

64 Darfeuille-Michaud A, Boudeau J, Bulois P *et al*. High prevalence of adherent-invasive Escherichia coli associated with ileal mucosa in Crohn's disease. *Gastroenterology* 2004; **127**: 412–421.

65 Sokol H, Lepage P, Seksik P *et al*. Temperature gradient gel electrophoresis of fecal 16S rRNA reveals active Escherichia coli in the microbiota of patients with ulcerative colitis. *J Clin Microbiol* 2006; **44**: 3172–3177.

66 Chassaing B, Darfeuille-Michaud A. The commensal microbiota and enteropathogens in the pathogenesis of inflammatory bowel diseases. *Gastroenterology* 2011; **140**: 1720–1728.

67 Benjamin JL, Hedin CR, Koutsoumpas A *et al*. Smokers with active Crohn's disease have a clinically relevant dysbiosis of the gastrointestinal microbiota. *Inflamm Bowel Dis* 2012; **18**: 1092–1100.

68 Johansson ME, Larsson JM, Hansson GC. The two mucus layers of colon are organized by the MUC2 mucin, whereas the outer layer is a legislator of host-microbial interactions. *Proc Natl Acad Sci USA* 2011; **108** (suppl 1): 4659–6465.

69 Cario E. Microbiota and innate immunity in intestinal inflammation and neoplasia. *Curr Opin Gastroenterol* 2013; **29**: 85–91.

70 Shanahan F. The colonic microbiota in health and disease. *Curr Opin Gastroenterol* 2013; **29**: 49–54.

71 Martinez-Medina M, Aldeguer X, Lopez-Siles M *et al*. Molecular diversity of Escherichia coli in the human gut: new ecological evidence supporting the role of adherent-invasive E. coli (AIEC) in Crohn's disease. *Inflamm Bowel Dis* 2009; **15**: 872–882.

72 Glasser AL, Boudeau J, Barnich N *et al*. Adherent invasive Escherichia coli strains from patients with Crohn's disease survive and replicate within macrophages without inducing host cell death. *Infect Immun* 2001; **69**: 5529–5537.

73 Meconi S, Vercellone A, Levillain F *et al*. Adherent-invasive Escherichia coli isolated from Crohn's disease patients induce granulomas in vitro. *Cell Microbiol* 2007; **9**: 1252–1261.

74 Ryan P, Kelly RG, Lee G *et al*. Bacterial DNA within granulomas of patients with Crohn's disease — detection by laser capture microdissection and PCR. *Am J Gastroenterol* 2004; **99**: 1539–1543.

75 Ohkusa T, Sato N, Ogihara T *et al*. Fusobacterium varium localized in the colonic mucosa of patients with ulcerative colitis stimulates species-specific antibody. *J Gastroenterol Hepatol* 2002; **17**: 849–853.

76 Ohkusa T, Okayasu I, Ogihara T *et al*. Induction of experimental ulcerative colitis by Fusobacterium varium isolated from colonic mucosa of patients with ulcerative colitis. *Gut* 2003; **52**: 79–83.

77 Strauss J, Kaplan GG, Beck PL *et al*. Invasive potential of gut mucosa-derived Fusobacterium nucleatum positively correlates with IBD status of the host. *Inflamm Bowel Dis* 2011; **17**: 1971–1978.

78 Kostic AD, Gevers D, Pedamallu CS *et al.* Genomic analysis identifies association of Fusobacterium with colorectal carcinoma. *Genome Res* 2012; **22**: 292–298.

79 Rubinstein MR, Wang X, Liu W *et al.* Fusobacterium nucleatum promotes colorectal carcinogenesis by modulating E-cadherin/beta-catenin signaling via its FadA adhesin. *Cell Host Microbe* 2013; **14**: 195–206.

80 Kitajima S, Morimoto M, Sagara E *et al.* Dextran sodium sulfate-induced colitis in germ-free IQI/Jic mice. *Exp Anim* 2001; **50**: 387–395.

81 Callaway TR, Edrington TS, Anderson RC *et al.* Probiotics, prebiotics and competitive exclusion for prophylaxis against bacterial disease. *Anim Health Res Rev* 2008; **9**: 217–225.

82 Kamada N, Chen G, Nunez G. A complex microworld in the gut: harnessing pathogen-commensal relations. *Nat Med* 2012; **18**: 1190–1191.

83 Medellin-Pena MJ, Wang H, Johnson R *et al.* Probiotics affect virulence-related gene expression in Escherichia coli O157:H7. *Appl Environ Microbiol* 2007; **73**: 4259–4267.

84 Atarashi K, Tanoue T, Oshima K *et al.* Treg induction by a rationally selected mixture of Clostridia strains from the human microbiota. *Nature* 2013; **500**: 232–236.

85 Kelly D, Campbell JI, King TP *et al.* Commensal anaerobic gut bacteria attenuate inflammation by regulating nuclear-cytoplasmic shuttling of PPAR-gamma and RelA. *Nat Immunol* 2004; **5**: 104–112.

86 Llopis M, Antolin M, Carol M *et al.* Lactobacillus casei downregulates commensals' inflammatory signals in Crohn's disease mucosa. *Inflamm Bowel Dis* 2009; **15**: 275–283.

87 Sokol H, Pigneur B, Watterlot L *et al.* Faecalibacterium prausnitzii is an anti-inflammatory commensal bacterium identified by gut microbiota analysis of Crohn disease patients. *Proc Natl Acad Sci USA* 2008; **105**: 16731–16736.

88 Sokol H, Seksik P, Furet JP *et al.* Low counts of Faecalibacterium prausnitzii in colitis microbiota. *Inflamm Bowel Dis* 2009; **15**: 1183–1189.

89 Willing B, Halfvarson J, Dicksved J *et al.* Twin studies reveal specific imbalances in the mucosa-associated microbiota of patients with ileal Crohn's disease. *Inflamm Bowel Dis* 2009; **15**: 653–660.

90 Varela E, Manichanh C, Gallart M *et al.* Colonisation by Faecalibacterium prausnitzii and maintenance of clinical remission in patients with ulcerative colitis. *Aliment Pharmacol Ther* 2013; **38**: 151–161.

91 Ahmad MS, Krishnan S, Ramakrishna BS *et al.* Butyrate and glucose metabolism by colonocytes in experimental colitis in mice. *Gut* 2000; **46**: 493–499.

92 Smith PM, Howitt MR, Panikov N *et al.* The microbial metabolites, short-chain fatty acids, regulate colonic Treg cell homeostasis. *Science* 2013; **341**: 569–573.

93 Duncan SH, Hold GL, Barcenilla A *et al.* Roseburia intestinalis sp. nov., a novel saccharolytic, butyrate- producing bacterium from human faeces. *Int J Syst Evol Microbiol* 2002; **52**: 1615–1620.

94 Sellon RK, Tonkonogy S, Schultz M *et al.* Resident enteric bacteria are necessary for development of spontaneous colitis and immune system activation in interleukin-10-deficient mice. *Infect Immun.* 1998; **66**: 5224–5231.

95 Erickson AR, Cantarel BL, Lamendella R *et al.* Integrated metagenomics/metaproteomics reveals human host-microbiota signatures of Crohn's disease. *PloS One.* 2012; **7**: e49138.

96 Human Microbiome Project Consortium. Structure, function and diversity of the healthy human microbiome. *Nature* 2012; **486**: 207–214.

97 Meyer F, Trimble WL, Chang EB *et al.* Functional predictions from inference and observation in sequencebased inflammatory bowel disease research. *Genome Biol* 2012; **13**: 169.

98 Morgan XC, Tickle TL, Sokol H *et al.* Dysfunction of the intestinal microbiome in inflammatory bowel disease and treatment. *Genome Biol* 2012; **13**: R79.

99 Erickson AR, Cantarel BL, Lamendella R *et al.* Integrated metagenomics/metaproteomics reveals human host- microbiota signatures of Crohn's disease. *PLoS One* 2012; **7**: e49138.

100 Rowan F, Docherty NG, Murphy M *et al.* Desulfovibrio bacterial species are increased in ulcerative colitis. *Dis Colon Rectum* 2010; **53**: 1530–1536.

101 Devkota S, Wang Y, Musch MW *et al.* Dietary-fat-induced taurocholic acid promotes pathobiont expansion and colitis in Il10-/- mice. *Nature* 2012; **487**: 104–108.

102 Mack DR, Neumann AW, Policova Z *et al.* Surface hydrophobicity of the intestinal tract. *Am J Physiol.* 1992; **262**: G171–G177.

103 Antoni L, Nuding S, Weller D *et al.* Human colonic mucus is a reservoir for antimicrobial peptides. *J Crohns Colitis.* 2013; **7**: e652–e664.

104 Troeger H, Richter JF, Beutin L *et al.* Escherichia coli alpha-haemolysin induces focal leaks in colonic epithelium: a novel mechanism of bacterial translocation. *Cell Microbiol.* 2007; **9**: 2530–2540.

105 Bucker R, Schulz E, Gunzel D, *et al.* Alpha-Haemolysin of Escherichia coli in IBD: a potentiator of inflammatory activity in the colon. *Gut* [published online ahead of print February 17, 2014]. doi: 10.1136/gutjnl- 2013-306099.

106 Steck N, Mueller K, Schemann M *et al.* Bacterial proteases in IBD and IBS. *Gut* 2012; **61**: 1610–1618.

107 Golinska E, Tomusiak A, Gosiewski T *et al.* Virulence factors of Enterococcus strains isolated from patients with inflammatory bowel disease. *World J Gastroenterol.* 2013; **19**: 3562–3572.

108 Young VB, Chien CC, Knox KA *et al.* Cytolethal distending toxin in avian and human isolates of Helicobacter pullorum. *J Infect Dis.* 2000; **182**: 620–623.

109 Smith JL, Bayles DO. The contribution of cytolethal distending toxin to bacterial pathogenesis. *Crit Rev Microbiol.* 2006; **32**: 227–248.

110 Bell SJ, Chisholm SA, Owen RJ *et al.* Evaluation of Helicobacter species in inflammatory bowel disease. *Aliment Pharmacol Ther.* 2003; **18**: 481–486.

111 Mack DR, Ahrne S, Hyde L *et al.* Extracellular MUC3 mucin secretion follows adherence of Lactobacillus strains to intestinal epithelial cells in vitro. *Gut.* 2003; **52**: 827–833.

112 Ananthakrishnan AN, Higuchi LM, Huang ES *et al.* Aspirin, nonsteroidal anti-inflammatory drug use, and risk for Crohn disease and ulcerative colitis: a cohort study. *Ann Intern Med.* 2012; **156**: 350–359.

CHAPTER 20

Ankylosing spondylitis, *Klebsiella* and the low-starch diet

Alan Ebringer[1], Taha Rashid[1] and Clyde Wilson[2]

[1] King's College London, London, United Kingdom
[2] King Edward VII Memorial Hospital, Bermuda

20.1 Introduction

Gut bacteria play an important role in the causation of many diseases. This concept is best illustrated by the crippling arthritic disease ankylosing spondylitis (AS). The discovery made in 1973 by two groups, one in London[1] and the other one in Los Angeles[2] that the histocompatibility antigen HLA-B27 was present in over 95% of AS patients whilst it is only found in 8% of the general population in the UK or USA raises important questions about the pathogenesis of this condition. This observation, which was confirmed by other rheumatology centres throughout the world, has provided an important clue on the cause of this disease. Any solution to the problem of AS must, at the same time, provide an explanation for why it is so frequent in HLA-B27-positive populations. Early studies showed that rabbits immunized with HLA-B27 cells made antibodies to the gut microbe *Klebsiella*[3]. Subsequent studies showed that *Klebsiella* microbes were isolated in fecal samples obtained from AS patients. Antibodies to the *Klebsiella* microbe have been reported from many different countries. The discovery that AS was caused by the gut microbe *Klebsiella* prompted the proposal, that if these gut bacteria could somehow be reduced, then this might provide a therapeutic benefit to AS patients. It is proposed to review the evidence that AS is caused by the gut microbe *Klebsiella*, a form of environmental dysbiosis, and how this observation could be exploited for therapeutic intervention.

20.2 Clinical features of AS

AS occurs more frequently in men than women, with a ratio of 3 males: 1 female. The great majority of AS patients are HLA-B27 positive and the disease starts in the late teens or early twenties. The disease onset is often insidious, when the patient complains of transient muscle pains, especially in the back and buttocks, with tiredness and generally feeling unwell. Sometimes the disease affects the rib

---

The Human Microbiota and Chronic Disease: Dysbiosis as a Cause of Human Pathology, First Edition.
Edited by Luigi Nibali and Brian Henderson.

cage, with severe pains on coughing. Eventually the disease is characterized by severe muscle pains in the back and generalized stiffness, usually more prominent in the mornings. Frequently the patient will say that it is so difficult to get up in the mornings that they "have to roll out of bed." This observation is almost a diagnostic sign of AS. Some 30% of AS patients will also develop eye inflammation or uveitis, which usually responds to steroid eye drops.

Eventually the AS patient will have characteristic radiographic changes of sacroiliitis, when the diagnosis of AS can be confirmed. Often there will be a characteristic "delay-to-diagnosis" of several years, when the patient complains of recurrent backache, until the diagnosis of AS can be confirmed by radiographic examination, magnetic resonance imaging and computerized tomography[4]. Furthermore, there is a characteristic inflammation at the point of insertion of muscles to bone at the entheses, and this is labeled as enthesitis. After several years the AS patient will develop a characteristic curvature of the spine or kyphosis, caused by fibrosis and calcification of intervertebral joints. These are irreversible changes and the kyphotic patient will require analgesic drugs to control the pains. Some patients complained that "there was not a single day in their life when they were free of pain," and this was a significant consideration in their outlook on life. Many complained of the unsympathetic way they were regarded by their doctors, with some of them almost accusing them of malingering, especially in the early stages of the disease.

20.3 Gut bacteria and total serum IgA

Wilson in chapter 1 described how the gut-associated lymphoid tissue (GALT) contains the largest collection of immunocompetent cells in the human body, with the aim to control bacterial and other antigens. Mucosal inflammation in the respiratory and gastrointestinal tract is controlled by the GALT system. Plasma cells in the GALT system are the major source of serum IgA[5]. Thus the measurement of total serum IgA provides a metric marker of mucosal immune activity and could be used to assess the antigenic load occurring in the gastrointestinal tract of AS patients.

Total serum IgA, IgG and IgM were measured in 122 AS patients from London. The AS patients were divided into two groups: those who had active inflammation with an erythrocyte sedimentation (inflammation) rate (ESR) above 15 mm/hour and those with an ESR below 15 mm/hour, who were deemed to be inactive. The mean serum IgA in AS patients having an ESR greater than 15 mm/hour was 307 mg/dl and this was significantly higher than in AS patients having an ESR less than 15 mm/hour ($p<0.001$) or normal blood donors ($p<0.001$). There was no significant difference in the levels of total serum IgG or IgM between the groups examined[6]. These results show that total serum IgA elevations are associated with active phases of disease activity in the gut of AS patients. These observations clearly incriminate a pathological process in the gastrointestinal tract, a form of bacterial dysbiosis in this disease. Extensive French studies have shown that pelvic and paraspinal lymphadenopathy precedes the development of radiological changes in the spine of AS patients[7]. Lymphangiographic studies indicate that regional lymph nodes draining the pelvis and pelvic colon, namely the presacral and para-aortic lymph nodes, become enlarged during active phases of AS before the development of sacroiliitis[8]. There is thus a clear link between gut pathology and subsequent paraspinal lymphatic abnormalities in the development of spinal disease in AS patients. Flemish researchers have pointed out that there is a high prevalence of clinically silent gut inflammatory lesions in over 50% of AS patients, thereby reinforcing the link between the gut and its bacterial contents in this disease[9].

The elevation in total serum IgA indicates that it is not part of the clinical phenotype of AS but it does correlate with the inflammatory profile in the patient. So an AS patient with severe kyphotic deformity due to long-standing disease but currently not inflamed will not have an elevation in serum IgA. Thus an elevation in total serum IgA strongly suggests that a bacterial agent is acting as a dysbiotic factor in the onset of AS. Discovering the source of this dysbiotic factor requires a molecular approach to this problem.

20.4 Molecular mimicry in AS

The concept that "molecular mimicry," or similarity between self-antigens and external bacteria, has pathological implications goes back to the discovery that streptococcal; infections are involved in rheumatic fever[10] and Sydenham's chorea[11]. The possibility arises that a similar process might operate in AS. The observation that HLA-B27 evokes antibodies to *Klebsiella* in rabbits supports this concept.

Studies from La Jolla, California showed that there is molecular mimicry between an enzyme found in *Klebsiella* microbes, namely nitrogenase reductase, and HLA-B27[12]. Other studies from London showed that another enzyme, *Klebsiella* pullulanase, also exhibited molecular mimicry with HLA-B27[13]. The sequence was a tetramer (DRDE) (aspartic acid-arginine-aspartic acid-glutamic acid) present in the pullulanase secretion protein pulD (residues 596–599) of a starch-debranching enzyme present in *Klebsiella* bacteria. This resembles the sequence (DRED) (residues 75–78) found in the HLA-B27 molecule. It is relevant to note that aspartic and glutamic acids are negatively charged whilst arginine is positively charged at biological pH. Charged amino acids form powerful antigenic determinants that evoke strong immune responses. Once the patient is exposed to the microbe *Klebsiella*, an immune response will be evoked that will target HLA-B27 molecules. The presence of the pullulanase system in *Klebsiella* allows the microorganism to cleave $\alpha(1-6)$ bonds present in branched starches such as amylopectin, which is found in potatoes and wheat- or flour-derived products such as bread, cakes, pasta and rice (Figure 1).This property of the pullulanase system suggests a possible therapeutic approach to the treatment of AS patients. Branched dietary starch molecules cannot be readily hydrolysed by human digestive enzymes, but the pullulanase system of *Klebsiella* microbes allows them to gain access to partially hydrolysed starch, to another carbon source, and thereby not only increase the fecal mass in the AS patients but also stimulate the immune GALT system to produce high levels of anti-*Klebsiella* antibodies.

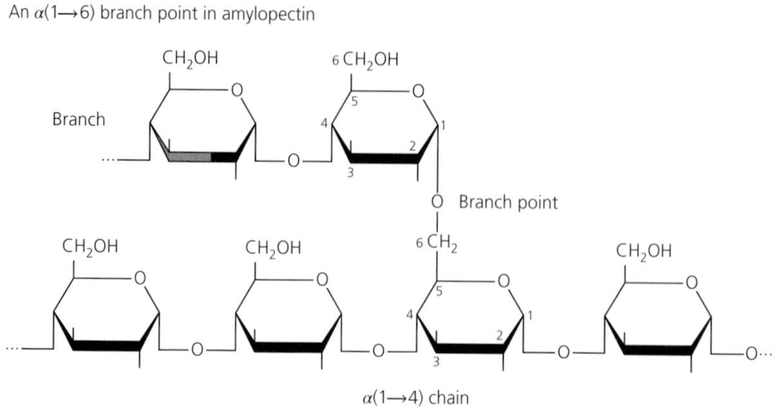

An α(1→6) branch point in amylopectin

Figure 1 Chemical structure showing the point of action by *Klebsiella* pullulanase enzyme on the α(1–6) links.

Table 1 Comparison of amino acid sequence homologies between Klebsiella pullulanase (pulA) and collagens I, III and IV (G=glycine, X=any other amino acid, P=proline, A=alanine, D=aspartic acid, E=glutamic acid).

Protein	Residues	Amino acid sequence
pulA	11–29	GXP-GXP-GXP-GXP-GXP-GXP
collagen I		GXP-GXP-GAD-GPA-GXP-GXP
collagen III		GXP-GXP-GXP-GXP-GXP
collagen IV		GAE-GXP-GXP-GXP

20.5 Pullulanase system and collagens

There is a second property of the pullulanase system that is relevant to the dysbiosis of *Klebsiella* in AS. The pulD fragment is the terminal secretion moiety of the pullulanase system and is known to have regions said to form strongly amphipathic β-pleated sheets that confer membrane-spanning properties[14]. The pullulanase enzyme appears to have, at the N-terminal end of the pulA protein in *Klebsiella* microbes, repeats of the collagen tri-peptide Gly-X-Pro, sequences that are also found in collagens type I, type III and type IV[15]. In the case of *Klebsiella* pullulanase, there is an unusual N-terminal amino acid sequence that includes six repeats of the tripeptide collagen Gly-X-Pro (Table 1). This type of sequence is characteristic of animal collagens, which are known to form triple helical structures. Type I collagen is found predominantly in tendons and bones and could readily explain the clinical features of the disease. Cross-reacting anti-*Klebsiella* antibodies may be involved in the deposition of fibrous tissue in the axial skeleton and involved in the inflammation at the entheses or enthesitis. Type III collagen is found in muscles and anti-*Klebsiella* antibodies could thus explain the muscle stiffness that affects AS patients. Type IV collagen is found in basement membranes of the gut, retina and uvea. Therefore this could explain the episodes of uveitis that occur in AS patients.

20.6 Specific antibodies to *Klebsiella* in AS patients

Extensive and specific immunological studies have been carried out over the last 30 years in 16 different countries that measure anti-*Klebsiella* antibodies by a variety of methods and have involved over 1000 AS patients in 40 different investigations (Table 2). All these investigations have been recently reviewed and gave the same results[16]. Active AS patients have elevated levels of antibodies to the bowel microbe *Klebsiella* and do not have elevated levels of antibodies to other microbes, such as *Salmonella, Shigella, Yersinia* or *E. coli*[16].

The specificity of the elevated anti-*Klebsiella* titres in AS patients was compared to antibody levels of another microbe that was involved in rheumatoid arthritis. In 1978, Stastny from Dallas showed that some 80% of rheumatoid arthritis patients were carriers of the histocompatibility antigen HLA-DR4 whilst it was present in only 35% of the U.S. general population[16]. Similar results were reported from the UK and other countries in Europe. Injection of HLA-DR4 lymphocytes from healthy blood donors into rabbits produced antibodies against the urinary microbe *Proteus*. A urinary tract infection by *Proteus* bacteria would readily explain the increased prevalence of rheumatoid arthritis in women compared to men. The suggestion was proposed that rheumatoid arthritis was caused by an upper urinary tract infection by *Proteus* bacteria. The hypothesis was tested by simultaneous measurements of antibodies to *Klebsiella* and *Proteus* in patients with AS or rheumatoid arthritis. The tests were carried out with patients from three different countries: UK[17], the Netherlands[18] and Japan[19]. All three studies

Table 2 Geographical distribution of anti-*Klebsiella* antibodies in patients with ankylosing spondylitis (AS) compared to those with other rheumatic diseases (ORD) or to healthy controls (HC), as reviewed by Rashid and Ebringer in *Curr Rheum Reviews*, 2012; **8**: 109–119. (With permission)

	Country and (No.) of the studies	No. of AS patients	No. of patients with (ORDs)*	No. of HCs	Years of the studies
1	ENGLAND (8)	546	149 (RA); 21 (PsA)	419	1983 to 2001
2	SCOTLAND (2)	73		49	1988; 1992
3	USA (1)	24		90	1987
4	RUSSIA (1)	25		28	1989
5	SLOVAKIA (1)	20		20	1992
6	CHINA (1)	60	28 (RA)	45	1993
7	CANADA (1)	31	18 (RA/OA)	15	1993
8	GERMANY (1)	41	24 (RA); 24 (PsA); 24 (SLE); 20 (ReA)	95	1994
9	SPAIN (1)	84	22 (RA); 41 (NIA)		1994
10	TURKEY (1)	40		40	1996
11	JAPAN (1)	52	50 (RA)	50	1997
12	NETHERLANDS (1)	34	25 (RA)	34	1998
13	MEXICO	44		40	1998
14	TAIWAN (1)	52		51	1998
15	FINLAND (3)	276	10 (RA)	200	1993; 1997; 1998
16	INDIA (1)	20		15	2002
	Total numbers	**1422**	**455**	**1191**	

gave similar results. AS patients had elevated levels of antibodies to *Klebsiella* but not to *Proteus*, whilst the reverse was observed with the rheumatoid arthritis patients who had elevated levels of antibodies to *Proteus* but not to *Klebsiella*. Each microbe assay was a specificity control for the other disease. Control sera from healthy blood donors had low levels of antibodies to both *Klebsiella* and *Proteus* microbes. The general conclusion was drawn that AS is caused by *Klebsiella* infection in the gut and rheumatoid arthritis is caused by an upper urinary tract infection by *Proteus* bacteria[20]. Since *Proteus* bacteria are found in the human gut and a retrograde infection of the upper urinary tract occurs, this could be considered another example of a dysbiotic condition[21] (covered in chapter 18 of this book).

The specific bowel dysbiosis in AS is based not only on molecular mimicry between *Klebsiella* and HLA-B27 but also in being able to explain why specific collagen tissues are targeted by the *Klebsiella* antibodies. Cytopathic studies using sheep red cells coated with HLA-B27 antigens clearly demonstrated that specific complement-dependent cytotoxicity occurs in AS patients[22]. The pathogenesis of AS involves the production of two types of anti-*Klebsiella* antibodies. One set has anti-HLA-B27 activity and the other has anti-collagen activity, thereby explaining the clinical features of the disease, namely association with HLA-B27 and muscle stiffness, backache and uveitis (Figure 2). The demonstration that anti-*Klebsiella* antibodies have cytopathic properties raises the issue that reduction of intra-luminal proliferation of *Klebsiella* microbes could have beneficial effects in the treatment of AS patients.

Sulphasalazine has been used in the treatment of AS patients over a period of 26 weeks in a Finnish study that showed a significant drop in the level of anti-*Klebsiella* antibodies[23]. Another study, using a different antibiotic, moxifloxacin, showed significant improvement in AS patients in a clinical trial lasting over a period of 12 weeks[24]. However, the long-term use of antibiotics raises the issue of development of resistant strains of bacteria in the treatment of this disease. A more rational approach would be to decrease the size of the *Klebsiella* bowel flora by reducing the substrates on which these bacteria grow. The demonstration that *Klebsiella* have a debranching enzyme, pullulanase that acts on partially digested dietary starch could provide a novel and simpler way of treating this disease[25].

20.7 The low-starch diet in AS

American researchers have demonstrated, using post-meal hydrogen studies, that starch in the form of bread or pasta is a significant component of bowel bacterial proliferation[26]. A "low starch diet" was developed for the patients attending the AS Research Clinic of the Middlesex Hospital. The main features of the "London Low-Starch Diet" were "No bread, no potatoes, no pasta, no cakes and no rice." The London Low-Starch Diet was tested in a group of 21 healthy subjects consisting of University students and hospital staff for a period of eight weeks. To compensate for the calorie loss in reducing starch intake, the control subjects were advised to increase their consumption of fish, meat, fruits and vegetables but to exclude potatoes. There were no restrictions on their drinking

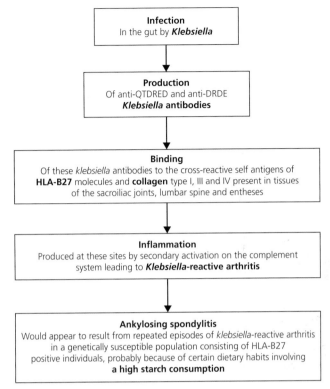

Figure 2 Pathogenesis of ankylosing spondylitis in a patient following exposure to *Klebsiella* bacteria in the gut.

habits, an important point with some of the University students. Total serum IgA, an estimate of the quantity of bowel flora, was measured before and after completion of the trial. The level of total serum IgA dropped from 255 mg/dl at the start of the study to 212 mg/dl (p < 0.001) by the end of eight weeks[27]. Since the control subjects appeared to tolerate the low-starch diet without any difficulties, it was prescribed to all the AS patients attending the AS Research Clinic of the Middlesex Hospital. The effect of the diet was examined over a period of 10 months in 43 AS patients who had an ESR above 15 mm/hour at the start of the study. There was a statistically significant drop in ESR, in those patients who had at the start of the study an ESR above 15 mm/hour: 40 individuals showed a drop in ESR, two patients worsened and one subject's ESR remained constant (Figure 3)[28]. Almost all AS patients claimed a significant reduction in pain, early morning stiffness and rigidity. It is important to control the placebo effect in clinical trials. Therefore dietary studies in AS patients should be monitored by objective measurements, such as ESR, C-reactive protein, total serum IgA, anti-*Klebsiella* antibodies and hemoglobin levels. The effect of the diet can be illustrated by the story of Mr. George McCaffery, quoted by Carol Sinclair, author of the book *IBS and the low starch diet*. Carol Sinclair's protocol is a "no-starch diet," a more radical form of a low-starch diet since foods reacting to iodine are excluded from consumption.

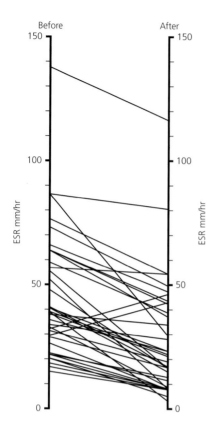

Figure 3 Individual erythrocyte sedimentation rate values before and after 10 months on low-starch diet in 43 ankylosing spondylitis patients, belonging to the group initially having an erythrocyte sedimentation rate level of 15 mm/hour or above at the start of the diet.

20.8 Conclusions

The first discovery of the link between AS and HLA-B27 by two groups, one from the UK and the other one from the USA, opened a new way of looking at this disease. The second discovery was the demonstration of "molecular mimicry" between *Klebsiella* and HLA-B27 by two groups, one from London and the other from San Diego. The third discovery by the London group was that *Klebsiella* microbes possessed pullulanase, a debranching enzyme that enhanced the quantity of substrate from dietary starch that could be used for bacterial proliferation in the gut. The fourth discovery was that elevated levels of antibodies to *Klebsiella* were present in active AS patients from 16 different countries. The fifth discovery was that starch restriction led to a reduction of objective parameters of inflammation in AS patients and could be considered a viable and useful addition to the treatment of AS. Other theories that have been proposed, such as labile behavior of the HLA-B27 molecule, do not account for the majority of clinical data observed in AS patients.

The general conclusion can be made that gut dysbiosis involving bacteria such as *Klebsiella* in AS can be dealt with dietary measures that could provide a significant therapeutic addition to the problem of this world-wide arthritic condition. A similar approach might be useful in other gut-related diseases. Rheumatoid arthritis with its link to *Proteus* could also be considered to fall within the ambit of dysbiotic disorders.

TAKE-HOME MESSAGE

- Evidence suggests that ankylosing spondylitis is caused by overgrowth of *Klebsiella* microbes in the gut as a form of dysbiosis.
- The disease can be treated by reducing the quantity of *Klebsiella* substrate when the patients go on a low-starch diet.

References

1   Caffrey MFP, James DCO. Human lymphocyte association in ankylosing spondylitis. *Nature* 1973; **242**: 121.

2   Schlosstein L, Terasaki PI, Bluestone R, Pearson CM. High association of an HLA antigen W27 with ankylosing spondylitis. *N Engl J Med.* 1973; **288**: 704–6.

3   Welsh J, Avalian H, Cowling P, Ebringer A, Wooley P, Panayi G, Ebringer R. Ankylosing spondylitis, HLA-B27 and *Klebsiella*. I. Cross-reactivity studies with rabbit antisera. *Br J Exp Pathol.* 1980; **61**: 85–91.

4   Sieper J, Braun J, Rudwaleit M, Boonen A, Zink A. Ankylosing spondylitis: an overview. *Ann Rheum Dis* 2002; **61** (suppl III): 8–18.

5   Lamm ME. Cellular aspects of immunoglobulin A. *Adv Immunol.* 1976; **22**: 223–90.

6   Cowling P, Ebringer R, Ebringer A. Association of inflammation with raised serum IgA in ankylosing spondylitis. *Ann Rheum Dis.* 1980; **39**: 545–9.

7   Fournier AM, Denizet D, Delagrange A. La lymphographie dans la spondylarthrite ankylosante. *J Radiol Electrl Med Nucl.* 1969; **50**: 7 73–84.

8   Langness U, Banavsky D, Alexopoulos J, Kriegel W, Burger R. Differentiation of inflammation and ossification in spondylitis ankylopoietica. *Acta Rheumatol Scand.* 1971; **17**: 15–22.

9   Mielants H, Veys EM, Cuvelier C, de Vos M. Ileocolonoscopic findings in seronergative spondyloarthropathies. *Br J Rheumatol.* 1988; **27** (Suppl II): 95–105.

10 Perricone C, Rinkevich S, Blank M, Launda-Rouben N, Alessandri C, Conti F, Leor J, Shoenfeld Y, Vatesini G. The autoimmune side of rheumatic fever. *IMAJ* 2014; **16**: 645–55.

11 Myers PJ, Kane KE, Porter BG, Mazzaccaro RJ. Sydenham's chorea: a rare consequence of rheumatic fever. *West J Emerg Med* 2014; **15**: 840.

12 Schwimmbeck PL, Yu DTY, Oldstone MBAS. Autoantibodies to HLA-B27 in sera of HLA-B27 patients with ankylosing spondylitis and Reiter's syndrome. *J Exp Med.* 1987; **166**: 173–81.

13 Fielder M, Pirt SJ, Tarpey I, Wilson C, Cunningham P, Ettelaie C, Binder A, Bansal S, Ebringer A. Molecular mimicry and ankylosing spondylitis: possible role of a novel sequence in pullulanase of *Klebsiella pneumonia*. *FEBS Lett.* 1995; **369**: 243–8.

14 Scofield RH, Warren WL, Koelsch G, Harley JB. A hypothesis for HLA-B27 dysregulation in spondyloarthropathy: contributions from enteric organisms, B27 structure, peptides bound by B27 and convergent evolution. *Proc Natl Acad Sci.* 1993; **90**: 9330–4.

15 Charalambous BM, Keen JN, McPherson MJ. Collagen-like sequences stabilize homotrimers of a bacterial hydrolase. *EMBO J.* 1988; **7**: 2903–9.

16 Stastny P. Asociation of the B-cell alloantigen DRW4 with rheumatoid arthritis. *N Engl J Med* 1978; **298**: 869–71.

17 Ebringer A, Ptaszynska T, Corbett M, Wilson C, Macafee Y, Avakian H, Baron P, James DCO. Antibodies to *Proteus* in rheumatoid arthritis. *Lancet* 1985; **2**: 305–7.

18 Blankenberg-Sprenkels SHD, Fielder M, Feltkamp TEW, Tiwana H, Wilson C, Ebringer A. Antibodies to *Klebsiella pneumonia* in Dutch patients with ankylosing spondylitis and acute anterior uveitis and to *Proteus mirabilis* in rheumatoid arthritis. *J Rheumatol.* 1998; **25**: 743–7.

19 Tani Y, Tiwana H, Hukuda S, Nishioka J, Fielder M, Wilson C, Bansal S, Ebringer A. Antibodies to *Klebsiella*, *Proteus* and HLA-B27 peptides in Japanese patients with ankylosing spondylitis and rheumatoid arthritis. *J Rheumatol.* 1997; **24**: 109–14.

20 Rashid T, Ebringer A. Detection of *Klebsiella* antibodies and HLA-B27 allotypes could be used in the early diagnosis of ankylosing spondylitis with a potential for the use of "low starch diet" in the treatment. *Curr Rheumatol Rev.* 2012; **8**: 109–19.

21 Ebringer A. *Rheumatoid arthritis and Proteus*. Springer, London 2012.

22 Wilson C, Rashid T, Tiwana H, Beyan H, Hughes L, Bansal S, Ebringer A, Binder A. Cytotoxicity responses to peptide antigens in rheumatoid arthritis and ankylosing spondylitis. *J Rheumatol.* 2003; **30**: 972–8.

23 Nissila M, Lehtinen K, Leirisalo-Repo M,Mutru O, Yli-Kerttula U. Sulfasalazine in the treatment of ankylosing spondylitis: a 26 week, placebo controlled clinical trial. *Arthritis Rheum.* 1988; **31**: 111–6.

24 Ogrendik M. Treatment of ankylosing spondylitis with moxifloxacine. *South Med J.* 2007; **100**: 366–70.

25 Ebringer A, Rashid T, Wilson C, Ptaszynska T, Fielder M. Ankylosing spondylitis, HLA-B27 and *Klebsiella*. An overview: proposal for early diagnosis and treatment. *Curr Rheumatol Rev.* 2006; **2**: 55–68.

26 Anderson LH, Levine AS, Levine MD. Incomplete absorption of carbohydrate in all- purpose wheat flour. *N Engl J Med.* 1981; **304**: 891–2.

27 Ebringer A, *Ankylosing spondylitis and Klebsiella*. Springer, London, 2013.

28 Ebringer A, Wilson C. The use of low starch diet in the treatment of patients with ankylosing spondylitis. *Clin Rheumatol.* 1996; **15** Suppl 1: 61–5.

CHAPTER 21

Microbiome of chronic plaque psoriasis

Lionel Fry

Imperial College, London, United Kingdom

21.1 Introduction

Psoriasis has been recognized for over 2,000 years. Hippocrates did recognize scaly eruptions, which he grouped under the term lopoi (from "lepo," to scale). Psoriasis as we know it today was grouped with leprosy until the mid-19th century. However, psoriasis comes from the Greek word psora, to itch, and was first used by Galen (133–200 AD), but from his description he was probably describing seborrhoeic eczema, "scaling of the eyelids and corners of the eyes". Thus it could be argued that the wrong names were given to clinical diseases we recognize today.

The pathogenesis of psoriasis has been debated for nearly two centuries, ever since psoriasis was recognized as a separate entity. The advances in our understanding of the pathogenesis have depended on technological advances in medicine. However, clinical observation has contributed to our understanding of the disease: in particular, family studies, which have shown that heredity is important in psoriasis.

The two main histological features of psoriasis are inflammation and epidermal hyperplasia, and it has been argued which comes first. In the 1960s it was proposed that psoriasis was due to an inherent defect in the cell cycle of the epidermal cells. It was hypothesized that there was a shortened cell cycle that resulted in an increased mitotic rate and hyperproliferation of the epidermis. However, in the 1980s, with the advent of techniques to investigate lymphocytes, it was shown that psoriasis was characterized by lymphocytic infiltration into the dermis and epidermis. It was proposed that the hyperproliferation of the epidermal cells was secondary to cytokines released from the lymphocytes[1]. That still left the question, what stimulated the lymphocytes? The most obvious choice was a bacterial antigen. In fact, bacteria as the cause of psoriasis was first suggested at the end of the 19th century. Radcliffe Crocker in 1893 reported "enormous numbers of minute circular bodies with a central dark spot which lie in clusters between the separated layers, but which also exist in dense masses below[2]. Their appearances certainly suggest they are organisms of some kind. As to whether they are *materies morbi* of some etiological significance or merely there because tissue is diseased is unclear".

The Human Microbiota and Chronic Disease: Dysbiosis as a Cause of Human Pathology, First Edition.
Edited by Luigi Nibali and Brian Henderson.

Unna in 1896, considering the parasitic nature of psoriasis, stated that "unfortunately the sparsity of the micrococci in the horny layers makes their recognition a serious undertaking[3]. However if one takes a squash preparation of the horny layer and preserves it aseptically in a nutrient medium a pure culture of micrococci grows abundantly".

More recently, it was shown in the 1950s that one form of psoriasis, "guttate psoriasis," could be precipitated by streptococcal throat infections[4]. Recently it has been shown that the T-cell receptor is the same in the tonsil and in the skin in streptococcal-precipitated psoriasis[5]. It has also been shown that T-cells in the skin respond to streptococcal peptidoglycans[6].

It has now been shown that CD4 T-cells in psoriatic lesions respond to streptococcal peptidoglycan and it was proposed that this is one of the antigens that may trigger psoriasis[6]. However, psoriasis has different clinical patterns and it is conceivable that different bacterial and fungal antigens may be responsible for triggering psoriasis at different sites. This implies that a specific antigen is responsible for triggering psoriasis. Over the last decade, innate immunity has been implicated in a number of diseases that are thought to be immune-mediated. There is now accumulating evidence that innate immunity is implicated in initiating psoriasis, as the majority of the genetic abnormalities are concerned with innate immunity[7].

There are currently two theories as to the pathogenesis of psoriasis. The first is that psoriasis is an autoimmune disease[8,9]. The two groups who have published these papers argue that there is molecular mimicry between streptococcal proteins and keratins. Psoriasis is initiated by streptococcal peptides and it is maintained by keratin peptides. Valdimarsson and colleagues[8] postulated that M protein was the streptococcal antigen and that there was homology with keratin 16 and 17. They synthesized keratin peptides which shared sequence homology with streptococcal M proteins. However, when they tested peripheral blood with these peptides they found it was only the CD8 but not the CD4 T cells that responded to them. However, Prinz and his colleagues[9] immunized rabbits with *Streptococcus pyogenes* and raised antibodies that were found to react with various keratinocyte proteins including keratin 6, ezrin, maspin, peroxiredoxin 2 and Hsp27. Sera from patients with psoriasis also reacted with these keratinocyte proteins when tested against them. When these proteins and synthetic peptides were tested against peripheral blood cells from patients with psoriasis, it was mainly the CD8 T cells that responded. Sequence alignment of the keratin peptides with the whole genome of the group A Streptococcus led to the identification of various homologous peptides. These homologies were not only with M protein but also with other streptococcal proteins, namely RopA, RecF and FcR. Thus the two groups who support autoimmunity in psoriasis have suggested different streptococcal and keratin antigens. In addition, both groups found only CD8 T cells to respond to these homologous peptides. However, it is only CD8 T lymphocytes that respond to these peptides, whereas it is CD4 T cells that are associated with the initiation of psoriasis. In addition, in the SC1D mouse model for studying psoriasis, when uninvolved psoriatic skin is engrafted onto the mice and injected with autologous CD4 and CD8 T lymphocytes, it is only the CD4 cells that convert the uninvolved skin into typical psoriasis[10]. So if psoriasis is not an autoimmune disease, what triggers the disease? It has recently been suggested that it is the microbiota in the skin that may be responsible[11,12]. Since it has been possible, with the advent of molecular techniques, to investigate the microbiome in tissues, there have been studies on both normal skin and psoriasis that suggest that the microbiota may well be involved.

21.2 Microbiota in psoriasis

21.2.1 Bacteria

We have previously seen (chapter 5) how metagenomic analyses reveal that most skin bacteria fall into four different phyla: Actinobacteria, Firmicutes, Bacteroidetes and Proteobacteria. There have been three studies on the specific microbiota in psoriasis[13–15]. Two have employed swabs for sampling[13,15] and one biopsies[14]. There was agreement in the findings in the three studies, but also differences. The three studies have shown less diversity in the microbiome in samples taken from psoriatic lesions compared to those from normal controls. All studies found the three most common phyla in psoriasis specimens and that from uninvolved skin and normal control skin to be Firmicutes, Actinobacteria and Proteobacteria. Firmicutes was the commonest phylum in both biopsies and swabs. The first study, which was based on swabs, gave an incidence of Firmicutes as 46% in psoriasis, 39% for uninvolved skin and 24% for controls[13]. In the study employing biopsies the incidence was 43% for normal skin and 39% for psoriasis[14]. However, a major difference in the two studies was the representation of Proteobacteria in psoriasis (38% v 11.4%) in biopsies and swabs respectively, but the representations were similar in controls (27% v 22%). Interestingly, the second paper from the Blaser group[15] proposed two cutaneotypes of the microbiome in psoriasis: the first Proteobacteria-associated microbiota and the second Firmicutes and Actinobacteria-associated. However, there is a difference between biopsies and swabs at the genus level. In the swabs the commonest genus was Corynebacterium in psoriasis and Propionibacteria in controls. In the biopsies the commonest were Streptococci in both psoriasis and controls. The incidence of Streptococci was significantly higher in both psoriasis and control groups in the biopsies compared to swabs, 33% v 14.3% in psoriasis and 27% v 7.1% in controls. Staphylococci were higher in swabs, being 18.8% in psoriasis, 15.7% in controls, whereas in biopsies it was 16% in controls and only 5% in psoriasis. This may reflect the depth of the sample. One expects more Staphylococci on the surface of the skin as this is the most common genus causing skin disease. Interestingly, in a microbiome study of the blood in psoriasis patients the commonest genus was streptococcus in guttate psoriasis (related to Streptococcal throat infections) and Staphylococcus in chronic plaque psoriasis[16].

One of the significant findings is the increased abundance of Streptococcus and decreased abundance of Propionibacteria in psoriasis in both swabs and biopsies. The ratio of Streptococcus/Propionibacteria was 5:1 for swabs and biopsies. The difference was mainly due to a decrease in Propionibacteria. It was suggested that the decrease in Propionibacteria may have an etiological role in psoriasis[13]. Alekeysenko and colleagues[15] reported no difference in the abundance of the genera Streptococcus, Staphylococci, Propionibacteria and Corynebacteria between the three groups (psoriasis, uninvolved and normal skin). However, when the abundance of the four genera was combined, there was a difference between psoriatic individuals and controls. They also found greater variability in the microbiota in the psoriasis group than the controls and uninvolved specimens. The same group[17] found that molecular signatures for psoriasis based on the microbiome studies are accurate enough for the diagnosis of psoriasis. However, until the microbiome of more skin diseases has been studied, this may be a premature conclusion. The same group also found no correlation of changes in the microbiome with disease activity. This implies that although the skin microbiota may play a role in initiating the disease, other factors control the severity of the disease.

21.2.2 Fungi

Studies on the fungal microbiome have found Malassezia species to be the commonest in normal skin and psoriasi[18-20]. Two groups[18,20] have studied psoriasis as well as normal skin. Paulino and colleagues studied five normal individuals and three patients with psoriasis. In the control subjects swabs were taken from the right and the left forearms. In two subjects swabs were taken from the same sites 10 months later. In four of the five individuals, Malassezia restricta and Malassezia globosa were the commonest species, whereas in one of the arms of one individual they were absent and this site had a high abundance of *Malassezia sympodialis*. This latter species was present in all samples except one, but was of a lower abundance compared to *M. restricta* and *M. globosa*. *Malassezia pachydermatis* was present in three subjects in one arm. *Malassezia furfur* was only present in one arm in one patient and in low abundance. The results of the two swabs taken 10 months later showed similar results to the first swab. In the three subjects with psoriasis, swabs were taken from a digital lesion and elbow in patient 1, from a leg, arm and forearm in patient 2 and an elbow and a leg in patient 3. In patient 1 samples were taken six months later from the same digit and elbow. In all three subjects a swab was taken from uninvolved skin. *M. restricta* was present in all the psoriasis lesions but absent in the uninvolved skin in two. *M. globosa* was present in all the lesions except one, and in the uninvolved skin. *M. sympodialis* was present in all psoriasis samples except one, but in only one of the three samples taken from uninvolved skin. Undetermined phyla were present in all samples except uninvolved skin in one subject. One of the samples taken six months later in the first patient showed similar fungal species to the previous one.

In the study by Jagielski and colleagues[20], both culture and molecular were used. They found the results from the two techniques were concordant in 65% and discordant in 35% of cases. They studied six normal individuals, six patients with atopic eczema and six with psoriasis. Each subject had four swabs taken: from the scalp, face, chest (between the clavicles) and the back (interscapular region). All these sites would be classified as sebaceous. *M. sympodialis* was the predominant species in all three groups. In psoriasis, in five of the six patients *M. sympodialis* was present (seven of nine samples) and *M. furfur* was present in four of the patients (from five samples). In the controls *M. sympodialis* was present in all subjects (10 of 13 samples). *M. restricta* and *M. globosa* were not found in the psoriasis subjects but were each found in one control subject.

Thus there are differences between the two studies. Although in both Malassezia was found to be the commonest genus in the skin, there were differences in the species. Although *M. sympodialis* was present in both, *M. restricta* and

M. globosa were not found in the study by Jagielski. *M. furfur* was found in four of the six psoriasis subjects in the Jagielski study. It was not found by Paulino and colleagues. However, the technological methods were not the same. An interesting finding was that temporally there was no alteration in the mycology: the same species were present over six to 10 months.

21.3 Variation of microbiota with site

As reported above, the four most common phyla found on normal skin are Actinobacteria 51.8%, Firmicutes 24.4%, Proteobacteria 16.5% and Bacteroides 6.3%[21]. Over the last five years it has been shown that the microbiota varies with the sites from which the specimens were taken. The first study by Grice et al divided the body into three sites: sebaceous (back, chest, scalp and face), moist (groins and axillae, sub-mammary, umbilicus, interdigital web spaces, antecubital and popliteal fossae) and dry (arms, legs and buttocks)[21]. Propionibacteria and Staphylococcus predominated in the sebaceous sites, Corneybacteria and Staphylococcus in the moist sites. In the dry sites there was a mixed population of bacteria but with a greater prevalence of B-Proteobacteria and Flavobacteriales. In the same study the authors found that there was variation not only in the microbiome with site, but also between individuals, and temporal variation in the same individual. The varying anatomical sites that show different microbioma have now been extended to eighteen[19].

The fungal microbiome also shows variation with site[22]. The greatest variation was seen in the foot, with differences between the heel, toe web and nail. Malassezia was the dominant species at the core body sites (external auditory and inguinal region, manubrium, behind the ear, the back, nares and occiput). Temporal studies from the same sites, however, showed little change in the fungi, with Malassezia predominating.

21.4 Swabs versus biopsies

All the studies showing variation in the microbiome with site have been based on swabs taken from the surface of the skin[19,21,22]. Two of the studies on psoriasis employed swabs[13,15] and one biopsies[14] and there was a difference in the findings. A recent study[23] showed that bacteria are present in all layers of the normal skin and the superficial subcutaneous fat. Thus if one wishes to study the skin microbiome and how it may cause disease one has to ask, is a swab sufficient? It would appear that a biopsy is the method of choice for studying the microbiome.

21.5 Psoriatic arthritis

A significant number of patients with psoriasis (10–30%) have an associated arthritis. Recent studies have suggested that microbiota in the gut[24,25] may be related to psoriatic arthritis. The study by Scher and colleagues[24] reported that the changes in the gut microbiota in psoriatic arthritis were similar to that in Crohn's disease, and they mentioned the clinical relationship of Crohn's disease to psoriatic arthritis. It would be interesting to investigate the microbiome in psoriasis without the arthropathy.

Another report suggested that the skin microbiota as well as that of the gut may play a part in psoriatic arthritis[26]. Traditionally, it was not thought that joints have their own microbiome. After all, nobody thought that bacteria were present deep in the skin. However, as discussed by Carter in chapter 17, PCR documented the presence of all the definite bacterial triggers of rheumatoid arthritis (ReA), or their bacterial products, in the synovial tissue or fluid of patients with ReA, and perhaps a synovial-based microbiota exists and can undergo a dysbiotic process.

21.6 Microbiome and immunity

Several authors have proposed a relationship between immunity and the microbes that inhabit the skin and gut[27–30]. It has been claimed that bacteria in the tissue are responsible for maintaining a normal immune system and that bacteria at one site may influence the immunity at another, e.g. that the gut microbiota influences immune responses in the lungs[31]. Thus if we discuss immune-mediated diseases and their relationship to the microbiome, it may not be sufficient just to investigate the microbiota at the site of pathology. The relationship among the disease state, the microbiota and immunity may prove to be more complex than we thought.

21.7 Evidence that the skin microbiome may be involved in the pathogenesis of psoriasis

21.7.1 Psoriasis and crohn's disease

It is well known that there is a clinical relationship between Crohn's disease (CD) and psoriasis. Patients with CD are five times more likely to develop psoriasis than a control population[32–34]. Furthermore, there is a significant increase in psoriasis in first-degree relatives of patients with CD[34]. Conversely, there is an increase of CD in patients with psoriasis[35,36]. Both disorders tend to commence at the same age, late teens and early adulthood, and they both respond to the same biological therapies, implying similar pathogenic pathways. Many gastroenterologists now accept that CD is due to a breakdown of immune tolerance to the microbiota (dysbiosis) of the intestine. This leads to chronic inflammation of the intestinal wall in genetically susceptible individuals. Patients with psoriasis also have an increased incidence of periodontitis[37], which is a disease thought to be due to an abnormal response to microbiota in the mouth (previously discussed by Nibali in chapter 13)[38,39]. Thus we propose that all three disorders, psoriasis, CD and periodontitis, may share a common genetic defect that allows an inflammatory response to bacteria inhabiting different sites, i.e. the skin, intestine and oral cavity respectively. The term "genetic dysbiosis" has recently been introduced to describe this phenomenon[38].

21.7.2 Genetic factors

Heredity was known to be important in psoriasis by the end of the nineteenth century and has since been confirmed by clinical studies. The most recent comprehensive study on the genetics of psoriasis found 36 susceptibility loci[7]. This report found 15 new susceptibility loci and emphasized the role of the innate immune system (IIS) in psoriasis. Ten of these genes overlap with those found in CD and are concerned with innate immunity. IL-23 is produced by cells of the IIS and four of the genes in psoriasis are concerned with IL-23 signaling [(L-23R, IL-12B, IL-23A, TYK2)[40] NFkB is an important part of the IIS and six genes are concerned with NFkB signaling (REL,TN1P1, TRAF31P1, TNFA1P1, NFkB1A, FBXL19)[40]. The newly identified genes include STAT3, which plays a role in mediating the innate immune response, and ZC3H12C and KLF4 are concerned with macrophage activation[7]. The mutation in the gene for NOD2 (CARD15) concerned with NFkB signaling is strongly associated with CD and its severity[41]. No association between genetic polymorphisms in the NOD2 gene and psoriasis has been detected[42], but recently a mutation in CARD14 has been found in psoriasis that is also involved in NFkB signaling[43].

21.7.3 Innate immunity

In addition to genetic polymorphisms in the innate immune genes reported in psoriasis, there is evidence of involvement of the innate immune system (IIS) in the disease process. Normally there is a symbiotic relationship between the bacteria and the skin. However, if there is a defect in the IIS, genetically determined, there may be an abnormal response of the IIS to the bacteria normally present in the skin or to bacteria unique to psoriasis. Cells of the IIS include keratinocytes, mast cell and neutrophils and are the main source of anti-microbial peptides (AMPs), which have a bacteriocidal-like action against pathogenic organisms[44]. There are two main groups of AMPs, β-defensins and cathelicidins, and there is evidence that both are involved in psoriasis[44-48].

Another component of the IIS is the pathogen recognition receptors (PRRs) that detect bacterial, viral or fungal components. The PRRs found on the surface of cells include Toll-like receptors (TLRs) and peptidoglycan recognition proteins (PGRPs). In psoriasis, TLR1 and TLR2 are enhanced on keratinocytes in the upper epidermis and TLR5 is down-regulated in basal keratinocytes compared with normal skin[47]. TLR2 is involved in the recognition of products of Gram-positive bacteria such as peptidoglycan[49] and lipoproteins[49,50]. In CD TLR4, which recognizes lipopolysaccharides of Gram-negative organisms, is upregulated compared with normal gut epithelium[51]. Thus, the expression of TLRs is different in psoriasis and CD, suggesting that the IIS may be triggered by different bacteria in each instance.

There are four peptidoglycan recognition proteins that function in antibacterial immunity by binding PG of Gram-positive bacteria, a known stimulator of the IIS[52]. In psoriasis there is polymorphism of the genes for PGRP-3 and PGRP-4 on chromosome Iq (PSORS4)[53,54]. This may lead to an abnormal response to bacterial PG in psoriasis that results in inflammation. Interestingly, it has been suggested that bacterial peptidoglycan may also play a role in activating the adaptive immune system in psoriasis. Streptococcal PG has been found in antigen presenting cells in psoriasis[6] and CD4 T-cell lines from psoriasis respond to both Streptococcal and Staphylococcal PG[6]. In addition, activation of IIS cells leads to the production of cytokines, such as IL-23, that are involved in the development and differentiation of Th1, Th2 and Th17 cells. IL-23 produced by cells of the IIS appears to be essential for the development of psoriasis[55].

21.8 New hypothesis on the pathogenesis of psoriasis

Thus we have put forward the concept that psoriasis is due to genetic abnormalities allowing an inflammatory response to microbiota in the skin[12]. This is supported by the genetic findings and activation of the IIS. In addition, the findings of bacteria on and in the skin suggest they may play a role. What must still be determined is whether one, two or several bacteria are responsible for the initiation of psoriasis? There is some evidence that supports a role for Streptococci and Staphylococci. First, in the biopsies Streptococcus was the genus with the highest abundance. Second, the CD4 cells respond to streptococcal and staphylococcal peptidoglycan[6]. Further evidence is that, from studying the microbiome in the blood, Staphylococci and Streptococci were the genera with the highest abundance[16]. Finally, another form of psoriasis, guttate, an acute form of the disease that appears to be triggered by a different pathway, is related to Streptococcal throat infections. However, the varying microbiome with different anatomical sites may imply that different bacteria or fungi at the different sites may be responsible for triggering psoriasis. However, it has to be asked if the altered microbiome in psoriasis is secondary to the altered structure of the skin in the disease and of no etiological importance. In addition, as mentioned above, the altered microbiome at another site, e.g. the gut, may be responsible for influencing the immune response seen in psoriasis.

A further question that needs to be answered is why psoriasis usually presents in the late second or third decade of life: is there a change in the microbiota with age or is there an alteration of the genes? Shenderov in chapter 11 discussed mechanisms of microbial-host epigenetic interactions. Indeed, epigenetics and methylation of the genes may be responsible for psoriasis predisposition and the known triggers for psoriasis — stress, alcohol, smoking and obesity — have all been associated with methylation of genes[56-60].

Thus the concept that psoriasis is an autoimmune disease is no longer sustainable. The recent suggestions that bacteria can influence our immune systems not only at the local but also distant sites make the microbiome likely to be involved in psoriasis, which is an immune-mediated inflammatory disease. One of the mechanisms thought to be involved in Crohn's disease is a breakdown to immune tolerance to the bacteria in the gut — so-called dysbiosis. This is likely to be due to genetic factors and the fact that there are common genes linked to the innate immune system in both psoriasis and CD make dysbiosis a possible trigger for psoriasis. The fact that there is a clinical relationship between psoriasis and CD makes a common pathogenetic role likely in the two diseases.

TAKE-HOME MESSAGE

- The skin microbiota is different in psoriasis compared to normal skin.
- There is evidence to suggest that the skin microbiota is involved in the pathogenesis of psoriasis.

References

1 Valdimarsson H, Baker BS, Jonsdottir I, Fry L. Psoriasis: a disease of abnormal keratinocyte proliferation induced by T lymphocytes. *Immunol To-day* 1986; **71**: 256–259.

2 Radcliffe-Crocker H. *Diseases of the Skin*. London, HK Lewis, 1893.

3 Unna PC. *Histopathology of the Skin*. New York, Macmillan & Co, 1896.

4 Norrlind R. Psoriasis following infections with haemolytic streptococci. *Acta Dermatovenerol* 1950; **30**: 64–72.

5 Diluvio L, Vollmer P, Besgen J. *et al*. Identical TCR beta chain rearrangements in streptococcal angina and skin lesions of patients with psoriasis vulgaris. *J Immunol* 2006; **176**: 7104–7111.

6 Baker BS, Powles A, Fry L. Peptidoglycans: a major factor for psoriasis. *Trends Immunol* 2006; **27**: 545–51.

7 Tsoi LC, Spain SL, Knight J *et al*. Identification of 15 new psoriasis susceptibility loci highlights the role of innate immunity. *Nat Genet* 2012; **44**: 1341–1348.

8 Valdimarsson H, Thorleifsdottir RH, Sigurdardottir SL *et al*. Psoriasis as an autoimmune disease caused by molecular mimicry. *Trends Immunol* 2009; **30**: 494–501.

9 Besgen P, Trommier P, Vollmer S, Prinz JC. Ezrin, maspin, peroxiredoxin 2, and heat shock protein 27: potential targets of a streptococcal-induced autoimmune response in psoriasis. *J Immunol* 2010; **184**: 5392–5402.

10 Nickoloff BJ, Wrone T. Injection of pre-psoriatic skin with CD4+ T cells induces psoriasis. *Am J Pathol* 1999; **155**: 145–158.

11 Fry L, Baker BS, Powles AV *et al*. Is chronic plaque psoriasis triggered by microbiota in the skin? *Brit J Dermatol* 2013; **169**: 47–52.

12 Fry L, Baker BS, Powles AV, Engstrand L. Psoriasis is not an autoimmune disease. *Exp Dermatol* doi:10.1111/exd. 12572.

13 Gao Z, Tseng C, Strober BE *et al*. Substantial alterations of the cutaneous bacterial biota in psoriatic lesions. *PLoS One* 2008; **3**: e2719.

14 Fahlen A, Engstrand L, Baker BS *et al*. Comparison of bacterial microbiota in skin biopsies from normal and psoriatic skin. *Arch Dermatol Res* 2012; **304**: 15–22.

15 Alekseyenko AV, Perez-Perez GI, DeSouzab A *et al*. Community differentiation of the cutaneous microbiota in psoriasis. *Microbiome* 2013; **1**(1): 31. doi: 10.1186/2049-2618-1-31.

16 Munz OH, Sela S, Baker BS *et al*. Evidence for the presence of bacteria in the blood of psoriasis patients. *Arch Dermatol Res* 2010; **302**: 495–498.

17 Stanikov A, Alekseyenko AV, Li Z *et al*. Microbiomic signatures of psoriasis: feasibility and methodology comparison. *Nature*.com/srep/2013//130910.

18 Paulino LC, Tseng CH, Strober BE, Blaser MJ. Molecular analysis of fungal microbiota in samples from healthy skin snd psoriatic lesions. *J Clin Microbiol* 2006; **44**: 2933–41.

19 Oh J, ByrdClay AL, Conlan DS *et al*. Biogeography and individuality shape function in human skin metagenome *Nature* **514**, 59–64 (02 October) doi: 10.1038/nature 13786.

20 Jagielski T, Rup R, Ziolkowska A *et al*. Distribution of Malassezia species on the skin of patients with atopic dermatitis, psoriasis and healthy volunteers, assessed by conventional and molecular identification methods. *BMC Dermatol* 2014 **14**: 3 March 7, 2014 doi:10 1186/1471-5945-14-.

21 Grice EA, Kong HH, Conlan S, Denning CB *et al*. Topographical and temporal diversity of the human skin microbiome. *Science* 2009; **324**: 1190–1192.

22 Findley K, Oh J, Yong J *et al*. Topographic diversity of fungal and bacterial communities in human skin. *Nature*, 2013, **498**: 367–70.

23 Nakatsuji T, Chiang HI, Jiang SB *et al*. The microbiome extends to subepidermal compartments of normal skin. *Nat Commun* 2013; **4**: 1431. doi: 10.1038/ncomms 2441.

24 Scher JU, Ubeda C, Artacho A *et al*. Decreased bacterial diversity characterises an altered gut microbiota in psoriatic arthritis and resemble dysbiosis of inflammatory bowel disease. *Arthritis Rheumatol* 2014 Oct 15 doi:10.1002/art 38892.

25 Eppinga H, Konstantinov SR, Peppelenbosch MF, Thio HB. *The microbiome and psoriatic arthritis*. Springer, New York 2014; **10**. 10075 2 Fs 11926-013-0407-2.

26 Castelino M, Eyre S, Upton M, *et al*. The bacterial microbiome in psoriatic arthritis, an unexplained link in pathogenesis: challenges and opportunities offered by recent technological advances. *Rheumatology* 2014, **53**: 777–784.

27 Belkaid Y, Segre JA. Dialogue between skin microbiota and immunity, *Science* 2014; **346**: 954–959.

28 Spasova DS, Surh CD. Blowing on embers: commensal microbiota and our immune system. *Front Immunol* 2014; **5**: 318.

29 Fung I, Garrett J P-D, Shahane A, Kwan M. Do bugs control our fate? The influence of the microbime on autoimmunity. *Current Allergy and Asthma Reports*.2012; **10**. 1007/s11882-012-0291-2.

30 Shapiro H, Thaiss CA, Levy M, Elinav E. The cross talk between microbiota and the immune system: metabolites take center stage. *Current Opinion Immunol* 2014; **30**: 54–62.

31 Holmes D, Haffnagle G. Rewriting the rules of the lung microbiome. *Lancet* 2014; **334**: 653.

32 Yates VM, Watkinson G, Kelman A. Further evidence of an association between psoriasis, Crohn's disease and ulcerative colitis. *Brit J Dermatol* 1982; **06**: 323–330.

33 Hughes S, Williams SE, Turnberg LA. Crohn's disease and psoriasis. *New England J Med* 1983; **303**: 101 (letter).

34 Lee FI, Bellamy SV, Francis C. Increased incidence of psoriasis in patients with Crohn's disease and their relatives. *Am J Gastroenterol* 1990; **85**: 962–963.

35 Nair RP, Hensler T, Jenisch S *et al*. Evidence for two psoriasis susceptibility loci (HLA and 17q) and two novel candidate regions (16q and 20p) by genome wide scan. *Hum Med Genet* 1997; **6**: 1349–1356.

36 Li WQ, Han JL, Chan AT, Qureshi AA. Psoriasis, psoriatic arthritis and increased incident of Crohn's disease in US women. *Annals Rheum Dis* 2013; doi:10.1136/annrheumdis-2012-202143.

37 Preus HR, Khanifam P, Kolltveit K *et al*. Periodonititis in psoriasis patients. A blinded, case-controlled study. *Acta Odontol Scand* 2010; **68**: 165–170.

38 Nibali L, Henderson B, Sadiq ST, Donos N. Genetic dysbiosis: the role of microbial insults in chronic inflammatory diseases. *J Oral Microbiol* 2014; **6**: 10.3402/jomv6.22962.

39 Ye Y, Carlsson S, Wondimu B *et al*. Mutations in the ELANE gene are associated with develoment of periodontitis in patients with severe congenital neutropenia. *J Clin Immunol* 2011; **31**: 936–945.

40 Capon F, Burden AD, Trembath RC, Barker JN. Psoriasis and other complex trait dermatoses from loci to functional pathways. *J Invest Dermatol* 2012; **132**: 915–922.

41 Bhullar M, Mavrae F, Brown G *et al*. Prediction of Crohn's disease aggression through NOD2/CARD15 gene sequencing in an Australian cohort. *World J Gastroenterol* 2014; **20**: 5008–5014.

42 Plant D, Lear J, Marsland A *et al*. CARD15/NOD2 single nucleotide polymorphisms do not confer susceptibility to type 1 psoriasis. *Brit J Dermatol* 2004; **151**: 675–678.

43 Jordan CT, Cao L, Roberson ED *et al*. PSORS2 is due to a mutation in CARD14. *Am J Hum Genet* 2012; **90**: 784–795.

44 Morizane S, Gallo RL. Antimicrobial peptides in the pathogenesis of psoriasis. *J Dermatol* 2012; **39**: 225–230.

45 Harder J, Schröder JME *et al*. Psoriatic scales: a promising source for the isolation of antimicrobial proteins. *J Leuk Biol* 2005; **77**: 476–486.

46 Ong PY, Ohtake T, Brandt T *et al*. Endogenous antimicrobial peptides in skin infections in atopic dermatitis. *New Engl J Med* 2002; **347**: 1151–1160.

47 Baker BS, Ovigne JM, Powles AV *et al*. Normal keratinocytes express Toll-like receptors (TLRs) 1, 2 and 5: modulation of TLR expression in chronic plaque psoriasis. *Brit J Dermatol* 2003; **148**: 670–679.

48 Yoshimura A, Lien E, Ingalls RR. Cutting edge: recognition of Gram-positive bacterial cell wall components by the innate immune system occurs via Toll-like receptor 2. *J Immunol* 1999; **163**: 1–5.

49 Lien E, Selati TJ, Yoshimura A *et al*. Toll like receptor functions as a pattern recognition receptor for diverse bacterial products. *J Biol Chem* 1999; **274**: 33419–33425.

50 Brightbull HD, Libray DH, Krutzik SR *et al*. Host defense mechanisms triggered by microbial lipoproteins through toll-like receptors. *Science* 1999; **285**: 732–736.

51 Cario E, Podolsky DK. Differential alteration in intestinal epithelial cell expression of toll-like receptor 2 (TLR3) and TLR4 in inflammatory bowel disease. *Infect Immunol* 2000; **68**: 7012–7.

52 Dzarski R, Gupta G. Review: mammalian peptidoglycan recognition proteins [PGRPS] in innate immunity. *Innate Immunol* 2010; **16**: 168–174.

53 Sun C, Mathur P, Dupuis J *et al*. Peptidoglycan recognition proteins Pglyrp3 and Pglyrp4 are encoded from the epidermal differentiation complex and are candidate genes for the PSORS4 locus on chromosome 1q. *Hum Genet* 2006; **119**: 113–125.

54 Kainu K, Kivinen K, Zucchelli M *et al*. Association of psoriasis to PGLYRP and SPRR genes at PSORS4 locus on 1q shows heterogeneity between Finnish, Swedish and Irish families. *Exp Dermatol* 2009; **18**: 109–115.

55 Tonel G, Conrad C, Laggna *et al*. Cutting edge: A critical functional role for IL-23 in psoriasis. *J Immunol* 2010; **185**: 5688–5691.

56 Elliott HR, Tillin T, McArdle WL *et al*. Differences in smoking associated DNA methylation patterns in south Asians and Europeans. *Clin Epigenetics* 2014; **6**(1): 4. doi:10.1186/1868-7083-6-4.

57 Zamas AS, West AE. Epigenetics and the regulation of stress vulnerability and resilience. *Neuroscience* 2014; **264**: 157–170.

58 Nieratschker V, Batra A, Fallgatter A. Genetics and epigenetics of alcohol dependence. *J Mol Psychiatry* 2013; **1**: 11.

59 Arora P. Obesity genetics and epigenetics: dissecting causality. *Circ Cardiovasc Genet* 2014; **7**: 395–396.

60 Rodriguez E, Baurecht H, Wahn AF, *et al*. An integrated epigenetic and transcription analysis reveals distinct tissue-specific patterns of DNA methylation associated with atopic dermatitis. *J Invest Dermatol* 2014; **134**: 1873–1883.

CHAPTER 22

Liver disease: interactions with the intestinal microbiota

Katharina Brandl[1] and Bernd Schnabl[2]

[1] Skaggs School of Pharmacy, University of California, San Diego, United States
[2] University of California, San Diego, United States

22.1 Introduction

Previous sections of this book have extensively discussed that the mammalian intestine contains trillions of microorganisms that live in a symbiotic relationship with their host. The commensal microorganisms contribute to digestion and absorption of nutrients and occupy a protected space to prevent pathogenic bacteria to invade and flourish. In return, they gain a secure nutrient-rich environment. This symbiotic relationship requires a tight control, as the exposure of the host to the vast amount of bacteria poses a serious risk for disease development[1].

Disturbances of this symbiosis can contribute to the pathogenesis of liver disease, including non-alcoholic fatty liver disease (NAFLD) and alcoholic liver disease[2,3]. Several studies have shown that quantitative and qualitative changes in the microbiota are linked to liver disease. Dietary factors such as Western diet and alcohol are inducing changes in the gut microbiota. Changes in the microbiota can lead to bacterial overgrowth, intestinal inflammation and translocation of bacterial products to distant sites. However, other products released from bacteria might also promote liver injury and lead to disease.

In this review we focus on the disturbances of the microbiota leading to NAFLD and alcoholic liver disease, which are two of the most common chronic liver diseases in industrialized countries.

22.2 Non-alcoholic fatty liver disease

NAFLD is a spectrum of diseases with the important feature of hepatic steatosis, in the absence of other causes for hepatic fat accumulation[4]. NAFLD is the most prevalent liver disease in the US and is therefore considered a major health problem. Insulin resistance is a risk factor for NAFLD and contributes to the significant emergence of NAFLD[5]. NAFL, defined as the presence of hepatic steatosis with no evidence of hepatocellular injury in the form of ballooning of the hepatocytes[6], is

The Human Microbiota and Chronic Disease: Dysbiosis as a Cause of Human Pathology, First Edition.
Edited by Luigi Nibali and Brian Henderson.

generally a benign disease; however, it can progress to non-alcoholic steatohepatitis (NASH). NASH is defined as the presence of hepatic steatosis and inflammation with hepatocyte injury (ballooning) with or without fibrosis[6], and can lead to cirrhosis and hepatocellular carcinoma. Because approximately 30% of patients with NAFL develop NASH, it is of particular interest to understand the factors that drive the progression from NAFL to NASH. Whereas the accumulation of fat in the liver is well recognized as "first hit," several factors are required as additional stressors to result in sustained hepatocellular injury[7]. Alterations in the microbiota may, directly or indirectly, serve as additional stressors.

22.3 Qualitative and quantitative changes in the intestinal microbiota

Several studies have linked changes in the intestinal microbiota to NAFLD and NASH[8]. Small intestinal bacterial overgrowth has been demonstrated to coexist with NAFLD and NASH[9–11]. Although the molecular consequences of quantitative changes in the gut microbiome are not completely understood, a sequence of overgrowth, alterations in gut permeability, followed by increased translocation of bacterial products and an inflammatory response in the liver, has been proposed[8]. In contrast, qualitative changes in the microbiome have not been conclusively linked to NAFLD and NASH[12–14].

A study comparing stool samples from healthy subjects with obese NAFLD patients found a dominance of Lactobacillus species in obese NAFLD patients compared to healthy controls. Other members of the phylum Firmicutes were increased as well (Lachnospiraceae: *Robinsoniella, Roseburia* and *Dorea*)[13]. In a pediatric population, both obese- and biopsy-proven NASH patients showed an increased level of Bacteroidetes in fecal samples when compared with healthy controls. The only genus that was able to differentiate NASH patients from obese and healthy patients was the genus *Escherichia*, which was significantly elevated in NASH patients. In addition, obese children and children with NASH showed a decrease in the phylum Firmicutes compared with healthy controls[14]. In contrast, a study using biopsy-proven NASH, found a lower percentage of the phylum Bacteroidetes compared to healthy controls and patients with simple steatosis. Fecal *Clostridium coccoides* was higher in NASH patients compared to those with NAFL. No differences were found for bifidobacteria, Bacteroidetes, *Clostridium leptum*, *E. coli*, and total bacteria between the groups[12].

Studies investigating qualitative alterations in the microbiota associated with NAFLD show large variations and discrepancies and are not yet conclusive. Cohort size, differences within the cohorts and different methodologies might explain the discrepancies[8]. Also, changes in bacterial species might not be as relevant as changes in bacterial genes that are associated with the pathogenesis of NAFLD and NASH. Indeed, a direct link between bacterial genes and fat deposition in the liver has been proposed for obesity[15,16]. In obese animals and human subjects, certain bacterial genes have the capacity to harvest more energy from the diet. Distinct bacterial enzymes are capable of digesting and extracting calories from otherwise indigestible dietary polysaccharides. The extracted energy is subsequently deposited in host fat depots[17]. Furthermore, bacteria suppress fasting-induced adipocyte factor (FIAF), a lipoprotein lipase inhibitor, leading to increased LPL activity and increased accumulation of lipids in the liver[17,18]. Therefore, metagenomics or metatranscriptomics might provide evidence of a causal relationship between the gut microbiota and liver disease, including NAFL and NASH.

As mentioned above, only 30% of patients with steatosis progress to NASH. It is believed that certain stressors are required to trigger disease progression[7]. The most important factors associated with the gut microbiota are discussed in the next section.

22.4 Endotoxin

NAFLD is associated with altered intestinal tight junctions and increased intestinal permeability[9,14]. Although the underlying mechanism is not clear, Miele *et al.* showed increased permeability as a result of tight junction disruption in patients with NAFLD[9]. Studies in a pediatric population demonstrated similar findings, as children with NAFLD showed increased permeability, which correlated with disease severity[14]. Secondary to a dysfunction of the mucosal barrier, microbial products of the type known as pathogen-associated molecular patterns or PAMPs translocate from the intestinal lumen to extraintestinal organs, including the liver. Several studies in human subjects and animal models support a role of endotoxin for progression of NAFLD.

Children with NAFLD have significant higher serum levels of endotoxin than healthy controls[19]. Patients with NASH also showed endotoxemia[20]. Elevated endotoxin levels are associated with intestinal bacterial overgrowth and most of the studies correlate small intestinal bacterial overgrowth (SIBO) with increased endotoxin levels[10,11].

Although these studies in human subjects provide a correlation between endotoxin levels and liver disease, mechanistic explanations come from animal studies causatively linking endotoxin levels to liver disease. Microbial PAMPs bind to cellular receptors of the innate immune system to initiate an intracellular signaling cascade resulting in an inflammatory response. For example, LPS and bacterial DNA bind to Toll-like receptor (TLR)-4 and TLR9, respectively[21]. Mice deficient in TLR4, the adaptor molecule MD2 and TLR9 are resistant to NASH[22,23]. Therefore, sensing PAMPs via TLRs triggers an inflammatory response in the liver that is involved in disease progression.[24,25].

Recent studies also suggest a role for the inflammasome in the development of NASH[26]. The inflammasome participates in an inflammatory response and can be activated by endogenous and exogenous PAMPs and danger-associated molecular patterns (DAMPs)[27]. It has been recently recognized that the inflammasome is also required to maintain a stable microflora composition (see also chapter 15 by Yamazaki on the role of the inflammasome). Mice deficient in inflammasome components showed dysbiosis of the microbiota, particularly expansion of Prevotellaceae and Porphyromonadaceae families[27,28]. Dysbiosis resulted in colonic inflammation that was dependent on the chemokine CCL5. Translocation of the bacterial products further lead to abnormal accumulation of PAMPs in the portal circulation, exposing the liver to high concentrations and triggering an inflammatory response. In conclusion, a genetic disposition to dysbiosis can lead to intestinal inflammation and increased intestinal permeability. The presence of a leaky gut facilitates the crosstalk between gut and liver, allowing microbial products to provoke an inflammatory response in the liver, which drives the progression of simple hepatic steatosis to NASH.

22.5 Ethanol

Another factor that has been correlated with NAFLD progression is ethanol. Ethanol is an endogenous product from the intestinal microflora. *Escherichia coli* are alcohol producing bacteria and the predominance of *E.coli* has been linked to NASH[14]. Children with NASH showed higher blood ethanol levels compared to pediatric patients with NAFL. Increased blood ethanol levels could serve as additional stressor in the liver. The mechanism most likely involves the production of reactive oxygen species triggering liver inflammation. Animal studies support the involvement of ethanol in NAFLD[29,30]. Alcohol is also known to increase gut permeability, leading to increased translocation of bacteria and inflammation[31]. Therefore, this data suggests that a microbiota rich in ethanol-producing bacteria might facilitate NAFLD progression.

22.6 Choline

Choline is another metabolite that has been linked to liver disease. Feeding a choline-deficient diet to rodents has been used as an animal model for NASH[32]. However, until recently it was not clear whether choline deficiency occurs under pathophysiological conditions. A dysbiotic microbiota can convert dietary choline into methylamines, which are eliminated via the urine. Reduced levels of choline lead to a decrease in phosphatidylcholine synthesis that is further necessary for very-low-density lipoprotein (VLDL) formation and secretion from the liver. Decreased assembly and secretion of VLDL induces accumulation of fat in the liver, which in turn exacerbates steatosis and NAFLD[33]. Methylamines induce hepatotoxicity and hepatocarcinogenesis in rats, thus providing an additional mechanism of liver toxicity[34]. Consuming a choline-deficient diet in humans also leads to the development of fatty liver. However, a single-nucleotide polymorphism in the promoter region of PEMT (rs12325817), which affects *de novo* synthesis of phosphatidylcholine, is required for the development of a fatty liver[35]. In conclusion, nutrient imbalance and the presence of certain microbiota can serve as an important trigger for NAFLD and NASH.

Taken together, dietary changes such as a Western diet can induce qualitative and quantitative changes in the intestinal microbiota. Once dysbiosis is established, altered gut bacteria contribute to NAFLD in several ways: 1. Increased intestinal permeability releases microbial products from the intestinal lumen to the liver; PAMPs then trigger an inflammatory response. 2. Bacteria produce ethanol that might change intestinal permeability and induce an inflammatory response via oxygen radicals in the liver. 3. Bacteria metabolize choline that is required for VLDL formation, subsequently inducing fat accumulation in the liver. However, one cannot exclude the possibility that liver disease changes the microbiota, which in turn affects and damages the liver.

22.7 Alcoholic liver disease

Alcoholic liver disease is characterized by steatosis, which can progress to steato-hepatitis, fibrosis and cirrhosis[36,37]. Once it has progressed to cirrhosis, effective treatment is not available and transplantation is advised. Although chronic alcohol abuse results in liver steatosis in the majority of cases, not every heavy drinker develops advanced alcoholic liver disease beyond simple steatosis. Risk factors for developing progressive alcoholic liver disease include the amount of alcohol consumed, drinking pattern, gender, obesity, genetic factors and the presence of coexisting liver diseases such as chronic hepatitis C infection[38].

In addition, one very important factor that contributes to the progression of alcoholic liver disease is the gut-liver axis[39]. Alterations in the microbiota can lead to intestinal inflammation and increased intestinal permeability. Translocation of bacterial products into the systemic circulation exposes the liver to these toxins, triggering liver inflammation and disease progression. In recent years, changes in the intestinal microbiota and their functional consequences for alcoholic liver disease have begun to be elucidated.

22.7.1 Qualitative and quantitative changes in the intestinal microbiome

Small intestinal overgrowth is one important pathogenic risk factor for alcoholic liver disease. Intestinal bacterial overgrowth has been shown in animal models and in patients with chronic alcohol abuse[40–42].

In our own studies, we used an intragastric infusion system of ethanol, the so-called Tsukamoto-French model. to detect changes in the microbiome. Mice receive the same liquid diet at the same feeding rate with identical caloric intake except that one group receives ethanol as part of the diet and control mice receive dextrose. Our studies demonstrate a higher abundance of aerobic and anaerobic bacteria in almost the entire intestinal tract, using conventional culture techniques. The increased amount of total bacteria was further confirmed by culture-independent PCR[43].

The important role of intestinal bacterial overgrowth is further demonstrated by the fact that intestinal decontamination using non-absorbable antibiotics ameliorates alcoholic liver disease in mice and rats[38,44]. Long-term administration of non-absorbable antibiotics together with ethanol significantly reduced AST levels and changes in all components of the liver pathology score, including steatosis, inflammation and necrosis[44]. These data identify products derived from the microbiota as causative factors for alcohol-induced liver disease.

Besides quantitative changes (bacterial overgrowth), we also observed qualitative changes, particularly a decrease in Firmicutes and an increase in Bacteriodetes. Probiotic bacteria such as *Lactobacillus, Pediococcus, Leuconostoc* and *Lactococcus* were significantly suppressed. In another study that investigated chronic alcohol administration in rats, dysbiosis was also characterized by a suppression of probiotic bacteria such as *Lactobacillus*[45]. This is of particular interest as several studies have shown a benefit in Lactobacillus supplementation in animal models[45–47] and patients with chronic alcohol abuse[48].

Very few human studies have demonstrated qualitative changes in the microbiota in alcoholics. One study analyzed the mucosa-associated colonic microbiome and revealed decreased numbers of Bacteroidetes and higher levels of

Proteobacteria[49]. The alterations correlated also with higher endotoxin levels. This study mainly focused on the microbiome composition and suggests that changes in the microbiome correlate with endotoxemia in alcoholics. Interestingly, not all alcoholics develop gut leakiness. A subgroup of chronic alcohol abusers with high intestinal permeability showed dysbiosis, which is characterized by a decreased abundance in the Ruminococcaceae family (*Ruminococcus, Faecalibacterium, Subdoligranulum, Oscillibacter*, and *Anaerofilum*), and an increased abundance in the Lachnospiraceae family (*Dorea*) and the genus *Blautia*[50].

However, studies on microbiota function might be more useful to differentiate alcoholics from healthy controls and help to understand the causal relationship between the microbiome and liver pathogenesis. Studies using metagenomic, metatranscriptomic and metabolomic techniques in patients with alcoholic liver disease are still limited. The profile of fecal metabolites is altered in chronic alcoholics with a barrier dysfunction, but the functional consequences of changes in these metabolites are currently unknown[50].

The exact mechanism for bacterial overgrowth and dysbiosis are unknown. Different factors contribute to intestinal homeostasis, the motility of the GI tract[51,52], changes in gastric acid secretion[53] and, of major importance, the intestinal immune system.

Our studies have suggested that chronic ethanol exposure to mice changes the gene and protein expression of Reg3b and Reg3g[43]. These molecules have been implicated in intestinal homeostasis[54–57]. Reg3g has bactericidal activity and targets gram-positive bacteria by binding to the peptidoglycan layer[58,59]. Decreased expression of Reg3g and Reg3b might be one factor contributing to the qualitative and quantitative changes of the microbiota observed in our study. Administration of prebiotics was able to restore the Reg3g levels and decrease bacterial overgrowth and finally ameliorate alcoholic steatohepatitis. Down-regulation of Reg3g has also been shown in duodenal biopsies of alcoholic patients[43]. Further functional studies are needed to elucidate the contribution of Reg3 molecules to intestinal dysbiosis and/or bacterial translocation.

Besides Reg3 proteins, there might be other factors contributing to changes in the gut microbiome. How much ethanol and its metabolites acetaldehyde and acetate directly affect bacteria has not conclusively been addressed. The consequences of dysbiosis following chronic alcohol exposure are not completely understood at a molecular level. However, several factors have been described to follow dysbiosis: intestinal inflammation, alterations in gastrointestinal permeability and increased endotoxin levels.

22.7.2 Contribution of dysbiosis to alcoholic liver disease

One consequence of dysbiosis is intestinal inflammation. Feeding mice with alcohol for eight weeks increases small intestinal TNF gene expression significantly compared to wild-type controls. TNF mRNA expression was also significantly higher in duodenal biopsies from alcoholics. These data suggest that chronic alcohol consumption leads to dysbiosis followed by increased TNF production in mice and men[60]. It is not clear how dysbiosis leads to intestinal inflammation and increased TNF production. We have recently demonstrated that alcohol-associated changes to the intestinal metagenome and metabolome are characterized by reduced bacterial synthesis of saturated long-chain fatty acids in mice. Since saturated long-chain fatty acids are metabolized by commensal *Lactobacillus* and promote their growth, reduced levels of saturated long-chain fatty acids might result in lower amounts of intestinal lactobacilli. Probiotic *Lactobacillus* is known to suppress intestinal inflammation. Thus, it is possible that lower enteric lactobacilli favor intestinal inflammation and disruption of the gut barrier[61]. Further studies in patients or mice with alcoholic liver disease are necessary to investigate whether this is mediated through which microbial metabolites or products.

What are the consequences of intestinal inflammation in the setting of chronic alcohol consumption? One could hypothesize that intestinal inflammation directly contributes to a leaky gut and causes translocation of bacterial products contributing to liver disease pathogenesis. TNF is a major pro-inflammatory cytokine and our studies recently identified TNF-receptor 1 on intestinal epithelial cells as master regulator of intestinal barrier function[60]. Following chronic ethanol administration in mice, TNF is secreted from monocytes and macrophages in the small intestine and binds to the TNF-receptor 1 on enterocytes. Myosin light-chain kinase (MLCK) is phosphorylated and activated downstream of the TNF-receptor 1. MLCK then redistributes tight junction proteins, which leads to increased gut permeability. Since MLCK-deficient mice are only partially protected from gut barrier dysfunction and alcoholic liver disease, other mediators likely contribute to this process. Alcohol and its metabolite acetaldehyde both can induce a direct disruption of intestinal epithelial tight junctions[62–64]. The mechanism most likely involves an inhibition of the regulation of the phosphorylating-dephosphorylating balance of tight junction proteins[31]. Ethanol can be metabolized to acetaldehyde not only by the liver but also by the intestinal microbiota. The relative contribution of each metabolic pathway requires further investigation. However, one might speculate that changes in the microbiota following alcohol administration could contribute to alterations in ethanol metabolism. Another contributor to changes in mucosal barrier are mast cells that can be stimulated by acetaldehyde and release several mediators that might contribute to epithelial cell pathophysiology[65].

Several studies have shown a correlation between changes in intestinal permeability and alcoholic liver damage in animal models and in humans[66–70]. Disruption of the intestinal barrier followed by translocation of bacterial products into the systemic circulation is thought to be central to the pathogenesis of liver disease. There is a direct relationship between the intraluminal load of enteric bacteria and the amount of translocated bacterial products in the setting of a leaky gut[36,43]. There is strong evidence for endotoxin as a marker for bacterial translocation triggering further liver damage. Elevated endotoxin levels have been found in animal models of alcoholic liver disease and in alcoholics. Endotoxemia can not only lead to severe systemic inflammation in patients with alcoholic liver disease, it is also implicated in the progression of alcoholic liver disease to fibrosis and cirrhosis. One mechanism is the activation of innate immune receptors by bacterial products (such as LPS) that result in the activation of several transcription factors and the production of various molecules involved in the inflammatory host response. Animal models using genetically manipulated mice support an important role for TLR4. TLR4 mutant mice (C3H/HeJ) show significant less ethanol-induced hepatic steatosis, inflammation and necrosis than wild-type mice[71].

The question remains, which cell type is responsible for mediating the TLR4 effects? TLR4 is expressed on several liver cells including hepatic stellate cells and

Kupffer cells, and both cells are involved in liver pathogenesis[72]. Mice with ablated Kupffer cells are protected from alcohol-induced liver injury[73]. Hepatic stellate cells are the major producer of extracellular matrix proteins in the fibrotic liver. In bone-marrow chimeric mice, both TLR4 expression on bone-marrow-derived (e.g. Kupffer cells) and on non-bone-marrow-derived (e.g. hepatic stellate cells) cells was required for liver disease development after chronic alcohol treatment[72]. However, one cannot exclude the importance of TLR4 expression on non-liver cells (e.g. intestinal cells) to contribute to disease progression. TLR4 is expressed on e.g. epithelial cells (non-bone-marrow-derived) and dendritic cells (bone-marrow derived) in the intestinal tract. As bacterial overgrowth is present in alcoholic liver disease, TLR4 on epithelial cells could directly sense bacterial products and contribute to the immune response. Dendritic cells in the lamina propria extend across the epithelium and can also sample bacteria in the lumen[74]. Therefore, the involvement of intestinal epithelial cells and dendritic cells in alcoholic liver disease requires future studies. Besides TLR4, other TLRs and ligands might be involved in alcoholic liver disease as well. TLR9, which recognizes bacterial DNA is a promising candidate, yet its role is not yet fully elucidated.

Taken together, in alcoholic liver disease, once dysbiosis has been established, the sequence of events most likely involves intestinal inflammation, increased gut permeability, translocation of bacterial products to the liver, activation of the hepatic innate immune system and progression of liver inflammation and pathology. Although the initial trigger for NAFLD/NASH is different from alcoholic liver disease (ethanol), the consequences of dysbiosis appear similar.

TAKE-HOME MESSAGE

- Several studies linked changes in the composition of the intestinal microbiota with liver disease.
- Metagenomics, metatranscriptomics and metabolomics provide functional links between the gut microbiome and liver disease.
- Once changes in the microbiota occur, inflammation can develop, followed by increased gut permeability, bacterial translocation and liver inflammation and leading to liver pathology.

References

1 Hooper LV, Littman DR, Macpherson AJ. Interactions between the microbiota and the immune system. *Science* 2012; **336**: 1268–73.
2 Schnabl B. Linking intestinal homeostasis and liver disease. *Curr Opin Gastroenterol* 2013; **29**: 264–70.
3 Abu-Shanab A, Quigley EM. The role of the gut microbiota in nonalcoholic fatty liver disease. *Nat Rev Gastroenterol Hepatol* 2010; **7**: 691–701.
4 Yeh MM, Brunt EM. Pathological features of fatty liver disease. *Gastroenterology* 2014.
5 Sun Z, Lazar MA. Dissociating fatty liver and diabetes. *Trends Endocrinol Metab* 2013; **24**: 4–12.
6 Chalasani N, Younossi Z, Lavine JE *et al.* The diagnosis and management of non-alcoholic fatty liver disease: practice Guideline by the American Association for the Study of Liver Diseases, American College of Gastroenterology, and the American Gastroenterological Association. *Hepatology* 2012; **55**: 2005–23.
7 Basaranoglu M, Basaranoglu G, Senturk H. From fatty liver to fibrosis: a tale of "second hit". *World J Gastroenterol* 2013; **19**: 1158–65.

8 Schnabl B, Brenner DA. Interactions between the intestinal microbiome and liver diseases. *Gastroenterology* 2014; **146**: 1513–24.

9 Miele L, Valenza V, La Torre G *et al.* Increased intestinal permeability and tight junction alterations in nonalcoholic fatty liver disease. *Hepatology* 2009; **49**: 1877–87.

10 Sabate JM, Jouet P, Harnois F *et al.* High prevalence of small intestinal bacterial overgrowth in patients with morbid obesity: a contributor to severe hepatic steatosis. *Obes Surg* 2008; **18**: 371–7.

11 Wigg AJ, Roberts-Thomson IC, Dymock RB *et al.* The role of small intestinal bacterial overgrowth, intestinal permeability, endotoxaemia, and tumour necrosis factor alpha in the pathogenesis of non-alcoholic steatohepatitis. *Gut* 2001; **48**: 206–11.

12 Mouzaki M, Comelli EM, Arendt BM *et al.* Intestinal microbiota in patients with nonalcoholic fatty liver disease. *Hepatology* 2013; **58**: 120–7.

13 Raman M, Ahmed I, Gillevet PM *et al.* Fecal microbiome and volatile organic compound metabolome in obese humans with nonalcoholic fatty liver disease. *Clin Gastroenterol Hepatol* 2013; **11**: 868–75 e3.

14 Zhu L, Baker SS, Gill C *et al.* Characterization of the gut microbiome in non-alcoholic steatohepatitis (NASH) patients: a connection between endogenous alcohol and NASH. *Hepatology* 2013; **57**: 601–9.

15 Ley RE, Backhed F, Turnbaugh P *et al.* Obesity alters gut microbial ecology. *Proc Natl Acad Sci U S A* 2005; **102**: 11070–5.

16 Ley RE, Turnbaugh PJ, Klein S *et al.* Microbial ecology: human gut microbes associated with obesity. *Nature* 2006; **444**: 1022–3.

17 Backhed F, Ding H, Wang T *et al.* The gut microbiota as an environmental factor that regulates fat storage. *Proc Natl Acad Sci U S A* 2004; **101**: 15718–23.

18 Backhed F, Manchester JK, Semenkovich CF *et al.* Mechanisms underlying the resistance to diet-induced obesity in germ-free mice. *Proc Natl Acad Sci USA* 2007; **104**: 979–84.

19 Alisi A, Manco M, Devito R *et al.* Endotoxin and plasminogen activator inhibitor-1 serum levels associated with nonalcoholic steatohepatitis in children. *J Pediatr Gastroenterol Nutr* 2010; **50**: 645–9.

20 Farhadi A, Gundlapalli S, Shaikh M *et al.* Susceptibility to gut leakiness: a possible mechanism for endotoxaemia in non-alcoholic steatohepatitis. *Liver Int* 2008; **28**: 1026–33.

21 Beutler B, Hoebe K, Du X *et al.* How we detect microbes and respond to them: the Toll-like receptors and their transducers. *J Leukoc Biol* 2003; **74**: 479–85.

22 Csak T, Velayudham A, Hritz I *et al.* Deficiency in myeloid differentiation factor-2 and toll-like receptor 4 expression attenuates nonalcoholic steatohepatitis and fibrosis in mice. *Am J Physiol Gastrointest Liver Physiol* 2011; **300**: G433–41.

23 Miura K, Kodama Y, Inokuchi S *et al.* Toll-like receptor 9 promotes steatohepatitis by induction of interleukin-1beta in mice. *Gastroenterology* 2010; **139**: 323–34 e7.

24 Rivera CA, Adegboyega P, van Rooijen N *et al.* Toll-like receptor-4 signaling and Kupffer cells play pivotal roles in the pathogenesis of non-alcoholic steatohepatitis. *J Hepatol* 2007; **47**: 571–9.

25 Roh YS, Seki E. Toll-like receptors in alcoholic liver disease, non-alcoholic steatohepatitis and carcinogenesis. *J Gastroenterol Hepatol* 2013; **28 Suppl 1**: 38–42.

26 Henao-Mejia J, Elinav E, Jin C *et al.* Inflammasome-mediated dysbiosis regulates progression of NAFLD and obesity. *Nature* 2012; **482**: 179–85.

27 Henao-Mejia J, Elinav E, Thaiss CA *et al.* Role of the intestinal microbiome in liver disease. *J Autoimmun* 2013; **46**: 66–73.

28 Elinav E, Strowig T, Kau AL *et al.* NLRP6 inflammasome regulates colonic microbial ecology and risk for colitis. *Cell* 2011; **145**: 745–57.

29 Cope K, Risby T, Diehl AM. Increased gastrointestinal ethanol production in obese mice: implications for fatty liver disease pathogenesis. *Gastroenterology* 2000; **119**: 1340–7.

30 Nair S, Cope K, Risby TH *et al.* Obesity and female gender increase breath ethanol concentration: potential implications for the pathogenesis of nonalcoholic steatohepatitis. *Am J Gastroenterol* 2001; **96**: 1200–4.

31 Rao RK, Seth A, Sheth P. Recent advances in alcoholic liver disease I. Role of intestinal permeability and endotoxemia in alcoholic liver disease. *Am J Physiol Gastrointest Liver Physiol* 2004; **286**: G881–4.

32 Blumberg H, McCollum EV. The prevention by choline of liver cirrhosis in rats on high fat, low protein diets. *Science* 1941; **93**: 598–9.

33 Dumas ME, Barton RH, Toye A *et al.* Metabolic profiling reveals a contribution of gut microbiota to fatty liver phenotype in insulin-resistant mice. *Proc Natl Acad Sci USA* 2006; **103**: 12511–6.

34 Lin JK, Ho YS. Hepatotoxicity and hepatocarcinogenicity in rats fed squid with or without exogenous nitrite. *Food Chem Toxicol* 1992; **30**: 695–702.

35 Spencer MD, Hamp TJ, Reid RW *et al.* Association between composition of the human gastrointestinal microbiome and development of fatty liver with choline deficiency. *Gastroenterology* 2011; **140**: 976–86.

36 Hartmann P, Chen WC, Schnabl B. The intestinal microbiome and the leaky gut as therapeutic targets in alcoholic liver disease. *Front Physiol* 2012; **3**: 402.

37 Yan AW, Schnabl B. Bacterial translocation and changes in the intestinal microbiome associated with alcoholic liver disease. *World J Hepatol* 2012; **4**: 110–8.

38 O'Shea RS, Dasarathy S, McCullough AJ. Alcoholic liver disease. *Hepatology* 2010; **51**: 307–28.

39 Chen P, Schnabl B. Host-microbiome interactions in alcoholic liver disease. *Gut Liver* 2014; **8**: 237–41.

40 Bode C, Kolepke R, Schafer K *et al.* Breath hydrogen excretion in patients with alcoholic liver disease — evidence of small intestinal bacterial overgrowth. *Z Gastroenterol* 1993; **31**: 3–7.

41 Bode JC, Bode C, Heidelbach R *et al.* Jejunal microflora in patients with chronic alcohol abuse. *Hepatogastroenterology* 1984; **31**: 30–4.

42 Casafont Morencos F, de las Heras Castano G, Martin Ramos L *et al.* Small bowel bacterial overgrowth in patients with alcoholic cirrhosis. *Dig Dis Sci* 1996; **41**: 552–6.

43 Yan AW, Fouts DE, Brandl J *et al.* Enteric dysbiosis associated with a mouse model of alcoholic liver disease. *Hepatology* 2011; **53**: 96–105.

44 Adachi Y, Moore LE, Bradford BU *et al.* Antibiotics prevent liver injury in rats following long-term exposure to ethanol. *Gastroenterology* 1995; **108**: 218–24.

45 Mutlu E, Keshavarzian A, Engen P *et al.* Intestinal dysbiosis: a possible mechanism of alcohol-induced endotoxemia and alcoholic steatohepatitis in rats. *Alcohol Clin Exp Res* 2009; **33**: 1836–46.

46 Forsyth CB, Farhadi A, Jakate SM *et al.* Lactobacillus GG treatment ameliorates alcohol-induced intestinal oxidative stress, gut leakiness, and liver injury in a rat model of alcoholic steatohepatitis. *Alcohol* 2009; **43**: 163–72.

47 Nanji AA, Khettry U, Sadrzadeh SM. Lactobacillus feeding reduces endotoxemia and severity of experimental alcoholic liver (disease). *Proc Soc Exp Biol Med* 1994; **205**: 243–7.

48 Kirpich IA, Solovieva NV, Leikhter SN *et al.* Probiotics restore bowel flora and improve liver enzymes in human alcohol-induced liver injury: a pilot study. *Alcohol* 2008; **42**: 675–82.

49 Mutlu EA, Gillevet PM, Rangwala H *et al.* Colonic microbiome is altered in alcoholism. *Am J Physiol Gastrointest Liver Physiol* 2012; **302**: G966–78.

50 Leclercq S, Matamoros S, Cani PD *et al.* Intestinal permeability, gut-bacterial dysbiosis, and behavioral markers of alcohol-dependence severity. *Proc Natl Acad Sci USA* 2014; **111**: E4485–93.

51 Madrid AM, Hurtado C, Venegas M *et al.* Long-term treatment with cisapride and antibiotics in liver cirrhosis: effect on small intestinal motility, bacterial overgrowth, and liver function. *Am J Gastroenterol* 2001; **96**: 1251–5.

52 Gupta A, Dhiman RK, Kumari S *et al.* Role of small intestinal bacterial overgrowth and delayed gastrointestinal transit time in cirrhotic patients with minimal hepatic encephalopathy. *J Hepatol* 2010; **53**: 849–55.

53 Shindo K, Machida M, Miyakawa K *et al.* A syndrome of cirrhosis, achlorhydria, small intestinal bacterial overgrowth, and fat malabsorption. *Am J Gastroenterol* 1993; **88**: 2084–91.

54 Brandl K, Plitas G, Mihu CN *et al.* Vancomycin-resistant enterococci exploit antibiotic-induced innate immune deficits. *Nature* 2008; **455**: 804–7.

55 Brandl K, Plitas G, Schnabl B *et al.* MyD88-mediated signals induce the bactericidal lectin RegIII gamma and protect mice against intestinal Listeria monocytogenes infection. *J Exp Med* 2007; **204**: 1891–900.

56 Vaishnava S, Behrendt CL, Ismail AS *et al.* Paneth cells directly sense gut commensals and maintain homeostasis at the intestinal host-microbial interface. *Proc Natl Acad Sci USA* 2008; **105**: 20858–63.

57 Vaishnava S, Yamamoto M, Severson KM *et al.* The antibacterial lectin RegIIIgamma promotes the spatial segregation of microbiota and host in the intestine. *Science* 2011; **334**: 255–8.

58 Cash HL, Whitham CV, Behrendt CL *et al.* Symbiotic bacteria direct expression of an intestinal bactericidal lectin. *Science* 2006; **313**: 1126–30.

59 Mukherjee S, Partch CL, Lehotzky RE *et al.* Regulation of C-type lectin antimicrobial activity by a flexible N-terminal prosegment. *J Biol Chem* 2009; **284**: 4881–8.

60 Chen P, Starkel P, Turner JR *et al.* Dysbiosis-induced intestinal inflammation activates TNFRI and mediates alcoholic liver disease in mice. *Hepatology* 2014.

61 Chen P, Torralba M, Tan J *et al.* Supplementation of saturated long-chain fatty acids maintains intestinal eubiosis and reduces ethanol-induced liver injury in mice. *Gastroenterology* 2014.

62 Wang Y, Kirpich I, Liu Y *et al.* Lactobacillus rhamnosus GG treatment potentiates intestinal hypoxia-inducible factor, promotes intestinal integrity and ameliorates alcohol-induced liver injury. *Am J Pathol* 2011; **179**: 2866–75.

63 Rao RK. Acetaldehyde-induced barrier disruption and paracellular permeability in Caco-2 cell monolayer. *Methods Mol Biol* 2008; **447**: 171–83.

64 Seth A, Basuroy S, Sheth P *et al.* L-Glutamine ameliorates acetaldehyde-induced increase in para-cellular permeability in Caco-2 cell monolayer. *Am J Physiol Gastrointest Liver Physiol* 2004; **287**: G510–7.

65 Ferrier L, Berard F, Debrauwer L *et al.* Impairment of the intestinal barrier by ethanol involves enteric microflora and mast cell activation in rodents. *Am J Pathol* 2006; **168**: 1148–54.

66 Choudhry MA, Fazal N, Goto M *et al.* Gut-associated lymphoid T cell suppression enhances bacterial translocation in alcohol and burn injury. *Am J Physiol Gastrointest Liver Physiol* 2002; **282**: G937–47.

67 Keshavarzian A, Choudhary S, Holmes EW *et al.* Preventing gut leakiness by oats supplementation ameliorates alcohol-induced liver damage in rats. *J Pharmacol Exp Ther* 2001; **299**: 442–8.

68 Mathurin P, Deng QG, Keshavarzian A *et al.* Exacerbation of alcoholic liver injury by enteral endo-toxin in rats. *Hepatology* 2000; **32**: 1008–17.

69 Lambert JC, Zhou Z, Wang L *et al.* Prevention of alterations in intestinal permeability is involved in zinc inhibition of acute ethanol-induced liver damage in mice. *J Pharmacol Exp Ther* 2003; **305**: 880–6.

70 Schmidt KL, Henagan JM, Smith GS *et al.* Effects of ethanol and prostaglandin on rat gastric mucosal tight junctions. *J Surg Res* 1987; **43**: 253–63.

71 Uesugi T, Froh M, Arteel GE *et al.* Toll-like receptor 4 is involved in the mechanism of early alcohol-induced liver injury in mice. *Hepatology* 2001; **34**: 101–8.

72 Inokuchi S, Tsukamoto H, Park E *et al.* Toll-like receptor 4 mediates alcohol-induced steatohepatitis through bone marrow-derived and endogenous liver cells in mice. *Alcohol Clin Exp Res* 2011; **35**: 1509–18.

73 Enomoto N, Ikejima K, Bradford BU *et al.* Role of Kupffer cells and gut-derived endotoxins in alcoholic liver injury. *J Gastroenterol Hepatol* 2000; **15 Suppl**: D20–5.

74 Chieppa M, Rescigno M, Huang AY *et al.* Dynamic imaging of dendritic cell extension into the small bowel lumen in response to epithelial cell TLR engagement. *J Exp Med* 2006; **203**: 2841–52.

CHAPTER 23

The gut microbiota: a predisposing factor in obesity, diabetes and atherosclerosis

Frida Fåk

Lund University, Lund, Sweden

23.1 Introduction

The cluster of pathologies comprising the metabolic syndrome (MetS) includes increased waist circumference, hyperglycemia, elevated blood pressure and hyperlipidemia. With time, these conditions present a major risk of developing obesity, type 2 diabetes and atherosclerosis. To date, treatment mainly includes symptom management, and effective prevention strategies are largely lacking. With the exciting discovery of an "obesogenic" microbiota in recent years, a new field of research has emerged, with attempts to use microbial manipulations to modulate our gut microbiome as a new preventive and therapeutic approach for different aspects of the MetS. In this chapter, the scientific rationale behind this exciting new research area is presented.

23.2 The "obesogenic" microbiota: evidence from animal models

In the years 2004–2009, Bäckhed, Ley, Turnbaugh, Gordon and colleagues presented ground-breaking studies on the role of the gut microbiota in host energy metabolism and proposed the hypothesis that obesity alters the composition of bacteria in the gut[1–4]. Secondly, with a series of elegant experiments using germ-free mice, they demonstrated that an "obesogenic" microbiota, i.e. bacteria from obese mice, can be transferred to a germ-free recipient, resulting in increased adiposity as compared to transfer of a "lean" microbiota from lean mice. Furthermore, they demonstrated the pivotal role of the gut microbiota for host metabolism by reporting that germ-free mice were protected from diet-induced obesity. Mechanisms involved included identification of microbiota-induced suppressed expression of angiopoietin-like factor 4 — a lipoprotein lipase inhibitor — in the intestine, leading to increased storage of fat in adipose tissue. Other mechanisms have since been reported, with gut bacteria affecting secretion of satiety hormones, liver lipogenesis and energy extraction from the diet.

23.3 The "obesogenic" microbiota in humans

In humans, clear shifts in the gut microbiome can be observed when comparing obese and lean individuals, with increased *Firmicutes* to *Bacteroides* ratio in obese individuals, at the phylum level of bacterial taxonomy[4]. However, such studies do not provide evidence of causality, i.e. a distinction between cause and consequence when observing gut microbiota differences between lean and obese people. Nutrient content in the diet directly affects the gut microbiome, where high-energy diets increase the ratio of *Firmicutes* to *Bacteroides* in humans[5]. Thus, a diet high in energy will contribute to development of obesity not only due to the energy content in the diet *per se*, but also through maintaining an "obesogenic" gut microbiota[6]. Researchers found that a 20% increase in the abundance of *Firmicutes* resulted in increased energy absorption of approximately 150 kcal, while 20% increase in *Bacteroides* led to 150 kcal less energy absorption from the diet in humans[5]. However, this association was observed only in lean individuals, indicating that obese individuals may respond differently to energy content in the diet or that the effects were dependent on the diet regularly ingested before the experiment.

It is clear that a high-fat diet induces an "obese" microbiota independent of body weight state and the altered gut microbiota in obese individuals is not merely a consequence of the obese state[6]. Together with mechanistic causality experiments using animal models, it is likely that the gut microbiota contributes to obese states. This is an important point, as it opens up possibilities for preventing and treating obesity using microbial-based strategies.

23.4 A leaky gut contributing to inflammation and adiposity

The monolayer of epithelial cells lining the gut mucosa has the delicate dual function of being an efficient absorptive layer to nutrients, whilst still maintaining a tight barrier to invading pathogens. This is accomplished through several specific features of the intestinal epithelium: first, the intestine hosts 70–90% of the body's immune cells, which together with the extensive enteric nervous system closely monitor events in the gut. Second, the junctions between epithelial cells consist of an intricate collection of proteins comprising the so-called tight junctions, which can be regulated by the cells themselves but are also affected by bacteria. We have covered the implications of a "leaky" gut with regards to hepatic disease (see chapter 22 by Brandl and Schnabl). Similarly, in obesity, these junctions appear leaky, as an increased influx of bacterial cell wall lipopolysaccharides (LPS) is observed as compared to lean individuals[7–10]. LPS is a component of Gram-negative bacteria cell walls and is pro-inflammatory through activation of Toll-like receptor 4 (TLR4) on antigen-presenting cells located both in the gut and in other tissues in the body. The low-grade systemic inflammation observed in obese individuals can be ascribed partly to increased LPS levels in the blood, but also to an increased secretion of pro-inflammatory cytokines from adipose tissue. Interestingly, a high-fat diet directly increases LPS blood levels, as LPS can form and bind to micelles during fat absorption and are therefore shuttled into the body from the gut[7]. Furthermore, systemic inflammation induced by LPS has been shown in mice to cause increased adiposity and body weight and insulin resistance[11,12]. As bacteria in the gut can affect tight junctions, gut hormones (glucacon-like peptide 2) as well as inflammatory signaling in the gut, it is not surprising that the microbial ecology in the gut affects host metabolism[12].

23.5 Obesity-proneness: mediated by the gut microbiota?

In both humans and rodents, resistance to diet-induced obesity is occasionally observed, but the reasons remain unclear why the same amount of energy ingested does not result in the same weight increase in all individuals. In obesity-prone and obesity-resistant mice fed a high-fat diet, fecal carbohydrate calories differed, whereas fecal fat or protein calories did not differ[13]. This surprisingly indicates that lean mice display decreased carbohydrate absorption from the diet as compared to obese mice, despite similar caloric intake[13]. Carbohydrate absorption is greatly affected by gut microbial activities, and the authors did indeed observe differences in microbial composition between obesity-prone and obesity-resistant mice. Other studies have also demonstrated differences in the gut microbiota between obesity-prone and obesity-resistant rodents, and have shown that development of inflammation via increased intestinal permeability was associated with both hyperphagia and obesity development[14]. Further, the *Enterobacteriales* order was increased in the obesity-prone rats[14]. For unknown reasons, obesity-prone rats respond to high-fat diets by developing inflammation, perhaps due to specific shifts in the gut microbiota, leading to increased energy intake. These differences in response to diet could be partly mediated by host genetics, but this issue requires further elucidation. The presence of obesity-prone and obesity-resistant individuals exemplifies yet another mechanism by which bacteria can affect energy metabolism and the complex nature of host-microbiome interactions.

23.6 Bacterial metabolites provide a link between bacteria and host metabolism

Chapter 6 by Bermudez described how gut bacteria degrade dietary components reaching the colon, giving rise to metabolites such as short-chain fatty acids (SCFA), which are formed during bacterial digestion of indigestible carbohydrates (e.g. dietary fiber). These bacterial metabolites may play an important role for host physiology, as they can be utilized as energy by colon cells, as well as be absorbed into the blood stream[15]. Thus, a diet low in dietary fiber may affect colonic health. Lately, the systemic effects of bacterial metabolites are also being explored. Lean mice resistant to diet-induced obesity display decreased amount of fecal short-chain fatty acids[13]. In addition, mice who underwent weight-loss surgery had a decreased amount of SCFA[16]. This was also found in humans, where fecal levels of SCFA were increased in obese individuals as compared to lean people[17]. It should be noted that not only the total level of SCFA matters for host metabolism, but also the proportion of individual fatty acids. Acetate, propionate and butyrate are the most important acids in colon. Acetate is involved in lipogenesis and propionate has been shown to inhibit liver lipogenesis involving inhibition of acetate conversion into lipids[16]. SCFA have also been shown to regulate satiety hormones[16] and intestinal gluconeogenesis[18]. Hence, bacterial metabolites provide an important mechanistic link between the activity of the gut microbiota and microbiota-induced effects on adiposity and body weight regulation.

23.7 Fecal microbiota transplants: can we change our gut bacterial profiles?

Fecal microbiota transplant (FMT) experiments performed in mice have given an intriguing foundation for a microbial-based therapy of obesity in humans. Initial trials with fecal microbiota transplants have been conducted in humans, with varying results (more details are given in chapter 31). A study in 18 males with the metabolic syndrome showed that FMT of feces from lean donors improved insulin sensitivity six weeks after transplantation but did not affect body weight[19]. In addition, the effects were present only short-term and dissipated over time. Fecal microbiota changes after FMT included 2.5-fold increases in *Roseburia intestinalis*, a bacterium that produces the short-chain fatty acid butyrate. Also in the small intestine, butyrate-producing bacteria increased, notably *Eubacterium halli*. As the FMT-induced changes in the gut microbiota of the recipients did not last, this strategy for obesity and insulin resistance treatment needs improvement. One strategy could be to provide a healthy, low-fat, high-fiber diet together with the FMT, as this would help the newly established microbiota be maintained[20]. At least in mouse experiments, this regimen has proven successful in maintenance of a "lean" microbiota[20].

23.8 What happens with the gut microbiota during weight loss?

Roux-en-Y gastric bypass (RYGB) surgery is a radical treatment against obesity in which the major part of the stomach is bypassed by linking a small portion of the upper stomach mucosa to the small intestine. Also, the first part of the small intestine, the duodenum, is bypassed, and ingested food processed in the stomach instead enters the jejunum. This results in dramatic weight-loss, but side effects include nausea, vomiting and stomach pain. Interestingly, these patients display improved insulin sensitivity rapidly after surgery, indicating that RYGB alters hormone signaling and glucose homeostasis[21]. Recent work now shows that the gut microbiota is altered by RYGB and that these changes include increased gut abundance of *Escherichia* and *Akkermansia* genera of bacteria[16]. These changes can be observed both in animal models and in humans. Further, transfer of fecal microbiota of RYGB-treated mice to germ-free, untreated mice resulted in decreased weight and adiposity as compared to mice given feces from sham-operated mice[16]. *Escherichia* is a genus of bacteria that includes several opportunistic and pathogenic bacteria, with pro-inflammatory LPS in their cell walls. Hence, it is difficult to explain why this genus increases in RYGB mice and patients. *Akkermansia*, on the other hand, is a mucin-degrader previously implicated in metabolic improvement after e.g. prebiotic intake[22]. As we learn more about microbiota changes after weight-loss surgery, it may be possible to develop microbial-based therapies mimicking the surgery-induced alterations in the gut microbiota, with associated improvement in glucose and body weight regulation.

23.9 The "diabetic" microbiota

Diabetes mellitus (DM) presents in two major forms: type I and type II, characterized by vastly different molecular events leading up to malfunction of glucose homeostasis. In type I diabetes, autoimmune reactions to insulin-producing beta-cells in the pancreas results in cell death and a gradual loss of insulin-production capacity, often starting early in life. The disease requires careful monitoring of blood glucose levels and life-long administration of insulin. Type II diabetes, on the other hand, is usually described as life-style-related, as obesity, little exercise and an unhealthy diet are key risk factors. Type II diabetics can often reduce their need for insulin by adopting a more healthy life style. Another important difference between the two forms of diabetes is the phenomenon of "insulin resistance" in type II diabetics. These patients have no lack of insulin; instead, they become insensitive to insulin and require higher doses of insulin to be able to transport glucose from the blood stream into the cells.

23.9.1 Type I diabetes and the gut microbiota

It is currently unknown what triggers the autoimmune reactions in type I diabetes, but a study in 2015 using genetically engineered diabetic mice (non-obese diabetic, NOD-mice) showed that bacteria in the phylum *Bacteroidetes* (including S24-7, *Prevotella* and unclassified *Bacteriodales*) may protect against development of type I diabetes, while increased levels of members of the *Firmicutes* phylum, i.e. *Ruminococcus*, *Oscillospira* and *Lachnospiraceae*, promoted the disease[23]. Further, by re-deriving NOD-mice as germ-free, the impact of microbiota on diabetes development was evaluated[24]. Results indicated that absence of microbiota increased inflammation in the islets of Langerhans in the pancreas (insulitis) and reduced glycemic control, while diabetes incidence was not affected[24].

Only a few studies on the gut microbiota have so far been conducted in humans with type I diabetes. In 2014 researchers compared the gut microbiota of children aged 1–5 years with age-matched healthy controls and found that children with diabetes had higher abundance of the combined levels of the phylum *Bacteroidetes* and the class Bacilli[25]. Healthy children had higher abundance of *Clostridium* cluster IV and XIVa[25]. This is in contrast to the NOD mouse experiments, and shows important species-differences in host-microbiome interactions. Regarding bacterial metabolites, healthy children had increased fractions of butyrate-producing bacteria[25]. Type I diabetic children were also characterized by increased gut microbial diversity[25]. This is interesting, as obese and type II diabetic adults generally have lower gut microbial diversity, which may point towards a complex age-related dynamic of the gut microbiota development.

23.9.2 Type II diabetes

It has been observed that not all obese people develop type II diabetes — and not all type II diabetics are obese. Hence, factors other than obesity have been proposed to play a role for insulin resistance, such as the gut microbiota. In recent years, DNA sequencing techniques have revealed that the gut microbiota composition in type II diabetic patients is distinctly different from that found in healthy, age-matched counterparts[26-29]. Mechanistic insight from animal models has further showed that the gut microbiota indeed can cause insulin resistance[3,11,30]. Pathways involved include those described above for obesity, with LPS from gut microbes entering the blood stream from a leaky gut barrier, causing low-grade systemic inflammation and, in turn, insulin resistance and obesity.

In general, the type II diabetic microbiota has lower species diversity and a lower abundance of butyrate-producing bacteria, such as *Faecalibacterium prausnitzii*[26,28,31]. Also, *Clostridium* cluster IV and subcluster XIVa have been found to be lower in type II diabetics as compared to lean non-diabetic individuals[32]. In addition, a higher abundance of lactic acid bacteria and bifidobacteria has been observed[31,32]. Lactic acid bacteria are usually regarded as beneficial, and the relevance of this finding is presently not known. One might speculate whether type II diabetics consume a diet with higher content of sugar, which may favor the growth of carbohydrate-utilizing bacteria, such as lactobacilli and bifidobacteria. This illustrates the need for causative studies to delineate which bacterial species in the gut cause insulin resistance and which bacteria are merely found as a consequence of unhealthy diet. When performing comparative microbiota studies of healthy versus diseased states in humans, careful documentation of dietary patterns is vital for interpretation of results.

23.10 The "atherosclerotic" microbiota

Bacterial infections (e.g. *Chlamydia pneumonia*) and pathogenic periodontal bacteria have long been known to be associated with increased risk of cardiovascular disease[33-36]. Only in the last few years has a direct role of the non-pathogenic commensal microbiota been demonstrated. A key finding in this aspect was the discovery of a microbial-dependent degradation of dietary choline and carnitine, found in e.g. meat and eggs, into pro-atherogenic metabolites called trimethyl-amine (TMA) and trimethylamine *N*-oxide (TMAO)[37,38]. By suppressing the gut microbiota with antibiotics in mice fed choline-diet, atherosclerosis was inhibited[39]. In addition, humans with high TMA and TMAO levels displayed increased risk of cardiovascular disease. An "atherosclerotic" microbiota can even be transferred between mice if they are fed a diet containing choline[40]. Also, the gut microbial metagenome, i.e. the functional capacity of the gut microbiota, was altered in patients with symptomatic atherosclerosis, who had reduced levels of the bacterial enzyme phytoene dehydrogenase, which is involved in β-carotene metabolism[41]. This corresponded with decreased blood levels of β-carotene in patients. The gut microbial composition in healthy individuals was characterized by increased levels of *Roseburia* and *Eubacterium*, while patients with symptomatic atherosclerosis had increased abundance of *Collinsella*[41]. Additional studies have also shown alterations in the gut microbiota of atherosclerosis patients, as well as the presence of bacterial DNA in atherosclerotic plaques[42]. Hence, bacteria may affect athero-genesis in several ways, including production of pro- or anti-atherogenic metabolites, inflammation in plaques and through modulation of cholesterol metabolism via bile acids[43]. Interestingly, the use of probiotic bacteria has shown efficacy in prevention of atherosclerosis in animal models[44] and can reduce biomarkers for cardiovascular risk in humans[45]. It remains to be elucidated whether it will be possible to prevent cardiovascular events in humans using targeted microbial modulations.

23.11 Conclusions

Exploring the important interplay between bacteria and its host — us — is a growing field of research and may unravel new basic aspects of biology as well as new prevention strategies and therapeutics aimed at the gut microbiome. The use of animal models is essential in this aspect to prove causality, but comparison between the human and mouse/rat microbiomes needs to be done in order to ensure successful translation of results from animal models to human physiology. A recent study set out to explore the human and mouse microbiomes and found that the major groups of bacteria were shared, while the relative abundance of bacteria differed[46]. Analysing three common laboratory mice (BALB/c, B6.V-*Lep*[ob]/J and NOD-mice), the authors could see differences in the gut microbiota composition due to the genetic background (all mice were kept on the same, standard mouse chow), but also concluded that the diet is an additional important factor for shaping the gut microbiota. When performing clinical trials with microbial manipulations, it is advised to standardize the diet to reduce variability of results. However, this poses a great challenge in long-term bacterial interventions in humans. The taxonomic level that is reported in gut microbiota studies should be emphasized. Analysing higher levels (phylum, order) may not be sufficient to draw conclusions; instead, genus and species levels may give better insight into bacterial shifts and provide more relevance for physiology[46,47]. In this respect, sequencing depth is important, as genus-level comparisons require greater sequencing effort than phylum level[46].

TAKE-HOME MESSAGE

- The gut microbiota composition plays a pivotal role in host metabolism and is altered in patients with obesity, diabetes and atherosclerosis.
- Butyrate-producing bacteria are associated with health and may protect against development of metabolic diseases.
- Diet is an important factor shaping the gut microbiota and can be used to alter aberrant gut microbial composition.

References

1 Bäckhed F, Manchester J, Semenkovich C, Gordon J. Mechanisms underlying the resistance to diet-induced obesity in germ-free mice. *PNAS* 2007; **104**: 979–984.
2 Bäckhed F *et al.* The gut microbiota as an environmental factor that regulates fat storage. *Proc Natl Acad Sci USA* 2004; **101**: 15718–15723.

3 Turnbaugh P *et al.* An obesity-associated gut microbiome with increased capacity for energy harvest. *Nature* 2006; **444**: 1027–1031.

4 Ley R, Turnbaugh P, Klein S, Gordon J. Microbial ecology: human gut microbes associated with obesity. *Nature* 2006; **444**: 1022–1023.

5 Jumpertz R *et al.* Energy-balance studies reveal associations between gut microbes, caloric load, and nutrient absorption in humans. *The American Journal of Clinical Nutrition* 2011; **94**; 58–65, doi:10.3945/ajcn.110.010132.

6 Hildebrandt MA *et al.* High-fat diet determines the composition of the murine gut microbiome independently of obesity. *Gastroenterology* 2009; **137**: 1716–1724 e1711–1712, doi:10.1053/j. gastro.2009.08.042.

7 Amar J *et al.* Energy intake is associated with endotoxemia in apparently healthy men. *The American Journal of Clinical Nutrition* 2008; **87**: 1219–1223.

8 Clemente-Postigo M *et al.* Endotoxin increase after fat overload is related to postprandial hypertriglyceridemia in morbidly obese patients. *Journal of Lipid Research* 2012; **53**: 973–978, doi:10.1194/jlr.P020909.

9 Manco M, Putignani L, Bottazzo GF. Gut microbiota, lipopolysaccharides, and innate immunity in the pathogenesis of obesity and cardiovascular risk. *Endocr Rev* 2010; **31**: 817–844, doi:10.1210/er.2009-0030.

10 Kheirandish-Gozal L *et al.* Lipopolysaccharide-binding protein plasma levels in children: effects of obstructive sleep apnea and obesity. *The Journal of Clinical Endocrinology and Metabolism* 2014; **99**: 656–663, doi:10.1210/jc.2013-3327.

11 Cani P *et al.* Metabolic endotoxemia initiates obesity and insulin resistance. *Diabetes* 2007; **56**: 1761–1772.

12 Cani PD *et al.* Changes in gut microbiota control inflammation in obese mice through a mechanism involving GLP-2-driven improvement of gut permeability. *Gut* 2009; **58**: 1091–1103, doi:10.1136/gut.2008.165886.

13 Li M *et al.* Gut carbohydrate metabolism instead of fat metabolism regulated by gut microbes mediates high-fat diet-induced obesity. *Beneficial Microbes* 2014; **5**: 335–344, doi:10.3920/bm2013.0071.

14 de La Serre CB *et al.* Propensity to high-fat diet-induced obesity in rats is associated with changes in the gut microbiota and gut inflammation. *American journal of physiology. Gastrointestinal and Liver Physiology* 2010; **299**: G440–448, doi:10.1152/ajpgi.00098.2010.

15 Hijova E, Chmelarova A. Short chain fatty acids and colonic health. *Bratisl Lek Listy* 2007; **108**: 354–358.

16 Liou AP *et al.* Conserved shifts in the gut microbiota due to gastric bypass reduce host weight and adiposity. *Sci Transl Med* 2013; **5**: 178ra141, doi:10.1126/scitranslmed.3005687.

17 Patil DP *et al.* Molecular analysis of gut microbiota in obesity among Indian individuals. *J Biosci* 2012; **37**: 647–657.

18 De Vadder F *et al.* Microbiota-generated metabolites promote metabolic benefits via gut-brain neural circuits. *Cell* 2014; **156**: 84–96, doi:10.1016/j.cell.2013.12.016.

19 Vrieze A *et al.* Transfer of intestinal microbiota from lean donors increases insulin sensitivity in individuals with metabolic syndrome. *Gastroenterology* 2012; **143**: 913–916 e917, doi:10.1053/j. gastro.2012.06.031.

20 Ridaura VK *et al.* Gut microbiota from twins discordant for obesity modulate metabolism in mice. *Science* 2013; **341**: 1241214, doi:10.1126/science.1241214.

21 Madsbad S, Dirksen C, Holst JJ. Mechanisms of changes in glucose metabolism and bodyweight after bariatric surgery. *The Lancet. Diabetes & Endocrinology* 2014; **2**: 152–164, doi:10.1016/s2213-8587(13)70218-3.

22 Everard A *et al.* Cross-talk between Akkermansia muciniphila and intestinal epithelium controls diet-induced obesity. *Proc Natl Acad Sci USA* 2013; **110**: 9066–9071, doi:10.1073/pnas.1219451110.

23 Krych L, Nielsen DS, Hansen AK, Hansen CH. Gut microbial markers are associated with diabetes onset, regulatory imbalance, and IFN-gamma level in NOD mice. *Gut Microbes* 2015: 0, doi:10.1080/19490976.2015.1011876.

24 Greiner TU, Hyotylainen T, Knip M, Backhed F, Oresic M. The gut microbiota modulates glycaemic control and serum metabolite profiles in non-obese diabetic mice. *PloS One* 2014; **9**: e110359, doi:10.1371/journal.pone.0110359.

25 de Goffau MC *et al.* Aberrant gut microbiota composition at the onset of type 1 diabetes in young children. *Diabetologia* 2014; **57**: 1569–1577, doi:10.1007/s00125-014-3274-0.

26 Qin J *et al.* A metagenome-wide association study of gut microbiota in type 2 diabetes. *Nature* 2012; **490**: 55–60, doi:10.1038/nature11450.

27 Karlsson FH *et al.* Gut metagenome in European women with normal, impaired and diabetic glucose control. *Nature* 2013; **498**: 99–103, doi:10.1038/nature12198.

28 Le Chatelier E *et al.* Richness of human gut microbiome correlates with metabolic markers. *Nature* 2013; **500**: 541–546, doi:10.1038/nature12506.

29 Larsen N *et al.* Gut microbiota in human adults with type 2 diabetes differs from non-diabetic adults. *PloS One* 2010; **5**: e9085, doi:10.1371/journal.pone.0009085.

30 Cani P *et al.* Changes in gut microbiota control metabolic endotoxemia-induced inflammation in high-fat diet-induced obesity and diabetes in mice. *Diabetes* 2008; **57**: 1470–1481.

31 Remely M *et al.* Microbiota and epigenetic regulation of inflammatory mediators in type 2 diabetes and obesity. *Beneficial Microbes* 2014; **5**: 33–43, doi:10.3920/bm2013.006.

32 Sasaki M *et al.* Transglucosidase improves the gut microbiota profile of type 2 diabetes mellitus patients: a randomized double-blind, placebo-controlled study. *BMC Gastroenterology* 2013; **13**: 81, doi:10.1186/1471-230x-13-81.

33 Desvarieux M *et al.* Periodontal microbiota and carotid intima-media thickness: the Oral Infections and Vascular Disease Epidemiology Study (INVEST). *Circulation* 2005; **111**: 576–582.

34 Hyvärinen K *et al.* A common periodontal pathogen has an adverse association with both acute and stable coronary artery disease. *Atherosclerosis* 2012; **223**: 478–484.

35 Nakano K *et al.* Detection of cariogenic Streptococcus mutans in extirpated heart valve and atheromatous plaque specimens. *J Clin Microbiol* 2006; **44**: 3313–3317, doi:10.1128/jcm.00377-06.

36 Pinho MM, Faria-Almeida R, Azevedo E, Manso MC, Martins L. Periodontitis and atherosclerosis: an observational study. *J Periodontal Res* 2013; **48**: 452–457, doi:10.1111/jre.12026.

37 Koeth RA *et al.* Intestinal microbiota metabolism of L-carnitine, a nutrient in red meat, promotes atherosclerosis. *Nat Med* 2013; **19**: 576–585, doi:10.1038/nm.3145.

38 Tang WH *et al.* Intestinal microbial metabolism of phosphatidylcholine and cardiovascular risk. *N Engl J Med* 2013; **368**: 1575–1584, doi:10.1056/NEJMoa1109400.

39 Wang Z *et al.* Gut flora metabolism of phosphatidylcholine promotes cardiovascular disease. *Nature* 2011; **472**: 57–63, doi:10.1038/nature09922.

40 Gregory JC *et al.* Transmission of atherosclerosis susceptibility with gut microbial transplantation. *Journal of Biological Chemistry* 2014, doi:10.1074/jbc.M114.618249 (2014).

41 Karlsson FH *et al.* Symptomatic atherosclerosis is associated with an altered gut metagenome. *Nat Commun* 2012; **3**: 1245.

42 Koren O *et al.* Microbes and Health Sackler Colloquium: Human oral, gut, and plaque microbiota in patients with atherosclerosis. *Proc Natl Acad Sci USA* 2011; **108 Suppl 1**: 4592–4598.

43 Sayin SI *et al.* Gut microbiota regulates bile acid metabolism by reducing the levels of tauro-beta-muricholic acid, a naturally occurring FXR antagonist. *Cell Metab* 2013; **17**: 225–235, doi:10.1016/j.cmet.2013.01.003.

44 Chen L *et al.* Lactobacillus acidophilus ATCC 4356 attenuates the atherosclerotic progression through modulation of oxidative stress and inflammatory process. *International Immunopharmacology* 2013; **17**; 108–115: doi:10.1016/j.intimp.2013.05.018.

45 Naruszewicz M, Johansson ML, Zapolska-Downar D, Bukowska H. Effect of Lactobacillus plantarum 299v on cardiovascular disease risk factors in smokers. *The American Journal of Clinical Nutrition* 2002; **76**: 1249–1255.

46 Krych L, Hansen CH, Hansen AK, van den Berg FW, Nielsen DS. Quantitatively different, yet qualitatively alike: a meta-analysis of the mouse core gut microbiome with a view towards the human gut microbiome. *PloS One* 2013; **8**: e62578, doi:10.1371/journal.pone.0062578.

47 Fak F, Backhed F. Lactobacillus reuteri prevents diet-induced obesity, but not atherosclerosis, in a strain dependent fashion in Apoe-/- mice. *PloS One* 2012; **7**: e46837, doi:10.1371/journal.pone.0046837.

CHAPTER 24

The microbiota and susceptibility to asthma

Olawale Salami and Benjamin J. Marsland

Service de pneumologie, CHUV, Faculty of Biology and Medicine, University of Lausanne, Lausanne, Switzerland

24.1 Introduction

Asthma is a common chronic respiratory disease that affects millions of people of all ages in every part of the world[1]. The global burden of asthma remains high, with current estimates of worldwide prevalence put at over 300 million people[1,2]. Importantly, the last few decades have witnessed a dramatic increase in asthma prevalence, a situation observed in both high- and low-income countries[3]. A growing body of epidemiological evidence suggests that early-life exposure to environment-derived microbes and farm milk consumption — prominent features of life in agrarian communities — confer protection against allergic disorders[4–6]. Conversely, atopic disorders including asthma correlate with several factors, such as increased antibiotic use in infancy[7–9], urban dwelling[5], and western diet-induced changes in the intestinal microbiota[10,11]. Recently, culture-independent analyses performed on the human microbiome have provided an overview of microbial communities in various tissue habitats[12]. Early-life airway microbiota could impact directly on asthma modulation via the outgrowth of pro-inflammatory, asthma-inducing lung microbiota[13] or the induction of tolerogenic mechanisms[14] and, through indirect mechanisms, via gastrointestinal tract-derived anti-inflammatory signals[15].

Today, culture-independent molecular techniques of microbial identification have disproved the traditional paradigm that the lower airways are a sterile environment, facilitating a new understanding of the microbial constituents in healthy and diseased lungs. This chapter provides a synopsis of current knowledge of the airway microbiota and how dysbiosis may underlie the susceptibility to the severity and persistence of asthma and other chronic lung diseases.

24.2 The microenvironment of the lower airways

The lower airways consist of distinct anatomical regions of which the broncho-alveolar space remains the most well characterized[16]. Sampling of the bronchoalveolar space by bronchoalveolar lavage (BAL) remains an important tool for

The Human Microbiota and Chronic Disease: Dysbiosis as a Cause of Human Pathology, First Edition.
Edited by Luigi Nibali and Brian Henderson.

clinical diagnoses as well as experimental models of lung diseases[17]. However, the global picture provided by the BAL fluid masks the regional differences in the airways. In this regard, it has been shown that differences in lung airway micro-architecture result in spatially distinct bacterial communities[18]. In this study, the authors sampled multiple discrete sites in explanted lungs of patients with COPD[18]. Interestingly, bacterial communities in the upper lobes of the lungs were dominated by *Haemophilus* spp., which support a previously observed correlation between lung upper lobe Haemophilus colonization and COPD[19]. Of note, up to 50% of patients with COPD manifest emphysematous changes in the upper lobe[20]. By virtue of its oxygen richness[21], the microenvironment of the upper lobe may support the outgrowth of aerophilic bacteria like haemophilus[19], creating a niche that could potentially promote COPD.

Bacteria that can thrive in the airways must be able to harness nutrients from their environment for energy generation and cellular biosynthesis. Gut microbiota have evolved to utilize dietary components like glycerides, cholesterol and short chain fatty acids for their metabolism[22]. Currently, our knowledge of nutrient sources required for optimal bacterial growth remains limited. Studies examining biochemical profiles of low-molecular-weight endogenous metabolites present in exhaled breath[23], sputum[24] and BAL fluid samples[25] in diseased and healthy lungs could provide clues on possible energy sources present in the airways. Notably, the bioavailability of lactate, valine, taurine and alanine has been reported to be higher in patients with cystic fibrosis as compared to healthy controls[25]. Other possible energy sources in the airways are bronchial mucus, which is rich in glycolipids and glycoproteins[26] and lung surfactant[26]. Currently, there is paucity of data on how airway metabolites shape lung microbiota, but this is an important area of future research.

24.3 Development of the airway microbiota in the neonate

The development of the airway microbiota in the neonate may be influenced by the following (see Figure 1):
1 Intra-uterine microbial exposure
2 Perinatal events
3 Breast-milk as a source of airway microbiota
4 Airborne microbiota
5 Microbiota of the upper airways

24.3.1 Intrauterine microbial exposure and airway microbiota
Birth heralds a transition from an intrauterine milieu to an external, microbe-laden environment. Recent emerging evidence suggests that the intrauterine microenvironment, previously thought to be sterile[27], contains a unique microbiota, raising questions whether the development of the infant microbiota predates the process of birth. Amniotic fluid, which bathes foetal epithelial surfaces and plays an essential role in pulmonary and intestinal organogenesis[28], has been found to contain diverse microbial communities whose constituents are important determinants of perinatal outcomes.[29–31] The isolation of Mycoplasma and Ureaplasma

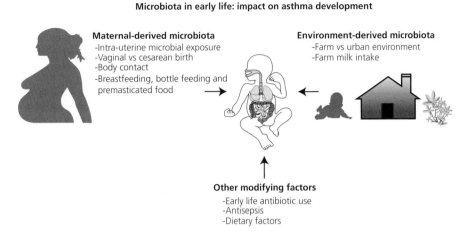

Figure 1 The microbiota in early life. An interplay of several early life factors shape the development of the microbiota in infants and risk of asthma later in life. Early host-microbe interactions commence in utero and may influence the programming of developing immune cells. Vaginal delivery signals the first exposure of neonatal body surfaces to the maternal cervico-vaginal microbiota. Antibiotic use in infancy may be related to increased risk of asthma childhood. Breastfeeding is associated with direct ingestion by the neonate of maternal skin-resident bacteria, and maternal gut bacteria via the entero-mammary pathway. Maternal transfer of skin and oral microbiota occurs by direct contact and pre-mastication of food. Growing up in a rural environment leads to early-life exposure to diverse, environment-derived microbes and is associated with protection against the development of allergy. Fibre-rich diets increase short-chain fatty acid levels, which have a protective effect on airway inflammation.

from amniotic fluid has been shown to be associated with detrimental foetal outcomes like necrotizing enterocolitis (NEC)[32], suggesting that intra-uterine microbial composition can impact post-natal life. Furthermore, there are intriguing reports that point to the existence of a unique, low-abundance placental microbiota[33,34], foetal exposure to which may hypothetically have significant implications for foetal lung immune development. The source of these intrauterine microbes remains unknown, but the maternal oral cavity[34], as well as the maternal gut via intestinal translocation[35] have been suggested. Indeed, a recent study elegantly showed the microbiome of pre-term and full-term placentas and defined a unique microbiota, consisting of low numbers of organisms that belong primarily to the phyla Tenericutes, Firmicutes, Proteobacteria, Fusobacteria and Bacteroidetes[34]. Furthermore, upon close comparison with other tissue sites, the placental microbiome showed a striking resemblance to that of the oral cavity[34]. The cross-talk between bacteria and the foetal immune system at such early time-points in development and possible effects on disease susceptibility or protection in later life remain unknown.

24.3.2 Perinatal events and airway microbiota

Following birth, environment-derived bacteria, a significant proportion of which are of maternal origin, colonize neonatal body surfaces[36]. The impact of the mode of delivery on microbial colonization of foetal surfaces in early life remains intriguing and was previously discussed in chapters 6 (Bermudez) and 16 (Sadiq and Hay). Several reports suggest that Cesarean and vaginal routes of delivery result in different patterns of microbial colonization[36,37]. Vaginal delivery leads to neonatal gut colonization by resident cervico-vaginal microbiota[38], dominated to a large extent by Lactobacilli[39], within the first 48 hours of life, which display striking similarities across the skin, oral cavity and naso-pharynx[40]. Furthermore, vaginal delivery has been found to be associated with a higher diversity of the Bacteroidetes phylum in infant gut as well as increased circulating levels of Th1-associated chemokines CXCL10 and CXCL-11[41]. The microbiota in neonates born via Cesarean section bears a strong resemblance to that of the maternal skin[36]. However, the sequence of events that determines the evolution of a stable microbiota in the airway remains an intriguing subject awaiting further scrutiny. Collectively, current evidence suggests that the pioneer microbiota within body surfaces, including the lung epithelia, may be key players in the calibration of neonatal immunity in preparation for future host-microbial encounters.

24.3.3 Breast milk as a source of airway microbiota

Human breast milk, which serves numerous important nutritional and immuno-logical functions in infancy, has been shown, using both culture-based and molec-ular techniques, to contain a variety of skin- and gut-associated bacteria[42-44]. Infant positioning for breastfeeding entails sustained contact between the infant oral cavity and the maternal periareolar skin[45], meaning that the observed similarities between neonatal gut and maternal skin microbiota are not completely unex-pected[46]. However, maternal gut-derived bacteria, via a process of entero-mammary trafficking by maternal intestinal dendritic cells[47,48], may also be important components of the gut microbiota in the neonate[48]. Following breastfeeding, gastroeosophageal reflux of ingested breast milk occurs frequently in the majority of infants, and may be associated with micro-aspiration into the lower airways[49]. This process may contribute to the earliest seeding of the airways in the neonate with breastmilk-derived microbiota.

24.3.4 Airborne microbiota and airway microbiota

The surface area of the lungs, measuring in the range of $75\,m^2$, is suffused daily with inspired air replete with airborne particulate matter, aerosolized bacteria and fungi. Recent reports utilizing culture-independent techniques such as 16S ribosomal RNA gene coding DNA (16S rDNA) analysis and metagenomic sequencing have shown that the air harbors its own unique microbiota[50,51] reaching up to 1800 bacterial types, a richness approaching that of some soil microbial communities[51]. Some of these aerosolized airborne bacteria display active replication and metabo-lism, suggesting a unique ability to survive in the air[52]. It is therefore plausible that some of these airborne environment-derived microbes may, over time, become adapted to the airway microenvironment and lead to the establishment of stable communities. Further work is needed to understand the impact of environmentally derived microbes on airway microbial communities and chronic lung diseases.

24.4 Upper airway microbiota

Functionally, the upper airway consists of the nasal cavity, the pharynx, larynx and trachea[53]. The continuous flow of inspired and expired air over these passages that characterizes normal respiratory activity therefore provides immense opportunities for lower-airway colonization by upper-airway-derived microbes. Asymptomatic colonization of the upper airway by *Streptococcus pneumoniae* has been recognized as a strong risk factor for invasive pneumococcal pneumonia, thus further highlighting the possible critical role of the upper-airway microbial environment in regulating the balance of microbial communities downstream. Currently, there is a paucity of knowledge of the temporal events that culminate in a stable upper airway microbiota. It is known, however, that colonization of the gut starts to occur within the first 24–48 hours of postnatal life[54] and as pharyngeal reflux and micro-aspiration of gastro-intestinal and breastmilk contents occur frequently in neonates[49], gut-derived bacteria could be the colonizers of the upper airways and subsequently the lower airways in humans. A recent report suggests that distinct upper airway microbiota profiles begin to emerge in healthy infants from about 1.5 months[55] and specific profiles were associated with dynamic microbiota stability and change over the first two years of life[55]. Specifically, Corynebacterium/Dolosigranulum-dominated microbial communities proved stable over the first two years of life and were associated with lower rates of respiratory tract infections[55]. Collectively, these studies outline a possible role for upper-airway-derived bacteria in determining the final constituents of lower-airway microbiota.

24.5 What constitutes a healthy airway microbiota

It is now established that the airway harbors a microbiota whose constituents as well as impact on respiratory health we are only beginning to elucidate. Using culture-independent techniques of bacterial identification, healthy airways in the adult have been shown to contain up to 2000/cm^3 of tissue[13], similar to what is obtained in the upper gastrointestinal tract[56]. The constitution of airway microbial communities have been characterized in patients with cystic fibrosis, chronic obstructive airway disease (COPD), asthma, and lung-transplant recipients[56]. The lung microbiota in patients with COPD shows striking differences from that of "normal" control subjects[57,58]. Similar observations in children with cystic fibrosis indicate the existence of significantly different microbial communities as compared with children without cystic fibrosis (further discussed by Bruzzese and co-workers in chapter 32)[59,60]. The most prevalent phyla identified in "normal" airways from these studies include Bacteroidetes, Firmicutes, Proteobacteria and Actinobacteria (similarly to the skin microbiota discussed in chapter 5)[61]. This "core" airway microbiota consists mainly of the genera *Pseudomonas* (Proteobacteria), *Streptococcus* (Firmicutes), *Prevotella* (Bacteroidetes), *Fusobacteria* (Fusobacteria), *Veillonella* (Firmicutes), *Haemophilus* (Proteobacteria), *Neisseria* (Proteobacteria) and *Porphyromonas* (Bacteroidetes)[61]. A prominent and recurrent feature in most recent studies is the finding that an outgrowth of certain pathogenic bacteria occurs that perturbs existing community diversity and is associated with diseased states in the lungs[56].

24.6 Microbiota and asthma

Evidence from large epidemiological studies points to an association between higher microbial exposure and decreased incidence of allergy[4–6]. A persuasive argument has been proposed on the importance of not only microbial exposure, but also the diversity and timing of such exposures in reducing airway inflammation[6]. In a major study, the airway microbiota in patients with asthma was shown to be distinctly different from non-asthmatic controls[19]. The question whether a dysbiotic microbiota confers increased susceptibility to asthma, or the possibility that airway inflammation could induce observed changes in airway microbiota, remain unanswered. Such hypotheses invoked by observations from clinical and epidemiological work prompted investigations in mice bred under completely germ-free conditions, in attempts aimed at elucidating the possible role of the microbiota in regulating chronic airway inflammation. The absence of lung microbiota has been shown to result in a dyregulated maturation of airway dendritic cells and an overall exaggeration of allergic airway inflammation[62]. Furthermore, direct administration of innocuous bacteria or bacterial products into the airway can abrogate airway inflammation[63,64]. In the same vein, experimental treatment of mice with antibiotics led to an increase in Th2-mediated allergic airway inflammation[65].

The airway microbiota may also play a significant role in immune maturation in the lung. Lung microbiota development was recently shown to be associated with decreased allergic airway responsiveness, congruent with the emergence of Helios – T-regulatory cells[14]. These T-regulatory cells required PDL-1 for their development, and the absence of lung microbiota or PDL-1 blockade maintained baseline neonatal exaggerated inflammatory responses through to adulthood[14,62]. Emerging evidence suggests that there is a developmental window in early life within which the beneficial effects of airway microbial colonization on airway inflammation are most pronounced. Thus, in line with observations in both large cohort studies[66] and murine experimental models[67], lung colonization within an early-life tight developmental window may result in protection against allergic airway inflammation in early adulthood.

24.7 Dietary metabolites and asthma

Our knowledge of the critical role of the gut microbiota in intestinal immune homoestasis and the pathophysiological consequences of its perturbation has improved over the past decades. What remains limited, however, is our understanding of the mechanisms by which intestinal microbial colonization regulates homeostatic processes occurring at remote extra-intestinal sites. Clues to this are beginning to emerge. Attention in the field has focused on potentially important immune regulatory signals provided by metabolites derived from fermentation of intestinal contents by microbiota, notably short-chain fatty acids (SCFAs)[68,69,70]. Recently, the SCFA propionate has been shown to dampen Th2-mediated allergic airway inflammation via a GPR-41-dependent induction of bone marrow DC and macrophage precursors with impaired ability to drive TH2 mediated inflammation[71]. The increased SCFA levels observed in mice fed on high-fibre diets were associated with profound changes in the gut microbial phyla, most notably in the ratio of Firmicutes/Bacteroidetes[71]. Collectively, the current body of evidence points to a critical role of the microbiota, via its fermentation specific dietary components, in the regulation of both local and systemic immune homeostasis.

24.8 Conclusion, future perspectives and clinical implications

The pathophysiologic features of asthma encompass a spectrum of changes in immune cell compartments of the airways, culminating in structural changes in the airway epithelium and smooth muscles. Current evidence strongly supports a crucial role of the microbiota in regulating airway inflammation. However, our knowledge of the key factors in early life that lead up to the establishment of stable microbiota in the lungs and the cross-talk with developing stromal and immune cell compartments remains limited. In the gut, resident bacteria have evolved to utilize host dietary components as energy sources, with release of metabolic by-products capable of modulating local and systemic immune responses. The lung microenvironment is devoid of such rich energy sources, and thus resident microbiota are likely to have evolved to utilize alternative sources of energy present in the airways. Currently, evidence of such alternative energy sources in the airways, as well as metabolic pathways required to utilize them, is lacking. This is an intriguing area of further research that may resolve what key energy sources in the airways are needed during health or disease. Finally, emerging evidence from preclinical in-vitro and in-vivo studies could be the basis for novel treatment strategies for asthma. However, first, studies utilizing engineered microbiota, microbial products with immuno-modulatory effects or energy sources that promote outgrowth of specific airway-resident bacteria are needed in order to bring the field forward.

TAKE-HOME MESSAGE

- The airway harbors a unique microbiota in homoeostasis.
- Host-microbiota cross-talk occurs within tight developmental windows in early life which influence airway inflammation in later life.
- Dysbiosis could underlie susceptibility to severity and progression of asthma.

References

1 Braman S. The global burden of asthma. *Chest* 2006; **130**: 4S–12S.
2 To T, Stanojevic S, Moores G *et al.* Global asthma prevalence in adults: findings from the cross-sectional world health survey. *BMC Public Health* 2012; **12**: 204.
3 Masoli M, Fabian D, Holt S *et al.* The global burden of asthma: executive summary of the GINA Dissemination Committee report. *Allergy* 2004; **59**: 469–78.
4 Fuchs O, Genuneit J, Latzin P *et al.* Farming environments and childhood atopy, wheeze, lung function, and exhaled nitric oxide. *Journal of Allergy and Clinical Immunology* 2012; **130**: 382–8.e6.
5 Solis Soto MT, Patino A, Nowak D *et al.* Prevalence of asthma, rhinitis and eczema symptoms in rural and urban school-aged children from Oropeza Province, Bolivia: a cross-sectional study. *BMC Pulmonary Medicine* 2014; **14**: 40.
6 Ege MJ, Mayer M, Normand A-C *et al.* Exposure to environmental microorganisms and childhood asthma. *New England Journal of Medicine* 2011; **364**: 701–9.
7 Wickens K, Pearce N, Crane J *et al.* Antibiotic use in early childhood and the development of asthma. *Clinical and Experimental Allergy: Journal of the British Society for Allergy and Clinical Immunology* 1999; **29**: 766–71.

8 Droste JH, Wieringa MH, Weyler JJ *et al.* Does the use of antibiotics in early childhood increase the risk of asthma and allergic disease? *Clinical and Experimental Allergy: Journal of the British Society for Allergy and Clinical Immunology* 2000; **30**: 1547–53.

9 McKeever TM, Lewis SA, Smith C *et al.* Early exposure to infections and antibiotics and the incidence of allergic disease: a birth cohort study with the West Midlands General Practice Research Database. *Journal of Allergy and Clinical Immunology* 2002; **109**: 43–50.

10 Kirjavainen PV, Apostolou E, Arvola T *et al.* Characterizing the composition of intestinal microflora as a prospective treatment target in infant allergic disease. *FEMS Immunology and Medical Microbiology* 2001; **32**: 1–7.

11 Bottcher MF, Nordin EK, Sandin A *et al.* Microflora-associated characteristics in faeces from allergic and nonallergic infants. *Clinical and Experimental Allergy: Journal of the British Society for Allergy and Clinical Immunology* 2000; **30**: 1590–6.

12 Human Microbiome Project Consortium. Structure, function and diversity of the healthy human microbiome. *Nature* 2012; **486**: 207–14.

13 Hilty M, Burke C, Pedro H *et al.* Disordered Microbial communities in asthmatic airways. *PloS One* 2010; **5**: e8578.

14 Gollwitzer ES, Saglani S, Trompette A *et al.* Lung microbiota promotes tolerance to allergens in neonates via PD-L1. *Nature Medicine* 2014; **20**: 642–7.

15 Noverr MC, Huffnagle GB. The 'microflora hypothesis' of allergic diseases. *Clinical & Experimental Allergy* 2005; **35**: 1511–20.

16 Tschernig T, Pabst R. What is the clinical relevance of different lung compartments? *BMC Pulmonary Medicine* 2009; **9**: 39.

17 Meyer KC. Bronchoalveolar lavage as a diagnostic tool. *Semin Respir Crit Care Med* 2007; **28**: 546–60.

18 Erb-Downward JR, Thompson DL, Han MK *et al.* Analysis of the lung microbiome in the "healthy" smoker and in COPD. *PloS One* 2011; **6**: e16384.

19 Hilty M, Burke C, Pedro H *et al.* Disordered microbial communities in asthmatic airways. *PloS One* 2010; **5**: e8578.

20 Foster WL, Jr., Gimenez EI, Roubidoux MA *et al.* The emphysemas: radiologic-pathologic correlations. *Radiographics: a Review Publication of the Radiological Society of North America, Inc* 1993; **13**: 311–28.

21 Martin CJ, Cline F, Jr., Marshall H. Lobar alveolar gas concentrations; effect of body position. *The Journal of Clinical Investigation* 1953; **32**: 617–21.

22 Neish AS. Microbes in gastrointestinal health and disease. *Gastroenterology* 2009; **136**: 65–80.

23 Carraro S, Rezzi S, Reniero F *et al.* Metabolomics applied to exhaled breath condensate in childhood asthma. *American Journal of Respiratory and Critical Care Medicine* 2007; **175**: 986–90.

24 Yang J, Eiserich JP, Cross CE *et al.* Metabolomic profiling of regulatory lipid mediators in sputum from adult cystic fibrosis patients. *Free Radical Biology & Medicine* 2012; **53**: 160–71.

25 Wolak JE, Esther CR, O'Connell TM. Metabolomic analysis of bronchoalveolar lavage fluid from cystic fibrosis patients. *Biomarkers: biochemical Indicators of Exposure, Response, and Susceptibility to Chemicals* 2009; **14**: 10.1080/13547500802688194.

26 Reid LM, Bhaskar KR. Macromolecular and lipid constituents of bronchial epithelial mucus. *Symposia of the Society for Experimental Biology* 1989; **43**: 201–19.

27 Moller BR, Kristiansen FV, Thorsen P *et al.* Sterility of the uterine cavity. *Acta Obstetricia et Gynecologica Scandinavica* 1995; **74**: 216–9.

28 Underwood MA, Gilbert WM, Sherman MP. Amniotic fluid: not just fetal urine anymore. *J Perinatol* 2005; **25**: 341–8.

29 Yoon BH, Romero R, Kim M *et al.* Clinical implications of detection of Ureaplasma urealyticum in the amniotic cavity with the polymerase chain reaction. *American Journal of Obstetrics and Gynecology* 2000; **183**: 1130–7.

30 Payne MS, Bayatibojakhi S. Exploring preterm birth as a polymicrobial disease: An overview of the uterine microbiome. *Frontiers in Immunology* 2014; **5**.

31 Mysorekar IU, Cao B. Microbiome in parturition and preterm birth. *Semin Reprod Med* 2014; **32**: 50–5.

32 DiGiulio DB. Diversity of microbes in amniotic fluid. *Seminars in Fetal and Neonatal Medicine* 2012; **17**: 2–11.

33 Aagaard K, Ganu R, Ma J *et al.* Whole metagenomic shotgun sequencing reveals a vibrant placental microbiome harboring metabolic function. *American Journal of Obstetrics and Gynecology* 2013; **208**: S5.

34 Aagaard K, Ma J, Antony KM *et al.* The placenta harbors a unique microbiome. *Science Translational Medicine* 2014; **6**: 237ra65.
35 Funkhouser LJ, Bordenstein SR. Mom knows best: the universality of maternal microbial transmission. *PLoS Biol* 2013; **11**: e1001631.
36 Dominguez-Bello MG, Costello EK, Contreras M *et al.* Delivery mode shapes the acquisition and structure of the initial microbiota across multiple body habitats in newborns. *Proceedings of the National Academy of Sciences* 2010; **107**: 11971–5.
37 Matamoros S, Gras-Leguen C, Le Vacon F *et al.* Development of intestinal microbiota in infants and its impact on health. *Trends in Microbiology* 2013; **21**: 167–73.
38 Rampersaud R, Randis TM, Ratner AJ. Microbiota of the upper and lower genital tract. *Seminars in fetal & neonatal medicine* 2012; **17**: 51–7.
39 Redondo-Lopez V, Cook RL, Sobel JD. Emerging role of lactobacilli in the control and maintenance of the vaginal bacterial microflora. *Review of Infectious Diseases* 1990; **12**: 856–72.
40 Karlsson CLJ, Molin G, Cilio CM *et al.* The pioneer gut microbiota in human neonates vaginally born at term — A Pilot Study. *Pediatr Res* 2011; **70**: 282–6.
41 Jakobsson HE, Abrahamsson TR, Jenmalm MC *et al.* Decreased gut microbiota diversity, delayed Bacteroidetes colonisation and reduced Th1 responses in infants delivered by Caesarean section. *Gut* 2014; **63**: 559–66.
42 Heikkilä MP, Saris PE. Inhibition of Staphylococcus aureus by the commensal bacteria of human milk. *J Appl Microbiol* 2003; **95**: 471–8.
43 Gueimonde M, Laitinen K, Salminen S *et al.* Breast milk: a source of bifidobacteria for infant gut development and maturation? *Neonatology* 2007; **92**: 64–6.
44 Hunt KM, Foster JA, Forney LJ *et al.* Characterization of the diversity and temporal stability of bacterial communities in human milk. *PloS One* 2011; **6**: e21313.
45 Holmes AV. Establishing successful breastfeeding in the newborn period. *Pediatric Clinics of North America* 2013; **60**: 147–68.
46 Palmer C, Bik EM, DiGiulio DB *et al.* Development of the human infant intestinal microbiota. *PLoS Biol* 2007; **5**: e177.
47 Macpherson AJ, Uhr T. Induction of protective IgA by intestinal dendritic cells carrying commensal bacteria. *Science* 2004; **303**: 1662–5.
48 Jost T, Lacroix C, Braegger CP *et al.* Vertical mother-neonate transfer of maternal gut bacteria via breastfeeding. *Environmental Microbiology* 2014; **16**: 2891–904.
49 Jadcherla SR. Pathophysiology of aerodigestive pulmonary disorders in the neonate. *Clinics in Perinatology* 2012; **39**: 639–54.
50 Tringe SG, Zhang T, Liu X *et al.* The airborne metagenome in an indoor urban environment. *PloS One* 2008; **3**: e1862.
51 Brodie EL, DeSantis TZ, Parker JPM *et al.* Urban aerosols harbor diverse and dynamic bacterial populations. *Proc. National Academy of Sciences USA* 2007; **104**: 299–304.
52 Womack AM, Bohannan BJM, Green JL. *Biodiversity and biogeography of the atmosphere*, Vol. **365**. 2010.
53 Hasleton PS. The internal surface area of the adult human lung. *Journal of Anatomy* 1972; **112**: 391–400.
54 Fanaro S, Chierici R, Guerrini P *et al.* Intestinal microflora in early infancy: composition and development. *Acta Pediatrica (Oslo, Norway: 1992). Supplement* 2003; **91**: 48–55.
55 Biesbroek G, Tsivtsivadze E, Sanders EAM *et al.* Early respiratory microbiota composition determines bacterial succession patterns and respiratory health in children. *American Journal of Respiratory and Critical Care Medicine* 2014; **190**: 1283–92.
56 Marsland BJ, Yadava K, Nicod LP. The airway microbiome and disease. *Chest* 2013; **144**: 632–7.
57 Sze MA, Dimitriu PA, Hayashi S *et al.* The lung tissue microbiome in chronic obstructive pulmonary disease. *American Journal of Respiratory and Critical Care Medicine* 2012; **185**: 1073–80.
58 Huang YJ, Kim E, Cox MJ *et al.* A persistent and diverse airway microbiota present during chronic obstructive pulmonary disease exacerbations. *Omics: a Journal of Integrative Biology* 2010; **14**: 9–59.
59 Renwick J, McNally P, John B *et al.* The microbial community of the cystic fibrosis airway is disrupted in early life. *PloS One* 2014; **9**: e109798.
60 Zemanick ET, Sagel SD, Harris JK. The airway microbiome in cystic fibrosis and implications for treatment. *Current Opinion in Pediatrics* 2011; **23**: 319–24.

61 Marsland BJ, Gollwitzer ES. Host-microorganism interactions in lung diseases. *Nat Rev Immunol* 2014; **14**: 827–35.

62 Herbst T, Sichelstiel A, Schär C *et al.* Dysregulation of allergic airway inflammation in the absence of microbial colonization. *American Journal of Respiratory and Critical Care Medicine* 2011; **184**: 198–205.

63 Nembrini C, Sichelstiel A, Kisielow J *et al.* Bacterial-induced protection against allergic inflammation through a multicomponent immunoregulatory mechanism. *Thorax* 2011; **66**: 755–63.

64 Smits HH, Gloudemans AK, van Nimwegen M *et al.* Cholera toxin B suppresses allergic inflammation through induction of secretory IgA. *Mucosal Immunology* 2009; **2**: 331–9.

65 Russell SL, Gold MJ, Hartmann M *et al.* Early life antibiotic-driven changes in microbiota enhance susceptibility to allergic asthma. *EMBO Reports* 2012; **13**: 440–7.

66 Riedler J, Braun-Fahrländer C, Eder W *et al.* Exposure to farming in early life and development of asthma and allergy: a cross-sectional survey. *Lancet* 2001; **358**: 1129–33.

67 Olszak T, An D, Zeissig S *et al.* Microbial exposure during early life has persistent effects on natural killer T cell function. *Science* 2012; **336**: 489–93.

68 Park J, Kim M, Kang SG *et al.* Short-chain fatty acids induce both effector and regulatory T cells by suppression of histone deacetylases and regulation of the mTOR-S6K pathway. *Mucosal Immunology* 2015; **8**: 80–93.

69 Smith PM, Howitt MR, Panikov N *et al.* The microbial metabolites, short-chain fatty acids, regulate colonic Treg cell homeostasis. *Science* 2013; **341**: 569–73.

70 Furusawa Y, Obata Y, Fukuda S *et al.* Commensal microbe-derived butyrate induces the differentiation of colonic regulatory T cells. *Nature* 2013; **504**: 446–50.

71 Trompette A, Gollwitzer ES, Yadava K *et al.* Gut microbiota metabolism of dietary fiber influences allergic airway disease and hematopoiesis. *Nature medicine* 2014; **20**: 159–66.

CHAPTER 25

Microbiome and cancer

Ralph Francescone[1] and Débora B. Vendramini-Costa[2]

[1] Fox Chase Cancer Center, Cancer Prevention and Control, Philadelphia, United States
[2] Institute of Chemistry, University of Campinas, Campinas-SP, Brazil

25.1 Introduction

The microbiome is becoming increasingly recognized for its important roles in the homeostasis and wellbeing of mammals. When the delicate balance between host and microbiota is disrupted, the consequences can be deleterious to the host and can lead to (or exacerbate) a wide variety of diseases, including cancer. In this chapter, we explore the general concepts regarding how the microbiome prevents or promotes cancer. Some key concepts that will be discussed include; how the microbiome becomes "unbalanced" in cancer, the interplay between microbes and the host immune system in the tumor microenvironment (inflammation), how microbiota can drive tumorigenesis, and how diet and metabolism alter the microbiome in the setting of cancer. Finally, we will take a brief look at potential therapeutic avenues and where the field is heading in the future.

Cancer is a leading cause of death worldwide and despite the advances in the field of anticancer drug discovery, the statistics are noteworthy; in 2012, 14.1 million new cases of cancer were diagnosed worldwide, with 8.2 million deaths[1]. The increased cancer incidence can be a result of demographics (population aging) and the adoption of "risky" lifestyle choices such as smoking, high-fat diets and physical inactivity[2].

The emergence of a cancer (carcinogenesis) is a complex and multistep process where normal cells progressively acquire a neoplastic phenotype. It can be divided into three phases: tumor initiation, tumor promotion and tumor progression. Tumor initiation involves irreversible changes on DNA, by chemical or physical carcinogens, leading to the activation of oncogenes and/or inactivation of tumor suppressor genes. At the tumor promotion stage, stimulation by certain agents or process known as promoters lead mutated cells to expand by increasing cell proliferation and/or reducing the processes of cell death[3]. Finally, increase in size, invasion of adjacent tissues and metastasis characterize the progression stage, when tumors can accumulate other mutations, thus exacerbating their phenotype.

The Human Microbiota and Chronic Disease: Dysbiosis as a Cause of Human Pathology, First Edition.
Edited by Luigi Nibali and Brian Henderson.
© 2016 John Wiley & Sons, Inc. Published 2016 by John Wiley & Sons, Inc.

Each genetic modification confers to tumor cells a type of advantage, such as self-sustained proliferation, evasion of growth signals suppressors, resistance to cell death, limitless replication, induction of angiogenesis and activation of invasion and metastasis processes. Besides cancer hallmarks, the tumor microenvironment also influences cancer development, and one prominent microenvironmental stimulus in carcinogenesis is inflammation[4,5]. Evidence for the involvement of inflammation on tumor development has been demonstrated in studies on the regular use of aspirin, which is associated with decreased risk for colorectal cancer development[6–8]. Also, some randomized trials have shown that nonsteroidal anti-inflammatories (NSAIDs) have protective action against colon, breast, prostate and lung cancers[9,10].

Inflammation plays an important role in cancer development, either by increasing cancer risk (as in many cases of chronic inflammation) or being induced by growing cancers, the so-called tumor-elicited inflammation event[11]. Products released by inflammatory cells during chronic inflammation are potentially genotoxic and can damage DNA. This excessive production of pro-inflammatory mediators leads to the establishment of oxidative stress, which functions as a chemical effector of carcinogenesis. These chemical effectors, when not inducing cell death, can cause damage to the DNA of stromal and epithelial cells. In this case, mutations may activate oncogenesis and/or inactivate tumor suppressor genes, resulting in genomic instability[12]. Besides the production of reactive oxygen species (ROS), there is activation of tumor necrosis factor (TNF), interleukins 1 and 6 (IL-1 and IL-6) and chemokines, mediators known to be amplifiers of the inflammatory response, thus leading to activation of transcription factors that modulate and reinforce the pro-inflammatory, survival and antiapoptotic profiles. At this stage, initiated cells depend on a "promotion environment", where there is exacerbation of proliferative signals, to the detriment of cell death processes — both of which can be driven by inflammation. The tumor is now able to recruit more inflammatory cells, which will contribute to its growth, nutrition and expansion. Many are the stimuli for cancer development, from environmental factors to pollutants, diet, physical stress, infections and hereditary factors[13]. Recently, the internal microbiota was recognized as an environmental factor that also contributes to carcinogenesis[14]. The mechanisms underlying the pro-carcinogenic profile of microbiota are still under investigation, but so far it has been demonstrated to have major roles in cellular metabolism, immune and inflammatory responses. There are two categories of microbiota driving carcinogenesis: those that are acquired by infectious processes (external environment) and those that rely on the internal commensal microbiota and its "corrupted" function.

The relationship between microbial infections and disease is already known for some types of cancer, where the infection promotes cancer development (Table 1). Up to 20% of cancers can be attributed to chronic infections, and besides the ability of some pathogens to directly induce cell transformation, the majority of them induce chronic inflammation at the site of infection[15]. Despite the well-known link between infectious processes and the development of some types of cancer, there is evidence suggesting that global changes in internal microbiome also may promote human diseases, including cancer[16].

Humans have around 10 times more microbial cells in their body than their own cells[17]. In general, these organisms are confined to epithelial tissues, especially

Table 1 Microorganisms associated with cancer development: well established associations between a microbe and the development of a particular type of cancer.

Microbial infection	Cancer type
Helicobacter pylori	gastric, esophageal
Human papillomavirus (HPV)	anogenital, oropharyngeal
Epstein-Barr virus (EBV)	lymphomas, nasopharyngeal
Human immunodeficiency virus (HIV)	lymphomas, Kaposi's sarcoma
Hepatitis B virus	hepatocellular
Hepatitis C virus	hepatocellular, lymphomas
Human T-cell lymphotropic virus type I (HTLV-1)	adult T-cell leukemia, lymphoma
Human herpesvirus 8 (HHV-8)	Kaposi's sarcoma

in the intestinal mucosa, skin and genital tract; interestingly, their taxonomic composition varies not only between different individuals, but also between different organs and intra-organ compartments inside the same individual. This gives a very "personalized" character to the microbiome-oriented disease. But if the internal microbiome can have a pivotal role on carcinogenesis, why doesn't everybody have cancer? The key concept for this question is homeostasis. Under normal conditions, the microbiome is in a symbiotic interaction with the host, what means that both have biological benefits from this interaction (eubiosis). On one side, we provide a nutrient-rich environment for these microorganisms, and on the other, we benefit from their metabolic capacities to harvest essential non-nutrient factors and also from the protective barrier that they promote by limiting available resources to potential external pathogens[18].

The homeostatic coexistence between the microbiome and host is possible because they are physically separated by multi-level barriers; disruption of these barriers may be one of the factors leading to the "corrupted" role of microbiome in promoting diseases[16]. So, the host immune system tolerates the normal microbiota, but provides immunosurveillance against external pathogens[19]. Another key factor that defines the pro-carcinogenic effect of the microbiome is changes in its composition, abundance and function, which can lead to changes in microbial composition: under inflammatory conditions, there is a decrease of obligate anaerobic bacteria and increase of facultative anaerobic (mainly bacteria belonging to the family Enterobacteriaceae), which can use nitrate as a source of energy[20].

The microbiome-driven disease can occur not only by direct exposure to the microbiota. A fitting example was provided by Brandl and Schnabl in chapter 22, who highlighted that the liver, for example, does not have its own microbiome, but still is influenced by gut microbiota. This occurs because these organisms can either migrate or release subproducts and metabolites, the microorganisms' associated molecular patterns (MAMP), which can travel to other regions nearby and be recognized by immune cells present in the environment (also discussed by Bermudez in chapter 6). In fact, translocation of intestinal bacteria and MAMPs is common in chronic liver disease, and is a risk factor for hepatocellular carcinoma (HCC) development[21,22]. Besides the liver, the breast is also known to be influenced by translocation of bacteria and metabolites. This fact amplifies the "limited" view of microbiome-host organ: it can act on a systemic level.

25.2 Microbiome and cancer: where is the link?

Microbes and their metabolites can be sensed by specialized receptors called Toll-like receptors (TLR) and their adaptor MyD88 on macrophages, myofibroblasts, epithelial cells and tumor cells and downstream its activation is the innate immune response, with pro-inflammatory stimuli[23]. In an interesting way, these receptors can sense not only microbiota-derived patterns, but also endogenous damage-associated molecular patterns (DAMPs), which are released by stressed and dying cells. So, in a stress situation, where barriers disruption occurs or there is an infection, a pro-inflammatory process occurs that can lead to more reactive species release and aggravation of tissue injury. The chronic injury results in more infiltration of microbes and microbial products. In the case of colon cancer, the rapid tumor growth and expansion can lead to barrier disruption, which will attract more bacteria and result in amplification of the inflammatory response. Chemokines and cytokines released on this very compromised tissue will serve as growth factors, mitogens, and activators of wound healing, migration and angiogenesis, thus acting as extra signals for tumor promotion[24].

In the molecular view, activation of nuclear factor-κB (NF-κB) and signal transducer and activator of transcription 3 (STAT3) transcription factors by Th1 and Th17 immune profiles are the major players in the cancer — the link to inflammation. Recently it was demonstrated that infiltration of commensal microbiota in the tissue due to barrier defects induced IL-23/IL-17, IL-22 and IL-6 activation in colon adenoma models[25]. In this study, genetic inactivation of *Tlr2*, *Tlr4* and *Tlr9*, as well as antibiotic treatment decreased IL-23/IL-17 expression, thus showing the importance of the microbiota and its recognition for the pro-inflammatory signaling[25].

In a more direct way, bacteria can interact with host cells and induce DNA damage by producing genotoxic toxins. Bacterial toxins known to have genotoxic potential are cytolethal distending toxin (CDT) and colibactin, both affecting cellular responses to DNA damage. CDTs constitute a family of bacterial protein toxins that are produced by Gram-negative bacteria including *Escherichia coli* and they act by generating imbalance between microbiota and host (dysbiosis), but the mechanisms leading to these modifications are not well understood. These variations can be due to host genetics and environmental factors, such as age, exposure to different diets, stress, therapies and daily environment[14]. Internal factors as well can drive these changes and in this aspect inflammation may have an important function, as the production of specific mediators can favor the survival of different sets of microorganisms by providing sources of energy or by promoting selective pressures[16]. As an example, the inflammatory process in the large bowel is accompanied by induction of DNA strand breaks. Similarly, colibactin is a genotoxin encoded by the *pks* genomic island of *E. coli* strains that can induce double-strand DNA breaks and cell cycle arrest, thus promoting mutations and chromosomal instability[26]. Bacterial products can also be genotoxic, as hydrogen sulphide, superoxide radicals and oxide nitric, or ROS and reactive nitrogen species

(RNS) that are powerful mutagens causing point mutations, DNA breaks and protein-DNA crosslinking, also leading to chromosomal instability[27].

But if on one side studies have shown that under depletion of TLRs, cytokines, germ-free environments and antibiotic therapy animals show a decreased incidence of tumors, on the other hand optimal responses to cancer therapy require an intact commensal microbiota[28]. One example is therapy with oxaliplatin, which "uses" gut microbiota as Trojan horses, as they will promote tumor infiltration of myeloid cells, which will produce ROS as one of the antineoplastic effects[19]. Moreover, commensal microbiota act as guardians, protecting the host from invading pathogens[29]. Thus, this is a two-way street: under infectious and inflammatory processes microbiota can promote cancer but commensal microbiota may be important for the effectiveness of anticancer therapy. The general concepts regarding tumor progression and the microbiome are summarized in Figure 1.

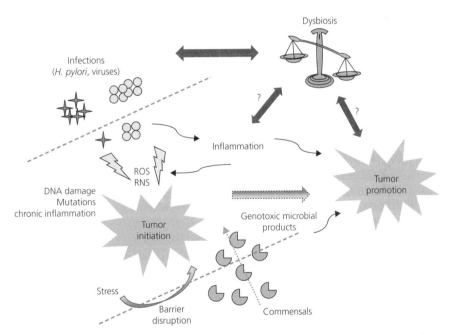

Figure 1 Overview of the core concepts: microbiome and cancer progression. The microbiome is a complex collection of microorganisms that inhabit the human body. In cancer, the microbiome can influence cancer initiation and promotion in a wide variety of ways. First, pathogenic microbes can induce damage to cells directly through induction of DNA damage, ROS/RNS, and chronic inflammation, which promotes the initiation of transformation. Pathogenic microbes can also alter the resident microbiota in the host, leading to dysbiosis. Commensal microbes can also enhance tumorigenesis by overamplifying inflammation when mislocalized, due to barrier disruption, to compartments of tissue they normally do not reside in. The microbiome can also generate genotoxic products through metabolism, further advancing tumor growth. Question marks (?) signify unanswered questions pertaining to dysbiosis in the field. Is dysbiosis a result of cancer, or is it a two-way street? What are the mechanisms involved with inflammation and dysbiosis?

25.3 Microbiome and barrier disruption

The location of resident bacteria in an organ, especially the colon, is crucial to the proper function of that organ. These commensals are generally harmless and are often necessary for proper immune system development and intestinal homeostasis[30]. However, the disruption of the epithelial layer of organs, either by injury, genetic predisposition, or tumor overgrowth, can allow microbes and microbial products to move into previously uninhabited areas, eliciting a robust immune response. If the damage to the epithelial barrier is left unchecked, chronic inflammation ensues, and can lead to DNA damage and even cell death[11].

When cell death occurs, contents of the cell that would not normally be exposed to the extracellular environment, such as heat shock proteins, danger signals (HMGB-1 and S100A8/A9), and DNA, spill out into the local microenvironment and act as pro-inflammatory cues, further exacerbating the damaging inflammatory response. This can promote pre-existing tumors or may help to initiate cancer, as roughly 20% of all cancers are preceded by inflammation. This phenomenon has been demonstrated to be a key driving force behind the progression of inflammatory bowel diseases (IBD) and colon cancer[25,31]. Additionally, the functionality of the mucus layer is critical for maintaining organ homeostasis and preventing bacteria from penetrating the epithelial layer. The colonic mucus is bi-layered and its main component, known as mucin-2 (MUC-2), forms a gel-like structure that can slow and trap bacteria in the less structurally restricted outer layer[32,33]. The inner layer is directly attached to the epithelium and generally remains microbe-free. Bacteria have many proteolytic enzymes that can degrade the outer layer, and this forms a habitat for many species to thrive. Mucus layers differ slightly in other organs, such as a single MUC-2 layer in the small intestine, consisting of larger pores, or a different major mucin component, MUC5AC, in the stomach.

Recent studies in mice have begun to elucidate some of the mechanisms behind the roles of the mucus layer. Muc-2 KO mice developed spontaneous colitis, had increased CD3+ immune cell infiltration, had elevated TNF-α and IL-1β expression, and had increased disease scores when treated with the chemical irritant dextran sodium sulfate (DSS), indicating the importance of the mucus layer in the protection of epithelial cells[34]. MUC-2 defects have been found in mouse models and in patients with IBD, allowing bacteria to adhere and penetrate the epithelial layer, causing disease-promoting inflammation[35]. MUC-2 is also critical for immune tolerance towards commensals, keeping potentially harmful inflammation in check[36]. It was shown that glycans attached to MUC-2 conditioned dendritic cells with an anti-inflammatory program, through increased IL-10 expression and inhibition of inflammatory NF-κB signaling, allowing immune tolerance towards commensal bacteria. Overall, the mucus layer is a vital element in the architecture and function of organ systems, providing a protective covering for epithelial cells and maintaining a healthy balance between the immune system and resident microbes. Figure 2 summarizes the impact of the barrier integrity on the progression of cancer.

Interestingly, antibiotic depletion of microbes was shown to reduce inflammation and tumorigenesis in mouse models of colon cancer and in patients with active IBD[25,31,33,37], suggesting that the flux of microbiota past the protective mucus and epithelial layers plays a key role in eliciting disease causing inflammation. Nevertheless, non-specific elimination of bacteria may not be the most efficacious strategy to

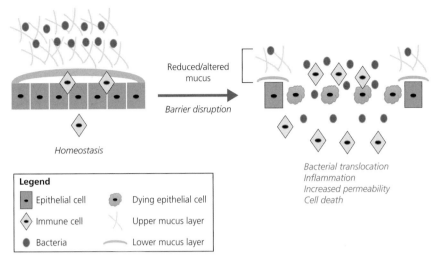

Figure 2 Barrier disruption in mucosal tissue. Under homeostatic conditions a mucosal tissue, such as the colon, has an intact dual mucus layer, with the upper layer serving as an ecosystem for resident microbes and the bottom layer attached to the epithelium as barrier to microbes. The epithelial layer maintains tight contacts and has basal amounts of immune cells. When the tissue barrier integrity is disrupted, as in cancer, the mucus layers are diminished and tight junctional contacts between epithelial cells deteriorate, increasing permeability. Cells undergo death, releasing inflammatory factors that recruit and activate the immune system. Bacterial translocation then occurs as commensals come into contact with the epithelium and penetrate deep into the tissue, further exacerbating inflammation. This leads to perpetual inflammatory conditions, promoting tumor progression.

combat microbial-driven inflammation in cancer. This is due to the fact that, although pathogenic organisms can lead to tumor development, the vast majority of microbes that penetrate into tissue during disease are normal residents of the body and are often beneficial to the host. Thus, broad-spectrum antibiotic therapy may ameliorate symptoms in the short term but cause dysbiosis in the long term, as opportunistic microbes could occupy the niche of beneficial bacteria after therapy. Moreover, it has been demonstrated that pre-operative antibiotic therapy is not effective in humans with colon cancer and actually leads to an increased incidence of *Clostridium difficile* infections[38]. Thus, targeting the tumor, the tumor-promoting inflammation, or repairing the underlying barrier defects would most likely be the best course of action clinically, as it would not be feasible to stay on antibiotics for the course of the disease, due to toxic side effects of the drugs and the lack of positive microbes.

25.4 Microbiome and different types of cancer

25.4.1 Colon cancer

The effect of the microbiome on tumorigenesis has been most extensively studied in colorectal cancer. The colon is an excellent model to study the influence of the microbiota on cancer for a few reasons. First, trillions of bacteria reside in the colon, and consequently, a large proportion of the host immune system is contained there as well, offering researchers an opportunity to investigate how bacteria and inflammation affect tumor development. Second, there are a wide variety of excellent mouse models of colitis associated and spontaneous colon cancer[11], allowing a much more thorough examination of the mechanisms involved in tumorigenesis relating to the microbiome (genetic manipulation, fecal transplants, bone marrow transfers, microbial depletion and colonization, pre-clinical therapies, etc.), when compared to human studies alone. Because colon cancer and the microbiome are discussed extensively in the next chapter, we will not cover them in depth here; instead, we highlight recent findings of the effects the microbiome has on cancers of the skin, breast, and liver and how the gut bacteria influence cancer at distant sites.

25.4.2 Skin cancer

The skin is an organ system with a large surface area that serves as our primary barrier to the outside world. Thus, it comes into contact with a diverse range of microbes and serves as an ecosystem for these bacteria[39]. Characteristics of the skin microbiota have been covered by Zeeuwen in chapter 5. As with the colon, the skin has developed a symbiotic relationship with these microorganisms and selects for a certain microbiota composition through a complex interaction of immune tolerance and beneficial characteristics of the resident microbes. One recent study demonstrated the importance of the host adaptive immune system in maintaining skin barrier function and restricting commensals from invading farther through the skin layers in mice[40]. This group uncovered, using a RAG-1 KO mouse model to eliminate all adaptive immune cells in the mice, that a subset of T-cells was responsible for limiting bacterial expansion on the surface of the skin (especially mycobacterium), as well as limiting the influx of bacteria into the draining lymph nodes of mice, partially through IL-17A and IFN-γ mediated mechanisms. Along the same lines, another group found that germ-free mice contained lower amounts of IL-17A producing cells, and a concomitant increase in Treg cells in the skin[41]. Moreover, they showed that mono-colonization of the skin with *Staphylococcus epidermidis* restored T-cell effector function, signified by increased levels of IL-17A, the ability to resist *L. major* infection, and that this was dependent on the IL-1 arm of the microbial sensing pathway.

These studies underscore the importance of bacterial colonization in the proper functioning of the host immune system and maintenance of the skin ecosystem. When this delicate balance is disrupted, diseases of the skin can arise, including psoriasis, dermatitis, and perhaps cancer, although this field is still in its infancy. In cancer cachexia (wasting syndrome) patients, the skin microbiome profile of secondary phyla shifts towards more *Staphylococcus* spp. and fewer *Corynebacterium* spp., suggesting that the disease state alters the habitats for resident microbes[42]. In relation to skin cancer, more work must be done first to define bacterial population shifts in healthy vs. skin cancer patients, followed by dissecting the underlying causes for the shifts. A very interesting study done using a two-step skin carcinogenesis model (7,12-dimethylbenz(a)anthracene [DMBA] and croton oil [CO]), demonstrated that High Mobility Group Box 1, HMGB-1 (a danger signal), through TLR-4 signaling was critical for inflammation and tumorigenesis, and that lipopolysaccharide (LPS) treatment actually inhibited tumorigenesis, suggesting that in this model, commensal bacterial products may impede cancer development[43]. It is tempting to speculate that LPS, which also binds TLR-4, has some negative/tolerizing effect on tumor-promoting inflammation in this model. This lends credence to the idea that resident bacteria can have protective functions in the skin. Bacteria can also promote tumorigenesis, as microbial flagellin, through its receptor TLR-5 on host cells, exacerbated wound-induced skin cancer by increasing immune cell infiltration, inflammation, and HMGB-1 release from immune cells[44]. Moreover, antibiotic depletion of skin microbes reduced carcinogenesis, implying that when commensals can bypass the protective skin barrier (as in a wound or extreme inflammatory condition), they have the ability to promote tumor development. This is consistent with colon cancer studies, as barrier disruption leads to chronic inflammation and exacerbation of tumorigenesis, as discussed earlier. Therefore, depending on the context, commensals can ameliorate or exacerbate cancer.

25.4.3 Breast cancer

While no concrete examples have determined the effect of microbiota in breast tissue on breast cancer, because breast tissue was thought to be sterile, a recent report has confirmed a unique microbiota composition in the breast distinct from other organs[45]. Here, the authors determined that normal breast tissue is inhabited by large numbers of *Proteobacteria* and *Firmicutes* groups, and they suggest this is owed to their ability to effectively process fatty acids, as the breast is a high-fat tissue. Moreover, there is a clear shift in a few species of bacteria in breast cancer tissue compared with normal breast tissue[45,46]. This first group observed an increased abundance of *E. coli* in tumor tissue and the second group found that tumor tissue from breast cancer patients was enriched with the bacterium *Methylobacterium radiotolerans*, while paired healthy tissue was enriched with *Sphingomonas yanoikuyae*. Certain strains of *E. coli*, such as NC101, have been previously associated with tumor progression in colon cancer[37], which is consistent with this finding in breast cancer. As for the latter study, two noteworthy results were observed. First, unlike in colon cancer and wound-induced skin cancer, the microbial sensing network, and its downstream effectors, were significantly down-regulated in breast cancer tissue. Second, overall levels of the bacterial 16S gene were lower in tumor tissue compared to normal breast tissue, indicative of less bacterial load in tumor tissue, and the bacterial load also decreased with tumor stage. This would seem to suggest that in breast cancer the local microbiota may play a protective role, possibly through modulation of the immune system, but further work must be done to explore the role of the breast microbiota in tumor development and progression.

25.4.4 Liver cancer

As previously mentioned, Hepatitis B and C infections are classic examples of microbes inducing chronic inflammation, which eventually initiates cancer in the liver. In addition, cirrhosis from heavy alcohol intake or a high-fat diet leading to fatty liver disease can also cause chronic inflammation in the liver and can spark tumor development. How the microbiome shapes the development of liver cancer has not been fully elucidated, however microbial sensing once again comes to the forefront as a potential regulator of tumorigenesis. But how do microbes reach the liver? Brandl and Schnabl in chapter 22 introduced us to the concept that the presence of a leaky gut facilitates the crosstalk between gut and liver, allowing microbial products to provoke an inflammatory response in the liver. One major route for bacteria to travel to and interact with the liver is through the portal vein[47]. The portal vein contains metabolic and microbial products that are constantly flowing from the intestine, and thus the liver, like the colon, must be tolerant of microbes but also limit their expansion towards the epithelia. Therefore, the balance between immune tolerance versus immune reaction to the microbiome is critical for tumor progression, as the combination of inflammation and immune suppression shapes the microenvironment of the tissue.

In a chemically induced and injury-sustained HCC model, diethylnitrosamine/carbon tetrachloride (DEN/CCl$_4$), TLR-4 deficient mice had reduced tumor progression, as hepatic proliferation and healing markers were markedly lower, resulting in slower-growing tumors[21]. To see whether or not intestinal microbiota had an effect on liver tumor development, either antibiotic depletion or LPS infusion was performed. Antibiotic depletion proved to limit tumorigenesis, while LPS infusion exacerbated tumorigenesis. Lastly, it was further shown that the effect of intestinal microbiota on cancer progression was most important at later stages of development, as late-stage antibiotic depletion was more effective in limiting tumor numbers than early-stage antibiotics. All in all, intestinal bacteria, or their products, promoted liver cancer in a chemically induced chronic-injury model. Along the same lines, mice lacking the TLR-4 adaptor protein, MyD88, also had reduced tumorigenesis in chemically induced liver cancer[48], signifying the importance of microbial sensing in liver cancer progression. On the other hand, another group used the same TLR-4 deficient mice but only injected the carcinogen DEN, and did not add a chronic irritant, such as CCl$_4$[49]. Surprisingly, they found exactly the opposite results: TLR-4 was protective against cancer in this model, as loss of TLR-4 activity was associated with decreased inflammatory signature, reduced DNA repair through Ku70, and increased susceptibility to ROS-induced DNA damage. Thus, the authors conclude that TLR-4 is important for liver homeostasis and effective immune clearance of damaged/senescent cells, preventing potential pre-neoplastic cells from developing into cancer cells.

These seemingly conflicting results actually may be complementary, due to the fact that one model relied on repeated injury and the other did not. In the context of other cancers, the repeated damage induced by CCl$_4$ would greatly perturb the integrity of the epithelial layer, and may result in a situation similar to barrier disruption seen in colon cancer, causing microbes to enter regions of the organ not normally exposed to bacteria. Therefore, a massive influx of microbes would generate an overwhelming immune response, exacerbating cancer. On the other hand, in the second model, the liver architecture is more or less intact, and in this situation TLR-4 could be necessary for proper immune homeostasis and control of microbes, limiting their impact on tumorigenesis. Similar contrasting findings were seen with TLR-2 as well in liver cancer[21,50]. Overall, these studies underscore the complexity of the microbiome on overall organ health, and more work must be put forth to understand if dysbiosis occurs in the gut/liver during liver cancer and if specific commensals promote/impede carcinogenesis, in order to pinpoint targeted therapies to restore balance. Figure 3 illustrates the inflammatory/

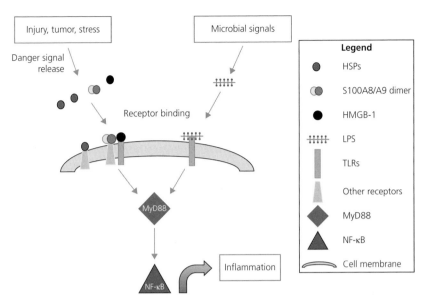

Figure 3 Microbial sensing in the progression of cancer. In many cancers, microbial sensing through TLRs and MyD88-dependent signaling generally promotes cancer. The microbial sensing pathway can be activated in two ways, directly through the binding of microbial products (as in barrier disruption), such as LPS, to TLRs or indirectly through the production of danger signals by tumor cells or immune cells in the tumor microenvironment and then subsequent binding of danger signals to TLRs or other pattern-recognition receptors. In both cases, the adaptor protein MyD88 is activated, which in turn activates the master inflammatory transcription factor NFκB, generating tumor promoting inflammation. Tumor cells or immune cells can respond to danger signals or microbial products.

pro-tumorigenic side of microbial sensing, which is a reoccurring theme in the microbiome and cancer relationship.

25.4.5 Local microbes affecting distant cancers

While it is obvious that the local microbial communities of a tissue would induce responses from the host at that particular site, some exciting findings implicating the effects of the gut microbiome on distant tissues, as in breast cancer, have been recently uncovered. For example, in the APC[Min/+] mouse model of colon tumorigenesis, colonization of mice with the intestinal pathogen *Helicobacter hepaticus* amazingly promoted cancer in the breast, and further enhanced tumorigenesis in the Rag2 KO background[51]. In addition to breast cancer, *H. hepaticus* in the intestine also accelerated aflatoxin B1-induced liver tumorigenesis in mice[52]. The model by which *H. hepaticus* promoted liver cancer was presumably through upregulated TNF-α and IFN-γ in the liver, which increased inflammation[53]. Furthermore, induction of metabolic stress and oncogenic signaling was seen in the liver as well, and this combination resulted in abnormal proliferation and cancer. Although *H. hepaticus* acted as a pathogen in these models, it nevertheless demonstrates that the microbiome of one tissue can influence the homeostasis of another tissue, illustrating that in a case of dysbiosis, one may expect similar results when commensal populations become altered or if the niche for a pathogenic microbe became available.

Results from clinical studies have implied a correlation between pancreatic cancer risk and infectious microbes[54]. In particular, *Helicobacter pylori* in the stomach, and *Porphyromonas gingivalis* in the oral cavity during periodontal disease, have been shown to be associated with pancreatic cancer risk, as *H. pylori* increased risk by up to 38% and periodontal disease up to 64%[55,56]. While evidence is lacking, the proposed mechanism of infection-induced pancreatic cancer relies on systemic inflammation and stress, thus activating a pro-tumorigenic microenvironment in the pancreas. Why these specific microbes induce pancreatic cancer preferentially remains unknown, but dysbiosis leading to opportunistic colonization of these bugs may cause overamplification of microbial- sensing/inflammatory pathways (TLRs and NF-κB), eventually inducing oncogenesis after prolonged inflammation.

Metabolic products from bacteria also influence cancer progression in distant organs. An excellent example of key metabolites of microbes is in the processing of estrogens[17]. Depending on the modifications to estrogen, different cellular effects can occur, as well as drastically altered pharmacodynamics/kinetics. In relation to the clinic, higher circulating estrogen levels correlate with higher risk of endometrial and breast cancers[57,58]. As one can imagine, estrogen metabolism by microbes is only one of the potentially thousands of metabolites generated by the microbiota of the gut, and the variability of bacterial populations among individuals only further increases the sheer volume of byproducts that could affect cancer initiation and progression. The characterization of metabolites, and their assignment to specific bacterial species, could afford opportunities for therapeutic intervention, diagnosis, and prognosis.

In sum, the influence of the intestinal microbiota is immense on the host as a whole. Tumorigenesis in areas far from the intestine is affected in a number of ways by the microbiota there, including modulation of the immune system, systemic inflammation, oncogenic signaling, and metabolite production. As it is well established that pathogenic microbes can cause systemic inflammation and promote cancer development, it will be interesting to dissect the contribution of resident intestinal commensals on distant tumor development.

25.5 Microbiota and metabolism: the good and the bad sides

The microbiome and the host interact at the metabolic level, as the human metabolism is a combination of microbial and human enzymatic machinery[59]. It is well known that the microbiota plays an important role on carbohydrate and protein fermentation, enzymatic processing of polyphenols, metabolism of hormones and xenobiotic metabolism, but many other processes as well are probably microbiota-dependent. The way to be processed depends on diet consumption (if there are more complex carbohydrates or more proteins) and on the exposure to different xenobiotics; these choices will also define the microbiota composition. In the case of colorectal cancer, diet contributes as a risk factor and it is already known that high intakes of red and processed meat and high alcohol consumption have a causative role, whereas intake of dietary fiber is suggested to have a protective effect[60].

Human enzymes cannot degrade complex carbohydrates and many plant-derived polysaccharides as fibers, which will be done by microbiota fermentation in the colon. This process generates energy for microbial essential functions and products such as short-chain fatty acids (SCFAs), mainly acetate, propionate and butyrate, which have beneficial effects for the host[61]. Butyrate is a SCFA that is selectively transported to the colonic epithelium, where it is the primary source of energy for colonocytes (70%). Basically all sources of butyrate come from the microbial metabolism and without butyrate, colonocytes would undergo cell cycle arrest and autophagy[62]. Interestingly, butyrate has controversial roles on normal and tumor tissue: in normal tissue it promotes colonocyte growth, but it decreases colorectal tumor cell proliferation and increases their apoptosis rate. This "selectivity" occurs because tumor colonocytes undergo the Warburg effect, increasing glucose intake, which results in a decrease on both oxidative metabolism and butyrate metabolization[14]. The accumulation of butyrate in the cell nucleus leads to histone deacetylation inhibition, thus controlling antitumor gene expression. Butyrate also has anti-inflammatory activities — there histone deacetylase inhibition activity leads to a decrease in the ability of intestinal macrophages to produce pro-inflammatory mediators like NO, IL-6 and IL-12[63]. Indeed, butyrate produced by microbial metabolism upregulates Foxp3 gene and thus colonic Treg cells function, which have a central role in the suppression of inflammatory responses[64]. On the other hand, protein fermentation generates toxic metabolites such as ammonia, phenols, nitrosamines and sulphides, which can be cancer promoters. A good example is the excessive consumption of red meat, which is among the top ten factors associated with incidence and progression of cancer in a World Cancer Research Foundation report[65]. In a high-protein diet, particularly red meat, there is a high activity of sulfur-reducing microbiota that generates hydrogen sulfide, a genotoxic metabolite. Studies show that patients with colon cancer and IBDs have higher levels of sulfur-producing bacteria such as *Desulfovibrio vulgaris*[66]. The hydrogen sulfide is generated from L-cysteine and can activate potassium channels, stimulate kinases and inhibits phosphodiesterase, thus playing an important role in proliferative and energetic pathways. A recent study showed that there is an overproduction of hydrogen sulphide in colon cancer that is correlated to tumor's cellular bioenergetics and angiogenesis maintenance[67].

Many studies show the antioxidant and anti-inflammatory activities of plant polyphenols, which can have a protective effect against cancer. As an example, phenolic compounds (especially high in coffee, tea, berries, nuts, vegetables, and cereals) are poorly absorbed and may accumulate in the colon, where they can be subjected to conversion by microorganisms into metabolites with potential biological activity[68]. The same is true for phytoestrogens as genistein, which have antiproliferative and pro-apoptotic effects on colon cancer[68,69].

The microbiome also has a role in metabolism of xenobiotics, either by detoxifying carcinogens or contributing to side effects, as secondary metabolites can be toxic. The gut bacterial enzymes β-glucuronidases can contribute to the severe diarrhea that is associated with a commonly used colon-cancer chemotherapy drug called irinotecan. The balance between circulating and excreted estrogens levels is also controlled by microbiota through its ability of deconjugate estrogens, and it is already known that estrogen levels has a direct effect on breast cancer risk, thus connecting microbiota metabolism to cancer risk[17]. The intestinal microbiota is also important for the deconjugation of taurine and glycine conjugates bile acids, which are precursors of secondary bile acids. High concentrations of secondary bile acids in various biological fluids have been associated with diseases, by modulating signaling pathways related to lipid peroxidation, triglyceride storage and production of hepatic fatty acids. One of these secondary bile acids, deoxycholic acid, is associated with several models of carcinogenesis and the enzymatic activity of 7a-dehydroxylating bacteria (some *Clostridia*), which may be a target in the study of risk factors for gastrointestinal tumors[70].

Another important factor for the risk of many diseases including cancer is obesity. Studies have shown a correlation between the composition of the microbiome and consumption of high-fat products, which reduces the total volume of the intestinal microbiota and induces the growth of Gram-negative bacteria, thus contributing to dysbiosis[71]. Chapter 23 by Fak described the presence of an "obesogenic" microbiota. Obese mice exhibited a decreased frequency of *Bacteroidetes* and an increased proportion of *Firmicutes*[72]. Obesity also contributes to increased prevalence of *Clostridia*, leading to deoxycholic acid production and augmented risk for HCC development[73,74]. In summary, the relationship among microbiome, metabolism and cancer prevention or promotion depends on our daily decisions, which will drive microbial growth.

25.6 Chemotherapy, the microbiome and the immune system

As previously discussed, the microbiome has a profound effect on the function of the immune system, and it is therefore not surprising that resident microbes can regulate the efficacy of chemotherapy through antitumor immunity. Two significant studies confirmed the role of the intestinal microbiota in the ability of chemotherapeutics to fight cancer[28,29]. The first group determined that commensal microbiota were essential in mediating the anti-cancer effects of the chemotherapy treatments[28]. In the first treatment regimen, which consisted of CpG-oligodeoxynucleotides, a TLR-9 ligand, and an anti-IL-10R antibody, the tumor-killing mechanism was found to be through CD11b+ or Ly6C+ myeloid cell production of TNF-α, and this was dependent on TLR-4 signaling, suggesting that sensing of the microbiome is essential for drug effectiveness. Additionally, antibiotic depletion of commensals inhibited therapy. Similarly, in the next chemo regimen, the efficacy of oxaliplatin to impede tumor growth in mice was demonstrated to be regulated by commensal-induced infiltration of ROS-producing Gr1+ myeloid cells, which stunted tumor growth. In the second study, treatment of tumor bearing mice with cyclophosphamide, an alkylating chemotherapeutic, induced dysbiosis and increased intestinal permeability. Nevertheless, the resulting microbiome, which shifted to contain more Gram-positive bacteria, was necessary for the efficacy of cyclophosphamide in killing tumors, which was through a "pathogenic" Th17 cell response[29]. Antibiotic depletion of the intestinal microbiota or of Gram-positive bacteria disrupted the pathogenic Th17 cell response against the tumor after cyclophosphamide treatment. These studies establish a key role for the microbiome in the interplay between anti-cancer immunity and chemotherapeutic efficiency.

The relationship between chemotherapeutic agents and the microbiome could have extensive clinical implications, and consideration of the general dysbiosis and tissue damage caused by drugs, as well as the immunomodulatory properties of commensal microbiota, will be vital for the success of therapy.

25.7 Therapeutic avenues

Modulation of the microbiome has limitless potential in therapeutic endeavors. Possible interventions include but are not limited to: 1) Modification of the activity of bacterial enzymes that interfere with host homeostasis; 2) Development of extremely selective antibiotics to eliminate problematic species and restore balance to microbial populations; 3) Pre/probiotics; 4) fecal transplantation (Figure 4).

25.7.1 Modulation of bacterial enzyme activity

Targeted alteration of bacterial enzyme activity is an attractive therapeutic option, as it maintains microbial population balance, while removing a disease-causing characteristic of the microbiome. Some exciting proof-of-principle studies undertaken have demonstrated effective modulation of specific enzymes in bacteria without inducing dysbiosis. Deletion of the pks island from the NC101 strain of *E. coli* reduced the severity of colitis-associated cancer in IL-10 KO mice[37]. In another study, administration of lipoteichoic-acid–deficient *Lactobacillus acidophilus* strain limited colitis and cancer burden in mice[75]. One research group generated specific inhibitors against the bacterial enzyme β-glucuronidase, present in many commensal species, which ameliorated the toxic side effects of the chemotherapy drug irinotecan in cancer patients without causing dysbiosis in the gut[76]. These are

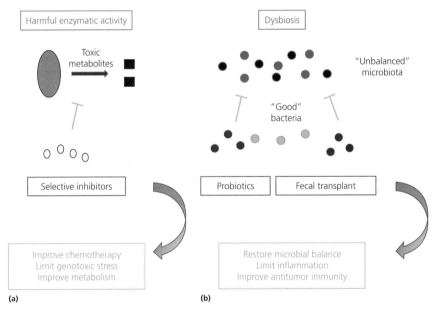

(a)   (b)

Figure 4 Therapeutic opportunities: selective targeting is the best medicine. Inhibition of bacterial enzymes with selective inhibitors can be a minimally disruptive approach to reduce harmful effects of commensal microbes without inducing dysbiosis. Therapeutically, chemotherapy and metabolism could be improved, reducing pro-tumorigenic capacity of commensals. Administration of "good" bacteria, through probiotic intake or fecal microbiota transplant, to patients suffering from dysbiosis could help to restore balance to the host microbiome, limiting pro-tumorigenic inflammation and enhancing antitumor immunity.

some of the noteworthy recent findings that will undoubtedly pave the way for the development of a wide variety of targeted bacterial therapies applied to a number of conditions, including cancer.

25.7.2 Antibiotics

There is substantial evidence that elimination of cancer-causing pathogenic bacteria, such as *H. pylori*, through the use of antibiotics, is a clinically effective way of ameliorating damaging inflammation and cancer risk associated with a particular pathogen[77]. However, indiscriminate targeting of the microbiome, which has been shown to be effective in many mouse models of inflammation and cancer, can also induce dysbios and thus be detrimental to cancer therapy or the future health of the host. Therefore, it will be critical to develop extremely selective antibiotics in order to minimally perturb microbiome homeostasis while rectifying any problems induced by a particular microbe. Alternatively, dysbiosis is often present in cancer, and thus one could imagine using an extremely selective antibiotic to eliminate one set of microbes from a niche that could be occupied by a more beneficial microorganism. This may prove to be a difficult approach, as the generation of novel antibiotics is slow, and much more needs to be understood about the interaction between microbial species and what niches they occupy.

25.7.3 Pre- and probiotics

The use of dietary supplements that favor the growth of a particular group of beneficial bacteria, known as prebiotics, and the consumption of potentially beneficial microbes to the host, known as probiotics, has dramatically increased in recent years. The idea behind using pre/probiotics is to alter the microbiota existing in a host in order to either maintain current health or improve some unwanted symptoms, as in the use of yogurt to alleviate constipation. One major prebiotic is fiber. The degradation of fiber into SCFAs, such as butyrate, by commensals is known to create a niche for more beneficial microbes (such as *Faecalibacterium prausnitzii*) and has been shown to be anti-inflammatory, which would limit cancer development[78]. Two of the most common probiotic microorganisms belong to the *Lactobacillus* and *Bifidibacterium* genera. Members of these genera have been shown to impart beneficial effects on the host, such as aiding in the removal of toxin metabolites and carcinogens in the colon, stimulating host antitumor immunity, and producing compounds that limit carcinogenesis by directly killing tumor cells or protecting normal cells against damage[79–82]. All in all, pre/probiotics can be an excellent preventive measure to limit systemic inflammation, combat dysbiosis, and maintain organ homeostasis, ultimately protecting the host against tumor development. Prebiotics and probiotics will be covered extensively in chapter 30.

25.7.4 Fecal transplantation

IBDs, which predispose patients to increased risk of colon cancer, are greatly influenced by dysbiosis. Fecal microbial transplantation (FMT) is the method of moving the microbiome of a healthy individual into an afflicted individual and has been suggested as a therapy for IBDs, although it is still controversial[83]. In some cases, FMT improved IBD symptoms in children but exacerbated them in an adult treated for *C. difficile* infection, where FMT is known to eliminate *C. difficile* and restore microbial balance[84,85]. Essentially, FMT is an interesting therapeutic intervention, but great care must be taken in characterizing the donor microbiome, as it may inadvertently worsen a certain disease. Conversely, the recipient microbiota should be known as well in order to predict potential positive or negative interactions with the donor.

25.8 Unresolved questions and future work

While we are beginning to uncover the basic mechanisms relating to the effects of the microbiome on cancer, there are still many outstanding questions. First, while we have a fairly strong understanding of the relative abundance and ratios of individual microbe families in healthy versus cancerous tissue across different organ systems, the reasons behind the variations in the microbiome remain elusive. How does dysbiosis occur? Is it a result of having cancer itself? Do certain external factors select for pro-tumorigenic bacteria, and thus, in turn, help initiate tumorigenesis? How do barrier defects arise and how does subsequent misplacement of microbiota affect tumor promoting inflammation? The scientific community will need carefully to address the contexts in which experiments are performed and compare their model systems against each other, and data from the clinic, in order truly to unveil how certain microbial communities affect inflammation and tumor development. This is an exciting time to study the human microbiome, and discoveries in the near future will certainly lead to treatment of cancer and other dysbiosis-related conditions.

TAKE-HOME MESSAGE

- The microbiome undoubtedly plays vital roles in carcinogenesis.
- Microbes exert their effects on tumors by modulating the immune response, maintaining organ homeostasis and structural integrity, and producing metabolites.
- Microbial populations in the human body can be altered by genetic predisposition or environmental factors, resulting in dysbiosis.
- Dysbiosis is an important phenomenon in cancer, but it is not known if it precedes cancer initiation or is a result of malignant transformation.
- Commensals and pathogenic bacteria can promote or inhibit cancer development, depending on the context.
- The study of the microbiome's effects on cancer is in the early stages, and we should expect great advances in our knowledge, and therapeutics utilizing microbes, in the next decade.

References

1 Ferlay J, Soerjomataram I, Dikshit R *et al.* Cancer incidence and mortality worldwide: sources, methods and major patterns in GLOBOCAN 2012. *Int J Cancer* 2015; **136**: E359–386.
2 Jemal A, Bray F, Center MM *et al.* Global cancer statistics. *CA Cancer J Clin* 2011; **61**: 69–90.
3 Weinberg R. *The biology of cancer*. 2013, Garland Science.
4 Hanahan D, Weinberg RA. Hallmarks of cancer: the next generation. *Cell* 2011; **144**: 646–674.
5 Grivennikov SI, Greten FR, Karin M. Immunity, inflammation, and cancer. *Cell* 2010; **140**: 883–899.

6 Rothwell PM, Wilson M, Price JF *et al.* Effect of daily aspirin on risk of cancer metastasis: a study of incident cancers during randomised controlled trials. *Lancet* 2012; **379**: 1591–1601.

7 Thun MJ, Namboodiri MM, Heath CW. Aspirin use and reduced risk of fatal colon cancer. *New England Journal of Medicine* 1991; **325**: 1593–1596.

8 Ruder EH, Laiyemo AO, Graubard BI *et al.* Non-steroidal anti-inflammatory drugs and colorectal cancer risk in a large, prospective cohort. *Am J Gastroenterol* 2011; **106**: 1340–1350.

9 Sandler RS, Halabi S, Baron JA *et al.* A randomized trial of aspirin to prevent colorectal adenomas in patients with previous colorectal cancer. *N Engl J Med* 2003; **348**: 883–890.

10 Wang D, Dubois RN. Eicosanoids and cancer. *Nat Rev Cancer* 2010; **10**: 181–193.

11 Grivennikov SI. Inflammation and colorectal cancer: colitis-associated neoplasia. *Semin Immunopathol* 2013; **35**: 229–244.

12 Vendramini-Costa DB, Carvalho JE. Molecular link mechanisms between inflammation and cancer. *Curr Pharm Des* 2012; **18**: 3831–3852.

13 Aggarwal BB, Shishodia S, Sandur SK *et al.* Inflammation and cancer: how hot is the link? *Biochem Pharmacol* 2006; **72**: 1605–1621.

14 Bultman SJ. Emerging roles of the microbiome in cancer. *Carcinogenesis* 2014; **35**: 249–255.

15 Trinchieri G. Cancer and inflammation: an old intuition with rapidly evolving new concepts. *Annu Rev Immunol* 2012; **30**: 677–706.

16 Schwabe RF, Jobin C. The microbiome and cancer. *Nat Rev Cancer* 2013; **13**: 800–812.

17 Plottel CS, Blaser MJ. Microbiome and malignancy. *Cell Host Microbe* 2011; **10**: 324–335.

18 Maynard CL, Elson CO, Hatton RD *et al.* Reciprocal interactions of the intestinal microbiota and immune system. *Nature* 2012; **489**: 231–241.

19 Zitvogel L, Galluzzi L, Viaud S *et al.* Cancer and the gut microbiota: An unexpected link. *Sci Transl Med* 2015; **7**: 271ps271.

20 Winter SE, Winter MG, Xavier MN *et al.* Host-derived nitrate boosts growth of *E. coli* in the inflamed gut. *Science* 2013; **339**: 708–711.

21 Dapito DH, Mencin A, Gwak GY *et al.* Promotion of hepatocellular carcinoma by the intestinal microbiota and TLR4. *Cancer Cell* 2012; **21**: 504–516.

22 Luedde T, Schwabe RF. NF-kappaB in the liver — linking injury, fibrosis and hepatocellular carcinoma. *Nat Rev Gastroenterol Hepatol* 2011; **8**: 108–118.

23 Rakoff-Nahoum S, Medzhitov R. Regulation of spontaneous intestinal tumorigenesis through the adaptor protein MyD88. *Science* 2007; **317**: 124–127.

24 Francescone R, Hou V, Grivennikov SI. Microbiome, inflammation, and cancer. *Cancer J* 2014; **20**: 181–189.

25 Grivennikov SI, Wang K, Mucida D *et al.* Adenoma-linked barrier defects and microbial products drive IL-23/IL-17-mediated tumour growth. *Nature* 2012; **491**: 254–258.

26 Cougnoux A, Delmas J, Gibold L *et al.* Small-molecule inhibitors prevent the genotoxic and protumoural effects induced by colibactin-producing bacteria. *Gut* 2015.

27 Irrazabal T, Belcheva A, Girardin SE *et al.* The multifaceted role of the intestinal microbiota in colon cancer. *Mol Cell* 2014; **54**: 309–320.

28 Iida N, Dzutsev A, Stewart CA *et al.* Commensal bacteria control cancer response to therapy by modulating the tumor microenvironment. *Science* 2013; **342**: 967–970.

29 Viaud S, Daillere R, Boneca IG *et al.* Gut microbiome and anticancer immune response: really hot sh*t! *Cell Death Differ* 2015; **22**: 199–214.

30 Artis D. Epithelial-cell recognition of commensal bacteria and maintenance of immune homeostasis in the gut. *Nat Rev Immunol* 2008; **8**: 411–420.

31 Salim SY, Soderholm JD. Importance of disrupted intestinal barrier in inflammatory bowel diseases. *Inflamm Bowel Dis* 2011; **17**: 362–381.

32 Hansson GC. Role of mucus layers in gut infection and inflammation. *Curr Opin Microbiol* 2012; **15**: 57–62.

33 Wang SL, Wang ZR, Yang CQ. Meta-analysis of broad-spectrum antibiotic therapy in patients with active inflammatory bowel disease. *Exp Ther Med* 2012; **4**: 1051–1056.

34 Van der Sluis M, De Koning BA, De Bruijn AC *et al.* Muc2-deficient mice spontaneously develop colitis, indicating that MUC2 is critical for colonic protection. *Gastroenterology* 2006; **131**: 117–129.

35 Johansson ME, Gustafsson JK, Holmen-Larsson J *et al.* Bacteria penetrate the normally impenetrable inner colon mucus layer in both murine colitis models and patients with ulcerative colitis. *Gut* 2014; **63**: 281–291.

36 Shan M, Gentile M, Yeiser JR *et al.* Mucus enhances gut homeostasis and oral tolerance by delivering immunoregulatory signals. *Science* 2013; **342**: 447–453.

37 Arthur JC, Perez-Chanona E, Muhlbauer M *et al.* Intestinal inflammation targets cancer-inducing activity of the microbiota. *Science* 2012; **338**: 120–123.

38 Wren SM, Ahmed N, Jamal A *et al.* Preoperative oral antibiotics in colorectal surgery increase the rate of *Clostridium difficile* colitis. *Arch Surg* 2005; **140**: 752–756.

39 Grice EA, Segre JA. The skin microbiome. *Nat Rev Microbiol* 2011; **9**: 244–253.

40 Shen W, Li W, Hixon JA *et al.* Adaptive immunity to murine skin commensals. *Proc Natl Acad Sci U S A* 2014; **111**: E2977–2986.

41 Naik S, Bouladoux N, Wilhelm C *et al.* Compartmentalized control of skin immunity by resident commensals. *Science* 2012; **337**: 1115–1119.

42 Li W, Han L, Yu P *et al.* Molecular characterization of skin microbiota between cancer cachexia patients and healthy volunteers. *Microb Ecol* 2014; **67**: 679–689.

43 Mittal D, Saccheri F, Venereau E *et al.* TLR4-mediated skin carcinogenesis is dependent on immune and radioresistant cells. *EMBO J* 2010; **29**: 2242–2252.

44 Hoste E, Arwert EN, Lal R *et al.* Innate sensing of microbial products promotes wound-induced skin cancer. *Nat Commun* 2015; **6**: 5932.

45 Urbaniak C, Cummins J, Brackstone M *et al.* Microbiota of human breast tissue. *Appl Environ Microbiol* 2014; **80**: 3007–3014.

46 Xuan C, Shamonki JM, Chung A *et al.* Microbial dysbiosis is associated with human breast cancer. *PLoS One* 2014; **9**: e83744.

47 Kern M, Popov A, Kurts C *et al.* Taking off the brakes: T cell immunity in the liver. *Trends Immunol* 2010; **31**: 311–317.

48 Naugler WE, Sakurai T, Kim S *et al.* Gender disparity in liver cancer due to sex differences in MyD88-dependent IL-6 production. *Science* 2007; **317**: 121–124.

49 Wang Z, Yan J, Lin H *et al.* Toll-like receptor 4 activity protects against hepatocellular tumorigenesis and progression by regulating expression of DNA repair protein Ku70 in mice. *Hepatology* 2013; **57**: 1869–1881.

50 Lin H, Yan J, Wang Z *et al.* Loss of immunity-supported senescence enhances susceptibility to hepatocellular carcinogenesis and progression in Toll-like receptor 2-deficient mice. *Hepatology* 2013; **57**: 171–182.

51 Rao VP, Poutahidis T, Ge Z *et al.* Innate immune inflammatory response against enteric bacteria Helicobacter hepaticus induces mammary adenocarcinoma in mice. *Cancer Res* 2006; **66**: 7395–7400.

52 Fox JG, Feng Y, Theve EJ *et al.* Gut microbes define liver cancer risk in mice exposed to chemical and viral transgenic hepatocarcinogens. *Gut* 2010; **59**: 88–97.

53 Rogers AB. Distance burning: how gut microbes promote extraintestinal cancers. *Gut Microbes* 2011; **2**: 52–57.

54 Michaud DS. Role of bacterial infections in pancreatic cancer. *Carcinogenesis* 2013; **34**: 2193–2197.

55 Michaud DS, Joshipura K, Giovannucci E *et al.* A prospective study of periodontal disease and pancreatic cancer in US male health professionals. *Journal of the National Cancer Institute* 2007; **99**: 171–175.

56 Trikudanathan G, Philip A, Dasanu C *et al.* Association Between helicobacter pylori infection and pancreatic cancer: a meta-analysis. *American Journal of Gastroenterology* 2010; **105**: S48–S48.

57 Lukanova A, Lundin E, Micheli A *et al.* Circulating levels of sex steroid hormones and risk of endometrial cancer in postmenopausal women. *Int J Cancer* 2004; **108**: 425–432.

58 Hankinson SE, Willett WC, Manson JE *et al.* Plasma sex steroid hormone levels and risk of breast cancer in postmenopausal women. *J Natl Cancer Inst* 1998; **90**: 1292–1299.

59 Gill SR, Pop M, Deboy RT *et al.* Metagenomic analysis of the human distal gut microbiome. *Science* 2006; **312**: 1355–1359.

60 Huxley RR, Woodward M, Clifton P. The epidemiologic evidence and potential biological mechanisms for a protective effect of dietary fiber on the risk of colorectal cancer. *Current Nutrition Reports* 2013; **2**: 63–70.

61 Tremaroli V, Backhed F. Functional interactions between the gut microbiota and host metabolism. *Nature* 2012; **489**: 242–249.

62 Donohoe DR, Wali A, Brylawski BP *et al.* Microbial regulation of glucose metabolism and cell-cycle progression in mammalian colonocytes. *PloS One* 2012; **7**.

63 Chang PV, Hao L, Offermanns S *et al.* The microbial metabolite butyrate regulates intestinal macrophage function via histone deacetylase inhibition. *Proc Natl Acad Sci U S A* 2014; **111**: 2247–2252.

64 Furusawa Y, Obata Y, Fukuda S *et al.* Commensal microbe-derived butyrate induces the differentiation of colonic regulatory T cells. *Nature* 2013; **504**: 446–450.

65 Wiseman M. The second World Cancer Research Fund/American Institute for Cancer Research expert report. Food, nutrition, physical activity, and the prevention of cancer: a global perspective. *Proc Nutr Soc* 2008; **67**: 253–256.

66 Rooks MG, Garrett WS. Bacteria, food, and cancer. *F1000 Biol Rep* 2011; **3**: 12.

67 Szabo C, Coletta C, Chao C *et al.* Tumor-derived hydrogen sulfide, produced by cystathionine-beta-synthase, stimulates bioenergetics, cell proliferation, and angiogenesis in colon cancer. *Proc Natl Acad Sci U S A* 2013; **110**: 12474–12479.

68 Macdonald RS, Wagner K. Influence of dietary phytochemicals and microbiota on colon cancer risk. *J Agric Food Chem* 2012; **60**: 6728–6735.

69 Guo JY, Li X, Browning JD, Jr *et al.* Dietary soy isoflavones and estrone protect ovariectomized ERalphaKO and wild-type mice from carcinogen-induced colon cancer. *J Nutr* 2004; **134**: 179–182.

70 Holmes E, Li JV, Athanasiou T *et al.* Understanding the role of gut microbiome-host metabolic signal disruption in health and disease. *Trends Microbiol* 2011; **19**: 349–359.

71 Muszer M, Noszczyńska M, Kasperkiewicz K *et al.* Human Microbiome: When a Friend Becomes an Enemy. *Archivum immunologiae et therapiae experimentalis* 2015; 1–12.

72 Ley RE, Turnbaugh PJ, Klein S *et al.* Microbial ecology: human gut microbes associated with obesity. *Nature* 2006; **444**: 1022–1023.

73 Yoshimoto S, Loo TM, Atarashi K *et al.* Obesity-induced gut microbial metabolite promotes liver cancer through senescence secretome. *Nature* 2013; **499**: 97–.

74 Ridlon JM, Kang DJ, Hylemon PB. Bile salt biotransformations by human intestinal bacteria. *Journal of Lipid Research* 2006; **47**: 241–259.

75 Lightfoot YL, Mohamadzadeh M. Tailoring gut immune responses with lipoteichoic acid-deficient *Lactobacillus acidophilus*. *Front Immunol* 2013; **4**: 25.

76 Wallace BD, Wang H, Lane KT *et al.* Alleviating cancer drug toxicity by inhibiting a bacterial enzyme. *Science* 2010; **330**: 831–835.

77 Wang F, Meng W, Wang B *et al. Helicobacter pylori*-induced gastric inflammation and gastric cancer. *Cancer Lett* 2014; **345**: 196–202.

78 Hooda S, Boler BM, Serao MC *et al.* 454 pyrosequencing reveals a shift in fecal microbiota of healthy adult men consuming polydextrose or soluble corn fiber. *J Nutr* 2012; **142**: 1259–1265.

79 Challa A, Rao DR, Chawan CB *et al. Bifidobacterium longum* and lactulose suppress azoxymethane-induced colonic aberrant crypt foci in rats. *Carcinogenesis* 1997; **18**: 517–521.

80 Femia AP, Luceri C, Dolara P *et al.* Antitumorigenic activity of the prebiotic inulin enriched with oligofructose in combination with the probiotics *Lactobacillus rhamnosus* and *Bifidobacterium lactis* on azoxymethane-induced colon carcinogenesis in rats. *Carcinogenesis* 2002; **23**: 1953–1960.

81 Han W, Mercenier A, Ait-Belgnaoui A *et al.* Improvement of an experimental colitis in rats by lactic acid bacteria producing superoxide dismutase. *Inflamm Bowel Dis* 2006; **12**: 1044–1052.

82 Carroll IM, Andrus JM, Bruno-Barcena JM *et al.* Anti-inflammatory properties of Lactobacillus gasseri expressing manganese superoxide dismutase using the interleukin 10-deficient mouse model of colitis. *American Journal of Physiology-Gastrointestinal and Liver Physiology* 2007; **293**: G729–G738.

83 de LeBlanc AD, LeBlanc JG. Effect of probiotic administration on the intestinal microbiota, current knowledge and potential applications. *World Journal of Gastroenterology* 2014; **20**: 16518–16528.

84 Kunde S, Pham A, Bonczyk S *et al.* Safety, tolerability, and clinical response after fecal transplantation in children and young adults with ulcerative colitis. *Journal of Pediatric Gastroenterology and Nutrition* 2013; **56**: 597–601.

85 De Leon LM, Watson JB, Kelly CR. Transient flare of ulcerative colitis after fecal microbiota transplantation for recurrent *Clostridium difficile* infection. *Clinical Gastroenterology and Hepatology* 2013; **11**: 1036–1038.

CHAPTER 26

Colorectal cancer and the microbiota

Iradj Sobhani[1] and Séverine Couffin[2]

[1] Centre Hospitalier Universitaire Henri Mondor-Assistance Publique Hôpitaux, de Paris, Paris, France
[2] UPEC, Université Paris Est Créteil Val de Marne-Equipe Universitaire EC2M3, Paris, France

26.1 Introduction

Colorectal cancer (CRC) is a common disease in the developed countries, with incidence steadily increasing over the last five decades and with the mortality remaining at very high levels. Thus the disease incurs high economic costs. A good understanding of the factors promoting carcinogenesis in CRC should make possible a better prevention policy and a more efficient screening protocol. Except for germline DNA mutations which have been attributed to less than 5% of patients, little is known about the main causes of CRC. The majority of CRC occurrences are considered sporadic, meaning they are due to environment rather than constitutional genetic alteration. Among factors from the environment, the Western lifestyle is considered to favor the occurrences of CRC.

Microbes are major actors in biological environments: about 16% of cancers around the world have been estimated to be caused by microbes and several cancers in the liver and gastro-intestinal (GI) tract have been clearly identified as microbe-related[1]. Among these, *Helicobacter pylori* has been considered by WHO's Agency for Research on Cancer (IARC) to be associated with causation of gastric adenocarcinoma and Mucosa-associated Lymphoid Tissue (MALT) lymphoma. Evidence on such associations has been obtained from epidemiological data in populations, experimental results in animal models and intervention trials by eradication of the bacterium in humans. Studying physiology and pathophysiology of the gastric function in relation to a single microorganism has made it possible to better understand human gastric carcinogenesis.

However, the colon microflora is a very complex system hosting several trillions of bacteria and multitudes of functions. Furthermore, due to the complexity of the current disease model used for colon carcinogenesis, the first step was to verify whether specific dysbioses could be considered as related to the disease. Consequently, first phylogenetic studies were undertaken followed by metagenomic analyses in well documented CRC patients compared to matched individuals with normal colonoscopy. Now mechanisms that could be involved are under investigation in humans as well in animal models.

The Human Microbiota and Chronic Disease: Dysbiosis as a Cause of Human Pathology, First Edition.
Edited by Luigi Nibali and Brian Henderson.

26.2 Colon carcinogenesis and epidemiological data

The incidence of CRCs is increasing in all Western countries and in many developing countries, suggesting that the main cause might be related to the Western life-style. Mortality due to CRC and its associated global cost to health care systems justify deployment of screening programs based on biological testing prior to colonoscopy or direct access to colonoscopy in average-risk and higher-risk populations, respectively[2].

26.2.1 Human carcinogenesis model

In the original model of colon carcinogenesis, it was proposed[3] that only polyps displaying tubular and/or tubulovillous adenoma patterns had the potential to progress to invasive adenocarcinoma (Figure 1). Serrated polyps including sessile serrated adenomas (SSAs) and traditional serrated adenomas (TSAs) are now recognized as having the potential for malignant transformation as well. Subset of hyperplastic polyps that are not usually recognized as precancerous lesions may progress to serrated neoplasms (SSAs or TSAs) and a fraction of these SSAs progress to cancer. Various alterations in tumor genes are observed. A few genes, such as the adenomatous polyposis coli (APC) and the β-catenin (CTNNB1) genes, P53 genes, Kirsten rat sarcoma (KRAS) and myelocytomatosis (MYC), are identi-fied in a large majority of tumors. In addition to gene mutations, abnormalities such as chromosome instability and microsatellite instability, and aberrant DNA methylation at CpG islands[4] are sometimes detected. These gene alterations can be induced by chemical and/or environmental factors. The occurrence of tumors is due to DNA alterations in the stem cells at the base of the villus crypt. These changes

Colon Carcinogenesis

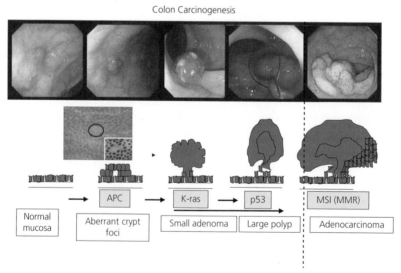

Figure 1 Colorectal carcinogenesis from normal to invasive cancer. At top are slides from colonoscopy after cleaning preparation showing precancerous lesions such as polyps at various diameters; genes involved in each steps are indicated supporting the theory of accumulation of gene mutations from aberrant crypt foci (ACF) (high magnification) up to the invasive cancer through precancerous lesions at various size. (*see color plate section for color details*).

appear to be permanent and prone to the accumulation of additional mutations[5]. Overexpression of specific bacteria in each of these steps, either in individuals' stools or in their colonic or rectal mucosa-adherent compartment, brings evidence of a link between cancer and bacteria and opens new investigative windows on CRC carcinogenesis.

26.2.2 Age-related risk in the general population
About 3% to 6% of the population aged 50 to 75 years old present with adenomatous polyps of varying sizes or cancer[6]. Time of exposure is very important, since 10 to 15 years are probably needed to make polyps evolve into cancer if they are not removed; thus tools for mass screening are considered regarding the age and mucosal polyp phenotypes. The alteration of microbial populations with aging may reflect the waning of immunity.

26.2.3 Gene- and familial-related risks
Although no more than a dozen genes are involved in early colon carcinogenesis, less than 5% of CRC result from a constitutional mutation. In these cases, the relatives who carry the mutated gene remain at very high risk of developing cancer (Table 1). The risk for this group to develop a cancer varies from 25% up to 100% depending on the mutated gene, age and location of the tumor or pre-neoplastic lesions. Even though the hereditary germline mutation constitutes the main risk factor, the actual development of cancer among the carriers in these family syndromes, probably depends on environmental factors. In addition, about 10% to 15% of individuals with CRC and/or adenomas have other affected family members whose conditions do not fulfil the criteria for a constitutional gene-related carcinoma. A simple family history of CRC (defined as one or more close relatives with CRC) confers a twofold to sixfold increase in CRC risk. A personal history of adenomatous polyps confers a 15% to 20% risk of subsequently developing polyps and increases the risk of CRC in relatives with diagnosed adenomas before the age of 60 (Figure 2). The estimated relative risk for this population is 2.59 (95% CI, 1.46–4.58)[7]. Interestingly, the relative risk (RR) of CRCs for spouses also appears to be higher than the RR in general population[8].

Indeed, putative roles of genetic and environmental factors have been approached by two studies performed in twin pairs and in spouses. After analysis

Table 1 Risk of CRCs in gene mutation carriers.

Syndrome	Risk in mutation carriers
FAP	90% by age 45 y
Attenuated FAP	69% by age 80 y
Lynch	40% to 80% by age 75 y
MYH-associated polyposis	35% to 53%
Peutz-Jeghers	39% by age 70 y
Juvenile polyposis	17% to 68% by age 60 y

Although absolute risk is clearly higher than average risk in the general population, note the impact of age and range of risk depending on time of exposure to the environmental factors; FAP: Familial adenomatous polyposis.

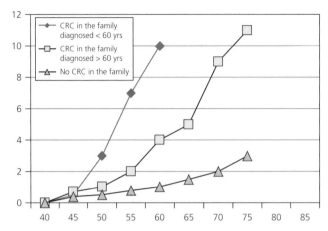

Figure 2 Familial risk of CRC. Cumulative incidence (%) in the population is indicated at the age (yr) of individuals according to the age of CRC diagnosis in the family; stratification is done on the age of CRC cases: no CRC in the family vs CRC diagnosed at 60 yrs old or less, vs CRC diagnosed after 60 yrs. (Adapted from Winawer SJ. *N Engl J Med*. 1996; 334(2): 82–7.) Guidelines of National Cancer Institute at the National Institutes of Health http://www.cancer.gov/cancertopics/pdq/genetics/colorectal/HealthProfessional.

of environment factors, they do explain the degree to which hereditary factors contribute to familial CRCs[9,10] and support intra-familial transmission of cancer. Briefly, familial clusters account for approximately 20% of all CRC cases in developed countries, while highly penetrant Mendelian CRC diseases contribute only up to 5% of familial cases. Overall shared environmental factors may contribute to 90% or more of the CRCs.

26.2.4 Environment-related risk

Although CRC remains one of the most common cancers, its rates of incidence vary up to tenfold in both sexes worldwide, the highest rate being estimated in the Western Europe, Australia/New Zealand and the lowest in Africa and India. Elevated intake of red meat and animal fat increases CRC risk, while greater consumption of fibre reduces this risk[11]. The link between Western diet and elevated risk of colon cancer has been established for quite some time[12]. The effect of diet on colonic mucosa can either maintain colonic health or promote chronic inflammation[11]. The first attractive explanation of this observation was that dietary constituents might influence colon cancer prevalence and it was hypothesized that these might be involved in the CRC carcinogenesis through the microbial flora of the gut. Changes in the microbiota and dysbiosis have been documented in colon cancer or adenomatous polyp patients as compared to control groups[11,13–15,16].

26.3 The microbiota

The intestinal microbial population known as the "intestinal microbiota" is a heterogeneous and complex entity composed of possibly more than 1000 different bacterial species in each individual. The analysis of the microbiota makes it possible to discuss "factors from the environment" (e.g. nutrients) and interactions with the host. This complex ecosystem contributes to the maturation of the immune system, provides a direct barrier against colonization of the intestine by pathogens[17] and seems able to metabolize pro-carcinogens and carcinogens from the environment[18]. Although more than 80% of intestinal bacteria cannot be cultured, identification of all bacteria has become possible by using high technology to perform whole DNA genome sequencing. This metagenomic approach has allowed characterization of healthy individuals' microbiota[19] which is now used as a reference, although the concept of a universal healthy microbiota still remains controversial. The bacterial community in an individual host is relatively stable in the distal digestive tract throughout adult life, although there is much variation from person to person[20]. The "core microbiome" hypothesis, which states that a similar set of microbial functions can be found at the gene level in all individuals, despite variations at the species level, may be a possible explanation for this phenomenon. Nevertheless, it seems plausible to expect that certain individuals (as determined by genetic make-up and/or lifestyle) are predisposed to bacteria-related diseases such as obesity, asthma, autoimmunity, or inflammatory bowel diseases[21,22] and CRCs. The main challenge is how to describe healthy conditions. In spite of large variations in healthy individuals' microflora[20], it has been shown that intestinal microbiota variation is generally stratified, not continuous and might respond differently to diet and drug intake based on healthy individuals' microbiota analyses through different continents and countries[23]. The latter group identified three major enterotypes suggesting that the impact of host-environment can be categorized through the microbiota. Further, data-driven marker genes or functional modules could be identified for host properties. For example, twelve genes significantly correlate with age and three functional modules with the body mass index, hinting at a diagnostic potential of microbial markers[23].

26.4 Bacteria and CRCs links

The bacterial density in the large intestine ($\sim 10^{12}$ cells per ml) is much greater than that in the small intestine ($\sim 10^2$ cells per ml), and this is paralleled by an approximately 12-fold increase in cancer risk for the large intestine compared to the small intestine. These two observations combined point towards the hypothesis that colon cancer might be linked with bacteria.

26.4.1 Historical data

Streptococcus bovis/gallolyticus was traditionally considered a lower-grade pathogen involved in endocarditis. Although McCoy and Mason[24] suggested a relationship between colonic carcinoma and the presence of infectious endocarditis in 1951, it was only in 1974 that the association of *S. bovis* and colorectal neoplasia was recognized.

In a general way, a study intended to identify associations between human microbiota composition and colorectal carcinogenesis without any specific hypothesis concerning the nature of the bacteria was conducted in 1971. However, it had to be abandoned because of technical difficulties in recovering all the bacteria in the stool samples particularly the large majority that are anaerobic. Later on, 13 bacterial species were shown to be associated significantly with a high risk of colon cancer and the Western diet[25]. However, these results were somewhat unconvincing because of the small number of subjects studied without performing any intestinal investigation (e.g. radiology or colonoscopy).

26.4.2 Clinical data

The phylogenic core of the microbiota, based on metagenomic analyses, is now characterized in healthy individuals, and therefore a comparison with other individual human groups has become possible. We and others have reported that the phylogenic core of the human microbiota was significantly different in colon cancer patients' stools (Figure 3) or mucosa as compared to matched normal individuals[26,27]. Although these studies did not reveal whether phylogenic core changes in colon cancer patients were a cause or consequence of the disease, quantification of major bacterial dominant and subdominant groups allowed identification of *Bacteroidetes*, as a predominant phylum in colon cancer patients.

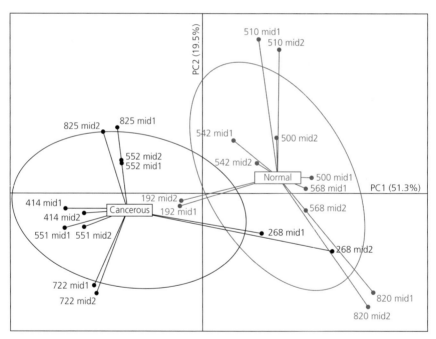

Figure 3 Composition of bacteria in colon cancer patients' stools. Fresh samples were collected prior to the colonoscopy, DNA was extracted and submitted to pyrosequencing analyses of 16S rRNA V3-V4 region and comparisons of bacteria sequences were performed between normal-colonoscopy individuals and colon cancer patients on the basis of bacterial species abundance; this belonged to the phylogenetic core and differentiates cancer patients and healthy individuals. Principal component analysis, based on the 16S rRNA gene sequence abundance of 10 discriminates phylogenetic core species, was carried out with six normal individuals (red points) and six cancer patients (black points) with two replicates. Two first components (PC1 and PC2) were plotted and represented 57.95% of whole inertia. Individuals (shown by their sample ID) were clustered and center of gravity computed for each class. This is the first comparison of stool composition in healthy individuals and colon cancer patients showing significant dysbiosis in CRC.

Bacterial genera such as *Bacteroides, Shigella, Citrobacter* and species such as *Salmonella* spp., or *E. coli*, are more abundant in the early stages of CRC, including adenomas, and disappear from cancerous tissue while *Fusobacterium* spp., *Streptococcus gallolyticus subsp. gallolyticus, Clostridium septicum* and *Coriobacteriaceae* (*Slackia and Collinsella* spp.), the genus *Roseburia* and the genus *Faecalibacterium* are more abundant in advanced cancer[28]. In terms of phyla, more *Proteobacteria* and less *Bacteroidetes* and in terms of genera more *Bacteroides*, while in terms of species, more *Dora* spp. *Faecalibacterium* spp. *Shigella* spp. and *Rhuminococcus* spp. and less *Bactreroides* spp. and *Coprococcus* spp., have been reported as adherent to the normal mucosa of those patients who display adenomas at colonoscopy[15]. Very recently, a panel of 22 bacteria species has been characterized in a large series of more than 200 patients undergoing colonoscopy as a discriminative tool for identifying those displaying CRC[29]; these bacteria have been found underexpressed in a large sample of healthy control individuals[23]. All these clinical and translational research data show clearly that microbiota heterogeneity in stools is significantly reduced in CRC patients conducting to the specific biomarker identification.

Many cancers arise from sites of infection, chronic irritation, and inflammation. The strongest association of chronic inflammation with malignant diseases is found in inflammatory bowel diseases (IBD) of the colon[30] with a lifetime incidence of 10%. Thus, those patients suffering from chronic inflammatory bowel disease (Crohn's disease, ulcerative colitis) are considered at high risk of CRC occurrence. They should undergo colonoscopy every two to three years even in case of clinical remission and special zoom endoscopy and/or coloration process should be used to enhance the sensitivity of the colonoscopy for detecting precancerous lesions such as aberrant crypt foci (ACF) and/or high-grade dysplasia. In patients with Crohn's disease, *Enterobacteria* were observed significantly more frequently than in health, and more than 30% of the dominant flora belonged to yet undefined phylogenetic groups[31]. The microbiota has been found to be both phylogenetically and functionally disturbed in these conditions and could contribute to the increasing CRC risk.

26.4.3 Experimental data and mechanisms involved
26.4.3.1 Inflammation
A recent study performed in animals has clearly shown that colitis can promote tumorigenesis by altering microbial composition and inducing the expansion of microorganisms with genotoxic capabilities[32]. This involves two different groups of bacteria, those inducing chronic inflammation and those favoring carcinogenesis. Monocolonization with the commensal *Escherichia coli NC101* promoted invasive carcinoma in azoxymethane (AOM)-treated IL10(−/−) mice. Deletion of the polyketide synthase (pks) genotoxic island from *E. coli* NC101 decreases tumor multiplicity and invasion in AOM-IL10(−/−) mice, without altering intestinal inflammation. The role of bacteria altering host colonocyte DNA by direct effect and/or stimulating pro-inflammatory mediators is therefore important.

26.4.3.2 Bacteria-induced DNA alteration
Whether some bacteria can induce DNA damage is a crucial question. It has been suggested that enteropathogenic *E. coli* possesses the ability to downregulate DNA mismatch repair proteins and, therefore, it could promote colonic

tumorigenesis by favoring directly DNA damage[33]. *E. faecalis* induces aneuploidy in colonic epithelial cells and aggressive colitis in mono-associated IL10(−/−) mice. Furthermore, inhibitors of reactive oxygen and nitrogen species (RONS) prevent *E. faecalis*-induced aneuploidy[32]. Thus the unique ability of this bacterium to induce RONS can lead to chromosomal instability in a susceptible host. Consequently, one needs to distinguish between indigenous intestinal bacteria able to drive the epithelial DNA damage, thereby contributing to the initiation of CRC and intestinal niche alterations that favor the proliferation or inhibition of tumor growth.

26.4.3.3 Mucosa-adherent bacteria
Epithelium-adherent bacteria appear to be different from stool microbiota in terms of abundance and heterogeneity. *Streptococcus gallolyticus and Fusobacterium* are probably passenger species because in pre-cancerous adenomatous polyps, analysis of the 16S rRNA genes from adherent bacteria has not shown these two species to be significant[15,16]. Overall, *Firmicutes* (62%), *Bacteroidetes* (26%) and *Proteobacteria* (11%) are most dominant phyla in adherent bacteria to these pre-cancerous lesions. Although significant differences in bacterial composition between cases and controls have surprisingly been identified in one of these studies[16], biopsies from adenomatous polyps showed a 20-fold relative reduction of mucosa adherent bacteria compared with normal tissue. This finding can be due to technical bias since quantification is performed on formalin-fixed materials. *Fusobacterium nucleatum*, although not a dominant species in the stools, has been detected from tumor biopsies in colon cancer patients by two independent groups; these have suggested that this bacterium might be the cause of CRC occurrence[34,35]. *Fusobacterium nucleatum* is commonly found with other species in the oral cavity. This bacterium adheres to the host tissue cells as an invasion and modulates the host's immune response[36,37]. The bacterium could be suggested as a co-factor of yet unidentified cause.

26.4.3.4 Focal and diffuse alterations in the intestinal mucosa
Stool versus adherent bacteria composition may impact diffuse and focal injuries, respectively. The transplant of human microbiota in the intestine of conventional animals stimulates cell proliferation in the intestine as compared to germ-free animals[38]. It is likely that this effect is direct and in need of bacteria attachment to the epithelial cells. In the mid 1970s and early 80s, Deschner and colleagues identified focal histologic lesions in the colonic mucosa in experimental animals as the earliest morphologic alteration preceding tumor development[39]. Later they showed that this phenomenon was associated with a diffuse hyperproliferation in the intestinal mucosa[40]. The pitfalls in the use of colonic mucosal cell proliferation in assessing the potential risk of developing a colonic neoplasia[41] and neoplastic paradigm (e.g., ACF) based on focal areas along the colon have been reviewed[42]. Nevertheless, an increase in the epithelial cell proliferation rate in the normal human mucosa of those patients with CRC[43], associated with a loss of the epithelium's ability to inhibit altered (gene-mutated) progenitor cells into differentiation, increases the risk of neoplastic development[44,45]. In this scheme, the human microbiota plays an important role particularly via progenitor or stem cells.

26.4.3.5 Stem cell as pivotal target from normal to neoplastic lesion

The stem cells and their descendant proliferating crypt precursors[46] with a large number of differentiated cells (enterocytes, enteroendocrine cells and goblet cells) occupy the crypt. They self-renew to regenerate the epithelium after injury while progenitor cells arrest their cell cycle and differentiate when they reach the tip of the crypt and when the number of cells produced by the crypt compartment is compensated by apoptosis at the tip of the crypt[47–49]. Thus, it is presumed that the first carcinogenic event occurs in stem and/or progenitor cells and CRC is considered as the consequence of a succession of steps with identifiable specific gene mutations in cells[3]. Here, the adherent bacteria could be considered as a crucial link with the environment offering interactions with stem cells to yield focal lesions; in this regard mucus layer[50] and crypt-specific core microbiota[51] are considered as facilitators. In the two main hereditary syndromes involving no lethal germline mutations (APC, MMR system), the delay for achieving critical accumulation of mutations in the stem cells up to the development of cancer is significantly shorter (20 to 45 years of age) than in sporadic cancers (65 years of age). The mosaic gene mutation mechanism[52] in stem and/or progenitor cells can possibly explain how accumulation of gene mutations (i.e. KRAS, β–catenin) necessary at very early colon carcinogenic steps might occur in lesions such as ACFs, serrated polyps or adenomas. Such DNA alterations, if due to bacteria, require direct contact and interaction with environment mutagenic co-factors. Therefore, stool versus mucosa bacteria analyses as well as focal versus diffuse colonic mucosa should be taken in consideration. Nevertheless, all these data support the hypothesis that CRC can be initiated by some bacteria, called "driver" and promoted by others called "passenger"[28]. For a driver bacterium, adherence to the mucosa is a pre-requisite needing recognition by the host.

26.4.3.6 Recognition of microbial species by the host and responses

Humans, mice, and other eukaryotes are equipped with an elegant repertoire of receptors, with each receptor recognizing specific conserved microbial patterns, such as components of bacterial cell walls or nucleic acids. These microbial sensors include several RNA helicases, lectin receptors, and Toll-like receptors (TLR)[21]. These receptors are also expressed on intestinal epithelial cells including endocrine cells, as well as on various mucosal immune cells. They initiate signaling cascades involving mitogen-activated protein kinases (MAPK), nuclear factor of kappa B (NF-κB), and interferon regulatory factors that in turn modulate apoptosis, proliferation, and cell migration directly or via cytokines and/or hormones through a paracrine pathway. These mechanisms are in agreement with several clinical trials that have shown the importance of hyperproliferation in the normal mucosa of individuals with high risk of CRCs[53,54]. The substantial role of hyperproliferation induced by microbiota as driving colorectal cancer has been best demonstrated in gene-mutated animals. For instance, the APC mouse, a model for human familial adenomatous polyposis, develops tens to hundreds of intestinal adenomas when raised in conventional or specific pathogen-free (SPF) conditions while in germ-free mice conditions they exhibit less than ≈50% of intestinal tumors[55]. The susceptibility of animals to developing cancer depends also on whether bacteria can stimulate nuclear targets through specific cell receptors (NOD, NLR-Nod like, Toll-like receptors), all involved in inflammatory and immune responses in the mucosa. In the newly developed AOM-IL10(-/-) model in which intestinal inflammation occurs spontaneously from the lack of immuno-suppressive IL10, tumorigenesis has been initiated with the colon-specific carcin-ogen azoxymethane (AOM) through extensive intestinal inflammation in response to bacteria[32] since those animals raised in germ-free conditions are found to be devoid of intestinal inflammation and tumors. In all animal experimental models in germ-free conditions, lower numbers of tumors are observed than when the conditions involve a conventional microbiota irrespective of whatever tumors are induced by chemical agents or due to germline mutations[38]. *Bacteroides fragilis*, a common intestinal commensal species from this bacteria family, induces spontaneous colonic tumorigenesis in APC Min/+ mice[56] as compared to germ-free animals. Intestinal flora may promote colon tumor formation in these APC knocked mice model via inflammatory and immunologic mechanisms[56]: indeed, the *enterotoxigenic B. fragilis strain*, induces more ACFs and cancer as compared to the control by a TH17-dependent pathway; this effect could be suppressed by using an anti- IL17 antibody. It is likely that adherent bacteria activate regulatory T cells (Tregs) in the colonic mucosa, which can promote TH17-inflammation. These observations are consistent with IL17 immune reactive pro-inflammatory and FoxP3 immune cells infiltration in the normal mucosa of cancer patients in our human colonoscopy series as compared to normal mucosa in individuals with normal colonoscopy[26]. Lymphocytes TH17 are known to be regulated by retinoic acid-related orpholen receptor gamma T (RORγT), which is also a stimulator of Wnt/β-catenin pathway. This group of receptors makes balance between protector Tregs and pro-inflammatory Tregs[57].

26.4.3.7 Indirect bacteria effects, energy balance and metabolism

An alternative pathway by which microbiota may drive carcinogenesis effects would be through hormone and metabolic changes[21]. Very recently, it has been shown that more than half of metabolites in the lumen of colon are due to the bacterial activity alone, one quarter to the colonocytes alone and the rest requires interaction of colonocytes and bacteria. As an example, 11 out of 23 metabolites produced from L-tryptophan in the colon of mice result from microbiota activity and 8 out them to the interaction of colonocytes and microbiota[58]. Interest for metabolite activity in the colon carcinogenesis has been highlighted by the link between resistance to the insulin effect and the risk of CRC and colonic adeno-mas[59]. These results are consistent with those from prospective and interventional studies performed in obese individuals that have shown higher colonic cell prolif-eration and elevated number of ACFs in the colonic mucosa. A standard low-fat (SLF) diet changing to a high-fat diet (HFD) is associated with a shift in the bal-ance of the two dominant phyla, the *Bacteroidetes* and *Firmicutes* due to the expan-sion of *Mollicutes* in *Firmicutes* phylum. This trait can be transferred by microbiota transplant into lean recipients. Similar changes in the proportion of *Firmicutes* and *Bacteroidetes* have been identified in overweight and obese humans, in genetically obese mice, and in obesity-resistant mice fed HFD diet (discussed by Fak in chap-ter 23)[21,60]. These highlight the critical role of specific members of the commensal microbiota in the development of metabolic-associated colorectal cancer. For instance, men of all ages and postmenopausal women are at increased risk for CRC if they are obese[59]. Visceral associated fat (VAF), stigmata of various hormone disorders, is considered the source of the presumptive metabolic risk factors for colon cancer in mice. These develop hyperinsulinemia. Pre-neoplastic colonic mucosal changes are observed more often in mice with than those without visceral abdominal obesity. Insulin-like growth factors are mitogens and regulate energy-dependent growth processes, cell proliferation and inhibit apoptosis. Serum leptin level is elevated in obese individuals. Hyperleptinemia and hypeinsulinemia in db/db rat or in APC mice under exogenous leptin infusion, are shown to increase the numbers of pre-cancerous lesions induced by AOM [61,62]. In the first model, AOM induces somatic gene alteration (i.e. β–catenin) including stem cells and endogenous hyperleptinemia which is a result of lack of feed-back regulatory signal due to leptin receptor alteration. In the second model, APC gene alteration is constitutional and affects all cells including stem cells within crypts and subjected to pharmacologically obtained hyperleptinemia. In both models, leptin plays the role of epithelial cell stimulation.

26.4.3.8 Bacterial enzyme activity

The production of bioactive carcinogenic compounds from environmental factors (diet, chemical agents) may occur through enzyme activities (β-glucuronidase, β-glucosidase, azoreductase, and nitroreductase). For instance, AOM is first hydrolyzed in the liver to methylazoxymethanol and conjugated with glucuronic acid before transport to the intestine through bile secretion. Similarly, bacterial β-glucuronidase spontaneously yields the highly reactive methyl carbonium ion, a carcinogenic from AOM, while inhibition of β-glucuronidase significantly reduces the ability of AOM to induce tumors in rats[63].

26.4.3.9 Anti-cancer bacterial effect

Exclusion of opportunistic pathogens by commensal bacteria may represent a natural defense against gastrointestinal diseases, including colorectal cancer. Probiotic bacteria *(Lactobacillus sp.* and *Bifidobacterium sp.)* exert anti-carcinogenic effects, likely in an indirect way, e.g. by inactivating microbial enzymes. For example, probiotic lactic acid bacteria including *L. casei* and *L. acidophilus* can decrease the activity of β-glucuronidase, azoreductase, and nitroreductase. *Bifidobacterium longum* reduces AOM-induced aberrant crypt formation, which correlates with a decrease in AOM-activating β-glucuronidase activity. *Lactobacillus sp.* and *Bifidobacterium sp.* can inhibit DNA damage and tumorigenesis induced by N-methyl-N'-nitro-N-nitrosoguanidine (MNNG), 1,2-dimethylhydrazine (DMH)[63]. In general, microbial fermentation of dietary fibers leads to short-chain fatty acids (SCFA) acetate, propionate, and butyrate productions. These are absorbed by colonocytes and used as a primary source of energy, and provide protection during the early stages of tumorigenesis[64]. Whether the use of probiotics may help preventing mild inflammation in the colonic mucosa is the main future therapeutic challenge.

26.4.3.10 Time-related effects

Age remains the best indicator of the time of exposure to the environment factors. Over time, the events due to host and environment interactions are linked in a spiral of alteration, diminished immunity, more alteration, and so forth with chronic interactions (e.g. inflammation) induced by the bacteria at particular locations (e.g., stomach, lung, colon, skin) leading to tissue destruction (e.g., atrophic gastritis), which leads to malignancy; this hypothesis is considered as the main explanation of the occurrence of malignant diseases in the elderly[65]. This is consistent with chronic inflammation as well as immune system alteration, as the main clinical situations increasing risk of cancer development in association with microorganisms[37]. We should note that those bacteria presumed to initiate carcinogenesis may disappear from cancerous tissue as they are outcompeted by passenger bacteria more involved in the promotion step with a growth advantage in the tumor microenvironment. This is somehow in line to the keystone pathogen hypothesis reported by Hajishengallis and Lamont in chapter 14 for the pathogenesis of periodontitis. These observations make necessary longitudinal studies in human to establish putative and relatively long-term oncogenic and anti-oncogenic effects.

26.5 Hypotheses and perspectives

We propose that the composition of the microbiota can shape a healthy immune response or predispose to disease. Many factors can contribute to dysbiosis, including lifestyle, and medical practices and probably host genotype. An individual with mutations in genes involved in immune regulatory mechanisms or pro-inflammatory pathways could undergo unrestrained inflammation in the intestine. It is possible that inflammation alone influences the composition of the microbiota, skewing it in favor of pathobionts. Alternatively, a host could "select" or exclude the colonization of particular organisms.

This selection can be either active (as in the case of an organism recognizing a particular receptor on the host) or passive (where the host environment is more conducive to fostering the growth of select organisms). Hypotheses on this "genetic dysbiosis" concept are discussed by Nibali in chapter 28. The composition of microbiota can now be fully analysed by using metagenomic sequencing on whole bacteria genes and a limited panel of bacteria was found to be significantly associated with CRC[29]. Several of these bacteria are known as pro-inflammatory and others as anti-inflammatory. A more general dysbiosis common to multiple disease conditions cannot be ruled out. We applied our recently published metagenomes classifier to 21 IBD patients and found an increased prediction rate of up to 25% in these patients. This probably reflects common alterations in the microbiota in these diseases, which is consistent with IBD patients being at greater risk of developing CRC. We monitored the five most discriminative CRC marker species for significant changes in abundance in IBD patients relative to controls and showed that *Fusobacterium* species as well as *Porphyromonas asaccharolytica,* are specific to CRC when *Peptostreptococcus stomatis* was found at elevated levels in IBD compared to controls, although its expansion in CRC was more pronounced. *Eubacterium ventriosum,* which is negatively associated with CRC, showed an even stronger decrease in IBD patients consistent with a study that reported reduced colonization of the intestinal mucus with *Eubacterium* species in IBD (Figure 4). In summary, even though the four most important CRC marker species are significantly more abundant in CRC patients than in IBD patients, the metagenomic classifier predicted CRC for a proportion of IBD patients that is higher than its overall false positive rate. Furthermore, functions differ significantly between healthy and cancer-afflicted

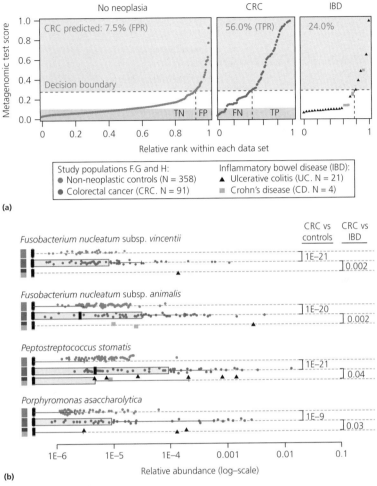

Figure 4 Colorectal cancer and inflammatory bowel diseases microbial signature. (a) Ranked predictions of the metagenomic classifier are plotted for each patient subgroup with the percentage of positive CRC predictions annotated in red. Proportions of true negatives (TN), false positives (FP), false negatives (FN) and true positives (TP) are shown at bottom for a decision boundary of 0.65. Application of the CRC classifier to metagenomes from Ulcerative colitis (UC) and Crohn's disease (CD) patients suggests moderately increased positive prediction rates for inflammatory bowel disease (IBD) patients. (b) Relative abundance distributions of the five most important species for CRC classification are plotted for each patient subgroup, including UC and CD. While in this data set the marker species *Fusobacterium* species and *P. asaccharolytica* appeared specific for CRC, others, most prominently *E. ventriosum*, were also significantly (negatively) associated with CD or IBD (UC and CD tested together). Source: Zeller, http://onlinelibrary.wiley.com/doi/10.15252/msb.20145645/full. Used under CC-BY-4.0http://creativecommons.org/licenses/by/4.0/

individuals, with a global metabolic shift from predominant utilization of dietary fibre in the tumor-free colon to host-derived energy sources in CRC. In the healthy gut, fibre-degrading enzymes and fibre binding domains were enriched more than sixfold. Along the progression of CRC, the microbiota functions *i.,g.* amino acids gradually shift to growth substrates derived from host cells reaching 10-fold

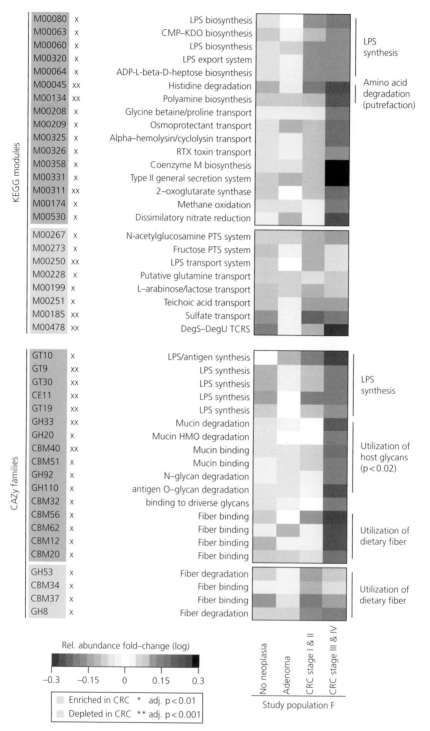

Figure 5 Functional changes in the CRC-associated metagenome. A. Significant changes in relative abundance of genes summarized by KEGG module annotations between cancer and non-cancer metagenomes are shown for cases with a >1.5-fold change and FDR-adjusted p-value<0.1. General trends in functional potential, such as enrichment of secretion and

enrichment; amino acids are more abundant in the tumor microenvironment as reported by metabolomics approaches in CRC patients. These suggest an increased capacity of the CRC-associated microbiota for amino acid uptake and metabolism via putrefaction pathways (Figure 5). Thus, it is crucial to analyse prospectively both metagenomic as well as metabolomic changes after fecal microbiota transfer in those human individuals undergoing such a therapy and correlate them to clinical and endoscopic outcomes. Finally, "Does the shift in the microbiota directly alter the course of disease?" is a very important question and requires animal studies.

TAKE-HOME MESSAGE

- The putative roles of gene alteration(s)field and environment factors in colon cancer genesis are beginning to be elucidated. Data are now available on the involvement of nutriment, hormone and metabolic disorders favoring colon cancer genesis.
- Colon microbiota should be considered as a novel part of colon carcinogenesis.
- Targets for colon cancer biomarkers are now well identified and their validations in large series are needed in screening and in those patients in medical therapy to predict tumor response.

References

1 de MC, Ferlay J, Franceschi S, Vignat J, Bray F, Forman D et al. Global burden of cancers attributable to infections in 2008: a review and synthetic analysis. *Lancet Oncol* 2012; **13**(6): 607–615.
2 Sobhani I, Alzahouri K, Ghout I, Charles DJ, Durand-Zaleski I. Cost-effectiveness of mass screening for colorectal cancer: choice of fecal occult blood test and screening strategy. *Dis Colon Rectum* 2011; **54**(7): 876–886.
3 Fearon ER, Vogelstein B. A genetic model for colorectal tumorigenesis. *Cell* 1990; **61**(5): 759–767.
4 Ogino S, Nosho K, Kirkner GJ, Kawasaki T, Meyerhardt JA, Loda M et al. CpG island methylator phenotype, microsatellite instability, BRAF mutation and clinical outcome in colon cancer. *Gut* 2009; **58**(1): 90–96.
5 Baker M. Stem cells: fast and furious. *Nature* 2009; **458** (7241): 962–965.
6 Winawer S, Fletcher R, Rex D, Bond J, Burt R, Ferrucci J et al. Colorectal cancer screening and surveillance: clinical guidelines and rationale: update based on new evidence. *Gastroenterology* 2003; **124**(2): 544–560.
7 Winawer SJ, Zauber AG, Gerdes H, O'Brien MJ, Gottlieb LS, Sternberg SS et al. Risk of colorectal cancer in the families of patients with adenomatous polyps. National Polyp Study Workgroup. *N Engl J Med* 1996; **334**(2): 82–87.
8 Quintero E, Castells A, Bujanda L, Cubiella J, Salas D, Lanas A et al. Colonoscopy versus fecal immunochemical testing in colorectal-cancer screening. *N Engl J Med* 2012; **366**(8): 697–706.

Figure 5 (*Continued*) transport systems, two-component regulatory systems (TCRS), iron (Fe) manganese (Mn) transport, and putrefaction in the CRC microbiome are summarized to the right of the heatmap (biosynth., biosynthesis; ascorbate-sp., ascorbate-specific). B. The heatmap displays significant abundance changes of genes summarized by CAZy family annotation with a >1.33 fold change and an FDR-adjusted p-value<0.1. A metabolic switch from utilization of dietary fibre to degradation of host carbohydrates, e.g. mucins, in CRC metagenomes, as well as an CRC-associated increase in metabolism of potentially pro-inflammatory bacterial cell wall components, such as lipopolysaccharide (LPS) and peptidoglycan (PG) is annotated to the right. (From Zeller et al. Mol Syst Biol 2014).

9 Lichtenstein P, Holm NV, Verkasalo PK, Iliadou A, Kaprio J, Koskenvuo M *et al.* Environmental and heritable factors in the causation of cancer — analyses of cohorts of twins from Sweden, Denmark, and Finland. *N Engl J Med* 2000; **343**(2): 78–85.

10 Hemminki K, Chen B. Familial risk for colorectal cancers are mainly due to heritable causes. *Cancer Epidemiol Biomarkers Prev* 2004; **13**(7): 1253–1256.

11 Greer JB, O'Keefe SJ. Microbial induction of immunity, inflammation, and cancer. *Front Physiol* 2011; **1**: 168.

12 O'Keefe SJ. Nutrition and colonic health: the critical role of the microbiota. *Curr Opin Gastroenterol* 2008; **24**(1): 51–58.

13 McIllmurray MB, Langman MJ. Large bowel cancer: causation and management. *Gut* 1975; **16**(10): 815–820.

14 de Filippo C, Cavalieri D, Di Paola M, Ramazzotti M, Poullet JB, Massart S *et al.* Impact of diet in shaping gut microbiota revealed by a comparative study in children from Europe and rural Africa. *Proc Natl Acad Sci USA* 2010; **107**(33):14691–14696.

15 Shen XJ, Rawls JF, Randall T, Burcal L, Mpande CN, Jenkins N *et al.* Molecular characterization of mucosal adherent bacteria and associations with colorectal adenomas. *Gut Microbes* 2010; **1**(3): 138–147.

16 Pagnini C, Corleto VD, Mangoni ML, Pilozzi E, Torre MS, Marchese R *et al.* Alteration of local microflora and alpha-defensins hyper-production in colonic adenoma mucosa. *J Clin Gastroenterol* 2011; **45**(7): 602–610.

17 Gaboriau-Routhiau V, Lecuyer E, Cerf-Bensussan N. Role of microbiota in postnatal maturation of intestinal T-cell responses. *Curr Opin Gastroenterol* 2011; **27**(6): 502–508.

18 Bordonaro M, Lazarova DL, Sartorelli AC. Butyrate and Wnt signaling: a possible solution to the puzzle of dietary fiber and colon cancer risk? *Cell Cycle* 2008; **7**(9): 1178–1183.

19 Qin J, Li R, Raes J, Arumugam M, Burgdorf KS, Manichanh C *et al.* A human gut microbial gene catalogue established by metagenomic sequencing. *Nature* 2010; **464** (7285): 59–65.

20 Eckburg PB, Bik EM, Bernstein CN, Purdom E, Dethlefsen L, Sargent M *et al.* Diversity of the human intestinal microbial flora. *Science* 2005; **308**(5728):1635–1638.

21 Vijay-Kumar M, Aitken JD, Carvalho FA, Cullender TC, Mwangi S, Srinivasan S *et al.* Metabolic syndrome and altered gut microbiota in mice lacking Toll-like receptor 5. *Science* 2010; **328**(5975): 228–231.

22 Bach JF. The effect of infections on susceptibility to autoimmune and allergic diseases. *N Engl J Med* 2002; **347**(12): 911–920.

23 Arumugam M, Raes J, Pelletier E, Le PD, Yamada T, Mende DR *et al.* Enterotypes of the human gut microbiome. *Nature* 2011; **473**(7346):174–180.

24 McCit WC, Mason JM, III. Enterococcal endocarditis associated with carcinoma of the sigmoid; report of a case. *J Med Assoc State Ala* 1951; **21**(6): 162–166.

25 Savage DC. Microbial ecology of the gastrointestinal tract. *Annu Rev Microbiol* 1977; **31**: 107–133.

26 Sobhani I, Tap J, Roudot-Thoraval F, Roperch JP, Letulle S, Langella P *et al.* Microbial dysbiosis in colorectal cancer (CRC) patients. *PLoS One* 2011; **6**(1):e16393.

27 Marchesi JR, Dutilh BE, Hall N, Peters WH, Roelofs R, Boleij A *et al.* Towards the human colorectal cancer microbiome. *PLoS One* 2011; **6**(5): e20447.

28 Tjalsma H, Boleij A, Marchesi JR, Dutilh BE. A bacterial driver-passenger model for colorectal cancer: beyond the usual suspects. *Nat Rev Microbiol* 2012; **10**(8): 575–582.

29 Zeller G, Tap J, Voigt AY, Sunagawa S, Kultima JR, Costea PI *et al.* Potential of fecal microbiota for early-stage detection of colorectal cancer. *Mol Syst Biol* 2014; **10**: 766.

30 Balkwill F, Mantovani A. Cancer and inflammation: implications for pharmacology and therapeutics. *Clin Pharmacol Ther* 2010; **87**(4): 401–406.

31 Seksik P, Rigottier-Gois L, Gramet G, Sutren M, Pochart P, Marteau P *et al.* Alterations of the dominant faecal bacterial groups in patients with Crohn;s disease of the colon. *Gut* 2003; **52**(2): 237–242.

32 Arthur JC, Perez-Chanona E, Muhlbauer M, Tomkovich S, Uronis JM, Fan TJ *et al.* Intestinal inflammation targets cancer-inducing activity of the microbiota. *Science* 2012; **338**(6103): 120–3.

33 Cuevas-Ramos G, Petit CR, Marcq I, Boury M, Oswald E, Nougayrede JP. Escherichia coli induces DNA damage in vivo and triggers genomic instability in mammalian cells. *Proc Natl Acad Sci U S A* 2010; **107**(25):11537–11542.

34 Castellarin M, Warren RL, Freeman JD, Dreolini L, Krzywinski M, Strauss J *et al.* Fusobacterium nucleatum infection is prevalent in human colorectal carcinoma. *Genome Res* 2012; **22**(2): 299–306.

35 Kostic AD, Gevers D, Pedamallu CS, Michaud M, Duke F, Earl AM *et al.* Genomic analysis identifies association of Fusobacterium with colorectal carcinoma. *Genome Res* 2012; **22**(2): 292–298.

36 Bolstad AI, Jensen HB, Bakken V. Taxonomy, biology, and periodontal aspects of Fusobacterium nucleatum. *Clin Microbiol Rev* 1996; **9**(1): 55–71.

37 Kostic AD, Chun E, Robertson L, Glickman JN, Gallini CA, Michaud M *et al.* Fusobacterium nucleatum potentiates intestinal tumorigenesis and modulates the tumor-immune microenvironment. *Cell Host Microbe* 2013; **14**(2): 207–215.

38 Sobhani I, Amiot A, Le BY, Levy M, Auriault ML, Van Nhieu JT *et al.* Microbial dysbiosis and colon carcinogenesis: could colon cancer be considered a bacteria-related disease? *Therap Adv Gastroenterol* 2013; **6**(3): 215–229.

39 Deschner EE. Experimentally induced cancer of the colon. *Cancer* 1974; **34**(3): suppl-8.

40 Lipkin M. Biomarkers of increased susceptibility to gastrointestinal cancer: new application to studies of cancer prevention in human subjects. *Cancer Res* 1988; **48**(2): 235–245.

41 Whiteley LO, Klurfeld DM. Are dietary fiber-induced alterations in colonic epithelial cell proliferation predictive of fiber's effect on colon cancer? *Nutr Cancer* 2000; **36**(2): 131–149.

42 Kinzler KW, Vogelstein B. Lessons from hereditary colorectal cancer. *Cell* 1996; **87**(2): 159–170.

43 Dejea CM, Wick EC, Hechenbleikner EM, White JR, Mark Welch JL, Rossetti BJ *et al.* Microbiota organization is a distinct feature of proximal colorectal cancers. *Proc Natl Acad Sci U S A* 2014; **111**(51): 18321–18326.

44 Wilcox DK, Higgins J, Bertram TA. Colonic epithelial cell proliferation in a rat model of non-genotoxin-induced colonic neoplasia. *Lab Invest* 1992; **67**(3): 405–411.

45 Zheng Y, Kramer PM, Lubet RA, Steele VE, Kelloff GJ, Pereira MA. Effect of retinoids on AOM-induced colon cancer in rats: modulation of cell proliferation, apoptosis and aberrant crypt foci. *Carcinogenesis* 1999; **20**(2): 255–260.

46 Reya T, Clevers H. Wnt signalling in stem cells and cancer. *Nature* 2005; **434** (7035): 843–850.

47 Vandesompele J, De PK, Pattyn F, Poppe B, Van RN, De PA *et al.* Accurate normalization of real-time quantitative RT-PCR data by geometric averaging of multiple internal control genes. *Genome Biol* 2002; **3**(7): RESEARCH0034.

48 Clevers H. Wnt/beta-catenin signaling in development and disease. *Cell* 2006; **127** (3):469–480.

49 Vogelmann R, Amieva MR. The role of bacterial pathogens in cancer. *Curr Opin Microbiol* 2007; **10**(1): 76–81.

50 Coic YM, Baleux F, Poyraz O, Thibeaux R, Labruyere E, Chretien F *et al.* Design of a specific colonic mucus marker using a human commensal bacterium cell surface domain. *J Biol Chem* 2012; **287**(19): 15916–15922.

51 Pedron T, Mulet C, Dauga C, Frangeul L, Chervaux C, Grompone G *et al.* A crypt-specific core microbiota resides in the mouse colon. *MBio* 2012; **3**(3).

52 Lindhurst MJ, Sapp JC, Teer JK, Johnston JJ, Finn EM, Peters K *et al.* A mosaic activating mutation in AKT1 associated with the Proteus syndrome. *N Engl J Med* 2011; **365**(7): 611–619.

53 Ponz de LM, Roncucci L, Di DP, Tassi L, Smerieri O, Amorico MG *et al.* Pattern of epithelial cell proliferation in colorectal mucosa of normal subjects and of patients with adenomatous polyps or cancer of the large bowel. *Cancer Res* 1988; **48** (14): 4121–4126.

54 Mills SJ, Mathers JC, Chapman PD, Burn J, Gunn A. Colonic crypt cell proliferation state assessed by whole crypt microdissection in sporadic neoplasia and familial adenomatous polyposis. *Gut* 2001; **48**(1): 41–46.

55 Zhu Y, Michelle LT, Jobin C, Young HA. Gut microbiota and probiotics in colon tumorigenesis. *Cancer Lett* 2011; **309**(2): 119–127.

56 Wu S, Rhee KJ, Albesiano E, Rabizadeh S, Wu X, Yen HR *et al.* A human colonic commensal promotes colon tumorigenesis via activation of T helper type 17 T cell responses. *Nat Med* 2009; **15**(9): 1016–1022.

57 Keerthivasan S, Aghajani K, Dose M, Molinero L, Khan MW, Venkateswaran V *et al.* beta-Catenin promotes colitis and colon cancer through imprinting of proinflammatory properties in T cells. *Sci Transl Med* 2014; **6**(225): 225ra28.

58 Sridharan GV, Choi K, Klemashevich C, Wu C, Prabakaran D, Pan LB *et al.* Prediction and quantification of bioactive microbiota metabolites in the mouse gut. *Nat Commun* 2014; **5**: 5492.

59 Frezza EE, Wachtel MS, Chiriva-Internati M. Influence of obesity on the risk of developing colon cancer. *Gut* 2006; **55**(2): 285–291.

60 Turnbaugh PJ, Ley RE, Mahowald MA, Magrini V, Mardis ER, Gordon JI. An obesity-associated gut microbiome with increased capacity for energy harvest. *Nature* 2006; **444**(7122): 1027–1031.

61 Hirose Y, Hata K, Kuno T, Yoshida K, Sakata K, Yamada Y *et al.* Enhancement of development of azoxymethane-induced colonic premalignant lesions in C57BL/KsJ-db/db mice. *Carcinogenesis* 2004; **25**(5): 821–825.

62 Aparicio T, Kotelevets L, Tsocas A, Laigneau JP, Sobhani I, Chastre E *et al.* Leptin stimulates the proliferation of human colon cancer cells in vitro but does not promote the growth of colon cancer xenografts in nude mice or intestinal tumorigenesis in Apc(Min/+) mice. *Gut* 2005; **54**(8): 1136–1145.

63 Arthur JC, Jobin C. The struggle within: microbial influences on colorectal cancer. *Inflamm Bowel Dis* 2011; **17**(1): 396–409.

64 Scheppach W. Effects of short chain fatty acids on gut morphology and function. *Gut* 1994; **35** (1 Suppl): S35–S38.

65 Blaser MJ, Webb GF. Host demise as a beneficial function of indigenous microbiota in human hosts. *MBio* 2014; **5**(6).

CHAPTER 27

The gut microbiota and the CNS: an old story with a new beginning

Aadil Bharwani[1] and Paul Forsythe[2]

[1] The Brain-body Institute and Firestone Institute for Respiratory Health, Ontario, Canada
[2] McMaster University, Hamilton, Ontario, Canada

27.1 Introduction

Most individuals become aware of brain-gut communication when disturbed gastrointestinal function is communicated to the brain, leading to the perception of visceral events such as nausea, satiety and pain, or, conversely, when stressful experiences lead to altered gastrointestinal secretions and motility[1]. However, bi-directional communication between the gut and brain occurs constantly and largely at a subconcious level, playing a critical role in maintaining homeostasis. Indeed, it is suggested that defects in gut-brain axis communication are an underlying cause of functional bowel disorders including irritable bowel syndrome (IBS)[2] and, potentially, inflammatory bowel disease[3]. This brain-gut axis consists of "hard-wired" anatomical connections involving vagal, spinal and enteric nerves and humoral components provided by the endocrine and immune systems (Figure 1).

The gastrointestinal tract is a nexus for the largest concentration of immune cells in the body, a network of 200-600 million neurons and the approximately 100 trillion bacteria[4] that constitute the human gut microbiota. It therefore seems almost intuitive that intestinal bacteria could influence gut-to-brain communication and potentially lead to modulation of central nervous system (CNS) function. However, for much of the last century the idea that the gut and attendant microbes had influence over mood and behavior was consigned to the realms of pseudoscience. Only in the past decade, with advances in sequencing technology, metabolomics and neurophysiology, has the concept of the microbiota-gut-brain axis gained serious attention and consideration; but what is seen by many as a novel approach to our understanding of brain development and mental health is, in many respects, the resurgence of an old idea.

The Human Microbiota and Chronic Disease: Dysbiosis as a Cause of Human Pathology, First Edition.
Edited by Luigi Nibali and Brian Henderson.
© 2016 John Wiley & Sons, Inc. Published 2016 by John Wiley & Sons, Inc.

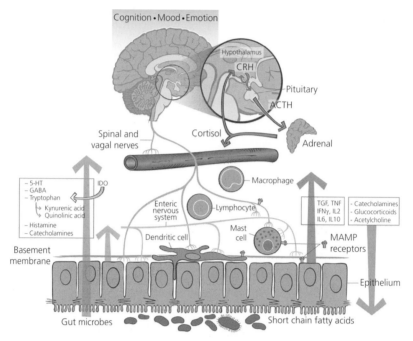

Figure 1 The multiple pathways that enable bidirectional communication along the microbiota-gut-brain axis. Adapted from Forsythe 2010[175].

27.2 The microbiota-gut-brain axis: a historical framework

The inception of the idea that microorganisms and other contents of the gut can influence behavior and the brain predates our understanding of certain fundamental concepts of neuroscience and microbiology. More than 150 years ago, the elegant work of Louis Pasteur detailing the role of microorganisms in the pathogenesis of various diseases set the stage for hypotheses on the role of the gut in mental illnesses[5]. The 19th century saw the advent of the most popular of such hypotheses: autointoxication. Briefly, autointoxication was a process describing the influence of intestinal contents, especially toxins, on systemic health[6]. The colon was viewed as "a receptacle and laboratory of poisons", as described in *Lectures on autointoxication in disease*, which garnered attention worldwide[7]. While the precise toxins themselves remained unknown — metabolites, bacterial antigens, or substances secreted by such organisms — these were the early origins of gut-brain communication: viewing melancholic depression and other psychological changes as a consequence of diet and gut-derived substances[8]. Pathologist John G. Adami argued that autointoxication was, in fact, a low-grade immune response directed against gut bacteria, which could consequentially induce symptoms of depression — a phenomenon well described today as the adaptive "sickness behavior"[9-11]. These suggestions were bolstered by observations of psychological stress-induced increase in intestinal permeability, resulting in increased bacterial translocation across the mucosal barrier[12]. Aligning with such thinking, Elie Metchnikoff proposed fighting "microbe with microbe" — today referred to as "probiotic" therapy[6]. This was extended to the treatment of melancholia using Lactobacillus bacteria, given the association of depression with symptoms of

constipation and disturbed digestion[13]. Unconcerned with whether these disturbances were the cause or consequence of depression, Phillips proposed the ingestion of such bacterial cultures with a strictly regimented diet to enable the
proliferation of the beneficial microbe in the gut. Lactobacillus treatment, Phillips
noted, was associated with reduced depression and lethargy, weight recovery,
and amelioration of constipation, all the while the patient "wears a happier
expression".

In what would be hailed as a ground-breaking discovery a century later,
Michael Cohendy successfully raised the first germ-free model at the Pasteur
Institute: chickens that lacked an intestinal microbiome[6]. Although this breakthrough was hailed by contemporaries for enabling research into the consequences of microbiota reconstitution, it was not until 2004 that this model would
be adopted in seminal work investigating the development of the brain and
behavior in such animals, and studying the effects of post-natal microbial colonization[14]. In fact, by the mid-1900s, the concept of gut-brain communication and
the role of autointoxication in mental health had been dismissed as mere pseudo-
scientific efforts. This was largely owing to the commercial and marketing over-
reach of bacterial products, lack of sufficient evidence for the benefits of lactic acid
bacteria administration, and the unsubstantiated nature of the vague and poorly
defined concept of autointoxication, which came to be associated with unsafe,
unregulated, and dubious therapies: colon purges, colonic contraptions, incor-
rectly labelled bacterial preparations[6]. These factors contributed to a shift in perception that ultimately undermined how scientifically sound research on the
subject was received, including the insightful "unifying theory" of Stokes and
Pillsbury that posited microbiota changes as the mechanism underlying similarities between inflammatory skin conditions and emotional states[6,15,16].

The concept of autointoxication and related hypotheses were consequentially
dismissed, and this area of research — investigating the bottom-up influence of
the gut on the brain — remained dormant for several decades. In the early 2000s,
however, these concepts were raised anew and incorporated into a scientifically
sound hypothesis describing how probiotics may serve as an adjuvant in the treatment of depression and chronic fatigue syndrome[17,18]. Moreover, with the dawn
of the 21st century, there was a growing recognition of the multifaceted nature of
mood and psychiatric conditions. Indeed, this line of thinking coincided with
observations of the effects of the immune system and inflammation on behavior
and mental health[11]. Evidence also emerged demonstrating the critical role of the
microbiome in the development of the host immune system and mucosal architecture[19], as well as its ability to influence host behavior, both through and independent of the immune system[20,21]. These preliminary yet pivotal lines of evidence
set the stage for the resurgence of a seemingly outdated concept in our understanding of the brain and, particularly, mental health.

27.3 The microbiota-gut-brain axis: an evolutionary perspective

Although data and observations derived from mammalian models possess greater
generalizability to human health and potential for translation to clinical models,
evolutionary analysis of the host-microbiome relationship in a less complex system

can yield tremendous insight. *Drosophila melanogaster* is such a system that has been thoroughly studied in the past, thus providing a less obfuscated view of host-microbiome interplay. For instance, proliferation of even a single bacterial species — in this case, *Lactobacillus plantarum* — can influence fruit-fly mating preference due to changes in salient odor cues produced by this bacterium[22]. Treatment with antibiotics abolished mating preference and was only restored upon infection with *L. plantarum*. In a recent study, researchers examined the phenomenon of larval and adult fruit-fly preference for food previously occupied by larvae[23]. No such attraction was observed to axenic (bacteria-free) food occupied by axenic larvae. However, upon the addition of the fruit fly gut commensals *L. brevis* and *L. plantarum*, or volatile compounds from *L. brevis*, the food preference was restored, suggesting that the volatile compounds produced by the microbiome serve as cues that influence social behavior. A similar role of the microbiome in the collective social behavior of hosts has also been demonstrated in locusts. Aggregation pheromones, which are responsible for locust-congregation behavior, are derived from metabolites, such as guaiacol, produced by members of the gut microbial community[24,25]. These observations from insect models demonstrate remarkable parallels to those stemming from rodent studies, despite 782.7 million years separating the divergence of the two groups[26]. It is also proposed such observations support a hologenome theory of evolution[27]. The hologenome is defined as the sum of the genetic material of the host and its microbiota. It is posited that the holobiont (host plus its associated microorganisms) acts as a unit of selection in evolutionary change and that variation, an important factor in evolution, can occur through modification in either the host or the microbiota genomes[27]. Ryan discussed this concept in chapter 10 and described how microorganisms may have helped shape the human genome.

The co-evolution of microbes and host is also evident is common signaling pathways that facilitate interkingdom communication. It is not generally appreciated that gut bacteria "speak the language" of the nervous system, producing a range of biological mediators, such as GABA, serotonin, melatonin, histamine and acetylcholine, normally associated with neurotransmission[28]. Indeed, it has been proposed that many of the enzymes involved in the synthetic and metabolic pathways of catecholamines, histamine, acetylcholine and GABA are a result of late horizontal gene transfer from bacteria. Evidence that neurotransmitters produced by the host can influence the function of components of the microbiota supports the concept of shared signaling pathways. For example, in the QseC sensor kinase of *Escherichia coli* O157:H7 is a receptor for host-derived epinephrine/norepinephrine, activation of which leads to transcription of virulence genes in the bacteria[29]. Conversely, there is increasing evidence that signaling molecules of quorum sensing systems, used by bacterial to communicate and coordinate their actions[30], can also bind to mammalian receptors, including taste receptors[31], and directly influence the host[31-33]. Indeed, the level of integration between animals and microbes has led to the suggestion that animals can no longer be considered as individuals but must be considered holobionts "whose anatomical, physiological, immunological and developmental functions evolved in shared relationships of different species"[34]. Within the context of this holobiont paradigm, the influence of the gut microbiota over brain development, mood, motivation and behavior is a natural consequence of the co-evolution of a multi-species organism.

To provide clearer insight into the implications of this new idea of "self" for health and disease, we need greater understanding of the mechanisms, pathways and consequences of communication along the microbiota-gut-brain axis.

27.4 The gut microbiota influence on brain and behavior

Clear evidence that the gut microbiota plays a direct role in development and function of the CNS comes from the study of germ-free animals. The absence of a microbiota results in an exaggerated hypothalamic–pituitary–adrenal (HPA) axis response to stress[14] and altered behavior, particularly reduced anxiety-related behavior, in rodents[35,36]. Restoration of a gut microbiota in germ-free mice, early in life, can normalize HPA axis activity[14] and several germ- free behavioral patterns[35], while similar conventionalization of adult mice fails to change the germ-free phenotype[14,35]. This indicates that the gut microbiota contributes to developmental programming; a process whereby an environmental factor acting during a developmental window can have a sustained impact on physiological function[37]. Reduced anxiety-related behaviors in germ-free animals are associated with changes in neurochemistry and gene expression in the brain. Enhanced turnover of noradrenaline, dopamine, and 5-HT has been demonstrated in the striatum of germ-free compared to specific pathogen-free mice[35]. Changes have also been reported in NMDA receptor subunit expression[14,36]. Sudo *et al.* identified a reduction in brain derived neurotrophic factor (BDNF) expression levels in the cortex and hippocampus of germ- free relative to conventional mice[14]. BDNF is involved in the regulation of multiple aspects of cognitive and emotional behaviors, being a key promoter of neuronal survival and growth as well as differentiation of new neurons and synapses[38–40]. Heijtz *et al.* also noted lower levels of BDNF mRNA expression in germ-free animals[35], specifically in the hippocampus, amygdala and cingulate cortex, all important components of the neural circuitry underlying anxiety and fear[41,42]. In contrast, Neufeld *et al.*[36] identified that reduced anxiety in germ-free mice was associated with an upregulation in the expression of BDNF mRNA in the dentate gyrus of the hippocampus. The conflicting observations regarding hippocampal BDNF in germ-free animals are in keeping with the lack of clarity in the association between anxiety and BDNF; with studies noting positive, negative or no correlation between hippocampal levels and anxiety[43–46] but may also be related to sex differences in mice used in these experiments. There is obviously much more scope for experiments designed to identify the role of the microbiota in driving the development of specific brain circuitry.

Recent evidence suggests that the gut microbiota plays a role in another aspect of CNS development- the formation of the blood-brain barrier (BBB)[47]. An intact BBB ensures optimal neuronal growth and protect against colonizing microbes during the neonatal period of brain development. *In utero*, germ-free mice display increased BBB permeability compared to mice with a normal gut microbiota. The increased BBB permeability is maintained in germ-free mice into adulthood and is associated with reduced expression of the tight junction proteins occludin and claudin-5, which are known to regulate barrier function in endothelial tissues. Notably, unlike other aspects of CNS development and behavior, the germ-free BBB phenotype can be normalized by exposure of mice to a pathogen-free gut microbiota even in adulthood[47]. While studies in germ-free mice identify a role for the gut microbiota in modulating CNS development, the germ-free state is artificial and these studies tell us nothing about the effect of physiologically relevant, qualitative changes in the gut microbiota. However, a number of approaches have been used to address this issue, including use of gnotobiotic mice, fecal transplants and treatments with antibiotics and probiotics.

Alterations in diet can lead to marked shifts in gut microbial populations[48,49] and mice fed a diet containing 50% lean ground beef were found to have a greater diversity of gut bacteria than those receiving standard rodent chow[50]. This increased bacterial diversity was associated with improvements in working and reference memory[50] and decreased anxiety-like behavior in response to the novel testing environment. While direct effects of dietary components likely contribute to the behavioral changes observed, this study did provide early support for the suggestion that diet-induced changes in bacterial diversity could influence the CNS. Associations between diet and microbial dysbiosis are discussed in detail in chapter 29.

The implications of gut microbe disruption for brain and behavior have been addressed more directly through the use of antimicrobial drugs to induce experimental dysbiosis. Bercik *et al.* reported that oral administration of neomycin and bacitracin along with the antifungal agent primaricin[51] leads not to quantitative changes in culturable bacteria but to a significant change in composition with increased Actinobacteria and Lactobacilli species and decreased γ-proteobacteria and bacteroidetes. The antibiotic treatment also resulted in behavioral changes. In the step-down and light/dark preference tests, treated animals demonstrated increased exploratory drive and decreased anxiety. Behavioral changes in antibiotic-treated animals were associated with decreased BDNF levels in the amygdala and increased levels in the hippocampus[51]. Similar changes in BDNF and associated behaviors have also been observed upon the adoptive transfer of gut microbiota[51]. Development of the anxious phenotype of Balb/c strain following microbiota transfer into germ-free NIH Swiss mice was associated with a reduction in BDNF levels. Conversely, greater levels of BDNF and lower levels of anxiety were observed upon the transfer of the NIH Swiss mice microbiota into germ-free Balb/c mice.

The ability of certain bacteria in the gut directly to modulate central neural pathways was first indentfied in the context of responses to pathogens[52]. Orally administered *Camphylobacter jejuni*, in doses too low to elicit immune activation have anxiety-provoking effects in mice. These same low doses of *C. jejuni* also activate visceral sensory nuclei in the brainstem, including the nucleus tractus solitarii (NTS) and lateral parabrachial nucleus, which participate in the neural information processing underlying autonomic neuroendocrine and behavioral responses[52]. Non-pathogenic gut bacteria also activate central neural pathways. Here the majority of data comes from studies of the effect of exposure to Lactobacillus and Bifidobacteria species. Intraduodenal injection of *Lactobacillus johnsonii* La1 was demonstrated to reduce renal sympathetic nerve activity while enhancing gastric vagal nerve activity[53] through a centrally mediated pathway invoving the hypothalamic suprachiasmatic nucleus, a major regulator of circadian rhythm[53]. Oral administration of a *L. rhamnosus* strain was demonstrated to reduce anxiety-like behavior and stress-induced plasma corticosterone levels in mice[54]. Mice that received *L. rhamnosus* also demonstrated alterations in central

GABA receptor subunit mRNA expression. Long-term *L. rhamnosus* administration decreased expression of GABA type B (GABAB) subunit 1 isoform b (GABAB1b) mRNA in the amygdala and hippocampus, while increasing expression in cortical areas. Expression of GABAAα2 receptor mRNA was reduced in the amygdala and cortical areas and increased in the hippocampus[54]. Additional studies in rodent models indicate that oral treatment with certain candidate probiotic bacteria can prevent the anxiety-like behaviors induced by infection, inflammation or maternal separation[55–57].

As detailed above, much of the research on microbiota effects on behavior has focused on anxiety- and depression-related responses. However, exposure to bacteria in the gut can also modulate the central circuitry underlying cognition and memory. The age-related deficit in long-term potentiation (LTP) in the rat hippocampus, a process closely associated with learning memory and fear processing, was attenuated by feeding a mixture of eight Gram-positive bacterial strains[58]. The effects on LTP were associated with a decrease in markers of microglial activation and an increase in expression of BDNF together with evidence of a change in genes that impact on neuronal plasticity. As indicated in studies outlined above, while we are still in the early stages, we are beginning to develop a picture of the changes in the neural circuitry and physiology that underlie the influence of the microbiome on and behavior and cognition.

To date studies on central effects of modulating gut microbiota on central pathways of humans have been limited. However, long-term ingestion of a fermented milk product with probiotic by healthy women lead to marked alterations in the response of a widely distributed brain network to a task probing attention to negative context[59]. The alterations included reductions in activity and changes in connectivity in interoceptive, somatosensory and affective brain regions. These changes were not observed if women consumed a non-fermented milk product of identical taste, so that the findings appear to be related to the ingested bacteria strains and their effects on the host.

27.5 Microbes and the hardwired gut brain axis

At the heart of understanding how neural signaling between the gut and brain can modulate mood and behavior lies the concept of interoception. Interoception is the sensing of the physiological condition of the body[60], as well as the representation of the internal state[61] within the context of ongoing activities. Interoception is closely associated with emotional awareness and well-being[62] and informs motivated actions that homeostatically regulate the internal state[61]. Interoceptive signals include sensations such as pain, temperature, itch, tickle, sensual touch, muscle tension, air hunger and intestinal tension[60]. These sensations are transmitted to the brain by vagal and glossopharyngeal afferents synapsing with the nucleus of the solitary tract (NTS) and via small-diameter primary sympathetic afferent fibers to a specific thalamo-cortical relay nucleus, and are integrated to provide a sense of the physiological condition[60]. It has been suggested that altered interoceptive signals can influence our attitude to the outside world and that pathological changes in visceral sensory inputs increase the risk of affective behavioral disorders[63].

It is clear that gut microbiota can modulate the function of the nervous system responsible for visceral perception. Feeding of a Lactobacillus Rhamnousus strain to rats completely suppressed colonic distension, induced pseudo-affective cardiac responses and reduced electrical discharge in single fibres of the dorsal root ganglia, reflecting inhibition of the perception of visceral pain[64]. Similar effects have been observed with other lactobacillus species. Rousseaux *et al.*[65] demonstrated that oral administration of specific Lactobacillus strains promotes morphine-like analgesic functions by inducing the expression of μ-opioid and cannabinoid receptors in intestinal epithelial cells. It therefore appears that at least some of the influence of the gut microbiota on mood and behavior occurs as a consequence of altered interoceptive signaling. In order to establish the full implications and theraputic oppurtunities arising from microbial modulation of interoception, it will be important to identify and better understand mechanisms and neural pathways involved.

27.5.1 The vagus

The vagus nerve is the major hardwired pathway for communication between the gut and brain. The vast majority (80%) of vagal nerve fibres supplying the gut are afferents[66,67]. Vagal afferents innervate the muscular and mucosal layers of the entire gut[68] and while vagal innervation is densest proximally, it is still substantial for the colon. Sensory vagal inputs arrive in the nucleus of the solitary tract (NTS, and from there are transmitted to widespread areas of the CNS, many of which[69] are associated with stress-related behavior and affective disorders. Vagal afferent fibres with terminals lying in the mucosa[66] do not cross the basal membrane to innervate the epithelial layer of the gut[68] and are therefore not in a position to sense luminal nutrients directly. However, the mucosal vagal fibres are in close anatomical apposition to the basal membrane of enteroendocrine cells[70] and express a large variety of chemosensitive receptors that are targets of gut hormones and regulatory peptides such as ghrelin, CCK, GLP-1 and PYY that influence the control of food intake and regulation of energy balance[67]. Vagal afferent fibers also sense the immune environment of the intestine and are activated by inflammatory cytokines, which can lead to the characteristic lethargy, depression, anxiety, loss of appetite, sleepiness and hyperalgesia associated with sickness behavior[11,71–75]. Furthermore, intraganglionic laminar vagal afferent endings (IGLEs), located in the connective tissue capsule of myenteric plexus ganglia, respond to tension associated with passive stretch and active contraction of the muscle layers[76] and are thought to be important for generating vagal afferent tone that contributes to balanced interoceptive awareness and thus emotional well-being.

In keeping with this, vagal stimulation is an FDA-accepted alternative treatment for intractable depression. While controversial due to a lack of positive sham treatment controlled clinical trials, there have been reports that vagal nerve stimulation is beneficial in at least some patients with depression and may be particularly effective with chronic treatment[77,78].

There is now good evidence, from rodent studies, that elements of the gut microbiota can activate the vagus nerve and that such activation plays a critical role in mediating effects on the brain and subsequently behavior. Gut pathogens provided the first evidence for vagus nerve mediated communication between intestinal bacteria and the CNS. Subdiaphragmatic vagotomy attenuated c-fos expression in the PVN of rats inoculated with Salmonella typhimurium[79]. Although

S. typhimurium infection was accompanied by intestinal inflammation, subsequent studies have indicated that microbes in the gastrontential tract can directly activate vagal pathways even in the absence of an identified immune response[52]. The anxiogenic effect of orally administered subclinical doses of Camphylobacter jejuni in mice was associated with a significant increase in c-Fos expression in neurons bilaterally in the vagal ganglia and activated visceral sensory nuclei in the brainstem. The areas of brainstem activation, the NTS and lateral parabrachial nucleus, participate in neural information processing that ultimately lead to autonomic neuroendocrine and behavioral responses[52]. Many of the effects of nonpathogenic bacteria on the CNS also appear to be mediated via activation of vagal signaling between gut and brain. The ability of intraduodenal *Lactobacillus johnsonii* La1 to reduce renal sympathetic nerve activity and blood pressure[53] was eliminated by denervation of vagal nerve fibers surrounding the oesophagus, indicating that the vagal signaling is required for at least some of the effects of this bacteria on autonomic nerve responses[53]. Intraluminal infusion of *Lactobacillus rhamnosus* elicits rapid single- and multi-unit firing of vagus nerve afferents in the mesenteric nerve bundle[80] of the mouse within minutes of application. The observed effect was not a general response to bacterial exposure, as another Lactobacillus species, *L. salivarius*, did not evoke changes in vagas firing. Correspondinlgly, subdiaphragmatic vagotomy blocks the anxiolytic and antidepressant effects of chronic *L. rhamnosus* ingestion, along with the assocaited alterations in GABA receptor expression in the amygdala[54]. Similarly, Bercik *et al.* demonstrated that the ability of *B. longum* to attenuate anxiety associated with experimental colitis was abolished by vagotomy[56]. That the contrasting anxiogenic and anxiolytic effects of pathogens and probiotics, respectively, can be driven by vagal signaling suggests the nerve endings can distinguish between elements of the gut microbiota. Although the mechanism underlying the ability of the vagus to distinguish between microbial stimuli remain unclear, it does appear that information regarding the composition of the microbiota can be transmitted along ascending vagal pathways. These signals accordingly elicit differential neural activity and structural changes in the brain that are reflected by distinct behavioral phenotypes.

Until very recently it was the widely held view that intestinal vagal sensory neurons, like those elsewhere in the body, are entirely primary in nature: that is, they respond non-synaptically to luminal stimuli. Correspondingly, it was also believed that enteric intrinsic primary afferent neurons did not contribute synaptically to impulses that reach the brain[81,82]. However, Perez-Burgos *et al.*[83] identified that gut-intrinsic primary afferent neurons can relay signals originating from the lumen to the vagal sensory ganglia. Furthermore, greater than 50% of vagal afferent stimulation observed following exposure to a psychoactive *L.rhamnousus* strain was a result of receiving input from this "sensory synapse." This finding expands our understanding of the anatomical gut-brain connection and significantly increases the significance of the enteric nervous system in the microbiota-gut-brain axis.

27.5.2 The enteric nervous system

The ENS can control intestinal function independently from the CNS while still allowing for input from the central and autonomic nervous systems. The chemo- and mechanosensory intrinsic primary afferent neurons (IPANs)[84] are necessary for normal peristalsis and viability. Intrinsic primary afferent neurons project to other IPANs, inter- and motor neurons, and send long processes into the gut mucosa, terminating near the epithelial layer[84–86]. Indeed, by far the richest innervation of mucosal epithelial layer cells derives from the myenteric plexus, which provides more than 90% of sensory neuropeptide containing fibres to the mucosal layer[87,88]. Myenteric IPANs are thus in a favorable position to transmit information from the intestinal lumen to the nervous system as a whole.

Recent evidence indicates that the gut microbiota is critical to the normal development of the ENS[89,90]. Germ-free mice were observed to have a decrease in nerve density, a decrease in the number of neurons per ganglion, and an increase in the proportion of myenteric nitrergic neurons in the jejunum and ileum[89]. The changes in neural density were associated with decreased frequency and amplitude of muscle contractions in the jejunum and ileum of germ-free mice. In a separate study a decrease in excitability of myenteric, IPANS was also noted in germ-free animals[90]. Notably, exposing the germ-free mice to a normal gut microbiota (conventionalization) normalized both density and activity of enteric neurons.[89,90].

While it is not clear how the gut microbiota contribute to normal ENS development and function, alterations in gut microbiota can also impact intestinal glial cells that provide support and nourishment for neurons as well as regulating synaptic transmission[91]. Kabouridis et al. demonstrated that antibiotic treatment, in addition to reducing the number of microbes within the gut lumen, also reduced the number of glial cells within the mucosa. Thus while there are no reports of glial cells being examined in germ-free mice, reduced glial cells numbers may underlie at least some of the ENS deficits in these animals.

That IPANs are a cellular target of specific bacteria has been demonstrated by whole-cell patch clamp recording experiments using rats that were fed L.rhamnosus. Myenteric IPANs but not motor- or interneurons within colon segments of probiotic-treated animals were more excitable than were those from controls.[92]. This increase in excitability was accompanied by a reduction in the post-action potential slow afterhyperpolarization, which mediates discharge accommodation in IPANs[92]. It was proposed that the underlying molecular mechanism involved inhibition of an intermediate conductance calcium-dependent potassium (IK_{Ca})[92,93,94].

The human enteric nervous system (ENS) contains up to 100 million IPANs, in contrast to the 50,000 extrinsic sensory vagal and spinal sensory neurons[67]. Based purely on their large number and dense innervation of mucosal villi[81], it seems likely that that IPANs are the initial neural responders to certain luminal stimuli, including microbes. This role appears reified by the observation that the majority of vagal signaling from gut lumen to brain is instigated by the ENS. As such, the ENS is a potential origin of gut-brain axis pathology, with enteric neural disruption leading to brain and limbic system dysfunction by altering afferent vagal impulses. On the other hand, the ENS, as a gateway to the brain, may offer unique therapeutic oppurtunities, including those involving modulation of the gut microbiota.

27.6 Hormonal pathways to the brain

In addition to direct neural pathways, hormonal signaling pathways play a key role in gut bran communication. Gut peptides from enteroendocrine cells can act directly on the brain at the area postrema (which lies outside the blood-brain barrier).

These gut peptides include orexin, galanin, ghrelin, gastrin and leptin. The gut peptides are most widely studied for their role in modulating feeding behavior and energy homeostasis but have also been associated with changes in sleep-wake cycle, sexual behavior, arousal and anxiety[95,96].

Studies in germ-free animals suggest that the gut microbes influence release of biologically active peptides and participate in the regulation of enteroendocrine cells[97], but to date little is known about the effect of qualitative changes in the gut microbiota on the hormonal components of gut-brain communication. However, given the ability of gut microbiota to alter nutrient availability[98] and the close relationship between nutrient sensing and peptide secretion by enteroendocrine cells[99], microbial modulation of hormonal signaling by the gut is highly plausible. Galanin can activate the central branch of the major adaptive endocrine response, the HPA axis. This peptide can also directly stimulate glucocorticoid secretion from adrenocortical cells and norepinephrine release from the adrenal medulla[100,101] and appears to play a role in modulating the HPA-axis response to stress. Given the established deleterious effects of galanin on cognitive function, the hormone may act as a link between stress, anxiety and memory[102,103] and it has been suggested that galaninergic drugs could provide a novel therapeutic option for psychopathologies such as post-traumatic stress syndrome[101]. Another major gut hormone, ghrelin, also induces ACTH/cortisol-release in humans and is likely involved in the modulation of the HPA response to stress[104–106]. In keeping with this, studies in gastrin-deficient mice indicate that normal circulating levels of gastrins play a role in the regulation of locomotor activity and anxiety-like behavior[107,108].

The pancreatic polypeptide–fold (PP-fold) family includes pancreatic polypeptide (PP) and peptide YY (PYY), and neuropeptide Y (NPY). These peptides have broad peripheral actions on a number of organs. Both NPY and PYY have anxiolytic effects in rats and NPY has been implicated in feeding and obesity, neuronal excitability, memory retention, anxiety and depression[109]. Moreover, intracerebrovetricular injection of NPY to rats has anti-depressive effects that are antagonized by NPY receptor blockers[110]. Lesniewska *et al.*[111] demonstrated that treatment with a mixture of *Lactobacillus rhamnosus GG*, *Bifidobacterium lactis* and inulin, in addition to altering intestinal microbiota, increased plasma levels of NPY and PYY in adult rats. In aged animals, the PYY concentration was unchanged and NPY levels were decreased by treatment. This study both supports the concept that changes in composition of the microbiota can alter gut hormone and suggests that the effects are dependent on age and presumably initial gut physiology and microbiome of the host. Fetissov *et al.* have proposed a novel means of gut bacterial influence over hormonal communication to the brain[112,113]. Autoantibodies directed against leptin, ghrelin, peptide YY, neuropeptide Y, and other gut regulatory peptides are present in normal human and rat sera, suggesting that the immune system may interfere with peptidergic systems involved in appetite and emotional control[112]. This concept is supported by the demonstration of autoantibodies directed against two melanocortin peptides, α-MSH and adrenocorticotropic hormone (ACTH) in subjects with eating disorders, with correlation between autoantibody levels and the core psychopathologic traits in these patients[114,115]. It is suggested that production of gut peptide autoantibodies is modulated by the gut microbiota as a consequence of molecular mimicry. There have been numerous cases of sequence homology with gut hormones and protein components of commensal and pathogenic micro-organisms including Lactobacilli, bacteroides, Helicobacter pylori, Escherichia coli, and Candida species. In a specific example, the ClpB heat-shock disaggregation chaperone protein of commensal gut bacteria Escherichia coli was identified as a conformational mimetic of α-MSH[116] and it is suggested the resultant antibodies to α-MSH may contribute to the development of eating disorders.

27.7 Microbes and immune pathways to the brain

The gut microbiota is critical to normal development of the mammalian immune system[14,117–120]. Balance in microbiota composition, together with the influence of pivotal species that induce specific responses, are important determinants of immunity. Studies involving the selective reconstitution of microbiota in germ-free mice have identified that classes of bacteria, such as clostridia and segmented filamentous bacteria, can influence the development of specific immune cells[118–120]. There is also clear evidence for the direct modulatory effects of certain gut bacteria on immune cells, particularly the epithelial and dendritic cells that are likely to be the first host cells to come into contact with microbes in the intestine.

Commensals and their associated antigens can be detected by the extending processes of lamina propria dendritic cells or by their transport across the epithelial layer by specialized M cells in Peyer's patches. Through such interactions certain commensals and other non-pathogenic bacteria have been identified as inducing immunoregulatory or anti-inflammatory effects in the host including enhanced NK cell activity, altered antigen presentation by dendritic cells and subsequent decrease in IgE responses[121], a skewing of T-cell polarization towards Th1 responses[122,123], the induction of regulatory T-cells[124–126] and inhibition of mast cell responses to antigen[127–130].

It is now widely acknowledged that cytokines and other mediators of the immune system can interact with the nervous system to bring about changes in CNS physiology and behavior[11]. Additionally, the adaptive arm of the immune system has been observed to play a supportive role in learning behavior and cognition[131]. Recent studies also suggest the peripheral adaptive immune system may be involved in the mechanisms underlying stress resilience. Mice receiving lymphocytes from chronically stressed donors appear less anxious, more sociable and have increased hippocampal cell proliferation compared with those receiving no cells or cells from unstressed donors[132]. In accordance with such observations, there exists evidence of immune-mediated bidirectional signaling between the CNS and the microbiome. For example, oral administration of specific lactobacilli normalizes the aberrant cytokine profile and behavioral deficits elicited by maternal separation stress in rats[133,134]. In addition to cytokines, there may other regulatory elements related to the immune response to gut bacteria that can influence both the peripheral and central nervous system. Indoleamine 2,3-dioxygenase 1 (IDO) is a rate-limiting enzyme in tryptophan metabolism that plays a crucial role in the induction of immune tolerance during infection, pregnancy, transplantation and autoimmunity[135–139]. Dendritic cells expressing IDO have an immunosuppressive function[138,140] and studies have identified IDO up-regulation in dendritic cells as key to the immunoregulatory and anti-inflammatory effects of potentially beneficial gut microbes[126,141,142]. However, the regulatory influence of IDO is not limited to the immune response. IDO activity regulates tryptophan availability for conversion to serotonin[143] and the tryptophan metabolites kynurenic and quinolinic acid, which are respectively antagonists and agonists of the NMDA receptor. These tryptophan-derived neuroactive compounds have been identified as playing critical roles in the regulation of neuronal excitation and mechanisms underlying pain and depression[144–146].

Another important immunoregulatory enzyme is heme oxygenase, the rate-limiting enzyme for heme metabolism that catalyzes heme to bilverdin, carbon monoxide (CO) and free iron $(Fe2^+)$[147–149]. Increased HO-1 expression in dendritic cells promotes IL-10 secretion and reduced capacity of the cells to trigger effector immune responses and reduced inflammation[150–153]. Feeding mice with an *L. rhamnosus* strain was demonstrated to increase HO-1 expression by dendritic cells[154]. As with IDO, there is evidence to suggest also plays a role in regulation of the nervous system, specifically that HO-derived CO influences a number of neurological processes including central regulation of nociceptive transmission, inflammation-associated peripheral pain[155–157], olfactory signal transduction[158] and long-term potentiation (LTP)[159,160]. Activation of IDO and HO-1 are critical to the immunoregulatory effects of certain gut bacteria, but further studies will be required; it is not yet clear to what extent they might also contribute to effects on the CNS.

27.8 Metabolites of the microbiota: short-chain fatty acids

Several chapters of this book have described that the fermentation of dietary carbohydrates by intestinal bacteria results in the production of short-chain fatty acids (SCFAs) such as acetic, propionic, and butyric acid. The SCFAs are, to varying degrees, histone deacetylase (HDAC) inhibitors[161] and have been associated with epigenetic regulation of components of the immune and nervous system through changes in histone acetylation. Epigenetic mechanisms have for long been thought to underlie the stable and chronic nature of depression[162]. Chronic social defeat in rodents induces symptoms of depression along with altered expression of *Bdnf* transcripts due to changes in methylation levels[163]. Administration of sodium butyrate in this model induced antidepressant-like effects by increasing levels of histone acetylation in the frontal cortex and the hippocampus, resulting in an elevation of Bdnf transcript levels[163,164]. Similar epigenetic changes also appear to underlie the effects of early-life maternal care on programming of the stress response[165]. It is not yet clear whether the systemic levels achieved by these fermentation metabolites are sufficient to produce behavioral changes. Nevertheless, it has been suggested[161] that the chronic effects of SCFAs produced by the microbiota may result in subtle, stable changes in chromatin regulation and gene expression by altering levels of histone acetylation in the CNS. In addition, SCFAs have been identified as being critical to the immunoregulatory capacity of certain commensals, such as non-pathogenic clostridia[166]. Therefore, as an alternative or an addition to direct effects on gene expression in the nervous system, SCFAs may regulate brain chemistry and behavior indirectly though immune-mediated mechanisms.

27.9 Clinical implications of the microbiota-gut-brain axis

The consequences of microbiome-gut-brain communication for health and disease are far from being fully understood. Certainly, based on existing evidence, there may be potential for microbe-based therapeutic approaches in mood disorders and stress-related conditions such as post-traumatic stress disorder. However, given the role of the microbiota in brain development, it is suggested that understanding microbiota-CNS interactions may also have important implications for the pathophysiology of neurodevelopmental conditions including autism spectrum disorder (ASD). GI symptoms, often correlating with the severity of ASD, have been frequently reported and are hypothesized to be associated with changes in the activity and diversity of the microbiome[167,168]. Prenatal valproic acid (VPA) challenge and maternal immune activation produce animal models that mimic the core ASD symptoms: communication deficits, social impairment, and repetitive and stereotypical behavior[169,170]. Notably, both these models also produce offspring with an altered intestinal microbiome[171,172]. Conversely, germ-free mice exhibit symptoms resembling neurodevelopmental disorders, including deficits in social cognition, changes in motivation and engagement in stereotypical and repetitive behavior[173]. Colonization of germ-free mice abolished certain social deficits and repetitive behavior[173]. Animal models also produce the GI symptoms reported in human ASD studies, including a compromised intestinal barrier. Hsiao *et al.* provided direct evidence for specific components of the microbiota playing a role in neurodevelopmental disorders[171]. Human *B. fragilis* treatment normalized the serum metabolite profile of MIA mice and restored gut barrier integrity, while reversing changes in microbiome and certain ASD-like behavioral abnormalities — communicative, repetitive, and anxiety-like behavior. Conversely, treatment of normal mice with 4-ethylphenylsulfate (4EPS), whose serum levels are modulated by the gut microbiota and are elevated in the MIA model, induced anxiety-like behavior resembling that exhibited by MIA mice[171]. The data outlined above clearly implicates the microbiota-gut-brain axis not only in CNS development but also in the pathogenesis of prevalent neurodevelopmental disorders.

27.10 Conclusion

The resurgence of interest in the gut microbiota as a potentially major contributor to normal brain development and the pathophysiology of affective and neurodevelopmental disorders continues apace. The level of enthusiasm for this "new" development in biological psychiatry is clear from high-profile media coverage and many scientific reviews and editorials produced in the past few years. However, we must learn from the past and not let our zeal for the potential implications of the microbiota-gut-brain axis outstrip empirical evidence regarding causality and the therapeutic efficacy of modulating the microbiota.

To date there is good evidence that the gut microbiota plays a role in normal CNS development and, in particular, influences systems associated with stress response, anxiety[14,35,36] and memory[174]. Exposure to certain key commensals can also attenuate the effects of early-life stress on CNS development[57,134]. It is also clear that disruption of the microbiota or exposure to specific gut bacteria can modulate brain chemistry and behavior in adult animals. Nevertheless, we are at the very early stages of understanding the complex communication systems of the holobiont. There are major gaps in our knowledge regarding how the brain perceives signals from gut bacteria. What, for example, is the commensal-induced chemical code triggering ENS and vagal discharge associated with an anxiolytic response? Furthermore, how does the brain process and respond to the multiple signaling pathways relaying information from the microbiota? Consequently, to what extent does disruption of the microbiota or microbe-generated signals contribute to the development and severity of pathology? There are now intense research efforts being made in these areas and it is certain that such studies will help us better understand mental health and the biological underpinnings of mood and neurodevelopmental disorders.

TAKE-HOME MESSAGE

- The gut microbiota is involved in developmental programming of the brain and stress response systems.
- There is now good evidence that gut bacteria can modulate the brain chemistry and behavior of adult animals.
- The enteric nervous system and vagus nerve play an important role in relaying signals from gut bacteria to the brain.
- Humoral signaling involving gut hormones and the immune system is also likely involved in mediating effects of the gut microbiota on the CNS.
- Moving forward, the holobiont paradigm is likely to provide novel insight into mental health and neurodevelopmental disorders.

References

1 Drossman DA. Presidential address: gastrointestinal illness and the biopsychosocial model. *Psychosom Med* 1998; **60**: 258–267.
2 Wood JD. Neuropathophysiology of irritable bowel syndrome. *J Clin Gastroenterol* 2002; **35**: S11–22 (2002).
3 Bonaz BL, Bernstein CN. Brain-gut interactions in inflammatory bowel diseases. *Gastroenterology* (2012).
4 Frank DN, Pace NR. Gastrointestinal microbiology enters the metagenomics era. *Curr Opin Gastroenterol* 2008; **24**: 4–10.
5 Pasteur L, Joubert C, Chamberland C. La théorie des germes et ses applications à la medicine et à la chirurgie. *CR Acad Sci Hebd Séances Acad. Sci* 86, 1037–1043 (1878).
6 Bested AC, Logan AC, Selhub, EM. Intestinal microbiota, probiotics and mental health: from Metchnikoff to modern advances: Part I: autointoxication revisited. *Gut Pathog* 2013; **5**: 5.
7 Bouchard C. *Lectures on auto-intoxication in disease, or, Self-poisoning of the individual.* 1906, F.A. Davis Company, Philadelphia.
8 Brower DR, Bannister HM. *A practical manual of insanity for the medical student and general practitioner.* 1902, WB Saunders & Company.
9 Adami JG. Chronic intestinal stasis: autointoxication and subinfection. *Br J Med* 1914; **1**: 177–183.
10 Menzies WF. The mechanism of involutionary melancholia. *J Ment Sci* 1920; **66**: 355–414.
11 Dantzer R, Konsman JP, Bluthe RM, Kelley KW. Neural and humoral pathways of communication from the immune system to the brain: parallel or convergent? *Auton Neurosci* **85**, 60–65 (2000).
12 Turck FB. The diffusion of bacteria into the intestinal wall. *Trans Am Gastroenterol Assoc*, 1914; **17**: 198–219.

13 Phillips JGP. The treatment of melancholia by the lactic acid bacillus. *The British Journal of Psychiatry* 1910; **56**: 422.

14 Sudo N *et al.* Postnatal microbial colonization programs the hypothalamic-pituitary-adrenal system for stress response in mice. *J Physiol* 2004; **558**: 263–275.

15 Bowe WP, Logan AC. Acne vulgaris, probiotics and the gut-brain-skin axis — back to the future? *Gut Pathog* 2011; **3**: 1.

16 Stokes JH, Pillsbury DM. The effect on the skin of emotional and nervous states: iii. theoretical and practical consideration of a gastro-intestinal mechanism. *Archives of Dermatology and Syphilology* 1930; **22**: 962–993.

17 Logan AC, Katzman M. Major depressive disorder: probiotics may be an adjuvant therapy. *Medical hypotheses* 2005; **64**: 533–538.

18 Rao AV *et al.* A randomized, double-blind, placebo-controlled pilot study of a probiotic in emotional symptoms of chronic fatigue syndrome. *Gut Pathog* 2009; **1**: 6.

19 Umesaki Y, Setoyama H. Structure of the intestinal flora responsible for development of the gut immune system in a rodent model. *Microbes Infect* 2000; **2**: 1343–1351.

20 Goehler LE, Lyte M, Gaykema RP. Infection-induced viscerosensory signals from the gut enhance anxiety: implications for psychoneuroimmunology. *Brain Behav Immun* 2007; **21**: 721–726.

21 Lyte M, Varcoe J J, Bailey MT. Anxiogenic effect of subclinical bacterial infection in mice in the absence of overt immune activation. *Physiol Behav* 1998; **65**: 63–68.

22 Sharon G, Segal D, Zilber-Rosenberg I, Rosenberg E. Symbiotic bacteria are responsible for diet-induced mating preference in Drosophila melanogaster, providing support for the hologenome concept of evolution. *Gut Microbes* 2011; **2**: 190–192.

23 Venu I, Durisko Z, Xu J, Dukas R. Social attraction mediated by fruit flies' microbiome. *J Exp Biol* 2014; **217**: 1346–1352.

24 Dillon RJ, Vennard CT, Charnley A. Exploitation of gut bacteria in the locust. *Nature* 2000; **403**: 851.

25 Dillon RJ, Vennard CT, Charnley A. A note: gut bacteria produce components of a locust cohesion pheromone. *J Appl Microbiol* 2002; **92**: 759–763.

26 Hedges SB, Dudley J, Kumar S. TimeTree: a public knowledge-base of divergence times among organisms. *Bioinformatics* 2006; **22**: 2971–2972.

27 Rosenberg E, Koren O, Reshef L, Efrony R, Zilber-Rosenberg I. The role of microorganisms in coral health, disease and evolution. *Nat Rev Microbiol* 2007; **5**: 355–362.

28 Iyer LM, Aravind L, Coon SL, Klein DC, Koonin EV. Evolution of cell-cell signaling in animals: did late horizontal gene transfer from bacteria have a role? *Trends Genet* 2004; **20**: 292–299.

29 Clarke MB, Hughes DT, Zhu C, Boedeker EC, Sperandio V. The QseC sensor kinase: a bacterial adrenergic receptor. *Proc Natl Acad Sci USA* **103**, 10420–10425 (2006).

30 Hughes DT, Sperandio V. Inter-kingdom signalling: communication between bacteria and their hosts. *Nat Rev Microbiol* 2008; **6**: 111–120.

31 Saunders CJ, Christensen M, Finger TE, Tizzano M. Cholinergic neurotransmission links solitary chemosensory cells to nasal inflammation. *Proc Natl Acad Sci USA* 2014; **111**: 6075–6080.

32 Boontham P *et al.* Significant immunomodulatory effects of Pseudomonas aeruginosa quorum-sensing signal molecules: possible link in human sepsis. *Clin Sci (Lond)* 2008; **115**: 343–351.

33 Telford G *et al.* The Pseudomonas aeruginosa quorum-sensing signal molecule N-(3-oxododecanoyl)-L-homoserine lactone has immunomodulatory activity. *Infect Immun* 1998; **66**: 36–42.

34 Gilbert SF, Sapp J, Tauber AI. A symbiotic view of life: we have never been individuals. *Q Rev Biol* 2012; **87**: 325–341.

35 Heijtz RD *et al.* Normal gut microbiota modulates brain development and behavior. *Proc Natl Acad Sci USA* **108**, 3047–3052 (2011).

36 Neufeld KM, Kang N, Bienenstock J, Foster JA. Reduced anxiety-like behavior and central neurochemical change in germ-free mice. *Neurogastroenterol Motil* 2011; **23**: 255–64, e119.

37 Lucas A. Programming by early nutrition in man. *Ciba Found Symp* 1991; **156**: 38–50; discussion 50-5.

38 Deng YS, Zhong JH, Zhou XF. Effects of endogenous neurotrophins on sympathetic sprouting in the dorsal root ganglia and allodynia following spinal nerve injury. *Exp Neurol* 2000; **164**: 344–350.

39 Garraway SM, Petruska JC, Mendell L. M. BDNF sensitizes the response of lamina II neurons to high threshold primary afferent inputs. *Eur J Neurosci* 2003; **18**: 2467–2476.

40 Nguyen N, Lee, SB, Lee YS, Lee KH, Ahn JY. Neuroprotection by NGF and BDNF against neurotoxin-exerted apoptotic death in neural stem cells are mediated through Trk receptors, activating PI3-kinase and MAPK pathways. *Neurochem Res* 2009; **34**: 942–951.

41 Cannistraro PA, Rauch SL. Neural circuitry of anxiety: evidence from structural and functional neuroimaging studies. *Psychopharmacol Bull* 2003; **37**: 8–25.

42 Sah P, Faber ES, Lopez De Armentia M, Power J. The amygdaloid complex: anatomy and physiology. *Physiol Rev* 2003; **83**: 803–834.

43 Fuss J *et al.* Deletion of running-induced hippocampal neurogenesis by irradiation prevents development of an anxious phenotype in mice. *PLoS One* 2010; **5**(9), e12769.

44 Martinowich K, Manji H, Lu B. New insights into BDNF function in depression and anxiety. *Nat Neurosci* 2007; **10**: 1089–1093.

45 Yee BK, Zhu SW, Mohammed AH, Feldon J. Levels of neurotrophic factors in the hippocampus and amygdala correlate with anxiety- and fear-related behaviour in C57BL6 mice. *J Neural Transm* 2007; **114**: 431–444.

46 Bergami M *et al.* Deletion of TrkB in adult progenitors alters newborn neuron integration into hippocampal circuits and increases anxiety-like behavior. *Proc Natl Acad Sci USA* 2008; **105**: 15570–15575.

47 Braniste V *et al.* The gut microbiota influences blood-brain barrier permeability in mice. *Sci Transl Med* 2014; **6**: 263ra158.

48 Crowther JS, Drasar BS, Goddard P, Hill MJ, Johnson K. The effect of a chemically defined diet on the faecal flora and faecal steroid concentration. *Gut* 1973; **14**: 790–793.

49 Zentek J, Marquart B, Pietrzak T, Ballevre O, Rochat F. Dietary effects on bifidobacteria and Clostridium perfringens in the canine intestinal tract. *J Anim Physiol Anim Nutr (Berl)* 2003; **87**: 397–407.

50 Li W, Dowd SE, Scurlock B, Acosta-Martinez V, Lyte M. Memory and learning behavior in mice is temporally associated with diet-induced alterations in gut bacteria. *Physiol Behav* 2009; **96**: 557–567.

51 Bercik P *et al.* The intestinal microbiota affect central levels of brain-derived neurotropic factor and behavior in mice. *Gastroenterology* 2011; **141**: 599–609, 609.e1–3.

52 Goehler LE *et al.* Activation in vagal afferents and central autonomic pathways: early responses to intestinal infection with Campylobacter jejuni. *Brain Behav Immun* 2005; **19**: 334–344.

53 Tanida M *et al.* Effects of intraduodenal injection of Lactobacillus johnsonii La1 on renal sympathetic nerve activity and blood pressure in urethane-anesthetized rats. *Neurosci Lett.* 2005; **389**: 109–14.

54 Bravo JA. *et al.* Ingestion of Lactobacillus strain regulates emotional behavior and central GABA receptor expression in a mouse via the vagus nerve. *Proc Natl Acad Sci USA* 2011; **108**: 16050–16055.

55 Bercik P *et al.* Chronic gastrointestinal inflammation induces anxiety-like behavior and alters central nervous system biochemistry in mice. *Gastroenterology* 2010; **139**: 2102–2112.e1.

56 Bercik P *et al.* The anxiolytic effect of Bifidobacterium longum NCC3001 involves vagal pathways for gut-brain communication. *Neurogastroenterol Motil* 2011; **23**: 1132–1139.

57 Desbonnet L *et al.* Effects of the probiotic Bifidobacterium infantis in the maternal separation model of depression. *Neuroscience* 2010; **170**: 1179–1188.

58 Distrutti E *et al.* Modulation of intestinal microbiota by the probiotic VSL#3 resets brain gene expression and ameliorates the age-related deficit in LTP. *PLoS One* 2014; **9**: e106503.

59 Tillisch K *et al.* Consumption of fermented milk product with probiotic modulates brain activity. *Gastroenterology* 2013; **144**: 1394–401, 1401.e1–4.

60 Craig AD. How do you feel? Interoception: the sense of the physiological condition of the body. *Nat Rev Neurosci* 2002; **3**: 655–666.

61 Craig AD. How do you feel — now? The anterior insula and human awareness. *Nat Rev Neurosci* 2008; **10**: 59–70.

62 Craig AD. Interoception: the sense of the physiological condition of the body. *Curr Opin Neurobiol* 2003; **13**: 500–505.

63 Zagon A. Does the vagus nerve mediate the sixth sense? *Trends Neurosci* 2001; **24**: 671–673.

64 Kamiya T *et al.* Inhibitory effects of Lactobacillus reuteri on visceral pain induced by colorectal distension in Sprague-Dawley rats. *Gut* 2006; **55** 191–196.

65 Rousseaux C *et al.* Lactobacillus acidophilus modulates intestinal pain and induces opioid and cannabinoid receptors. *Nat Med* 2007; **13**: 35–37.

66 Mei N. Recent studies on intestinal vagal afferent innervation. Functional implications. *J Auton Nerv Syst* 1983; **9**: 199–206.

67 Blackshaw LA, Brookes SJ, Grundy D Schemann M. Sensory transmission in the gastrointestinal tract. *Neurogastroenterol Motil* 2007; **19**: 1–19.

68 Wang FB, Powley TL. Vagal innervation of intestines: afferent pathways mapped with new en bloc horseradish peroxidase adaptation. *Cell Tissue Res* 2007; **329**,:221–230.

69 Aston-Jones G, Ennis M, Pieribone VA, Nickell WT, Shipley MT. The brain nucleus locus coeruleus: restricted afferent control of a broad efferent network. *Science* 1986; **234**: 734–737.

70 Li Y. Sensory signal transduction in the vagal primary afferent neurons. *Curr Med Chem* 2007; **14**: 2554–2563.

71 Luheshi GN *et al.* Vagotomy attenuates the behavioural but not the pyrogenic effects of interleukin-1 in rats. *Auton Neurosci* 2000; **85**: 127–132.

72 Konsman JP, Luheshi GN, Bluthe RM, Dantzer R. The vagus nerve mediates behavioural depression, but not fever, in response to peripheral immune signals; a functional anatomical analysis. *Eur J Neurosci* 2000; **12**: 4434–4446.

73 Bluthe RM *et al.* Lipopolysaccharide induces sickness behaviour in rats by a vagal mediated mechanism. *C R Acad Sci III* 1994; **317**: 499–503.

74 Laye, S. *et al.* Subdiaphragmatic vagotomy blocks induction of IL-1 beta mRNA in mice brain in response to peripheral LPS. *Am J Physiol* 1995; **268**: R1327–31.

75 Bret-Dibat JL, Bluthe RM, Kent S, Kelley KW,Dantzer R. Lipopolysaccharide and interleukin-1 depress food-motivated behavior in mice by a vagal-mediated mechanism. *Brain Behav Immun* 1995; **9**: 242–246.

76 Berthoud HR, Lynn PA, Blackshaw LA. Vagal and spinal mechanosensors in the rat stomach and colon have multiple receptive fields. *Am J Physiol Regul Integr Comp Physiol* 2001; **280**: R1371–81.

77 Martin, JL, Martin-Sanchez E. Systematic review and meta-analysis of vagus nerve stimulation in the treatment of depression: variable results based on study designs. *Eur Psychiatry* 2012; **27**: 147–155.

78 Rizvi SJ *et al.* Neurostimulation therapies for treatment resistant depression: a focus on vagus nerve stimulation and deep brain stimulation. *Int Rev Psychiatry* 2011; **23:** 424–436.

79 Wang X *et al.* Evidences for vagus nerve in maintenance of immune balance and transmission of immune information from gut to brain in STM-infected rats. *World J Gastroenterol* 2012; **8**: 540–545.

80 Perez-Burgos A *et al.* Psychoactive bacteria Lactobacillus rhamnosus (JB-1) elicits rapid frequency facilitation in vagal afferents. *Am J Physiol Gastrointest Liver Physiol* 2013; **304**: G211–20.

81 Furness JB. *The enteric nervous system*. Blackwell Publishing, Massachusetts, Oxford, Carlton, 2006.

82 Mayer EA. Gut feelings: the emerging biology of gut-brain communication. *Nat Rev Neurosci* 2011; **12**: 453–466.

83 Perez-Burgos A, Mao YK, Bienenstock J, Kunze WA. The gut-brain axis rewired: adding a functional vagal nicotinic "sensory synapse". *FASEB J* 2014; **28**: 3064–3074.

84 Furness JB, Jones C, Nurgali K, Clerc N. Intrinsic primary afferent neurons and nerve circuits within the intestine. *Prog Neurobiol* 2004; **72**: 143–164.

85 Clerc N, Gola M, Vogalis F, Furness JB. Controlling the excitability of IPANs: a possible route to therapeutics. *Current Opinion in Pharmacology* 2002; **2**: 657–664.

86 Kunze WA, Furness JB. The enteric nervous system and regulation of intestinal motility. *Annu Rev Physiol* 1999; **61**: 117–142.

87 Keast JR, Furness JB, Costa M. Somatostatin in human enteric nerves: distribution and characterization. *Cell Tissue Res* 1984; **237**, 299–308.

88 Ekblad E, Winther C, Ekman R, Hakanson R, Sundler F. Projections of peptide-containing neurons in rat small intestine. *Neuroscience* 1987; **20**: 169–188 (1987).

89 Collins J, Borojevic R, Verdu EF, Huizinga JD, Ratcliffe EM. Intestinal microbiota influence the early postnatal development of the enteric nervous system. *Neurogastroenterol Motil* 2014; **26**: 98–107.

90 McVey Neufeld KA, Mao YK, Bienenstock J, Foster JA, Kunze WA. The microbiome is essential for normal gut intrinsic primary afferent neuron excitability in the mouse. *Neurogastroenterol Motil* 2013; **25**: 183–e88.

91 Kabouridis PS *et al.* Microbiota controls the homeostasis of glial cells in the gut lamina propria. *Neuron* 2015; **85**: 289–295.

92 Kunze WA *et al.* Lactobacillus reuteri enhances excitability of colonic AH neurons by inhibiting calcium-dependent potassium channel opening. *J Cell Mol Med* 2009; **13**: 2261–2270.

93 Ishii TM *et al.* A human intermediate conductance calcium-activated potassium channel. *Proc Natl Acad Sci USA* 1997; **94**: 11651–11656.

94 Wang B *et al.* Luminal administration ex vivo of a live Lactobacillus species moderates mouse jejunal motility within minutes. *FASEB J* 2010; **24**: 4078–4088.

95 Cameron J, Doucet E. Getting to the bottom of feeding behaviour: who's on top? *Appl Physiol Nutr Metab* 2007; **32**: 177–189.

96 Wren AM, Bloom SR. Gut hormones and appetite control. *Gastroenterology* 2007; **132**: 2116–2130.

97 Uribe A, Alam M, Johansson O, Midtvedt T, Theodorsson E. Microflora modulates endocrine cells in the gastrointestinal mucosa of the rat. *Gastroenterology* 1994; **107**: 1259–1269.

98 Hsiao WW, Metz C, Singh DP, Roth J. The microbes of the intestine: an introduction to their metabolic and signaling capabilities. *Endocrinol Metab Clin North Am* 2008; **37**: 857–871.

99 Moran-Ramos S, Tovar AR, Torres N. Diet: friend or foe of enteroendocrine cells — how it interacts with enteroendocrine cells. *Adv Nutr* 2012; **3**: 8–20.

100 Tortorella C, Neri G, Nussdorfer GG. Galanin in the regulation of the hypothalamic-pituitary-adrenal axis (review). *Int J Mol Med* 2007; **19**: 639–647.

101 Wrenn CC, Holmes A. The role of galanin in modulating stress-related neural pathways. *Drug News Perspect* 2006; **19**: 461–467.

102 Rustay NR *et al.* Galanin impairs performance on learning and memory tasks: findings from galanin transgenic and GAL-R1 knockout mice. *Neuropeptides* 2005; **39**: 239–243.

103 Wrenn CC *et al.* Learning and memory performance in mice lacking the GAL-R1 subtype of galanin receptor. *Eur J Neurosci* 2004; **19**,:1384–1396.

104 Giordano R *et al.* Neuroregulation of the hypothalamus-pituitary-adrenal (HPA) axis in humans: effects of GABA-, mineralocorticoid-, and GH-Secretagogue-receptor modulation. *Scientific World Journal* 2006; **6**: 1–11.

105 Jaszberenyi M, Bujdoso E, Bagosi Z, Telegdy G. Mediation of the behavioral, endocrine and thermoregulatory actions of ghrelin. *Horm Behav* 2006; **50**: 266–273.

106 Carlini VP *et al.* Ghrelin induced memory facilitation implicates nitric oxide synthase activation and decrease in the threshold to promote LTP in hippocampal dentate gyrus. *Physiol Behav* 2010; **101**: 117–123.

107 Yamada K, Wada E, Wada K. Male mice lacking the gastrin-releasing peptide receptor (GRP-R) display elevated preference for conspecific odors and increased social investigatory behaviors. *Brain Res* 2000; **870**: 20–26.

108 Yamada K, Wada E, Wada, K. Female gastrin-releasing peptide receptor (GRP-R)-deficient mice exhibit altered social preference for male conspecifics: implications for GRP/GRP-R modulation of GABAergic function. *Brain Res* 2001; **894**: 281–287.

109 Berglund MM. Hipskind PA, Gehlert DR. Recent developments in our understanding of the physiological role of PP-fold peptide receptor subtypes. *Exp Biol Med (Maywood)* 2003; **228**: 217–244.

110 Ishida H *et al.* Infusion of neuropeptide Y into CA3 region of hippocampus produces antidepressant-like effect via Y1 receptor. *Hippocampus* 2007; **17**: 271–280.

111 Lesniewska V *et al.* Effect on components of the intestinal microflora and plasma neuropeptide levels of feeding Lactobacillus delbrueckii, Bifidobacterium lactis, and inulin to adult and elderly rats. *Appl Environ Microbiol* 2006; **72**: 6533–6538.

112 Fetissov SO *et al.* Autoantibodies against appetite-regulating peptide hormones and neuropeptides: putative modulation by gut microflora. *Nutrition* 2008; **24**: 348–359.

113 Fetissov SO *et al.* Emerging role of autoantibodies against appetite-regulating neuropeptides in eating disorders. *Nutrition* 2008; **24**: 854–859.

114 Fetissov SO *et al.* Autoantibodies against alpha-MSH, ACTH, and LHRH in anorexia and bulimia nervosa patients. *Proc Natl Acad Sci USA* 2002; **99**: 17155–17160.

115 Fetissov SO *et al.* Autoantibodies against neuropeptides are associated with psychological traits in eating disorders. *Proc Natl Acad Sci USA* 2005; **102**: 14865–14870.

116 Tennoune N *et al.* Bacterial ClpB heat-shock protein, an antigen-mimetic of the anorexigenic peptide alpha-MSH, at the origin of eating disorders. *Transl Psychiatry* 2014; **4**.

117 Kalliomaki M, Isolauri E. Role of intestinal flora in the development of allergy. *Curr Opin Allergy Clin Immunol* 2003; **3**: 15–20.

118 Atarashi K *et al.* Induction of colonic regulatory T cells by indigenous Clostridium species. *Science* 2011; **331**: 337–341.

119 Geuking MB *et al.* Intestinal bacterial colonization induces mutualistic regulatory T cell responses. *Immunity* 2011; **34**: 794–806.

120 Round JL Mazmanian SK. Inducible Foxp3+ regulatory T-cell development by a commensal bacterium of the intestinal microbiota. *Proc Natl Acad Sci USA* 2010; **107**: 12204–12209.

121 Hisbergues M *et al.* In vivo and in vitro immunomodulation of Der p 1 allergen-specific response by Lactobacillus plantarum bacteria. *Clin Exp Allergy* 2007; **37**: 1286–1295.

122 Baba N, Samson S, Bourdet-Sicard R, Rubio M, Sarfati M. Selected commensal-related bacteria and Toll-like receptor 3 agonist combinatorial codes synergistically induce interleukin-12 production by dendritic cells to trigger a T helper type 1 polarizing programme. *Immunology* 2009; **128**: e523–31.

123 Iwabuchi N *et al.* Suppressive effects of Bifidobacterium longum on the production of Th2-attracting chemokines induced with T cell-antigen-presenting cell interactions. *FEMS Immunol Med Microbiol* 2009; **55**: 324–334.

124 Karimi K, Inman MD, Bienenstock J, Forsythe P. Lactobacillus reuteri-induced regulatory T cells protect against an allergic airway response in mice. *Am J Respir Crit Care Med* 2009; **179**: 186–193.

125 Lyons A *et al.* Bacterial strain-specific induction of Foxp3+ T regulatory cells is protective in murine allergy models. *Clin Exp Allergy* **40**, 811–819 (May).

126 Kwon HK *et al.* Generation of regulatory dendritic cells and CD4+Foxp3+ T cells by probiotics administration suppresses immune disorders. *Proc Natl Acad Sci USA* 2010; **107**: 2159–2164.

127 Forsythe P, Wang B, Khambati I, Kunze WA. Systemic effects of ingested lactobacillus rhamnosus: inhibition of mast cell membrane potassium (ikca) current and degranulation. *PLoS One* 2012; **7**: e41234.

128 Kim JY, Choi YO, Ji GE. Effect of oral probiotics (Bifidobacterium lactis AD011 and Lactobacillus acidophilus AD031) administration on ovalbumin-induced food allergy mouse model. *J Microbiol Biotechnol* 2008; **18**: 1393–1400.

129 Magerl M *et al.* Non-pathogenic commensal Escherichia coli bacteria can inhibit degranulation of mast cells. *Exp Dermatol* 2008; **17**: 427–435.

130 Oksaharju A *et al.* Probiotic Lactobacillus rhamnosus downregulates FCER1 and HRH4 expression in human mast cells. *World J Gastroenterol* 2011; **17**: 750–759.

131 Brynskikh A, Warren T, Zhu J, Kipnis J. Adaptive immunity affects learning behavior in mice. *Brain Behav Immun* 2008; **22**: 861–869.

132 Brachman RA, Lehmann ML, Maric D, Herkenham M. Lymphocytes from chronically stressed mice confer antidepressant-like effects to naive mice. *J Neurosci* 2015; **35**: 1530–1538.

133 Bailey MT, Coe CL. Maternal separation disrupts the integrity of the intestinal microflora in infant rhesus monkeys. *Dev Psychobiol* 1999; **35**,:146–155.

134 Gareau MG. Jury J, Macqueen G, Sherman PM, Perdue MH. Probiotic treatment of rat pups normalizes corticosterone release and ameliorates colonic dysfunction induced by maternal separation. *Gut* (2007); **56**(11): 1522–1528.

135 Fallarino F *et al.* T cell apoptosis by kynurenines. *Adv Exp Med Biol* 2003; **527**:183–190.

136 Jaen O *et al.* Dendritic cells modulated by innate immunity improve collagen-induced arthritis and induce regulatory T cells in vivo. *Immunology* 2009; **126**: 35–44.

137 Kahler DJ, Mellor AL. T cell regulatory plasmacytoid dendritic cells expressing indoleamine 2,3 dioxygenase. *Handb Exp Pharmacol* 2009; 165–196.

138 King NJ, Thomas SR. Molecules in focus: indoleamine 2,3-dioxygenase. *Int J Biochem Cell Biol* 2007; **39**: 2167–2172.

139 Lopez AS, Alegre E. LeMaoult J, Carosella E, Gonzalez A. Regulatory role of tryptophan degradation pathway in HLA-G expression by human monocyte-derived dendritic cells. *Mol Immunol* 2006; **43**: 2151–2160.

140 Popov A, Schultze JL. IDO-expressing regulatory dendritic cells in cancer and chronic infection. *J Mol Med* 2008; **86**: 145–160.

141 Hayashi T *et al.* Enhancement of innate immunity against Mycobacterium avium infection by immunostimulatory DNA is mediated by indoleamine 2,3-dioxygenase. *Infect Immun* 2001; **69**: 6156–6164.

142 Forsythe P, Inman MD, Bienenstock J. Oral treatment with live Lactobacillus reuteri inhibits the allergic airway response in mice. *Am J Respir Crit Care Med* 2007; **175**: 561–569.

143 Leonard BE. The HPA and immune axes in stress: the involvement of the serotonergic system. *Eur Psychiatry* 2005; **20 Suppl** 3: S302–6.

144 Kim H *et al.* Brain indoleamine 2,3-dioxygenase contributes to the comorbidity of pain and depression. *J Clin Invest* 2012; **122**: 2940–2954.

145 Salazar A, Gonzalez-Rivera BL, Redus L, Parrott JM, O'Connor JC. Indoleamine 2,3-dioxygenase mediates anhedonia and anxiety-like behaviors caused by peripheral lipopolysaccharide immune challenge. *Horm Behav* 2012; **62**: 202–209.

146 Kohl C, Sperner-Unterweger B. IDO and clinical conditions associated with depressive symptoms. *Curr Drug Metab* 2007; **8**: 283–287.

147 Otterbein LE, Soares MP, Yamashita K, Bach FH. Heme oxygenase-1: unleashing the protective properties of heme. *Trends Immunol* 2003; **24**: 449–455.

148 George JF *et al.* Suppression by CD4+CD25+ regulatory T cells is dependent on expression of heme oxygenase-1 in antigen-presenting cells. *Am J Pathol* 2008; **173**: 154–160.

149 Moreau A *et al.* Tolerogenic dendritic cells actively inhibit T cells through heme oxygenase-1 in rodents and in nonhuman primates. *FASEB J* 2009; **23**: 3070–3077.

150 Chauveau C *et al.* Heme oxygenase-1 expression inhibits dendritic cell maturation and proinflammatory function but conserves IL-10 expression. *Blood* 2005; **106**: 1694–1702.

151 Carter EP, Garat C, Imamura M. Continual emerging roles of HO-1: protection against airway inflammation. *Am J Physiol Lung Cell Mol Physiol* 2004; **287**: L24–5.

152 Pae HO, Lee YC, Chung HT. Heme oxygenase-1 and carbon monoxide: emerging therapeutic targets in inflammation and allergy. *Recent Pat Inflamm Allergy Drug Discov* 2008; **2**: 159–165.

153 Xia ZW *et al.* Heme oxygenase-1-mediated CD4+CD25high regulatory T cells suppress allergic airway inflammation. *J Immunol* 2006; **177**: 5936–5945.

154 Karimi K, Kandiah N, Chau J, Bienenstock J, Forsythe P. A Lactobacillus rhamnosus strain induces a heme oxygenase dependent increase in Foxp3+ regulatory T cells. *PLoS One* 2012; **7**: e47556.

155 Bijjem KR, Padi SS, Lal Sharma P. Pharmacological activation of heme oxygenase (HO)-1/carbon monoxide pathway prevents the development of peripheral neuropathic pain in Wistar rats. *Naunyn Schmiedebergs Arch Pharmacol* (2012); **386**(1), 79–90.

156 Fan W *et al.* Carbon monoxide: a gas that modulates nociception. *J Neurosci Res* 2011; **89**: 802–807.

157 Hervera A, Leanez S, Negrete R, Motterlini R, Pol O. Carbon monoxide reduces neuropathic pain and spinal microglial activation by inhibiting nitric oxide synthesis in mice. *PLoS One* 2012; **7**: e43693.

158 Zufall F, Leinders-Zufall T. Identification of a long-lasting form of odor adaptation that depends on the carbon Monoxide/cGMP second-messenger system. *J Neurosci* 1997; **17**: 2703–2712.

159 Hawkins RD, Zhuo M, Arancio O. Nitric oxide and carbon monoxide as possible retrograde messengers in hippocampal long-term potentiation. *J Neurobiol* 1994; **25**: 652–665.

160 Stevens CF, Wang Y. Reversal of long-term potentiation by inhibitors of haem oxygenase. *Nature* 1993; **364**: 147–149.

161 Stilling RM, Dinan TG, Cryan JF. Microbial genes, brain & behaviour — epigenetic regulation of the gut-brain axis. *Genes Brain Behav* 2014; **13**: 69–86.

162 Tsankova N, Renthal W, Kumar A, Nestler EJ. Epigenetic regulation in psychiatric disorders. *Nat Rev Neurosci* 2007; **8**, 355–367.

163 Tsankova NM *et al.* Sustained hippocampal chromatin regulation in a mouse model of depression and antidepressant action. *Nat Neurosci* 2006; **9**: 519–525.

164 Schroeder FA, Lin CL, Crusio WE, Akbarian S. Antidepressant-like effects of the histone deacetylase inhibitor, sodium butyrate, in the mouse. *Biol Psychiatry* 2007; **62**: 55–64.

165 Weaver IC *et al.* Epigenetic programming by maternal behavior. *Nat Neurosci* 2004; **7**; 847–854.

166 Furusawa Y *et al.* Commensal microbe-derived butyrate induces the differentiation of colonic regulatory T cells. *Nature* 2013; **504**: 446–450.

167 Adams JB, Johansen LJ, Powell LD, Quig D, Rubin RA. Gastrointestinal flora and gastrointestinal status in children with autism — comparisons to typical children and correlation with autism severity. *BMC Gastroenterol* 2011; **11**: 22.

168 de Theije CG *et al.* Pathways underlying the gut-to-brain connection in autism spectrum disorders as future targets for disease management. *Eur J Pharmacol* 2011; **668 Suppl 1**: S70–80.

169 Iwata K, Matsuzaki H, Takei N, Manabe T, Mori N. Animal models of autism: an epigenetic and environmental viewpoint. *J Cent Nerv Syst Dis* 2010; **2**: 37–44.

170 Malkova NV, Yu CZ, Hsiao EY, Moore MJ, Patterson PH. Maternal immune activation yields offspring displaying mouse versions of the three core symptoms of autism. *Brain Behav Immun* 2012; **26**: 607–616.

171 Hsiao EY *et al.* Microbiota modulate behavioral and physiological abnormalities associated with neurodevelopmental disorders. *Cell* 2013; **155**: 1451–1463.

172 de Theije CG *et al.* Altered gut microbiota and activity in a murine model of autism spectrum disorders. *Brain Behav Immun* 2014; **37**: 197–206.

173 Desbonnet L, Clarke G, Shanahan F, Dinan TG, Cryan JF. Microbiota is essential for social development in the mouse. *Mol Psychiatry* 2014; **19**(2): 146–8.

174 Gareau MG *et al.* Bacterial infection causes stress-induced memory dysfunction in mice. *Gut* 2011; **60**: 307–317.

175 Forsythe P, Sudo N, Dinan T, Taylor VH, Bienenstock J. Mood and gut feelings. *Brain Behav Immun* 2010; **24**: 9–16.

CHAPTER 28

Genetic dysbiosis: how host genetic variants may affect microbial biofilms

Luigi Nibali

Centre for Oral Clinical Research, Queen Mary University of London, London, United Kingdom

28.1 The holobiont: humans as supra-organisms

Previous chapters have highlighted the importance of the human microbiome in health and disease. It is now clear that humans are heavily colonized by bacteria[1], and as such are considered supra-organisms[2] with bacteria representing 90% of the cells in the human body. It is estimated that in the human gut, the microbiome outnumbers the human genome by 150-fold[3]. The term "holobiont" refers to the human host plus its associated microorganisms.

In several human mucosal surfaces, bacteria adhere to each other forming aggregates termed biofilms (discussed by Snowden in chapter 8). Although only a small proportion of bacteria have ever been cultivated, thousands of bacterial phylotypes are now recognized as colonizing humans[4]. Each individual will harbor a subset of microbes, with variations among individuals affecting microbial biofilms qualitatively and quantitatively. These inter-individual differences and the different host responses to the microbial challenge will affect health and predisposition to disease states[5]. Site-specific factors have the potential to affect bacterial colonization. But which other factors are responsible for the "selection" of each individual's microbiome? A sensible hypothesis is that host genetic variants affect not only the individual ability to respond to microbial challenges, thus predisposing to the onset of acute microbial diseases[6], but also the normal composition of microbial biofilms, thus predisposing to a variation in the normal composition of microbial biofilms named "dysbiosis." This in turn could lead to what could be defined as "chronic microbial diseases," or diseases originated from the host response to a dysbiotic microbial biofilm[7]. This chapter will review and discuss the role of common human genetic variants in affecting microbial responses.

The Human Microbiota and Chronic Disease: Dysbiosis as a Cause of Human Pathology, First Edition.
Edited by Luigi Nibali and Brian Henderson.
© 2016 John Wiley & Sons, Inc. Published 2016 by John Wiley & Sons, Inc.

28.2 Genetic variants in the host response to microbes

The heritability of human physical and behavioral traits has long been recognized. With the discovery of DNA and the completion of the Human Genome Project we have learned a lot more on the human genetic makeup. We now know that the human genome consists of 19,000–25,000 genes located in 23 pairs of chromosomes[8]. Six billion basepairs of DNA make the human genome and over 60 million common genetic variants, responsible for determining differences among individuals including in the susceptibility to disease, are listed in the Single Nucleotide Polymorphism Database (dbSNP) by the National Center for Biotechnology in collaboration with National Human Genome Research Institute[9]. Different individuals are thought to be 99.4% identical in chromosomal structure and 99.9% identical at sequence level[10].

Functional SNPs could affect disease predisposition by major or subtle changes in gene activity (when located in the gene promoter) or in the protein produced by the gene (when located in the coding region of the gene). Single-gene defects can cause rare diseases such as hemophilia A, an X-linked recessive disease, where the presence of a specific mutation in the F8 gene causes defects in coagulation factor VIII, resulting in serious, often life-threatening blood-clotting disturbances[11]. However, most diseases are characterized by a complex susceptibility profile in which several common genetic variants (single nucleotide polymorphisms, SNPs) contribute to the disease risk by modifying the effect of the respective gene. For example, this is the case for SNPs involved in inflammation (able to perpetrate or amplify inflammatory cascades), in DNA repair (associated with a reduction in the ability to repair damaged DNA) and microbial recognition (determining aberrant responses to the normal microbiota).

Furthermore, the genome is not completely stable and can have structural variations which can alter gene dosage (microdeletions, microduplications). Error rates in DNA replication, damage to bases (depurination, cytosine deamination, oxidation, methylation, UV) and exogenous double-stranded breaks (for example due to radiations) are sources of genetic instability[12]. These structural variations can predispose to genetic diseases such as obesity, epilepsy or autism[13–15]. However, genomic instability is also incredibly useful in the immune system (programmed genomic instability to create antigen diversity). For example, going back to evolution, genomes have intentionally rearranged and mutated (e.g. IgG molecules) and lymphocytes have developed diverse antigen receptors to recognize and combat different pathogens.

Recent evidence suggests that host genetic factors play a major role in deciding which bacteria colonize which host and therefore in the composition of the natural human biofilms[16]. The concept of infectogenomics suggests that inherited genetic variants predispose to the colonization and proliferation of selected bacteria in the human body[6]. As a result of this, a group of human diseases originate from a genetically-determined failure to properly recognize or respond to members of the normal human microbiota. The two distinct infectogenomics pathways that can be recognized[6] are described in the following sections.

28.2.1 Bacterial recognition pathway
Upon exposure of the human body to a microbe, mechanisms are activated for its recognition. Mammals have a very wide variety of pattern-recognition receptors

(PRRs) that recognize evolutionarily-conserved constituents of microbes called pathogen-associated molecular patterns (PAMPs)[17]. Among normally cell-bound proteins are the toll-like receptors (TLRs), NOD-like receptors (NLRs), RIG-I-like receptors (RLRs), C-type-lectin like receptors (CLRs), scavenger receptors (SCs), innate DNA receptor proteins termed AIM2-like receptors (ALRs), members of the complement pathways and peptidoglycan-recognition proteins (PRPs). A range of soluble PRRs include collectins, ficolins, pentraxins, galectins, sCD (cluster of differentiation)[14] and natural IgM (Immunoglobulin M) (reviewed in[7]). These various receptor-based and soluble PRRs generally interact with various accessory proteins to allow selective cell signaling. Upon microbial binding to these PRRs, the target cell reacts with the generation of pro- and/or anti-inflammatory proteins such as cytokines. Mutations in the promoter regions and coding segments of the individual PRR genes may result in either altered expression of PRRs or differences in the ability to recognize the microbial constituents they bind to, with downstream consequences in the reaction against the invading microbes. Changes in the interaction between the PRR and obligatory accessory proteins may also be a factor in this "binding/recognition" process.

Initial evidence has recently been produced for the effect of microbial recognition genes on microbial presence in periodontal[18] and vaginal biofilms[19,20]. In other words, genetic factors that may determine an aberrant epithelial barrier (through defects in pattern-recognition receptors and innate immune signaling pathways) may induce microbial shifts which could in theory lead to disease[21].

A compelling example of how microbial recognition genes could affect the response to microbial challenge is given by the HIV. This virus is recognized by the human body thanks to a CCR5 chemokine receptor, coded for by a gene named CCR5 located on chromosome 3 codes. This receptor binds the HIV, allowing it to enter its target cells (macrophages, T cells). A genetic variant in the CCR5 gene, namely a deletion of a 32-bp segment (CCR5-Δ32), results in a non-functional receptor, thus preventing HIV R5 entry. Subjects homozygous for this variant are resistant to macrophage-tropic HIV and heterozygous subjects have a slower progression[22,23].

28.2.2 Bacterial proliferation

Several human surfaces in contact with the outside environment such as skin and oral cavity are inevitably colonized by a microbial *biofilm*. In parallel with microbial recognition, other host genetic variants may directly affect the proliferation of certain species within a biofilm by different mechanisms. For instance, genetic variants predisposing to an excessive inflammatory response create a favorable environment for the selective growth of specific bacteria within the human biofilms that, due to specific characteristics in their metabolism, grow well in more inflamed environments. It has been hypothesized that, in periodontitis, diseased periodontal pockets can become a "breeding ground" for selected periodontal bacteria despite the presence of active leukocytes and other defense mechanisms[24] (see chapter 15 of this book). Initial evidence in periodontitis suggests that cytokine gene polymorphisms may select and favor the growth of certain components of the subgingival biofilm[18,25]. Similarly to periodontitis, other human diseases might be affected by the overgrowth of certain components of the biofilms upon stimulation by a more or less "inflamed" environment in "hyper-inflammatory" individuals. It has been speculated that mutations in genes involved in immune regulatory mechanisms or pro-inflammatory pathways could lead to unrestrained inflammation in the intestine and that inflammation can influence the composition of the microbiota, skewing it in favor of pathological microorganisms[26]. However, it is not clear whether the inflammatory deregulation is the cause or in fact the consequence of a microbial shift in such cases. Sickle-cell anemia is an example of a condition where a specific human genetic variant has developed by selective pressure because it inhibits the growth of a pathogen (*Plasmodium falciparum*) in erythrocytes making subjects resistant to malaria[27]. Hence, *P. falciparum* is a clear example of a bacterium whose proliferation is severely affected by the environment determined by a single host genetic variant.

28.3 Genetic dysbiosis

We previously introduced the term "genetic dysbiosis" to define chronic human diseases due to an imbalance between the integrity of barrier organs and their colonizing microorganisms[7]. These diseases include periodontal disease, inflammatory bowel disease, psoriasis and bacterial vaginosis. The common underlying mechanism involves the presence of triggering microbial events acting on a genetically susceptible individual, in the presence of predisposing environmental factors. The epithelia of barrier organs such as skin, oral cavity, gastrointestinal tract and oro-genital mucosa are able to regulate the normal host-microbial homeostasis[28] by acting as a mechanical barrier and as a first line of host defense against invading pathogens, recognizing microorganisms and producing cytokines and antimicrobial peptides. Therefore, these epithelia are physiologically colonized by a number of opportunistic bacteria, with the potential to cause disease. Genetic regulation of microbial recognition and proliferation may be at the basis for the common pathogenesis and co-morbidity between these genetic dysbioses[29–33]. For example, TLR-4 genetic variants have been implicated in BV, PD and IBD pathogenesis[34–36]. Among other functional genes important in the barrier organ-microbe homeostasis, NLR (nucleotide-binding domain and leucine-rich repeat containing) genes, encoding mediators of innate immunity and providing the first line of defense against pathogens, may provide a link between these diseases[28]. Physiological bacterial translocation from gut and oral biofilms, by the intra-epithelial route and then via the mesenteric lymph nodes (or directly to the portal circulation in case of damage to the epithelium) or by diseased periodontium[37–39] could contribute to generalised immune activation and subsequent disease progression. In other words, onset of one genetic dysbiosis might alter the biodiversity of the microbiota, provoking a loss of immunological tolerance to commensal bacteria, hence predisposing to other dysbioses[28]. An additional pathogenic mechanism for these conditions is a host genome-driven misrecognition of the normal microbiota, which might overlap or in some cases represent a different disease-initiating pathway. Although the different epithelial and mucosal surfaces of the body have their own distinct microbiotas with little overlap between species in health[5], this may change in the presence of a genetically driven dysbiosis. The next sections review the evidence for a role of host genetic variants in influencing the composition of biofilms and hence in disease predisposition.

28.3.1 Genetic dysbiosis of oral biofilm

A biofilm is formed on the tooth surface upon its eruption in the oral cavity. Characteristics of the oral biofilm have been discussed in chapter 4. Chapter 13 covered the basic principles of periodontitis, an inflammatory disease of the supporting apparatus of the tooth due to an aberrant response to members of the subgingival biofilm, such as Gram-negative bacteria *Aggregatibacter actinomycetemcomitans*, *Porphyromonas gingivalis* and *Tannerella forsythia*[40]. Our group and others have shown in recent years that specific genetic variants affecting the inflammatory response (e.g. interleukin-1 and interleukin-6 genes) are associated with detection of periodontopathogenic bacteria (such as *A.actinomycetemcomitans* and *P.gingivalis*) below the gingival margin[18,42,43]. A recent study using a genome-wide approach in 1020 subjects identified a moderate association between locus 1q42 and periodontopathogenic bacteria (belonging to the so-called "red complex") in subgingival pockets and confirmed a moderate association our group previously reported between *IL6* genetic variants and presence of pathogenic bacteria subgingivally[25]. Therefore, periodontal "dysbiosis" or a shift towards a more pathogenic microbiota (including pathogenic bacteria which grow well in inflamed environments, such as *A.actinomycetemcomitans*) may be due to specific genetic variants in the host[44]. In a study on 12 patients with periodontitis selected based on their *IL6* genotypes, we detected higher *A. actinomycetemcomitans* counts in their gingival pockets in *IL6* "haplotype positive" subjects before treatment. Despite a reduction after treatment, these subjects tended to have a sharp increase in counts of *A. actinomycetemcomitans* again at the last study follow-up three months after periodontal treatment, suggesting a strong genetic influence on gingival pocket re-colonization[45]. Further evidence for possible host gene influence on microbial colonization in the periodontium is represented by the striking tropism of the JP2 clone of *A. actinomycetemcomitans* for specific ethnic groups, mainly of North African ancestry[41]. The associations of *IRF5* and *PRDM1* with IBD and AgP[46,47] may suggest that the risk variants at these loci contribute to a disturbance of the immunological barrier in different environments, which may promote dysbiosis of the local microflora. Evidence is still lacking on how host genetic variants may affect the composition of oral biofilms in supra-gingival plaque or in other niches such as tongue and tonsils. However, it is conceivable that, as well as periodontitis, other microbially driven oral diseases such as caries may be initiated by a host genome-driven dysbiotic process, enhanced by dietary factors[48].

28.3.2 Genetic dysbiosis of gut biofilm

Previous chapters have discussed the clinical characteristics and pathogenesis of inflammatory bowel disease (IBD) encompassing Crohn's disease (CD) and ulcerative colitis (UC). IBDs are associated with changes in intestinal microbial biofilms, even if a clear definition for a "healthy biofilm" does not actually exist and no bacterium has so far been singled out as causative for IBD[49,50]. Childhood exposure to antibiotics has been implicated in such intestinal dysbiosis, leading to IBD[51,52]. There is evidence that some key members of the gut microbiota can modulate the expression of genes involved in different intestinal activities[53].

Bacterial type VI secretion systems (T6SSs) of pathobiont *Helicobacter hepaticus*, a Gram-negative bacterium of the intestinal microbiota[54], have been shown to direct an anti-inflammatory gene expression profile in intestinal epithelial cells thus limiting colonization and intestinal inflammation. On the other hand, host genetic variants have been associated with gut microbial detection. Genome-wide association studies (GWAS) have made considerable progress in the identification of genetic variants predisposing to IBD. There is evidence for a role of genetic variants in genes coding for bacterial recognition receptors in disease pathology. Genetic variants in the NOD2 gene, which codes for an intracellular pattern recognition receptor, able to recognize molecules containing bacterial muramyl dipeptide[55], are associated with a higher risk of developing CD[56]. Other genes linked to epithelial barrier function seem to be specifically associated with UC, while bacterial recognition genes such as NOD2 and autophagy genes seem to be associated with CD[52]. This is in line with the involvement of the more superficial epithelial layers in UC and the deeper transmural inflammation of CD (supposedly caused by defects in cellular innate immunity and bacterial handling in the deeper layers of the lamina propria and beyond)[52].

Circumstantial evidence for a role of innate host factors in gut microbiota composition derive from studies showing that microbial profiles of fecal samples collected at various times from a given individual are more similar to each other than to the intestinal microbial communities in a different individual[57,58]. Furthermore, monozygotic twins frequently have more similar gut microbiomes than non-twin siblings[59]. Evidence for a direct role of genetic variants in intestinal dysbiosis in CD comes from an experimental ileal inflammation model in mice[60]. When a CD susceptible genotype (NOD2 mutations) was superimposed on *Toxoplasma gondii*-induced ileitis, an increase in inflammation and dysbiosis was noticed (measured by pyrosequencing as a shift from mainly Gram-positive to Gram-negative bacteria, associated with invasive *E. coli*). This suggests that inflammation may drive a progressive decrease in the microbial diversity, potentially leading to perturbations in the microenvironment such as increased availability of substrates for growth of Gram-negative bacteria (e.g. iron and serum, dead or dying cells) and loss of niche and substrates for Gram-positive flora (e.g. mucus, goblet cells)[7]. Furthermore, genetic variants in the NOD2 and autophagy-related 16-like 1 protein (ATG16L1) have been associated with gut microbiota structure alterations, including decreased *Faecalibacterium* levels and increased *Escherichia* levels[61]. Genetic variants could impact the threshold for dysbiosis and the individual ability to resolve the self-perpetuating cycle of dysbiosis and inflammation generated by an acute trigger[60]. Even if IBD-associated dysbiosis is not seen as an initiator of disease, it could still be important in perpetuating the disease[7].

28.3.3 Genetic dysbiosis of skin biofilm

The skin microbiota has a high degree of interpersonal variation[62,63]. It has been suggested that SNPs in innate or adaptive immune elements can lead to excessive inflammation causing significant microbial shifts[64,65]. Previous chapters have described the characteristics of microbial biofilms on the human skin and covered the definition of the common skin condition termed psoriasis (chapter 21 by Fry). 16S rRNA gene analysis using swabbing of the skin to recover bacteria, has revealed differences in bacterial colonization compared to healthy skin, with an increase in *Firmicutes* and a decrease in *Actinobacteria*[66]. Similarly, a study on biopsy specimens[67] using massive parallel sequencing of the 16S rRNA genes revealed differences in microbial colonization between psoriatic and healthy skin. Host genetic variants predisposing to the psoriasis trait include *PSORS1* (psoriasis susceptibility 1 on chromosome 6p21.3)[68,69] and other genes affecting adaptive immune response and epidermal barrier function[70]. More recently, a GWAS study identified 36 potential loci suspected to be associated with psoriasis[71]. Most genes involved are associated with innate immune system including NFkB activation and Interleukin (IL)-23 signaling. Mutations in genes encoding epidermal proteins (filaggrin and LCE3B/C) can result in variation of the skin barrier with cutaneous sensitization by penetration of microbes and bacterial products, causing inflammation leading to atopic dermatitis (AD) and psoriasis[72,73]. Mast cell degranulation and production of IgE and IL-4 are both hallmarks of AD. It has been suggested that *S. aureus*-derived deltatoxin is at the basis of these processes. However, deltatoxin might only penetrate the skin in the setting of genetic defects associated with disease (*e.g. filaggrin* mutations in AD)[74]. Thus it is possible that psoriasis and AD are due to dysbiosis of bacteria colonizing the skin[75] and that genetic variants predispose to this dysbiosis. However, further research in this field is needed to substantiate this hypothesis.

28.3.4 Genetic dysbiosis of vaginal biofilm

Host genetic variants have been postulated to affect the composition of the vaginal biofilm, along with oestrogen levels, environmental and behavioral factors[20,76]. In most human populations, the healthy vagina is mainly populated by *Lactobacillus* species, which have a symbiotic relationship with their female host. Lactobacilli metabolize the glycogen produced by the vaginal epithelia leading to lactic acid production, largely responsible for the normal acidic vaginal pH of <4.5. The increased acidity and the production of antimicrobials by different lactobacilli species are associated with inhibition of growth of potential pathogens, such as *Gardnerella vaginalis*, *Prevotella bivia*, *Mobiluncus* spp. and Group B Streptococcus[77]. A very common vaginal disease characterized by vaginal discharge, bacterial vaginosis, was discussed in chapter 16. In BV, there is a decrease in lactobacilli load and in the vaginal acidity, associated with an overgrowth of vaginal anaerobes. Bacterial species associated with BV include *G. vaginalis*, *Mycoplasma hominis*, *Mobiluncus species*, *Prevotella species*, *Leptotrichia sanguinegens/ amnionii* and *Atopobium vaginae*[78]. Genetic factors affecting microbial recognition have been implicated in the predisposition to the carriage of organisms associated with BV and host immunity may be key to why some women who are affected by BV have recurrent or relapsing infections and complications whilst others do not[20]. Genetic polymorphisms in genes encoding bacterial receptors such as TLR and inflammatory mediators such as Interleukin-1 receptor antagonist (IL-1RA) and tumor necrosis factor (TNF)-α may account for differences in BV and complications between women[76,79]. Therefore, BV could be considered a dysbiosis in which genetic factors affecting bacterial recognition and host response may predispose to a shift in the vaginal microbial biofilm, predisposing to disease initiation.

28.4 Summary and conclusions

Host genetic variants have a strong role in determining the response to microbial colonization. Both human association studies[25,61,74] and animal-model studies[60] are bringing forward new evidence suggesting that host genetic variants may affect the normal composition of human microbial biofilms, hence predisposing to genetic dysbioses such as inflammatory bowel disease, psoriasis, atopic dermatitis, periodontitis and bacterial vaginosis[7].

TAKE-HOME MESSAGE

- Host genetic variants have a role in determining the response to microbial challenge.

- Initial evidence is starting to show that the composition of human biofilms and dysbiotic processes may be modulated by host genetic variants.

- Through an effect on dysbiosis, host genetic variants may predispose to a range of chronic human diseases.

References

1 McFall-Ngai M, Henderson B, Ruby N (Editors). *The Influence of Cooperative Bacteria on Animal Host Biology*. 2005. New York: Cambridge University Press.
2 Ruby E, Henderson B, McFall-Ngai M. Microbiology — we get by with a little help from our (little) friends. *Science* 2004; **303**:1305–1307.
3 Wu GD, Lewis JD. Analysis of the human gut microbiome and association with disease. *Clin Gastroenterol Hepatol* 2013; **11**: 774–7.
4 Zoetendal EG, Rajilic-Stojanovic M, de Vos WM. High-throughput diversity and functionality analysis of the gastrointestinal tract microbiota. *Gut* 2008; **57**: 1605–15.
5 The Human Microbiome Project Consortium. Structure, function and diversity of the healthy human microbiome. *Nature* 2012; **486**: 207–14.
6 Kellam P, Weiss RA. Infectogenomics: insights from the host genome into infectious diseases. *Cell* 2006; **124**: 695–7.
7 Nibali L, Henderson B, Sadiq ST *et al*. Genetic dysbiosis: the role of microbial insults in chronic inflammatory diseases. *J Oral Microbiol* 2014; **6**.
8 Ezkurdia I, Juan D, Rodriguez JM *et al*. Multiple evidence strands suggest that there may be as few as 19,000 human protein-coding genes. *Hum Mol Genet* 2014; **23**: 5866–78.
9 Sherry ST, Ward MH, Kholodov M *et al*. dbSNP: the NCBI database of genetic variation. *Nucleic Acids Res* 2001; **29**: 308–11.
10 Girirajan S, Campbell CD, Eichler EE. Human copy number variation and complex genetic disease. *Annu Rev Genet* 2011; **45**: 203–26.
11 Fomin ME, Togarrati PP, Muench MO. Progress and challenges in the development of a cell-based therapy for hemophilia A. *J Thromb Haemost* 2014; **12**: 1954–65.
12 Ciccia A, Elledge SJ. The DNA damage response: making it safe to play with knives. *Mol Cell* 2010; **40**: 179–204.
13 Albuquerque D, Stice E, Rodriguez-Lopez R *et al*. Current review of genetics of human obesity: from molecular mechanisms to an evolutionary perspective. *Mol Genet Genomics* 2015.
14 Noebels J. Pathway-driven discovery of epilepsy genes. *Nat Neurosci* 2015; **18**: 344–50.
15 Higdon R, Earl RK, Stanberry L *et al*. The promise of multi-omics and clinical data integration to identify and target personalized healthcare approaches in autism spectrum disorders. *OMICS* 2015; **19**: 197–208.
16 Cooke GS, Hill AV. Genetics of susceptibility to human infectious disease. *Nat Rev Genet* 2001; **2**: 967–77.

17 Janeway CA, Jr., Medzhitov R. Innate immune recognition. *Annu Rev Immunol* 2002; **20**: 197–216.

18 Nibali L, Ready DR, Parkar M *et al.* Gene polymorphisms and the prevalence of key periodontal pathogens. *J Dent Res* 2007; **86**: 416–20.

19 Genc MR, Vardhana S, Delaney ML *et al.* Relationship between a toll-like receptor-4 gene polymorphism, bacterial vaginosis-related flora and vaginal cytokine responses in pregnant women. *Eur J Obstet Gynecol Reprod Biol* 2004; **116**: 152–6.

20 Verstraelen H, Verhelst R, Nuytinck L *et al.* Gene polymorphisms of Toll-like and related recognition receptors in relation to the vaginal carriage of Gardnerella vaginalis and Atopobium vaginae. *J Reprod Immunol* 2009; **79**: 163–73.

21 Karin M, Lawrence T, Nizet V. Innate immunity gone awry: linking microbial infections to chronic inflammation and cancer. *Cell* 2006; **124**: 823–35.

22 Carrington M, Dean M, Martin MP *et al.* Genetics of HIV-1 infection: chemokine receptor CCR5 polymorphism and its consequences. *Hum Mol Genet* 1999; **8**: 1939–45.

23 Tebas P, Stein D, Tang WW *et al.* Gene editing of CCR5 in autologous CD4 T cells of persons infected with HIV. *N Engl J Med* 2014; **370**: 901–10.

24 Hajishengallis G, Liang S, Payne MA *et al.* Low-abundance biofilm species orchestrates inflammatory periodontal disease through the commensal microbiota and complement. *Cell Host Microbe* 2011; **10**: 497–506.

25 Divaris K, Monda KL, North KE *et al.* Genome-wide association study of periodontal pathogen colonization. *J Dent Res* 2012; **91**: 21S–8S.

26 Sobhani I, Tap J, Roudot-Thoraval F *et al.* Microbial dysbiosis in colorectal cancer (CRC) patients. *PLoS One* 2011; **6**: e16393.

27 Kwiatkowski DP. How malaria has affected the human genome and what human genetics can teach us about malaria. *Am J Hum Genet* 2005; **77**: 171–92.

28 Mattozzi C, Richetta AG, Cantisani C *et al.* Psoriasis: new insight about pathogenesis, role of barrier organ integrity, NLR/CATERPILLER family genes and microbial flora. *J Dermatol* 2012; **39**: 752–60.

29 Brito F, de Barros FC, Zaltman C *et al.* Prevalence of periodontitis and DMFT index in patients with Crohn's disease and ulcerative colitis. *J Clin Periodontol* 2008; **35**: 555–60.

30 Srinivasan U, Misra D, Marazita ML *et al.* Vaginal and oral microbes, host genotype and preterm birth. *Med Hypotheses* 2009; **73**: 963–75.

31 Keller JJ, Lin HC. The effects of chronic periodontitis and its treatment on the subsequent risk of psoriasis. *Br J Dermatol* 2012; **167**: 1338–44.

32 Li WQ, Han JL, Chan AT *et al.* Psoriasis, psoriatic arthritis and increased risk of incident Crohn's disease in US women. *Ann Rheum Dis* 2013; **72**: 1200–5.

33 Persson R, Hitti J, Verhelst R *et al.* The vaginal microflora in relation to gingivitis. *BMC Infect Dis* 2009; **9**: 6.

34 Henckaerts L, Pierik M, Joossens M *et al.* Mutations in pattern recognition receptor genes modulate seroreactivity to microbial antigens in patients with inflammatory bowel disease. *Gut* 2007; **56**: 1536–42.

35 Genc MR, Schantz-Dunn J. The role of gene-environment interaction in predicting adverse pregnancy outcome. *Best Pract Res Clin Obstet Gynaecol* 2007; **21**: 491–504.

36 Schulz S, Zissler N, Altermann W *et al.* Impact of genetic variants of CD14 and TLR4 on subgingival periodontopathogens. *Int J Immunogenet* 2008; **35**: 457–64.

37 O'Boyle CJ, MacFie J, Dave K *et al.* Alterations in intestinal barrier function do not predispose to translocation of enteric bacteria in gastroenterologic patients. *Nutrition* 1998; **14**: 358–62.

38 Shanahan F. The host-microbe interface within the gut. *Best Pract Res Clin Gastroenterol* 2002; **16**: 915–31.

39 Naaber P, Smidt I, Tamme K *et al.* Translocation of indigenous microflora in an experimental model of sepsis. *J Med Microbiol* 2000; **49**: 431–9.

40 Socransky S, Haffajee A. Periodontal infections. 2012. Chapter 9 in Clinical Implantology and Implant Dentistry, Lindhe J, Lang N, Karring P. Blackwell Munksgaard, Oxford, UK.

41 Haubek D, Ennibi OK, Poulsen K *et al.* Risk of aggressive periodontitis in adolescent carriers of the JP2 clone of Aggregatibacter (Actinobacillus) actinomycetemcomitans in Morocco: a prospective longitudinal cohort study. *Lancet* 2008; **371**: 237–42.

42 Socransky SS, Haffajee AD, Smith C *et al.* Microbiological parameters associated with IL-1 gene polymorphisms in periodontitis patients. *J Clin Periodontol* 2000; **27**: 810–8.

43 Nibali L, Tonetti MS, Ready D *et al.* Interleukin-6 polymorphisms are associated with pathogenic bacteria in subjects with periodontitis. *J Periodontol* 2008; **79**: 677–83.

44 Nibali L, Donos N, Henderson B. Periodontal infectogenomics. *J Med Microbiol* 2009; **58**: 1269–74.

45 Nibali L, Pelekos G, Habeeb R *et al.* Influence of IL-6 haplotypes on clinical and inflammatory response in aggressive periodontitis. *Clinical Oral Investigations* 2013; **17**: 1235–1242.

46 Anderson CA, Boucher G, Lees CW *et al.* Meta-analysis identifies 29 additional ulcerative colitis risk loci, increasing the number of confirmed associations to 47. *Nat Genet.* 2011; **43**: 246–52.

47 Schaefer AS, Jochens A, Dommisch H *et al.* A large candidate-gene association study suggests genetic variants at IRF5 and PRDM1 to be associated with aggressive periodontitis. *J Clin Periodontol.* 2014; **41**: 1122–31.

48 Takahashi N, Nyvad B. The role of bacteria in the caries process: ecological perspectives. *J Dent Res* 2011; **90**: 294–303.

49 Indriolo A, Greco S, Ravelli P *et al.* What can we learn about biofilm/host interactions from the study of inflammatory bowel disease. *J Clin Periodontol* 2011; **38 Suppl** 11: 36–43.

50 Robles A, V, Guarner F. Linking the gut microbiota to human health. *Br J Nutr* 2013; **109 Suppl 2**: S21–S26.

51 Isaacs KL, Sartor RB. Treatment of inflammatory bowel disease with antibiotics. *Gastroenterol Clin North Am* 2004; **33**: 335–45, x.

52 Parkes M. Evidence from genetics for a role of autophagy and innate immunity in IBD pathogenesis. *Dig Dis* 2012; **30**: 330–3.

53 Bibiloni R, Mangold M, Madsen KL *et al.* The bacteriology of biopsies differs between newly diagnosed, untreated, Crohn's disease and ulcerative colitis patients. *J Med Microbiol* 2006; **55**: 1141–9.

54 Chow J, Mazmanian SK. A pathobiont of the microbiota balances host colonization and intestinal inflammation. *Cell Host Microbe* 2010; **7**: 265–76.

55 Inohara, Chamaillard, McDonald C *et al.* NOD-LRR proteins: role in host-microbial interactions and inflammatory disease. *Annu Rev Biochem* 2005; **74**: 355–83.

56 Ogura Y, Bonen DK, Inohara N *et al.* A frameshift mutation in NOD2 associated with susceptibility to Crohn's disease. *Nature* 2001; **411**: 603–6.

57 Eckburg PB, Relman DA. The role of microbes in Crohn's disease. *Clin Infect Dis* 2007; **44**: 256–62.

58 Matsuki T, Watanabe K, Fujimoto J *et al.* Quantitative PCR with 16S rRNA-gene-targeted species-specific primers for analysis of human intestinal bifidobacteria. *Appl Environ Microbiol* 2004; **70**: 167–73.

59 Turnbaugh PJ, Hamady M, Yatsunenko T *et al.* A core gut microbiome in obese and lean twins. *Nature* 2009; **457**: 480–4.

60 Craven M, Egan CE, Dowd SE *et al.* Inflammation drives dysbiosis and bacterial invasion in murine models of ileal Crohn's disease. *PLoS One* 2012; **7**: e41594.

61 Frank DN, Amand ALS, Feldman RA, Boedeker EC, Harpaz N, Pace NR. Molecular-phylogenetic characterization of microbial community imbalances in human inflammatory bowel diseases. *Proc Natl Acad Sci U S A* 2007; **104**: 13780.

62 Grice EA, Kong HH, Conlan S *et al.* Topographical and temporal diversity of the human skin microbiome. *Science* 2009; **324**: 1190–2.

63 Hollox EJ, Huffmeier U, Zeeuwen PL *et al.* Psoriasis is associated with increased beta-defensin genomic copy number. *Nat Genet* 2008; **40**: 23–5.

64 Boguniewicz M, Leung DY. Atopic dermatitis: a disease of altered skin barrier and immune dysregulation. *Immunol Rev* 2011; **242**: 233–46.

65 Gao Z, Tseng CH, Strober BE *et al.* Substantial alterations of the cutaneous bacterial biota in psoriatic lesions. *PLoS One* 2008; **3**: e2719.

66 Fahlen A, Engstrand L, Baker BS *et al.* Comparison of bacterial microbiota in skin biopsies from normal and psoriatic skin. *Arch Dermatol Res* 2012; **304**: 15–22.

67 Trembath RC, Clough RL, Rosbotham JL *et al.* Identification of a major susceptibility locus on chromosome 6p and evidence for further disease loci revealed by a two stage genome-wide search in psoriasis. *Hum Mol Genet* 1997; **6**: 813–20.

68 Nair RP, Stuart P, Henseler T *et al.* Localization of psoriasis-susceptibility locus PSORS1 to a 60-kb interval telomeric to HLA-C. *Am J Hum Genet* 2000; **66**: 1833–44.

69 Oka A, Mabuchi T, Ozawa A *et al.* Current understanding of human genetics and genetic analysis of psoriasis. *J Dermatol* 2012; **39**: 231–41.

70 Tsoi LC, Spain SL, Knight J *et al.* Identification of 15 new psoriasis susceptibility loci highlights the role of innate immunity. *Nat Genet* 2012; **44**: 1341–8.

71 Palmer CN, Irvine AD, Terron-Kwiatkowski A *et al*. Common loss-of-function variants of the epidermal barrier protein filaggrin are a major predisposing factor for atopic dermatitis. *Nat Genet* 2006; **38**: 441–6.

72 de CR, Riveira-Munoz E, Zeeuwen PL *et al*. Deletion of the late cornified envelope LCE3B and LCE3C genes as a susceptibility factor for psoriasis. *Nat Genet* 2009; **41**: 211–5.

73 Nakamura Y, Oscherwitz J, Cease KB *et al*. Staphylococcus delta-toxin induces allergic skin disease by activating mast cells. *Nature* 2013; **503**: 397–401.

74 Fry L, Baker BS, Powles AV *et al*. Is chronic plaque psoriasis triggered by microbiota in the skin? *Br J Dermatol* 2013; **169**: 47–52.

75 Turovskiy Y, Sutyak NK, Chikindas ML. The aetiology of bacterial vaginosis. *J Appl Microbiol* 2011; **110**: 1105–28.

76 Matu MN, Orinda GO, Njagi EN *et al*. In vitro inhibitory activity of human vaginal lactobacilli against pathogenic bacteria associated with bacterial vaginosis in Kenyan women. *Anaerobe* 2010; **16**: 210–5.

77 Hay P. Life in the littoral zone: lactobacilli losing the plot. *Sex Transm Infect* 2005; **81**: 100–2.

78 Cauci S, Di SM, Casabellata G *et al*. Association of interleukin-1beta and interleukin-1 receptor antagonist polymorphisms with bacterial vaginosis in non-pregnant Italian women. *Mol Hum Reprod* 2007; **13**: 243–50.

Mirroring the future: dysbiosis therapy

CHAPTER 29

Diet and dysbiosis

Mehrbod Estaki, Candice Quin and Deanna L. Gibson
The University of British Columbia Kelowna, Canada

29.1 Introduction

The symbiotic relationship between the mammalian host and the trillions of bacteria that colonize their intestinal tract tells a lengthy and complex tale of coevolution. Mammals provide food substrates for the bacteria, which they ferment into digestible dietary components. In return, the host allows resident bacteria to exist within the confines of the intestinal lumen without disruptions from the immune system. Since the moment of birth and possibly even during gestation[1,2], this mutually beneficial relationship strongly influences host immune development and health. We now know that a disturbance in this delicate balance of the host-microbiota relationship, termed dysbiosis, is associated with a multitude of diseases including inflammatory bowel disease (IBD), metabolic syndrome, obesity, and diabetes (covered in chapters 17–27). Consequently, the last decade has seen a drastic increase in research aimed at defining and altering the intestinal microbiota in an attempt to promote health, prevent the development and reverse the course of dysbiosis-associated diseases. Some of these methods include development of: antibiotics to eradicate dysbiosis-inducing pathogens, functional foods such as probiotics and prebiotics, agents targeted at promoting intestinal health, and customized diets. The last method may be the most viable tactic in promoting microbial health and is the focus of this section.

29.2 Coevolution of the host-microbiota super-organism

The intestinal microbiota of our early hunter-gatherer ancestors most likely differed greatly from those of today's industrialized city dwellers. Phylogenetic diversity analyses of Hadza tribesmen of Tanzania, who live nomadic lifestyles closely resembling those of the Paleolithic era, show remarkable differences in their gut microbiota compared to those from urban settings[3]. With no domestication or cultivation of plants and livestock, the Hadza diet consists of wild game and plants

The Human Microbiota and Chronic Disease: Dysbiosis as a Cause of Human Pathology, First Edition.
Edited by Luigi Nibali and Brian Henderson.
© 2016 John Wiley & Sons, Inc. Published 2016 by John Wiley & Sons, Inc.

such as berries and fibrous tubers and is devoid of dairy, processed, and grain-based foods. The fecal bacterial community from these mobile tribesmen shows a highly diverse and unique configuration. Their microbiome is enriched with the genus *Treponema* (encompassing pathobiont subspecies that cause syphilis, yaws, bejel, and pinta infections)[4] and *Prevotella*, and is low in *Bifidobacterium*, often considered a health-promoting genus[5]. Interestingly, the Hadza experience very low to no incidences of autoimmune diseases[6,7], suggesting that the current trends and generalization of bacterial groups may not be meaningful across all populations but rather are dependent on environmental and genetic contexts. With constant exposure to and dependency on their surrounding biosphere, the Hadza microbiome has likely coevolved to consist of bacteria more efficient in nutritional acquisition specific to available food sources. The *Prevotella* and *Treponema* genus for example allow for efficient hydrolysis of cellulose and xylan from fibre-rich foods[8], allowing the Hadza optimal nutrient extraction from their typical diets. The Hadza themselves, in return, have become tolerant to these bacterial species, despite their apparent pathogenic association in Westernized settings[4,9].

29.3 Gut microbiota in personalized diets

Evidence of sex-related microbiome differences in the Hadza also suggests an evolutionary divergence corresponding to the sexual division of labor within traditional social roles and structures. The customary gatherer role of the female Hadza and the evolutionary pressures of pregnancy and lactation demand different microbial adaptation than their male counterparts. It is widely accepted that nutritional needs of individuals within the same population differ depending on age, sex, and health status; however, the crucial role of microbiota in personalized diets is largely overlooked. For example, within Asian populations, soy-rich diets are protective against prostate cancer, osteoporosis, and cardiovascular disease as a result of the production of equol from soy isoflavone diadzein[10]. Equol is produced exclusively by certain gut bacteria found in over half of adult population from Japan, Korea, or China but only in 25–30% of adults from Western countries[11]. Due to the presence of these key gut bacteria, only the "equol-producing" populations profit from these health benefits of soy-rich diets.

According to existing DNA records, early humans migrated out of Africa some 60,000–70,000 years ago, finally reaching South America some 15,000 thousand years after (Genographic Project Consortium). As our early ancestors settled in various corners around the world, their gut microbiota also adapted to new environments, pathogen loads, and available food sources to optimize survival. Populations in Japan, for example, as a result of exposure to marine microbes, acquired gut microbiota enriched with genes encoding enzymes for digestion of porphyran, a complex carbohydrate. Porphyran is ubiquitous in traditional Japanese diets, which contain nori (edible seaweed) due to its geographic abundance[12]. Rapid movement of humans out of their ancestral lands without sufficient microbial adaptations may pose a disadvantageous biological strain leading to nutritional deficiencies and development of chronic diseases. In this regard, the adage "eat local" may hold more significance than its common "economically responsible" connotations. These findings suggest that the microbial composition of individuals should be considered a critical component in personalized diet prescriptions.

29.4 The evolution of diet

Around 10,000 years ago, the transition from the hunter-gatherer lifestyles of the Paleolithic period to the Neolithic era, characterized by domestication of wild animals and cultivation of carbohydrate-rich farming, marked one of the greatest dietary shifts in human evolution. Calcified dental plaques from various periods of history show a drastic shift in oral microbes at the onset of Neolithic period towards a more cariogenic or disease-associated configuration[13]. Another major shift occurred around 160 years ago corresponding to the industrial revolution with the introduction of processed flour and sugar, and urban living. At this point, the oral microbial ecosystem became significantly less diverse, and this is thought to have contributed to the rise in dental and possibly other chronic diseases of the post-industrial lifestyles[13]. Biodiversity boosts productivity in any biological ecosystem, while its decrease is correlated with negative events and loss of ecological equilibrium. The transition from farming lifestyles to industrial urban living was accompanied by drastic shifts in eating habits in which over-sanitized, pasteurized, and processed products including high-sugar, calorie-dense, low-fibre foods became conventional diets. It is thought that these changes reduced the adaptive potential of the gut microbiota by reducing their biodiversity and functionality, leading to increased incidences of chronic diseases, food sensitivities, and atopy[8,14]. In support of this, evidence suggest that children raised in farms who are in contact with animals, stables, and consume raw/unpasteurized milk have lower prevalence of hay fever, asthma, wheeze, and atopic sensitization compared to non-farmers' children[15,16]. One key feature common to these studies is the need for exposure to farm environments at an early age. This suggests that early colonization of the intestinal microbiome may be critical for balanced immune development and protection against disease susceptibilities later in life.

29.5 Plasticity of the microbiota and diet

The role of diet in shaping the human microbiome is still poorly understood due to the complex and confounding nature of its interaction with environmental factors. The human microbiota may be continuously reconfiguring its composition throughout the host's lifetime to adapt to environmental stressors such as lifestyle, psychological health, diet, physical activity, and host physiology. These changes appear to occur down the taxonomical line at the species level, while the overall deviations appear to anchor around a core set of bacteria groups belonging to the Bacteroidetes and Firmicutes phyla, and to a lesser extent the Actionbacteria, Proteobacteria, and Verrucomicrobia[17]. Furthermore, interpersonal variations in gut microbiota within populations appear to surround a functional core rather than a microbial core[18]. Once established (usually within the first 2–4 years of life), a healthy gut microbiota has substantial stability over time[19]. Alterations in some microbial groups can occur rapidly as quickly as 24 hours during controlled feeding[20]; however, long-lasting changes of the "core" microbiota require long-term dietary interventions and are influenced by the original native microbiota[21]. Agrarian populations from Burkina Faso, Malawi, and Venezuela, whose diets are dominated by plant-derived polysaccharides, have a microbiome enriched by

Prevotella[8,9,22]. In the US and Europe, however, long-term diets of animal protein, several amino acids, and saturated fats is associated with an abundance of *Bacteroides* taxa[21]. In humans, five days of animal-based diets was associated with increases in bile-tolerant microorganisms such as *Bilophila wadsworthia* and *Alistipes* spp., in addition to increases in expression of genes involved in vitamin biosynthesis[23]. Understanding the mechanisms by which the gut microbiota resists change and maintains stable states would allow us to develop novel strategies and diets to revert dysbiosis-associated diseases to healthy states. The following sections highlight the tripartite relationship among gut microbiota, host and food as well as the role of specific macronutrients in conferring changes to gut microbiota.

29.6 Interaction among gut microbiota, host and food

The intestinal bacteria reside at the interface of the external environment and the host, allowing for a tightly regulated crosstalk between the three (Figure 1). The bacteria reside within the lumen of the intestinal tract and interact directly with dietary antigens where they aid in digestion of partially digested food, vitamin biosynthesis, and absorption. The bacteria also act as a first line of defense against pathogens in the gut through competition for food and space. As discussed by Ding and Hart in chapter 19, separating the microbiota from the sterile sub-mucosal area is a highly selective monolayer made up of intestinal epithelial cells (IEC). Goblet cells found within the IEC produce glycoprotein polymers such as mucin that form the mucus layer in the gut. In the colon, the innermost layer of the mucus is dense and prevents the entry of bacteria, while the less dense outer layer acts as a habitat for bacteria[24]. In addition, released mucin monosaccharides from the outer layer can also be recycled as a food source for the commensal bacteria. Interestingly, the gut microbiota also regulates goblet cell mucus secretion, suggesting the involvement of microbiota-goblet cell crosstalk in dysbiosis. Occasionally a commensal bacteria or pathogen can reach the IEC surface by circumventing the lumenal defenses and mucus layer. This stimulates immune responses within the host by downstream activation of innate and adaptive immunity. Beyond the IEC, antigen-presenting cells (APCs) such as dendritic cells and macrophages act as the next line of defense against bacteria that may have traversed the IEC layer. In the gut, APCs are tolerogenic to the commensal bacteria but they become inflammatory in response to pathogens[25]. This is critical in preventing a state of chronic inflammation due to the vast bacterial exposure in the gut. APCs activate the involvement of the adaptive immune system by recruiting T helper (Th) cells, including Th1, Th2 and Th17 subpopulations. Another subset of T-cells, the regulatory T cells (Treg), play an important role in tolerance to self-antigens as well as commensal bacteria by suppressing an overly pro-inflammatory immune response abrogating autoimmunity. The antibody-producing B cells are activated by Th cells and secretory IgA are particularly important in guarding the mucosal surface from pathogenic microbes. Overall, the complex tripartite interactions among the intestinal microbiota, host, and the environment regulate intestinal health and by extension the host. Disturbances in any of these factors can have deleterious effects on the host, in part by inducing dysbiosis.

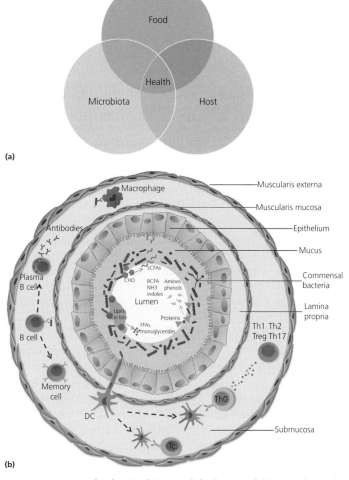

Figure 1 (a) Interactions among food, microbiota, and the host. A delicate and complex relationship exists among the microbiota, food and the host, and even slight deviations in this tripartite relationship can have deleterious effects on the host. (b) **Food-microbiota:** Dietary intake can directly influence microbial composition by promoting blooms of certain groups while inhibiting the growth of others. Partially digested food such as complex carbohydrates are broken down in the gut by the bacteria via fermentation and produce short-chain fatty acids (SCFAs). Lipid digestion in the gut requires the secretion of liver-derived bile from the gallbladder. High-lipid diets demand higher bile release, which could induce dysbiosis in part by changing intestinal pH levels. Fermentation of proteins in the gut can result in production of branched-chain fatty acids (BCFA) as well as potentially toxic products such as amines, ammonia (NH3), phenols, and indoles[26]. Gut bacteria are essential in synthesizing various vitamins such as riboflavin, folate, B12, and vitamin K[26]. **Microbiota-host:** The microbiota acts as a first line of defense against enteropathogens by competition for food and niche. Host immunity is modulated by the intestinal microbiota starting at birth. Antigen-presenting cells (APC) such as dendritic cells (DC) or macrophages phagocytose invading bacteria and subsequently initiate differentiation of naïve T cells (Th0) to subset of T-helper cells (Th): Th1, Th2, Th17, or regulatory T-cells (Treg). B-cells presented by bacterial antigen differentiate into antibody-producing effector B cells or memory cells. Antibodies produced by effector B-cells neutralize and tag the specific pathogen for removal. Bacteria can regulate mucus production by stimulation of goblet cells. Paneth cells of the intestinal epithelium layer sense bacteria and in response produce antimicrobial peptides (AMPs) that protect the host from bacterial invasion. Evidence has emerged showing that gut microbiota interact with central and peripheral neural processes, suggesting their involvement in hormone signaling as well as psychological health (see chapter 27 by Barwani and Forsythe). **Host-Food:** Food antigens stimulate endocrine secretion, which initiates downstream digestion cascades such as bile and amylase production by the liver and pancreas, respectively. SCFAs produced by bacteria in the colon are used by colonocytes as a major energy source. Ingestion of contaminated food is a major source of pathogen entry into the body. Hormonal regulations by the host control appetite and satiety. (*see color plate section for color details*).

29.7 Consequences of diet-induced dysbiosis on host health

We know from controlled experiments that disturbances in any of the three factors mentioned above (host, intestinal microbiota, or diet) can induce dysbiosis. In clinical settings, however, despite strong evidence suggesting the involvement of dysbiosis in various chronic inflammatory diseases such as IBD, irritable bowel syndrome (IBS), diabetes, colorectal cancer (CRC), atherosclerosis, and non-alcoholic fatty liver disease, a causative relationship has not yet been established. It is likely that under certain circumstances dysbiosis is a product of the disease, exaggerating symptoms, while in other cases it may instigate pathogenesis. As in any ecosystem, increased diversity and resistance to change are hallmarks of a healthy microbiota. The onset of dysbiosis may be asymptomatic; over time, however, imbalanced nutrition and poor dietary habits can significantly alter the microbiota towards a dysbiotic state that leaves the host vulnerable to various infections and chronic diseases. For example, introduction of high-fat, high-sugar Western diets to African-Americans have been linked to altered microbial profiles leading to increases incidences of CRC and associated mortality[27]. Diet and microbiota have also been implicated in the pathogenesis of IBS, a functional disorder of the gastrointestinal tract. Dysbiosis can lead to improper digestion of carbohydrates and prolonged hydrogen production in the intestines, causing increased methane levels, which correspond to increased IBS severity[28]. Despite the absence so far of a proven causative link between dysbiosis and IBD (encompassing ulcerative colitis and Crohn's disease), it is generally thought that in clinical settings, a customized diet can alleviate symptoms in some. In animal models this has been shown to act through changes in the microbiota. For example, a diet high in milk fat induced dysbiosis by proliferating sulfide-reducing pathobionts, leading to increased incidences of colitis in susceptible IL-10 knock-out mice[29]. Similarly, in other studies, excessive omega-6 PUFA consumption has been linked to increased incidences of ulcerative colitis in humans[30] and promotion of dysbiosis in mice[31]. In obesity, dysbiosis can favor increases in energy-harvesting bacteria[32] and modulation in leptin sensitivity[33], leading to increased food intake.

While diet may play a role in the pathogenesis of various dysbiosis-associated diseases, it is unknown if manipulation alone can reverse these conditions. Once a state of dysbiosis has been established, opportunistic bacteria and/or pathogens can eradicate various native species by outcompeting them for food and habitable niches. Only by reintroduction of new bacteria can the extinct have the opportunity to recolonize; this process of treating diseases using bacteria is referred to as bacteriotherapy. Today extensive laboratory and clinical research is exploring the use of bacteriotherapy in various diseases. The use of healthy whole microbiome communities, for example from stool, has successfully been used in combating dysbiosis-associated diseases such as antibiotic-resistant *Clostridium difficile* infection (see also chapter 31)[34]. The use of probiotics, by far the most intensely studied branch of bacteriotherapy, is covered in chapter 30. While many experimental animal models suggest probiotics have a beneficial effect, clinical studies have not been conclusive.[35-37] Some studies provide evidence that probiotics do not colonize or persist in the mammalian gut efficiently[38]. While bacteriotherapy has tremendous potential as a novel strategy in combating dysbiosis-associated disease, additional research is needed to determine its safety and efficacy.

29.8 The role of gut microbes on the digestion of macronutrients

Mammals, on their own, are inefficient at digesting food. To utilize food energy, the host-microbiome superorganism has evolved to work as a unit customized to the benefit of both. Up to 10% of the dietary energy in humans can be due to the activities of their intestinal microbiota[39]. This may be beneficial or detrimental depending on the needs of the host. For example, in the context of obesity, having microbes that extract more energy is not favorable but in the context of cachexia this would be of benefit.

29.8.1 Carbohydrates

Mammalian genomes do not encode for most enzymes required for the degradation of insoluble structural polysaccharides. Undigested food particles arriving at the large intestine are typically comprised of insoluble materials derived from plant cell walls and resistant starch[40]. Specialized bacteria species such as *Roseburia bromii*, belonging to the Firmicutes, act as primary degraders of insoluble polysaccharides, in turn producing substrates for other amylolytic bacteria[41]. The primary result of carbohydrate fermentation under anaerobic conditions in the gut is the production of SCFAs such as acetate, lactate, propionate, and butyrate, of which the last plays a particularly important role in gut homeostasis and health. While lactate and acetate become available to the host systemically[42], butyrate directly becomes the primary food source of the colonocytes. Butyrate is known to possess anti-cancer and anti-inflammatory properties[43] and is involved in gut motility[44], energy expenditure[45], and appetite control[46]. Butyrate can also be produced through conversion of other available acids such as acetate and lactate by members of the Firmicutes phyla, including *Eubacterium hallii* and *Anaerostipes* spp[47].

29.8.2 Proteins

Partially digested proteins, mucus secretions, and amino acids from shed IEC are fermented in the large intestine and serve as a source of carbon and nitrogen for intestinal microorganisms[48,49]. The amount of protein entering the large intestine depends on the total intake as well as the source of the protein. The digestibility of animal-derived protein is higher (94–99%) than that derived from plant sources (70–90%)[50]. Protein catabolism by intestinal microorganisms yields a variety of end products, many which are toxic to the host[51]. Degradation of undigested or endogenous protein, referred to as putrefaction, is particularly prominent in the etiology of CRC and ulcerative colitis (UC)[52,53]. Bacterial proteases and peptidases can initiate the degradation process by means of hydrolysis to convert proteins into smaller peptides and amino acids. This process is optimal in the distal colon where pH is higher. Fermentation of amino acids by reductive deamination can, albeit to a much lesser extent, produce SCFA similar to those found during carbohydrate fermentation[54]. They also produce ammonia, amines, thiols, phenols, and indoles, which are exclusive to amino acid fermentation. The accumulation of these by-products is pathogenic to the host. For example, phenols and indoles are considered carcinogens, ammonia a mutagen, and thiols a cellular toxin[50]. Another product of protein fermentation is branched-chain fatty acids (BCFA). The physiological significance of BCFA is not well understood; however, they appear to be important in developmental stages during gestation and immediately following birth[55]. Further, sulfate-reducing bacteria such as those in the genus *Desulfovibrio* instigate fermentation of dietary and mucinous sulfate and sulfur amino acids, leading to increased production of hydrogen sulfide (H_2S), which has been associated with pathogenesis of various intestinal diseases including UC[56,57]. As H_2S is involved in many normal processes within the colon, it is likely that its association with etiology of intestinal diseases is relevant only in conditions where H_2S levels reach higher than subtoxic levels[56].

29.8.3 Lipids

Different members of the gut microbiota are thought to regulate fat absorption via distinct mechanisms. Several roles of microbes in the harvest and storage of dietary lipids have been postulated; however, the exact mechanisms in play remain to be elucidated. In zebrafish, the presence of members of the Firmicutes phyla such as *Bacillus* spp. and *Clostridia* spp. stimulates fatty acid uptake and lipid droplet formation in the gut epithelium and liver[58]. This leads to an increased rate of lipid absorption by the intestinal cells (enterocytes), resulting in a greater accumulation of fatty acids in various tissues. Once dietary lipids are absorbed into the enterocytes, they are repackaged into lipoprotein particles known as chylomicrons. These chylomicrons transport lipids in the form of triglycerides to adipocytes, where lipoprotein lipases breakdown triglycerides into free fatty acids that can then be taken into the tissue for storage. Interestingly, a study in mice found that the presence of gut microbes reduced serum levels of chylomicrons[59]. This outcome could have been due to decreased lipid absorption in the gut or an increased lipid clearance in the periphery. It was found that the presence of microbiota did not affect the absorbance rate of lipids in mice, but rather affected the clearance of chylomicrons. Another study found that intestinal microbiota can induce lipoprotein lipase activity by suppressing its inhibitor, fasting-induced adipocyte factor (FIAF)[60]. Suppression of FIAF leads to increases in lipid clearance, subsequent reduction in serum triglycerides and an increase in lipid storage in adipocytes[59].

29.9 Diet induces dysbiosis in the host

Long-term dietary intake has been changing worldwide due to marked changes in agricultural practices, population growth, lifestyle, and socio-economic status[61]. This has wide-ranging implications in terms of gut microbiota and health. Traditionally, different cultures exhibit different gut microbiota. A clinical study examining the differences in the intestinal microbiota of children raised in Africa and children raised in Italy found that those of African children consuming vegetarian diets rich in fiber were markedly different from those of Italian children consuming Western diets[8]. The African children's microbiome was abundant in Gram-negative bacteria (58.5%) compared to children from Italy, which were abundant in Gram-positive bacteria (~70%). It has been reported that modification of diet can cause significant and rapid changes to human gut microbiome in as few as five days[23]. In mice, changes to the Bacteroidetes: Firmicutes ratio can occur within 24 hours, a change that stabilizes within several days. In humans, however, the change caused by short-term dietary interventions is not sufficient to change enterotype clustering and long-term dietary interventions are required[21]. Enterotypes are a recently introduced classification of gut microbiomes based on specific patterns of bacteriological communities. Enterotype clusters vary by genus composition; type 1 is defined by high levels of *Bacteroides*, type 2 by high levels of *Prevotella* and low *Bacteroides*, and type 3 is associated with high levels of *Ruminococcus*[62]. This classification system, however, is based on very discrete separations and current evidence suggests that enterotypes are more continuous than previously thought[63]. Regardless, alterations to bacterial clusters through worldwide changes in dietary practices could have dire epidemiological consequences without sufficient understanding of diet's role in gut microbial regulation. Although the research is limited, studies are now being conducted to examine the effects of macronutrients (protein, carbohydrates, and lipids) in modulating gut microbes.

29.9.1 Protein

Dietary proteins can have profound effects on gut microbial composition. For example, in obese humans, high animal-derived protein intake decreases total microbial abundance as well as the proportion of butyrate-producing *Roseburia/Eubacterium rectale* groups of bacteria[64]. Butyrate is essential for colonic mucosal health, while reduced levels of this SCFA reportedly contribute to the etiology of ulcerative colitis[65]. Therefore, alterations to gut microbial communities through management of protein intake could serve as a therapy for treatment and management of ulcerative colitis. Additionally, in humans, high-protein diets decrease the abundance of anti-inflammatory *Bifidobacteria* genus[66]. High animal-based-protein consumption in Western diets is associated with *Bacteroides* enterotypes. This observation echoes the previously mentioned comparative study of African and Italian children gut microbiota. Although it is evident that dietary protein affects gut microbial composition, the consequences of these changes are not well understood.

29.9.2 Carbohydrates

Diets rich in carbohydrates (polysaccharides, oligosaccharides, and resistant starch) and simple sugars are commonly seen in agrarian societies and are associated with a *Prevotella*-dominated enterotype[21]. Oligosaccharide polymers act as prebiotics and promote microbial diversity in the gut[67]. On the other hand, a reduced consumption of carbohydrates can also alter gut microbiota by reducing the butyrate-producing *E. rectale*, *Roseburia* spp. and *Bifidobacteria* spp[20]. *Bifidobacteria* spp. are desirable for a number of reasons: they act as immune modulators, they produce vitamins, and their metabolic end products (acetate and lactate) inhibit the growth of opportunistic pathogens[5]. It has been shown that the microbiomes found in obese individuals have an increased capacity to harvest energy from diet[68] and a Bacteroidetes:Firmicutes ratio that favors the Firmicutes[69]. A recent study suggests that the increased energy harvested is derived from dietary carbohydrate intake, rather than proteins and lipids, and that this increased carbohydrate absorption could be due to differences in the gut microbial composition[70].

29.9.3 Lipids

The relationship between gut microbes and lipids has been a relatively active area of research. In particular, high-fat Western diets are typically compared to low-fat controls in regards to their interaction with the gut microbiota. Recent evidence, however, suggests that the type of fat and not only the amount of fat ingested may be critical in inducing dysbiosis[31].

29.9.3.1 Saturated fatty acid

Conflicting evidence exists regarding the link between gut microbes and saturated fatty acids (SFA) such as animal-fat products like milk and certain vegetable products like coconut oil. Some evidence suggests that diets rich in SFA alter microbial diversity by reducing the abundance of bacteria in the Bacteroidetes phylum, such as *Bacteroides* spp. This in turn increases the Firmicutes:Bacteroidetes ratio associated with obesity[71,72]. In contrast, one study found SFA to increase *Bacteroides* spp. in the human fecal microbiota[73]. These conflicting results may be attributed to the different sources of SFA, which are often not identified in studies. For example, the increase in the Firmicutes:Bacteroidetes ratio was seen in a study using palm oil as the source of SFA, while the increase in *Bacteroides* was seen in a study using SFA derived from animal sources. SFA also promotes the growth of sulphite-reducing *Bilophila wadsworthia*, a pathobiont that promotes pro-inflammatory Th1 responses through hepatic bile conjugation[29]. Bile acids further facilitate fat absorption and disrupt bacterial membrane integrity leading to antibacterial activity[74]. However, some bacteria are bile-resistant, such as *B. wadsworthia, H. hepaticus and L. monocytogenes*[29]. The by-products of these microbes, such as secondary bile and hydrogen sulfide, can compromise mucosal barrier function and allow the infiltration of immune cells that induce tissue damage. This suggests a mechanism by which high-fat diets may direct immune-mediated diseases. Interestingly, diets supplemented with bile acids mirror microbial patterns found in high-fat diets: there is a reduction in Bacteroidetes and an increase in prominent Firmicutes classes such as *Clostridia* and *Erysipelotrichi* spp[75]. In addition, laurate and palmitate, two SFAs, have been shown to activate both toll-like receptor (TLR)-4 and TLR-2[76]. These innate immune cell surface receptors result in the activation of immunity critical for host defense against pathogens and aid in wound healing but also can perpetuate chronic inflammation. Overall, different sources of SFA with comparable degrees of saturation may elicit different results.

29.9.3.2 Mono-unsaturated fatty acid

Traditional Mediterranean diets are high in monounsaturated fatty acids (MUFA) due to their large olive oil intake rich in oleic acid. Diets rich in MUFA are thought to have some health benefits such as optimizing cardiac energy metabolism in obese individuals[77], and are protective against ulcerative colitis-associated colorectal cancer[78]. MUFA decreases total bacteria numbers, which could be important in protecting against bacterial overgrowth, but do not alter the relative abundance of individual bacterial populations[79]. However, it has been shown that MUFA-rich diets are associated with higher numbers of *Bacteroides* spp. and lower numbers of

Bifidobacteria spp.[73]. Curiously, a high intake of MUFA as well as SFA has also been associated with bacterial vaginosis, a mild bacterial infection of the vagina (covered in chapter 16)[80]. This association was found by comparing the diets of pregnant women and adjusting for covariates to see which dietary factors appear to contribute to vaginosis. Additional studies on MUFA-rich diets are needed to understand how this anti-inflammatory diet alters the microbiome.

29.9.3.3 Polyunsaturated fatty acid

- n-6 polyunsaturated fatty acids: N-6 PUFA (omega-6) are found in many vegetable oils such as corn, sunflower, and canola oils. There are two main types of n-6 PUFAs: linoleic acid and arachidonic acid. Diets rich in n-6 PUFA, particularly linoleic acid, change gut microbiota in mice, promoting increased numbers of pathobionts and leading to increased intestinal mucosal damage[81] and exacerbation of colitic symptoms[31]. Additionally, PUFA-rich diets decrease the Firmicutes:Bacteroidetes ratio and negatively correlated with the production of SCFA essential to overall gut health[82]. However, human studies are lacking and need to be conducted to compare whether linoleic acid and arachidonic acid have similar effects on gut microbial health.

- n-3 polyunsaturated fatty acids: N-3 PUFA (omega-3) such as docosahexaenoic acid (DHA), eicosapentenic acid (EPA) and their metabolite, eicosanoids, are known for their anti-inflammatory actions in humans[83]. As a result, the food industry has taken to supplementing conventional foods with DHA as a means to combat the prevalence of chronic inflammatory diseases. For instance, baby formula enriched with DHA and EPA is commonly marketed as having various health benefits despite contradictory evidence regarding their efficacy. While fish-oil supplementation in formula has no effect on vision, cognition, growth, or motor development[84], it appears to change the composition of an infant's gut microbiota[85,86]. The magnitude of these changes depends on the background diet being consumed with the n-3 PUFA supplements. For example, fish-oil supplementation in infants consuming cow milk leads to a different gut microbial composition from that of infants consuming fish-oil-supplemented formula[85]. Breast-feeding can also influence the degree to which fish-oil supplements affect the infant microbiome[86]. In mice, n-3 PUFA supplements fed postnatally cause a decreased ratio of Firmicutes:Bacteroidetes with an increase in anti-inflammatory microbes such as *Lactobacillus* spp. and *Bifidobacteria* spp[31]., the latter of which plays an important role in improving mucosal barrier function[87]. Additionally, n-3 PUFA supplementation promotes the growth of beneficial *Enterococcus faecium* while simultaneously preventing blooms of pathogenic members of the Enterobacteriaceae family[31]. Furthermore, n-3 PUFA supplementation down-regulates lymphocyte proliferation, antigen presentation, and pro-inflammatory cytokine expression[88]. These findings correspond to the discovery that n-3 PUFA from fish-oil-supplemented mice on an n-6 PUFA-rich diet, from corn oil, are unable to mount an appropriate immune response when challenged with a *Citrobacter rodentium* infection model of colitis as a result of impaired immunity resulting in sepsis[31]. Likewise, mice supplemented with n-3 PUFA from fish oil have an exacerbated DSS-induced colitis in addition to a further decrease in adiponectin expression, which is involved in regulating glucose levels and lipid metabolism[89]. α-linolenic acid (ALA), a less commonly studied n-3 PUFA, also appears to exacerbate colitis. ALA is a precursor to the long-chain n-3 PUFA; EPA and DHA. As seen with the consumption of these long-chain fatty acids, an enriched diet of flaxseed oil (rich in α-linolenic acid) increased colitis severity in a chemically induced model of murine colitis[90]. While the flaxseed oil diet resulted in an increased availability of anti-inflammatory metabolites such as DHA and EPA, there were no anti-inflammatory effects in the infected mice. Overall, n-3 PUFA supplementation appears to decrease inflammation at the expense of impaired immune responses. There is also evidence that long-chain n-3 PUFA such as EPA, DHA, and arachidonic acid (AA) have a more pronounced effect on the physiology of the host than short-chain n-3 PUFA like ALA[71]. Therefore, further investigations might uncover different effects of the various types of n-3 PUFA postnatally.

29.10 The effect of maternal diet on offspring microbiota

Maternal dietary intake influences the ecology of the offspring gut microbiota and, by extension, their health[91]. The mother's genital tract contains microbes that are altered by dietary intake. Dysbiosis of vaginal microbiota, termed bacterial vaginosis, occurs when facultative and anaerobic bacteria such as *Gardnerella vaginalis*, *Atopobium vaginalis*, and species of *Leptotrichia, Prevotella,* and *Clostridales* replace the dominating *Lactobacillus* spp[92]. The loss of *Lactobacillus* spp. is associated with a 75% higher risk of preterm delivery[93], suggesting that diet-induced dysbiosis plays a part in preterm births. Dietary intake of alliums and dried fruits, which contain antimicrobial components, are associated with a reduced risk of spontaneous preterm delivery, possibly due to the promotion of beneficial microbes and reduction of microbes associated with preterm delivery such as *Streptococcus mutans*[94]. Maternal dietary habits during gestation and lactation can also structure and persistently alter the intestinal microbiota in the offspring as maternal microbes are transferred from the mother[95]. Maternal n-3 PUFA rich diets fed to pregnant rat dams during gestation and lactation result in higher concentrations of opportunistic pathogens associated with sepsis such as *Bilophila wadsworthia, Enterococcus faecium,* and *Bacteroides fragilis* in their pups[88]. Interestingly, however, these mice exhibit an overall compromised immune system including reduced levels of: CD8+ T cells, important in defending the host against pathogens; T regulatory cells, which modulate self-tolerance; and M2 macrophages, which are pivotal in resolution of acute inflammation. Similar to their mothers, these offspring also suffered from increased susceptibility to chemically induced colitis as adults since their crypt structures in addition to intestinal permeability were impaired[96]. These studies reveal that maternal-driven changes have long-lasting effects on the offspring's ability to respond to disease susceptibility later in life.

Dietary interventions in infants post-weaning can only partially correct maternal diet-driven dysbiosis during gestation and lactation. For example, primate maternal consumption of high-fat diets during gestation and lactation depletes Proteobacteria such as the *Epsilonproteobacteria Campylobacter* and *Helicobacter* spp. and enriches Firmicutes members *Dialister* spp. and *Ruminococcus* spp. of the Clostridia class[97]. The resulting microbial dysbiosis is only partially reversed in the offspring following a low-fat post-weaning diet. Overall, maternal high-fat diets during pregnancy and post-weaning both cause depletion in species richness and diversity in their offspring.

It is apparent that offspring are susceptible to metabolic programming *in utero* and during early postnatal periods[98]. This process occurs predominantly *in utero* and to a lesser extent during the developmental stages of infancy. Therefore, the influences of maternal diet and microbiota on neonatal microbial composition and immunity constitute a crucial dimension of development that warrants further investigation.

29.11 The effects of post-natal diet on the developing microbiota of neonates

Neonatal nutrition is essential during the initial development of microbial ecology[99]. Current research shows that breast-fed and formula-fed infants have distinct gut colonization patterns as well as susceptibility to intestinal diseases[100].

29.11.1 Breast milk

Breast milk is one of the most important elements in programming a neonate's immune system. Breastfeeding helps protect newborns from infection[100], and the effects from breastfeeding on a neonate's gut microbiota extends beyond cessation. This is exemplified in the fact that breast-fed babies have lower incidences of infection from diarrheal diseases than those formula-fed[101]. It is further believed that breastfeeding can protect against the development of allergies[102]. Breast milk acts as an intermediary between the maternal and offspring microbiota. During lactation, gut lymphocytes travel to the lactating mammary glands by means of peripheral blood and lymphatic system[103], giving rise to breast-milk leukocytes. During mouse pregnancy and lactation there is an increase in bacterial translocation from the gut to the breast milk, suggesting that changes to maternal microbial components by means of diet may affect breast milk composition. Breast milk provides infants with a myriad of factors that influence the neonate's microbial composition, such as lysozymes and lactoferrin[103]. It contains maternally generated antibodies and other immune-active compounds, including cytokines and antimicrobial enzymes, that are utilized by the infant[61]. The antimicrobial components in breast milk inhibit the growth of certain pathogenic bacteria such as *Streptococcus mutans* in the offspring[104]. Breast milk further contains growth factors that stimulate the maturation and development of the intestinal mucosa and the growth of beneficial bacteria[105]. In addition to these factors, breast milk also contains various bacterial species associated with the mother's microbiota, predominantly *Streptococcus* spp., *Enterococcus* spp., *Peptostreptococcus* spp., *Staphylococcus* spp., *Corynebacterium* spp., *Bifidobacteria* spp., and *Lactobacillus* spp[103,106]. The lactic acid bacteria such as *Lactobacillus gasseri*, *Lactobacillus rhamnosus*, and *Enterococus faecium* inhibit pathogenic bacteria in the infant through competitive exclusion and production of antimicrobial compounds such as hydrogen peroxide, bacteriocins, and organic acids[107–109]. *Bifidobacterium* spp. such as *B. breve*, *B. longum*, and *B. infantis* are also found in breast milk,[110] and may prime the immune system[111]. *B. longum* and *B. infantis* have been shown *in vitro* to down regulate growth of pathogenic bacteria such as *E. coli*, *C. perfringens*, *Campylobacter jejuni*, and *Entamoeba histolytica*[111,112].

Another important component of breast milk, oligosaccharides, which are digested in the distal gut of the newborn by microbes[113], have been shown to improve microbial diversity[67]. Oligosaccharides also promote initial colonization of specific subsets of beneficial microbes that protect against pathogenic bacteria[114]. It is also known that the composition of maternal breast milk changes over the course of lactation. For example, fat concentration increases with both duration of lactation and volume of milk delivered[115]. In humans there is considerable variation in breast-milk fat composition, which can also be influenced by the maternal diet. The consumption of foods containing trans-fatty acid (TFA) during lactation results in the incorporation of TFA in breast milk[116] and has adverse effects on infant development[117]. The highest rates of breast-milk TFA in the world are seen in southwestern regions of the US, an alarming 7-18%[116]. Canadian women in comparison have significantly less TFA content in breast milk as a result of differences in dietary intake[118].

Immediately following birth, *Weisella, Staphylococcus, Streptococcus, Lactococcus,* and *Leuconostoc* initially dominate the neonates' gut microbiota. *Veillonella, Leptotrichia,* and *Prevotella* spp. replace these bacteria after a month[103,119], corresponding to changes to the mother's milk composition. Thus, an increasingly popular area of nutrition is developing strategies to alter maternal gut microbiota through dietary intervention in order to optimize offspring health.

29.11.2 Formula

Despite the myriad of benefits of breast milk to early immune and microbial development, it is common in Western societies to replace breastfeeding with formula at an early age. Infant formula manufacturers aim to produce products that mimic the functional properties of breast milk and seek ways to improve the protective properties of infant formulas. At present, prebiotics (covered in Chapter 30) are being investigated as supplements to promote microbial health in formula-fed infants and make it resemble those in breast-fed infants[120,121]. These prebiotics, including galactooligosaccharides, lactulose, inulin, fructooligosaccharides, and polydextrose, typically contain combinations of non-digestible oligosaccharides similar to those found in breast milk. During the first day of life, formula-fed and breast-fed infants have similar microbial communities. They both have prevalent members of the family Enterobacteriacea, as well as *Streptococcus, Enterococcus* and *Saphylococcus* spp[122]. In contrast, the presence of anaerobic bacteria such as *Lactobacillus* spp., *Bifidobacteria* spp. and *Bacteroides* spp. is rare. Throughout the first week of life, there appear to be no dominating organisms in formula-fed infants, but they exhibit significantly less *Bifidobacteria* spp. than breast-fed infants. One study found 60−91% of the bacterial community in breast-fed infants to be comprised of *Bifidobacteria* spp., whereas formula-fed infants only harbored approximately 50% abundance[123–126]. As *Bifidobacteria* spp. can inhibit pathogen growth, this may explain the higher rates of morbidity seen in formula-fed infants[127]. In contrast, another study found only minor differences in the abundance of *Bifidobacteria* spp. between breast-fed and formula-fed infants, especially with the introduction of prebiotics, which significantly increased the presence of *Bifidobacteria* spp[128]. Additionally, while *Bifidobacteria* spp. are less in formula-fed infants, they remain the dominating species during the second month of life, even though the abundance of other bacteria such Enterobacteria, as well as *Enterococci* spp., *Lactobacilli* spp. *Bacteroides* spp. and members of the Clostridia class is significantly greater in formula-fed infants than those who are breast-fed[129].

Interestingly, formula-fed infants ted to have increased species richness and diversity[130] and have intestinal microbiota that closely resembles that of an adult, including increased levels of *B. longum* and *B. adolescentis,* compared to breast-fed infants, who have dominant *B. breve* and develop an adult microbiome over the course of the first two or three years[105]. For example, compared to breast-fed infants, formula-fed infants have gut microbial communities with significantly greater abundances of the family Verrucomicrobiaceae and Peptostreptococcacea, such as *Akkermansia* spp. and *C. difficile* spp. respectively[130], as well as *Bacteroides* spp., and pathogenic *Klebsiella* spp. and *Enterobacter* spp[128]. In addition, formula-fed mice pups show dominant Proteobacteria and Bacteroidetes phyla compared to maternally fed pups whose guts are dominated by Firmicutes[131,132]. At the genus level, formula-fed mice have an increased abundance of *Serratia* spp. and *Lactococcus* spp., while breast-fed mice have an increase in *Lactobacillus* spp. These microbial communities, specifically the decreased abundance of *Lactobacillus* spp., could presumably increase vulnerability to disease in formula-fed mice[131]. Dietary choices during fetal life and early infancy modify immune and metabolic programming, which has important implications for later health and risk of susceptibility to immune-mediated inflammatory diseases[98].

29.12 Conclusion

The tripartite relationship among humans, their environment, and the plethora of microorganisms residing in their gut has significant implications for host health. Of all the exogenous factors implicated in structuring the phylogenetic makeup of the microbiota, diet is the most extensively studied[131]. Currently, long-term dietary intake has been changing worldwide due to alterations in agricultural practices, population growth, lifestyle, and socioeconomic status[61]. This has wide-ranging implications for gut microbiota and health as new research consistently links dysbiosis to various chronic diseases and infections. As such, specialized diets aimed at reversing dysbiosis have become an increasingly popular topic of research. With advances in technology and reduced costs of metagenomics sequencing techniques, the concept of personalized diets that embrace the unique microbial makeup of individuals is within reach. It is conceivable that in the not-so-distant future, determining the microbial composition of individuals will become common practice in a medical context. Just as blood types are invaluable in patient medical profiles, routine microbiota analysis will aid in predicting various chronic diseases as well as determining proper dietary prescriptions optimized for restoring and maintaining health.

TAKE-HOME MESSAGE

- Human gut microbiota composition is reflective of evolutionary adaptations in response to the surrounding biosphere and environmental challenges.
- Dietary components play an extremely important role in shaping gut microbial communities and may be an essential tool in treating dysbiosis-associated diseases.
- Maternal dietary habits have long-lasting impacts on offspring's microbiota and health.
- Evidence has emerged showing that gut microbiota interacts with central and peripheral neural processes, suggesting their role in hormone signaling as well as psychological health.

References

1 Mshvildadze M *et al.* Intestinal microbial ecology in premature infants assessed with non-culture-based techniques. *The Journal of Pediatrics* 2010; **156**: 20–25.

2 Hu J *et al.* Diversified microbiota of meconium is affected by maternal diabetes status. *PLoS One* 2013; **8**: e78257.

3 Schnorr SL *et al.* Gut microbiome of the Hadza hunter-gatherers. *Nature Communications* 2014; **5**: 3654.

4 Giacani L, Lukehart SA. The endemic treponematoses. *Clin Microbiol Rev* 2014; **27**: 89–115.

5 Gibson GL, Roberfroid MB. Dietary modulation of the human colonic microbiota: introducing the concept of prebiotics. *J Nutrition* 1995; **125**: 1401–12.

6 Blurton Jones NG, Smith LC, O'Connell JF, Hawkes K, Kamuzora CL. Demography of the Hadza, an increasing and high density population of Savanna foragers. *Am J Phys Anthropol* 1992; **89**: 159–181.

7 Bennett FJ, Barnicot NA, Woodburn JC, Pereira MS, Henderson BE. Studies on viral, bacterial, rickettsial and treponemal diseases in the Hadza of Tanzania and a note on injuries. *Hum Biol* 1973; **45**: 243–272.

8 De Filippo C *et al.* Impact of diet in shaping gut microbiota revealed by a comparative study in children from Europe and rural Africa. *Proceedings of the National Academy of Sciences USA* 2010; **107**: 14691–14696.

9 Sobhani I *et al.* Microbial dysbiosis in colorectal cancer (CRC) patients. *PLoS One* 2011; **6**: e16393.

10 Jackson RL, Greiwe JS, Schwen RJ. Emerging evidence of the health benefits of S-equol, an estrogen receptor beta agonist. *Nutr Rev* 2011; **69**: 432–448.

11 Setchell KD, Brown NM, Lydeking-Olsen E. The clinical importance of the metabolite equol-a clue to the effectiveness of soy and its isoflavones. *J Nutr* 2002; **132**: 3577–3584.

12 Hehemann JH *et al.* Transfer of carbohydrate-active enzymes from marine bacteria to Japanese gut microbiota. *Nature* 2010: **464**: 908–912.

13 Adler CJ *et al.* Sequencing ancient calcified dental plaque shows changes in oral microbiota with dietary shifts of the Neolithic and Industrial revolutions. *Nat Genet* 2013; **45**: 450–455, 455e451.

14 Bloomfield SF, Stanwell-Smith R, Crevel RW, Pickup J. Too clean, or not too clean: the hygiene hypothesis and home hygiene. *Clin Exp Allergy* 2006; **36**: 402–425.

15 Riedler J *et al.* Exposure to farming in early life and development of asthma and allergy: a cross-sectional survey. *Lancet* 2001; **358**: 1129–1133.

16 Von Ehrenstein, OS *et al.* Reduced risk of hay fever and asthma among children of farmers. *Clin Exp Allergy* **30**, 187–193 (2000).

17 Eckburg PB *et al.* Diversity of the human intestinal microbial flora. *Science* **308**, 1635-1638 (2005).

18 Lozupone CA, Stombaugh JI, Gordon JI, Jansson JK, Knight R. Diversity, stability and resilience of the human gut microbiota. *Nature* 2012; **49**: 220–230.

19 Costello EK *et al.* Bacterial community variation in human body habitats across space and time. *Science* 2009; **326**: 1694–1697.

20 Walker AW *et al.* Dominant and diet-responsive groups of bacteria within the human colonic microbiota. *ISME J* 2011; **5**: 220–230.

21 Wu GD *et al.* Linking long-term dietary patterns with gut microbial enterotypes. *Science* 2011; **334**: 105–109 (2011).

22 Yatsunenko T *et al.* Human gut microbiome viewed across age and geography. *Nature* 2012; **486**: 222–227.

23 David LA *et al.* Diet rapidly and reproducibly alters the human gut microbiome. *Nature* 2014; **505**: 559–563.

24 Johansson ME, Larsson JM, Hansson GC. The two mucus layers of colon are organized by the MUC2 mucin, whereas the outer layer is a legislator of host-microbial interactions. *Proceedings of the National Academy of Sciences USA* 2011; **108 Suppl 1**: 4659–4665.

25 Franchi L *et al.* NLRC4-driven production of IL-1beta discriminates between pathogenic and commensal bacteria and promotes host intestinal defense. *Nat Immunol* 2012; **13**: 449–456.

26 Rist VT, Weiss E, Eklund M, Mosenthin R. Impact of dietary protein on microbiota composition and activity in the gastrointestinal tract of piglets in relation to gut health: a review. *Animal* 2013; **7**: 1067–1078.

27 Mai V, McCrary QM, Sinha R, Glei M. Associations between dietary habits and body mass index with gut microbiota composition and fecal water genotoxicity: an observational study in African American and Caucasian American volunteers. *Nutr J* 2009; **8**: 49.

28 Ong DK *et al.* Manipulation of dietary short chain carbohydrates alters the pattern of gas production and genesis of symptoms in irritable bowel syndrome. *J Gastroenterol Hepatol* 2010; **25**: 1366–1373.

29 Devkota S *et al.* Dietary-fat-induced taurocholic acid promotes pathobiont expansion and colitis in Il10-/- mice. *Nature* 2012; **487**: 104–108.

30 Investigators, I.B.D.i.E.S. *et al.* Linoleic acid, a dietary n-6 polyunsaturated fatty acid, and the aetiology of ulcerative colitis: a nested case-control study within a European prospective cohort study. *Gut* 2009; **58**: 1606–1611.

31 Ghosh S *et al.* Fish oil attenuates omega-6 polyunsaturated fatty acid-induced dysbiosis and infectious colitis but impairs LPS dephosphorylation activity causing sepsis. *PloS One* 2013; **8**: e55468–e55468.

32 Jumpertz R *et al.* Energy-balance studies reveal associations between gut microbes, caloric load, and nutrient absorption in humans. *Am J Clin Nutr* 2011; **94**: 58–65.

33 Everard A *et al.* Responses of gut microbiota and glucose and lipid metabolism to prebiotics in genetic obese and diet-induced leptin-resistant mice. *Diabetes* 2011; **60**: 2775–2786.

34 van Nood E *et al.* Duodenal infusion of donor feces for recurrent Clostridium difficile. *N Engl J Med* 2013; **368**: 407–415.

35 Jonkers D, Penders J, Masclee A, Pierik M. Probiotics in the management of inflammatory bowel disease: a systematic review of intervention studies in adult patients. *Drugs* 2012; **72**: 803–823.

36 Naidoo K, Gordon M, Fagbemi AO, Thomas AG, Akobeng AK. Probiotics for maintenance of remission in ulcerative colitis. *Cochrane Database Syst Rev* 2011; CD007443.

37 Sinagra E *et al.* Probiotics, prebiotics and symbiotics in inflammatory bowel diseases: state-of-the-art and new insights. *J Biol Regul Homeost Agents* 2013; **27**: 919–933.

38 Gibson MK, Pesesky MW, Dantas G. The yin and yang of bacterial resilience in the human gut microbiota. *J Mol Biol* 2014; **426**: 3866–3876.

39 McNeil NI. The contribution of the large intestine to energy supplies in man. *Am J Clin Nutr* 1984; **39**: 338–342.

40 Van Wey AS,*et al.* Bacterial biofilms associated with food particles in the human large bowel. *Mol Nutr Food Res* 2011; **55**: 969–978.

41 Ze X, Duncan SH, Louis P, Flint HJ. Ruminococcus bromii is a keystone species for the degradation of resistant starch in the human colon. *ISME J* 2012; **6**: 1535–1543.

42 Pomare EW, Branch WJ, Cummings JH. Carbohydrate fermentation in the human colon and its relation to acetate concentrations in venous blood. *J Clin Invest* 1985; **75**: 1448–1454.

43 Hamer HM *et al.* Review article: the role of butyrate on colonic function. *Aliment Pharmacol Ther* 2008; **27**: 104–119.

44 Scheppach W. Effects of short chain fatty acids on gut morphology and function. *Gut* 1994; **35**: S35–38.

45 Gao Z *et al.* Butyrate improves insulin sensitivity and increases energy expenditure in mice. *Diabetes* 2009; **58**: 1509–1517.

46 Sleeth ML, Thompson L, Ford HE, Zac-Varghese SE, Frost G. Free fatty acid receptor 2 and nutrient sensing: a proposed role for fibre, fermentable carbohydrates and short-chain fatty acids in appetite regulation. *Nutr Res Rev* **23**, 135–145 (2010).

47 Duncan SH, Louis P, Flint HJ. Lactate-utilizing bacteria, isolated from human feces, that produce butyrate as a major fermentation product. *Applied and Environmental Microbiology* 2004; **70**: 5810–5817.

48 Windey K, De Preter V, Verbeke K. Relevance of protein fermentation to gut health. *Mol Nutr Food Res* 2012; **56**: 184–196.

49 Smith EA, Macfarlane GT. Enumeration of amino acid fermenting bacteria in the human large intestine: effects of pH and starch on peptide metabolism and dissimilation of amino acids. *FEMS Microbiology Ecology* 1998; **25**: 355–368.

50 Gilbert JA, Bendsen NT, Tremblay A, Astrup A. Effect of proteins from different sources on body composition. *Nutrition, Metabolism, and Cardiovascular Diseases: NMCD* 2011; **21 Suppl 2**: B16–31.

51 Macfarlane GT, Macfarlane S. Bacteria, colonic fermentation, and gastrointestinal health. *J AOAC Int* 2012; **95**: 0–60.

52 Silvester KR, Cummings JH. Does digestibility of meat protein help explain large bowel cancer risk? *Nutr Cancer* 1995; **24**: 279–288.

53 Davis CD, Milner JA. Gastrointestinal microflora, food components and colon cancer prevention. *J Nutr Biochem* 2009; **20**: 743–752.

54 Blachier F, Mariotti F, Huneau JF, Tome D. Effects of amino acid-derived luminal metabolites on the colonic epithelium and physiopathological consequences. *Amino Acids* 2007; **33**: 547–562.

55 Ran-Ressler RR, Devapatla S, Lawrence P, Brenna JT. Branched chain fatty acids are constituents of the normal healthy newborn gastrointestinal tract. *Pediatr Res* 2008; **64**: 605–609.

56 Lewis S, Cochrane S. Alteration of sulfate and hydrogen metabolism in the human colon by changing intestinal transit rate. *The American Journal of Gastroenterology* 2007; **102**: 624–633.

57 Roediger WE, Moore J, Babidge W. Colonic sulfide in pathogenesis and treatment of ulcerative colitis. *Dig Dis Sci* 1997; **42**: 1571–1579.

58 Semova I *et al*. Microbiota regulate intestinal absorption and metabolism of fatty acids in the zebrafish. *Cell Host & Microbe* 2012; **12**: 259–261.

59 Velagapudi VR *et al*. The gut microbiota modulates host energy and lipid metabolism in mice. *Journal of Lipid Research* 2010; **51**: 1101–1112.

60 Bäckhed F, Manchester JK, Semenkovich CF, Gordon JI. Mechanisms underlying the resistance to diet-induced obesity in germ-free mice. *Proceedings of the National Academy of Sciences USA* 2007; **104**: 979–984.

61 Kau AL, Ahern PP, Griffin NW, Goodman AL, Gordon JI. Human nutrition, the gut microbiome and the immune system. *Nature* 2011; **474**: 327–336.

62 Arumugam M *et al*. Enterotypes of the human gut microbiome. *Nature* 2011; **473**: 174–180.

63 Knights D *et al*. Rethinking "enterotypes". *Cell Host Microbe* 2014; **16**: 433–437.

64 Russell W *et al*. High-protein, reduced-carbohydrate weight-loss diets promote metabolite profiles likely to be detrimental to colonic health. *The American Journal of Clinical Nutrition* 2011; **93**: 1062–1072.

65 Kumari R, Ahuja V, Paul J. Fluctuations in butyrate-producing bacteria in ulcerative colitis patients of North India. *World Journal of Gastroenterology* 2013; **19**: 3404–3414.

66 Hentges DJ, Maier BR, Burton GC, Flynn A, Tsutakawa RK. Effect of a high-beef diet on the fecal bacterial flora of humans. *Cancer Res* 1977; **37**: 568–571.

67 Thum C *et al*. Can nutritional modulation of maternal intestinal microbiota influence the development of the infant gastrointestinal tract? *The Journal of Nutrition* 2012; **142**: 1921–1928.

68 Turnbaugh PJ *et al*. An obesity-associated gut microbiome with increased capacity for energy harvest. *Nature* 2006; **444**: 1027–1031.

69 Ley RE, Turnbaugh PJ, Klein S, Gordon JI. Microbial ecology: human gut microbes associated with obesity. *Nature* 2006; **444**: 1022–1023.

70 Li M *et al*. Gut carbohydrate metabolism instead of fat metabolism regulated by gut microbes mediates high-fat diet-induced obesity. *Beneficial Microbes* 2014; **5**: 335–344.

71 Liu T, Hougen H, Vollmer AC, Hiebert SM. Gut bacteria profiles of Mus musculus at the phylum and family levels are influenced by saturation of dietary fatty acids. *Anaerobe* 2012; **18**: 331–337.

72 de Wit N *et al*. Saturated fat stimulates obesity and hepatic steatosis and affects gut microbiota composition by an enhanced overflow of dietary fat to the distal intestine. *American Journal of Physiology. Gastrointestinal and Liver Physiology* 2012; **303**: G589–599.

73 Simoes CD, Maukonen J, Kaprio J, Rissanen A, Pietilainen, KH. Habitual dietary intake is associated with stool microbiota composition in monozygotic twins 1–3. *J Nutr* 2013; **143**: 417–423.

74 Kurdi P, Kawanishi K, Mizutani K, Yokota A. Mechanism of growth inhibition by free bile acids in lactobacilli and bifidobacteria. *Journal of Bacteriology* 2006; **188**: 1979–1986.

75 Islam KB *et al*. Bile acid is a host factor that regulates the composition of the cecal microbiota in rats. *Gastroenterology* 2011; **141**: 1773–1781.

76 Huang S *et al.* Saturated fatty acids activate TLR-mediated proinflammatory signaling pathways. *Journal of Lipid Research* 2012; **53**: 2002–2013.

77 Ebaid GMX, Seiva FRF, Rocha KKHR, Souza GA, Novelli ELB. Effects of olive oil and its minor phenolic constituents on obesity-induced cardiac metabolic changes. *Nutrition Journal* 2010; **9**: 46–46.

78 Sánchez-Fidalgo S *et al.* Extra-virgin olive oil-enriched diet modulates DSS-colitis-associated colon carcinogenesis in mice. *Clinical Nutrition* 2010; **29**: 663–673.

79 Fava F *et al.* The type and quantity of dietary fat and carbohydrate alter faecal microbiome and short-chain fatty acid excretion in a metabolic syndrome 'at-risk' population. *International Journal of Obesity (2005)* 2013; **37**: 216–223.

80 Neggers YH *et al.* Dietary intake of selected nutrients affects bacterial vaginosis in women. *J Nutr* 2007; **137**: 2128–33.

81 Hekmatdoost, A, Wu X, Morampudi V, Innis SM, Jacobson K. Dietary oils modify the host immune response and colonic tissue damage following Citrobacter rodentium infection in mice. *American Journal of Physiology. Gastrointestinal and Liver Physiology* 2013; **304**: G917–928.

82 Feehley T, Stefka AT. Cao S, Nagler CR. Microbial regulation of allergic responses to food. *Seminars in Immunopathology* 2012; **34**: 671–688.

83 Calder PC. n−3 Polyunsaturated fatty acids and inflammation: from molecular biology to the clinic. *Lipids* 2003; **38**: 343–352.

84 Simmer, K. Longchain polyunsaturated fatty acid supplementation in infants born at term (review). Cochrane Neonatal Group 2001.

85 Nielsen, S, Nielsen DS, Lauritzen L, Jakobsen M, Michaelsen KF. Impact of diet on the intestinal microbiota in 10-month-old infants. *Journal of Pediatric Gastroenterology and Nutrition* 2007; **44**: 613–618.

86 Andersen AD, Mølbak L, Michaelsen KF, Lauritzen L. Molecular fingerprints of the human fecal microbiota from 9 to 18 months old and the effect of fish oil supplementation. *Journal of Pediatric Gastroenterology and Nutrition* 2011; **53**: 303–309.

87 Cani PD, Bibiloni R, Knauf C, Neyrinck AM. Changes in gut microbiota control metabolic endotoxemia-induced inflammation in high-fat diet-induced obesity and diabetes in mice. *Diabetes* 2008; **57**: 1470–81.

88 Gibson, D *et al.* Maternal exposure to fish oil primes offspring to harbor intestinal pathobionts associated with altered immune cell balance. *Gut Microbes* 2015; 1–9 (2015).

89 Matsunaga, H *et al.* Omega-3 fatty acids exacerbate DSS-induced colitis through decreased adiponectin in colonic subepithelial myofibroblasts. *Inflammatory Bowel Diseases* 2008; **14**: 1348–1357.

90 Zarepoor L *et al.* Dietary flaxseed intake exacerbates acute colonic mucosal injury and inflammation induced by dextran sodium sulfate. *American Journal of Physiology. Gastrointestinal and Liver Physiology* 2014; **306**: G1042–1055

91 Palmer C, Bik EM, DiGiulio DB, Relman DA, Brown PO. Development of the human infant intestinal microbiota. *PLoS Biology* 2007; **5**: e177–e177.

92 de Andrade Ramos B, Kanninen TT, Sisti G, Witkin SS. Microorganisms in the female genital tract during pregnancy: tolerance versus pathogenesis. *American Journal of Reproductive Immunology (New York, N.Y.:1989)* 2014; 1–7.

93 Donders GG *et al.* Predictive value for preterm birth of abnormal vaginal flora, bacterial vaginosis and aerobic vaginitis during the first trimester of pregnancy. *BJOG: an International Journal of Obstetrics and Gynaecology* 2009; **116**: 1315–1324.

94 Myhre R *et al.* Intakes of garlic and dried fruits are associated with lower risk of spontaneous preterm delivery. *J Nutr* 2013; **143**:1100–1108.

95 Becker N, Kunath J, Loh G, Blaut M. Human intestinal microbiota: characterization of a simplified and stable gnotobiotic rat model. *Gut Microbes* 2011; **2**: 25–33.

96 Innis SM, Dai C, Wu X, Buchan AMJ, Jacobson K. Perinatal lipid nutrition alters early intestinal development and programs the response to experimental colitis in young adult rats. *American Journal of Physiology. Gastrointestinal and Liver Physiology* 2010; **299**: G1376–1385.

97 Ma J *et al.* High-fat maternal diet during pregnancy persistently alters the offspring microbiome in a primate model. *Nature Communications* 2014; **5**: 3889–3889.

98 Collado MC, Rautava S, Isolauri E, Salminen S. Gut microbiota: source of novel tools to reduce the risk of human disease? *Pediatric Research* 2014; 1–7 (2014).

99 Chan YK, Estaki M, Gibson DL. Clinical consequences of diet-induced dysbiosis. *Annals of nutrition & metabolism* 2013; **63 Suppl 2**, 28–40.

100 Wright AL, Bauer M, Naylor A, Sutcliffe E, Clark L. Increasing breastfeeding rates to reduce infant illness at the community level. *Pediatrics* 1998; **101**: 837–44.

101 Long KZ *et al.* Proportional hazards analysis of diarrhea due to enterotoxigenic Escherichia coli and breast feeding in a cohort of urban Mexican children. *American Journal of Epidemiology* **139**: 193–205.

102 Kramer MS. Breastfeeding and allergy: the evidence. *Annals of Nutrition & Metabolism* 2011; **59 Suppl 1**: 20–26.

103 Donnet-Hughes A *et al.* Potential role of the intestinal microbiota of the mother in neonatal immune education. *The Proceedings of the Nutrition Society* 2010; **69**: 407–415 (2010).

104 Holgerson P *et al.* Oral microbial profile discriminates breast-fed from formula-fed infants. *J Pediatr Gastroenterol Nutr* 2013; **56**: 127–136.

105 Mackie RI, Sghir A, Gaskins HR. Developmental microbial ecology of the neonatal gastrointestinal tract. *American Journal of Clinical Nutrition* **69** (1999).

106 Solís G, de Los Reyes-Gavilan CG, Fernández N, Margolles A, Gueimonde M. Establishment and development of lactic acid bacteria and bifidobacteria microbiota in breast-milk and the infant gut. *Anaerobe* 2010; **16**: 307–310.

107 Heikkila MP, Saris PE. Inhibition of Staphylococcus aureus by the commensal bacteria of human milk. *Journal of Applied Microbiology* 2008; **95**: 471–478.

108 Beasley SS, Saris PE. Nisin-producing Lactococcus lactis strains isolated from human milk. *Applied and Environmental Microbiology* 2004; **70**: 5051–5053.

109 Martin R *et al.* Human milk is a source of lactic acid bacteria for the infant gut. *The Journal of Pediatrics* 2003; **143**: 754–758.

110 Favier CF, de Vos WM, Akkermans, ADL. Development of bacterial and bifidobacterial communities in feces of newborn babies. *Anaerobe* 2003; **9**: 219–229.

111 Newburg DS, Morelli L. Human milk and infant intestinal mucosal glycans guide succession of the neonatal intestinal microbiota. *Pediatric Research* 2014; **1–6**.

112 Gibson GR, Wang X. Regulatory effects of bifidobacteria on the growth of other colonic bacteria. *Journal of Applied Bacteriology* 1994; **77**: 412–420.

113 Marcobal, A *et al.* Bacteroides in the infant gut consume milk oligosaccharides via mucus-utilization pathways. *Cell Host Microbe* 2011; **10**: 507–514.

114 Hopkins MJ, Macfarlane GT. Nondigestible oligosaccharides enhance bacterial colonization resistance against Clostridium difficile in vitro. *Applied and Environmental Microbiology* 2003; **69**: 1920–1927.

115 Michaelsen KF, Skafte L, Badsberg JH, Jorgensen M. Variation in macronutrients in human bank milk: influencing factors and implications for human milk banking. *J Pediatr Gastroenterol Nutr* 1990; **11**: 229–39.

116 Mosley EE, Wright AL, McGuire MK, McGuire MA. Trans fatty acids in milk produced by women in the United States 1–3. *Am J Clin Nutr* 2005; **82**: 1292–1297.

117 Hornstra G, van Eijsden M, Dirix C, Bonsel G. Trans fatty acids and birth outcome: some first results of the MEFAB and ABCD cohorts. *Atherosclerosis. Supplements* 2006; **7**, 21–23.

118 Friesen R, Innis SM. Trans fatty acids in human milk in Canada declined with the introduction of trans fat food labeling. *J Nutr* 2006; **136**: 2558–2561.

119 Cabrera-Rubio R *et al.* The human milk microbiome changes over lactation and is shaped by maternal weight and mode of delivery 1–4. *Am J Clin Nutr* 2012; **96**: 544–551.

120 Ghisolfi J. Dietary fibre and prebiotics in infant formulas. *Proceedings of the Nutrition Society* 2007; **62**: 183–185.

121 Nakamura N *et al.* Molecular ecological analysis of fecal bacterial populations from term infants fed formula supplemented with selected blends of prebiotics. *Applied and Environmental Microbiology* 2009; **75**: 1121–1128.

122 Fanaro S, Chierici R, Guerrini P, Vigi V. Intestinal microflora in early infancy: composition and development. *Acta Paediatrica* 2003; **91**: 48–55.

123 Donovan SM *et al.* Host-microbe interactions in the neonatal intestine: role of human milk oligosaccharides. *Advances in Nutrition (Bethesda, Md.)* 2012; **3**: 450S–455S.

124 Harmsen HJ *et al.* Analysis of intestinal flora development in breast-fed and formula-fed infants by using molecular identification and detection methods. *Journal of Pediatric Gastroenterology and Nutrition* 2000; **30**: 61–67.

125 Koenig JE *et al.* Succession of microbial consortia in the developing infant gut microbiome. *Proceedings of the National Academy of Sciences USA* 2011; **108 Suppl 1**: 4578–4585.

126 Boesten R *et al.* Bifidobacterium population analysis in the infant gut by direct mapping of genomic hybridization patterns: potential for monitoring temporal development and effects of dietary regimens. *Microbial Biotechnology* 2011; **4**: 417–427.

127 Dewey KG, Heinig MJ, Nommsen-Rivers LA. Differences in morbidity between breast-fed and formula-fed infants. *The Journal of Pediatrics* 1995; **126**: 696–702.

128 Adlerberth I, Wold AE. Establishment of the gut microbiota in Western infants. *Acta Paediatrica (Oslo, Norway: 1992)* 2009; **98**; 229–238.

129 Benno Y, Sawada K, Mitsuoka T. The intestinal microflora of infants: composition of fecal flora in breast-fed and bottle-fed infants. *Microbiology and Immunology* 1984; **28**: 975–986.

130 Azad MB *et al.* Gut microbiota of healthy Canadian infants: profiles by mode of delivery and infant diet at 4 months. *CMAJ : Canadian Medical Association journal = journal de l'Association medicale canadienne* 2013; **185**: 385–394.

131 Carlisle EM *et al.* Murine gut microbiota and transcriptome are diet dependent. *Annals of surgery* 2013; **257**: 287–294.

132 LeBlanc JG. *et al.* Bacteria as vitamin suppliers to their host: a gut microbiota perspective. *Current opinion in biotechnology* 2013; **24**: 160–168.

CHAPTER 30

Probiotics and prebiotics: what are they and what can they do for us?

Marie-José Butel, Anne-Judith Waligora-Dupriet

Université Paris Descartes, Sorbonne Paris, Paris, France

30.1 The gut microbiota, a partnership with the host

Previous sections of this book have highlighted that the human gut is an extremely complex ecosystem where microbiota, nutrients, and host cells interact extensively. Many studies have reported the health benefits of the commensal gut microbiota, which is an active element of the gut physiology that displays many functions. Hence, bacterial imbalance, called dysbiosis, is currently involved in many diseases, included allergy, inflammatory bowel diseases, obesity, and metabolic diseases. This relationship between gut microbiota and health has raised the interest in modulation of this ecosystem through the administration of pre- or probiotics for prevention, and even cure, of some diseases.

30.2 Probiotics

30.2.1 Probiotics, a story that began a long time ago

Probiotics are currently defined as live microorganisms that when consumed in adequate amounts confer a health benefit on the host[1]. The idea of beneficial bacteria is very old; several texts mentioning the use of fermented milk — therefore containing bacteria — to treat gastroenteritis. In Genghis Khan's era, in the 12th century, fermented milk was considered as a source of strength and health, and Mongolian women sprayed horses and riders with it to protect them during battles. However, the era of probiotics really started at the beginning of the 20th century with Elie Metchnikoff, who correlated the longevity of Bulgarians with their high consumption of fermented milk[2]. At the same time, Henry Tissier, a French pediatrician, observed that infants suffering from diarrhea had few Y-shaped Gram-positive rods in their stools whereas these bacteria were dominant in healthy infants' stools. His observations led him to suggest giving to the diarrheic children "1 to 2 glasses of a culture of *Bacillus acidiparalactici* or a synbiosis of this species with *Bacillus bifidus*," so as to favor a flora with preventive functions

The Human Microbiota and Chronic Disease: Dysbiosis as a Cause of Human Pathology, First Edition.
Edited by Luigi Nibali and Brian Henderson.
© 2016 John Wiley & Sons, Inc. Published 2016 by John Wiley & Sons, Inc.

against pathogens[3]. Metchnikoff's observation of the health benefit of fermented milk and its bioactive component, i.e. the bacteria, led him to argue that "all microbes are not dangerous for health" and that it could be beneficial to replace the "putrid flora" by enriching the normal flora with bacteria able to ferment glucose and weakly proteolytic, such as lactic bacteria[2]. This concept of beneficial bacteria was forgotten during the era of antibiotics and vaccination; however, studies on the commensal microbiota and its relationship with health have renewed interest in them.

30.2.2 What are probiotics?

The word *probiotic* was first used in 1965 by Lilly and Stillwell, as opposed to the word antibiotic, to qualify a "microbial substance able to stimulate the growth of another microorganism"[4]. At that time the notion of probiotics was restricted to live microorganisms and defined as "live microorganisms with health benefits on the host by modifying the equilibrium of the gut microbiota." Currently, the modulation of the gut microbiota is no longer part of the definition, probiotics being able to act through various mechanisms. They are therefore defined as "live microorganisms which, when consumed in adequate amounts, confer a health effect on the host"[1]. Numerous studies, either *in vitro*, or in animal models, or through clinical trials, have proven their potential in terms of health benefits regarding prevention of many diseases.

The characteristics of bacterial strains to be considered as probiotics have been defined[1,5]. First, these strains must reach their site of action, usually the gut. For this, they have to survive the physiological stresses met during ingestion, which include gastric and gut acidity and the presence of biliary salts. Second, they must have proven beneficial effect. Third, they must be without any risk for the host. Finally, they must keep all their characteristics and remain stable throughout the manufacturing process and storage in the matrix in which they are incorporated.

The most frequent bacterial genera used as probiotics are lactic bacteria, mainly *Lactobacillus* and *Bifidobacterium*. Probiotic microorganisms were mainly isolated from fermented milk, non-fermented food such as meat, sausage, or vegetables, and the gut microbiota of human or animals. Some probiotic strains are not lactic bacteria, such as for example *Escherichia coli*, *Propionibacterium*, or *Enterococcus*. However, they are less frequently used, in particular due to their greater potential risk of adverse effects. Besides bacteria, the yeast *Saccharomyces boulardii* has also been used for several decades.

30.2.3 How do probiotics work?

The mechanisms of action of probiotics are not always well established. This is a point considered by the European Food Safety Authority (EFSA), which recently rejected the health claims of marketed probiotics for lack of sufficient evidence[6]. The mode of action of probiotics can be related to the modulation of the host's microbiota. One of the modes of action is the improvement of the "barrier effect," also called resistance to colonization, exerted against exogenous bacteria, thus preventing or limiting colonization by pathogenic bacteria. The bacterial inhibition can be due to the production of either bacteriocins or metabolites such as short-chain fatty acids leading to a decrease of pH unfavorable to the bacterial growth, or even the presence of biosurfactants with an antimicrobial activity.

This barrier effect can also act via other mechanisms such as competition with pathogens for binding sites, inhibition of adhesion, and so on.

The second mode of action involves the improvement of the intestinal mucosal barrier. This function is related to the quality of the tight junctions between the intestinal epithelial cells. Other elements also participate in this barrier function, such as Paneth cells, by producing antimicrobial peptides (defensins, lysozyme) and mucus cells by secretion of mucus, providing a protective layer that prevents any direct contact with the luminal gut bacteria. Probiotics can activate the signaling pathways leading to an increase in the mucus layer, production of defensins, as well as the proteins of the tight junctions.

The third mode of action is modulation of the immune system. In chapter 23 we learned that more than 70% of immune cells are located in the gut, especially in the mucosa of the small bowel, making up the gut associated lymphoid tissue (GALT). Peyer patches — specific sites consisting of a follicular center and covered by M cells associated with the epithelium — are a major route for passage of antigens. The activation of the immune response requires the recognition of specific receptors of innate immune cells (dendritic cells, macrophages and epithelial cells). These receptors, called pattern recognition receptors (PRR), include mostly Toll-like receptors (TLRs) and nucleotide-binding oligomerization domains (NODs). They are recognized by bacterial components called microbial associated molecular patterns (MAMPs), and therefore interact with the GALT[7]. The consequences are activation of regulatory T cells and T helper (Th) lymphocyte differentiation into Th1 and Th2 cells, leading to production of either pro- and/or anti-inflammatory cytokines. Bacteria with probiotic potential may have different effects depending on the cytokines profile stimulated[8]. The effects can be local because they are limited to stimulation at the gut level (induction of secretory IgA production, for instance) or may be systemic. Probiotic strains can also have beneficial effects by providing enzymes such as beta-galactosidase and are thus able to improve gastrointestinal symptoms after lactose ingestion in lactose-intolerant patients[9].

Whatever the mechanisms of action, probiotic strains act *via* (i) their components such as DNA (especially CpG motifs), peptidoglycan, LPS, flagellin, and/or (ii) their metabolites, especially short-chain fatty acids. Their mode of action can be linked to their capacity to colonize the gut or their capacity to modulate the gut microbiota, leading to an increase level in beneficial bacteria. The probiotic effects depend on the strain. Hence, probiotic strains must be clearly identified and characterized based on both phenotypic and genotypic characteristics. Furthermore, they must be safe for the human population.

30.2.4 Safety of probiotics

Probiotics, especially bifidobacteria and lactobacilli, have a long history in terms of safety through their historical use in food and fermented milk. Moreover, these bacteria are frequently encountered in the environment, where they act as commensal bacteria in plants, animals, and humans. Lactobacilli, bifidobacteria, lactococci, and yeasts are classified in the category of organisms "Generally Regarded as Safe (GRAS)." However, a few cases of infections have been reported, but mostly in immunocompromised patients[10]. Not all probiotics belong to the GRAS category, examples being enterobacteria or enterococci, although some strains of these organisms are used as probiotics[11]. There are different theoretical risks related to the use of probiotics. First, probiotics may be responsible for infections due to a bacterial translocation, defined as the passage from the gut lumen to extra-intestinal sites, a first step leading to infection. Few authors have analyzed this risk. In fact, they have instead focused on the effect of probiotic strains, some strains having the capacity to decrease their translocation, leading to a protective effect. Few probiotic strains have been a source of infection. Bacteremia, endocarditis, or abscesses have been reported with the probiotic strain *Lactobacillus* GG[12]. However, these cases were only observed in patients with risks factors such as short-bowel syndrome or patients harboring a central venous catheter. Fungemia with the probiotic yeast *Saccharomyces boulardii* has been reported[13], but it has been linked to catheter colonization when opening the probiotic pack rather than to translocation from the gut[14].

Other negative effects of probiotics could be related to the production of potentially toxic metabolites, such as D-lactate responsible for lactic acidosis. Indeed, in infants suffering from short-bowel syndrome, either a gut microbiota enriched in lactobacilli or administration of probiotic lactobacilli was associated with an acidosis that could lead to hyperventilation or encephalopathy. Nevertheless, no lactic acidosis has been reported in healthy children[15].

30.3 Prebiotics

30.3.1 What are prebiotics?

These non-viable food components were first defined as non-digestible food ingredients. However, like probiotics, this definition was refined by the Food and Agriculture Organization of the United Nations (FAO) as "a non-viable food component that confers a health benefit on the host associated with modulation of the microbiota"[16]. This definition emphasized the impact of prebiotic on a limited number of microbial genus/species in the gut microbiota, bifidobacteria and lactobacilli being the usual target. Prebiotics must therefore resist gastric acidity, mammalian enzymes and absorption in the upper gastrointestinal tract, and reach the colon to be selectively fermented by the gut microbiota, stimulating the growth and/or activities of bacteria that result in specific changes in the composition and/or activity of the gastrointestinal microbiota, thus conferring benefit(s) upon host health. This selectivity is a major characteristic of prebiotics, prebiotics being fibre but not all dietary fibre being prebiotics[17].

Prebiotics are generally regarded as safe because of their long history of safe use. The most frequently used prebiotics are inulin, fructo-oligosaccharides (FOS), galacto-oligosaccharides (GOS) and trans-galacto-oligosaccharides (TOS), alone or in combination. The mixture GOS/FOS with 90% short-chain galactoologosaccharide and 10% long-chain frutooligosaccharides mimics the molecular size distribution of human milk oligosaccharides[18] and is therefore frequently encountered in infant formulas (see also chapter 29 by Gibson et al). Inulin-type fructan including inulin/oligofructose 50/50 are more likely used in adults. Oligosaccharides are naturally present in our diet from birth. Indeed, human breast milk contains about 5–10 g/l of complex oligosaccharides, reaching up to 20 g/l in colostrum[19]. Their diversity is huge, with more than 100 structurally distinct molecules thus far identified[20]. The structural diversity of these human milk oligosaccharides varies from mother to mother and even varies during lactation for each mother. Inulin and oligosaccharides are also found naturally in foods, including bananas, onions, garlic, chicory, asparagus, artichokes, soybeans, wheat, and oats. Inulin is also used as a fat substitute or as texturing agent in dairy products, dressings and some desserts. Oligofructose has properties comparable to sucrose but is less sweet and provides less energy, and hence is used in food as a substitute for sucrose. Prebiotics are commonly found in biscuits, cereals, chocolate and chocolate. Other molecules such as lactulose, resistant starch, polydextrose, wheat dextrin, acacia gum, psyllium, whole-grain wheat and whole corn grain also have prebiotic properties.

30.3.2 How do prebiotics work?

An important mechanism of prebiotics is fermentation in the colon, resulting in changes in gut microbiota composition. Generally, growth of saccharolytic bacteria such as lactobacilli and bifidobacteria is favored by prebiotics. However, other bacteria or bacterial groups can also be affected, such as *Bacteroides* whose numbers increased by wheat dextrin supplementation[21], and *Faecalibacterium prausnitzii* increased with mixed lc-inulin/scFOS and GOS supplementation[22]. The response of *F. prausnitzii* to FOS stimulation *in vitro* and *in vivo* is potentially important, given the ubiquity of this strain in healthy individuals and its reduced abundance in ileal Crohn's disease patients.

Modulation of the gut microbiota leads to modulation of its functionality. For instance, prebiotics can stimulate certain butyrate-producing Firmicutes species and butyrate has the potential to benefit colonic health[23]. Other genera, such as *Eubacterium* and *Roseburia*, can also be modulated. Prebiotics can also improve the barrier effect of the microbiota. Moreover, they can have a direct inhibitory impact on gut pathogens. Many pathogens use monosaccharide or short oligosaccharide sequences as receptors. Binding of these pathogens to these receptors is the first step in the colonization process. Prebiotic are derivatives of sugar and thus can act as blocking factors for pathogens such as against enteropathogenic *E. coli* (EPEC) and *Listeria*, attenuating their virulence[20,24]. Prebiotics may also influence the immune system directly, for instance through activation of pattern-recognition receptors, or indirectly, as a result of intestinal fermentation and promotion of the growth of certain members of the gut microbiota.

30.4 Synbiotics

The concept of synbiotics is to combine probiotics and prebiotics. This concept is highly relevant since it combines the bacteria and its growth substrate, acting therefore synergistically.

30.5 Pro-, pre-, and synbiotics in human medicine today

30.5.1 Pro- and prebiotics and infectious diarrhea

One of the first recommendations for probiotic use was the treatment and/or the prevention of diarrhea. A recent meta-analysis of 63 clinical trials reports a significant benefit of probiotics for acute diarrhea, with a shortened duration of one day for diarrhea lasting more than four days, and a decreased number of stools[25]. Nevertheless, the effects are variable, depending on the study. One of the reasons may be that the specific effect of a given strain on a given enteropathogen makes comparison between studies difficult. Another recent meta-analysis reported no effect of probiotics on diarrhea in adults in comparison with children[26]. In children, probiotics decrease the duration of diarrhea and the fever but had no effect on the duration of hospitalization and on the number of stools[26] and a limited effect in cases of persistent diarrhea[27]. One of the most interesting effects was the decrease of nosocomial rotavirus diarrhea. In this indication, the probiotic *Lactobacillus* GG, especially in three randomized controlled trials including more than 1,000 children, induced a significant decrease of rotavirus diarrhea[28]. Another indication for probiotics is diarrhea associated to antibiotic intake, most often due to *Clostridium difficile*, the colonization of which is promoted by the dysbiosis induced by antibiotic treatments. Despite adequate antibiotherapy, recurrences are frequent, related partly to the resistance of *C. difficile* spores to antibiotics. Administration of probiotics clearly allows rebuilding a "barrier microbiota." The results of many meta-analyses and reviews suggest their effectiveness when administrated with antibiotics; however, benefits depend on the probiotic strain[29]. The most effective species or strains seem to be *Bifidobacterium lactis, Lactobacillus* GG, as well as the yeast *Saccharomyces boulardii*. The latest meta-analysis from the Cochrane Library demonstrates effectiveness in the prevention of *C. difficile* diarrhea but not in decreasing their incidence[30]. A systematic review and meta-analysis showed the benefit of a co-administration of antibiotics and probiotics[31].

Data are available from a few clinical studies comparing prebiotics with probiotics or synbiotics. A meta-analysis including five randomized controlled trials revealed statistically significant decreases in the number of infectious episodes requiring antibiotic therapy in infants and in the rate of overall infections in infants and children 0–24 months of age[32]. Among the prebiotics used, scGOS/LcFOS in an infant formula lowered the incidence and severity of gastroenteritis[33] and the number of infections during the first six months of life in healthy infants[34]. This prebiotic mixture associated with LCPUFAs in a growing-up milk (GUM) that also contains iron and vitamin D seems to have similar impact[35]. In adults, in a double-blind placebo-controlled randomized study, a galacto-olisaccharide mixture decreased the incidence and duration of traveler's diarrhea, with decreased abdominal pain and an increase in overall quality of life[36]. By contrast, a synbiotic that associates two probiotic strains, i.e. *Enterococcus faecium* and *Saccharomyces cerevisiae,* and a fructooligosaccharide did not reduce the risk of developing traveler's diarrhea among travelers, nor its duration or the use of antibiotics when diarrhea occurred in comparison with the placebo[37]. Fructo-oligosaccharide appeared to be effective at preventing relapse of *C. difficile*–associated diarrhea in a controlled trial of 142 patients[38]. This effect was associated with an increase in fecal bifidobacterial concentrations, but not with a reduction of fecal *C. difficile* carriage.

30.5.2 Pro- and prebiotics and inflammatory bowel diseases

Inflammatory bowel diseases (IBD), i.e. Crohn's disease (CD) and ulcerative colitis (UC), are chronic diseases of unknown etiology (discussed in detail in chapter 19 of this book). Patients present with ulcerations of the small bowel or colon mucosa, alternating periods of activity with periods of remission. Their incidence has been increasing for a few decades in developed countries. Some authors have reported, besides genetic predisposition, dysbiosis and decrease of microbial diversity in patients suffering from IBD, but the causal link remains to be determined. Ding and Hart in chapter 19 report that the best-defined microbial change noted in IBD patients is a reduced phylum *Firmicutes* population and increased *Gammaproteo-bacteria* and that *Faecalibacterium prausnitzii* seems to be underrepresented in IBD[39]. The likely relationship between microbiota and mucosal inflammation has occasioned growing interest in using probiotics in IBD, as a complement to the normal treatment to decrease recurrence, but their effectiveness has not been proved. The clinical trials were mostly conducted with the aim of inducing or prolonging remission[40]. The results were rather disappointing for CD and did not allow recommending probiotics in this indication[41,42]. Nevertheless, some positive effects observed with either *Escherichia coli* Nissle 1917 or *Saccharomyces boulardii* led to investigating this approach through larger studies. The conclusions were almost identical for UC; few authors reported either a prolongation of remission or effectiveness greater than the usual treatment[40,43,44]. The effects of probiotics seemed more significant for pouchitis. A systematic review concluded that there is insufficient data to recommend the use of probiotics in CD, but sufficient evidence to recommend them for induction and maintenance of remission in UC and pouchitis[45]. The complex relationships among genetic, microbial, and environmental factors, leading to various phenotypes in patients can explain the disparity in the response to this therapeutic approach. Likewise, despite a strong rational proof of concept in animal studies and hopes from open studies, the few studies using prebiotics alone do not provide evidence on their efficacy in IBD. Lactulose, in a pilot randomized controlled study with 14 patients presenting UC and 17 with CD, had no beneficial effect on clinical index and immunohistological parameters, despite a significant improvement of quality of life in UC patients[46]. Similarly, Benjamin et al. did not observe clinical benefits testing FOS supplementation in a randomized controlled trial enrolling 103 patients with active CD[47]. Trials using synbiotics showed more promising results. Indeed, an oligofructose/inulin mix was shown to reduce fecal calprotectin in a small randomized controlled study involving patient with active UC[48]. This prebiotic mix associated with a *Bifidobacterium longum* strain showed a trend toward improved UC symptomology in a trial including 18 patients with a reduced sigmoidoscopy scores and an increase level of bifidobacteria[49]. The same synbiotic showed significant improvements in clinical outcomes, with reductions in both activity indices and histological scores in CD in a double-blind placebo-controlled trial involving 35 patients[50].

30.5.3 Pro- and prebiotics and irritable bowel syndrome

The irritable bowel syndrome (IBS) or functional colopathy is an intestinal disorder characterized by abdominal pain and a chronic disorder of the intestinal transit such as diarrhea and/or constipation. The etiology of this disease, which is very frequent and disturbing to the patient, remains badly documented, making management difficult. Several factors have been implicated in the occurrence of this syndrome such as stress, low-grade inflammation of the intestinal mucosa, abnormal colic motility, and gastro-intestinal infectious diseases. Furthermore, the current data seems to prove a relationship with dysbiosis, even if no bacterial marker has yet been demonstrated. This has opened the way to probiotic use for functional colopathy[51], and meta-analyses of case controlled clinical trials have demonstrated their potential contribution[52,53]. However, some differences in effectiveness have been observed, depending on the strains and doses administered and on the experimental protocol. Furthermore, the placebo effect is known to be beneficial in this disease, and the better results come from the smallest studies. This would tend to overestimate the beneficial effect of probiotics, allowing use of this treatment only as a second step for the management of this type of functional colopathy[54]. A recent meta-analysis including 43 randomized clinical trials concluded that probiotics are effective for IBS, although which strains are most beneficial remains unclear, likely due to the heterogeneity of symptoms of IBS[5]. Nevertheless, probiotics are an interesting alternative treatment, even if complementary clinical trials are needed.

Data on prebiotic and synbiotics on IBS are sparse. The impact of a transgalactooligosaccharide mixture produced from the activity of galactosyltransferases from *B. bifidum* on lactose was studied on 44 patients with Rome II positive IBS in a 12-week parallel crossover controlled clinical trial[56]. The galactooligosaccharide specifically stimulated gut bifidobacteria in IBS patients and was effective in alleviating symptoms, significantly changing stool consistency, improving flatulence, bloating, a composite score of symptoms and subjective global assessment, as well as anxiety scores[55]. Constipation is very common in children and in adults as well. The success of ingestion of large amounts of fiber emphasizes the likely role of microbiota in its treatment or prevention. Some data suggest the potential interest of ingestion of *Lactobacillus casei* strain Shirota[57] or *Bifidobacterium lactis*.

30.5.4 Pro- and prebiotics and allergy

Several epidemiological studies have demonstrated differences of microbiota between allergic and non-allergic children[58]. The use of probiotics allowing an adequate stimulation of the immune system is once again attractive, despite the discordances in study results for the relationship between microbiota and allergy. Many authors of clinical trials have reported promising results[58]. Unfortunately, contradictory results among studies do not currently allow recommending probiotic use in the prevention of allergy[58]. Several points may explain the intra-study variations: (i) the type of population studied in terms of age, type of allergic disease, disease stage, number of patients included, genetic background, environment; and (ii) the probiotic(s) tested in terms of bacterial species, mixture of strains use, doses administered, and administration algorithm, with consequent great heterogeneity among studies[59]. Nevertheless, the results of some studies clearly prove the effectiveness of probiotics for the prevention of some allergic diseases, most frequently in high-risk families, justifying further experimental research and clinical trials, focusing on the selection of the strain(s) to be used and the time of administration[58,60].

Similarly to probiotics, further research is needed before routine use of prebiotics can be recommended for prevention of allergy. Indeed, the meta-analysis of Osborn and Sinn[61] and the review of Moura and Filho[18] concluded that there is some evidence that a prebiotic supplement added to infants' food may prevent eczema. However, it is unclear whether the use of prebiotics should be restricted to infants at high risk of allergy or may have an effect in low-risk populations, or whether it may have an effect on other allergic diseases including asthma[61].

By contrast, another meta-analysis published the same year showed that while administration of probiotics and synbiotics may reduce the incidence of infant eczema, prebiotic administration alone had no effect[62]. There is still little evidence available on prebiotic adjuvant treatment for moderate to severe atopic dermatitis mediated by IgE; individual studies showed in general a trend toward fewer allergic reactions in children receiving active therapy as compared with placebo[18].

30.5.5 Pro- and prebiotics and obesity and diabetes

Chapter 23 of this book discussed the evidence for a relationship between microbiota and obesity[63] as well as one with type 2 diabetes[64]. One of the explanations is related to the colic fermentation of food not digested by the microbiota in the small bowel, which induces the production of metabolites such as short-chain fatty acids that are absorbed by the mucosa, thus allowing recovery of energy[65,66]. Currently, the authors of interventional studies have not proved any beneficial effect of probiotics on the BMI, perhaps because no genera and/or species could be associated with the pathological state[65,66]. Nevertheless, some clear results, such as those reported by Qin et al. proving that fecal metagenomic markers differentiated obese from non-obese patients more easily than markers of the human genome, are reasons to continue working on microbiota modulation in the management of these diseases[64].

Dietary prebiotics might also represent a potential strategy in prevention and also in treatment of metabolic abnormalities. Kellow et al. included 36 randomized clinical trials involving 831 patients in their systematic review, all dealing with inulin, oligofructose, fructo-oligosaccharides or galactosaccharides[67]. Short-term human trials support the use of dietary prebiotics as a potential therapeutic intervention for regulation of appetite and reduction of postprandial glucose and insulin concentration. However, the effects of prebiotics on total energy intakes, body weight, peptides YY and GLP-1 concentration, insulin sensitivity, lipids and gastric emptying time were contradictory and do not allow conclusions on the impact of prebiotic on overweight/obesity, glucose intolerance, dyslipidemia and non-alcoholic steatohepatitis. One of the reasons is the lack of large-scale, long-term prospective clinical trials.

30.5.6 Other indications

Other indications have been suggested, but currently with insufficient proof. Some relationships have been found among dysbiosis, dysfunctions and/or alterations of the metagenome, and diseases, supporting a probiotic approach[68]. The gut microbiota, via the production of toxic metabolites, has been implicated in the genesis of gastrointestinal (GI) cancers (see chapter 26 by Sobhani and Couffin), and some bacterial genera, such as lactobacilli, have shown protective effects against carcinogenesis. It is thus justified to believe that manipulation of the microbiota may have beneficial effects in the prevention of at least some GI cancers, also based on data from animal models. The effects appear to depend on the probiotic strains administered and to be more effective for prevention. There are much less data in human models, partly because of lack of large cohorts and of long-term follow-ups. Using probiotic strains genetically modified for the production of antiproliferative or pro-apoptotic factors could be very helpful. Elena Serban, in an excellent and very recent review, makes a complete update of breakthroughs in this domain[69].

The authors of some studies have suggested a contribution of probiotics in the prevention of upper-respiratory-tract infections. The Cochrane Library recently published a meta-analysis including the results of 10 randomized controlled clinical trials. The administration of probiotics, mostly lactobacilli and bifidobacteria, was beneficial, resulting in fewer upper-respiratory-tract infections[70]. Nevertheless, the level of evidence for the use of probiotics in this indication remains low because of the great heterogeneity among studies, especially in their main objective, and for some types of infections because of the small number of patients included. Once again, complementary trials must be made before recommending probiotics in this indication[71].

30.5.7 Pre- and probiotics in pediatrics

Microbial colonization of neonates in the first weeks of life is a crucial step for the implementation of different functions: the immune system is not mature at birth and the sequential colonization of the digestive tract during the neonatal period has an impact on restoring of the Th1/Th2 balance. The neonate is born with a Th2 profile and a dysbiosis may prevent the switch from Th2 to Th1 that potentially affects health, a Th2 orientation being a risk factor for presenting with an allergic disease. This concept is the basis for what was called the "hygiene theory," "the excess of hygiene" in developed countries, especially during delivery, being responsible for a modification of the microbiota implementation that results in an inadequate maturation of the immune system[72]. Such early dysbiosis may have a prolonged impact for many months. For example, the dysbiosis induced by a high hygiene level in developed countries has been associated with an increased incidence of allergies in these countries[73], with C-section delivery and very premature birth to an increased risk of being allergic[74] or obesity[75]. Other diseases have been associated with an alteration of early colonization such as IBD[76], autism[77], or type 1 diabetes[78]. Many clinical trials focusing on an early probiotic supplementation in full-term neonates have been conducted. The European Society of Pediatric Gastroenterology, Hepatology, and Nutrition (ESPGHAN) recently published a review of these trials[79]. The authors did not find any significant clinical benefit of early supplementation, which is why it is not currently recommended. Nevertheless, some authors have reported beneficial effects such as decreased GI infections, decreased use of antibiotics, and less incidence of diarrhea. The prevention of allergy seems to be an interesting indication for children at high risk of allergy[80], even if this effect was not reported in every study. The Cochrane Library published a meta-analysis agreeing on the potential interest of probiotic use in the prevention of allergy[81]. Furthermore, a study testing the effectiveness of probiotic supplementation for obesity preventionreported a trend of body mass index (BMI) decrease at four years of age[82]. Currently, the clearest contribution of early probiotic administration for very premature neonates is to decrease the incidence of necrotizing enterocolitis (NEC), the first GI emergency in neonates born before 33 weeks of gestation. The very abnormal microbiota established in these infants is a risk factor for NEC. The authors of all the reviews and meta-analyses concluded on a decreased incidence and severity of NEC in premature babies supplemented with one or several probiotic strains, but questioned its effectiveness in neonates with very low birth weight (<1,000 g)[83,84]. Nevertheless, this supplementation is still not routinely used by neonatologists, perhaps because of the lack of documentation on the mechanism of action, as well as because of the potential risk induced by this bacterial supplementation in very premature neonates. Some studies are currently ongoing to determine the ideal probiotic, the dose and the optimal administration algorithm.

In parallel, early prebiotic supplementation through supplemented infant formulas has been investigated. The committee of nutrition of ESPGHAN did not find sufficient evidence to recommend this supplementation, but recognized its safety[79]. However, infant formula containing a specific mixture of GOS and FOS has been found to affect the development of the infant gut microbiota with equilibrium close to that observed in human-milk-fed infants, with similar SCFA composition suggesting a similarity in functionalities[85]. A recent systematic review indicates that prebiotics may be effective in decreasing the rate of overall infections in infants and children up to 24 months of age[86]. However, no impact on the gut microbiota has been observed in premature infants, suggesting a lack of any health benefits[87,88]. Further evidence for the treatment of pediatric diseases by modifications of the microbiota is discussed in chapter 31.

30.6 Concluding remarks

It is currently acknowledged that the gut microbiota interacts with human health and that its modulation by probiotics and/prebiotics is an interesting way to prevent some diseases. However, their effectiveness has not been proven in many conditions, limiting current recommendations for their use. Several reasons may be given: too many low-quality studies, variability of microbiota and response to modulation attempts and diversity of probiotic strains used. Nevertheless, the rationale and the encouraging results reported in some studies support further research on probiotics. Several approaches should be taken into account. It is necessary to work on large cohorts to take into account inter-patient variability, as well to determine the bacterial markers of diseases related to the microbiota so as to obtain clinical trials of adequate power. The new methods, i.e. next-generation sequencing (NGS), used to analyze the microbiota and its functions should make it much easier to work on a large number of samples and to take a very global approach to the microbiota and its relationships with the host. Furthermore, it is important to work on choosing the right probiotic strain for a given application, the effect being strain-dependent. The use of probiotic strains genetically modified to improve their health benefits is an appealing approach. Besides, the use of complex microbiota, such as the successful attempts at fecal transplantation to manage recurrent *C. difficile* diarrhea as shown in the first published randomized controlled trial[89], is also an alternative of growing interest. The benefits in *C. difficile* recurrence treatment led gastroenterologists to think about fecal transplantation in patients suffering from dysbiosis as in IBD, obesity, diabetes, or IBS, and there have already been some ongoing clinical trials[90] and reports of some health benefits[91]. However, even if so far no deleterious effects have been reported, fecal transplantation raises several questions on safety with the risk of the transmission of pathogenic bacteria or deleterious dysbiosis, although all precautions are currently taken before administration. This approach could lead to the definition of a complex mixture of beneficial bacteria.

The success encountered with fecal transfer argues for the great interest in microbiota modulation as a preventive approach for diseases linked to dysbiosis, whose number is continuously increasing. All potential indications of prebiotics and probiotics open the way to an impressive field of research focused on this commensal microbiota (pathogens having been considered only for so long), its relationship to the host's health, and the benefit of its modulation.

TAKE-HOME MESSAGE

- The relationships demonstrated between gut microbiota and health have aroused great interest in gut microbiota modulation through administration of pre- or probiotics for prevention and treatment of some diseases.

- Although recommendations for pre/probiotic administration are currently limited, many clinical trials have reported benefits of their use, particularly in prevention of allergic diseases and improvement of clinical symptoms in irritable bowel syndrome. In pediatrics, the main interest is in early administration to prevent long-term diseases such as allergies, obesity; and in preterm infants to prevent neonatal necrotizing enterocolitis.

- The increased evidence for relationships between microbiota and health and the very encouraging results on gut microbiota modulation by pre/probiotics justify going on with basic research and clinical trials in this field to determine the best pre/probiotics to be used, at what doses, and for which indications.

References

1 FAO/WHO Working Group. Guidelines for the evaluation of probiotics in food. London: 2002.
2 Metchnikoff E. *The prolongation of life: optimistic studies*, G.P. Putnam's Sons, New York and London: 1908.
3 Tissier MH. Traitement des infections intestinales par la méthode de transformation de la flore bactérienne de l'intestin. *Compte-rendus de la Société de Biologie* 1906; **60**: 359–361.
4 Lilly DM, Stillwell RH. Probiotics: growth-promoting factors produced by microorganisms. *Science* 1965; **147**: 747–8.
5 FAO/WHO. Health and nutritional properties of probiotics in food including powdered milk with live lactic acid bacteria. **30**(suppl 2), S23–S33. 2001. Argentina.
6 EFSA Panel on biological hazards (BIOHAZ). Scientific opinion on the maintenance of the list of QPS biological agents intentionally added to food and feed (203 update). *EFSA Journal* 2013; **11**: 3449–3557.
7 Round JL, Mazmanian SK. The gut microbiota shapes intestinal immune responses during health and disease. *Nat Rev Immunol* 2009; **9**: 313–23.
8 Menard O, Butel MJ, Gaboriau-Routhiau V *et al.* Gnotobiotic mouse immune response induced by *Bifidobacterium* sp. strains isolated from infants. *Appl Environ Microbiol* 2008; **74**: 660–6.
9 Ojetti V, Gigante G, Gabrielli M *et al.* The effect of oral supplementation with *Lactobacillus reuteri* or tilactase in lactose intolerant patients: randomized trial. *Eur Rev Med Pharmacol Sci* 2010; **14**: 163–70.
10 Cannon JP, Lee TA, Bolanos JT *et al.* Pathogenic relevance of *Lactobacillus*: a retrospective review of over 200 cases. *Eur J Clin Microbiol Infect Dis* 2005; **24**: 31–40.
11 Franz CM, Huch M, Abriouel H *et al.* Enterococci as probiotics and their implications in food safety. *Int J Food Microbiol* 2011; **151**: 125–40.
12 Snydman DR. The safety of probiotics. *Clin Infect Dis* 2008; **46 Suppl 2**: S104–S111.
13 Lherm T, Monet C, Nougiere B *et al.* Seven cases of fungemia with *Saccharomyces boulardii* in critically ill patients. *Intensive Care Med* 2002; **28**: 797–801.
14 Hennequin C, Kauffmann-Lacroix C, Jobert A *et al.* Possible role of catheters in *Saccharomyces boulardii* fungemia. *Eur J Clin Microbiol Infect Dis* 2000; **19**: 16–20.

15 Connolly E, Abrahamsson T, Bjorksten B. Safety of D(-)-lactic acid producing bacteria in the human infant. *J Pediatr Gastroenterol Nutr* 2005; **41**: 489–92.

16 Pineiro M, Asp NG, Reid G *et al.* FAO Technical meeting on prebiotics. *J Clin Gastroenterol* 2008; **42 Suppl 3 Pt 2**: S156–S159.

17 Slavin J. Fiber and prebiotics: mechanisms and health benefits. *Nutrients* 2013; **5**: 1417–35.

18 de Moura PN, Rosario Filho NA. The use of prebiotics during the first year of life for atopy prevention and treatment. *Immun Inflamm Dis* 2013; **1**: 63–9.

19 Hennet T, Weiss A, Borsig L. Decoding breast milk oligosaccharides. *Swiss Med Wkly* 2014; **144**: w13927.

20 Bode L, Jantscher-Krenn E. Structure-function relationships of human milk oligosaccharides. *Adv Nutr* 2012; **3**: 383S–91S.

21 Lefranc-Millot C, Guerin-Deremaux L, Wils D *et al.* Impact of a resistant dextrin on intestinal ecology: how altering the digestive ecosystem with NUTRIOSE(R), a soluble fibre with prebiotic properties, may be beneficial for health. *J Int Med Res* 2012; **40**: 211–24.

22 Ramirez-Farias C, Slezak K, Fuller Z *et al.* Effect of inulin on the human gut microbiota: stimulation of *Bifidobacterium adolescentis* and *Faecalibacterium prausnitzii. Br J Nutr* 2009; **101**: 541–50.

23 Scott KP, Martin JC, Duncan SH *et al.* Prebiotic stimulation of human colonic butyrate-producing bacteria and bifidobacteria, in vitro. *FEMS Microbiol Ecol* 2014; **87**: 30–40.

24 Gibson GR, McCartney AL, Rastall RA. Prebiotics and resistance to gastrointestinal infections. *Br J Nutr* 2005; **93 Suppl 1**: S31–S34.

25 Allen SJ, Martinez EG, Gregorio GV *et al.* Probiotics for treating acute infectious diarrhoea. *Cochrane Database Syst Rev* 2010; CD003048.

26 Salari P, Nikfar S, Abdollahi M. A meta-analysis and systematic review on the effect of probiotics in acute diarrhea. *Inflamm Allergy Drug Targets* 2012; **11**: 3–14.

27 Bernaola AG, Bada Mancilla CA, Carreazo Pariasca NY *et al.* Probiotics for treating persistent diarrhoea in children. *Cochrane Database Syst Rev* 2010; CD007401.

28 Szajewska H, Wanke M, Patro B. Meta-analysis: the effects of *Lactobacillus rhamnosus* GG supplementation for the prevention of healthcare-associated diarrhoea in children. *Aliment Pharmacol Ther* 2011; **34**: 1079–87.

29 Hempel S, Newberry SJ, Maher AR *et al.* Probiotics for the prevention and treatment of antibiotic-associated diarrhea: a systematic review and meta-analysis. *JAMA* 2012; **307**: 1959–69.

30 Goldenberg JZ, Ma SS, Saxton JD *et al.* Probiotics for the prevention of Clostridium difficile-associated diarrhea in adults and children. *Cochrane Database Syst Rev* 2013; **5**: CD006095.

31 Pattani R, Palda VA, Hwang SW *et al.* Probiotics for the prevention of antibiotic-associated diarrhea and *Clostridium difficile* infection among hospitalized patients: systematic review and meta-analysis. *Open Med* 2013; **7**: e56–e67.

32 Lohner S, Kullenberg D, Antes G *et al.* Prebiotics in healthy infants and children for prevention of acute infectious diseases: a systematic review and meta-analysis. *Nutr Rev* 2014; **72**: 523–31.

33 Bruzzese E, Volpicelli M, Squeglia V *et al.* A formula containing galacto- and fructo-oligosaccharides prevents intestinal and extra-intestinal infections: an observational study. *Clin Nutr* 2009; **28**: 156–61.

34 Arslanoglu S, Moro GE, Boehm G. Early supplementation of prebiotic oligosaccharides protects formula-fed infants against infections during the first 6 months of life. *J Nutr* 2007; **137**: 2420–4.

35 Chatchatee P, Lee WS, Carrilho E *et al.* Effects of growing-up milk supplemented with prebiotics and LCPUFAs on infections in young children. *J Pediatr Gastroenterol Nutr* 2014; **58**: 428–37.

36 Drakoularakou A, Tzortzis G, Rastall RA *et al.* A double-blind, placebo-controlled, randomized human study assessing the capacity of a novel galacto-oligosaccharide mixture in reducing travellers' diarrhoea. *Eur J Clin Nutr* 2010; **64**: 146–52.

37 Virk A, Mandrekar J, Berbari EF *et al.* A randomized, double blind, placebo-controlled trial of an oral synbiotic (AKSB) for prevention of travelers' diarrhea. *J Travel Med* 2013; **20**: 88–94.

38 Lewis S, Burmeister S, Brazier J. Effect of the prebiotic oligofructose on relapse of *Clostridium difficile*-associated diarrhea: a randomized, controlled study. *Clin Gastroenterol Hepatol* 2005; **3**: 442–8.

39 Sokol H, Seksik P, Furet JP *et al.* Low counts of *Faecalibacterium prausnitzii* in colitis microbiota. *Inflamm Bowel Dis* 2009; **15**: 1183–9.

40 Meijer BJ, Dieleman LA. Probiotics in the treatment of human inflammatory bowel diseases: update 2011. *J Clin Gastroenterol* 2011; **45 Suppl**: S139–S144.

41 Butterworth AD, Thomas AG, Akobeng AK. Probiotics for induction of remission in Crohn's disease. *Cochrane Database Syst Rev* 2008; CD006634.

42 Rahimi R, Nikfar S, Rahimi F *et al*. A meta-analysis on the efficacy of probiotics for maintenance of remission and prevention of clinical and endoscopic relapse in Crohn's disease. *Dig Dis Sci* 2008; **53**: 2524–31.

43 Kato K, Mizuno S, Umesaki Y *et al*. Randomized placebo-controlled trial assessing the effect of bifidobacteria-fermented milk on active ulcerative colitis. *Aliment Pharmacol Ther* 2004; **20**: 1133–41.

44 Kruis W, Fric P, Pokrotnieks J *et al*. Maintaining remission of ulcerative colitis with the probiotic *Escherichia coli* Nissle 1917 is as effective as with standard mesalazine. *Gut* 2004; **53**: 1617–23.

45 Ghouri YA, Richards DM, Rahimi EF *et al*. Systematic review of randomized controlled trials of probiotics, prebiotics, and synbiotics in inflammatory bowel disease. *Clin Exp Gastroenterol* 2014; **7**: 473–87.

46 Hafer A, Kramer S, Duncker S *et al*. Effect of oral lactulose on clinical and immunohistochemical parameters in patients with inflammatory bowel disease: a pilot study. *BMC Gastroenterol* 2007; **7**: 36.

47 Benjamin JL, Hedin CR, Koutsoumpas A *et al*. Randomised, double-blind, placebo-controlled trial of fructo-oligosaccharides in active Crohn's disease. *Gut* 2011; **60**: 923–9.

48 Casellas F, Borruel N, Torrejon A *et al*. Oral oligofructose-enriched inulin supplementation in acute ulcerative colitis is well tolerated and associated with lowered faecal calprotectin. *Aliment Pharmacol Ther* 2007; **25**: 1061–7.

49 Furrie E, Macfarlane S, Thomson G *et al*. Toll-like receptors-2, -3 and -4 expression patterns on human colon and their regulation by mucosal-associated bacteria. *Immunology* 2005; **115**: 565–74.

50 Steed H, Macfarlane GT, Blackett KL *et al*. Clinical trial: the microbiological and immunological effects of synbiotic consumption — a randomized double-blind placebo-controlled study in active Crohn's disease. *Aliment Pharmacol Ther* 2010; **32**: 872–83.

51 Almansa C, Agrawal A, Houghton LA. Intestinal microbiota, pathophysiology and translation to probiotic use in patients with irritable bowel syndrome. *Expert Rev Gastroenterol Hepatol* 2012; **6**: 383–98.

52 Moayyedi P, Ford AC, Talley NJ *et al*. The efficacy of probiotics in the treatment of irritable bowel syndrome: a systematic review. *Gut* 2010; **59**: 325–32.

53 Hoveyda N, Heneghan C, Mahtani KR *et al*. A systematic review and meta-analysis: probiotics in the treatment of irritable bowel syndrome. *BMC Gastroenterol* 2009; **9**: 15.

54 McKenzie YA, Alder A, Anderson W *et al*. British Dietetic Association evidence-based guidelines for the dietary management of irritable bowel syndrome in adults. *J Hum Nutr Diet* 2012; **25**: 260–74.

55 Ford AC, Quigley EM, Lacy BE *et al*. Efficacy of prebiotics, probiotics, and synbiotics in irritable bowel syndrome and chronic idiopathic constipation: systematic review and meta-analysis. *Am J Gastroenterol* 2014; **109**: 1547–61.

56 Silk DB, Davis A, Vulevic J *et al*. Clinical trial: the effects of a trans-galactooligosaccharide prebiotic on faecal microbiota and symptoms in irritable bowel syndrome. *Aliment Pharmacol Ther* 2009; **29**: 508–18.

57 Koebnick C, Wagner I, Leitzmann P *et al*. Probiotic beverage containing *Lactobacillus casei* Shirota improves gastrointestinal symptoms in patients with chronic constipation. *Can J Gastroenterol* 2003; **17**: 655–9.

58 Waligora-Dupriet AJ, Butel MJ. Microbiota and allergy: from dysbiosis to probiotics. In: *Allergic Diseases – Highlights in the Clinic, Mechanisms and Treatment* (Pereira, C, ed). Rijeka: Intech, 2012: 413–34.

59 Ozdemir O. Various effects of different probiotic strains in allergic disorders: an update from laboratory and clinical data. *Clin Exp Immunol* 2010; **160**: 295–304.

60 Pelucchi C, Chatenoud L, Turati F *et al*. Probiotics supplementation during pregnancy or infancy for the prevention of atopic dermatitis: a meta-analysis. *Epidemiology* 2012; **23**: 402–14.

61 Osborn DA, Sinn JK. Prebiotics in infants for prevention of allergy. *Cochrane Database Syst Rev* 2013; **3**: CD006474.

62 Dang D, Zhou W, Lun ZJ *et al*. Meta-analysis of probiotics and/or prebiotics for the prevention of eczema. *J Int Med Res* 2013; **41**: 1426–36.

63 Turnbaugh PJ, Ley RE, Mahowald MA *et al*. An obesity-associated gut microbiome with increased capacity for energy harvest. *Nature* 2006; **444**: 1027–31.

64 Qin J, Li Y, Cai Z *et al*. A metagenome-wide association study of gut microbiota in type 2 diabetes. *Nature* 2012; **490**: 55–60.

65 Shen J, Obin MS, Zhao L. The gut microbiota, obesity and insulin resistance. *Mol Aspects Med* 2013; **34**: 39–58.

66 Sanz Y, Rastmanesh R, Agostonic C. Understanding the role of gut microbes and probiotics in obesity: how far are we? *Pharmacol Res* 2013; **69**: 144–55.

67 Kellow NJ, Coughlan MT, Reid CM. Metabolic benefits of dietary prebiotics in human subjects: a systematic review of randomised controlled trials. *Br J Nutr* 2014; **111**: 1147–61.

68 de Vos WM, Nieuwdorp M. Genomics: A gut prediction. *Nature* 2013; **498**: 48–9.

69 Serban DE. Gastrointestinal cancers: influence of gut microbiota, probiotics and prebiotics. *Cancer Lett* 2014; **345**: 258–70.

70 Hao Q, Lu Z, Dong BR *et al*. Probiotics for preventing acute upper respiratory tract infections. *Cochrane Database Syst Rev* 2011; CD006895.

71 Williams K, Tang M, Williams K. Probiotics may prevent upper respiratory tract infections, but should we recommend them? *J Paediatr Child Health* 2012; **48**: 942–3.

72 Okada H, Kuhn C, Feillet H *et al*. The 'hygiene hypothesis' for autoimmune and allergic diseases: an update. *Clin Exp Immunol* 2010; **160**: 1–9.

73 Rautava S, Ruuskanen O, Ouwehand A *et al*. The hygiene hypothesis of atopic disease--an extended version. *J Pediatr Gastroenterol Nutr* 2004; **38**: 378–88.

74 Laubereau B, Filipiak-Pittroff B, von BA *et al*. Caesarean section and gastrointestinal symptoms, atopic dermatitis, and sensitisation during the first year of life. *Arch Dis Child* 2004; **89**: 993–7.

75 Huh SY, Rifas-Shiman SL, Zera CA *et al*. Delivery by caesarean section and risk of obesity in preschool age children: a prospective cohort study. *Arch Dis Child* 2012; **97**: 610–6.

76 Schippa S, Conte MP, Borrelli O *et al*. Dominant genotypes in mucosa-associated *Escherichia coli* strains from pediatric patients with inflammatory bowel disease. *Inflamm Bowel Dis* 2009; **15**: 661–72.

77 Finegold SM, Molitoris D, Song Y *et al*. Gastrointestinal microflora studies in late-onset autism. *Clin Infect Dis* 2002; **35**: S6–S16.

78 Cardwell CR, Stene LC, Joner G *et al*. Caesarean section is associated with an increased risk of childhood-onset type 1 diabetes mellitus: a meta-analysis of observational studies. *Diabetologia* 2008; **51**: 726–35.

79 Braegger C, Chmielewska A, Decsi T *et al*. Supplementation of infant formula with probiotics and/or prebiotics: a systematic review and comment by the ESPGHAN committee on nutrition. *J Pediatr Gastroenterol Nutr* 2011; **52**: 238–50.

80 Foolad N, Brezinski EA, Chase EP *et al*. Effect of nutrient supplementation on atopic dermatitis in children: a systematic review of probiotics, prebiotics, formula, and fatty acids. *JAMA Dermatol* 2013; **149**: 350–5.

81 Boyle RJ, Bath-Hextall FJ, Leonardi-Bee J *et al*. Probiotics for treating eczema. *Cochrane Database Syst Rev* 2008; CD006135.

82 Luoto R, Kalliomaki M, Laitinen K *et al*. The impact of perinatal probiotic intervention on the development of overweight and obesity: follow-up study from birth to 10 years. *Int J Obes (Lond)* 2010; **34**: 1531–7.

83 Alfaleh K, Anabrees J, Bassler D *et al*. Probiotics for prevention of necrotizing enterocolitis in preterm infants. *Cochrane Database Syst Rev* 2011; CD005496.

84 Deshpande GC, Rao SC, Keil AD *et al*. Evidence-based guidelines for use of probiotics in preterm neonates. *BMC Med* 2011; **9**: 92.

85 Oozeer R, van LK, Ludwig T *et al*. Intestinal microbiology in early life: specific prebiotics can have similar functionalities as human-milk oligosaccharides. *Am J Clin Nutr* 2013; **98**: 561S–71S.

86 Lohner S, Kullenberg D, Antes G *et al*. Prebiotics in healthy infants and children for prevention of acute infectious diseases: a systematic review and meta-analysis. *Nutr Rev* 2014; **72**: 523–31.

87 Campeotto F, Suau A, Kapel N *et al*. A fermented formula in preterm infants: clinical tolerance, gut microbiota, down regulation of fecal calprotectin, and up regulation of fecal secretory IgA. *Br J Nutr* 2011; **105**: 1843–51.

88 Underwood MA, Kalanetra KM, Bokulich NA *et al*. Prebiotic oligosaccharides in premature infants. *J Pediatr Gastroenterol Nutr* 2014; **58**: 352–60.

89 Van NE, Vrieze A, Nieuwdorp M *et al*. Duodenal infusion of donor feces for recurrent *Clostridium difficile*. *N Engl J Med* 2013; **368**: 407–15.

90 Borody TJ, Brandt LJ, Paramsothy S. Therapeutic faecal microbiota transplantation: current status and future developments. *Curr Opin Gastroenterol* 2014; **30**: 97–105.

91 Rossen NG, MacDonald JK, de Vries EM *et al*. Fecal microbiota transplantation as novel therapy in gastroenterology: A systematic review. *World J Gastroenterol* 2015; **21**: 5359–71.

CHAPTER 31

The microbiota as target for therapeutic intervention in pediatric intestinal diseases

Andrea Lo Vecchio and Alfredo Guarino

University of Naples, Naples, Italy

31.1 Introduction

This book has so far emphasized the increasingly clear and stimulating role that the gut microbiota exerts on human life and its potential involvement in the pathophysiology of several diseases. The human microbiota exerts important immune, metabolic, trophic, and protective functions that are currently interpreted with a model of symbiosis between the host and intestinal microbes[1]. It is well established that the commensal microflora is able to inhibit colonization by pathogenic bacteria through a variety of local mechanisms and may interact with the immune system at the local and systemic level[2,3].

The gut microbiota may be seen as a "dynamic hidden organ" whose structure varies according to many factors including genetic factors, age, nutrition, disease and related treatments. Many of the factors that affect and regulate microbiota composition develop during infancy and early childhood and the progressive age periods are an ideal setting to study and understand the functional roles of gut microbiota. During childhood the gut microbiota is more exposed to factors that may change its composition than in adulthood and, in addition, it possesses a relatively simple structure at this age and its composition is intrinsically plastic and rather unstable over the time. In contrast, the adult microbiota is more complex, but relatively stable and similar among individuals[4].

Intestinal colonization begins at birth with amniotic membrane rupture or perhaps even before, according to a very recent experimental hypothesis[5]. Bacteria from the mother's intestinal and vaginal sites and from the outer environment colonize the neonatal gut within a few hours of birth with significant differences between children born from vaginal delivery and Cesarian section[6,7]. Subsequently the intestinal microbiota is significantly affected by child age, type of feeding, hygiene status, diet and other family and environmental exposures, the presence of intestinal and extra-intestinal diseases and the use of antibiotics and foods.

The Human Microbiota and Chronic Disease: Dysbiosis as a Cause of Human Pathology, First Edition.
Edited by Luigi Nibali and Brian Henderson.

Specific aberrations of the microbiota with a predominance of harmful bacterial species and/or a reduction in variability (dysbiosis) have been detected in populations of children with specific diseases and have been defined as "microbiological signatures"[1]. The concept of microbiological signature has a double meaning: selected bacterial species may have a key role in causing — or contributing to — specific diseases and/or the microflora composition may be a hallmark of selected disease and may help to recognize it or monitor its course. Microbiological signatures have been proposed for many intestinal diseases in children and adults, including celiac disease, inflammatory bowel diseases, and irritable bowel syndrome[8-11].

If the gut microbiota has such a determinant role in intestinal diseases, it seems logical to set up interventions that could modify its composition and hence have an effect on disease-related symptoms or, ideally, affect disease onset and progression. The overall hypothesis is to replace the "ill microbiota" with a "healthy" one that may reduce dysbiosis and mucosal inflammation, and restore the intestinal immune response and its regular activities. Theoretically, this may be obtained in three different ways:

1 Providing nutritional ingredients or non-digestible foods able selectively to stimulate the growth/activity of a limited number of colonic bacteria; this may be done with the so-called *prebiotics*;

2 Providing live bacteria (in some cases yeast) that may colonize the bowel, modify the composition of colonic microflora and counteract enteric pathogens; i.e., administration of *probiotics*;

3 Replace the gut microbiota of a patient with intestinal illness with that of healthy donor: this has been attempted with *fecal microbiota transplantation.*

Some of these approaches have been tested in adults and children and are included in other chapters of this book (see chapters 30 and 32). We focus now on the last two interventions with a specific pediatric perspective.

31.2 Use of probiotics in pediatric intestinal diseases

Probiotics are live micro-organisms that act both at intestinal level competing with pathogens for nutrients and receptors and regulating absorption, permeability and local inflammation, and at systemic level by modulating the immune response. More detailed information on their mechanisms of action is reported in chapter 30. Their ease of administration and excellent safety profile make them an ideal intervention in pediatric diseases. Probiotic bacteria may act differently in various diseases and the effect in selected conditions significantly varies depending on strains and doses. Evidence in support of their use changes significantly for different diseases (Table 1).

31.2.1 Acute diarrhea

The use of probiotics in children with acute diarrhea has been extensively studied in the last 20 years. As seen in the previous chapter, the evidence that supports their use for the treatment of acute diarrhea of infectious etiology (acute gastroenteritis) is conclusive. This is the only condition for which evidence-based guidelines specifically recommend the use of probiotics. However, only two strains are

Table 1 Use of probiotics in pediatric intestinal diseases.

Disease	Available evidence	Probiotic strain with more evidence in support
Treatment		
Acute infectious diarrhea	Conclusive	LGG, *S. boulardii*
Infant colic	Strong	*L. reuteri*
Protracted diarrhea	Weak	—
Functional abdominal pain	Weak	LGG, VSL#3
Prevention		
Antibiotic-associated diarrhea	Strong	LGG; *S. boulardii*
Necrotizing enterocolitis	Promising	—
Acute infectious diarrhea	Limited	LGG
Ulcerative colitis recurrence	Weak	VSL#3, *L. reuteri*
Crohn's recurrence	Absent	—

LGG: *Lactobacillus rhamnosus GG*

recommended based on compelling data for the treatment of children with acute gastroenteritis: *Lactobacillus rhamnosus GG* (LGG) and *Saccharomyces boulardii*[12]. High doses of LGG (>10^{10} CFU) are more effective than low doses in reducing the duration of diarrhea in children with acute intestinal infections, demonstrating a clear dose-related effects[13].

Despite the strong clinical evidence in support, acute infectious diarrhea is an exception to the hypothesis that correlates the effects of probiotics with the modification of gut microbiota. Usually diarrhea starts some hours or a few days after the contact with the causative pathogen (in most cases viruses), and the early administration of probiotic strains rapidly restores a normal stool pattern, reducing the duration of symptoms in a few hours. It is unlikely that such a rapid effect is induced by a modification of intestinal microbiota; other mechanisms have been proposed, including local anti-inflammatory and ion pro-absorptive actions (Figure 1). In addition, a long-term use of the same strains, which might induce a more significant modification of gut microflora, has inconsistent effects in preventing new episodes of infectious diarrhea in children[14].

A further common cause of acute onset diarrhea in childhood is antibiotic-associated diarrhea. In this situation the diarrhea is not the expression of an acute infection but is rather the direct consequence of a disruption of gut microbiota equilibrium with a reduction in bacterial richness and a prevalence/overgrowth of pathogen species (such as *Clostridium difficile*, CD). Antibiotic-associated diarrhea occurs in about 10–40% of children upon administration of broad-spectrum antibiotics taken for other infectious illnesses (respiratory infections in most cases). Despite a relevant risk of bias, available evidence included in a recent meta-analysis (2874 participants) demonstrated a reduction of the incidence of antibiotic-associated diarrhea in children receiving probiotics (RR 0.52; 95% CI 0.38 to 0.72). Again, effective strains were LGG or *Saccharomyces boulardii* and, as demonstrated for acute gastroenteritis, high probiotic doses (≥5 billion CFUs/day) are more effective than low doses (<5 billion CFUs/day) (p = 0.010) and the number needed to treat to prevent one case of diarrhea is as low as 7 (95% CI 6 to 10)[15]. A recent Cochrane Review that specifically analyzed the effect of probiotics in

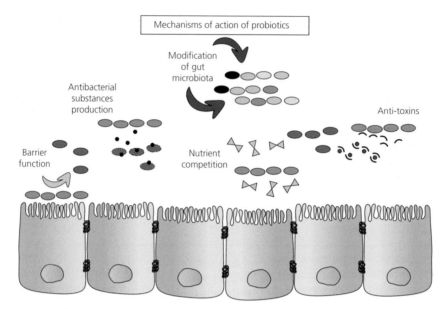

Figure 1 Antimicrobial mechanisms of probiotics. The known mechanisms whereby probiotics exert their antimicrobial effects include: 1) modification of gut microbiota in a healthy state; 2) secretion of antimicrobial substances such as bacteriocins, hydrogen peroxide and short-chain fatty acids; 3) competition for nutrients; 4) antitoxin effect; 5) inhibition of pathogen adhesion to the intestinal epithelium (barrier function).

preventing *Clostridium difficile*-associated diarrhea included three studies on children (*n* = 605) and reported a relevant effect in children receiving probiotics (RR 0.37; 95% CI 0.23 to 0.60)[16].

31.2.2 Inflammatory bowel diseases

Many studies showed that the intestinal microbiota profoundly differs between patients with inflammatory bowel diseases (IBD) and healthy individuals, and intestinal dysbiosis may contribute to the risk of IBD or its relapses. Schwiertz *et al.* showed that *Faecalibacterium prausnitzii* (a component of the Firmicutes phylum) which probably has an anti-inflammatory effect, is less represented in IBD patients than in healthy individuals[9]. In adults, probiotics may potentially prevent IBD relapses by stabilizing the gut microbiota and reduce mucosal inflammation. However, this does not appear entirely true in pediatric patients.

Evidence in children with Crohn's disease is very limited and inconsistent and it is also limited in ulcerative colitis, where two studies support the use of selected probiotics in children. The first published trial demonstrated the efficacy of VSL#3 in maintaining remission in children with ulcerative colitis[17]. A second trial proposed the endorectal administration of *L. reuteri* in children with active distal ulcerative colitis, and reported an improvement in mucosal inflammation and a modification of cytokines involved in the mechanisms of inflammatory bowel disease at a mucosal level[18]. The sensitive equilibrium between mucosal inflammation and gut microbiota provides a stimulating background to study the effects of probiotics on microbiota in patients with IBD. However, further studies also including other possible interventions are warranted (see below).

31.2.3 Irritable bowel syndrome

Irritable bowel syndrome (IBS) is a common gastrointestinal complaint in developed countries associated with poor quality of life and substantial costs both in adults and children[19–21]. IBS is diagnosed according to the Rome criteria and to the patient age[22,23]. However, the clinical picture is usually characterized by the presence of abdominal pain, flatulence and bloating as major symptoms and/or by changes in frequency and/or consistency of stools, discomfort or pain improved by defecation, in the absence of specific endoscopic, biochemical or radiological abnormalities. Recent evidence indicates that an alteration in the composition of intestinal microbiota may contribute to IBS[24]. A study showed specific and consistent changes in fecal microbiota of IBS patients, and specific changes in microbiota composition were associated with different IBS subtypes. The microbiota of subjects with IBS was characterized by a significant decrease of *Bacteroides*, *Bifidobacterium* and *Faecalibacterium* spp. and an increase in *Dorea*, *Ruminococcus*, and *Clostridium* spp.[25]. Interestingly, a close relationship between a specific bacterial profile with the severity of symptoms has been shown in pediatric IBS[11].

The direct effect on the pathophysiology of symptoms, the good safety profile and the possibility of a long-term administration show probiotics to be an ideal candidate for the treatment of IBS-related symptoms. A recent meta-analysis explored their role in childhood functional intestinal disorders and found a significant effect in reducing abdominal pain, mainly as part of IBS syndrome (RR 1.50 95% CI 1.22–1.84), but not in functional constipation[26]. However, different strains and doses have been studied (and included together in the analysis), and to date it's still difficult to identify a specific intervention to be routinely applied in clinical practice.

31.2.4 Infant colic

Infant colic, or excessive crying of unknown cause, affects up to 20% of infants and is a major burden on families and health services[27]. Recent studies suggest a possible link between infant colic and gut microbiota, indicating probiotics as a promising treatment. However, only few strains have been tested, and results from randomised controlled trials are conflicting. A meta-analysis of data available up to 2014 showed a significant effect of *Lactobacillus reuteri* in reducing crying and fussing at 21 days after the end of intervention (mean difference −66.72 95%CI −99.79 to −33.64); however, the same research group tested this intervention in a large infant population and had inconsistent results, showing that the treatment with *L. reuteri* did not reduce crying or fussing, nor was it effective in improving infant sleep, maternal mental health, or family quality of life[28]. A more recent clinical trial performed in Canada further confirmed the efficacy of *L.reuteri* DSM 17938, demonstrating a significant reduction in daily crying (60 min IQR 64 min/day vs 102 min IQR, 87 min/day; p = 0 .045) and fussing times at the end of treatment period[29]. The data available up to date that support the use of *L.reuterii* are very promising. A large European trial has been presented and is currently ongoing (PROSPERO CRD42014013210) and might potentially solve this issue, supporting the use of a selected probiotic strain to be used in infants with colic[30].

31.2.5 Necrotizing enterocolitis

Necrotizing enterocolitis (NEC) is a severe and potentially fatal disease that selectively affects preterm neonates. The role of gut microbiota in the pathogenesis of NEC is not well defined and the data currently available are conflicting. Some authors reported a decrease in microbiota diversity (even more evident than that reported in preterm infants without NEC), with a predominance of Gammaproteobacteria and a reduction of other bacterial species[31]. However, others did not find relevant difference in diversity[32] or even found an opposite pattern, with a higher bacterial diversity, expressed as band richness, in infants with NEC than in controls[33].

The administration of probiotics to preterm infants at high risk of developing NEC is probably the most challenging and discussed intervention in neonatal gastroenterology in recent years. This intervention may have dramatic consequences for child health and survival; however, the evidence, even if strongly supports the use of probiotics, is still affected by methodological biases and seen with skepticism by many neonatologists. The beneficial effects have been confirmed in more than one meta-analysis[34-37]. However, other authors raised the suspicion of a misleading effect of probiotics in this fragile population due to relevant biases in the published studies in terms of heterogeneity, inclusion criteria and major difference in incidence of NEC strains and doses used. All these observations could change the claimed benefit of probiotics in the prevention of NEC, sepsis and mortality[38]. More recent evidence confirmed a reduction of the incidence of NEC and — even more interestingly — of mortality for all causes in preterm infants treated with probiotics[39]. However, the efficacy of probiotics in preventing NEC, if confirmed, would be the most spectacular and important clinical effect obtained with probiotics in pediatrics, and should be translated in a clinical recommendation.

31.3 Fecal microbiota transplantation for treatment of intestinal diseases

Fak in chapter 23 introduced the concept of fecal microbiota transplantation (FMT) for the treatment of subjects with the metabolic syndrome. FMT is aimed at re-establishing the normal composition and restoring the wide diversity of gut microbiota through the instillation of healthy donor feces into the gastrointestinal tract. This approach, which comes historically from veterinary medicine (treatment of diarrhea in horses), is able to restore phylogenetic richness and colonization resistance and promote normal bowel function in animal models and humans[40].

31.3.1 Preparation and administration

Fecal microbiota is usually obtained from household members or unrelated/anonymous donors, screened for infectious diseases, factors that may affect microbiota composition (e.g. use of antibiotics), sexual at-risk contacts or recent hospitalizations. In addition, before FMT, patients should avoid intake of substances that may potentially cause allergies[41]. Preparation of the recipient varies slightly depending on the instillation method, but usually antibiotic treatment is stopped

24–72 hours before FMT and a bowel preparation with polyethylene glycol is suggested. Stools should be accurately prepared hours before FMT (Table 2) and may be instilled through either nasogastric tube or enema or during upper and lower endoscopy (Table 3). More recently, the advantages of banking frozen, ready-to-use fecal material obtained from healthy volunteers has been explored with good efficacy and with no reported adverse effects[42]. This approach may have potential

Table 2 Preparation of donor stools and recipient.

Universal precautions and devices
Hand hygiene, Gloves, Gown, Mask, Eye shield

Preparation of stools
- Timing depends from the route of administration: < 6 hours (upper route) to < 24 hours (lower route)
- Prepare stools in a dedicated room (under a hood) and use a dedicated blender;
- Homogenize stool sample with 0.9% saline, or alternatively 4% milk, in a conventional household blender;
- Blend the stools for 2–4 minutes to obtain a homogenized, liquid slurry;
- Filter the mixture through gauze pads to remove large particulate matter;
- Prepare the stool suspension for upper or lower gastrointestinal route.

Pre-treatment before FMT
Patients receiving FMT through upper gastrointestinal route should be treated with proton pump inhibitor the evening before and the morning of the FMT. In contrast, those receiving FMT through lower route should be treated with Loperamide before receiving donor stools.

Table 3 Advantages and barriers in different routes of FMT administration.

	Lower gastrointestinal route	Upper gastrointestinal route
Route of administration	• Colonoscopy • Enema	• Nasogastric tube • Nasojejunal tube • Gastroscopy
Advantages	Lower endoscopy allows instillation in focused and difficult-to-reach areas (e.g. diverticuli) Retention enema may be performed at home	Easy to perform Low risk of intestinal perforation Early discharge Low cost
Disadvantages	Enemas reaches only the splenic flexure and usually needs multiple instillation Risk of colon perforation in case of colonoscopy Refrain from defecation for 30–45 minutes after transplantation	Stools may not reach distal intestinal tract Risk of short bowel bacterial overgrowth
Side effects	Abdominal cramps or pain	• Nausea • Vomiting • Belching • Abdominal cramps or pain • Diarrhea • Constipation

benefits in standardizing the FMT procedure, facilitating prompt treatment, reducing laboratory screening time and costs, and potentially resetting the microbiota by instilling bacterial assets completely different from those partially shared in households[43].

31.3.2 Advantages and barriers

Fecal microbiota transplantation is usually seen as a safe procedure. However, adverse reactions to FMT have not been actively searched for, but merely spontaneously reported in series of cases, and data on long-term adverse events are very limited. Mild and temporary gastrointestinal symptoms, including diarrhea, cramping, belching or constipation, have been reported hours to days after the procedure[44]. The potential risk of transmitting pathogens from the donor and the risk of small intestinal bacterial overgrowth after duodenal installation of feces have been hypothesized; however, evidence of transmission of infection has never been reported. Studies that specifically explore long-term consequences of FMT are lacking. The only available study with a long follow-up reported four autoimmune diseases including peripheral neuropathy, Sjogren's syndrome, idiopathic thrombocytopenic purpura, rheumatoid arthritis in 77 adult patients[45].

Current protocols do not require testing of donors and recipients for autoimmune diseases, and the true risk of subsequent autoimmune diseases can only be derived from paired follow-up studies of donors and recipients. These potential long-term consequences may be of greater importance in children and adolescents who have a longer time to develop side effects than adults. It's general opinion that acceptability of FMT might be the major barrier to the large dissemination of this treatment for intestinal diseases. However, some recent surveys reported that both physicians and patients seems to be prone to consider FMT as a optional treatment for some recurrent gastrointestinal diseases[47], and a large percentage of patients who experienced it for *Clostridium difficile* infection (CDI) even choose FMT as first-line treatment instead of antibiotics[45]. The recent dissemination of this practice in different countries and its increasing acceptance suggest that physicians and patients may be more prone to exploit FMT than expected[46].

31.3.3 The use of FMT in specific intestinal diseases
31.3.3.1 Clostridium difficile infection (CDI)

The incidence and severity of CDI increased worldwide in the last two decades, possibly linked with an excess of antibiotics and other treatment responsible for gut microflora disruption and with the concomitant emergence of hyper-virulent epidemic strains (eg. NAP1/BI/027) and the increase in potentially highly susceptible individuals[47,48]. According to age and the presence of risk factors and underlying conditions, CD may be responsible for a broad spectrum of diseases ranging from a self-limiting diarrhea to life-threatening conditions such as pseudomembranous colitis, toxic megacolon, intestinal perforation and septic shock.

The risk of CD colonization, virulence and recurrence is strongly related to disruption of the host microbiota. The fecal microbiome of patients with recurrent CDI displayed a marked and consistent decrease in diversity (phylotype richness) compared with healthy controls and those with initial CDI; a marked decrease in the phyla *Bacteroidetes* and *Firmicutes* was evident in recurrent CDI patients[49]. Relevant changes in short-chain fatty acids (acetate and butyrate) that usually represent the main nutrients of colonic flora have been reported[50,51]. The reported increase in treatment failure, ranging between 15 and 40%[52] according to risk factors and number of recurrences, is leading to the development of new therapeutic strategies including different antibiotics, active and passive immunotherapy and different attempts to reach a stable microflora.

Restoration of host microbiota is a key step in the treatment of CDI and prevention of relapses: the use of probiotics has been proposed in pediatrics, as already used for adults and for prevention of CDI. However, current evidence (mainly based on the use of *S. boulardii*) does not support this approach for the treatment of CDI and its recurrence and probiotics are not included in recent pediatric guidelines for the management of CDI[53] (see section 31.2.1 for the use of probiotics in the prevention of CDI). FMT is a reasonable therapeutic option that is rapidly gaining acceptance among new treatments for recurrent and severe CDI in adults, and more recently in children as well[54]. To date, only a few pediatric cases of recurrent CDI treated with FMT have been reported[55], and most of the supporting evidence comes from adult populations in which the efficacy was higher than 90%, with even better results in a subgroup analysis of high-quality studies[56]. In 2013, a three-arm, open-label, randomized controlled trial reported a significantly higher resolution rate in elderly patients with recurrent CDI allocated to FMT administered through the nasogastric route, in comparison to controls receiving standard vancomycin treatment with or without bowel lavage (81% vs 31% vs 23% respectively, p<0.001)[44]. Three out of the 16 patients in the FMT group required a second infusion with feces of different donors, and two of them had resolution of CD-associated diarrhea (with a cumulative resolution rate of 94%). More recently, Youngster and colleagues reported an overall 90% (95% CI, 68%–98%) efficacy in resolution of CD-associated diarrhea following administration of FMT using frozen encapsulated inoculum from unrelated donors[57]. This year Kronman and colleagues reported a case series of 10 children with recurrent CDI successfully treated with FMT administered through nasogastric tube, as done in the adult trial. The response rate was 90% and only one patient developed a recurrence after receiving a new treatment with antibiotics[58].

Administration of healthy donor stools restores patients' microbiota by increasing diversity in intestinal flora, making it similar to that of donors[44,59]. This modification in microflora composition seems to be long-lasting and fecal samples of recipients may demonstrate an increase in microbiota diversity and richness for as long as a year post-FMT[55]. Although the evidence of efficacy and the good acceptability by physicians and patients has led some authors to propose FMT as "first-line therapy" for CDI, fecal transplant with donor feces infusion is currently considered an alternative treatment for severe and recurrent cases and still needs close long-term follow-up and 'pharmaco-vigilance to determine long-term effects[44].

31.3.3.2 Inflammatory bowel diseases
As elegantly presented in chapter 19 by Ding and Hart, the etiology of inflammatory bowel diseases (IBD) is still unclear, and both genetic and environmental factors play a role in these diseases, although there is growing evidence that the

alteration of intestinal microbiota has a pivotal role in its pathogenesis. Whether the dysbiosis has an etiological role or is an epiphenomenon is not yet clear, but significant alterations of gut microbiota are related to drug treatment and the risk of relapses. Based on these principles, the use of FMT has been attempted in patients with IBD since 1989. when Justin Bennet self-administered a fecal infusion through enema to treat his active and refractory ulcerative colitis[60]. To date. many IBD patients treated with FMT have been described, most of them adults with ulcerative colitis[61]. However, Kunde *et al.* treated 10 children with refractory ulcerative colitis with a success rate of 78%[62]. Although some patients received FMT as treatment of a concomitant *Clostridium difficile* infection (they were IBD patients at relevant risk for CDI), an improvement of symptoms, and in some cases of the endoscopic picture, was reported in patients with active IBD in about 70% of cases[62].

Although the hypothesis that a modification of the gut microbiota may indirectly regulate the bowel and systemic inflammation is fascinating, the available evidence is limited. In addition, patients with such a chronic/relapsing conditions as IBDs would probably require several infusion of feces in order to maintain a stable microbiota.

31.3.3.3 Irritable bowel syndrome

Like probiotics, FMT has been proposed and tested in adults with IBS to reduce intestinal symptoms. Data available up to date are very limited, but show a potential role of microbiota transplantation in patients with functional constipation[63,64]. However, experience in children and adolescents is still lacking and, although some reports describe a 20-month follow-up, the long-term effects of this treatment in functional disorders are still unexplored.

Although the scientific evidence and popularity of FMT has grown rapidly in recent years, it should be noted that safety and efficacy of FMT still need attention before it should be included in the pipeline. The potential long-term consequences should be better (and specifically) explored, mainly in children and young adults, who have a longer time to develop potential side effects. In addition, new applications have been recently proposed for the treatment of extra-intestinal diseases with a potential microbiota involvement; in any case, benefits should always be balanced against risks before exposing patients with non-life-threatening conditions (such as IBS or constipation) to FMT until the consequences are better explored and defined.

31.4 Conclusion

The gut microbiota structure varies according to many factors including genetic factors, age, nutrition, disease and related treatments. Many of these factors develop during infancy and early childhood. Overall, increasing evidence supports the use of strategies targeting intestinal microecology in several intestinal diseases, including the administration of probiotics, prebiotics and instillation of a "healthy" microflora through the use of fecal microbiota transplantation; all these interventions have been tested in children. Strong evidence supports the use of selected probiotic strains (*Lactobacillus GG and S. boulardii*) for the treatment of acute infectious diarrhea and the prevention of antibiotic-associated diarrhea in children. *Lactobacillus reuteri* is effective in reducing crying and fussing in children with infant colic. The use of probiotics for the prevention of necrotizing enterocolitis is still debated; however, this intervention may have dramatic consequences in clinical practice by having a relevant impact on neonatal mortality. The mechanisms of efficacy of probiotics may involve local effects in the intestine such as the production of antibiotics or bacteriocines or interaction with receptors. Immunomodulation as a consequence of changes of intestinal bacterial populations has been implicated in the beneficial effects in inflammatory states. It is somehow surprising that the effects in gastroenteritis, which are seen unequivocally and rapidly early after probiotic administration and hence unlikely to be immune-mediated, are the more convincing ones based on clinical evidence. Fecal microbiota transplantation (FMT) is a promising treatment option for serious and recurrent *Clostridium difficile* infection, and current (although weak) evidence demonstrates consistent and excellent efficacy in clinical outcomes. It has been proposed also for children with inflammatory bowel diseases and irritable bowel syndrome, with conflicting results. However, more data on safety and long-term effects are needed before promoting this intervention for chronic, non-life-threatening conditions.

TAKE-HOME MESSAGE

- The gut microbiota exerts an active role in several intestinal diseases and could a target for therapy or prevention.

- Three different approaches have been proposed to change microbial composition: probiotics, prebiotics and fecal microbiota transplantation.

- Strong evidence (described as "conclusive") supports the use of selected probiotic strains (*Lactobacillus GG and S. boulardii*) for treatment of acute infectious diarrhea and the prevention of antibiotic-associated diarrhea in children. Less compelling evidence is available for the use of probiotics in adjunct to standard treatment in functional intestinal disorders and inflammatory bowel diseases. A major issue is the prevention of mortality in pre-term babies, in particular necrotizing enterocolitis.

- Another option to change microbial structure is the use of prebiotics. However, their efficacy is less clear than that of probiotics and the main use is prevention rather than treatment.

- A third and more recent option is fecal microbiota transplantation (FMT), which has been successfully used for serious and recurrent *Clostridium difficile* infection in children with chronic conditions, but whose long-term side effects have not been studied.

References

1 Buccigrossi V, Nicastro E, Guarino A. Functions of intestinal microflora in children. *Curr Opin Gastroenterol* 2013; **29**: 31–8.
2 Stecher B, Hardt WD. The role of microbiota in infectious disease. *Trends Microbiol* 2008; **16**: 107–14.
3 Stecher B, Hardt WD. Mechanisms controlling pathogen colonization of the gut. *Curr Opin Microbiol* 2011; **14**: 82–91.
4 Jalanka-Tuovinen J, Salonen A, Nikkila J *et al*. Intestinal microbiota in healthy adults: temporal analysis reveals individual and common core and relation to intestinal symptoms. *PloS One* 2011; **6**: e23035.

5 Aagaard K, Ma J, Antony KM, Ganu R, Petrosino J, Versalovic J. The placenta harbors a unique microbiome. *Sci Transl Med* 2014; **6**(237): 237ra65.

6 Penders J, Thijs C, Vink C *et al.* Factors influencing the composition of the intestinal microbiota in early infancy. *Pediatrics* 2006; **118**: 511–521.

7 van Nimwegen FA, Penders J, Stobberingh EE *et al.* Mode and place of delivery, gastrointestinal microbiota, and their influence on asthma and atopy. *J Allergy Clin Immunol* 2011; **128**: 948–955.

8 Sellitto M, Bai G, Serena G *et al.* Proof of concept of microbiome-metabolome analysis and delayed gluten exposure on celiac disease autoimmunity in genetically at-risk infants. *PloS One* 2012; **7**: e33387.

9 Schwiertz A, Jacobi M, Frick J-S *et al.* Microbiota in pediatric inflammatory bowel disease. *J Pediatr* 2010; **157**: 240–244.e1.

10 Hooda S, Boler BMV, Serao MCR *et al.* 454 pyrosequencing reveals a shift in fecal microbiota of healthy adult men consuming polydextrose or soluble corn fiber. *J Nutr* 2012; **142**: 1259–1265.

11 Saulnier DM, Riehle K, Mistretta TA *et al.* Gastrointestinal microbiome signatures of pediatric patients with irritable bowel syndrome. *Gastroenterology* 2011; **141**: 1782–1791.

12 Guarino A, Ashkenazi S, Gendrel D *et al.* European Society for Pediatric Gastroenterology, Hepatology, and Nutrition/European Society for Pediatric Infectious Diseases. Evidence-based guidelines for the management of acute gastroenteritis in children in Europe: update 2014. *J Pediatr Gastroenterol Nutr* 2014; **59**(1): 132–52.

13 Szajewska H, Skorka A, Ruszczynski M *et al.* Meta-analysis: Lactobacillus GG for treating acute gastroenteritis in children — updated analysis of randomised controlled trials. *Aliment Pharmacol Ther* 2013; **38**: 467–76.

14 Guarino A, Lo Vecchio A, Canani RB. Probiotics as prevention and treatment for diarrhea. *Curr Opin Gastroenterol* 2008; **25**: 18–23.

15 Johnston BC, Goldenberg JZ, Vandvik PO, Sun X, Guyatt GH. Probiotics for the prevention of pediatric antibiotic-associated diarrhea. *Cochrane Database Syst Rev.* 2011; **11**: CD004827.

16 Goldenberg JZ, Ma SS, Saxton JD, Martzen MR *et al.* Probiotics for the prevention of Clostridium difficile-associated diarrhea in adults and children. *Cochrane Database Syst Rev.* 2013 May 31; **5**: CD006095.

17 Miele E, Pascarella F, Giannetti E *et al.* Effect of a probiotic preparation (VSL#3) on induction and maintenance of remission in children with ulcerative colitis. *Am J Gastroenterol* 2009; **104**(2): 437–43.

18 Oliva S, Di Nardo G, Ferrari F *et al.* Randomised clinical trial: the effectiveness of Lactobacillus reuteri ATCC 55730 rectal enema in children with active distal ulcerative colitis. *Aliment Pharmacol Ther.* 2012; **35**(3): 327–34.

19 Hillila MT, Farkkila NJFM. Societal costs for irritable bowel syndrome: a population based study. *Scand J Gastroenterol* 2010; **45**: 582–591.

20 Simrén, M, Svedlund J, Posserud I *et al.* Health-related quality of life in patients attending a gastroenterology outpatient clinic: functional disorders versus organic diseases. *Clin Gastroenterol Hepatol Journal* 2006; **4**: 187–195.

21 Hyams J, Burke G, Davis P. Abdominal pain and irritable bowel syndrome in adolescents: a community-based study. *Journal of Pediatrics* 1996; **129**: 220–226.

22 Rasquin A, Di Lorenzo C, Forbes D *et al.* Childhood functional gastrointestinal disorders: child/adolescent. *Gastroenterology* 2006; **130**: 1527–1537.

23 Hyman PE, Milla P, Benninga MA *et al.* Childhood functional gastrointestinal disorders: Neonate/toddler. *Gastroenterology* 2006; **130**: 1519–1526.

24 Jeffery I, O'Toole P, Öhman L *et al.* An irritable bowel syndrome subtype defined by species-specific alterations in faecal microbiota. *Gut* 2012; **61**: 997–1006.

25 Rajilić-Stojanović M, Biagi E, Heilig HGHJ *et al.* Global and deep molecular analysis of microbiota signatures in fecal samples from patients with irritable bowel syndrome. *Gastroenterology* 2011; **141**: 1792–1801.

26 Korterink JJ, Ockeloen L, Benninga MA *et al.* Probiotics for childhood functional gastrointestinal disorders: a systematic review and meta-analysis. *Acta Paediatr* 2014; **103**(4): 365–72.

27 Wake M, Morton-Allen E, Poulakis Z *et al.* Prevalence, stability, and outcomes of cry-fuss and sleep problems in the first 2 years of life: prospective community-based study. *Pediatrics* 2006; **117**: 836–42.

28 Sung V, Hiscock H, Tang ML *et al.* Treating infant colic with the probiotic Lactobacillus reuteri: double blind, placebo controlled randomised trial. *BMJ* 2014; **348**: g2107.

29 Chau K, Lau E, Greenberg S *et al.* Probiotics for infantile colic: a randomized, double-blind, placebo-controlled trial investigating Lactobacillus reuteri DSM 17938. *J Pediatr* 2015; **166**(1): 74–78.

30 Sung V, Cabana MD, D'Amico F *et al.* Lactobacillus reuteri DSM 17938 for managing infant colic: protocol for an individual participant data meta-analysis. *BMJ Open* 2014; **4**(12): e006475.

31 Wang Y, Hoenig JD, Malin KJ *et al.* 16S rRNA gene-based analysis of fecal microbiota from preterm infants with and without necrotizing enterocolitis. *ISME J* 2009; **3**: 944–954.

32 Mai V, Young CM, Ukhanova M *et al.* Fecal microbiota in preterm infants prior to necrotizing enterocolitis. *PloS One* 2011; **6**: e20647.

33 Smith B, Bode S, Skov TH *et al.* Investigation of the early intestinal microflora in premature infants with/without necrotizing enterocolitis using two different methods. *Pediatr Res* 2012; **71**: 115–120.

34 Deshpande G, Rao S, Patole S, Bulsara M. Updated meta-analysis of probiotics for preventing necrotizing enterocolitis in preterm neonates. *Pediatrics* 2010; **125**: 921–930.

35 Alfaleh K1, Anabrees J, Bassler D, Al-Kharfi T. Probiotics for prevention of necrotizing enterocolitis in preterm infants. *Cochrane Database Syst Rev* 2011; (3): CD005496.

36 Wang Q, Dong J, Zhu Y. Probiotic supplement reduces risk of necrotizing enterocolitis and mortality in preterm very low-birth-weight infants: an updated meta-analysis of 20 randomized, controlled trials. *J Pediatr Surg* 2012; **47**(1): 241–8.

37 Bernardo WM1, Aires FT, Carneiro RM *et al.* Effectiveness of probiotics in the prophylaxis of necrotizing enterocolitis in preterm neonates: a systematic review and meta-analysis. *J Pediatr (Rio J)* 2013; **89**(1): 18–24.

38 Mihatsch WA, Braegger CP, Decsi T *et al.* Critical systematic review of the level of evidence for routine use of probiotics for reduction of mortality and prevention of necrotizing enterocolitis and sepsis in preterm infants. *Clin Nutr* 2012; **31**(1): 6–15.

39 Bonsante F, Iacobelli S, Gouyon JB. Routine probiotic use in very preterm infants: retrospective comparison of two cohorts. *Am J Perinatol* 2013; **30**(1): 41–6.

40 Brandt LJ. American Journal of Gastroenterology Lecture: Intestinal Microbiota and the role of fecal microbiota transplant (FMT) in the treatment of C. difficile infection. *Am J Gastroenterol* 2013; **108**: 177–185.

41 Bakken JS, Borody T, Brandt LJ *et al.* and the Fecal Microbiota Transplantation Workgroup. Treating Clostridium difficile infection with fecal microbiota transplantation. *Clin Gastroenterol Hepatol* 2011; **9**: 1044–9.

42 Hamilton MJ, Weingarden AR, Sadowsky MJ, Khoruts A. Standardized frozen preparation for transplantation of fecal microbiota for recurrent Clostridium difficile infection. *Am J Gastroenterol* 2012; **107**: 761–7.

43 Lo Vecchio A, Cohen MB. Fecal microbiota transplantation for Clostridium difficile infection: benefits and barriers. *Curr Opin Gastroenterol.* 2014; **30**(1): 47–53.

44 van Nood E, Vrieze A, Nieuwdorp M *et al.* Duodenal infusion of donor feces for recurrent Clostridium difficile. *N Engl J Med* 2013; **368**: 407–15.

45 Brandt LJ, Aroniadis OC, Mellow M *et al.* Long-term follow-up of colonoscopic fecal microbiotia transplant for recurrent Clostridium difficile infection. *Am J Gastroenterol* 2012; **107**: 1079–87.

46 Jiang ZD, Hoang LN, Lasco TM *et al.* Physician attitudes toward the use of fecal transplantation for recurrent Clostridium difficile infection in a metropolitan area. *Clin Infect Dis* 2013; **56**: 1059–60.

47 Walia R, Garg S, Song Y. Efficacy of fecal microbiota transplantation in 2 children with recurrent Clostridium difficile infection and its impact on their growth and gut microbiome. *J Ped Gastroenterol Nutr* 2014; **59**: 565–570.

48 Lo Vecchio A, Cohen MB. Fecal microbiota transplantation for Clostridium difficile infection: benefits and barriers. *Curr Opin Gastroenterol* 2014; **30**(1): 47–53.

49 Chang JY, Antonopoulos DA, Kalra A *et al.* Decreased diversity of the fecal microbiome in recurrent Clostridium difficile-associated diarrhea. *J Infect Dis* 2008; **197**: 435–8.

50 Lawley TD, Clare S, Walker AW *et al.* Targeted restoration of the intestinal microbiota with a simple, defined bacteriotherapy resolves relapsing Clostridium difficile disease in mice. *PLoS Pathog* 2012; **8**: e1002995.

51 Landy J, Al-Hassi HO, McLaughlin SD *et al.* Review article. Faecal transplantation therapy for gastrointestinal disease. *Alim Pharmacol Ther* 2011; **34**: 409–415.

52 Cohen SH, Gerding DN, Johnson S *et al*. Clinical practice guidelines for Clostridium difficile infection in adults: 2010 update by the Society for Healthcare Epidemiology of America (SHEA) and the Infectious Diseases Society of America (IDSA). *Infect Control Hosp Epidemiol* 2010; **31**: 431–55.

53 Schutze GE, Willoughby RE. Committee on Infectious Diseases, American Academy of Pediatrics. Clostridium difficile infection in infants and children. *Pediatrics* 2013; **131**(1): 196–200.

54 Russell G, Kaplan J, Ferraro M, Michelow IC. Fecal bacteriotherapy for relapsing Clostridium difficile infection in a child: a proposed treatment protocol. *Pediatrics* 2010; **126**: e239–42.

55 Walia R, Garg S, Song Y. Efficacy of fecal microbiotia transplantation in 2 children with recurrent Clostridium difficile nfection and its impact on their growth and gut microbiome. *J Ped Gastroenterol Nutr* 2014; **59**: 565–570.

56 Kassam Z, Lee CH, Yuan Y, Hunt RH. Fecal microbiota transplantation for Clostridium difficile infection: systematic review and meta-analysis. *Am J Gastroenterol* 2013; **108**: 500–8.

57 Youngster I, Russell GH, Pindar C *et al*. Oral, capsulized, frozen fecal microbiota transplantation for relapsing Clostridium difficile infection. *JAMA*. 2014; **312**(17): 1772–8.

58 Kronman MP, Nielson HJ, Adler AL *et al*. Fecal microbiota transplantation via nasogastric tube for recurrent Clostridium difficile infection in pediatric patients. *JPGN* 2015; **60**: 23–26.

59 Khoruts A, Dicksved J, Jansson JK, Sadowsky MJ. Changes in the composition of the human fecal microbiome after bacteriotherapy for recurrent Clostridium difficile-associated diarrhea. *J Clin Gastroenterol* 2010; **44**: 354–60.

60 Bennet JD, Brinkman M. Treatment of ulcerative colitis by implantation of normal colonic flora. *Lancet* 1989; **1**: 164.

61 Ianiro G, Bibbò S, Scaldaferri F *et al*. Fecal microbiota transplantation in inflammatory bowel disease: beyond the excitement. *Medicine* 2014; **19**: 1–11.

62 Kunde S, Pham A, Bonczyk *et al*. Safety, tolerability and clinical response after fecal microbiota transplantation in children and young adults with ulcerative coltiis. *J Pediatr Gastroenterol Nutr* 2013; **56**: 597–601.

63 Andrews P, Borody T. Bacteriotherapy for chronic constipation — a long term follow-up. *Gastroenterology* 1995; **108**: A563.

64 Smits LP, Bouter KEC, De Vos WM *et al*. Therapeutic potential of fecal microbiota transplantation. *Gastroenterology* 2013; **145**: 946–953.

CHAPTER 32

Microbial therapy for cystic fibrosis

Eugenia Bruzzese, Vittoria Buccigrossi, Giusy Ranucci and Alfredo Guarino
University of Naples, Naples, Italy

32.1 Introduction: pathophysiology of cystic fibrosis

Cystic fibrosis (CF) is an autosomal recessive inherited disease in Caucasians, caused by mutations in the cystic fibrosis transmembrane regulator (CFTR) gene, which leads to a deficiency or absence of functional CFTR proteins at the apical membrane of epithelial cells in several body systems[1]. CFTR protein is an ion channel that controls Cl⁻ and H_2O transepithelial flux. CFTR loss results in electrolyte abnormalities and an acidic pH that has consequences on mucus consistency in the extracellular milieu. As a result of CFTR impairment, epithelial cells in lungs, pancreas and intestine produce abnormally thick, sticky mucus. In addition to its role as a chloride channel, CFTR protein also affects bicarbonate transport, inducing the formation of dehydrated and viscous mucus in the lungs and digestive tract, with an increased risk of recurrent and chronic pulmonary infections and inflammation, pancreatic insufficiency (PI), CF-related liver disease (CFRLD) and diabetes (CFRD). The prevalence of CF-related chronic comorbidities has increased in line with the substantial extension of life expectancy of CF subjects[2].

The clinical hallmark of CF is recurrent severe pulmonary inflammation and infection, beginning in early childhood and responsible for progressive respiratory failure defined as pulmonary exacerbation. The most common respiratory pathogen in CF patients is *Pseudomonas aeruginosa*. Other bacteria originating from the lungs or gastrointestinal tract may act as pathogens in CF. Patients colonized with *P. aeruginosa* are at increased risk for pulmonary infections and persistent inflammation and frequently require multiple antibiotic treatment. Pulmonary exacerbations are the leading cause of morbidity with a progressive decline in lung function. The pathogenesis of intestinal involvement in CF is multi-factorial and several mechanisms are involved[3]: (i) loss of the functional CFTR chloride channel with poorly hydrated and acidic luminal fluid; (ii) reduced digestive enzyme activity from the exocrine pancreas, which impairs digestion and absorption; (iii) increased mucus production with viscous mucus, and (iv) impaired mucus clearance secondary to disturbed motility.

The Human Microbiota and Chronic Disease: Dysbiosis as a Cause of Human Pathology, First Edition.
Edited by Luigi Nibali and Brian Henderson.
© 2016 John Wiley & Sons, Inc. Published 2016 by John Wiley & Sons, Inc.

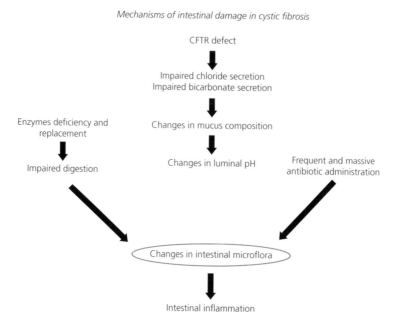

Figure 1 The intestine is a target organ in CF and a major role may be played by the disrupted microflora. This contributes to the intestinal inflammation and may affect the rates and severity of respiratory involvement. Probiotic administration may in part restore intestinal microbiota, reduce intestinal inflammation and, together with other treatment, reduce the risk of pulmunary exacerbations, ultimately improving the long-term outcome of this progressive disease.

Although CF is typically considered a disease targeting the respiratory tract, CFTR defects also affect organs such as the intestine[3]. The intestinal epithelium shows a proximal-to-distal gradient with the highest CFTR expression in the duodenum and decreasing distally along the small intestine to the ileum[4]. The gastrointestinal tract may be strongly affected by CF, with negative impact on nutritional status and poor nutritional status, associated with pulmonary deterioration and predicting a fatal outcome[5]. Mouse models of CF have greatly contributed to understanding of the physiopathology of gut involvement. In all affected epithelia, including the CF intestine, mutations in the *CFTR* gene result in total or partial loss of chloride channel function. Combined loss of fluid volume and abnormal acidity in the intestinal lumen[6] lead to the accumulation of mucus, which will affect digestion, nutrient absorption, motility, the gut microbiota and gut inflammation (Figure 1).

32.2 Intestinal inflammation in CF

Chronic intestinal inflammation has been demonstrated in cystic fibrosis patients, even in the absence of gastrointestinal symptoms. In a study comparing duodenal endoscopy, specimens from 14 pancreatic-insufficient patients and 20 healthy controls, the mucosal morphology appeared normal, but increased infiltration of the lamina propria by mononuclear cells expressing inflammatory markers, such as the intercellular adhesion molecules ICAM-1, CD-25, IL-2 and IFN-gamma, was described[7]. Considering that patients with chronic pancreatitis but not CF do not show duodenal inflammation, the inflammation in CF could be the direct result of an abnormal immune response or might be linked to the CFTR defect. Gut lavage collected from 21 asymptomatic pancreatic-insufficient children with CF showed significant increases in the concentrations of albumin, IgG and IgM, eosinophilic cationic protein, neutrophil elastase, IL-8 and IL-1 compared to 12 controls[8]. This data supports the hypothesis that intestinal immune activation could be linked to the CFTR defect. More recently, using wireless capsule endoscopy in 41 patients with CF, including 13 pancreatic-sufficient patients, signs of small-bowel inflammation were identified in 60% of the CF cohort[9]. In a study of unselected CF children the authors evaluated fecal calprotectin, a non-specific marker of intestinal inflammation that is increased in inflammatory bowel diseases, and found that 29 out of 30 pancreatic-insufficient CF patients showed an increased fecal calprotectin concentration, significantly higher than that found in healthy controls,[10] raising the hypothesis that intestinal inflammation could be a major feature of CF, independent of other features. However, fecal calprotectin was increased in pancreatic-insufficient patients but was in the normal range in pancreatic-sufficient patients in a subset of the Werlin *et al.* study[9].

More recently, an increase in fecal calprotectin concentration was confirmed in CF patients, although the authors were unable to demonstrate a correlation between increased fecal calprotectin concentration and pancreatic insufficiency[11]. The authors found that 10 of 16 CF patients with pancreatic sufficiency had elevated fecal calprotectin concentration, with increase comparable to those of patients with pancreatic insufficiency. This supports the hypothesis that intestinal inflammation is a feature of CF regardless of pancreatic status and that "CF enteropathy" is independent of pancreatic involvement[11].

The course of CF-related intestinal inflammation is as yet unclear. Patients with lower forced expiratory volume in 1 second (FEV1) showed higher fecal calprotectin concentrations. Lower FEV1 is associated with worse pulmonary status. The increase in respiratory secretions probably contributes to increased intestinal inflammation[8]. Bacteria and other contents of sputum may be involved in direct stimulation of the intestinal mucosa and calprotectin in the sputum may itself increase the fecal levels. However, another hypothesis to explain intestinal inflammation in CF is that inflammatory triggers such as cytokines or bacterial products (e.g. lipopolysaccharide (LPS)) activate intestinal monocytes and epithelial cells. Increased levels of fecal calprotectin concentration were found in children with CF and with aberrant intestinal microbial composition. CF children showed an abundance of *E. coli* in fecal microbiota compared to healthy controls. In particular, a significant correlation was found between *E. coli* abundance and fecal calprotectin and the fractional amount of fat in the stools, suggesting that *E. coli* may directly contribute to cystic fibrosis gastrointestinal dysfunction[12].

32.3 Dysbiosis in CF

CF patients have multiple risk factors for altered intestinal microbiota, including thick intestinal mucus, constipation and slow intestinal motility. In addition, they must frequently undergo heavy courses of antibiotics[13]. Fecal microbiota population structure showed a temporal instability and reduced species richness in CF children compared with healthy controls[14,15]. Dysbiosis in CF patients included a reduced presence of Bifidobacteria, Clostridium cluster IV and XIVa and Firmicutes and increased concentrations of Proteobacteria[16,17]. In CF children, dysbiosis is an aberrant early event due to a very unstable gut microbial community. An aberrant microbial structure was observed in CF children compared with healthy controls. In particular, the levels of *Eubacterium rectale, Bacteroides uniformis, Bacteroides vulgatus, Bifidobacterium adolescentis, Bifidobacterium catenulatum*, and *Faecalibacterium prausnitzii* were decreased in children with CF[16,17].

Salami and Marsland in chapter 24 suggested that through pharyngeal reflux and microaspiration of gastro-intestinal and breastmilk contents in neonates, gut-derived bacteria could be the colonizers of airways in humans. Along these lines, the existence of a gut-lung axis in CF has recently been hypothesized (Figure 2). However, experimental support for a link between intestinal microbiota and bacterial colonization of the respiratory tract is far from being obtained. A very interesting aspect is the early development of gut and airway microbiota in CF infants. A recent study analyzed gut and airway microbiota in CF newborns approximately every three months up to the first 21 months of age. Interestingly, the gut and respiratory tract microbiota showed similar results with a clear association of a large number of genera that were present in the gut prior to colonization of the respiratory tract[18]. Also, the severity of the disease correlates with intestinal dysbiosis in CF[19]. The severity of CF depends on host genetic variation. There are several genetic mutations responsible for the maturation process and transfer of CFTR to the apical compartment of cells. One of the most common mutations, the

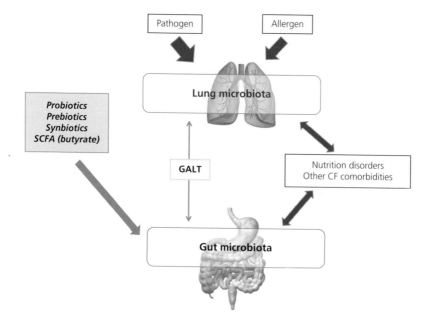

Figure 2 Gut-lung axis in CF. A link between intestinal microbiota and bacterial colonization of the respiratory tract has been hypothesized. Manipulation of intestinal microbiota through the administration of probiotics, prebiotics and synbiotics may influence lung function and nutritional status in patients with CF (modified by reference 25). SCFA: short-chain fatty acids; GALT: gut-associated lymphatic tissue.

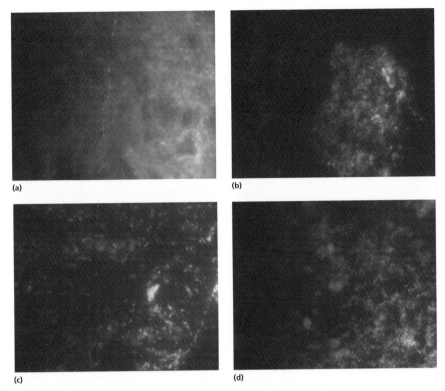

Figure 3 Habitually constitutive bacteria intestinal microflora evaluated by FISH (fluorescence in situ hybridization). Top: Total bacteria and Faecalibacterium prausnitzii were evaluted by EUB (green) and Fpra (red) probes in healthy (a) and cystic fibrosis (b) children. Bottom: Total bacteria and Eubacterium rectale were evaluted by EUB (green) and Erec (red) probes in healthy (c) and cystic fibrosis (d) children. Images kindly provided by Dr. V. Buccigrossi. (*see color plate section for color details*).

F508del, is associated with the most severe phenotypes of CF. It was reported that dysbiosis is enhanced in homozygous F508del patients as shown by the increase in harmful bacteria, such as *Escherichia coli* and *Eubacterium biforme*, and a decrease in beneficial species such as *Faecalibacterium prausnitzii*, *Bifidobacterium spp* and *Eubacterium limosum*[19].

CF is characterized by an increased susceptibility to respiratory infections, and as a consequence CF patients frequently undergo antibiotic therapy. Hence, dysbiosis also depends on the massive antibiotic use (Figure 3). The pressure of high doses of multiple antimicrobial agents to which CF patients are frequently exposed has a dramatic impact on the bacterial colonization, especially in the intestines[20]. Antibiotics induce the loss of *Oxalobacterformigenes*, an important commensal agent that metabolizes oxalate[21]. As a consequence, CF patients are at risk of hyperoxaluria and formation of calcium-oxalate kidney stones. On the other hand, *Clostridium difficile* frequently colonizes CF patients[22]. A reduced concentration of *Bacteroides uniformis*, *Bacteroides vulgates*, *Eubacterium rectale*, *Bacteroidetes* and *Faecalibacterium prausnitzii* was observed compared to healthy controls[23]. In addition, *Escherichia coli* was more abundant in children with CF and was correlated with intestinal inflammation and with intestinal dysfunction in children with CF[12].

Several bacterial species are associated with mucosal inflammation. For example, butyrate is an important energy source for the colonic mucosa, protecting it from colitis and colorectal cancer and promoting the normal development of colonic epithelial cells (discussed in chapter 27 by Sobhani). *Eubacterium rectale* may have a protective role against intestinal inflammation through production of butyrate. In addition, the reduction of *Faecalibacterium prausnitzii* is considered a hallmark of chronic inflammatory disease[24]. A significant inverse correlation between the richness of CF microbiota and intestinal inflammation was observed[23] and children with fecal calprotectin concentrations greater than 200 mg/g showed a significantly lower number of Bacteroidetes in DGGE analysis[23].

In summary, several lines of evidence support the concept that the intestinal environment is abnormal in CF and, together with massive antibiotic therapy, promotes the development of an aberrant microbiota. This is associated with intestinal and respiratory inflammation.

32.4 Microbial therapy in CF

Dietary approaches (mainly manipulation of fat and indigestible carbohydrates) or probiotic and/or prebiotic administration may provide an effective early intervention to restore the abnormal intestinal microbiota in CF, namely to enhance gut microbiota diversity in order to modify the course of lung colonization, thereby ultimately improving patient outcomes. In particular, oral administration of probiotics and prebiotics or a combination of both (the so called "synbiotic approach" described in chapter 30) could influence the composition of the airway microbiota, either indirectly, through the release of bacterial products or metabolites that reach the lung and favor the outgrowth of probiotic bacteria, or directly, via microaspiration of the probiotic strain from the intestinal tract to the airways. These theoretical mechanisms may restore a health-promoting microbiota and have a beneficial effect on the course of the disease[25]. Given the continuous increase of bacterial resistance to antibiotics and the lack of novel antibiotics, the interest in these bacteria for management of broncho-pulmonary exacerbations in CF is increasing. Neither breastfeeding nor the introduction of non-solid food had effects on biodiversity of gut or airway microbiota in CF over time. On the other hand, statistical clustering of the gut and airway microbiota samples demonstrated that specific groups of bacteria are associated with solid-food introduction and breast-feeding[18]. There is increasing evidence that probiotics modulate immune response in the lung; in particular, gut microbial stimulation can enhance T regulatory response in the airway through an interaction with the gut-associated lymphatic tissue (GALT), such as Peyer's patch cells, influencing pulmonary inflammatory response[25,26]. The mechanisms underlying these effects of probiotics are far from being understood. Intranasal and oral probiotics induced up-regulation of natural killer cells and macrophage activity in the airway mucosa, expansion of T-regulatory cells, increase of the IgA-secretory cells in the bronchial mucosa, production of antibacterial compounds, and inhibition of virulence factors[27,28].

Probiotics have an anti-inflammatory effect activating specific microbe-derived ligands signaling pathways[26]. However, few probiotic strains have a role in modifying the course in lung involvement. In animal studies lactobacilli have profound immunoregulatory effects on the lung, but results of clinical trials in humans have been highly variable. Strain differences may in part explain the observed variability. In humans administration of Lactobacillus GG (LGG) negatively influences the incidence of ventilator-associated pneumonia[29] and reduced respiratory infections in healthy[30] and hospitalized children[31]. The first evidence of the potential benefits of probiotic administration in CF came from a prospective randomized placebo-controlled cross-over trial performed in two groups of patients with CF chronically colonized by *Pseudomonas aeruginosa* (PA). Nineteen children were given LGG for six months followed by placebo (oral rehydration solution) for the subsequent six months. At the same time, 19 children were given the placebo for six months and then the probiotic for the same period of time. The patients on LGG had a significant reduction in intestinal inflammation and of episodes of pulmonary exacerbations and hospitalization rates, with a decrease in IgG, suggesting that there is a relationship between intestinal and pulmonary inflammation. The intake of this probiotic was associated to a significant increase of the maximal FEV1 compared to the placebo as well as to a significant increase in body weight[32].

Those important findings were in part confirmed by a study with a commercially available mixture of probiotics (*Lactobacilllus, Bifidobacterium and Streptococcus thermophiles* spp.) given to 10 patients chronically infected with PA for six months with a significant reduction in the pulmonary exacerbation rate compared to the two previous years[33]. In addition, administration of *Lactobacillus reuteri* (LR) was effective in reducing the number of pulmonary exacerbations and of upper-respiratory-tract infection in patients with CF[34]. In this prospective randomized, double-blind, placebo-controlled study, 61 patients with CF were randomly assigned to receive 10^{10} colony-forming units LR in drops per day or placebo for six months. However, probiotics did not change hospitalization rates, FEV1 or fecal calprotectin and cytokines[34]. In a population of non-selected children and adolescents with CF, LGG supplementation reduced intestinal inflammation as assessed by fecal calprotectin concentration and rectal nitric oxide[10,23]. LGG also restored intestinal microbiota as shown by an increase in *Bacteroides* counts[23]. Therefore LGG significantly decreases intestinal inflammation (in particular calprotectin levels) and increases digestive comfort, restoring in part the normal microbial homeostasis as shown by the total bacterial density and the increase in microbial diversity. Similar results were reported with *Lactobacillus reuteri* [17]. However, the effects of probiotics are likely to be time-, dose- as well as strain-dependent, indicating the need for comparative clinical trials on probiotic therapy in CF. Future work is required in order to identify reliable biomarkers of intestinal inflammation to be used in long term clinical trials on targeted anti-inflammatory therapies in the gut[35].

Less is known about the potential use of prebiotics in CF management. In one study in rats, administration of prebiotics such as the non-digestible carbohydrate fructo-oligosaccharide increased cecal active GLP-1 levels acting via modulation of gut hormones induced by gut microbiota variations[36,37]. Active GLP-1 levels are influenced by pro-inflammatory cytokines and reduced levels of active GLP-1 may be associated with CFRD onset[36]. Modification of gut microbiota by changing dietary content of indigestible carbohydrate may improve under-nutrition in human CF[38]. Prebiotic supplementation of probiotic products is proposed to prolong the intestinal survival of the probiotic strain and therefore enhance immunomodulatory capacity. Human breast milk contains oligosaccharides as well as lactic acid bacteria, which makes it a natural synbiotic. Synbiotics have produced some positive results in the treatment of conditions associated with chronic inflammation[39]. Thus, synbiotic approaches might be more effective in controlling CF than probiotic or prebiotic treatment alone and might therefore be an effective prophylactic or therapeutic intervention. Finally, manipulating gut microbiota by changing dietary content of indigestible carbohydrate and short-chain fatty acids (butyrate in particular) may both improve undernutrition and have an anti-inflammatory effect in gut microbiota in CF and indirectly on the lung[25].

32.5 Conclusion

In conclusion, the gastrointestinal tract may be strongly affected by CF with a negative impact on nutritional status. CFTR dysfunction results in chronic inflammation and in an altered microbial composition in lung and intestine. Some evidence for the gut-lung axis in CF comes from the link between intestinal microbiota and bacterial colonization of the respiratory tract. In addition, the effect of a restructure of intestinal microflora on intestinal inflammation and respiratory lung function provides new and compelling proof. Probiotics and/or prebiotics in CF have shown some promise, including potential benefits in nutritional status, energy intake and respiratory function. However, their effect requires further clarification before therapeutic implementation.

TAKE-HOME MESSAGE

- The intestine is a target organ in cystic fibrosis and intestinal inflammation has been demonstrated in patients even in the absence of gastrointestinal symptoms.

- The pathogenesis of intestinal involvement in CF is multifactorial and several mechanisms are involved.

- Dysbiosis is a feature of CF children as a consequence of CFTR dysfunction, frequent use of antibiotics, pancreatic enzyme supplementation and modification of intraluminal pH.

- Intestinal inflammation and respiratory lung function are associated with aberrant gut microbiota composition.

- A restructure of gut microbiota, through probiotics and/or probiotics, might improve nutritional status, energy intake and respiratory function.

References

1 Garcia MAS, Quinton PM, Yang N. Normal mouse intestinal mucus release requires cystic fibrosis transmembrane regulator-dependent bicarbonate secretion. *J Clin Invest* 2009; **119**: 2613e22.
2 Parkins MD, Parkins VM, Rendall JC, Elborn S. Changing epidemiology and clinical issues arising in an ageing cystic fibrosis population. *Ther Adv Respir Dis* 2011; **5**: 105 e19.
3 De Lisle RC, Borowitz D. The cystic fibrosis intestine. *Cold Spring Harb Perspect Med* 2013; **3**: a009753.
4 Strong TV, Boehm K, Collins FS. Localization of cystic fibrosis transmembrane conductance regulator mRNA in the human gastrointestinal tract by in situ hybridization. *J Clin Invest* 1994; **93**: 347–54.
5 Stallings VA, Stark LJ, Robinson KA, Feranchak AP, Quinton H; Clinical Practice Guidelines on Growth and Nutrition Subcommittee; Ad Hoc Working Group. Evidence-based practice recommendations for nutrition-related management of children and adults with cystic fibrosis and pancreatic insufficiency: results of a systematic review. *J Am Diet Assoc* 2008; **108**(5): 832–9.

6 De Lisle RC, Isom KS, Ziemer D, Cotton CU. Changes in the exocrine pancreas secondary to altered small intestinal function in the CF mouse. *Am J Physiol Gastrointest Liver Physiol* 2001; **281**(4): G899–906.

7 Raia V, Maiuri L, De Ritis G, De Vizia B, Vacca L, Conte R, Auricchio S, Londei M.. Evidence of chronic inflammation in morphologically normal small intestine of cystic fibrosis patients. *Pediatr Res* 2000; **47**(3): 344–350.

8 Smyth RL, Croft NM, O'Hea U, Marshall TG, Ferguson A. Intestinal inflammation in cystic fibrosis. *Arch Dis Child* 2000; **82**(5): 394–399.

9 Werlin SL, Benuri-Silbiger I, Kerem E, Adler SN, Goldin E, Zimmerman J, Malka N, Cohen L, Armoni S, Yatzkan-Israelit Y, Bergwerk A, Aviram M, Bentur L, Mussaffi H, Bjarnasson I, Wilschanski M. Evidence of intestinal inflammation in patients with cystic fibrosis. *J Pediatr Gastroenterol Nutr* 2010; **51**(3): 304–8.

10 Bruzzese E, Raia V, Gaudiello G, Polito G, Buccigrossi V, Formicola V, Guarino A. Intestinal inflammation is a frequent feature of cystic fibrosis and is reduced by prebiotic administration. *Aliment Pharmacol Ther* 2004; **20**: 813.819.

11 Rumman N, Sultan M, El-Chammas K, Goh V, Salzman N, Quintero D, Werlin S. Calprotectin in cystic fibrosis. *BMC Pediatr.* 2014; **29**; 14: 133. doi: 10.1186/1471-2431-14-133.

12 Hoffman LR, Pope CE, Hayden HS, Heltshe S, Levy R, McNamara S, Jacobs MA, Rohmer L, Radey M, Ramsey BW, Brittnacher MJ, Borenstein E, Miller SI. Escherichia coli dysbiosis correlates with gastrointestinal dysfunction in children with cystic fibrosis. *Clin Infect Dis* 2014; **58**: 396–9.

13 Fridge JL, Conrad C, Gerson L, Castillo RO, Cox K. Risk factors for small bowel bacterial overgrowth in cystic fibrosis. *J Pediatr Gastroenterol Nutr* 2007; **44**(2): 212–218.

14 Duytschaever G, Huys G, Bekaert M, Boulanger L, De Boeck K, Vandamme P. Cross-sectional and longitudinal comparisons of the predominant fecal microbiota compositions of a group of pediatric patients with cystic fibrosis and their healthy siblings. *Appl Environ Microbiol* 2011; **77**: 8015–24.

15 Scanlan PD, Buckling A, Kong W, Wild Y, Lynch SV, Harrison F. Gut dysbiosis in cystic fibrosis. *J Cyst Fibros* 2012; **11**: 454–5.

16 Duytschaever G, Huys G, Bekaert M, Boulanger L, De Boeck K, Vandamme P. Dysbiosis of bifidobacteria and Clostridium cluster XIVa in the cystic fibrosis fecal microbiota. *J Cyst Fibros* 2013; **12**: 206–15.

17 Del Campo R, Garriga M, Pérez-Aragón A, Guallarte P, Lamas A, Máiz L, Bayón C, Roy G, Cantón R, Zamora J, Baquero F, Suárez L. Improvement of digestive health and reduction in proteobacterial populations in the gut microbiota of cystic fibrosis patients using a Lactobacillus reuteri probiotic preparation: A double blind prospective study. *J Cyst Fibros.* 2014; **13**: 716–22.

18 Madan JC, Koestler DC, Stanton BA, Davidson L, Moulton LA, Housman ML, Moore JH, Guill MF, Morrison HG, Sogin ML, Hampton TH, Karagas MR, Palumbo PE, Foster JA, Hibberd PL, O'Toole GA. Serial analysis of the gut and respiratory microbiome in cystic fibrosis in infancy: interaction between intestinal and respiratory tracts and impact of nutritional exposures. *MBio* 2012; **3**. pii: e00251-12.

19 Schippa S, Iebba V, Santangelo F, Gagliardi A, De Biase RV, Stamato A, Bertasi S, Lucarelli M, Conte MP, Quattrucci S. Cystic fibrosis transmembrane conductance regulator (CFTR) allelic variants relate to shifts in faecal microbiota of cystic fibrosis patients. *PLoS One.* 2013; **8** :e61176

20 Duytschaever G, Huys G, Boulanger L, De Boeck K, Vandamme P. Amoxicillin-clavulanic acid resistance in fecal Enterobacteriaceae from patients with cystic fibrosis and healthy siblings. *J Cyst Fibros* 2013; **12**: 780–3.

21 Sidhu H, Hoppe B, Hesse A, Tenbrock K, Brömme S, Rietschel E, Peck AB. Absence of Oxalobacter formigenes in cystic fibrosis patients: a risk factor for hyperoxaluria. *Lancet* 1998; **352**: 1026–9.

22 Yahav J, Samra Z, Blau H, Dinari G, Chodick G, Shmuely H. Helicobacter pylori and Clostridium difficile in cystic fibrosis patients. *Dig Dis Sci* 2006; **51**: 2274–9.

23 Bruzzese E, Callegari ML, Raia V, Viscovo S, Scotto R, Ferrari S, Morelli L, Buccigrossi V, Lo Vecchio A, Ruberto E, Guarino A. Disrupted intestinal microbiota and intestinal inflammation in children with cystic fibrosis and its restoration with Lactobacillus GG: a randomised clinical trial. *PLoS One* 2014; **9**: e87796.

24 Miquel S, Martín R, Rossi O, Bermúdez-Humarán LG, Chatel JM, Sokol H, Thomas M, Wells JM, Langella P. Faecalibacterium prausnitzii and human intestinal health. *Curr Opin Microbiol* 2013; **16**: 255–61.

25 Li L, Somerset S. The clinical significance of the gut microbiota in cystic fibrosis and the potential for dietary therapies. *Clinical Nutrition* 2014; **33**: 571–580.

26 Forsythe P. Probiotics and lung disease. *Chest* 2011; **139**(4): 901–908.

27 Harata G, He F, Kawase M, Hosono A, Takahashi K, Kaminogawa S. Differentiated implication of Lactobacillus GG and L. gasseri TMC0356 to immune responses of murine Peyer's patch. *Microbiol Immunol* 2009; **53**(8): 475–480.

28 Fink LN, Zeuthen LH, Christensen HR, Morandi B, Frokiaer H, Ferlazzo G. Distinct gut-derived lactic acid bacteria elicit divergent dendritic cell-mediated NK cell responses. *Int Immunol* 2007; **19**: 1319–1327.

29 Morrow LE, Kollef MH, Casale TB. Probiotic prophylaxis of ventilator-associated pneumonia: a blinded, randomized, controlled trial. *Am J Respir Crit Care Med* 2010; **182**: 1058.1064.

30 Hatakka K, Savilahti E, Ponka A, Meurman JH, Poussa T, Nase L, Saxelin M, Korpela R. Effect of long term consumption of probiotic milk on infections in children attending day care centres: double blind, randomised trial. *BMJ* 2001; **322**: 1327.

31 Hojsak I, Abdovic S, Szajewska H, Milosevic M, Krzaric Z, Lolacek S. Lactobacillus GG in the prevention of nosocomial gastrointestinal and respiratory tract infections. *Pediatrics* 2010; **125**: e1171–7.

32 Bruzzese E, Raia V, Spagnuolo MI, Volpicelli M, De Marco G, Maiuri L, Guarino A. Effect of Lactobacillus GG supplementation on pulmonary exacerabtions in patients with cystic fibrosis: a pilot study. *Clinical Nutrition* 2007; **26**: 322–328.

33 Weiss B, Bujanover Y, Yahav Y, Vilozni D, Fireman E, Efrati O. Probiotic supplementation affects pulmonary exacerbations in patients with cystic fibrosis: a pilot study. *Pediatric Pulmonology* 2010; **45**: 536–540.

34 Di Nardo G, Oliva S, Menuchella A, Pistelli R, De Biase RV, Patriarchi F, Cucchiara S, Stronati L. Lactobacillus reuteri ATCC55730 in cystic fibrosis. *J Pediatr Gastroenterol Nutr* 2014; **58**: 81–86.

35 Munck A. Cystic fibrosis: evidence for gut inflammation. *Int J Biochem Cell Biol* 2014; **52**: 180–183.

36 Kok NN, Morgan LM, Williams CM, Roberfroid MB, Thissen J-P, Delzenne NM. Insulin, glucagon-like peptide 1, glucose-dependent insulin-tropic polypeptide and insulin-like growth factor I as putative mediators of the hypolipidemic effect of oligofructose in rats. *J Nutr* 1998; **128**: 1099e103.

37 Davis CD, Milner JA. Gastointestinal microflora, food components and colon cancer prevention. *J NutrBiochem* 2009; **20**: 743–752.

38 Vaisman N, Tabachnik E, Sklan D. Short-chain fatty acid absorption in patients with cystic fibrosis. *J Pediatr Gastroenterol Nutr* 1992; **15**: 146e9.

39 Kolida S, Gibson GR. Synbiotics in health and disease. *Annu Rev Food Sci Technol* 2011; **2**: 373–393.

Index

Page numbers in **bold** refer to figures and tables.

The Human Microbiota and Chronic Disease: Dysbiosis as a Cause of Human Pathology, First Edition.
Edited by Luigi Nibali and Brian Henderson.
© 2016 John Wiley & Sons, Inc. Published 2016 by John Wiley & Sons, Inc.

CHAPTER 1

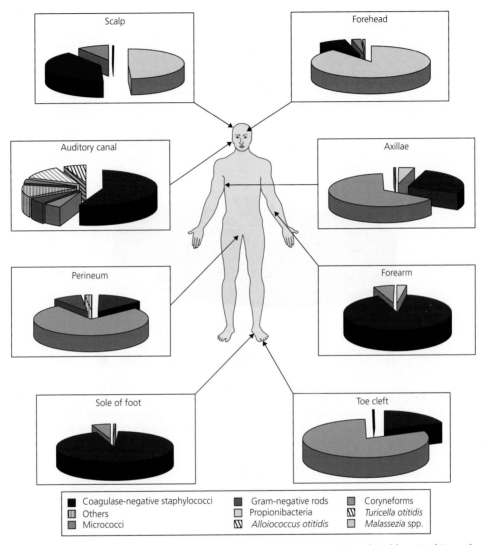

Figure 1 Relative proportions of the various organisms comprising the cultivable microbiota of a number of skin sites (reproduced with permission from Wilson M. *Bacteriology of humans: an ecological perspective*. Oxford: Wiley-Blackwell, 2008).

The Human Microbiota and Chronic Disease: Dysbiosis as a Cause of Human Pathology, First Edition.
Edited by Luigi Nibali and Brian Henderson.
© 2016 John Wiley & Sons, Inc. Published 2016 by John Wiley & Sons, Inc.

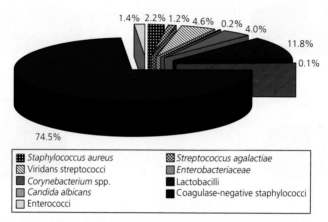

Figure 10 Relative proportions of the various microbes that comprise the cultivable microbiota of the labia majora of post-menarcheal/pre-menopausal females. Data represent the mean values obtained in a study involving 102 individuals (reproduced with permission from Wilson M. *Bacteriology of humans: an ecological perspective*. Oxford: Wiley-Blackwell, 2008).

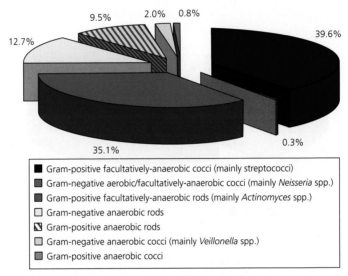

Figure 13 Relative proportions of organisms comprising the cultivable microbiota of the gingival crevice. Data are derived from a study involving seven healthy adults (reproduced with permission from Wilson M. *Bacteriology of humans: an ecological perspective*. Oxford: Wiley-Blackwell, 2008).

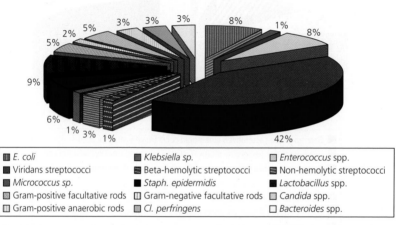

Figure 17 Relative proportions of organisms comprising the cultivable microbiota of the duodenal mucosa of 26 healthy adults (reproduced with permission from Wilson M. *Bacteriology of humans: an ecological perspective.* Oxford: Wiley-Blackwell, 2008).

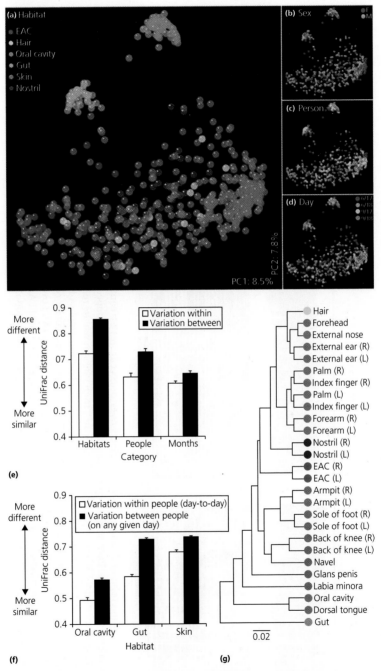

Figure 1 16S rRNA gene surveys reveal hierarchical partitioning of human-associated bacterial diversity. (a to d) Communities clustered using principal component analysis of the unweighted UniFrac distance matrix. Each point corresponds to a sample colored by (a) body habitat, (b) host sex, (c) host individual, or (d) collection date. The same plot is shown in each panel. F, female; M, male. (e and f) Mean (± SEM) unweighted UniFrac distance between communities. In (e) habitats are weighted equally and in (f) skin comparisons are within sites. (g) UPGMA (Unweighted Pair Group Method with Arithmetic Mean) clustering of composite communities from the indicated locales. Leaves are colored according to body habitat as in (a). R, right; L, left. Reproduced with permission from Costello EK, Lauber CL, Hamady M, Fierer N, Gordon JI, Knight R. Bacterial community variation in human body habitats across space and time. *Science* 2009; **326**: 1694–1697.

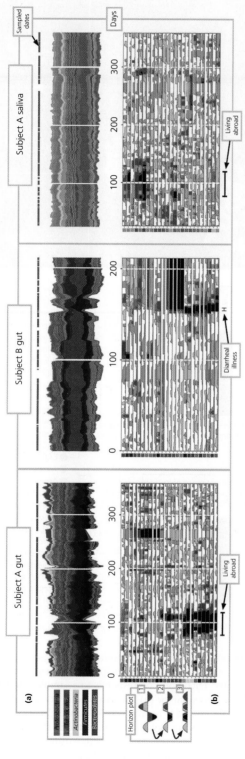

Figure 2 Gut and salivary microbiota dynamics in two subjects over one year. (a) Stream plots showing OTU (Operational Taxonomic Unit) fractional abundances over time. Each stream represents an OTU and streams are grouped by phylum: Firmicutes (purple), Bacteroidetes (blue), Proteobacteria (green), Actinobacteria (yellow), and Tenericutes (red). Stream widths reflect relative OTU abundances at a given time point. Sampled time points are indicated with gray dots over each stream plot. (b) Horizon graphs of most common OTUs abundance over time. Warmer regions indicate date ranges where a taxon exceeds its median abundance and cooler regions denote ranges where a taxon falls below its median abundance. Colored squares on the vertical axis correspond to stream colors in (A). Lower black bars span Subject A's travel abroad (days 71 to 122) and Subject B's Salmonella infection (days 151 to 159). For further details see David LA, Materna AC, Friedman J *et al.* Host lifestyle affects human microbiota on daily timescales. *Genome Biol* 2014; **15**: R89. Reproduced with permission.

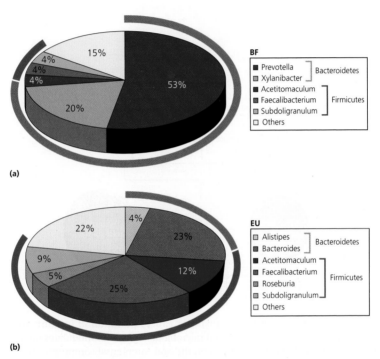

Figure 3 16S rRNA gene surveys reveal a clear separation of two children populations in Burkina Faso (a) and Europe (b). Pie charts of median values of bacterial genera present in fecal samples of Burkina Faso and European children (>3%). Rings represent corresponding phylum (Bacteroidetes in green and Firmicutes in red) for each of the most frequently represented genera. Adapted from De Filippo C, Cavalieri D, Di Paola M *et al.* Impact of diet in shaping gut microbiota revealed by a comparative study in children from Europe and rural Africa. *Proc Natl Acad Sci USA* 2010; **107**: 14691–14696. Reproduced with permission.

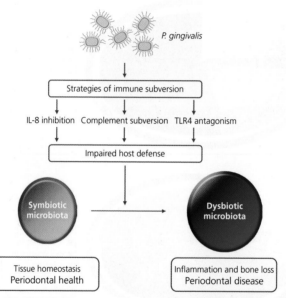

Figure 7 The red complex bacterium *P. gingivalis* causes periodontal inflammation and bone loss by remodeling the oral commensal microbiota. *P. gingivalis* modulates innate host defense functions that can have global effects on the oral commensal community. Immune subversion of IL-8 secretion, complement activity, or TLR4 activation can result in an impaired host defense. The inability of the host to control the oral commensal microbial community in turn results in an altered oral microbial composition and an increased microbial load. This alteration from a symbiotic to a dysbiotic microbiota is responsible for pathologic inflammation and bone loss. Reproduced with permission from Darveau RP, Hajishengallis G, Curtis MA. *Porphyromonas gingivalis* as a potential community activist for disease. *J Dent Res* 2012; **91**: 816–820.

CHAPTER 3

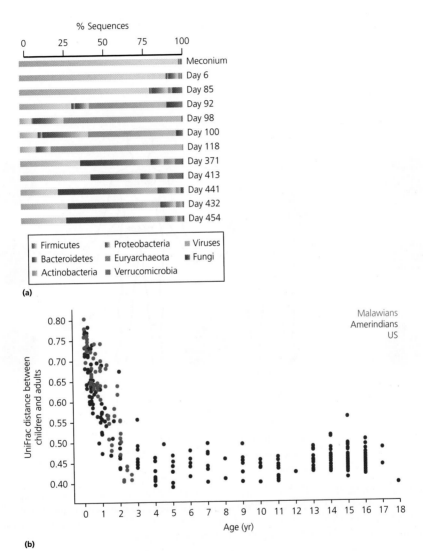

(a)

(b)

Figure 1 (a) Metataxonomic DNA analysis of an infant fecal microbiota in the first year and a half of life showing a chaotic evolution impacted by weaning and antibiotic consumption, from Koenig *et al*[13]. (b) Fecal microbial differences between children and adults from Amazonia (Amerindians: green dots), Malawi (red dots) and USA (blue dots), showing reduction of distance to adult microbiota during the first three years of life and then stability until adulthood, from Yatsunenko *et al*[14].

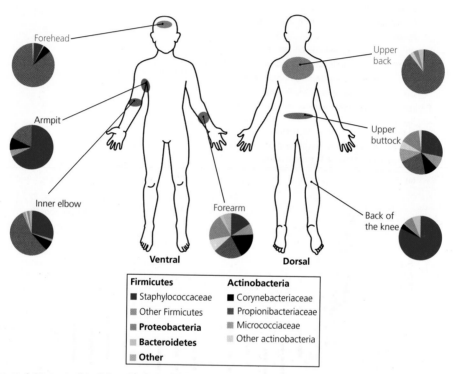

Figure 1 Topographical distribution of bacteria on specific skin sites. The microbial composition is shown at phylum and family level. Moist sites are labelled in green, sebaceous sites are labelled in blue, and dry surfaces in red. Data from Zeeuwen *et al.*[38] except for volar forearm, back of the knee, and the back (from Human Microbiome Project).

CHAPTER 9

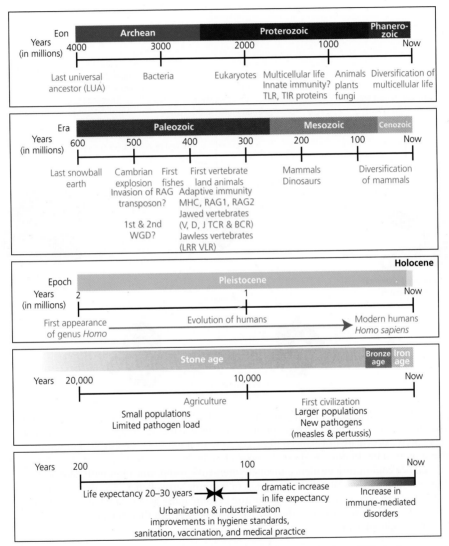

Figure 1 Time scale of immune-microbial co-evolution. The evolution of life is shown and important evolutionary (red), immunological (green), and modernization events (black) are indicated. The different eons, eras, epochs, and ages are indicated in the colored boxes.

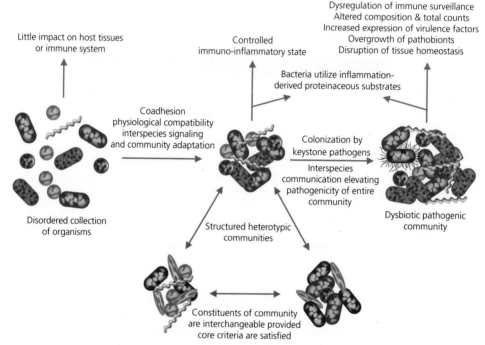

Figure 1 The polymicrobial synergy and dysbiosis (PSD) model of periodontal disease etiology. (a) Model overview. In health, communities assembled through co-adhesion and physiological compatibility participate in balanced interactions with the host in a controlled immuno-inflammatory state. In disease, there is colonization by keystone pathogens such as *P. gingivalis* that enhance community virulence through interactive communication with accessory pathogens such as *S. gordonii,* and disruption of immune surveillance. The dysbiotic community proliferates, pathobionts (green) overgrow and become more active, and tissue destruction ensues. The participation of individual species is less important than the presence of the appropriate suite of genes.

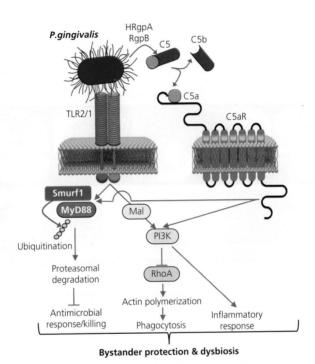

Figure 5 Subversion of neutrophil function and dysbiosis. *P. gingivalis* co-activates TLR2 and C5aR in neutrophils. and the resulting crosstalk leads to E3 ubiquitin ligase Smurf1-dependent ubiquitination and proteasomal degradation of MyD88, thereby inhibiting a host-protective antimicrobial response. Moreover, the C5aR-TLR2 crosstalk activates PI3K, which prevents phagocytosis through inhibition of RhoA activation and actin polymerization, while stimulating an inflammatory response. In contrast to MyD88, another TLR2 adaptor, Mal, is involved in the subversive pathway and acts upstream of PI3K. The integrated mechanism provides "bystander" protection to otherwise susceptible bacterial species and promotes polymicrobial dysbiotic inflammation *in vivo*. From Maekawa *et al.*, 2014 (Ref. 59) with permission.

CHAPTER 26

Colon Carcinogenesis

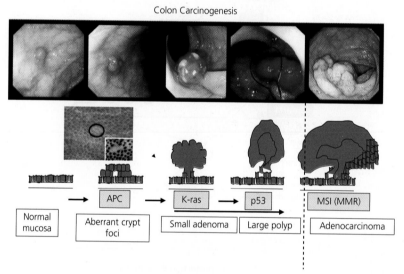

Figure 1 Colorectal carcinogenesis from normal to invasive cancer. At top are slides from colonoscopy after cleaning preparation showing precancerous lesions such as polyps at various diameters; genes involved in each steps are indicated supporting the theory of accumulation of gene mutations from aberrant crypt foci (ACF) (high magnification) up to the invasive cancer through precancerous lesions at various size.

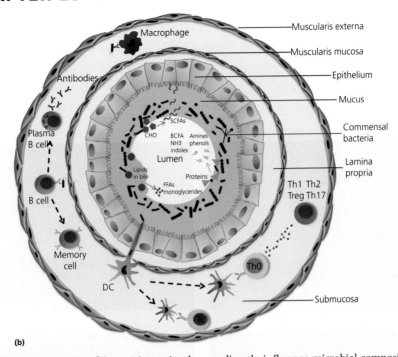

(b)

Figure 1 (b) Food-microbiota: Dietary intake can directly influence microbial composition by promoting blooms of certain groups while inhibiting the growth of others. Partially digested food such as complex carbohydrates are broken down in the gut by the bacteria via fermentation and produce short-chain fatty acids (SCFAs). Lipid digestion in the gut requires the secretion of liver-derived bile from the gallbladder. High-lipid diets demand higher bile release, which could induce dysbiosis in part by changing intestinal pH levels. Fermentation of proteins in the gut can result in production of branched-chain fatty acids (BCFA) as well as potentially toxic products such as amines, ammonia (NH3), phenols, and indoles[26]. Gut bacteria are essential in synthesizing various vitamins such as riboflavin, folate, B12, and vitamin K[26]. **Microbiota-host:** The microbiota acts as a first line of defense against enteropathogens by competition for food and niche. Host immunity is modulated by the intestinal microbiota starting at birth. Antigen-presenting cells (APC) such as dendritic cells (DC) or macrophages phagocytose invading bacteria and subsequently initiate differentiation of naïve T cells (Th0) to subset of T-helper cells (Th): Th1, Th2, Th17, or regulatory T-cells (Treg). B-cells presented by bacterial antigen differentiate into antibody-producing effector B cells or memory cells. Antibodies produced by effector B-cells neutralize and tag the specific pathogen for removal. Bacteria can regulate mucus production by stimulation of goblet cells. Paneth cells of the intestinal epithelium layer sense bacteria and in response produce antimicrobial peptides (AMPs) that protect the host from bacterial invasion. Evidence has emerged showing that gut microbiota interact with central and peripheral neural processes, suggesting their involvement in hormone signaling as well as psychological health (see chapter 27 by Barwani and Forsythe). **Host-Food:** Food antigens stimulate endocrine secretion, which initiates downstream digestion cascades such as bile and amylase production by the liver and pancreas, respectively. SCFAs produced by bacteria in the colon are used by colonocytes as a major energy source. Ingestion of contaminated food is a major source of pathogen entry into the body. Hormonal regulations by the host control appetite and satiety.

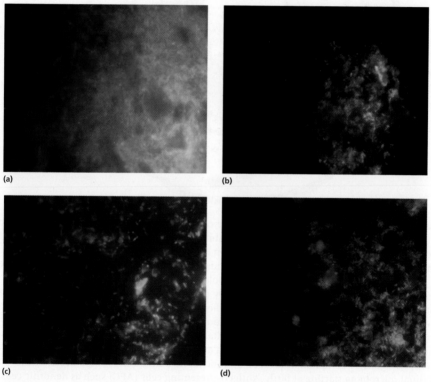

Figure 3 Habitually constitutive bacteria intestinal microflora evaluated by FISH (fluorescence in situ hybridization). Top: Total bacteria and Faecalibacterium prausnitzii were evaluted by EUB (green) and Fpra (red) probes in healthy (a) and cystic fibrosis (b) children. Bottom: Total bacteria and Eubacterium rectale were evaluted by EUB (green) and Erec (red) probes in healthy (c) and cystic fibrosis (d) children. Images kindly provided by Dr. V. Buccigrossi.